Flooding and Environmental Challenges for Venice and its Lagoon: State of Knowledge

Time may be running out for Venice. With rising average water levels, the frequency of city flooding is increasing and the threat of a repeat of the November 1966 event, when a violent storm surge took water levels nearly two metres higher than usual, remains. Surrounding the city is a severely degraded lagoon ecosystem. This timely scientific and technical volume synthesises the great wealth and diversity of recent interdisciplinary research on Venice and its Lagoon and the prospects for large engineering interventions to separate the lagoon and sea, as well as other measures in the built environment, discussed at an International Conference held at Churchill College, Cambridge in September 2003. The lessons and inferences reported here show how Venice, with its mix of challenges to protect its prestigious cultural heritage within one of the largest coastal wetlands in the Mediterranean, and against a background of pressures brought about by industry, port activities and tourism, shares many issues with other areas threatened by coastal flooding, including areas of the Netherlands, the USA and the cities of London and St Petersburg.

Caroline Fletcher is Senior Research Associate, Cambridge Coastal Research Unit, University of Cambridge, Venice in Peril Fellow, Churchill College, Cambridge and Principal Environmental Scientist, HR Wallingford Ltd. She specialises in sustainable development and the impacts of natural processes and human activities on estuarine and coastal systems. She has worked in the UK, Europe, USA, the Middle East and the Far East on projects concerned with aquatic pollution, habitat creation and the beneficial use of dredged material, sediments and port developments. She represents the Central Dredging Association (CEDA) at international conventions and is a member of two Permanent International Association for the Navigation Community (PIANC) international working groups on the management of dredged material.

Tom Spencer is University Senior Lecturer, Department of Geography, University of Cambridge, Director, Cambridge Coastal Research Unit and Official Fellow, Magdalene College, Cambridge. His research interests – wetland hydrodynamics and sedimentation, coral reef geomorphology, sea level rise and coastal management – have taken him to the Caribbean Sea, the Pacific and Indian Oceans and, closer to home, the coastline of Eastern England. He is author (with H. Viles) of *Coastal Problems: Geomorphology, Ecology and Society at the Coast* and is currently co-editing *Big Flood*, a collection of papers commemorating the fiftieth anniversary of the 1953 North Sea storm surge.

Flooding and Environmental Challenges for Venice and its Lagoon: State of Knowledge

EDITED BY

C. A. FLETCHER AND T. SPENCER

Cambridge Coastal Research Unit, Department of Geography,
University of Cambridge

C C R U

Venice in Peril

THE BRITISH COMMITTEE FOR THE PRESERVATION OF VENICE

CAMBRIDGE
UNIVERSITY PRESS

CAMBRIDGE UNIVERSITY PRESS
Cambridge, New York, Melbourne, Madrid, Cape Town, Singapore,
São Paulo, Delhi, Dubai, Tokyo

Cambridge University Press
The Edinburgh Building, Cambridge CB2 8RU, UK

Published in the United States of America by Cambridge University Press, New York

www.cambridge.org
Information on this title: www.cambridge.org/9780521124140

First published 2005
This digitally printed version 2009

A catalogue record for this publication is available from the British Library

ISBN 978-0-521-84046-0 Hardback
ISBN 978-0-521-12414-0 Paperback

Additional resources for this publication at www.cambridge.org/9780521124140

Contents

Flooding and Environmental Challenges for Venice and its Lagoon: State of Knowledge, ed. C. A. Fletcher and T. Spencer.
Published by Cambridge University Press, © Cambridge University Press 2005.

Contributors

Giovanni Abrami
Università IUAV di Venezia
Santa Croce 191
30135 Venice, Italy

Francesco Acri
Istituto di Scienze Marine/Biologia del Mare
Consiglio Nazionale delle Ricerche
Castello 1364/a
30122 Venice, Italy

Albert J. Ammerman
Department of Classics
Colgate University
13 Oak Drive
Hamilton NY 13346, USA

Carl L. Amos
Southampton Oceanography Centre
University of Southampton
Empress Dock
Southampton SO14 3ZH, UK

Fabrizio Bernardy Aubry
Istituto di Scienze Marine/Biologia del Mare
Consiglio Nazionale delle Ricerche
Castello 1364/a
30122 Venice, Italy

Marco Bajo
Istituto di Scienze Marine – Venezia
Consiglio Nazionale delle Ricerche
San Polo 1364
30125 Venice, Italy

Asterios Bakolas
Department of Chemical Engineering
National Technical University of Athens
Iroon Polytechniou St
Zografou Campus
157 73 Athens, Greece

Andrea Barbanti
Thetis S.p.A.
Castello 2737/f
30122 Venice, Italy

Adriano Barbi
Agenzia Regionale per la Prevenzione e protezione
Ambientale del Veneto (ARPAV)
Via Marconi 55
Teolo
35037 Padua, Italy

Alberto Giulio Bernstein
Consorzio Venezia Nuova
San Marco 2803
30124 Venice, Italy

Andrea Berton
Istituto di Scienze Marine/Biologia del Mare
Consiglio Nazionale delle Ricerche
Castello 1364/a
30122 Venice, Italy

Franco Bianchi
Istituto di Scienze Marine/Biologia del Mare
Consiglio Nazionale delle Ricerche
Castello 1364/a
30122 Venice, Italy

Guido Biscontin
Dipartimento di Scienze Ambientali
Università Ca' Foscari di Venezia
Calle Larga Santa Marta
Dorsoduro 2137
30123 Venice, Italy

Sandro Boato
Agenzia Regionale per la Prevenzione e protezione
Ambientale del Veneto (ARPAV)
Piazzale Stazione 1
35131 Padua, Italy

Flooding and Environmental Challenges for Venice and its Lagoon: State of Knowledge, ed. C. A. Fletcher and T. Spencer.
Published by Cambridge University Press, © Cambridge University Press 2005.

Rene Bol
Directorate-General for Public Works and Water
    Management
PO Box 556
3000 AN Rotterdam, The Netherlands

Michele Bolla Pittaluga
Dipartimento di Ingegneria Ambientale
Università degli studi di Genova
Via Montallegro 1
16145 Genoa, Italy

Lorenzo Bonometto
Osservatorio naturalistico della laguna e Centro
    informativo sulla salvaguardia e la
    manutenzione urbana
Comune di Venezia
S. Croce 1704
30135 Venice, Italy

John Boyd
Churchill College
Storey's Way
Cambridge CB3 0DS, UK

Margaretha Breil
Fondazione Eni Enrico Mattei
Castello 5252
30122 Venice, Italy

Maria Teresa Brotto
Consorzio Venezia Nuova
San Marco 2803
30124 Venice, Italy

Silvano Buogo
Istituto di Acustica 'O. M. Corbino' (IDAC)
Consiglio Nazionale delle Ricerche
Tor Vergata
Via del Fosso del Cavaliere 100
00133 Rome, Italy

Roy Butterfield
Department of Civil and Environmental Engineering
University of Southampton
3 Furzedown Road
Highfield
Southampton S017 1PN, UK

Warren R. L. Cairns
Istituto per la Dinamica dei Processi Ambientali
Consiglio Nazionale delle Ricerche
Calle Larga Santa Marta
Dorsoduro 2137
30123 Venice, Italy

Nicoletta Calace
Dipartimento di Chimica
Università 'La Sapienza' di Roma
P. le Aldo Moro 5
00185 Rome, Italy

Elisa Camatti
Istituto di Scienze Marine/Biologia del Mare
Consiglio Nazionale delle Ricerche
Castello 1364/a
30122 Venice, Italy

Pierpaolo Campostrini
CORILA
Palazzo Franchetti
San Marco 2847
30124 Venice, Italy

Dario Camuffo
Istituto di Scienze dell'Atmosfera
Consiglio Nazionale delle Ricerche
Corso Stati Uniti 4
35127 Padua, Italy

Ernesto Canal
Istituto di Scienze Marine – Venezia
Consiglio Nazionale delle Ricerche
San Polo 1364
30125 Venice, Italy

Paolo Canestrelli
Centro Previsioni e Segnalazioni Maree
Comune di Venezia
Palazzo Cavalli
San Marco 4090
30124 Venice, Italy

Giovanni Caniato
Archivio di Stato di Venezia
San Polo 3002
30125 Venice, Italy

Giovanni Caniglia
Dipartimento di Biologia
Università degli studi di Padova
Via Trieste 75
30135 Padua, Italy

Giovanni Bosco Cannelli
Istituto di Acustica 'O. M. Corbino' (IDAC)
Consiglio Nazionale delle Ricerche
Tor Vergata
Via del Fosso del Cavaliere 100
00133 Rome, Italy

Gabriele Capodaglio
Dipartimento di Scienze Ambientali
Università Ca' Foscari di Venezia
Calle Larga Santa Marta
Dorsoduro 2137
30123 Venice, Italy

Eugenio Carminati
Dipartimento di Scienze della Terra
Università 'La Sapienza' di Roma
P. le Aldo Moro 5
00185 Rome, Italy

Fabio Carrera
Worcester Polytechnic Institute / Massachusetts
    Institute of Technology
100 Institute Road
Worcester MA 01609, USA

Roberto Casarin
Segreteria regionale Ambiente e Lavori pubblici
    – Regione Veneto
Calle Priuli
Cannaregio 99
30121 Venice, Italy

Silvia Cavazzoni
Istituto di Scienze Marine – Venezia
Consiglio Nazionale delle Ricerche
San Polo 1364
30125 Venice, Italy

Giovanni Cecconi
Consorzio Venezia Nuova
San Marco 2803
30124 Venice, Italy

Massimiliano Cervelli
Istituto di Scienze Marine/Biologia del Mare
Consiglio Nazionale delle Ricerche
Castello 1364/a
30122 Venice, Italy

Flaviano Collavini
Istituto di Scienze Marine – Venezia
Consiglio Nazionale dele Ricerche
San Polo 1364
30125 Venice, Italy

Alessandra Comaschi
Istituto di Scienze Marine/Biologia del Mare
Consiglio Nazionale delle Ricerche
Castello 1364/a
30122 Venice, Italy

Giovanni Coppini
Istituto Nazionale di Geofisica e Vulcanologia
Sede di Bologna
Via Donato Creti 12
40129 Bologna, Italy

Elisa Coraci
Istituto di Scienze Marine – Venezia
Consiglio Nazionale delle Ricerche
San Polo 1364
30125 Venice, Italy

Simone Cosoli
Istituto di Scienze Marine – Venezia
Consiglio Nazionale delle Ricerche
San Polo 1364
30125 Venice, Italy

Franco Costa
Istituto di Scienze Marine – Venezia
Consiglio Nazionale dele Ricerche
San Polo 1364
30125 Venice, Italy

Andrea Critto
Dipartimento di Sicenze Ambientali
Università Ca' Foscari di Venezia
Calle Larga Santa Marta
Dorsoduro 2137
30123 Venice, Italy

Andrea Cucco
Istituto di Scienze Marine – Venezia
Consiglio Nazionale delle Ricerche
San Polo 1364
30125 Venice, Italy

Daniele Curiel
SELC scarl
Via dell'Elettricità 5/d
30174 Marghera-Venice, Italy

Luigi D'Alpaos
Dipartimento di Ingegneria Idraulica
Marittima, Ambientale e Geotecnica
Università degli Studi di Padova
Via Loredan 20
35131 Padua, Italy

Jane Da Mosto
CORILA
Palazzo Franchetti
San Marco 2847
30124 Venice, Italy

Luisa Da Ros
Istituto di Scienze Marine/Biologia del Mare
Consiglio Nazionale delle Ricerche
Castello 1364/a
30122 Venice, Italy

Caterina Dabalà
CORILA
Palazzo Franchetti
San Marco 2847
30124 Venice, Italy

Edoardo Danzi
Dipartimento di Storia dell'Architettura
Università IUAV di Venezia
San Polo 2468
30125 Venice, Italy

John.W Day Jr
Department of Oceanography and Coastal Sciences
School of the Coast and Environment
Louisiana State University
Baton Rouge
LA 70803, USA

Francesco De Biasio
Istituto di Scienze Marine – Venezia
Consiglio Nazionale dele Ricerche
San Polo 1364
30125 Venice, Italy

Stefania De Zorzi
CORILA
Palazzo Franchetti
San Marco 2847
30124 Venice, Italy

Maurizio Di Donato
Consorzio Venezia Nuova
San Marco 2803
30124 Venice, Italy

Giampaolo Di Silvio
Dipartimento di Ingegneria Idraulica
Marittima, Ambientale e Geotecnica
Università degli Studi di Padova
Via Loredan 20
35131 Padua, Italy

Carlo Doglioni
Dipartimento di Scienze della Terra
Università 'La Sapienza' di Roma
P. le Aldo Moro 5
00185 Rome, Italy

Bruno Dolcetta
Insula S.p.A.
Dorsoduro 2050
30123 Venice, Italy

Sandra Donnici
Istituto di Scienze Marine – Venezia
Consiglio Nazionale delle Ricerche
San Polo 1364
30125 Venice, Italy

Steve Dunthorne
Jacobs Gibb Ltd
Jacobs House
London Road
Reading RG6 1BL, UK

Yuil Eprim
Technital S.p.A.
Via Cassano d'Adda 27/1
20139 Milan, Italy

Edoardo Faganello
5D Coley Hill
Reading RG1 6AE, UK

Maurizio Ferla
Agenzia per la protezione dell'ambiente e per i
    servizi tecnici (APAT)
Palazzo X Savi
Ruga dei Oresi 50
30125 Venice, Italy

Vincenzo Ferrara
ENEA-Casaccia Research Center
Via Anguillarese 301
00060 S. Maria di Galeria (Rome), Italy

Giorgio Ferrari
Servizio Anti-inquinamento
Magistrato alle Acque di Venezia
Ministero dei Lavori Pubblici
Palazzo X Savi
San Polo 19
30125 Venice, Italy

Christian Ferrarin
Istituto di Scienze Marine – Venezia
Consiglio Nazionale delle Ricerche
San Polo 1364
30125 Venice, Italy

Alessandra Ferrighi
Dipartimento di Storia dell'Architettura
Università IUAV di Venezia
San Polo 2468
30125 Venice, Italy

Caroline A. Fletcher
Churchill College and Cambridge Coastal
    Research Unit
Department of Geography
University of Cambridge
Downing Place
Cambridge, UK

Marino Folin
Università IUAV di Venezia
Santa Croce 191
30135 Venice, Italy

Elisa Franzoni
Dipartimento di Chimica Applicata e Scienza dei
    Materiali
Facoltà di Ingegneria
Università di Bologna
V. le Risorgimento 2
Bologna 40136, Italy

Roberto Frassetto
Istituto di Scienze Marine – Venezia
Consiglio Nazionale delle Ricerche and IGBP
San Polo 1364
30125 Venice, Italy

Miro Gačić
Istituto Nazionale di Oceanografica di Geofisica
    Sperimentale
Borgo Grotta Gigante 42/c
34010 Sgonico
Trieste, Italy

Gretel Gambarelli
Fondazione Eni Enrico Mattei
Castello 5252
30122 Venice, Italy

Andrea Gambaro
Dipartimento di Scienze Ambientali
Università Ca' Foscari di Venezia
Istituto per la Dinamica dei Processi Ambientali –
    CNR
Calle Larga Santa Marta
Dorsoduro 2137
30123 Venice, Italy

Valeria Garotta
Dipartimento di Ingegneria Ambientale
Università degli studi di Genova
Via Montallegro 1
16145 Genoa, Italy

Domenico Gaudioso
Agenzia per la protezione dell'ambiente e per i
    servizi tecnici (APAT)
Via Vitaliano Brancati 48
00144 Rome, Italy

Herman Gerritsen
WL | Delft Hydraulics
PO Box 177
2600 MH Delft, The Netherlands

Joris Geurts Van Kessel
National Institute for Coastal and Marine
    Management/RIKZ
Kortenaerkade 1
PO Box 20907
2500 EX The Hague, The Netherlands

Antonio Gozzi
Technital S.p.A.
Via Carlo Cattaneo 20
37121 Verona, Italy

Anita Grezio
Istituto Nazionale di Geofisica e Vulcanologia
Sede di Bologna
Via Donato Creti 12
40129 Bologna, Italy

Peter M Guthrie
Department of Engineering
University of Cambridge
Trumpington Street
Cambridge CB2 1PZ, UK

Donald R F Harleman
Department of Civil and Environmental
    Engineering
Massachusetts Institute of Technology
Ralph M. Parsons Laboratory
77 Mass Ave
Cambridge MA 02139, USA

Marjukka Hiltunen
Finnish Environmental Institute
Mechelininkatu 34a
PO Box 140
FIN-0051
Helsinki, Finland

Peter Hunter
Jacobs Gibb Ltd
Jacobs House
London Road
Reading RG6 1BL, UK

Vedrana Kovačević
Istituto Nazionale di Oceanografia e di Geofisica
    Sperimentale
Borgo Grotta Gigante 42/c
34010 Sgonico
Trieste, Italy

Ralph Lasage
Institute for Environmental Studies
Vrije Universiteit, Amsterdam
De Boelelaan 1087
1081 HV Amsterdam, The Netherlands

Sarah Lavery
UK Environment Agency
Thames Barrier
Eastmoor Street
Charlton
London SE7 8LX, UK

Alberto Lezziero
Istituto di Scienze Marine – Venezia
Consiglio Nazionale delle Ricerche
San Polo 1364
30125 Venice, Italy

Angelo Libertini
Istituto di Scienze Marine/Biologia del Mare
Consiglio Nazionale delle Ricerche
Castello 1364/a
30122 Venice, Italy

Piero Lionello
Dipartimento di Scienza dei Materiali
Università degli Studi di Lecce
Via per Arnesano
73100 Lecce, Italy

Nicola Lonardoni
Dipartimento di Chimica Applicata e Scienza dei
    Materiali
Facoltà di Ingegneria
Università di Bologna
V. le Risorgimento 2
40136 Bologna, Italy

Isaac Mancero Mosquera
Istituto Nazionale di Oceanografia e di Geofisica
    Sperimentale
Borgo Grotta Gigante 42c
34010 Sgonico
Trieste, Italy

Andrea Mancuso
UNESCO office in Venice – Regional Bureau for
    Science in Europe (ROSTE)
Palazzo Zorzi
Castello 4930
30122 Venice, Italy

Antonio Marcomini
Dipartimento di Sicenze Ambientali
Università Ca' Foscari di Venezia
Calle Larga Santa Marta
Dorsoduro 2137
30123 Venice, Italy

Concepción Marcos
Departamento de Ecologica e Hidrologia
Universidad de Murcia
Facultad de Biologia
Campus de Espinardo 30151
Murcia, Spain

Paolo Martini
Dipartimento di Ingegneria Idraulica
Marittima
Ambientale e Geotecnica
Università degli Studi di Padova
Via Loredan
20, 35131 Padua, Italy

Andrea Mazzoldi
Istituto di Scienze Marine – Venezia
Consiglio Nazionale delle Ricerche
San Polo 1364
30125 Venice, Italy

Donata Melaku Canu
Istituto Nazionale di Oceanografia e di Geofisica
    Sperimentale
Borgo Grotta Gigante 42/c
34010 Sgonico
Trieste, Italy

Giselle Menel Lemos
Technital S.p.A.
Via C Cattaneo 20
37121 Verona, Italy

Christian Micheletti
Dipartimento di Sicenze Ambientali
Università Ca' Foscari di Venezia
Calle Larga Santa Marta
Dorsoduro 2137
30123 Venice, Italy

Rosa R Mikhailenko
Morzaschita
Department of St Petersburg City Administration
76 Moika Emb
St Petersburg 190000, Russia

Roberta Millini
Agenzia Regionale per la Prevenzione e protezione
    Ambientale del Veneto (ARPAV)
Via Marconi 55
Teolo
35037 Padua, Italy

Daniele Mion
SELC scarl
Via dell'Elettricità 5/d
30174 Marghera – Venice, Italy

William Mitsch
Olentangy River Wetland Research Park
School of Natural Resources
Ohio State University
Columbus OH 43210, USA

Marco Monai
Agenzia Regionale per la Prevenzione e protezione
    Ambientale del Veneto (ARPAV)
Via Marconi 55
Teolo
35037 Padua, Italy

Laura Montobbio
Consorzio Venezia Nuova
San Marco 2803
30124 Venice, Italy

Howard Moore
UNESCO Office in Venice – Regional Bureau for
    Science in Europe (ROSTE)
Palazzo Zorzi
Castello 4930
30122 Venice, Italy

Cristina Nasci
Istituto di Scienze Marine/Biologia del Mare
Consiglio Nazionale delle Ricerche
Castello 1364/a
30122 Venice, Italy

Paulo A. L. D. Nunes
Fondazione Eni Enrico Mattei
Università Ca' Foscari di Venezia and Faculty of
    Economics, Vrije Universiteit, Amsterdam
Castello 5252
30122 Venice, Italy

Paolo Oddo
Corso di Scienze Ambientali
Università di Bologna
Laboratorio FINCEM
Piazzale Kennedy 12
48100 Ravenna, Italy

Emanuela Pagan
Istituto di Scienze dell'Atmosfera
Consiglio Nazionale delle Ricerche
Corso Stati Uniti 4
35127 Padua, Italy

Daniela Pampanin
Istituto di Scienze Marine/Biologia del Mare
Consiglio Nazionale delle Ricerche
Castello 1364/a
30122 Venice, Italy

Paolo Parati
Centro di riferimento per il Bacino Scolante in
    laguna di Venezia – ARPAV
Via Lissa 6
30171 Mestre-Venice, Italy

Roberto Pastres
Dipartimento di Chimica Fisica
Università Ca' Foscari di Venezia
Calle Larga Santa Marta
Dorsoduro 2137
30123 Venice, Italy

Gisella Penna
Divisione Ambiente – Regione Veneto
Cannaregio 99
30121 Venice, Italy

Angel Peres-Ruzafa
Departamento de Ecologica e Hidrologia
Universidad de Murcia
Facultad de Biologia
Campus de Espinardo
30151 Murcia, Spain

Bianca M. Petronio
Dipartimento di Chimica
Università 'La Sapienza' di Roma
P. le Aldo Moro 5
00185 Rome, Italy

Mario Piana
Dipartimento di Storia dell'Architettura
Università IUAV di Venezia
San Polo 2468
30125 Venice, Italy

Andrea Pierini
SELC scarl
Via dell'Elettricità 5/d
30175 Marghera – Venice, Italy

Marco Pietroletti
Dipartimento di Chimica
Università 'La Sapienza' di Roma
P. le Aldo Moro 5
00185 Rome, Italy

Nadia Pinardi
Istituto Nazionale di Geofisica e Vulcanologia
Sede di Bologna
Via Donato Creti 12
40129 Bologna, Italy

Maria Giovanna Piva
Magistrato alle Acque di Venezia
Ministero dei Lavori Pubblici
Palazzo X Savi
San Polo 19
30125 Venice, Italy

Giulio Pojana
Dipartimento di Scienze Ambientali
Università Ca' Foscari di Venezia
Calle Larga Santa Marta
Dorsoduro 2137
30123 Venice, Italy

Philippe Pypaert
UNESCO office in Venice – Regional Bureau for
    Science in Europe (ROSTE)
Palazzo Zorzi
Castello 4930
30122 Venice, Italy

Sandro Rabitti
Istituto di Scienze Marine/Biologia del Mare
Consiglio Nazionale delle Ricerche
Castello 1364/a
30122 Venice, Italy

Astrid Raudner
Agenzia per la protezione dell'ambiente e per i
    servizi tecnici (APAT)
Via Vitaliano Brancati 48
00144 Rome, Italy

Andrea Razzini
Autorità Portuale di Venezia
Zattere 1401
30123 Venice, Italy

Paul Richens
Churchill College
Storey's Way
Cambridge CB3 0DS, UK

Enrico Rinaldi
CORILA
San Marco 2847
30124 Venice, Italy

Andrea Rismondo
SELC Scarl
Via dell'Elettricita 5/d
30175 Marghera – Venice, Italy

Pierluigi Rossetto
Thetis S.p.A.
Castello 2737/f
30122 Venice, Italy

John Rybczyk
Department of Environmental Sciences
Western Washington University
Bellingham WA 98225, USA

Henk L. F. Saejs
Erasmus Center for Sustainability and Management
Erasmus University Rotterdam
Europalaan 31
Middleburg 4334 EA, The Netherlands

Franco Sandrolini
Dipartimento di Chimica Applicata e Scienza dei
    Materiali
Università di Bologna
V. le Risorgimento 2
40136 Bologna, Italy

Alexander N. Savin
St Petersburg City Administration
76 Moika Emb
St Petersburg 190000, Russia

Francesco Scarton
SELC Scarl
Via dell'Elettricità 5/d
30175 Marghera – Venice, Italy

Mario Scattolin
Assessorato all'Ambiente
Comune di Venezia
San Marco 4136
30124 Venice, Italy

Alberto Scotti
Technital S.p.A.
Via Carlo Cattaneo 20
37121 Verona, Italy

Davide Scrocca
Istituto di Geologia Ambientale e Geoingegneria
    (IGAG)
Consiglio Nazionale delle Ricerche
P. le Aldo Moro 5
00185 Rome, Italy

Isabella Scroccaro
Istituto di Scienze Marine – Venezia
Consiglio Nazionale delle Ricerche
San Polo 1364
30125 Venice, Italy

Giovanni Seminara
Dipartimento di Ingegneria Ambientale
Università degli studi di Genova
Via Montallegro 1
16145 Genoa, Italy

Adriano Sfriso
Dipartimento di Sicenze Ambientali
Università Ca' Foscari di Venezia
Calle Larga Santa Marta
Dorsoduro 2137
30123 Venice, Italy

Giorgio Socal
Istituto di Scienze Marine/Biologia del Mare
Consiglio Nazionale delle Ricerche
Castello 1364/a
30122 Venice, Italy

Stefano Sofia
Agenzia Regionale per la Prevenzione e protezione
    Ambientale del Veneto (ARPAV)
Via Marconi 55
Teolo
35037 Padua, Italy

Cosimo Solidoro
Dipartimento di Oceanografia
Istituto Nazionale di Oceanografia e di Geofisica
    Sperimentale
Borgo Grotta Gigante 42/c
34010 Sgonico
Trieste, Italy

Anna Somers Cocks
Venice in Peril
Unit 4 Hurlingham Studios
Ranelagh Gardens
London SW6 3PA, UK

Robin J. S. Spence
The Martin Centre
Department of Architecture
University of Cambridge
6 Chaucer Road
Cambridge CB2 2EB, UK

Tom Spencer
Cambridge Coastal Research Unit
Department of Geography
University of Cambridge
Downing Place
Cambridge CB2 3EN, UK

Mario Spinelli
IUAV Studi e progetti srl
Dorsoduro 3900
30123 Venice, Italy

Peter K. Stansby
Dipartimento di Ingegneria Ambientale
Università degli studi di Genova
Via Montallegro 1
16145 Genoa, Italy

Fabio Strazzabosco
Direzione Regionale Geologia e Ciclo dell'acqua –
    Regione Veneto
Calle Priuli
Cannaregio 99
30121 Venice, Italy

Giovanni Sturaro
Istituto di Scienze dell'Atmosfera
Consiglio Nazionale delle Ricerche
Corso Stati Uniti 4
35127 Padua, Italy

Nicoletta Tambroni
Dipartimento di Ingegneria Ambientale
Università degli Studi di Genova
Via Montallegro 1
16145 Genoa, Italy

Alberto Tomasin
Dipartimento di Matematica Applicata
Università Ca' Foscari di Venezia
Cannaregio 3175
30121 Venice, Italy

Giuseppina Toscano
Dipartimento di Scienze Ambientali
Università Ca' Foscari di Venezia
Calle Larga Santa Marta
Dorsoduro 2137
30123 Venice, Italy

Renata Trisolini
Istituto di Scienze Marine/Biologia del Mare
Consiglio Nazionale delle Ricerche
Castello 1364/a
30122 Venice, Italy

Clara Turetta
Istituto per la Dinamica dei Processi Ambientali
Consiglio Nazionale delle Ricerche
Calle Larga Santa Marta
Dorsoduro 2137
30123 Venice, Italy

Ivano Turlon
Insula S.p.A.
Dorsoduro 2050
30123 Venice, Italy

Georg Umgiesser
Istituto di Scienze Marine – Venezia
Consiglio Nazionale delle Ricerche
San Polo 1364
30125 Venice, Italy

Hans Van Pagee
Rijkswaterstaat
Ministry of Transport Public Works and Water
    Management; National Institute for Coastal and
    Marine Management (RIKZ)
PO Box 8039
Grenadierweg 31
4330 EA Middelburg, The Netherlands

Marina Vazzoler
Agenzia Regionale per la Prevenzione e protezione
    Ambientale del Veneto (ARPAV)
Piazzale Stazione 1
35131 Padua, Italy

Pier Vellinga
Faculty of Earth and Life Sciences
Vrije Universiteit
De Boelelaan 1085/F148
Amsterdam 1081 HV, The Netherlands

Ettore Vio
Curia Patriarcale di Venezia
San Marco
30124 Venice, Italy

Rinus Vis
WL | Delft Hydraulics
P.O. Box 177
2600 MH Delft, The Netherlands

David J. Wilkes
Environment Agency
Phoenix House
Global Avenue
Leeds LS11 8PG, UK

Luca Zaggia
Istituto di Scienze Marine – Venezia
Consiglio Nazionale delle Ricerche
San Polo 1364
30125 Venice, Italy

Lucia Zampato
Centro Previsioni e Segnalazioni Maree
Comune di Venezia
Palazzo Cavalli
San Marco 4090
30124 Venice, Italy

Guido Zanovello
Stuido Altieri Srl
Via Colleoni 50
36016 Thiene, Italy

Stefano Zecchetto
Istituto di Scienze dell'Atmosfera e del Clima
Consiglio Nazionale delle Ricerche
Corso Stati Uniti 4
35127 Padua, Italy

Elisabetta Zendri
Dipartimento di Sicenze Ambientali
Università Ca' Foscari di Venezia
Via Torino 155/b
30174 Mestre-Venice, Italy

Alberto Zirino
Scripps Institution of Ocenography
University of California, San Diego
9500 Gilman Drive
La Jolla
CA 92093-0202, USA

Roberto Zonta
Istituto di Scienze Marine – Venezia
Consiglio Nazionale delle Ricerche
San Polo 1364
30125 Venice, Italy

Aleardo Zuliani
Istituto di Scienze Marine – Venezia
Consiglio Nazionale delle Ricerche
San Polo 1364
30125 Venice, Italy

Preface

CONVERTING RESEARCH INTO ACTION

The 16 million visitors who come to Venice annually see restoration going on everywhere. This is a city where property values are booming and monuments are expertly protected by the Superintendencies, the responsible government officials. But the deeper reality is less happy. Time is running out for this loveliest of cities. Even on a calm day, the water laps above the stone foundations of many of the buildings and attacks the brickwork above. The frequency of flooding is increasing, and Venice is essentially no better protected from an extreme weather event than it was at the time of the great flood of 1966. The future effects of climate change (difficult to predict precisely, but capable of being factored in nonetheless) are barely being considered. And yet large sums have been spent since the 1970s on scientific research into what should be done to defend Venice from the sea. The problem has been in converting that research into action. When the Venice in Peril, the British fund for the safeguarding of Venice, turned itself from an organization that mainly restored monuments into one that also looked into the chief peril facing Venice, the waters, we realized that the fundamental reason for this inaction was a lack of agreement about what science was actually telling us. The arguments about whether the proposed mobile barriers are vital or actually damaging have fiercely divided the citizens of Venice and Italian politicians and both sides invoke science. We discovered, also, that the reliable evidence was difficult to track down. Most of it was unpublished, some was not even listed in an accessible way and very little was in English, which was a loss to the international scientific community that has a lot to learn from the work done on Venice and its lagoon. In 2001, Venice in Peril began funding its collaboration with the Coastal Research

Unit of Cambridge University and CORILA in Venice with the aim of helping specialists link up with non-specialists; of encouraging free, open and informed debate among the scientists concerned; of making the world especially those who have the power to decide on the future for Venice, realize that Venice can be saved. A healthy, collaborative research environment may produce the knowledge that will save Venice, which is why this book is important. It is the broadest, most up-to-date published scientific survey of the question since research into this field began. It is a milestone in the endless process that is essential if we want our great-grand-children to see this incomparable achievement of man.

Anna Somers Cocks
Chairman
Venice in Peril

COLLABORATION BETWEEN THE WORLDS OF SCIENCE AND CULTURE

For many of my generation, Venice would be first encountered through those stark, black-and-white images of November 1966, and the harrowing scenes we witnessed of damage brought on by the devastating flooding. It was to be our first recognition of the fragile and vulnerable nature of that most exceptional of cities on the globe. The Campaign for the Safeguarding of Venice had its origins in the immediate reaction by UNESCO to that dramatic event, when an international campaign was launched by the then Director-General of the Organization, René Maheu. In response to the appeal of Maheu, many private bodies (some fifty-nine) were set up around the world to gather funds and use them in the

Flooding and Environmental Challenges for Venice and its Lagoon: State of Knowledge, ed. C. A. Fletcher and T. Spencer.
Published by Cambridge University Press, © Cambridge University Press 2005.

preservation and restoration of buildings, monuments and works of art in Venice. Even today, almost forty years on, there are well over twenty – in 11 countries – that are still active, and UNESCO, through its Office in the city and the Association of Private Committees, is pleased to play a part in their important work. UNESCO can also be satisfied to have quickly brought the attention of the international community to the need for extending the international campaign beyond the conservation of its monuments and works of art to more complex questions concerning the safeguarding of the entire lagoon system: questions that required close collaboration between the worlds of science and culture. (It is important to recall in this connection that it is 'Venice and its Lagoon' that appears on the World Heritage List.) UNESCO was thus to become one of the main actors in the Comitato Tecnico Scientifico (CTS), the scientific and technical committee for the safeguarding of Venice and its lagoon, which, during the 1980s, developed the idea of a major project to be later given life by the Italian Ministry for Universities and Scientific and Technological Research (MURST) under the title 'Venice Lagoon Ecosystem' Project. This provided a quantitative understanding of the major flows making up the coupling between pelagic and benthic systems within the lagoon waters, and related this understanding to the processes of eutrophication. Later, the UNESCO Office in Venice was to undertake a second initiative, the 'Venice Inner Canals' Project, which sought to develop, calibrate and validate a first water quality model for central Venice. The project included additional sampling and laboratory work on both the waters and sediments' bio-chemical and microbiological characteristics and led to the final calibration of a new hydrodynamic model, as well as to the development of a water quality model to be applied to the Venice inner canal system. It was against this backdrop that the UNESCO Office in Venice – Regional Bureau for Science in Europe (ROSTE) was pleased to be involved in, and support, the Cambridge conference of which this book provides the all-important written record. The organizers, Venice in Peril, the University of Cambridge, Churchill College and CORILA, are to be congratulated on having brought together, in a timely way,

scholars, scientists and those whose professional life is bound up with the city and Venice and its lagoon. The Italian Government has now set in train the MOSES Project for the protection of the City and its lagoon, although as the meeting in Cambridge showed, there is far from unanimity among the experts as to whether or not it will prove to be an effective means by which Venice and its lagoon can be protected against extreme environmental events in the future. Either way, the need for multidisciplinary research on this most complex of systems will be as great as ever, and I am convinced that this volume will prove to have been an important milestone on the road to our understanding it.

Howard Moore
Director
UNESCO Office in Venice
– Regional Bureau for Science in Europe (ROSTE)

THE FRAGILE CITY

'Sospesa sull'abisso, la vita degli abitanti d'Ottavia è meno incerta che in altre città. Sanno che più di tanto la rete non regge.'

('Suspended over the abyss, the life of Octavia's inhabitants is less uncertain than in other cities. They know the net will last only so long.')
(Italo Calvino, *Invisible Cities*)

In Calvino's book, Marco Polo describes 55 mysterious cities, each bearing a woman's name. They seem to expose the beauty and ugliness, the humanity and complexity, that can be found in any long-standing city; but slowly it dawns on us that each place is an aspect of Venice, and the overarching theme is that of fragility. There is a natural affinity between Cambridge and Venice, and Churchill College was more than pleased to support Venice in Peril and CORILA in the programme that has led to this book. One of our main functions as a College is to bring together researchers of different origins, interests and disciplines, and few causes can have a greater need of interdisciplinary understanding and cooperation than the survival of Venice and its lagoon. Our initial discussions with

Venice in Peril, particularly with Anna Somers Cocks, brought in architects, ecologists, engineers and geographers and led to the appointment of our Venice in Peril Research Fellow, Dr Caroline Fletcher (herself an environmental chemist), and an interdisciplinary steering committee chaired by Dr Tom Spencer (a coastal geographer). They built bridges to the Venetian researchers in CORILA, and we came to know Pierpaolo Campostrini and Jane Da Mosto rather well. The huge range of relevant issues was demonstrated in the first workshops held in 2002, which in turn set the agenda for the International Meeting, held in Churchill College in September 2003, which is recorded in this volume. The fundamental objective of the programme is to reinforce support for those working on the survival of Venice and its environment by exposing their work to the international scientific community. Though Venice is unique, its problems have many parallels across the world, and we believe that this book demonstrates both the potential for international engagement, and the benefit that it will yield.

Marco smiled. 'What else do you believe I have been talking about? ... Every time I describe a city I am saying something about Venice.'

Paul Richens, Vice-Master
Sir John Boyd KCMG, Master
Churchill College, Cambridge

EDITORIAL NOTE

All the substantive chapters in this volume have been subject to a full and thorough process of peer review, using at least two referees of international standing in their own particular specialism and with the editorial control of Dr Fletcher and Dr Spencer of Cambridge University. All authors have revised their contributions in the light of comments received. The chapters which appear here have been approved by Dr Fletcher, Dr Spencer and by the lead author of each chapter, on behalf of all authors of that chapter.

The editors wish to acknowledge here the exceptional contribution of Jane Da Mosto. She has played a major role to bring this large and complex work to fruition. Her assistance in liaising between the editors, individual authors and organizations in Venice and elsewhere, and her tireless contributions to the general editing of the volume, have been critical in maintaining the forward momentum of this publishing initiative.

Part I · Introduction

1 • Venice and the Venice Lagoon: creating a forum for international debate

T. SPENCER, J. DA MOSTO, C. A. FLETCHER AND P. CAMPOSTRINI

INTRODUCTION

'There really is no point in continuing to rescue and restore individual buildings in Venice if the city remains under increasing threat from flooding. Now what can be done about that wider issue?' With these words in late 2000, over lunch in the Master's Lodge at Churchill College, Cambridge, Anna Somers Cocks, the Chairman of Venice in Peril (The British Committee for the Preservation of Venice) set in train a broad series of research activities in Cambridge and Venice. One of several substantive outcomes of that process is this volume.

HOW IS VENICE TO BE SAVED?

It is now nearly forty years since the disastrous storm surge of 3–4 November 1966 which flooded many parts of Venice to water depths of nearly 2 m above mean sea level. As Howard Moore notes in the preface to this book, that event was the catalyst for an international gathering of scientists to discuss the 'Venice problem' in 1969. Since that time a great volume of detailed research has been undertaken in Venice, elsewhere in Italy and around the world, on the geology, hydrology, ecology and biogeochemistry of the Venetian built environment and the Venice Lagoon, and the linkages of these habitats to the Adriatic Sea and the hinterland of the Veneto. There have been considerable technical developments, including, for example, the monitoring and mathematical modelling of tidal inlet and lagoon hydrodynamics. New problems have emerged, covering such diverse issues as the anoxic crises and algal blooms of the 1980s, the ecological and morphological impacts of hydraulic dredging for clams since the 1990s, and the recent identification in the lagoon of a range of

chemicals which affect biological systems at very low concentrations (the endocrine disrupters). Furthermore, there has been a growing realization of the spatial and temporal scale of sampling needed to effectively define the complex dynamics of both lagoon and watershed. Most recently, these issues have come to be seen against the backdrop, and uncertainties, provided by what global environmental change might mean for Venice. All these developments, amongst many others, are covered in detail in the chapters of this volume. However, the overarching problem of rising water levels and associated serious city flooding and what to do about it still remains the most apparent key issue facing Venice. And there are many indications that the problem is worsening. Whilst the 1966 event has not been repeated, of the ten highest floods between 1900 and 2004, 8 occurred after 1960. In the first decade of the twentieth century the lowest part of the city, St Mark's Square, flooded ten times per year or less. By the 1980s, flooding was occurring 40 times per year. In the winter of 2002, there were 10 'exceptional' floods (when nearly 4 per cent or more of the city is inundated) in one period of three weeks.

Within this context, this broad review of the 'state of knowledge' comes at a particularly timely point in the long history of flood protection measures for the city. The 1966 event concentrated minds. In 1971 a 'competition of ideas' was held and won by a design for an underwater mobile barrier to close off the three lagoon inlets at times of storm surge. A deciding factor was the importance of aesthetics; the barrier would lie on the seabed and only be raised, and thus visible, at times of high water. In 1984 the Special Law for Venice (see Appendix A2, this volume) created the Consorzio Venezia Nuova (CVN), a consortium of large Italian engineering and construction

Flooding and Environmental Challenges for Venice and its Lagoon: State of Knowledge, ed. C. A. Fletcher and T. Spencer. Published by Cambridge University Press, © Cambridge University Press 2005.

companies, which was given sole responsibility for implementing the barrier solution, within a more general remit from the Ministry for Public Works for the planning, design and execution of all public works for the safeguarding of the lagoon. As part of a package of 'hard' and 'soft' engineering measures (now referred to singly as the 'MOSE system'), qualified approval for implementation of the mobile barrier system, and associated navigation locks, breakwaters and inlet modifications, was finally given in April 2003, at the halfway point of the Venice in Peril initiative that underpins this volume. These decisions have had a reinvigorating effect, both sharpening up some long-debated issues (the whole question of water, sediment and nutrient exchanges between the Adriatic Sea and the lagoon for instance) and generating some quite new questions (might the gates be used to control water circulation and thus water quality in the lagoon?) to which we have yet to apply existing knowledge or devise new, critical tests. Furthermore, one of the key questions concerning barrier impact, that of the frequency and duration of gate closures, and its seasonal variation, is crucially related to near-future changes in sea level and storminess consequent upon greenhouse-gas related environmental change.

WIDENING THE REMIT

The long delay between the original concept, completion of testing of the prototype mobile barrier in 1992 and the final decision to implement the 'MOSE' storm protection system reflects the intensity of scientific, social and political debate over what approach is best for Venice and its lagoon. Apart from the issue of cost (currently budgeted at €4bn or £2.7bn) and whether this expenditure should be a priority, serious and sustained challenges to this scheme have come from the engineering arena, when considering the risks associated with such an innovative barrier system and irreversibility of the structure, especially in the light of adaptations to the possible effects of climate change. Ecological concerns have also been repeatedly highlighted – when the floodgates are closed, the lagoon will become cut off from the benefits of tidal flooding. This has implications for both sediment exchange and water quality, the

former with consequences for the maintenance of lagoonal saltmarsh accretion in the face of sea level rise and the latter raising the spectre of a return to the algal blooms seen in the lagoon in the 1980s.

Over time numerous debates over the merits and demerits of a mobile barrier scheme have become highly polarized, resulting in oversimplification of the key issues governing the 'health' of the lagoon and safety of the city, notably lagoon ecosystem characterization, functional dynamics and evolutionary trends and the relations between the lagoon, the city and the large watershed of the Veneto that feeds into the lagoon. Tackling this level of questions requires a more sympathetic and more holistic view of Venice and its lagoon and a willingness to listen to, and draw information from, a wide range of specialist scientific disciplines into a more thorough synthesis.

CREATING AN INTERNATIONAL FORUM: PROCESS AND PRODUCT

With generous financial backing from Venice in Peril and its supporters in place, a research plan was formulated in 2001 by a group of scientists, engineers, architects and geographers, largely drawn from Cambridge University's Centre for Sustainable Engineering, Centre for Risk in the Built Environment and the Cambridge Coastal Research Unit, and developed in association with CORILA, the Venetian consortium for the coordination of research activities concerning Venice and the lagoon system. Within the broad aim of promoting the objective study and review of information concerning the 'Venice problem', it was agreed to develop a three-stranded collaborative research programme. This consisted of an information gathering process; the promulgation of a series of workshops; and the organization of an international discussion meeting.

By September 2002, the project was in a position to mount the scientific and technical workshops in a number of Cambridge Colleges. The workshops, each attended by 12-15 participants, covered the following five themes: urban flooding – architectural and structural issues; engineering solutions to storm surge flooding of Venice; physical and ecological processes of the Venice Lagoon; modelling the hydrodynamics, morphology and water quality of the Venice Lagoon;

and global environmental change, uncertainty, risk and sea-level rise in the North Adriatic Sea. The aim of the workshops was to identify and outline the key flooding and environmental issues; identify research requirements (in both the mid- and long-term); highlight areas of information need where current understanding of the complex Venice system requires further elaboration; raise matters concerning communication of science and data availability; and, not least, in a relatively informal setting with no formal recording of proceedings, allow the exploration of less mainstream and more innovative solutions to Venice's environmental issues. Participants at each workshop were a mixture of researchers working specifically in Venice and internationally recognised experts on the issues under consideration.

A prime aim of the workshops was to act as a sounding board and pump-priming operation for planning of the international discussion meeting to be held in Cambridge 12 months afterwards. This meeting duly took place in Churchill College, Cambridge in September 2003, under the title 'Flooding and Environmental Challenges for Venice and its Lagoon: State of Knowledge'. The papers given at that meeting, attended by 130 scientists and engineers from Italy, the UK, The Netherlands, Denmark, Germany, Spain, Lithuania, Russia and the USA, subsequently developed and revised, and with additional invited contributions and commentaries, are presented here. The aim of the collection is to show what is currently known about the city and lagoon; where debate remains; and where critical information and tests of key hypotheses are still lacking. The text reports the wide spectrum of scientific archival, field, laboratory and modelling studies on Venice and its lagoon; shows the considerable intellectual challenges involved in trying to understand the workings of a complex, highly managed and altered, and highly stressed set of ecological habitats (in the widest sense, from buildings to plants, both predominantly rooted in salty water); and identifies the need to overcome past shortcomings in research co-ordination and effective dissemination of research results in order to deliver key understandings of city and lagoon dynamics to opinion makers and decision takers.

ACKNOWLEDGEMENTS

This project is a contribution to the Cambridge Environmental Initiative of Cambridge University (formerly the Cambridge Committee for Interdisciplinary Environmental Studies). We are grateful to Peter Guthrie, Paul Richens and Robin Spence for their participation in the Project team, along with Frances Clarke and Anna Somers Cocks, distinguished by their vision and tenacity, and to Sir John Boyd and Janet Milne of Churchill College, who hosted and helped with the organization of the International Meeting. Venice in Peril, the Save Venice Boston Chapter, the UNESCO Office in Venice – Regional Bureau for Science in Europe (ROSTE) and the Regione Veneto provided material support for the Meeting. And above all, we recognize the many scientists who have applied, and continue to apply, their talents and energies to improving our understanding of this most remarkable of ecosystems and home to a unique cultural heritage.

2 • Between salt and fresh waters

G. CANIATO

Venice built her fortunes on the sea: a huge emporium of trade between the Orient and the West. A connecting point between the Middle East, the Mediterranean coasts and continental Europe, that reached the peak of its political and commercial power, between the fifteenth and sixteenth centuries, thanks to the control of maritime routes. But – this is the first question – was Venice (and is Venice) a city on the sea?

In fact Venice is not – as one would normally think – a maritime city provided with docks or waterfronts facing the sea, like Marseille, Genoa, Naples, Istanbul or most of the other major and minor sea ports on the Mediterranean and Atlantic coasts. Venice is in fact an island (or rather a series of islands, a crowded 'archipelago'), well inside a protected lagoon, capital of an independent Republic until 1797: the main island, amongst the dozens scattered all around the lagoon, the heart of a complex productive and residential environment densely populated until a few decades ago.

Today many parts of the lagoon are wastelands full of abandoned settlements formerly used for centuries, and even in the recent past, by fishermen, sailors, boatmen and people living on local agriculture and trade. More or less as in the twelfth century, when the ancient Rialto archipelago became definitively the capital of the Commune Venetiarum, cradle of the new-born Republic which developed as an international power after the Fourth Crusade, but was still located in a narrow marshy area spreading along the Adriatic coasts, compressed between the Adige and Isonzo river mouths.

Even today Venice and its lagoon should be seen as a single entity (as it used to be) and the ancient capital as the centre of a wider productive, commercial and residential system, in which water can be compared to the green belt commonly found in other major metropolitan areas. Not a liquid desert, but an essential part of a 'diffuse city', with the canals acting as boulevards and the lagoons as 'fields', to be frequented like Hyde Park in London or Central Park in New York. But the vast, shallow and placid water surface of the lagoon is today considered more as an obstacle, that can be criss-crossed by fast, and often huge, destructive motorboats racing from land to sea and back. Or maybe bypassed with rapid underground connections, as some local planners propose.

On the other hand, the lagoon is often simply considered as the natural 'frame' and background for the former amphibian 'Queen of the Sea', an ecosystem to be preserved – together with the stones of Venice – for its beauty and uniqueness. An ecosystem that should still be considered (as it was) a life-giving territory for the survival of the diffused city as a living organism and as a capital of European culture populated not only by hurried masses of day-trippers.

Comparisons between the present and the situation of the lagoon centuries ago is probably impossible, but Venetians were not interested in such 'aesthetic' questions concerning the lagoon where they lived: their primary, if not their only, concern, as we shall see, was with strategic security, maritime navigation and sanitary matters. The lagoon was modified over the centuries as the permanent site for a once Mediterranean-wide civilization. Individual needs were not allowed to prevail and always had to be subordinated to the general interest: full private property of water surfaces was not admitted and our forefathers managed to guarantee a precarious but lasting balance between land and sea, adapting and somehow taming an unstable environment, threatened both by freshwater floods and by sea-sand and river-silt deposits.

Flooding and Environmental Challenges for Venice and its Lagoon: State of Knowledge, ed. C. A. Fletcher and T. Spencer.
Published by Cambridge University Press, © Cambridge University Press 2005.

Venetians knew perfectly well – if they wanted to survive as an independent Nation – that they had to preserve a permanent balance between land and water: they understood that a coastal lagoon – left on its own – will naturally be transformed into open sea (if tides and currents prevail) or into land (if sediments brought by rivers prevail). Thus, water management slowly changed the first marshy location into an 'artificial' (or rather, 'anthropic-shaped') environment, permanently suitable for the safe settlement of a human community in constant expansion.

In fact, winding your way through the canals of the lagoon – far from the centre of Venice and beyond the major islands of Burano and Torcello – you will find yourself immersed in a flat landscape that can be fully understood only from the air, with little scattered islands emerging just a few inches above water level. A peculiar environment that – especially at low tide – often becomes a muddy plain: the only protection for the only capital in the world that never had walls. And truly the vast, shallow lagoon surfaces were – as our forefathers knew – the walls of Venice that enemy armies could not scale or destroy. A sort of dogma, recalled by the Renaissance humanist Giovan Battista Cipelli, better known as Egnatius:

Venetorum urbs divina disponente providentia aquis fundata, aquarum ambitu circumsepta, aquis pro muro munitur. Quisquid igitur, quoque modo, detrimentum publicis aquis inferre ausus fuerit, ut hostis patriae iudicetur, nec minori plectatur poena quam qui sanctos muros patriae violasset. Huius edicti ius ratum perpetuumque esto.

Fig. 2.1   Traditional fishing in the shallow waters of the lagoon near Sant'Erasmo sea port (photograph by Catullo and Lagomarsino: Venice, 1990).

Fig. 2.2   Late eighteenth-century engraving of the island of San Giorgio in Alga, along the canal connecting Venice with the Brenta river mouth.

The city of Venice, founded by divine providence on the waters, is surrounded by water and has water for its walls. Therefore, if anyone in any way should damage these waters, he shall be judged an enemy of the state and shall receive no less punishment than if he had violated the holy walls of the homeland. This ruling is given in perpetuity.

A muddy landscape, criss-crossed by a network of major channels and minor tortuous canals ramifying from the sea ports to the river deltas of the mainland. The 'veins and arteries of a living organism' as our forefathers used to say, the lagoon being the natural extension and indispensable support of the main island: Venice, fulcrum of a vast 'diffuse city' whose waterways were carefully maintained and used daily, giving vitality to the minor islands that provided important support for fishing (Fig. 2.1), transport, trade, agriculture and all those activities that used to make the lagoon a lively and populous environment.

The end of the Venetian Republic (and the widespread use of the islands for 'military' purposes during the following twenty or so years of recurrent warfare) marked the end of this ancient and well balanced interconnection between Venice and its liquid surroundings. A powerful 'protective belt' with massive fortifications – of which Venetians had never felt the need – was built well inside the lagoon and on the edge of the nearby mainland; insular convents and monasteries were abandoned or destroyed and the nearby canals gradually started to lose their importance (Figs. 2.2., 2.3). In more recent times people began to reach Venice across the railway bridge (built in 1846), followed in 1933 by a parallel road bridge, with adverse affects on the old waterways that used to serve the city linking it both to the mainland and the sea ports.

Fig. 2.3   Aerial photograph of San Giorgio in Alga. Still visible are the recently collapsed remains of the ancient monastery, used from 1810 until the Second World War as a military base (photograph by Giacomelli: Venice, 1960).

When examining how ancient Venetians acted in order to preserve their lagoon we can perceive apparent discrepancies. Sometimes, when the government finally decided to realize important hydraulic enterprises, such as massive excavations or river diversions, from the moment of the decision to the final execution a very short time elapsed. Just think – for example – of the diversion of the lower branch of the Po, the greatest of Italian rivers. Once it became clear that the easterly flows from the delta contributed to the silting of the southern portion of the lagoon, it was quickly decided to divert it to the south. The Senate took the final decision in the year 1600 and a huge excavation was completed in 1604, in spite of the delicate diplomatic problems that arose, the lower portion of the Po river being a political boundary between the Republic and the still powerful and certainly not always friendly State of the Roman Church.

Or just think of the huge sea-walls of Istrian stone along the shores of the Lido: the work was begun in 1740 and continued throughout the century with an enormous financial effort, in years considered, even by recent historians, to be ones of political and economic decadence. The two opposing parties, active in Venice even in the past, of those who wanted to reclaim land along the borders of the lagoon and those who wanted to leave the saltwater free to expand as far as possible on the mainland, were in perfect agreement instead about the need for adequate protection along the seaward side of the lagoon.

But – this is the next question – have we ever thought why it took centuries for the Venetians to complete not only other important hydraulic works (such as the diversion of the rivers Brenta, Sile and Piave outside the lagoon, directly into the Adriatic Sea) but also quite a simple operation such as the definition of the fiscal and judicial boundaries of the lagoon? The so called *conterminazione* ('boundary line') was planned in the late sixteenth century and finally concluded only in 1791, with different laws, regulations and prescriptions effective inside the line. Maybe one of the ancestral psychological reasons for the delay was that the early borders – facing often hostile rulers on the nearby mainland – were along the uncertain line, varying with the tide, that divided salt lagoons (first home to the Venetians) from the mainland and from the fresh water marshes occupied by the new barbarians after the fall of the Roman Empire. So the lagoons had to be maintained as wide as possible in order to keep the political and military boundaries far from the main inhabited islands. Yet at the same time ensuring that the tide could penetrate deep inside the lagoon, in and out twice a day, to wash sewage away, keep the waterways healthy and the canals and channels deep. Diseases such as malaria flourished in areas of prevailing fresh water – such as the surroundings of Torcello – that had to be abandoned. This long continued effort of the Venetians was encapsulated in the mid sixteenth century in a sonnet attributed to Cristoforo Sabbadino, the chief hydraulic engineer of the time:

> Quanto fur grandi le tue mura il sai,
> Venetia, hor come le s'attrova vedi,
> e s'al periglio lor tu non provedi
> deserta e senza mura rimarai.
>
> I fiumi e 'l mar e gli huomini tu hai
> per inimici, e 'l provi e non lo credi,
> non tardar, apri gli occhi e muovi i piedi,
> che volendolo poi far tu nol potrai.
>
> Scaccia i fiumi da te, le voglie ingorde
> de gl'huomini raffrena e poscia il mare,
> restato sol, sempre t'harà obedita.
>
> (…)

How great your walls were, you know Venice, now see how you find them.

and, if you don't provide against their ruin, deserted and wall-less you will remain.

The rivers, the sea, and men you have as enemies, and for proof and belief don't delay, open your eyes and get moving, (otherwise) what you wish to do, you will find impossible.

Drive the rivers away from you, and check the greedy desires of men and then, only the sea remaining, it has always obeyed you.

(…)

But why – this is the next question – were many other projects concerning the safeguarding of the lagoon proposed, discussed, analysed, sometimes experimentally started, but finally abandoned or never realized? Projects, that became numerous especially in the sixteenth and seventeenth centuries, such as those presented in 1672 to the Senate and to the Magistrato alle Acque by Agostino Martinelli (*Prodromus ad revocandam lacunam inclytae civitatis Venetiarum ad primam profonditatem*), who proposed building a stone wall across the sea-mouths of the lagoon, provided with huge sluice gates (…*per le quali a compiacenza si potessero aprire e serare, hora per introdur il mare et hora per negarli l'ingresso*) (which could be opened and closed at will, to let the sea in or to keep it out) and in 1673 by Alfonso Moscatelli who proposed closing two of the major port mouths (Chioggia and Sant'Erasmo) and separating with embankments the central lagoon from the marshes facing the mainland, in order to strengthen the vital natural action of the tides.

So our ancient patrician rulers and water management technicians planned, studied, proposed and discussed dozens and dozens of projects, often made mistakes, tried to correct them by learning from nature's reactions, or just changed their minds over time. Planning artificial diversions on the lower course of the Brenta river – as an example – went on for centuries, starting long before Venice achieved direct control of the nearby mainland. Padua ruled the whole area through which the river ran until the beginning of the fifteenth century but even in the

Fig. 2.4   A 1585 text painted on the inside wall of an anciemt warehouse on the Lazzaretto Novo island. Here wool and other goods were kept in quarantine (photograph by G. Caniato, 1978).

twelfth century rivers had been diverted, mainly for strategic reasons and not far from the boundaries of Venice. Here swamps and marshes, scattered amongst fields and agricultural land, were often compromised by river flooding. The river delta was important not only for trade between the sea, the lagoon and the densely populated plains of northern Italy, but also for fishing, hunting, brush-wood gathering, heating and mill-grinding. All these activities were needed for the survival of the thousands that could make a living from the lagoon and from the nearby mainland and marshes.

Health problems caused by natural or anthropic changes in the local environment forced the Venetians to act. In 1324, they decided to build a long earth bank in front of the Brenta lagoon delta, in order to divert the river to the south-west, as far as possible from the site of the capital. This enterprise soon caused serious environmental problems, both to the nearby mainland, where fresh water lakes and marshes expanded and forced peasants to abandon their lands and houses, and to the main sea-channels, which shallowed with the loss of the previously adequate and continuous river flow that had kept them naturally deep. At the beginning of the following century the 'filling up' of the lagoon (today we know it was probably caused by a general lowering of the average sea level, but our forefathers did not have such scientific data) and the reduced depth of the main canals and harbours – so vital for maritime navigation – forced Venetians to plan possible solutions.

But in order to decide first (and realize later) the most appropriate technical solutions, documents show us that Venetians were not so sure nor self-confident and relied on special enquiry commissions of patricians, supported by boards of experts. Not only hydraulic engineers and technicians but also fishermen, boatmen and other 'common people' who spent all their life in the lagoon were called: the *Savii ed esecutori alle acque* (water management authority) had by law to invite eight old (over the age of 35!) and expert fishermen to take part – once a month at least – in the general conference on the lagoon in the Ducal Palace, asking their opinions and involving them in the technical discussions (a law apparently no longer compulsory for the Venetian water management of our time). In 1415, for example, the Senate established a commission in order to verify the silting found along the borders of the lagoon (near the present day industrial site of Marghera) and in the nearby channels connecting with the sea-ports. They proposed to open some passages through the barrier built one century before, in order to encourage the natural deepening of the major canals with the flow of river water. In 1391 a special commission of patricians was sent by order of the Senate to decide how best to strengthen an embankment (*fortificatione ageris existenti*) with the advice of local common people (*proborum hominum*) verifying *cursus et motus aquarum* the water's behaviour under different tide conditions.

These practical solutions, when adopted, were supposed to be provisional and had to be verified in practice, on site (*sul campo*), showing us the doubts and hesitations of the Government in such delicate matters. Thus, for example, another special commission of *sapientes* was appointed to verify regularly, for one whole year, the variations in depth of the nearby canals and of sea-channels connecting lagoon and sea, both during low and high tide conditions. We could define it as an empirical or experimental method: verifying how nature reacted to human intervention, which had to be reversible, and had to be modified or abandoned if it was found ineffective or damaging. In practice, in the case of the Brenta delta, openings through earth barriers and embankments separating fresh from salt waters were repeatedly closed and reopened throughout the fifteenth century, then finally closed when unknown diseases

were discovered and supposed to be connected with the river flow. So the urgency to protect public health prevailed even over the needs of trade and navigation, and general dredgings were planned as an alternative, in order to assure an adequate depth for boats and ships, while major diversions of the river outside the boundaries of the lagoon started to be seriously planned and were realized first with the Brenta Nova channel, created in the first half of the fifteenth century, and finally in 1610 with the excavation of the Brenta Novissima leading directly into the Adriatic, south of Chioggia.

An 'empirical method', or 'experimental method', had been applied in this and other cases since the Middle Ages, and continued well after the fall of the Republic of Venice in 1797. In fact, in the 1830s, following heavy river floods that caused disasters in the countryside around Padua, the course of the Brenta was brought back once more to the south-west part of the lagoon causing, in a short time, serious environmental and health problems, as official reports in 1847 remark: rapid silting up of the fish-farms that provided a livelihood for the main part of the densely populated town of Chioggia, malaria caused by the prevalence of fresh over salt water, expansion of marshes and problems for the merchant and military navy, due to the silting up of the major sea channel. Exactly the same problems had been experienced three or four centuries before but with one little difference: the lagoon was no longer the cradle and protection of the capital of an independent republic and so problems and urgencies concerning the 'people of the lagoons' – transformed from rulers to subjects – could no longer prevail over those of the surrounding towns and territory of the mainland.

As a concluding question what should be the role of an archivist, who has the duty (and also the privilege) of analysing, preserving and transmitting to future generations the documents produced and deposited over the centuries? Certainly not to propose ancient drawings, projects and manuscripts as a scientific base for today's planning: knowledge, the environment, the social and economic life of the past are not comparable to the present. But studying documents at first hand can help us to understand the mentality and the 'sense of State' of the Venetians. This can be seen behind plans, actions and even the mistakes of our forefathers involved in water management: they felt the duty to serve the public interest, maybe because the private and public interests of the patrician ruling class coincided. The preservation of the lagoon environment, keeping it as wide as possible (*fin dove pol ascender le aque salse*) was essential for the preservation of an independent Republic. Without it Venice, long before the advent of Napoleon, would have been lost.

ACKNOWLEDGEMENTS

We are grateful to T. Spencer and M. A. Kershaw for their help in editing this paper; D. J. H. Murphy is thanked for his assistance with translations.

Part II · Geological and environmental context

3 • Introduction: geological and environmental context

T. SPENCER, J. DA MOSTO AND C. A. FLETCHER

The northern Adriatic basin is backed by the Alps to the north, the Apennines to the west and south and the Dinaric Alps to the east. The basin incorporates the remains of the Adriatic tectonic plate which is being subducted beneath the Apennines; the result is a long-term, regional subsidence rate in the Venice area of 1.0 mm a^{-1} (Carminati *et al.*, chapter 4). Mesozoic basement limestones in the basin are overlain by alternating marls and limestones of Eocene – Oligocene age and Pliocene clays. Above the Plio – Pleistocene boundary are *c.* 800 m of clayey sands, sandy silts and clays and silts, characteristic of shallow sea and deltaic environments. The upper 300 m of these sediments consist of continental floodplain and marsh deposits and localized aeolian deposits, indicating cycles of sea level regression and transgression, associated with low glacial and high interglacial sea level stands respectively. During the last glaciation, the Venice Lagoon was characterized by fluvial and lacustrine environments, followed by extensive alluviation at the beginning of the post-glacial period. These clays were consolidated in the period between *c.* 10 000 and 6000 years ago and now form the patchily distributed '*caranto*' beneath the city and the lagoon. The presence/absence of this deposit may help explain the variations in the degree of building subsidence seen across the city (Frassetto, chapter 5). The modern lagoon began to be formed *c.* 6000 years ago when rising sea levels flooded the easternmost part of the Po Plain, creating a coastline characterized by small cuspate deltas and the estuaries of the rivers Po, Adige, Brenta and Piave.

The base of the lagoon sediments lie at *c.* 5 m below modern mean sea level.[1] The earliest archaeological sites, at Torcello and San Francesco del deserto, lie at *c.* 2 m below modern sea level (and thus 3 m or more below presently occupied surfaces) (Ammerman,

chapter 13; Buogo *et al.*, chapter 14). Rates of sea-level rise were therefore *c.* 7 mm a^{-1} in the period between 4000 BC and AD 400, accelerating to *c.* 13 mm a^{-1} between AD 400 and 1897 (Ammerman, chapter 13). Comparison of rates of sea-level rise derived from archaeological evidence, extrapolated from the paintings of Canaletto and Belloto for the period 1727–58 and derived from the study of early photographs from the 1880s and 1890s (Camuffo *et al.*, chapter 16) and from tide gauge records after the 1930s (Ferla, chapter 12) are remarkably consistent, in the range 13–19 mm a^{-1}. The rate of sea-level rise for the last one hundred years has been *c.* 23 cm but some 12 cm of this rise can be attributed to the pumping of groundwater in the industrial complex of Port Marghera between 1930 and 1970 (Butterfield, chapter 15).

Venice's vulnerability to flood events comes from its position at the closed northern end of the Adriatic Sea and the interaction between this location and a large low pressure system above Northern Europe. Some 60% of surges relate to the maintenance of the main pressure minimum north of the Alps and such disturbances tend to lead to higher surge levels; the remaining *c.* 40% of surges are related to more

---

[1] Water level measurements in the city were formalized in 1440 when the Venice Water Authority (Magistrato alle Acque) began to record the upper level of wet algae growing on buildings facing canals with an engraved letter 'C' (the 'Comune Marino'), a measurement of the normal high tide level (Frassetto, chapter 5). Regular tidal measurements began in the 1870s, with the Santo Stefano tide-gauge, and the current reference point was established in 1897 with the Punta della Salute gauge, at the entrance to the Grand Canal. In Venice all water level measurements are related to the mean sea level established in 1897 and this convention is followed in this volume. It may be noted that, as a result of sea-level rise and land subsidence, modern (2004) sea level is now *c.* 25 cm higher than the 1897 reference level.

Flooding and Environmental Challenges for Venice and its Lagoon: State of Knowledge, ed. C. A. Fletcher and T. Spencer.
Published by Cambridge University Press, © Cambridge University Press 2005.

southerly tracks and intense cyclonic activity over Italy at the time of highest surge level. Although this short-term dynamic is well known, the mechanisms responsible for longer-term variations in surge frequency and intensity are poorly understood (records since the eighth century AD show intense flooding in the first half of the sixteenth and eighteenth centuries, more quiet periods, and a continuous positive trend during the second half of the twentieth century). It appears that periods with extreme surge events are characterized by a general circulation anomaly, represented by a pattern with a negative centre of action above the North Sea, according to which cyclones deviate southeastwards and penetrate into the Mediterranean Sea from the northwest. It should be noted, however, that this pattern is distinctively different from the North Atlantic Oscillation (NAO) index which is poorly correlated with the highest surges in Venice (Lionello, chapter 8). There is, however, a strong positive correlation between solar activity and the frequency of surges although quite what the nature of a connection between sunspots and some large-scale meteorological parameter might be is not known at present (Tomasin, chapter 9). Finally, additional, but equally poorly understood, complexities come from the connections between atmospheric forcing variability and shelf sea and ocean basin dynamics at seasonal, inter-annual and inter-decadal time scales. Thus the Otranto Strait provides the heat restoring mechanism for the large heat losses that take place in the northern Adriatic Sea in winter. Lateral exchange within the Adriatic is maintained by the balance between the Eastern Adriatic Current (EAC), flowing northward along the eastern side of the Adriatic Sea and partially determined by exchanges in the Otranto Strait, and the Western Adriatic Coastal Current (WACC), flowing southward and connected to the Po runoff and the wind forcing. These shelf sea exchanges are related to ocean – atmosphere processes within the wider Mediterranean Basin – which in turn have been found to be linked to both tropical (Indian monsoon and Sahel) and mid-latitude (NAO index) atmospheric variability (Pinardi *et al.*, chapter 6).

The detailed character of individual storm events depend upon the relations between disturbance characteristics (windspeed, duration and direction) and the astronomical tide (high or low, spring or neaps). Normal tides range between −50 and +80 cm around the 1897 'zero' reference level at Punta della Salute. A 'sustained' tide is one that exceeds +80 cm, an 'exceptional' event occurs when the water level reaches +110 cm or more and an 'extreme' event is one which is over +140 cm. The wind of greatest concern is the southeasterly *sirocco*, with a fetch over the entire length of the Adriatic Sea. These effects are exacerbated by the bathymetry of the Adriatic Sea, deeper in the south and shallower in the north, and the basin's shape – elongated, semi-enclosed and surrounded by mountain chains – funnelling wind and water towards the Gulf of Venice. Such events can produce water level differences inside the lagoon of more than 60 cm (Ferla, chapter 12; and see Umgiesser *et al.*, chapter 47). The northeasterly *bora* is frequently stronger but the fetch length is restricted and water can 'escape sideways' through the inlets to the Venice Lagoon. Nevertheless, *bora* winds can lead to water level set-up of *c.* 20 cm or more between the northern and southern parts of the lagoon (Ferla, chapter 12; and see also Gačić, chapter 49). In addition to these forcings, the Adriatic is close to a resonant condition which can result in free oscillations, or seiches, which give continued periodic high sea level events for several days subsequent to a surge peak, as occurred in November 1966; sometimes subsequent peaks may be higher than the initial surge height (Tomasin, chapter 9).

Prediction of water levels in Venice involves both correlation (black box) techniques, using linear autoregressive models and relying on historical databases, and various analytical and numerical methods which presume a knowledge of the underlying physical processes and the equations that describe them. Of the statistical models, the '*Semplificato*' model, which uses predictors of sea level in Venice and pressure in Venice, Genoa, Alghero and Bari has been found to be the most accurate model within the first ten hours. For time lags beyond 10 hours, the recursive '*Esteso*' model, which employs observed values of sea level in Venice, pressure from a number of Italian and Croatian stations and forecast pressures from the European Centre for Medium range Weather Forecasting (ECMWF), is the most accurate statistical model (Canestrelli and Zampato, chapter 11). The application of deterministic models raises questions as to the integration algorithm and the grid of depth data to be used, the area to be covered by the numerical scheme and the assimilation of data for model spin-up. Canestrelli and Zampato (chapter 11) describe the application of both the just-operational ISMAR-CNR Shallow water HYdrodynamic Finite Element Model (SHYFEM; and see also Umgiesser *et al.*, chapter 47) and the not-yet-operational University of Padua finite difference HYdrostatic Padua Sea Elevation model (HYPSE) to meteorological and sea level forecasting. The overall conclusion is that hydrodynamic and statistical models appear particularly efficient if used together; hydrodynamic models appear more sensitive at large forecast lags, of the order of two or three days, whereas statistical models are much more accurate at shorter forecast lags, of the order of 24 hours or less. For all kinds of models, however, a major difficulty remains the connection between sea and lagoon. Models that rely upon historical data are compromised by inlet modifications that post-date at least part of the dataset. The deterministic models typically simulate meteorological and water level conditions at the CNR oceanographic platform 'Acqua Alta' but this structure lies 15 km offshore. This shortcoming is recognised in current efforts to extend the computational grid of the finite element deterministic model to include the inner Venice Lagoon. Additional improvements are concerned with the introduction of data assimilation operational procedures into both types of deterministic model, to integrate observed sea level data into model simulations, and in the use of satellite wind fields – such as those measured by the QuikSCAT scatterometer on board the NASA satellite 'Seawind' (Zecchetto *et al.*, chapter 7) – alongside ECMWF wind fields to improve the quality of the meteorological forcing for the deterministic models. New statistical models, formulation of a neural network model and development of an ensemble prediction model, incorporating the results of the range of numerical models, are all in prospect.

For Frassetto (chapter 5), 'while in the 1960s the primary scientific concerns for the safeguarding of Venice were the local weather forecast of storm surges and the rate of subsidence, since the 1980s the major scientific concerns have become those of global climatic warming, with the risk of sea-level rise (SLR) and the occurrence of more frequent severe meteorological events'. Forecasts estimate an increase in temperature of 0.6–2.5 °C over the next 50 years and of 1.4–5.8 °C for the next 100 years, with significant local peaks (Barbi *et al.*, chapter 10). The Third Assessment Report of the Intergovernmental Panel on Climate Change estimated that global sea level rise to the year 2100 will be between 8 and 88 cm (Ferrara *et al.*, Appendix A1). However, it is not clear how this figure can be translated into a near-future regional sea-level rise for the Gulf of Venice. Mediterranean sea levels have fallen by as much as 20 mm relative to the Atlantic since 1960, probably as a result of declining freshwater input and consequent seawater density increase but also due to a strong negative correlation between sea level and the NAO index. Furthermore, syntheses of recent satellite altimetry time series for the Mediterranean Basin show a complex picture, with overall sea-level rise of *c.* 2.2 mm a^{-1}, masking more dramatic rises (up to 9.3 mm a^{-1}) in the eastern Mediterranean close to areas of sea level fall (reaching 11.9 mm a^{-1}) in the Ionian Sea.

No direct link has been found between the two warming periods which took place at a hemispheric scale from 1910 to 1945 and since 1976 and observed variations in extreme surge values. Rather, if the effect of regional sea-level rise is subtracted from the data, the record of extremes is dominated by large inter-decadal variability (Lionello, chapter 8). Climate modelling using two 30-year long time slice experiments to simulate present and 'doubled CO_2' scenarios, has produced very similar cyclonic regimes, with if anything a slightly higher number of cyclones under the present climate at the spatial scale of the Mediterranean Basin. However, some caution is necessary as the meso-scale features involved in the production of surges at Venice cannot be resolved in currently available global climate models. A downscaling procedure is needed and this introduces an underestimation of observed surge levels in Venice (Lionello, chapter 8). Clearly this area of research is one where predictions will be eagerly sought and evaluated over the next decade.

4 • Magnitude and causes of long-term subsidence of the Po Plain and Venetian region

E. CARMINATI, C. DOGLIONI AND D. SCROCCA

SUMMARY

In the Po Plain (Northern Italy), the natural component of subsidence can be split into: (1) a long-term component controlled by tectonics and geodynamics, active on time periods of about 10^6 yr; (2) a short-term component, controlled by climatic changes (glacial cycles), acting on periods of 10^3–10^4 yr, plus sediment compaction and loading of deltas. The magnitude of the long-term component of subsidence in the Venetian coastal area has been estimated using stratigraphic data from industrial wells. Thickness and absolute age of the Pleistocene base allow the calculation of long term subsidence rates of about 1.0 mm yr^{-1}. Long term subsidence in the Venice area and in the Po Plain is triggered by the flexure of the Adriatic plate subducting under the Apennines, as suggested by the interpretation of the CROP M-18 seismic line. Analysis of ^{14}C ages of recent sediments provides a constraint on total natural subsidence rates in the area of Venice, with average values of around 1.3 mm yr^{-1}. The deglaciation component is therefore limited to some 0.3 mm yr^{-1}. It is concluded that the most significant part of the natural subsidence of the Venetian coastal area is related to the north-eastward retreat of the Adriatic subduction, a process inducing subsidence in the whole Po Plain basin and part of the Alps.

INTRODUCTION

Recently, the study of present-day subsidence in the Po Plain (northern Italy; Fig. 4.1), a flat and fast subsiding area bounded to the north and to the south by mountain belts (the Alps and the Apennines; Fig. 4.1), has become progressively important due to the necessity to protect from sea level variations both his-
torical towns (such as Venice and Ravenna) and the fragile coastal ecosystems of the Po Delta (Carbognin and Marabini, 1989; Carbognin *et al.*, 2000; Pirazzoli, 2002; Bras *et al.*, 2002). Several datasets have been used to investigate subsidence in the Po Plain: stratigraphic data from deep boreholes (Pirazzoli, 1987; Doglioni, 1993; Gambolati, 1998; Carminati and Di Donato, 1999; Carminati *et al.*, 2003a; Massari *et al.*, 2004); seismic lines (Mariotti and Doglioni, 2000; Carminati *et al.*, 2003a); ^{14}C dates of recent sediments (Bonatti, 1968; Fontes and Bortolami, 1973; Bortolami *et al.*, 1977; Bortolami *et al.*, 1985; Serandrei Barbero *et al.*, 2001; Carminati *et al.* 2003b); geodetic surveys (Caputo *et al.*, 1970; Arca and Beretta, 1985; Bergamasco *et al.*, 1993; Carminati and Martinelli 2002); tide gauge measurement time series (Emery *et al.*, 1988; Carbognin and Taroni, 1996; Zerbini *et al.*, 1996; Carbognin and Tosi, 2002; Carbognin *et al.*, 2004a); archaeological and historical data (Flemming, 1992; Pirazzoli, 1996; Ammerman *et al.*, 1999; Camuffo and Sturaro, 2003), SAR Interferometry (Tosi *et al.*, 2002); and space geodesy (namely SRL, VLBI and GPS; Zerbini *et al.*, 1996).

The natural component of subsidence can be split into: (1) a long-term component controlled by tectonics and geodynamics, active on time periods of about 10^6 yr; and (2) a short term component, most likely controlled by climatic changes (glacial cycles), acting on periods of 10^3–10^4 yr. In this chapter we summarize knowledge on the magnitude and causes of long term natural subsidence, as this can be inferred from boreholes and seismic lines, with special reference to the area of Venice. The component related to deglaciation, inferred from ^{14}C ages on recent peat samples, is also briefly addressed.

Flooding and Environmental Challenges for Venice and its Lagoon: State of Knowledge, ed. C. A. Fletcher and T. Spencer.
Published by Cambridge University Press, © Cambridge University Press 2005.

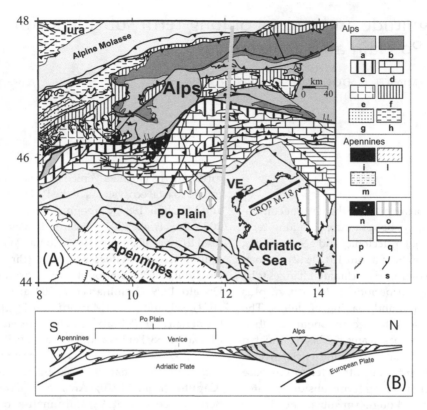

Fig. 4.1   (A) Geological sketch of northern Italy. Alps: (a) and (b) Australpine basement and cover units; (c) and (d) Southalpine basement and cover units; (e) and (f) Penninic basement and cover units; (g) and (h) Helvetic basement and cover units. Apennines: (i and l) Apenninic cover units; (m) post orogenic sediments; (n) Tertiary and Quaternary volcanic and plutonic bodies; (o) foreland units; (p) foreland basin units; (q) Dinaric units; (r) normal faults; (s) thrust faults. I.L.: Insubric lineament. VE: Venice. Redrawn and simplified after Bigi *et al.* (1990). The trace of CROP-M18 seismic line is also shown. (B) Schematic regional cross section through the Alps – Po Plain – Apennines system. The trace of the section is indicated by the grey line in panel (A).

GEODYNAMIC BACKGROUND

The Po Plain is the foreland basin of both the Alps and the Apennines (Fig. 4.1; Doglioni, 1993). The subsidence in this area is likely to be heavily influenced by the tectonic and geodynamic processes responsible for the formation of the two mountain belts. The geodynamics of the Alps – Po Plain – northern Apennines system are the result of the inversion of Mesozoic passive continental margins and intervening oceanic Tethys branches into subduction since at least 90 Myr.

During the Cretaceous-Paleocene, the consumption of the Tethys occurred by SE- to E-directed subduction of Tethyan oceanic lithosphere under the Adriatic continental lithosphere (e.g. Polino *et al.*, 1990; Schmid *et al.*, 1996). Some 50 Myr ago, collision took place, in the Alps, between the European and Adriatic continental crust (Schmid *et al.*, 1996). The subduction of European continental lithosphere under the Adriatic continental lithosphere has probably continued until present. In the Late Oligocene (c. 30-23 Myr ago), the Apennines W-directed subduction of the Adriatic plate started along the retrobelt of the Alps in the western Mediterranean region (Gueguen *et al.*, 1998). This process led to the development of the Apennines belt. The Apennines subduction hinge retreated rather continuously towards the east and north-east (Gueguen *et al.*, 1998; Carminati *et al.*, 1998). The roll-back of the Apennines subduction consumed the Adriatic lithosphere and resulted in a decrease in the distance between the Alpine and the Apenninic subduction. As shown in Fig. 4.1B, at present the Po Plain rests on the remains of the Adriatic plate and the distance between the two subduction zones is restricted to some 300 km, or less moving westward.

MAGNITUDE OF LONG-TERM SUBSIDENCE RATES

Tectonic and sedimentological processes control the long-term vertical velocities. These processes are active on typical time-scales of millions to hundreds thousands of years. Carminati and Di Donato (1999) estimated the long-term subsidence rates for the Po Plain (Fig. 4.2) using a backstripping procedure (Sclater and Christie, 1980), from the Quaternary stratigraphy of approximately 200 deep wells. Subsidence rates quoted below are referred to sea-level (relative land subsidence). The Pliocene-Quaternary boundary, marked in these wells by the first occurrence of *Hyalinea balthica*, has an absolute age of 1.43 Myr (e.g. Bortolami *et al.*, 1985). Averaging

subsidence rates over the Quaternary (i.e. a 1.43 Myr period) means that the subsidence components active on shorter time periods (e.g. related to deglaciation or to human activities) are filtered out. The main assumption adopted is that Quaternary sedimentation rates are equal to subsidence rates. The assumption is justified since the entire sequence was deposited under shallow marine to continental conditions (e.g., Massari *et al.*, 2004).

Figure. 4.2 shows in the north-eastern part of the Po Plain a southward slow increase of subsidence rates, followed in the Po Delta area by a rapid increase. South of the Po Delta, towards the Apennines front (buried under the Po Plain sediments) and the southern border of the Po Plain, subsidence rates decrease (see also Gambolati and Teatini, 1998).

In the north-western upper Adriatic coastal zone, deep industrial boreholes have encountered the base of the Pleistocene at depths of 960-1500 m. Decompacted thickness and absolute age of the Pleistocene base permit the calculation of Quaternary subsidence rates (Fig. 4.3). In the Lagoon and city of Venice, long-term subsidence rates are about 1 mm yr^{-1} (see also Bortolami *et al.*, 1985; Kent *et al.*, 2002; Brambati *et al.*, 2003;

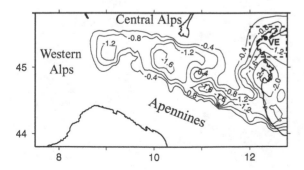

Fig. 4.2   Long-term subsidence rates (mm yr^{-1}) obtained from borehole stratigraphic data (modified after Carminati and Di Donato, 1999). The dashed box shows the area of Fig. 4.3.

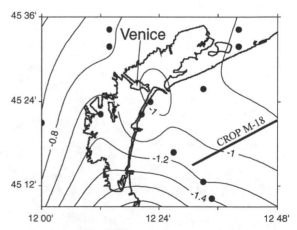

Fig. 4.3   Long-term subsidence rates (mm yr^{-1}) obtained for the area around Venice from industrial borehole stratigraphic data. The dots show the locations of considered wells. The position of the south-easternmost part of the CROP M-18 seismic line is also shown.

Carbognin *et al.*, 2004b). A paleobathymetry uncertainty of 100 m (from shallow marine to continental environments) gives an error of ±0.07 mm yr^{-1} in the subsidence rate estimates. Comparable subsidence rates occur along the foredeeps associated with the west-directed subduction zones (Doglioni, 1993).

CAUSES OF LONG-TERM SUBSIDENCE

The geometry of Quaternary sediments just offshore Venice is clearly shown on the CROP M-18 seismic reflection profile (Fig. 4.4), acquired by the deep seismic soundings Italian CROP project (Scrocca *et al.*, 2003). Figure 4.4 shows the Pliocene-Pleistocene boundary, based on deep industrial wells, and the trend of the regional monocline (Carminati *et al.*, 2003a). The Pleistocene sediments are almost horizontal, and their thickness decreases towards the northeast. The Pleistocene layers show progressive north-eastward migrating onlap and pinchout on the underlying sediments. The lower Pleistocene reflectors are tilted to the southwest (e.g., the reflector labelled 1 in Fig. 4.4) and are parallel to the Pliocene-Pleistocene boundary. The Upper Pleistocene reflectors are less inclined (e.g. the reflectors labelled 2 and 3 in Fig. 4.4). It is concluded that the foreland regional monocline is a growing Pleistocene structure. The

Pleistocene-Pliocene boundary shows undulations. Moving toward the Friuli plain or the north-easternmost Adriatic Sea, the upper Pleistocene sediments gradually thin and lie directly on Miocene sediments (Merlini *et al.*, 2002). This is related to a decrease of subsidence rates, moving north-eastward, i.e. far from the Apennines.

Figure 4.4 clearly suggests that the spacing for the Pleistocene sediments is accommodated by the downflexure towards the south of the Adriatic plate (Doglioni, 1993). Figure 4.5 shows the depth-migration (using 2000 m s^{-1} for Pleistocene and 2600 m s^{-1} for Pliocene sediments) of the reflectors shown in Figure 4.4. The dip of the regional monocline measured from Figure 4.5 is about 1.5°. The CROP M-18 section is not perfectly perpendicular to the Apennines front. The measured apparent dip corresponds to a real southwestward dip of 1.8°. In seismic reflection profiles of the Po Basin, the dip of the regional monocline gradually decreases from even more than 22° to close to 0° moving from the Apennines to the Alps. This feature has been interpreted as the effect of the flexure of the Adriatic plate subducting under the Apennines and a cause for the Venice subsidence (Doglioni, 1993; Mariotti and Doglioni, 2000; Doglioni and Carminati, 2002). Therefore, the curvature of the Apennines subduction reaches the area of Venice (and probably part

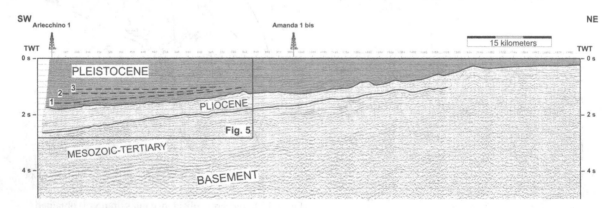

Fig. 4.4   Interpretation of the shallowmost portion of the CROP M-18 seismic line (TWT). The dashed lines labelled 1-3 mark three reflectors within the Pleistocene sediments. Notice that, although all the reflectors dip towards the SW, the dip of the reflectors decreases from the older (labelled 1) to the younger (labelled 3). For this reason, the foreland regional monocline is interpreted as a growing Pleistocene structure.

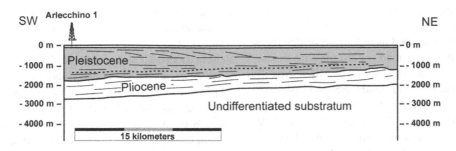

Fig. 4.5   Depth migration of CROP M-18 seismic line.

of the Alps) and triggers subsidence and sediment deposition. Sediment compaction and sedimentary loading further contribute to the subsidence. The subsidence rate estimates from Figs. 4.2 and 4.3 include the effects of these two processes.

SUBSIDENCE RELATED TO DEGLACIATION

The present-day natural subsidence of the Po Plain is still affected by the effects of deglaciation (Mitrovica and Davis, 1995; Di Donato *et al.*, 1999). Deglaciation-related sea-level rise started in the Adriatic Sea some 18 000 yr BP (Amorosi *et al.*, 1999). In coastal areas of the Po Plain, relative post-deglaciation subsidence rates can be inferred from ^{14}C ages of peats (Fontes and Bortolami, 1973; Bortolami *et al.*, 1977, 1985; Brunetti *et al.*, 1998; Serandrei Barbero *et al.*, 2001; Carminati *et al.*, 2003b). In the easternmost Po Plain, peat deposits formed in brackish environments, ranging in elevation between −3 and +3 m. Therefore, peat-rich layers are useful first order approximations of ancient sea-levels. For this reason, subsidence rates obtained considering depth and age of peat layers can be approximated to subsidence rates. Fontes and Bortolami (1973) and Bortolami *et al.* (1985) calculated for the area of Venice post-deglaciation subsidence rates of 1.3 mm yr[-1]. This value is the sum of the contribution of long-term tectonic processes with that of deglaciation. It has been shown that long-term subsidence rate is ca. 1 mm yr[-1]. This implies that the deglaciation component of subsidence rate in Venice is about 0.3 mm yr[-1].

DISCUSSION AND CONCLUSIONS

The natural subsidence rate for the area of Venice inferred from stratigraphic data is of the order of 1.3 mm yr[-1]. This value is remarkably consistent with the estimates obtained with completely different data sets. Ammerman *et al.* (1999) reconstructed a sea level curve using archaeological data and inferred, approximately for the last 2000 years, average natural subsidence rates of about 1.3 mm yr[-1]. A similar estimate is obtained from historical data (Camuffo and Sturaro, 2003; Camuffo *et al.*, this volume). Comparing maximum sea levels portrayed in Canaletto's paintings with present day levels, these authors obtain subsidence rates of 1.9 mm yr[-1]. Finally, these estimates are compatible with the rates of relative sea-level rise inferred from historical tide gauge measurements (Camuffo and Sturaro, 2003; Carbognin *et al.*, 2004a).

From the described stratigraphic data, it is concluded that the largest part (c. 1 mm yr[-1]) of natural subsidence rates for Venice are induced by long term tectonic processes. Long term subsidence is triggered by the flexure of the Adriatic plate subducting under the Apennines, as suggested by the interpretation of the CROP M-18 seismic line. The north-eastward retreat of the Adriatic subduction determines subsidence in the entire Po Plain and its effects reach Venice, located some 100 km north of the Apennines. The use of ^{14}C ages of peat samples constrains the deglaciation component of subsidence to ca. 0.3 mm yr[-1].

ACKNOWLEDGEMENTS

R. Frassetto kindly invited us to the Cambridge meeting. D. Camuffo is thanked for stimulating discussions. Research has been supported by MIUR Progetto Giovani Ricercatori (E. Carminati), ASI and Cofin 2001 (C. Doglioni). The CROP Project Committee is acknowledged for permission to publish the CROP M-18 seismic line. R. Fantoni is kindly thanked for providing well stratigraphies. Figs. 4.2 and 4.3 were plotted using GMT software (Wessel and Smith, 1995). Two anonymous reviewers and the editors are acknowledged for their suggestions.

REFERENCES

Ammerman, A. J., McClennen, C. E., De Min, M., and R. Housley. 1999. Sea-level change and the archaeology of early Venice. *Antiquity* **73**, 303–12.

Amorosi, A., Colalongo, M. L., Pasini, G., and D. Preti. 1999. Sedimentary response to Late Quaternary sea-level changes in the Romagna coastal plain (northern Italy). *Sedimentology* **46**, 99–121.

Arca, S., and G. P. Beretta. 1985, Prima sintesi geodetico-geologica sui movimenti verticali del suolo nell'Italia Settentrionale (1897–1957). *Boll. Geod. Sc. Aff.* **44**, 125–56.

Bergamasco, A., Teatini, P., and L. Carbognin. 1993. Confronto critico tra kriging e analisi oggettiva. *Il Nuovo Cimento* **16**, 289–302.

Bigi, G., Cosentino, D., Parotto, M., Sartori, R., and P. Scandone. 1990. Structural model of Italy. Scale 1:500.000. In *Progetto Finalizzato Geodinamica*, Florence: C.N.R.

Bonatti, E. 1968 Late-Pleistocene and postglacial stratigraphy of a sediment core from the Lagoon of Venice (Italy). *Mem. Biogeogr. Adriat.* **7**, 1–17.

Bortolami, G. C., Fontes, J. C., Markgraf, V., and J. F. Saliege. 1977. Land, sea and climate in the northern Adriatic region during late Pleistocene and Holocene. *Pal. Pal. Pal.* **21**, 139–56.

Bortolami, G. C., Carbognin, L., and P. Gatto. 1985. The natural subsidence in the Lagoon of Venice, Italy. In A. I. Johnson, L. Carbognin and L. Ubertini (eds.), *Land Subsidence*. IAHS 151, pp. 777–85.

Brambati, A., Carbognin, L., Quaia, T., Teatini, P., and L. Tosi. 2003. The Lagoon of Venice: geological setting, evolution and land subsidence. *Episodes* **26**, 264–8.

Bras, R. L., Harleman, D. R. F., Rinaldo, A., and P. Rizzoli. 2002. Obsolete? No. Necessary? Yes. The gates will save Venice. *EOS* **83**, 217–24.

Brunetti, A., Denèfle, M., Fontugne, M., Hatté, C., and P. A. Pirazzoli. 1998. Sea-level and subsidence data from a Late Holocene back-barrier lagoon (Valle Standiana, Ravenna, Italy). *Marine Geology* **150**, 29–37.

Camuffo, D., and G. Sturaro. 2003. Sixty-cm submersion of Venice discovered thanks to Canaletto's paintings. *Climatic Change* **58**, 333–43.

Caputo, M., Pieri, L., and M. Unguendoli. 1970. Geometric investigation of the subsidence in the Po Delta. Boll. *Geofis. Teor. Appl.* **13**, 187–207.

Carbognin, L., and F. Marabini. 1989. Evolutional trend of the Po River delta (Adriatic Sea, Italy). In *28th International Geological Congress*, Washington, DC, USA, 9–19 July. Washington DC: American Geophysical Union, vol. I, pp. 238–9.

Carbognin, L., and G. Taroni. 1996. Eustatismo a Venezia e Trieste nell'ultimo secolo. *Atti Istituto Veneto di Scienze e Lettere ed Arti* **165**, 281–98.

Carbognin, L., and L. Tosi. 2002. Interaction between climate changes, eustasy and land subsidence in the North Adriatic Region, Italy. *Marine Ecology* **23**, 38–50.

Carbognin, L., Cecconi, G., and V. Ardone. 2000. Intervention to the safeguard of the environment of the Venice Lagoon (Italy) against the effects of land elevation loss. In L. Carbognin, G. Gambolati, and A. I. Johnson (eds.), *Land Subsidence*. Vol. I, SISOLS Proceedings. Paduva: La Garangola, pp. 309–24.

Carbognin, L., Teatini, P., and L. Tosi. 2004a. Eustasy and land subsidence at the beginning of the new millennium. *Journal of Marine Systems* **51**, 1–4, 345–53.

Carbognin, L., Teatini, P., and L. Tosi. 2004b. Land subsidence in the Venetian area: known and recent aspects. In *Engineering Geology in Italy*, AIGA, Pitagora Editrice, in press, 250 pp.

Carminati, E., and G. Di Donato. 1999. Separating natural and anthropogenic vertical movements in fast subsiding areas: the Po Plain (N. Italy) case. *Geophysical Research Letters* **26**, 2291–4.

Carminati, E., and G. Martinelli. 2002. Subsidence rates in the Po Plain (Northern Italy): the relative impact of Natural and Anthropic causation. *Engineering Geology* **66**, 241–55.

Carminati, E., Wortel, M. J. R., Spakman, W., and R. Sabadini. 1998. A new model for the opening of the western-central Mediterranean basins: geological and geophysical constraints for a major role of slab detachment. *Earth and Planetary Science Letters* **160**, 651–65.

Carminati, E., Doglioni, C., and D. Scrocca. 2003a. Apennines subduction-related subsidence of Venice (Italy). *Geophysical Research Letters* **30**, doi:10.1029/2003GL017001.

Carminati, E., Martinelli, G., and P. Severi. 2003b. Influence of glacial cycles and tectonics on natural subsidence in the Po Plain (Northern Italy): insights from 14C ages. *G-cubed* **4**, doi:10.1029/2002GC000481.

Di Donato, G., Negredo, A. M., Sabadini, R. and L. L. A. Veermersen. 1999. Multiple processes causing sea-level rise in the central Mediterranean. *Geophysical Research Letters* **26**, 1769–72.

Doglioni, C. 1993. Some remarks on the origin of foredeeps. *Tectonophysics* **228**, 1–20.

Doglioni, C., and E. Carminati. 2002. The effects of four subductions in NE Italy. Transalpine Conference, *Mem. Scienze Geol.* **54**, 1–4.

Emery, K. O., Aubrey, D. G., and V. Goldsmith. 1988. Coastal neo-tectonics of the Mediterranean from tide-gauge records. *Marine Geology* **81**, 41–52.

Flemming, N. C. 1992. Predictions of relative coastal sea-level change in the Mediterranean based on archaeological, historical and tide-gauge data. In L. Jeftic *et al.* (eds.), *Climatic Change in the Mediterranean*. London: Edward Arnold, pp. 247–81.

Fontes, J. C. and G. Bortolami. 1973. Subsidence of the Venice area during the past 40,000 yr. *Nature* **244**, 339–41.

Gambolati, G. (ed.). 1998. *CENAS-Coastline evolution of the upper Adriatic Sea due to sea-level rise and natural and anthropogenic land subsidence*. Dordrecht, The Netherlands: Kluwer Academic Publ., 344 pp.

Gambolati, G., and P. Teatini. 1998. Natural land subsidence due to compaction coastal morphodynamics of the upper Adriatic Sea basin. *IGEA* **11**, 29–40.

Gueguen, E., Doglioni, C., and M. Fernandez. 1998. On the post 25 Ma geodynamic evolution of the western Mediterranean. *Tectonophysics* **298**, 259–69.

Kent, V. D., Rio, D., Massari, F., Kukla, G., and L. Lanci. 2002. Emergence of Venice during the Pleistocene. *Quaternary Science Review* **21**, 1719–27.

Mariotti, G., and C. Doglioni. 2000. The dip of the foreland monocline in the Alps and Apennines. *Earth and Planetary Science Letters* **181**, 191–202.

Massari, F., Rio, D., Serandrei Barbero, R., Asioli, A., Capraro, L., Fornaciari, E., Muellenders, W., Raffi, I., and P. P. Vergerio. 2004. The environment of Venice in the past two million years. *Paleogeography, Paleoclimatology and Paleoecology* **202**, 273–308.

Merlini, S., Doglioni, C., Fantoni, R., and M. Ponton. 2002. Analisi strutturale lungo un profilo geologico tra la linea Fella-Sava e l'avampaese adriatico (Friuli Venezia Giulia-Italia). *Mem. Soc. Geol. It.* **57**, 293–300.

Mitrovica, J. X., and J. Davis. 1995. Present-day post glacial sea level change far from the Late Pleistocene ice sheets: implications for recent analyses of tide gauge records. *Geophysical Research Letters* **22**, 2529–32.

Pirazzoli, P. A. 1987. Recent sea-level changes and related engineering problems in the Lagoon of Venice (Italy). *Progress in Oceanography* **18**, 323–46.

Pirazzoli, P. A. 1996. *Sea Level Changes. The Last 20000 Years*. Chichester: John Wiley & Sons, 211 pp.

Pirazzoli, P. A. 2002. Did the Italian government approve an obsolete project to save Venice? *EOS* **83**, 217–23.

Polino, R., Dal Piaz, G. V., and G. Gosso. 1990. Tectonic erosion at the Adria margin and accretionary processes for the Cretaceous orogeny of the Alps. *Mém. Soc. Géol. France* **156**, 345–67.

Schmid, S. M., Pfiffner, O. A., Froitzheim, N., Schönborn, G. and E. Kissling. 1996. Geophysical–geological transect and tectonic evolution of the Swiss-Italian Alps. *Tectonics* **15**, 1036–64.

Sclater, J. G. and P. A. F. Christie. 1980. Continental stretching: an explanation of the post-mid-Cretaceous subsidence of the central North Sea basin. *Journal of Geophysical Research* **85**, 3711–39.

Scrocca, D., Doglioni, C., Innocenti, F., Manetti, P., Mazzotti, A., Bertelli, L., Burbi, L., and S. D'Offizi (eds.). 2003. CROP Atlas: seismic reflection profiles of the Italian crust. *Memorie Descrittive della Carta Geologica d'Italia* **62**.

Serandrei Barbero, R., Lezziero, A., Albani, A., and U. Zoppi. 2001. Depositi tardo pleistocenici ed olocenici nel sottosuolo veneziano: paleoambienti e cronologia. *Il Quaternario* **14**, 1, 9–22.

Strozzi, T., Tosi, L., Teatini, P., Wegmuller, U., Werner, C., and L. Carbognin. 2003. Land subsidence monitoring service in the Lagoon of Venice. *Earth Observation Quarterly* **71**, ESA Publ., Div. Noordwijk, The Netherlands.

Tosi, L., Carbognin, L., Teatini, P., Strozzi, T., and U. Wegmuller. 2002. Evidence of the present relative land stability of Venice, Italy, from land, sea, and space observations. *Geophysical Research Letters* **29**, 10.1029/2001GL013211.

Wessel, P., and W. H. F. Smith. 1995. New version of the generic mapping tools released. *EOS* **76**, 329.

Zerbini, S. *et al.* (16 authors). 1996. Sea level in the
    Mediterranean: a first step towards separating crustal
    movements and absolute sea-level variations. *Global
    Planetary Change* **14**, 1–48.

5 • The facts of relative sea-level rise in Venice

R. FRASSETTO

INTRODUCTION

Through a synthetic review of 35 years of scientific research since the 1969 UNESCO Report on the Safeguarding of Venice, it is possible to demonstrate the useful results obtained so far in understanding, describing and predicting the major environmental risks menacing the conservation of the historical city, such as the rate of sea-level rise (RSLR), which includes subsidence and eustasy. New scientific challenges are being met by ongoing and planned local to global international research, the prime objective of which is to quantify the natural and anthropogenic rates of changes of the past and predict reliable future trends in climate. This review is focussed mostly on facts based on data and proxies.

While in the 1960s the primary scientific concerns for the safeguarding of Venice were the local weather forecast of storm surges and the rate of subsidence, since the 1980s the major scientific concerns have become those of global climatic warming, with the risk of sea-level rise (SLR) and the occurrence of more frequent severe meteorological events. In the dynamics of the Earth System, three main forcing agents drive relative sea level (RSL) changes.

1) crustal movements in the 10^6-time scale generating natural subsidence;
2) solar radiation and illumination of the planet earth in the 10^5 to 10^2-time scale with its glacial and interglacial cycles as cause of global sea level changes; and
3) anthropogenic forcing, with the use of land and natural resources causing soil compaction in the historical time scale.

For the case of Venice, we describe the regional and local lithosphere instability of deposited sediments and their vulnerability to compaction processes and we review the evidence for RSL changes.

GEOLOGICAL FORMATION AND EVOLUTION OF THE AREA IN THE 10^6–10^3 YEAR RANGE

From Friuli (N) to Emilia-Romagna (S), the NW lithological area of the Adriatic is a stratified, unstable geological system, which evolved through the glacial and interglacial climatological epochs in the 10^6 to 10^3-year time range. Several studies by oil companies interested in subsoil gas resources, by research institutes of the Italian National Research Council (CNR) and by universities of the region have reconstructed the geological formation of the area and its evolution. A schematic section of the Po valley, and the effect of tectonic stress between the Alps and the Apennines, from the Pliocene to the Eocene periods are shown in Fig. 5.1. Ravenna and Venice are included in this tectonically active and unstable area.

One of the most interesting theories of regional plate tectonics, which concerns this zone, is based on a mechanism of subduction (Fig. 5.2) (Doglioni, 1993). Supported by deep cores and seismic tomography, the theory opens a new vision of how, in the course of millions of years, endogenous forcings, induced also by the Earth's rotation, have forced the plates' movement and how connected vertical movements of the lithosphere have occurred at a rate of between 1.0 mm yr^{-1} to 0.5 mm yr^{-1}, with larger values of subsidence in the Romagna area (SW).

On the 10^5 to 10^3-year timescale in the NW Adriatic lithosphere area, a system of lagoons, beaches and estuaries underwent alternate shoreline retreats and advances following climatic changes

Flooding and Environmental Challenges for Venice and its Lagoon: State of Knowledge, ed. C. A. Fletcher and T. Spencer. Published by Cambridge University Press, © Cambridge University Press 2005.

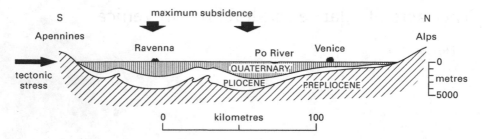

Fig. 5.1   A very schematic geological cross-section of the Po Valley between Venice and Ravenna. The pre-Quaternary substratum is characterized by folds and faulted overfolds running parallel to the main tectonic profiles of the Apennines (Agip Mineraria, 1969; redrawn by Carbognin *et al.*, 1981).

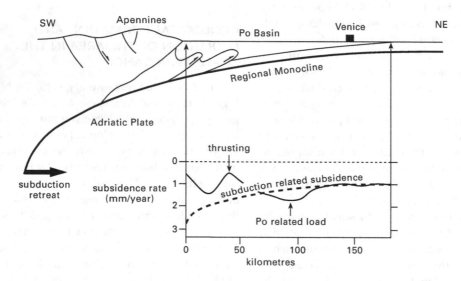

Fig. 5.2   The subduction of the Adriatic plate as cause of natural subsidence (after Doglioni, 1993).

forced by orbital variations of the Earth (solar radiation and illumination). Glacial and interglacial cycles strongly influenced different kinds of sedimentation, in terms of rate, type and extension. During continental epochs (low sea level), erosion and fluvial deposition complicated the distribution of sediments in the Venetian basin.

Three deep, undisturbed cores: the Venice 1 (950m), Venice 1 bis (~70m) at the Tronchetto and a larger diameter core Venice 2 (~ 100 m deep) at the Arsenale were made by CNR-ISDGM in 1970-3 to study the geological sequence and character of the

column of sediments below Venice. From the data, Serandrei Barbero (1972, 1975), amongst others, reconstructed the subsoil stratification on a plane traversing Venice and estimated the chronological succession. Sand, silt and clay are found more or less mixed in different proportions characterizing the continental and the marine littoral epochal conditions. All strata appear unconsolidated with the exception of a thin layer (0.1 to 10 m thick) of clay called 'caranto'. This clay was desiccated and consolidated, according to Gatto and Previatello (1974), after the last ocean regression of 6000 to 10 000 years

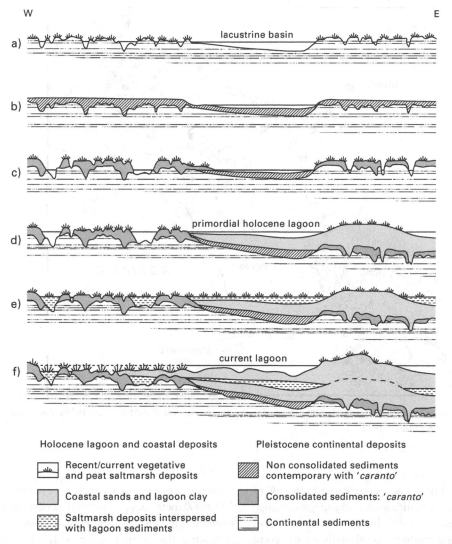

Fig. 5.3   The genetic phases of the 'caranto', the thin strata formed during a marine regression around 10 000 BP (Gatto and Previatello, 1974). (a) glacial maximum (*c.* 18 000 BP); (b) extensive flooding at the beginning of the post-glacial period; (c) return to a freshwater environment and consolidation of the 'caranto'; (d) marine transgression at 6000 BP; (e) marsh growth corresponding to phase of marine regression (3–4000 BP); (f) current status.

ago (Fig. 5.3). Its irregular horizontal distribution (Fig. 5.4) may have been one of the reasons why some Venetian palaces appear to have subsided more than others, due to the lack of support of the *caranto* layer on which to lay the foundation piles of heavy structures. New research by acoustic soundings and cores

is underway to describe and map with greater precision this interesting, anomalous consolidated layer.

The complex and diversified stratification makes the subsoil from Venice to Ravenna vulnerable to compaction by intense anthropogenic forcings, such as construction loads (land use) and water and gas

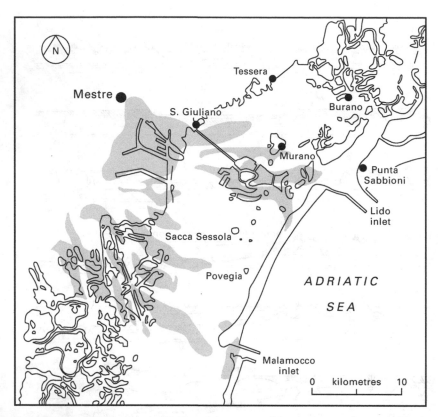

Fig. 5.4   A preliminary map of the irregular extension of the *caranto* with west to east gaps probably due to fluvial erosion (Gatto and Previatello, 1974).

pumping, which occurred in particular in the twentieth century. In the Venice area, anthropogenic subsidence has varied in time and space to such an extent that the analysis made professionally by different authors over different periods of time by means of land elevation surveys, geotechnical research, drawdown water levels of artesian wells and tide gauge data, are not comparable, demonstrating that the sinking process is very localized. Land settlement due to water table exploitation is clearly demonstrated by piezometric levels of a network of artesian wells established and observed by CNR (Serandrei Barbero, 1973; Mozzi, 2000).

Rowe (1975), and Ricceri and Butterfield (1974) through geotechnical analysis, particularly of the VE2 CNR core of 30 cm diameter, found a variable settlement to the drawdown ratio with an average of 20

millimetres per metre. Extraction of water from about 100 to 300 metre-deep aquifers described by Serandrei Barbero (Fig. 5.5, 1972) increased from 1930 to 1970 following the demands for water by developing industries and modern living, reaching a maximum in 1950-70. This 'anthropogenic jump' of subsidence, as Rowe defines it, is also clearly demonstrated by Carbognin *et al.* (1976) with the difference between Venice and Trieste tidal records, assuming Trieste to be stable.

Bortolami and Fontes (Fig. 5.6, 1974), through isotopic analysis, measured the velocity of the deep stored precipitation waters penetrating from the mountains, where the value reaches 50 metres per year, to the Venice coast, where the velocity is about 1 metre per year through sediments of finer granulometry. Despite the velocity of the water particles,

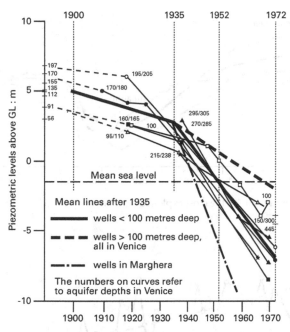

Fig. 5.5 Historical piezometric level of artesian wells in Venice and Marghera indicating the 'anthropogenic jump' from 1935 to 1972 (from Serandrei Barbero, 1972). The numbers on curves refer to aquifer depths.

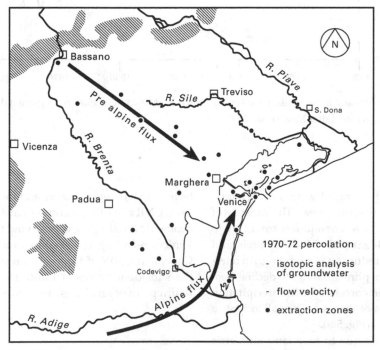

Fig. 5.6 The flow of deep water stored in the subsoil of the Venice catchment basin, with velocities reducing from 50 m per year to 1 m per year from the foot of the mountains to the coast respectively (after Bortolami and Fontes, 1974).

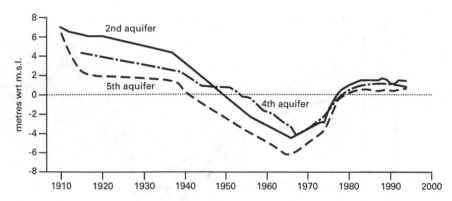

Fig. 5.7   Variation of artesian water pressure of the most exploited aquifers of Venice, showing evidence of the 1970 to 1980 rebound (Mozzi *et al.*, 2000).

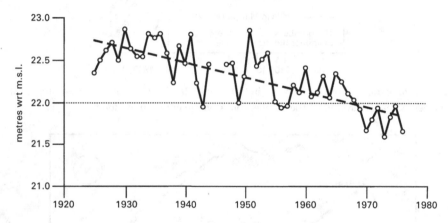

Fig. 5.8   The decline of phreatic water levels in the Venetian catchment basin from 1930 to the present day, an indication of decreasing precipitation rates (after Mozzi *et al.*, 2000).

the pressure varies quite rapidly, permitting a relatively quick rebound of settled areas. The variation of the pressure of three different aquifers from 1910 to 1990 is shown by Mozzi (Fig. 5.7) with a rebound starting in 1970 for a reduction of pumping. An interesting tendency of the phreatic water to decline from 1925 to 1975 has been correlated with precipitation and river runoff rates of the area. They all indicate a climatological decline (Fig. 5.8).

The common technique to map land altimetric variations is by means of precise geometric levelling. From the first Italian national survey of 1890, the technique has become more and more precise. Since 1971, CNR and other institutes have made a series of geometric levelling surveys around the lagoon and in Venice, referring the system to a bench mark in Conegliano, NW of Venice, taken as stable. Some of the subsidence rates at Punta Dogana (Salute), a return point of levellings, are shown in Table 5.1.

Table 5.1   *Review of different analyses on RSLR and components.*

Name	Date	Method	Yrs	Period	Yearly rates mm		
					RSLR	Subsidence	SLR
Zendrini	1981	C-Palazzo Ducale	170	1810-1980	1.84	0.74	
UIMA	1983	C-Palazzo Ducale + levelling 1873	107	1873-1980	3	2	
UIMA Rusconi	1983	C. Ponte Ferrovia	139	1841-1980	3.45		
Frassetto	2003	C. Ca' Corner	158	1841-1999	4.03		
UIMA	1983	Tide gauge S. Stefano	110	1871-1980	4.4	2.9	
UIMA	1983	Levelling Punta Dogana	80	1900-1980		1.73	
Polli	1965	Tide gauge + decadal var.	59	1881-1940	2.5		1.1
UIMA	1983	Tide gauge	110	1871-1980	2.5		
Mosetti	1969	Tide gauge VE	71	1896-1967	3.84	2.57	
Mosetti	1969	Tide gauge TS	55	1906-1961			1.27
Caputo	1971	Levelling Punta Dogana	61	1900-1961		2.0-3.0	
Carbognin	2003	Tide gauge VE	72	1908-1980	3.11		
Carbognin	2003	Tide gauge VE*	39	1931-1970	3.8		
Tosi *et al.*	2002	Tide gauge VE-TS	97	1896-1993	2.52	1.39-1.5	1.13
Serandrei	1974	Artesian heads*	35	1900-1935		1	
Serandrei	1974	Artesian heads*	17	1935-1952		4	
Ricceri	1974	Geotecn.*	20	1952-1972		6	
Ricceri	1974	Compressibility/Geotecn.	70	1900-1970		3	
Rowe	1975	Compressibility	32	1938-1970		4.0-6.0	
Carbognin	2002	Tides VE-TS	30	1970-2000		~ 0.0	
INSULA / Tosi *et al.*	2001	General estimates	100	20th century	2.3	1.2	1.1
IPCC TAR	2001	Tide gauges global	100	20th century			1.8 (1.0-2.5)
PAGES-IGBP	2002	Tide gauges global	100	20th century			1.5 (1.0-2.0)
Flemming & W.	1986	Medit. Eustasy	100	20th century			0.5-1.2
Wigley & Raper	1987-95	Gobal model	100	20th century			1.0-1.5
Pirazzoli	1980	Tides VE-TS	100	20th century		0.5-0.7	
' ' '		Trieste benchmark	50			0.1-0.2	
' ' '		Tides VE-TS	102	1872-1984	2.5	1.2	1.3
Pirazzoli	2000	Natural subsidence	Last interglacial			0.2/0.4	
Lambeck	1999	Global ocean volume	3000	6000-3000 bp	0.1/0.5		
Lambeck	1999	Geol. records	3000	Last 3000	0.1/0.2		
Ammerman	1990	Archeology	1500	400-1847		1.9	
Literature		Global ocean volume					1.0-1.5
Barnett	1983	Selected SL stations	Last century				1.5
Douglas	1997	Selected tide gauges	Last century				1.8±0.1
Min.LL.PP. I.		General view				1.4	1.1
INSULA / Tosi *et al.*	2003	Suggested estimates	100	21st century	1.66	natural 0.4	1.23

* Anthropogenic jump.

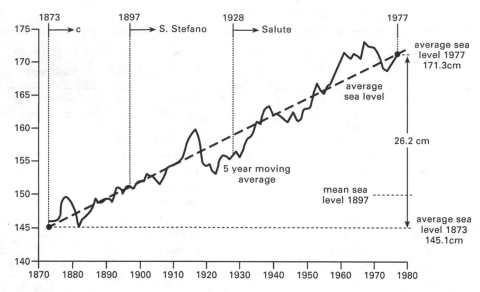

Fig. 5.9   The five-year smoothed trace of tide measurements in Venice, recorded through the *Comune Marino* from 1871, by the S. Stefano tide gauge from 1897 and by the Punta Salute tide gauge since 1925. The trend shows an average of 26 cm rate of change for the twentieth century (Rusconi, 1983).

THE RELATIVE MEAN SEA LEVEL VARIATIONS IN VENICE DURING THE LAST CENTURY

Since Venice's legendary foundation in AD 400, the morphology and water dynamics of the lagoon have always been of great concern for the growth of its naval power and for its defence. The awareness of the sinking city started early in the city's history. The Magistrato alle Acque and several offices were created to take charge of water management and enforce severe regulations on the population to grant the respect of the lagoon waters, artificially adapted to the needs of the city and its commercial activity. Water levels were therefore carefully observed and controlled, but only in 1440 did the archives of the Magistrato adopt the term *Comune Marino* to represent the upper level of wet algae growing on buildings facing canals to indicate the top mean level of tidal water levels. It was decided to carve on the stone a line of this level with a C (*Comune Marino*) above it. More precisely, this was the CAM (Comune Alte Maree, meaning normal high tide level). It was dated

to permit the measurement of the lagoon's water level variations over time. In 1983, the Hydrographic Office of the Magistrato alle Acque (Rusconi, 1983) made a survey finding 37 Cs still visible on aged stones, but only 8 of them could be dated and only 2 became usable for research on historical land elevation. In Fig. 5.9, Rusconi shows the five-year average tide data variation and the mean trend. From 1873 C data were used; then in sequence from 1897 the Santo Stefano tide gauge and from 1925 until today the Punta Dogana (Salute) tide gauge. The mean century trend (1873-1977) gives a RSL rise of 26.2 cm or a mean rise rate of 2.62 mm yr^{-1}. However, variability is evident and depends on the rate of use of aquifers.

Table 5.1 lists the different rates of relative sea-level rise and of its components resulting from the best known studies by different authors. They are the result of analyses carried out over different periods, varying lengths of time and using different methods. The Table shows that the results are not comparable and that land settlement has varied in time and location. The anthropogenic component has great variability, while the natural subsidence can be considered a constant over the century time scale. The eustasy component instead shows great variability, but has a definite rising trend. We give a special relevance to this global process, identified through analysis of geological archives and in the twentieth century through measurements.

GLOBAL AND REGIONAL EUSTATIC VARIATIONS

Global mean sea-level rise, with its local scale effects, is without doubt the major risk that Venice may face in future decades. It depends on the climatological global warming, swelling oceans and melting ice. The chaotic mechanism of climate, with its variation in time and space, makes precise predictions of future rates of warming impossible at the present time. This mechanism will represent for a long time one of the greatest challenges of internationally planned research and of numerous and proliferating programmes of national and joint researches. The present information on the historical courses of the climate can either be gathered from direct and indirect human records or from natural archives that we have continuously learned to decipher more precisely with the use of ever more sensitive and refined methods.

Since the 1980s, PAGES, a subprogramme of the IGBP, has been dealing with paleoclimate through proxy data and indicators gathered from ice, lake, ocean floor and land cores, analysis of tree rings and coral growth records. General estimates of variation and trends have been computed and the results of modelling and analysis using different approaches have been compared and assessed. Many consider that all this work is just at its beginning. The major processes and mechanisms of climate change, however, have been identified. Quality and quantity information are preliminary.

A great variety of evidence of the evolution and change of variables linked to climate and studies of indexes are now available for further investigation and for the reduction of declared uncertainties. To assess, synthesize, and inform governments and society on interesting and useful results of global research, the UN, with WMO and UNEP, created in 1988 the Intergovernmental Panel on Climate Change (IPCC). This is divided into three Working Groups on:

1) Science of Climate Change;
2) The Impacts, Adaptation and Vulnerability; and
3) Mitigation.

The IPCC produces, about every five years, a report on the review and assessment of the progress achieved internationally. Three assessment reports have been published so far and the fourth is being planned for presentation around 2007.

The general trend of temperature over the last two centuries shows a global increase of about 0.9 °C. Beer *et al.* (1995) have shown deviations of mean temperature of the Northern Hemisphere correlated with [10]Be cosmogenic isotope concentration in Greenland ice suggesting connections between solar variability and climate change. On a decade time scale, Tomasin (2002) has shown a correlation between solar sunspot number and storm surge frequency in Venice (see Tomasin, this volume). This is a suggestive indication of the influence of solar radiation variation on local meteorological forcings.

Since the announcement of catastrophic prediction of possible global sea-level rise of 50 to 345 cm for the year 2100, made by the US Environmental Protection Agency (EPA) in the 1970s, the predictions have become gradually more realistic. In 1990 the first IPCC Assessment Report (AR) estimated a global SLR from 31 to 110 cm for the twenty-first century with the best estimate 66 cm.

In 1992 Wigley and Raper, using the Integrated Scenario IS92a proposed at the Rio convention for the development of predictive models, estimated a SLR range of 15 to 90 cm with the best estimate 48 cm for the year 2100. In 1995 IPCC SAR estimated a range of 10 to 54 cm with a best estimate of 37 cm. In 2001 the TAR, using SRES scenarios, estimated a range of 10 to 50 cm with a best estimate of 30 cm. The Fourth Assessment Report will certainly provide more refined information, with a reduction of present uncertainties both in models and in data, and with better knowledge of the global ocean thermal circulation (THC) and of the acceleration of ocean warming. A recommendation has been made to give more emphasis to the chapter on the assessment of sea level variations, with more peer reviews.

Table 5.2 presents the estimated rates of global SLR averaged over the twentieth century. With posi-

Table 5.2   *Estimated rates of sea-level rise averaged over the twentieth century. For the ice-sheet and the total, the numbers in brackets are for the larger error bounds.*

	Minimum	Central value	Maximum
Thermal expansion	0.4	0.60	0.8
Glaciers	0.2	0.30	0.4
Greenland-Anthropogenic	0.0	0.05	0.1
Antarctica-Anthropogenic	-0.3	-0.20	-0.1
Ice sheets - Adjustment since LGM	-0.1 (-0.6)	0.30 (0.60)	0.7 (1.8)
Permafrost	0.0	0.10	0.2
Terrestrial storage terms	-0.45	-0.30	-0.15
Total	-0.25 (-0.75)	0.85 (1.15)	1.95 (3.05)
Observed	1.0	1.80	2.5

(From Church *et al.*, 2001.)

Table 5.3   *Recent estimates of sea-level rise from tide gauges. The standard error for these estimates is also given along with the method used to correct for vertical land movement (VLM).*

	Region	VLM	Rate ± s.e. (mm/yr)
Gornitz and Levedeff (1987)	Global	Geological	1.2 ± 0.3
Peltier and Tushingham (1989, 1991)	Global	ICE-3G/M1	2.4 ± 0.9
Trupin and Wahr (1990)	Global	ICE-3G/M1	1.7 ± 0.13
Nakiboglu and Lambeck (1991)	Global	PGR Model	1.2 ± 0.4
Douglas (1991)	Global	ICE-3G/M1	1.8 ± 0.1
Shennan and Woodworth (1992)	NW Europe	Geological	1.0 ± 0.15
Gornitz (1995a)	New Zealand	Geological	1.7 - 1.8
Gornitz (1995b)	N America E Coast	Geological	1.5 ± 0.7
Mitrovica and Davis (1995)	Global	PGR Model	1.4 ± 0.3
Peltier (1996)	US E Coast	ICE-4G/M2	1.9 ± 0.6
Peltier and Jiang (1997)	US E Coast	Geological	2.0 ± 0.6
Peltier and Jiang (1997)	Global	ICE-4G/M2	1.8 ± 0.6
Douglas (1997)	Global	ICE-3G/M1	1.8 ± 0.1
Lambeck *et al.* (1998)	Fennoscandia	PGR Model	1.1 ± 0.2
Woodworth *et al.* (1999)	British Isles	Geological	1.0

(From Church *et al.*, 2001.)

tive and negative values the total central values, amounts to 0.85 mm yr^{-1}. Observations give about twice the values obtained from models of ocean volume and mass increase; something to be verified. Table 5.3 shows the rates of SLR from more than 50 years of tide gauge data corrected for vertical land movement (VLM) taken globally or from limited geographical areas. The results from 15 different authors range from 1.0 to 2.4 mm yr^{-1}. The global average estimate for the twentieth century amounts to 1.8 mm yr^{-1} ± 0.1. Other authors (e.g. Peltier and Jung, 1997 (see Table 5.3)) suggest an uncertainty of ± 0.6 mm yr^{-1}.

The TAR presents also the results of numerous inter-comparison experiments on the thermal expansion of the oceans and on glaciers melting, along with their acceleration rates in the twentieth century. Predictions for the twenty-first century calculated through the AOGCM with IS92a and SRES scenarios are also presented and discussed. These results can be considered as the best and most reliable information available at present. Global sea level averages give basic indications of climatological variations and trends; however, sea level varies in time and over space as a function of local forcings such as temperature, wind, pressure and precipitation. Satellite altimeter observations demonstrate it but site verifications and observations are needed. A recent study finds coastal SLR greater than offshore SLR. A possible effect of boundary waves, this is another sign of variability that needs to be better understood.

CONCLUSIONS

Relative sea-level rise (RSLR) is becoming the major long-term menace to the conservation of Venice. This chapter has reviewed the best-known past and present data and information about the change in RSLR components. These are: local natural and anthropogenic subsidence and the global climatological sea-level rise.

Natural subsidence is due to tectonic and isostatic regional processes over the 10^6 years timescale. The present estimate of its rate of change is 5 to 10 cm per century. New plate tectonics research is planned to better quantify and describe this evidence. Collaboration with Italian oil companies, who own a great deal of data for the NW littoral of the Adriatic,

is recommended. Industrial data by now could be declassified. The local anthropogenic subsidence, if prevented with sound management and legal enforcements, can be considered practically negligible. Continuous piezometric controls and periodic geodetic levellings (at least every 10 years) are needed. A network of artesian wells with systematic piezometric level observations has been created by CNR for scientific demonstrations. It is ready to be transformed into a routine service by the regional government. Recent restoration and upkeep of monuments, dwellings and canal borders have upset the positions of several benchmarks. Land surveys and laser soundings indicate new minor elevation changes due to the instability of the Venice subsoil that we have described here. CNR is experimenting with deep moored benchmarks in the Arsenale area as reference points.

The global eustatic sea-level rise, with its yearly rate of change and acceleration is the climatological planetary process, which is under extensive international scientific research assessed about every five years by the IPCC WG1. A proliferation of group/activities (CLIVAR, MERCURE, PRUDENCE, etc.) is joining the larger programmes. The IPCC TAR estimate of the average global rate of change of MSL in the twentieth century is 1.8 ± 0.6 mm yr^{-1} with a calculated acceleration of 0.5 to 1.5 mm yr^{-1} per century for thermal expansion, which is the value posing the greatest uncertainties. The defence of Venice from the possible growth of frequency and amplitude of storm surges depends on sound scientific information being continuously updated. This is available through the IPCC Secretariat in Geneva.

ACKNOWLEDGEMENT

The author wishes to thank Jane Frankenfield Zanin, CNR-ISMAR, Venice for her effective editing of the paper.

ABBREVIATIONS AND ACRONYMS

AOGCM   Ocean-Atmosphere General Circulation Model
CNR   Consiglio Nazionale delle Ricerche (of Italy)

GHG   Greenhouse Gas
GL   Ground Level
IPCC   Intergovernmental Panel for Climate
         Change
MSL   Mean Sea Level
RSL   Relative Sea Level
RSLR   Relative Sea-level rise
SAI   Standard Anomaly Index
SLR   Sea-level rise
SRES   Special Report on Emissions Scenarios
TAR   Third Assessment Report (IPCC)
TS   Trieste
VE   Venice
VLM   Vertical Land Movement

REFERENCES

Beer, J. *et al*. 1995. Climate information from polar ice cores. *EAWAG News* **38 E**.

Bortolami, G., and J. C. Fontes. 1974. *Idrologia isotopica in un sistema multifalde – La pianura veneta*. Venice: CNR-ISDGM, TR 82.

Carbognin, L., Gatto, P., Mozzi, G., Gambolati, G. and G. Ricceri. 1976. New Trend in the subsidence of Venice. 2nd International Symposium on Land Subsidence, Anaheim, CA (USA), *IAHS* **121**, 65–81.

Carbognin, L., Gatto., G. and G. Mozzi. 1981. La riduzione altimetrica del territorio veneziano e le sue cause. *Istituto Veneto di Scienze Lettere ed Arti, Rapporti e Studi* **7**, 55–83.

Church, J. A., *et al*. (2001) Zool. Changes in sea-level. In J. T. Houghton *et al*. (eds.), *Climate Change 2001: The Scientific Basis. Contribution of Working Group I to the Third Assessment Report of the Intergovernmental Panel on Climate Change.* Cambridge and New York: Cambridge University Press.

Doglioni, C. 1993. Some remarks on the origin of foredeeps. *Tectonophysics* **228**, 1–2, 1–20.

Gatto., P. and P. Previatello. 1974. Significato stratigrafico, comportamento meccanico e distribuzione nella laguna di Venezia di una argilla sovraconsolidata nota come 'caranto'. CNR-ISDGM, TR 70.

Lindzen, R. S. 1994. Climate dynamics and global change. *Annu. Rev. Fluid Mech.* **26**, 353–78.

Mozzi, G. *et al*. 1997. *Venezia una città fragile*. Rome: Pres. Consiglio Ministri, Protezione Civile.

Mozzi G. *et al*. 2000. *CNR – Salvaguardia del Patrimonio idrico sotterraneo del Veneto*. Rome: Consiglio Ministri, Dip. Protezione Civile.

Ricceri, G., and B. Butterfield. 1974. *An Analysis of Compressibility Data from a Deep Borehole in Venice*. London: Institute of Civil Engineers.

Rowe, P. W. 1975. *Soil Mechanic Aspects of the Cores of Deep Borehole VE1 in Venice*. Venice: CNR-ISDGM, TR 57.

Rusconi, A. 1983. *Il Comune Marino a Venezia*. Publication 157. Venice: Uff. Idrogr. Magistrato alle Acque,

Serandrei Barbero, R. 1972. *Indagine sullo sfruttamento artesiano nel Comune di Venezia 1846–1970*. Venice: CNR-ISDGM, TR 31.

Serandrei Barbero, R. 1975. Sondaggio VE2 CNR – Stratigrafia e paleoecologia. *Giornale di Geologia Serie 2°* **40**, Fascicolo 1.

Tomasin, A. 2002. *The Frequency of Adriatic Surges and Solar Activity*. Venice: CNR-ISDGM, TN 194.

Wigley, T. M. L. and S. C. B. Raper. 1992. Implications for climate and sea level of revised IPCC emissions scenarios. *Nature* **357**, 293–300.

WMO. 2003. *Annual Review of the World Climate Research Programme and Report of the Twenty-third Session of the Joint Scientific Committee*. Hobart, Tasmania, Australia: 18–22 March 2002, WCRP JSC TD No. 1137, Jan. 2003.

UNESCO. 1969. *Rapporto su Venezia*. Milan: Mondadori.

6 • Ocean climate variability in the Mediterranean Sea: climate events and marine forecasting activities

N. PINARDI, G. COPPINI, A. GREZIO AND P. ODDO

INTRODUCTION

The long-term ocean variability of the Mediterranean Sea has been studied intensively in the past twenty years. Results illustrate the correlation between atmospheric forcing variability and ocean response at seasonal, interannual and interdecadal time scales. Major climate variability events have occurred in the 1980s and 1990s driven by long-term interannual variability of atmospheric forcing over the basin. The changes involve inversion of current direction in deep regions of the basin, strengthening and weakening sub-basin scale circulation structures. Moreover, shorter-term ocean variability, connected with the time scales from the seasonal to the mesoscale, has been thoroughly investigated. This has produced the implementation of a Mediterranean ocean Forecasting System (MFS) to predict ocean variability in the Mediterranean Sea from the global scale to the shelf areas. The MFS started operational activities in January 2000. Presently it produces daily analyses and weekly 10-day forecasts of currents and temperature and salinity fields for the entire Mediterranean at approximately 10 km resolution.

The main elements of the MFS – simultaneously operating in a near real time observational data network, a general circulation model and data assimilation scheme – were implemented and upgraded as part of two EU funded projects The main goal of the most recent research has been to advance monitoring technology to achieve maximum reliability of the near real time observing system, to demonstrate regional forecasting in several Mediterranean subregions (3 km resolution). One of these regions is the Adriatic Sea where forecast activities started in 2003 within a project called ADRIatic sea integrated COastal areaS and river basin Management system pilot project (ADRI-

COSM). The latter demonstrates that, given proper downscaling in the Adriatic Sea, realistic currents can be predicted down to the scale of 5 km and less. Products of MFS and ADRICOSM are available at www.bo.ingv.it/mfstep and www.ingv.it/adricosm.

OCEAN CLIMATE VARIABILITY IN THE MEDITERRANEAN SEA

The Mediterranean Sea is a semi-enclosed basin (Plate 2) with deep ocean areas and narrow continental shelves. Exceptions to this rule are given by the extended shelf areas of the Adriatic Sea and the Tunisian plateau. The Gibraltar Strait balances the water and heat losses at the air-sea interface of the basin, maintaining a long term steady state equilibrium. The complex morphological structure of the basin and its exchanges with the Atlantic Ocean make the Mediterranean Sea circulation particularly unsteady and variable, from the deep ocean areas to the shelves. A recent overview of the shelf scale circulation of the Mediterranean Sea has been given in Pinardi *et al.* (2005). The ocean climate variability of the Mediterranean Sea has been observed by several recent interdisciplinary international projects such as POEM (Physical Oceanography of the Eastern Mediterranean (Robinson *et al.*, 1991; Malanotte-Rizzoli *et al.*, 1999; Roether *et al.*, 1996)) and MATER (MAss Transfer and Ecosystem Response (Monaco and Peruzzi 2002)). The main findings show that the basin circulation is formed by sub-basin scale gyres and free and boundary current jets. Mesoscale processes are pervasive. Numerical modelling of the circulation has shown that the variability of all these structures is controlled by atmospheric forcing variability to a degree unknown up to few years ago (Pinardi *et al.*, 1997; Korres *et al.*, 2000; Molcard *et al.*, 2002; Demirov and Pinardi 2002).

Flooding and Environmental Challenges for Venice and its Lagoon: State of Knowledge, ed. C. A. Fletcher and T. Spencer.
Published by Cambridge University Press, © Cambridge University Press 2005.

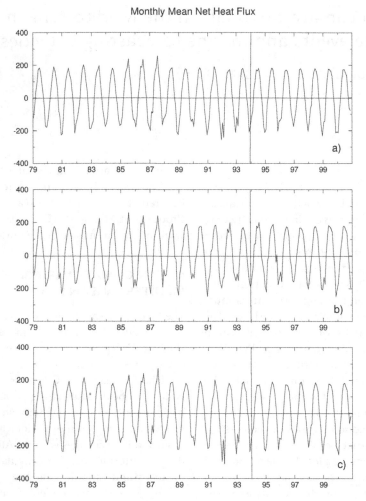

Fig. 6.1   Surface average heat fluxes (W/m^2) computed from a 1979-2001 model simulation forced by six hours ECMWF atmospheric forcing variables, as described in Castellari *et al.* (2000): (a) whole basin; (b) Western Mediterranean; (c) Eastern Mediterranean averages. The simulation was restarted at the end of 1993 to reduce the climate drift of the model.

It is worth mentioning two of the largest climate variability events documented in the literature. The first (Brankart and Pinardi, 2001) is concerned with an intermediate water cooling event that affected the entire Ionian and Levantine basins between 1981 and 1983. This event was due to a large anomaly in wind stress curl and heat fluxes over the Eastern Mediterranean that can be seen in Figs. 6.1 and 6.2. The wind stress curl anomaly was outstanding in the winter of 1981 while the heat flux anomalies in 1981-2 were only the third lowest minima. This forcing anomaly combination produced intermediate waters (at 200 metres) which were 0.45 °C cooler than water masses at the same depth in the eastern Mediterranean in the previous decade. Brankart and Pinardi (2001) show with different simulation experiments that this temperature anomaly was due to the atmospheric forcing anomaly of 1981-3.

Monthly Mean Wind stress Curl

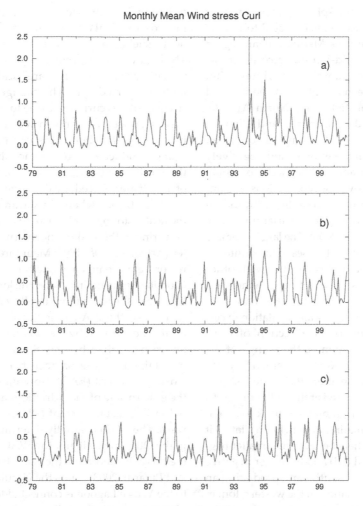

Fig. 6.2   Wind stress curl surface average time series (dyn/cm²) from a 1979-2001 model simulation forced by six hours winds from ECMWF analyses: (a) whole basin; (b) Western Mediterranean; and (c) Eastern Mediterranean averages. The simulation was restarted at the end of 1993 to reduce the climate drift of the model.

The second event is the so-called Eastern Mediterranean Transient (EMT; Roether *et al.*, 1996). This produced deep waters in the Cretan Sea that outflow through the Straits and replace the abyssal waters of the Eastern Mediterranean. The EMT started around 1988-9, reached maximum amplitude in 1992 and then relaxed down. Its remote effects are still continuing to appear in distant regions of the basin, such as the Adriatic Sea and the western basin. The Cretan Deep Waters are more saline and warmer than the previous deep waters of the Eastern Mediterranean, originating over the past fifty years from the southern Adriatic Sea. The Cretan Deep Waters have lifted the layer corresponding to Levantine Intermediate Waters (LIW) by several tens of metres so that more saline waters are entering into regions connected by sills and straits, such as the Otranto Channel and the Sicily Strait. Research is still underway to understand the effects of such a transient on the ecosystem dynamics of the basin.

The slowly varying atmospheric forcing variability of atmospheric surface fields is related to remote forcings that teleconnect the Mediterranean region with the tropical and mid-latitude global atmospheric variability. For example, the winter wind stress variability between the 1979-1987 period and the following is connected to the North Atlantic Oscillation (NAO) index (Demirov and Pinardi, 2002). The heat flux variability is instead correlated positively with NAO in the western and negatively in the Eastern Mediterranean (Marzocchi, personal communication). Moreover, teleconnections of summer Mediterranean climate variability with Indian Monsoon and Sahel atmospheric variability have been studied (Raicich *et al.*, 2003). The large anomaly events, such as the great heat losses in the winter of 1992-3, are found to be best correlated with other regional climate regimes, such as Etesian wind variability (Josey, 2003). In response to these changes in atmospheric forcing, the large-scale circulation in certain areas can also invert the direction of flow. Pinardi *et al.* (1997) documented the change of the cyclonic circulation in the Northern Ionian Sea into an anticyclonic circulation after 1987. Plate 3 shows the other change that occurred at the end of the 1990s that brought the northern Ionian Sea back to a generally cyclonically circulating flow field. In Plate 3 it is evident that the flow field between the smaller negative values (yellow and red) and the larger negative values (blue) in 1997 was southward, producing an overall anticyclonic circulation in the western Ionian Sea. In 2000 the situation totally changed and the overall circulation in the western Ionian was cyclonic. Satellite data have confirmed such changes (Lermusieaux and Robinson, 2001), together with *in situ* data (Manca *et al.*, 2003). This complex picture of the large scale basin circulation and water mass structure makes it impossible to assess the state of the system from only episodic surveys of the basin, collecting scattered data in different subregions. A large scale monitoring system should be set up that is capable of detecting changes in the circulation when they occur and follow them, from large scales to coastal areas.

The shelf area circulation is in fact extremely sensitive to changes in large scale current strength. An example is given by Echevin *et al.* (2003) in the Gulf of Lyon where the along-slope, westward flowing Liguro-Provencal current can intrude onto the continental shelf, changing the local circulation structure. The Liguro-Provencal current is part of the large scale Gulf of Lyon gyre that is a large scale, wind and thermohaline forced gyre. The changes in position and strength of this current are due to the large scale vorticity balance of the north-western basin and the impacts on the shelf areas can be as large as the changes induced by local atmospheric forcing over the specific continental shelf area. Another example of a shelf area 'forced' by the large scale flow field, is the Adriatic Sea where the Otranto Strait provides the heat restoring mechanism for the large heat losses occurring in the shallow northern areas during winter (Artegiani *et al.*, 1997; Maggiore *et al.*, 1998). This restoring mechanism is like the one working in the Gibraltar Strait and regulates the long term buoyancy balance of the basin (Zavatarelli *et al.*, 2002).

The Northern Adriatic Sea heat budget is then partially controlled by the lateral exchange of waters occurring during the year. Such lateral exchange is maintained by the balance between the Eastern Adriatic Current (EAC), flowing northward along the eastern side of the Adriatic Sea, and the Western Adriatic Coastal Current (WACC), flowing southward. The EAC is partially determined by exchanges in the Otranto Strait and the WACC in connected to the Po river runoff and the wind forcing (Zavatarelli and Pinardi, 2003). Thus the circulation at the inlets to the Venice Lagoon is connected to the local atmospheric forcings (heat losses) and the circulation related to the Otranto Strait transport. This provides the conceptual basis for a possible predictive system of the circulation at the inlets to the Venice Lagoon; monitoring should consider both large and shelf scale monitoring and downscaling from the open ocean to the shelf scale (Pinardi *et al.*, 2002).

THE MEDITERRANEAN FORECASTING SYSTEM

The large amplitude natural variability discussed above couples with man-induced changes in the Mediterranean regions. Anthropogenic effects are evident, especially on the hydrological cycle of the basin and particularly on runoff. The largest man-induced changes are occurring in the Nile Delta and generally in the northern shore rivers where agricultural activities have induces changes in the timing of runoff peaks and total runoff (Sanchez-Arcilla and Simpson, 2002; Hamza, 2003). Thus the coastal areas cannot be managed without a nowcasting / forecasting system that will allow for the continuous assessment of system evolution. Such a system is at the base of the most urgent societal concerns for the further exploitation and preservation of the world's natural resources in the oceans and coastal regions of the Mediterranean Sea.

Since September 1998, the Mediterranean Forecasting System (MFS) programme (Pinardi and Flemming, 1998) has begun to implement the backbone of a future fully integrated system for the protection of the Mediterranean Sea marine environment and the sustainable exploitation of its resources. The first project to implement elements of the MFS programme was the Mediterranean Forecasting System Pilot Project (MFSPP) that lasted from September 1998 until June 2001. The second project is the Mediterranean Forecasting System Toward Environmental Predictions (MFSTEP) which began in March 2003 and will end in 2005.

In parallel with these large scale monitoring and nowcasting/forecasting efforts, a regional and shelf scale forecasting system for the Adriatic Sea has been started that will provide the necessary downscaling to the shelf areas with appropriate horizontal and vertical resolution. The ADRIatic sea integrated COastal areaS and river basin Management system pilot project (ADRICOSM) started its activities in October 2001 and ended its first phase in October 2003. This system has established the first real time coastal monitoring and nowcasting/forecasting system for the Adriatic Sea nested within the large scale MFS forecasts.

In the following discussion we start first with the description of the MFS components and then consider the ADRICOSM implementation and results.

THE LARGE-SCALE OBSERVING AND DATA MANAGEMENT SYSTEM

The four observing system components of MFS are: (1) the satellite data, near real time data analysis and dissemination network; (2) the voluntary observing ship (VOS), vertical profiling system; (3) the Mediterranean moored multidisciplinary array (M3A) observing platforms; and (4) the autonomous drifting and profiling system.

Real time remotely sensed data collection and analysis is the backbone of nowcasting / forecasting in the ocean. The most important data sets used are: (1) sea surface height (SSH) or sea level anomaly (SLA) from satellite altimeters; (2) sea surface temperature (SST) from satellite radiometers; (3) sea surface chlorophyll (SSC) from radiometers and (4) surface winds from scatterometers. SST, SLA and surface winds are currently produced in real time by several centres and institutions. Retrieval algorithms have become very reliable and allow the accurate computation of derived quantities.

For the Mediterranean Sea, real time SST and SLA are assimilated in the forecasting ocean model. The algorithms developed for real time analysis of along track SLA seem to give reasonable data for the assimilation scheme. SST is produced as a weekly mean of daily nighttime images, taken from different passes over the Mediterranean and collected by two different data collection centres in France and Italy. The SST data set is used in the assimilation scheme as a heat flux correction term at the model surface. The scientific analysis of the satellite MFSPP data set has been done by intercomparing SST and SLA and SLA and XBT (eXpendable Bathy-Thermograph) data (Buongiorno Nardelli et al., 2003). SSC and surface winds are not yet analysed yet in real time. However, SSC is analysed in real time for the Adriatic Sea (see below) and surface wind analyses are evaluated with respect to operational analyses before starting the real time production.

The voluntary observing ships (VOS) for temperature data collection has been, and will continue to be, performed along several tracks across the Mediterranean deep ocean areas. The along track nominal resolution tracks is 12 nm and XBT are used down to 400 and 700 metres. The real time data collection system has been successfully organized and harmonization of the data collection exercise achieved. Results are presented in Fusco et al. (2003); the structure of the data management system is analysed in Manzella et al. (2003). There are two streams of data: one real time and without high level quality control; the other highly quality controlled and fully resolved in the vertical. The data collection is now done with full resolution profiles sent by portable phone teletransmission at the end of each track (i.e. with a maximum time delay of 24 hours after data collection). The VOS tracks will be used to deploy subsurface drifting buoys and to develop a new expandable instrument that will collect temperature and fluorescence data in order to reduce the cost-benefit ratio for such a monitoring system.

The Mediterranean moored multisensor array (M3A) design fulfils the requirements of MFS *in situ* multidisciplinary observations. The system is used to monitor the high temporal variability of the upper thermocline; euphotic zone field variables for the open ocean ecosystem; and air-sea interactions. The monitored state variables are: air temperature and dew point temperature, surface pressure, surface winds, precipitation, solar radiation for the surface, temperature and conductivity, oxygen, fluorescence, turbidity, and nitrates at selected depths. The basic idea of a modular system with acoustic links between different mooring lines is a promising one. The two to three month maintenance interval guarantees high data quality, with the exception of the turbidity and PAR (Photosynthetically Available Radiation) measurements. One year's physical, optical and biochemical data have been successfully collected at a location in the Cretan Sea from a water depth of 1050 m (Nittis *et al.*, 2003). The data have been used to calibrate ecosystem models in the area and the buoy system has proved to be a valuable data collection platform for that purpose (Triantafyllou *et al.*, 2003). The developments now being carried out consist of the adaptation of two more existing mooring lines to the M3A 'philosophy', one in the southern Adriatic and the other in the Ligurian Sea (north-western Mediterranean).

Current observing system developments are concerned with the definition of a autonomous drifting and profiling network for the Mediterranean Sea. This will consist of:

1) a high space-time resolution network of autonomous subsurface profiling floats (Array for Real-Time Geostrophic Oceanography (ARGO). The initial drifting depth will be fixed at 400 metres, with descent down to 700 m and surfacing every three-five days. Two different floats are already deployed in the western Mediterranean to check the functioning of the system, ultimately a total of 20-25 floats will be deployed;

2) a basin scale glider autonomous vehicle experiment. The target for the glider experiment is to carry out continuous unattended glider profiles down to 1 km along a 300 km section of the northern Ionian, approximately along the 23 °E longitude meridian. The repeat period should be approximately 30 days. Launching and recovery will be executed by small boats from a support base in Messina.

The main objective of the MFS data management system is to build a *regional data management* structure that will consider both real time data dissemination and archiving. This is necessary since regional data can be at higher resolution than required for the global ocean and will require adapted quality control procedures. The philosophy behind data management in regional/shelf areas is that quality control and data collection procedures should be validated and intercompared between the basin scale and the shelf scale systems and should have substantially the same protocol of data exchange.

The basin scale data management system considers the development of two streams of data in real time: one 'raw' and the other quality controlled and filtered. Such a data management structure is illustrated in Manzella *et al.* (2003) for the VOS subsystem. The regional data management structure will include a centralised archiving and dissemination data centre (ADDC) and several thematic expert data centres (TEDC) associated with the different data sources (Fig. 6.3). The communication between the data centres will be undertaken by WWW, ftp and email, with both automatic and manual retrieval modes. The ADDC system will provide the subsampling procedures necessary to send the collected data in real time to the World Meteorological Organization-WMO Global Teleconnection System (GTS), thus contributing to global ocean real time data exchange.

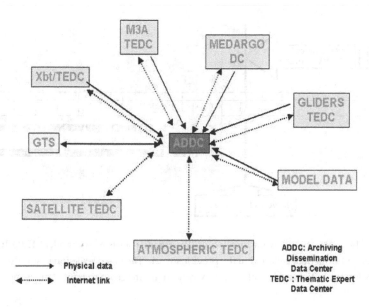

Fig. 6.3   MFS data management scheme. The centers are: (1) ENEA (IT) (www.santateresa.enea.it) for the VOS/XBT data (centre called Xbt/TEDC); (2) NCMR (GR) (www.ncmr.gr) for the M3A data (centre called M3A TEDC); (3) CLS (FR) (www.cls.fr) for the satellite data (centre called SATELLITE TEDC); (4) OGS (IT) (www.ogs.trieste.it) for the MEDARGO data (centre called MEDARGO DC); (4) IFM-KIEL (GE) (www.ifm.uni-kiel.de) for the gliders data (centre called GLIDERS TEDC); (5) IFREMER (FR) (www.ifremer.fr) for the ADDC; (6) IASA (GR) (www.iasa.gr) for the atmospheric data (ATMOSPHERIC TEDC); (7) INGV (IT) (www.bo.ingv.it) and UAT (GR) (www.oc.phys.uoa.gr) for the model data, analyses and forecasts.

THE BASIN SCALE ASSIMILATION AND FORECASTING ACTIVITIES

Ocean current forecasts are only possible if accurate initial and boundary conditions are known with sufficient accuracy and a numerical model is used to predict the future evolution of the flow field. In order to fulfill these requirements, an assimilation system and a numerical model have to be developed and coupled to the observing system collecting data in real time.

The ocean data assimilation algorithm and the impact of data strategy and accuracy is an important issue for MFSPP. Results have been achieved through the development of a new data assimilation scheme (scheme for ocean forecast and analysis (SOFA); De Mey and Benkiran, 2002). SOFA is an optimal interpolation reduced order scheme that uses multivariate

vertical Empirical Orthogonal Functions (EOF) as components of the reducing order projection operator. Vertical multivariate EOF have been computed for the basin, both from historical hydrological data and model results and intercomparison has been carried out successfully (Sparnocchia et al., 2003). SOFA has been implemented for combined SLA and XBT assimilation and tested under several different combinations of EOF and input data (Demirov et al., 2003).

Numerical forecasting was, and it still is, performed using an Ocean General Circulation Model (OGCM) that was implemented in the Mediterranean basin at $1/8 \times 1/8$ degrees resolution and 31 vertical levels. The pre-operational system started in January 2000 and is now continuing with a second version of the assimilation system, implemented on 1 January 2003. Real time atmospheric data collection is real-

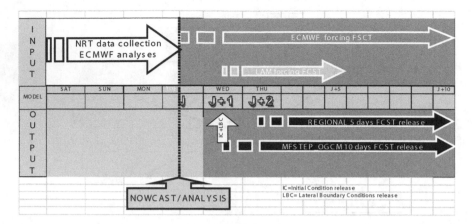

Fig. 6.4   The MFS weekly information flow for forecasting. The forecast starts every week at 12:00 Tuesday from an analysis/nowcast produced from data collected in the previous week. The basin scale forecast is done with ECMWF forcing and on Wednesday the initial and lateral boundary conditions are given to limited areas models, nested within the MFS OGCM.

ized for surface atmospheric field analyses and forecasts from the European Centre for Medium-range Weather Forecasts (ECMWF). Ten-day forecasts are produced once a week using an asynchronous coupling with atmospheric surface parameter forecasts. The forecasting system for MFSTEP is presented in Fig. 6.4. It includes NRT collection of ocean observations and meteorological forcing, production of analysis/nowcast on Tuesday of each week and the release of weekly 10-day forecasts. In addition, every week the forecast will be published on the website of MFSTEP (www.bo.ingv.it/mfstep). Analysis of forecast skill scores indicate that the rms (root mean square) surface temperature error is always less that 0.5 °C and that, at all levels, rms persistence error is always greater than the forecast error after the first to the tenth day (Fig. 6.5).

THE REGIONAL AND SHELF PREDICTIVE SYSTEM: A NESTED APPROACH

Regional and shelf models will approximately double and quadruple respectively the spatial resolution of the OGCM in critical areas. Regional models will run on a weekly basis, initialised from the MFSTEP OGCM fields (see Fig. 6.4), producing forecasting simulations for five days with high resolution atmos-

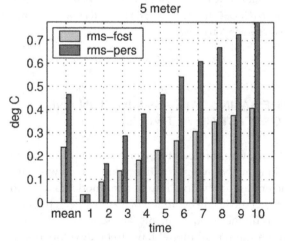

Fig. 6.5   The root mean square forecast error for the mean of the ten days and for each of the ten days of the forecast for sea surface temperature and for the enire basin and the whole 2002 year. The light bars indicate the forecast error computed as a difference between forecast and analyses, the dark bars indicate the persistence forecast error computed as a difference between the initial condition (nowcast or persistence) and the analyses for following ten days.

pheric fields. Lateral open boundary conditions for the regional and shelf models will be also provided by the MFSTEP-OGCM forecast fields.

During MFSTEP the meteorological community will be actively involved in producing the best estimates of atmospheric forcing fields at the best available temporal and horizontal resolutions. The Arpege (Deque and Piedelievre, 1995) Aladin (Radnoti *et al.*, 1995; Horanyi *et al.*, 1996) and Skiron/ETA (Kallos, 1997) limited area models will be run explicitly to produce 10 km hourly fields to force the basin scale and regional models. Together with the ECMWF fields, this will produce the best atmospheric dataset for future studies of ocean forcing and air-sea interaction coupling.

High resolution simulations in the coastal and shelf areas with nested model implementation have already been performed during MFSPP. Four 5 km intermediate models nested within the forecasting OGCM and nine shelf models (2-3 km resolution) nested into the intermediate, have been implemented and calibrated. Novel model implementation in areas previously lacking modelling experience has been successfully carried out (Korres, 2003). Understanding local shelf dynamics for forecasting needs has been accomplished in preparation for the next phase. All models have been able to capture known features of the seasonal circulation in the basin and different resolution models are consistent, even if the high resolution models are more realistic.

THE SHELF SCALE FORECASTING SYSTEM FOR THE ADRIATIC SEA

ADRICOSM represents the downscaling of the MFSTEP forecasting to the required scale for the shelf areas. This is the scale capable of resolving the physical flow field transport processes and the optimal scale to couple the physical flow field with land derived sources of pollutants and nutrients that normally determine the state of health of the coastal system.

Thus the major aims of this project are:

- to demonstrate the feasibility of Near Real Time (NRT) nested coastal current forecasts;
- to develop the integration of the river monitoring and modelling system with the coastal current forecasting.

On the basis of the MFS experience, the forecasting activities at weekly time scales have been demonstrated to be applicable also in this critical shelf area. An important aim of the project is also to develop a data assimilation scheme for both the large scale and the coastal datasets.

THE SHELF SCALE OBSERVING SYSTEM

The ADRICOSM observing system is composed of two parts: one for the open ocean, encompassing the Southern Adriatic deep areas, and the second one focussing on the shelf/coastal processes. The space and time resolution of the two components is quite different. The shelf scale is centered at weekly time scales and the deep ocean areas at monthly time scales. The effort here is to unify data collection protocols and data transmission procedures so that shelf scale data are also disseminated in real time.

The ADRICOSM shelf scale observing system is localized in four different coastal regions: Emilia Romagna, Gulf of Trieste and two areas along the Croatian coast. In these regions, traditional transects of temperature and salinity (from CTD) are realized and data transmitted in real time. The buoy network is dedicated to air-sea interaction monitoring and to the upper water column but the measurements are still not as interdisciplinary as in the M3A open ocean case. This shelf scale observing system is coupled with the two VOS-XBT tracks that are part of the deep ocean monitoring system and which run in the Southern Adriatic.

THE ADRIATIC AND SHELF NUMERICAL MODELLING AND FORECASTING

The Adriatic Sea regional model is at 5 km resolution and 21 sigma layers in the vertical. It is based on the POM (Princeton Ocean Model, Blumberg and Mellor, 1987) which was initially implemented for the Adriatic Sea by Zavatarelli and Pinardi (2003). The lateral open boundary conditions in the northern Ionian Sea are nested within the MFSTEP OGCM. The high frequency forcing is provided by ECMWF surface variables with interactive air-sea physics (Oddo *et al.*, 2005).

In the shelf area of the Croatian Islands that includes the major town of Split and the Neretva river outflow, a shelf model (POM at 1.5 km horizontal resolution) has been implemented and nested within the Adriatic regional model. In addition, the very near coastal waters of the same area have been modelled by a 2D hydro-dynamical model of variable resolution (with a finite elements grid), down to 20 m horizontal resolution. This very high resolution model receives lateral boundary conditions from the shelf model but in addition is coupled with a river basin modelling and forecasting system developed for the Cetina river and for the sewage system of Split.

In May 2003, a pre-operational system was set up with forecasts made in real time once a week, coupled with the MFSTEP forecasting cycle, as depicted in Fig. 6.4. The Po runoff is included with daily flow data while all the other rivers are represented by monthly mean runoff values. The Croatian shelf scale forecasting is also undertaken in real time. This ongoing real time experiment (www.ingv.it/adricosm) shows that nested forecasts are possible and under several circumstances are accurate. In the Adriatic Sea, the forecast accuracy is limited by the capability to predict the run off of the major rivers and by the unavailability of resolved atmospheric forcing fields (of the order of 5-7 km in the horizontal). It is in fact already well known that atmospheric forcing variability is high in this region and can generate jets and gyres due to wind stress curl structures of the order of several tens of kilometres.

CONCLUSIONS

A short-term forecasting system for the Mediterranean basin scale and the Adriatic coastal areas has been developed that provides continuous monitoring of the flow field evolution and its changes. Such a system is the backbone for more environmentally oriented systems that it should provide forecasts in the near future of coastal algal biomass variability, pollutant dispersion and indicators of ecosystem health and change.

The concept is that dynamical downscaling is required in order to reach the necessary resolution for the appropriate simulation of transport and coastal processes and for coupling with biochemical tracers.

This paper demonstrates that forecasting from the basin scales to the shelf areas is practical with present day technology. This system also provides the means to improve our understanding, and our capability to accurately model, the physical processes with an incremental approach and the optimal usage of all information. In the future it will be necessary to couple such a system with monitoring and modelling of biochemical fluxes. This will give rise to a complete, predictive system of ecosystem variability and change.

ACKNOWLEDGEMENTS

This work has been partially funded by the Italian Project ADRICOSM (supported by the Italian Ministry for the Environment and Territory) and MFSTEP (supported by the European Community, V Framework Programme – Energy, Environment and Sustainable Development; Contract no. EVK3-CT-2002-00075). All the partners of the two projects are thanked for their invaluable contributions to the material reported in this chapter.

REFERENCES

Artegiani, A., Bregant, D., Paschini, E., Pinardi, N., Raicich, F. and A. Russo. 1997. The Adriatic Sea general circulation. Part I: Air–Sea interactions and water mass structure. *Journal of Physical Oceanography* 27, 1492–514.

Brankart, J. M., and N. Pinardi. 2001. Abrupt cooling of the Mediterranean Levantine Intermediate Water at the beginning of the eighties: observational evidence and model simulation. *Journal of Physical Oceanography* 31, 8, 2307–20.

Blumberg, A. F., and G. L. Mellor. 1987. A description of a three dimensional coastal ocean circulation model. In Norman S. Hemps (ed.) *Three dimensional Ocean Models.* Washington DC: American Geophysical Union, pp. 1–16.

Buongiorno Nardelli, B., Larnicol, G., D'Acunzo, E., Santoleri, R., Marullo, S., and P.Y. Le Traon. 2003. Near Real Time SLA and SST products during 2 years of MFS pilot project: processing, analysis of the variability and of the coupled patterns. *Annales Geophysicae* 21, 1, 103–21.

Castellari, S., Pinardi, N., and K. D. Leaman. 2000. Simulation of water mass formation processes in the Mediterranean Sea: influence of the time frequency of the atmospheric forcing. *Journal Geophysical Research* 105, C10, 24,157–81.

De Mey, P., and M. Benkiran. 2002. A Multivariate reduced-order optimal interpolation method and its application to the Mediterranean basin-scale circulation. In N. Pinardi and J. D. Woods, (eds.), *Ocean Forecasting: Conceptual Basis and Application*. Berlin: Spinger-Verlag, 281–306.

Demirov, E., and N. Pinardi. 2002. Simulation of the Mediterranean Sea circulation from 1979–1993. Part I: The interannual variability. *Jnl. Mar. Syst.* **33/34**, 23–50.

Demirov, E., Pinardi, N., and C. Fratianni. 2003. Assimilation scheme of the Mediterranean Forecasting System: operational implementation. *Annales Geophysicae* **21**, 180–204.

Deque, M., and J. P. Piedelievre. 1995. High resolution climate simulation over Europe. *Climate Dyn.* **11**, 321–39.

Echevin, V., Crépon, M., and L. Mortier. 2003. Analysis of the mesoscale circulation in the North Western Mediterranean Sea simulated in the framework of the Mediterranean Forecast System Pilot Project. *Annales Geophysicae* **21**, 2, 281–97.

Fusco, G., Manzella, G. M. R., Cruzado, A., Gacic, M., Gasparini, G. P., Kovacevic, V., Millot, C., Tziavos, C., Velasquez, Z. R., Walne, A., Zervakis, V., and G. Zodiatis. 2003. Variability of mesoscale features in the Mediterranean Sea from XBT data analysis. *Annales Geophysicae* **21**, 1, 21–32.

Hamza, W., Ennet, P., Tamsalu, R., and V. Zalensky. 2003. The 3D physical-biological model study in the Egyptian Mediterranean coastal area. *Aquatic Ecology* **37**, 307–24.

Horanyi, A., Ihasz, I., and G. Radnoti. 1996. ARPEGE/ALADIN a numerical weather prediction model for Central-Europe with the participation of the Hungarian Meteorological Service. *IDÖJARAS* **100**, 4.

Josey, S. A. 2003. Changes in the heat and freshwater forcing of the Eastern Mediterranean and their influence on deep water formation. *Journal of Geophysical Research* **108** (C4), 3108–24.

Kallos, G. 1997. The Regional weather forecasting system SKIRON. In Kallos (ed.), *Proceedings of the Symposium on Regional Weather Prediction on Parallel Computer Environments, 15–17 October 1997, Athens, Greece*. Athens: University of Athens, pp. 1–9.

Korres, G., and A. Lascaratos. 2003. A one-way eddy resolving model of the Aegean and Levantine basins: implementation and climatological runs. *Annales Geophysicae* **21**, 205–20.

Korres, G., and A. Lascaratos. 2000. The ocean response to low frequency interannual atmospheric variability in the Mediterranean Sea, Part I: Sensitivity experiments and energy analysis. *Journal of Climate* **13**, 705–31.

Lermusieaux, P. F. J. and A. R. Robinson. 2001. Features of dominant mesoscale variability, circulation patterns and dynamics in the Strait of Sicily. *Deep Sea Research I* **48**, 1957–97.

Maggiore, A., Zavatarelli, M., Angelucci, M. G. and N. Pinardi. 1998. Surface heat and water fluxes in the Adriatic Sea: seasonal and interannual variability. *Phys. Chem. Earth* **23**, 561–7.

Malanotte-Rizzoli, P., Manca, B., d'Alcala, M. R., Theocharis, A., Brenner, S., Budillon, G., and E. Özsoy. 1999. The Eastern Mediterranean in the 80s and in the 90s, the big transition in the intermediate and deep circulations. *Dyn. Atm. Oceans* **29**, 365–95.

Manca, B., Budillon, G., Scarazzato, P. and L. Orsella. 2003. Evolution of dynamics in the Eastern Mediterranean affecting water mass structures and properties in the Ionian and Adriatic Seas. *Jour. Geophys. Res.* **108**, C9, 101029–46.

Manzella, G. M. R., Scoccimarro, E., Pinardi N. and M. Tonani. 2003. Improved near real-time data management procedures for the Mediterranean ocean Forecasting System – Voluntary Observing Ship program. *Annales Geophysicae* **21**, 1, 49–62.

Molcard, A., Pinardi, N., Iskandarani, M., and D. B. Haidvogel. 2002. Wind driven general circulation of the Mediterranean Sea simulated with a Spectral Element Ocean Model. *Dynamics of Atmospheres and Oceans* **35**, 2, 97–130.

Monaco, A., and S. Peruzzi. 2002. The Mediterranean Targeted Project MATER–a multiscale approach of the variability of a marine system–overview. *J. Mar. Syst.* **33/34**, 3–21.

Nittis, K., Tziavos, C., Thanos, I., Drakopoulos, P., Cardin, V., Gacic, M., Petihakis, G. and R. Basana. 2003. The Mediterranean Moored Multi-sensor Array (M3A): system development and initial results. *Annales Geophysicae* **21**, 1, 75–87.

Oddo, P., Pinardi, N., and M. Zavatarelli. 2005. A numerical study of the interannual variability of the Adriatic Sea (2000–2002). In press. *Journal of the Global Environment*.

Pinardi, N. and N. Flemming. 1998. *The Mediterranean Forecasting System Science Plan EuroGOOS*, Publication No. 11. Southampton: Southampton: Oceanography Centre.

Pinardi, N., Korres, G., Lascaratos, A., Russenov, V., and E. Stanev. 1997. Numerical simulation of the Mediterranean Sea upper ocean circulation. *Geophys. Res. Lett.* **24**, 425–8.

Pinardi, N., Auclair, F., Cesarini, C., Demirov, E., Fonda Umani, S., Giani, M., Montanari, G., Oddo, P., Tonani, M. and M. Zavatarelli. 2002. Toward marine environmental predictions in the mediterranean sea coastal areas: a monitoring approach. In Nadia Pinardi and John Woods (eds.), *Ocean Forecasting*. Berlin: Springer-Verlag, pp. 281–305.

Pinardi, N., Arneri, E., Crise, A., Ravaioli, M., and M. Zavatarelli. 2005. *The physical, sedimentary and ecological structure and variability of shelf areas in the Mediterranean Sea.* In press. Vol. XIV of The Sea.Cambridge, MA: Harvard University Press.

Radnoti, G., Ajjaji, R., Bubnova, R., Caian, M., Cordoneanu, E., Von Der Emde, K., Gril, J. D., Hoffman, J., Horanyi, A., Issara, S., Ivanovici, V., Janousek, M., Joly, A., Le Moigne, P., and S. Malardel. 1995. The spectral limited area model *ARPEGE/ALADIN. PWPR Report Series* **7**, WMO-TD 699, 111–17.

Raicich, F., Pinardi, N., and A. Avarra. 2003. Teleconnections between Indian monsoon and Sahel rainfall and the Mediterranean. *International Journal of Climatology* **23**, 2, 173–86.

Robinson, A. R., Golnaraghi, M., Leslie, W. G, Artegiani, A., Hecht, A., Lassoni, E., Michelato, A., Sansone, E., Theocaris, A., and U. Unluata. 1991. The Eastern Mediterranean general circulation: Features, structure and variability. *Dyn. Atmos. Oceans* **15**, 215–40.

Roether, W., Manca, B. B., Klein, B., Bregant, D., Georgopoulos, D., Beitzel, V., Kovacevic, V., and A. Luchetta. 1996. Recent changes in Eastern Mediterranean deep waters. *Science* **271**, 5247, 333–5.

Sanchez-Arcilla, A., and J. H. Simpson. 2002. The narrow shelf concept: coupling and fluxes. *Continental Shelf Research* **22**, 153–72.

Sparnocchia, S., Pinardi, N., and E. Demirov. 2003. Multivariate empirical orthogonal function analysis of the upper thermocline structure of the Mediterranean Sea from observations and model simulations. *Annales Geophysicae* **21**, 1, 167–87.

Triantafyllou, G., Petihakis, G., and I. J. Allen. 2003. Assessing the performance of the Cretan Sea ecosystem model with the use of high frequency M3A buoy data set. *Annales Geophysicae* **21**, 1, 365–75.

Zavatarelli, M., and N. Pinardi. 2003. The Adriatic Sea Modelling system: a nested approach. *Annales Geophysicae* **21**, 2, 345–64.

Zavatarelli, M., Pinardi, N., Kourafalou, V. H., and A. Maggiore. 2002. Diagnostic and prognostic model studies of the Adriatic Sea General Circulation: seasonal variability. *Journal Geophysics Research* **107**, 3004–25.

7 • Features of scatterometer wind observations in the Adriatic Sea

S. ZECCHETTO, F. DE BIASIO AND M. BAJO

INTRODUCTION

The Mediterranean Sea is a basin where the wind circulation at sub-synoptic scales may be strongly influenced by the coastal orography and by the presence of many islands. As a consequence, meteorological phenomena at spatial scales of up to a few kilometres are very common. They are important for the whole dynamic of the atmosphere but their simulation by atmospheric models is difficult. In the Adriatic Sea, for instance, the channelling effects due to the presence of the Apennines and Balkan Mountains chains occurs under the influence of both north-easterly and south-easterly winds. In this context, satellite wind observations from microwave scatterometers are of primary importance. With a spatial resolution of *c.* 25 km by 25 km, they provide a thorough description of many meso-scale phenomena, such as frontal systems and orographic winds, as well as of the structure of the regional wind systems (Guymer and Zecchetto, 1993; Zecchetto and Cappa, 2001; Zecchetto *et al.*, 2002).

This chapter aims to illustrate the potentialities offered by scatterometer wind observations in the characterization of wind fields in the Adriatic Sea, and to discuss their possible use as forcing fields in storm surge hindcasting. The second section describes the main characteristics of the scatterometer wind observations, their present and future availability, and their temporal sampling of basins like the

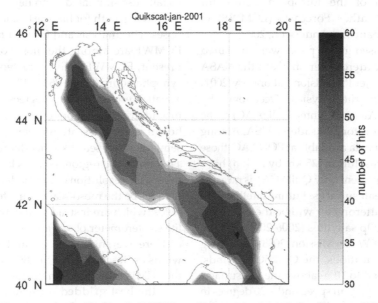

Fig 7.1   Typical monthly coverage provided by QuikSCAT satellite over the Adriatic Sea (January 2001).

Flooding and Environmental Challenges for Venice and its Lagoon: State of Knowledge, ed. C. A. Fletcher and T. Spencer.
Published by Cambridge University Press, © Cambridge University Press 2005.

Adriatic Sea. In the third, two case studies are presented, one of *bora* and one of *sirocco*, the two most important winds in the Adriatic Sea, and by the modelled wind fields. The fourth discusses the possible advantages in considering the scatterometer winds as meteorological forcing for storm surge hindcasting, while the last section is devoted to conclusions and recommendations for future work.

MAIN CHARACTERISTICS OF SATELLITE OBSERVATIONS

Microwave scatterometers sense the sea surface roughness, the development of which depends on wind speed and atmospheric stability. This is defined according to the difference, ΔT, between the air and sea temperatures i.e. $\Delta T < 0$ unstable, $\Delta T = 0$ neutral and $\Delta T > 0$ stable. Parametric algorithms are used by the space agencies to retrieve the wind vectors from the radar backscatter, yielding the wind at neutral stability conditions. Thus scatterometer winds need to be corrected for the actual atmospheric stability conditions. Satellite scatterometer data are used worldwide for scientific (determination of the wind forcing and of the air-sea fluxes) and operational (improvement of the weather and wave predictions) purposes. They have been assimilated since 1996 into the global model of the European Centre for Medium-Range Weather Forecasts (ECMWF) in Reading, U.K. (Trepaut and Andersson, 2003).

The wind data used in this paper were measured by the SeaWinds scatterometer on board the NASA QuikSCAT satellite (Jet Propulsion Laboratory, 2001) and distributed by the Physical Oceanography Distributed Active Archive Center (PODAAC) of the Jet Propulsion Laboratory, Pasadena, USA. Among the different kind of data available at PODAAC, those of highest spatial resolution (25 km by 25 km) have been used here. At present only QuikSCAT is measuring winds over ocean surfaces. Future satellite missions bearing a scatterometer will be the European Space Agency MetOp satellites (2006-2019) and the NASA Ocean Vector Winds Mission (2008). According to the sensor specifications, the QuikSCAT winds, which are referenced to 10 m above the sea surface, have ~ 2 m s^{-1} of accuracy in speed and ~ 20 degrees in direction. In this work they have been corrected by a

boundary layer model based on the theory of Kondo (1975), using the air temperatures and the atmospheric pressure of the analysis fields provided by ECMWF and the sea surface temperature observations provided by the Advanced Very High Resolution Radiometer (AVHRR) instruments on board the National Oceanographic and Atmospheric Administration (NOAA) satellites (Brown *et al.*, 1985; McClain *et al.*, 1985).

QuikSCAT observations, which span a period from July 1999 to the present, provide the best temporal sampling available in the Mediterranean Sea. Figure 7.1 reports a typical monthly coverage provided by QuikSCAT satellite over the Adriatic Sea, showing that the basin is not uniformly covered: the best coverage occurs in the North and in the South of the basin (~ 60 passes per month). Therefore, the Adriatic Sea is sampled at most two times per day, around 04:00 and 17:00 GMT. The lack of data close to coast is a consequence of the scatterometer spatial resolution of 25 km which prevents obtaining data not contaminated by the land closer than 12.5 km offshore.

SCATTEROMETER OBSERVATIONS AND ANALYSIS WIND FIELDS IN SMALL BASINS

Oceanographic models, like storm surge models, usually use simulated wind fields as the meteorological forcing, both for hindcasting and forecasting purposes. The analysis and forecast fields produced by ECMWF are among the most commonly used. At present, ECMWF delivers meteorological fields at synoptic hours (00:00, 06:00, 12:00, 18:00 GMT), with a spatial resolution of 0.5 degrees. The wind fields are provided at 10 m above the sea surface, the same height as for the scatterometer winds. However, in semi-enclosed seas like the Mediterranean Sea, and *a fortiori* in the regional basins like the Adriatic Sea, the spatial resolution of the model fields is too coarse to describe the meso-scale characteristics of the wind fields, which are instead well described by the satellite scatterometer observations.

Here we show two examples of Adriatic Sea winds as described by QuikSCAT observations and the ECMWF analysis wind fields, the latter representing 'the best gridded estimate of the state of the atmosphere (best fit to observations)' (ECMWF,

2003). Plate 4 shows a case of north-easterly *bora* (top panels) and one of south easterly *sirocco* (bottom panels), the two most stormy winds blowing across the Adriatic Sea. The left panels report the ECMWF analysis wind fields, the right panels the scatterometer observations. The ECMWF analysis wind fields, interpolated over a 25 km by 25 km grid for comparison with the satellite observations, are one (for the *sirocco* case) and two (for the *bora*) hours later than the scatterometer pass time. While this time difference is not seen as critical, it should be pointed out that different spatial resolution implicates different representations of the physical processes. In other words, a finer model resolution will not be able to describe some of the meso-scale features of the wind if the physical processes at that scale are not appropriately modelled and/or if the observations for the model initialization are too few in number. In our experience, limited area atmospheric model simulations in the Mediterranean Sea, initialized using the ECMWF analysis fields, do not offer particular improvements with respect to the ECMWF fields.

The example of *bora* (top panels) introduces well the main characteristics of the scatterometer winds: a wider range of speeds, higher spatial variability but lower spatial coverage than the analysis field. The observed *bora* wind field clearly shows the spatial structure of this wind, characterized by an alternate occurrence of high and low wind areas along the Adriatic major axis, which derives respectively from air flow funnelling through the Balkans mountains gaps of Trieste (~ 45.5 N, 13.5 E), Senj (~ 45 N, 15 E), Sibenik (~ 43.5 N, 16 E), Makarska (~ 43.3, N 17 E) and by mountain sheltering. Furthermore, the wind is stronger along the eastern Dalmatian coast. These features, which bring also strong spatial wind speed gradients and a characteristic pattern to the wind curl, are only weakly reproduced by the analysis field (top left panel), which shows a less pronounced spatial variability in the wind across the basin. The example of *sirocco* (bottom panels) has a tighter relationship with storm surge issues, as this wind is the main cause of the meteorological tides in the northern Adriatic Sea. The structure of the observed wind field shows two areas of high speed (~ 15 m s^{-1}) in the southern and north-eastern parts of the basin, along with a main wind direction almost aligned to the

Adriatic Sea major axis, probably due to the channelling mentioned above. The analysis field, instead, provides a less variable wind field, as in the case of the *bora*, with northward prevailing directions in the central and northern parts of the basin.

In both cases, the scatterometer wind observations results in wind speeds on average 3 m s^{-1} higher than the ECMWF analysis winds. This bias is slightly dependent on the model spatial resolution (as assessed in Zecchetto *et al.*, 2002 and recently verified by comparing satellite observations to wind fields produced by a regional atmospheric model). An important issue concerns the spatial distribution of the wind speed differences between the scatterometer and the analysis winds. Figure 7.2 reports the monthly mean field of this wind speed difference for January 2001, when the *sirocco* was the prevailing wind (top panel) and for February 2001, when the *bora* was the most frequent wind (bottom panel). In both cases, the largest differences occur close to the coasts where the orographic effects are not well accounted by the model, which tends to smooth out the fields. A substantial agreement is found in the southern and northern parts of the basin. In the case of prevailing *sirocco* winds, the central Adriatic shows the largest differences (~ 3m s^{-1}).

The bias of the analysis winds with respect to the scatterometer observations becomes even more crucial when it is considered that the real forcing exerted by the wind on the sea surface is expressed in terms of the wind stress, which is proportional to the square of the wind speed. An incorrect estimate of the wind speed by $\Delta u = 3$ m s^{-1} may result in a wind stress up to 4 times different, depending on the wind speed and on the atmospheric stability conditions.

The last consideration refers to the importance of the wind direction. In the *sirocco* case reported in Plate 4, there is a mean difference of about 45 degrees between the mean direction of the scatterometer (~ 150 degrees) and of the analysis wind fields (~ 105 degrees). Such a difference may have an important role when those fields are used in hindcasting the storm surge levels. Whilst the scatterometer winds are almost aligned with the Adriatic major axis, thus virtually blowing over the entire basin from south to north (~1000 km), the analysis winds have a fetch reduced to 200-300 km.

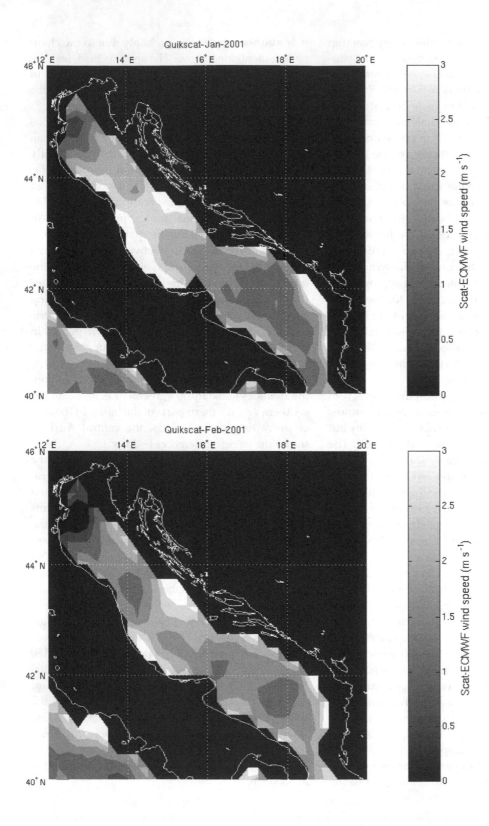

Fig. 7.2 Wind speed difference between QuikSCAT scatterometer observations and ECMWF analysis. Top panel: January 2001, when the *sirocco* wind was prevailing. Bottom panel: February 2001, when the *bora* was the most frequent wind.

DISCUSSION

In the topic of the storm surge modelling, the correct definition of the meteorological forcing, both in terms of wind stress intensity and direction, may be of crucial importance. For instance, the *sirocco* example shown above is evidence that, besides the wind strength, the wind direction may also play an important role as a result of the geographic configuration of the Adriatic Sea. Furthermore, since the meso-scale features of the wind field on a seasonal time scale have an important role in driving ocean circulation (Zecchetto and Cappa, 1998, 2001), one may speculate that they could have some impact on the determination of storm surge levels in the basin.

Scatterometer wind fields can be used only for hindcasting, which is an activity devoted to the understanding of the main factors influencing a physical event. They are essential to define the spatial structure of the meteorological forcing fields, since they are the only available wind data in open sea, and they are also useful in tuning the atmospheric model forecasts to be used as forcing fields in scatterometer winds. They can be also used directly to force the storm surge models to hindcast events of high water level (e.g. Canestrelli *et al.*, 2003). Despite the many problems brought about by the temporal sampling of the basin provided by the present scatterometer, the results seem promising. However, we are still far from having an exhaustive assessment about the benefits of using scatterometer winds in storm surge models, if any, and to understand the importance of both the spatial and temporal sampling in the hindcasting of water levels. It is clear that a better definition of the forcing fields represents one of the requirements of modelling.

Scatterometer winds are already assimilated by the ECMWF atmospheric model but the improvement in small basins seems at present negligible, due to its too coarse spatial resolution. It is also questionable if the scatterometer wind assimilation into regional atmospheric models could produce, by itself, remarkable improvements, since the surface winds represent only one of the datasets necessary to initialize the models, the other being the vertical structure of the atmosphere, which is often unknown at the regional scale. However, this is a matter for the atmospheric modelling research community and is beyond the aim of this chapter and our expertise. The development of methods for the integration of scatterometer winds with other kinds of data (*in situ* reports and model simulations), through objective analysis techniques, for instance, may provide some benefit in the definition of the initial conditions for storm surge forecasts.

CONCLUSIONS

Scatterometer wind observations from satellite provide a unique description of the meso-scale spatial features of the wind fields in small basins like the Adriatic Sea, not fully reproduced by the atmospheric models, either global or regional. In principle, they should be preferred to the model data when hindcasts of storm surge have to be produced. On the other hand, in the Adriatic Sea the present scatterometer provides wind fields two times per day 12 hours apart, making it doubtful that they are sufficient to improve the water level simulation when used as forcing. Present hindcasts and forecasts use ECMWF fields available four times per day 6 hours apart and therefore, to fully evaluate the impact of the observed winds into the storm surge models, scatterometer observations with a similar temporal sampling need to be available. Such a requirement has been almost achieved with ADEOS-II satellite. This satellite provided wind fields at 10:00 and 21:00 GMT, but unfortunately only for a few months in 2003. Satellite scatterometer wind observations cannot be exhaustive of the meteorological forcing, which is indeed represented by the wind stress, the calculation of which requires the knowledge of the air-sea precesses at local spatial scales and the availability of parameters from different sources (analysis, *in situ*, satellite, models). However, scatterometer winds may be extremely important to answer the following questions: what is the role of wind spatial structure, of the temporal evolution of atmospheric storms, of the air-sea stability conditions on water level set up? These represent the chief questions for future research.

ACKNOWLEDGEMENTS

The activity described in this chapter was realized with funding from the Agenzia Spaziale Italiana (ASI). The satellite wind data have been downloaded from the web site http://podaac.jpl.nasa.gov/. The ECMWF analysis data have been downloaded from the data delivery service of the ECMWF. The authors record their appreciation of the comments of the referees and editors, through which the paper has been improved.

BIBLIOGRAPHY

ADEOS-II User's Handbook, obtainable at
  http://sharaku.eorc.nasda.go.jp/ADEOS2/doc/document.html

Brown, O. B., Brown, J. B., and R. H. Evans. 1985. Calibration of advanced very high resolution radiometer infrared observations. *Journal of Geophysical Research* **90**, 11667–77.

Canestrelli, P., Cucco, A., De Biasio, F., Umgiesser, G., Zampato, L., and S. Zecchetto. 2003. The use of QuikSCAT wind fields in water level modeling of the Adriatic Sea. In *Proceedings of the Sixth European Conference on Applications in Meteorology ECAM 2003, Rome 15–19 September 2003.*

ECMWF. 2003. *MARS User Guide.* Reading: ECMWF.

Guymer, T. H., and S. Zecchetto. 1993. Applications of scatterometer winds in coastal areas. *International Journal of Remote Sensing* **14**, 9, 1787–812.

Jet Propulsion Laboratory. 2001. *QuikSCAT Science Data Product User's Manual*, Ver. 2.1 April 2001. JPL, D-18053.

Kondo, J. 1975. Air-sea bulk transfer coefficient in diabatic conditions. *Boundary Layer Meteorology* **9**, 91–112.

McClain, E. P., Pichel, W. G., and C. C. Walton. 1985. Comparative performance of AVHRR-based multichannel sea surface temperature. *Journal of Geophysical Research* **90**, 11587–601.

Trepaut, J. N., and E. Andersson. 2003. Assimilation of high-resolution satellite data. *ECMWF Newsletter* no. 97, 6–12.

Zecchetto, S., and C. Cappa. 1998. Oceanographic applications of scatterometer data in the Mediterranean Sea. In *Proceedings of a Joint ESA–Eutmetsat Workshop on Emerging Scatterometer Applications – From Research to Operations,* 5–7 October 1998. The Netherlands: ESA SP-424. ESTEC.

Zecchetto, S., and C. Cappa. 2001. The spatial structure of the Mediterranean Sea winds revealed by ERS-1 scatterometer. *International Journal of Remote Sensing* **22**, 1, 45–70.

Zecchetto, S., De Biasio, F., Music, S., Nickovic, S., and N. Pierdicca. 2002. Intercomparison of satellite observations and atmospheric model simulations of a meso-scale cyclone in the Mediterranean Sea. *Canadian Journal of Remote Sensing* **28**, 3, 413–23.

8 • Extreme storm surges in the Gulf of Venice: present and future climate

P. LIONELLO

INTRODUCTION

Exceptional floods of the lagoon islands are recorded in Venetian chronicles since the beginning of the permanent settlement of a local government. Occasional loss of human life, recurrent economic damage, and maximum level of the highest surges have been reported with a precision which is sufficient to reconstruct the frequency of past floods since the eighth century AD (Camuffo, 1993). The time series shows a succession of periods of recurrent floods, particularly intense in the first half of the sixteenth and of the eighteenth centuries, more quiet periods, and a continuous positive trend during the second half of the twentieth century. Although flooding of Venice has occurred many times during its history, the last 50 years represent an unprecedented period of frequent and intense events. In fact, instrumental records show clearly an increased frequency of floods and previous studies have already identified the corresponding positive trend of the extreme surges (e.g. Smith, 1986). There is a widespread concern that the intensity of such extreme events is increasing and that this might be, at least partially, associated with a climate change signal superimposed on subsidence of both natural and anthropic origin. The purpose of this study is to identify the atmospheric general circulation pattern responsible for the extreme floods of Venice, to evaluate their variability and trends in the present climate and their possible change in a 'doubled CO_2' scenario.

PRESENT VARIABILITY AND TRENDS OF EXTREME SEA LEVELS

The tide gauge PSal (Punta della Salute) provides hourly measurement of the sea level in the city centre and has done so almost continuously since 1940. This record represents the best source of information for analysing the behaviour of the sea-level oscillations inside the lagoon. These data have been post-processed to produce four different time series: 1) O, raw observed values; 2) ST, obtained from O by subtracting the astronomical tide; 3) SML, obtained from O by subtracting the mean sea level; 4) STML, obtained by subtracting both the astronomical tide and mean sea level. Therefore, the time series STML includes only the effect of storm surge and seiches. Mean sea level has been evaluated as the running mean value inside a 181-day wide time window. Single storms have been extracted from each of the four time series, considering events characterized with the highest sea level in a 72-hour time window; that is isolated peaks are attributed to a single storm if separated by less than 72 hours. The highest sea level in each storm is used to evaluate its intensity. The 72-hour length is not related to the duration of the synoptic forcing but to the requirement to include in a unique event the initial surge and the successive seiches triggered by it. For each year since 1940, this procedure results in a list of storms ranked in decreasing order of strength (from the peak sea level reached in each event).

The average value of the 10 highest peaks (one for each storm) is used to evaluate the overall intensity of the strongest events in each calendar year. The time series, smoothed with a three-year running mean, are shown in Fig. 8.1a. The mean sea level of the first analysed decade (1940–1950) has been subtracted from the time series O and ST, in order to remove the initial bias of these two time series. The time series show large inter-decadal variability and a systematic trend in O and ST, due to the local sea-level increase, or, more properly, to the local loss of ground

Flooding and Environmental Challenges for Venice and its Lagoon: State of Knowledge, ed. C. A. Fletcher and T. Spencer.
Published by Cambridge University Press, © Cambridge University Press 2005.

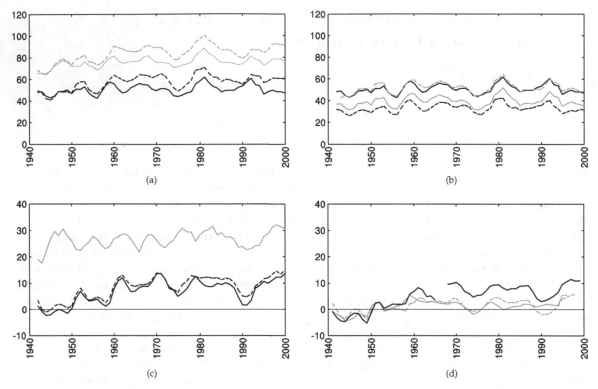

Fig. 8.1   Y-axis show sea level (cm) and x-axis shows the corresponding year in all panels. Panel (a): average value of the 10 highest storm surges in each year of the O (grey dashed line), ST (black dashed line), SML (grey continuous line), STML (black continuous line) time series. The average sea level in the decade 1940–1950 has been subtracted to the O and ST time series. Panel (b): values of the STML series estimate using a 31-day (black dashed line) 181-day (black continuous line), 365-day (grey continuous line), 1195-day (grey dashed line) window for the computation of the mean sea level. Panel (c): difference O-ST, which shows the effect of tide on storm surge (continuous grey), difference O-SML (dashed black) and ST-STML (continuous black), which both show the effect of local sea level rise on storm surge. Panel (d): sea level in Venice (black continuous line), Northern Adriatic (grey dashed line), North Western Mediterranean (grey continuous line). Sea levels are relative to the average value in the 1940–60 period. All lines in panels have been smoothed with a three-year running mean.

level. The two time series SML and STML, which are independent from ground level loss because of the subtraction of mean sea level, show a large inter-decadal variability but no appreciable trend.

Figure 8.1c shows the effect of local sea-level rise on the storm surge (difference between ST and STML) and on the superposition of tide and storm surge (difference between O and SML). In fact, there is a large increase in the peak values of the sea surface elevation in time series O and ST between the 1950s and the 1970s, parallel to the difference between sea level in Venice and the rest of the Mediterranean Sea in the same period. Figure 8.1d shows the three-year running mean of sea level relative to the 1940–1960 period in Venice, in the N Adriatic Sea (here represented as the average value of Trieste, Bakar, Split, Dubrovnik and Rovinij) and in the NW Mediterranean Sea (the average value of Genoa, Monaco and Marseille). The comparison between sea level in Venice and that in the rest of the

N Adriatic shows the increase associated with the well-known subsidence produced by soil water extraction (Carbognin *et al.*, 1976). Note that the initial part of the record presents some problem of homogeneity, as some data of coastal Croatian towns in the 1940s and early 1950s are missing. The effect of the tide inside the lagoon (shown by the difference between O – ST in Fig. 8.1c) on the peak level values is about 30 cm and shows a weak positive trend.

Since the main signal is associated with the presence of a positive trend in mean sea level, it is important to verify the importance of the time window used for its computation. In fact, long period (10 to 100-day period) oscillations, mainly due to planetary waves, are an important contribution to sea-level variability (Pasaric and Orlic, 2001). Figure 8.1b shows the STML time series computed using a 31, 181, 365, 1195-day long time window for the estimation of the reference mean sea level. Though the actual values are different (the longest time window gives the highest surge levels) the four lines show a similar behaviour, showing that the amount of inter-decadal variability, and the absence of trend, do not depend on the choice of the time window.

The extreme values statistics are estimated by fitting the GEV (Generalized Extreme Values) distribution, on the basis of the three highest events in each year (Smith, 1986; Tawn, 1988) of the STML time series computed using a 181-day long time window (Fig. 8.2a). The dashed horizontal line shows the 10-year return value, computed using the whole 60-year long period. The continuous grey and black lines show the 10-year return values, computed using only the highest surges within a 7 and 21-year wide time window respectively, and centred at each year of the 1940–2000 period. The smaller time window allows the resolution of shorter (inter-decadal) variability but has a larger statistical uncertainty, substantially because of the reduced sampling size. Average error is 18 cm, 11 cm and 6 cm for the estimate, based on the 7-year wide window, 21-year wide window and the whole period respectively. These uncertainties imply that, with the exception of the very first part of the record, differences are not statistically significant and that the values of the extremes can be considered constant over time. In other words, the single value computed on the basis of the whole record is compatible with the estimates based on the 7-year and 21-year windows, except for the first decade of the record during which extreme surges were lower. The presence in the time window of the 4 November 1966 surge, which with a height of 176 cm in the STML time series corresponds to a return period longer than 250 years, does not introduce statistically

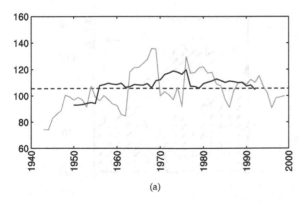

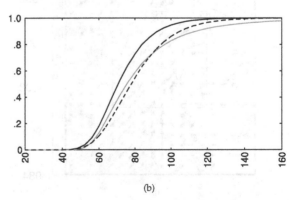

(a)   (b)

Fig. 8.2   Panel (a): variation of the 10-year return value (y-axis, cm.) during the 1940–2000 period. The horizontal dashed line shows the value computed on the basis of the whole 1040–2001 period. The grey and black continuous lines show the values computed using events inside a moving 7-year and 21-year long time window, respectively. Panel (b): GEV distributions resulting from three different time windows: 1940 to 1959 (black), 1960 to 1979 (grey), 1980 to 1999 (black-dashed). Each line shows the probability (Y-axis) of an annual maximum value lower than a given threshold (cm, X-axis).

significant difference in the GEV distribution. Figure 8.2b shows the 3 GEV distributions resulting from the division of the 1940-2000 period into three sub-periods, each 20-years long. The comparison of the three GEV distributions confirms that the first part of the record presents a different statistical distribution of the extremes. The GEV of the earlier period (1940–1959) is different from the two other distributions which almost overlap. Consequently, there is no direct link between the two warming periods which took place at an hemispheric scale from 1910 to 1945 and from 1976 onwards and the observed variations of the extreme surge values.

RELATION TO SYNOPTIC SCALE CIRCULATION AND TELECONNECTION PATTERNS

The meteorological situation which leads to surge in the N Adriatic Sea, and the consequent flooding of Venice, is well established (Robinson *et al.*, 1973). A deep low-pressure system induces a southeastward pressure gradient along the basin, which, because of the channelling due to the long coastal mountain ridges, produces a strong *sirocco* wind. Both wind

and inverse barometric effect contribute to the surge in the northern part of the basin. Figure 8.3 shows a model reconstruction of the highest ever recorded surge of 4 November 1966. Figure 8.3a shows the sea-level pressure and surface wind field. Figure 8.3b shows sea level and the significant wave height field. Both panels refer to the conditions at the peak of the storm. The model used is a regional coupled atmosphere-wave-ocean model called MIAO (Model of Interacting Atmosphere and Ocean; Lionello *et al.*, 2003b), which carried out this simulation on the basis of the NCEP (National Centre for Environmental Prediction) analysed wind and sea-level pressure fields (Kalnay *et al.*, 1996).

The synoptic situations responsible for surge events have already been shown to originate from a large low pressure system above Northern Europe (Trigo and Davies, 2002), which produces strong orographic cyclogenesis on the southern side of the Alps (Buzzi and Tibaldi, 1978). A detailed analysis (based on the NCEP geopotential height at 1000 hPa level) reveals that a further distinction is possible according to the location of the main minimum central pressure with respect to the Alps. The whole cyclone can take a southward path, resulting in an intense cyclone

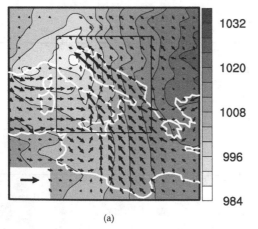

(a)

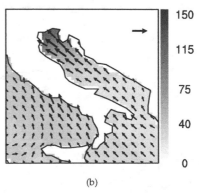

(b)

Fig. 8.3   Panel (a): meteorological situation at the peak of the 4 November 1966 flood event. The wind field (black arrows, a reference 25 m s⁻¹ arrow is shown at the lower-left corner) is superimposed to the contour lines of the sea level pressure (pressure values in hPa according to the greyscale bar). Panel (b): sea level (values in cm according to the greyscale bar) and wave field (arrows represent significant wave height and mean direction) in the area delimited by the rectangle in the left panel. A reference arrow corresponding to a 6m high wave is shown at the top-right corner.

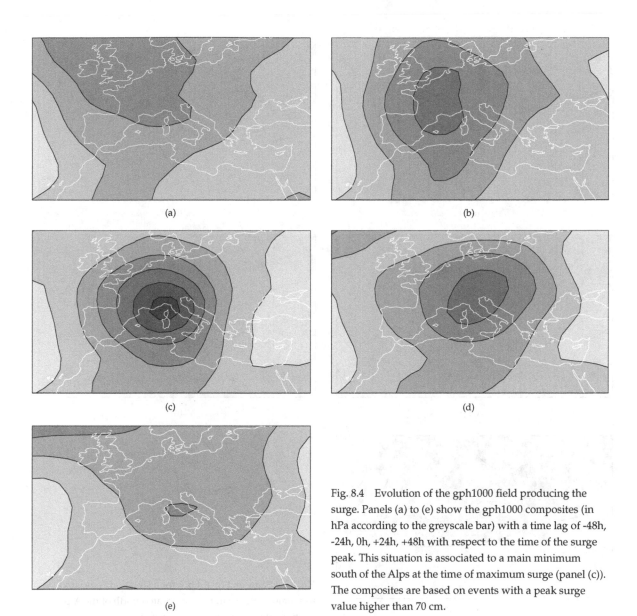

(a)

(b)

(c)

(d)

(e)

Fig. 8.4   Evolution of the gph1000 field producing the surge. Panels (a) to (e) show the gph1000 composites (in hPa according to the greyscale bar) with a time lag of -48h, -24h, 0h, +24h, +48h with respect to the time of the surge peak. This situation is associated to a main minimum south of the Alps at the time of maximum surge (panel (c)). The composites are based on events with a peak surge value higher than 70 cm.

over Italy at the time of the highest surge level; this subsequently attenuates, conserving a well defined identity throughout its evolution (Fig. 8.4). This meteorological pattern accounts for about 40 per cent of surges above 70 cm (18 events since 1958). Alternatively, the main pressure minimum can follow a northern path, remaining over central and northern Europe, but originating a secondary mini-

mum south of the Alps which is subsequently reabsorbed in the deeper system (Fig. 8.5). This second pattern accounts for about 60 per cent of surges above 70 cm (34 events since 1958) and is generally associated with higher surge levels. The average height of the 10 highest surges since 1958 is 96 cm and 105 cm for the first and the second patterns respectively. The occurrence of the surge in the Gulf

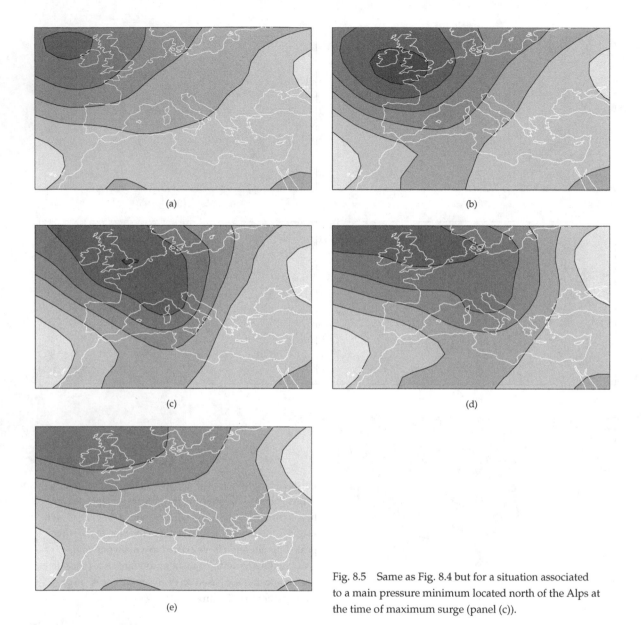

(a)

(b)

(c)

(d)

(e)

Fig. 8.5   Same as Fig. 8.4 but for a situation associated to a main pressure minimum located north of the Alps at the time of maximum surge (panel (c)).

of Venice is associated with the intensity and frequency of these two synoptic patterns.

These adverse conditions are triggered by the passage of a deep cyclone and their strength and frequency is related to the position and intensity of the storm track over western Europe. These are, in turn, associated with the spatial distribution of the average geopotential height, its meridional gradient and the presence of a trough over western Europe. This analysis considers the average geopotential height at 1000 hPa level during November and December, the time of year with most frequent floods. First, two composites, one for the periods with at least three surge events above 67 cm, another for those with no

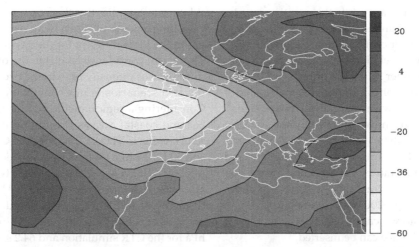

Fig. 8.6   Two-monthly average gph1000 pattern (in hPa according to the greyscale bar) associated with surge events during November and December for the period 1958–2001. The pattern shows the difference between composite representing periods with at least three high surge events and without any high surge.

surge higher than 63 cm, have been computed (Fig. 8.6). Each composite is based on 8 samples out of the 44 available for the period 1958–2001. The two composites are representative of stormy and calm periods, as far as storm surge is concerned. The pattern has been obtained by computing the difference between these two composites. It reveals a large circulation feature associated with the penetration of the storm track in the Mediterranean region. The large-scale pattern does not resemble the NAO (North Atlantic Oscillation) dipole, otherwise important for large-scale variability of precipitation and temperature over Europe (Hurrell, 1995). This shows that a large-scale pattern with a centre of action located above central and Northern Europe, different from the NAO, is associated with the variability of the floods of Venice and of storminess in the NW Mediterranean region. In fact, the NAO index and the values of extreme surges in Venice are poorly correlated. On the other hand, although there is no clear physical explanation, the sunspot number plays a role, which has already been noticed in the literature (Smith, 1986; Tomasin, 2002). Figure 8.7 shows the time series of the NAO index, the average value of five highest surges and the sunspot number. These last two quantities have been transformed into a dimen-

sionless index, subtracting the mean and normalizing with the standard deviation of the relative time series. Only November and December are considered for the NAO index and surge levels. The correlation between sun spot number and surge ($r = 0.58$ with the 90 per cent confidence interval from 0.26 to 0.79)

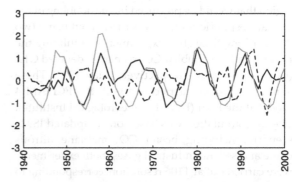

Fig. 8.7   Indexes of the STML time series (black continuous, same as in Fig. 8.1a), sunspots number (grey continuous), NAO (black dashed) during the 1940–2000 period (years on the x-axis). NAO and STML data include November and December only. All time series have been smoothed using a three-year running mean.

is statistically significant while the negative correlation between NAO and surge is not significant (r = –0.18 with the 90 per cent confidence interval from -0.52 to 0.22). The correlation between surge extremes and sun spot number raises the unresolved issue of the identification of physical mechanisms capable of amplifying an otherwise energetically weak signal. Other studies have shown an effect of the sun spot cycle on the troposphere resulting in a negative correlation with low-cloud amount and a broadening and weakening of the Hadley cell and northward displacement of the storm track at solar maximum (Gleisner and Thejll, 2003), therefore suggesting a global pattern into which the effect of the solar cycle on the floods of Venice can be inserted.

EVALUATION OF A FUTURE SCENARIO

The analysis above shows that the 0.6 K warming characteristic of the twentieth century at the hemispheric scale has no corresponding signal in the height of surge in the Gulf of Venice. However, a future trend could be expected to modify the present surge climate because it might, eventually, be associated with an increase in radiative forcing of much greater magnitude than that which has been experienced until now. This issue is strongly related to the effect of global warming on storminess in the Mediterranean area. A previous investigation of the variations of the cyclonic regime in the Mediterranean region that would be produced by the doubling of the CO_2 atmospheric content was carried out using two 30-year long time slice experiments simulating the present and the doubled CO_2 scenarios, denoted CTR (ConTRol) and CO_2 respectively (May, 1999). These simulations were carried out with the ECHAM 4 model at the DMI (Danish Meteorological Institute). The CO_2 simulation was based on an updated IS92a scenario, including, beside CO_2, methane, nitrous oxide and several industrial gases. Both experiments were carried out at T106 resolution, corresponding to a 160×320 global Gaussian grid, with 19 vertical levels. The results show that the cyclonic regime in the Mediterranean region is similar in the two climate scenarios (Lionello et al., 2002). The present climate is characterized by a slightly, but statistically significant, higher overall number of cyclones. The Mann-

Whitney test implies that the frequency of all cyclones exceeding a depth of 15hPa (with respect to the background field) is different for the two scenarios, at the 95 per cent confidence level. On the other hand, the difference of intensity of extreme cyclones between the two scenarios is not convincingly significant. Considering a region including the Mediterranean Sea and coastal areas surrounding it, the 100-year return value is 54.0 ± 2.3 hPa for the CTR simulation and 50.1 ± 2.1 hPa for the CO_2 simulation, with no significant difference between the two scenarios. However,, if a larger region from 10° W to 40° E and from 25° N to 55° N is considered, the situation reverses and the 100-year return value is 57.4 ± 2.5 hPa for the CTR simulation and 64.2 ± 3.4 hPa for the CO_2 simulation. Details are described in Lionello et al. (2002).

The CTR and CO_2 simulation have been carried out with global spectral model at T106 resolution, which is transferred to a regular lat-long grid with 1.1° resolution. This is inadequate for the reproduction of the storm surge in the Adriatic Sea, and it results in a gross under-estimate of extreme events. A statistical downscaling procedure was used to compensate for this shortcoming and to derive the regional surface wind fields (on a 0.25° lat-long grid) from the sea-level pressure fields (on a 1.1° lat-long grid) computed in the two CTR and CO_2 time slice experiments. The procedure is based on the CCA (Canonical Correlation Analysis) of PCA (Principal Component Analysis) pre-filtered sea-level pressure and wind fields. The procedure was established on the basis of a set of hindcast studies for which both high resolution wind fields and low resolution pressure fields were available and subsequently applied to the whole duration of the two time slice experiments. The downscaled wind fields have been used to force a shallow water model during the two 30-year long simulations of both present and doubled CO_2 climate scenarios. The downscaled wind fields produce a large improvement with respect to the original fields at 1.1° resolution but a systematic underestimation with respect to the observed surge levels remains present. This shortcoming of the analysis may prevent the identification of very high events. Consequently, the extreme value analysis of the results of the CTR simulation produces values

Table 8.1. *Levels and relative errors for 10, 50, 100, 250, 500 return periods. The columns refer to Observations (2nd and 3rd column), CTR (4th and 5th column) and the CO$_2$ (6th and 7th column) simulations.*

Return period	Observations					
	PdSalute		CTR		CO$_2$	
(years)	level (cm)	stantard error (cm)	level (cm)	standard error (cm)	level (cm)	standard error (cm)
10	106	6	92	6	94	9
50	137	13	112	12	128	22
100	152	18	120	16	144	31
250	172	26	130	21	167	45
500	189	33	138	25	186	59

lower than observed. The same systematic bias can be expected in the evaluation of future climate. Some caution is, therefore, necessary in the interpretation of the results of the downscaling. Nonetheless, the comparison between the present and future climate simulations shows no significant change of extreme surge levels as, although the CO$_2$ simulation produces higher extremes, the difference in the 100-year surge return value is 24 ± 47 cm (see Table 8.1). Details on the downscaling procedure and the results of the analysis can be found in Lionello *et al.* (2003a).

Figure 8.8 shows the average value of the three highest storms in each year of the CTR (dashed line) and CO$_2$ (continuous line) simulations. The two time series show an inter-decadal variability similar to that of the observations. Actually, the CO$_2$ simulation contains isolated higher surge values which are hidden by the three-year running mean used in Fig. 8.8. The GEV distribution resulting from the STML time series (observations in Venice at the PSal station, grey continuous line) is higher than those resulting for the two scenarios. This difference is only partially explained

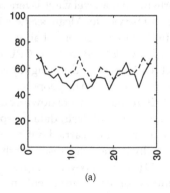

(a)

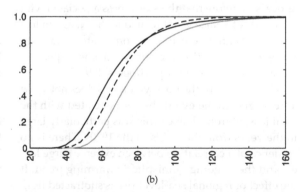

(b)

Fig. 8.8   Panel (a): average value (y-axis, cm) of the three highest storm in each year (x-axis) in the CTR (dashed line) and CO$_2$ (continuous line) simulations. Values have been smoothed with a three-year running mean. Panel (b): GEV distributions resulting from the STML time series (observations in Venice, PSal station, grey continuous line), the CTR (black dashed) and the CO$_2$ (light grey) simulation at the CNR platform. Each line in the panel shows the probability (Y-axis) of an annual maximum value lower than a given threshold (cm, X-axis).

by the different location, which implies that the surge level at the PSal station (in the city centre) is higher than that at the CNR platform (15 km offshore), and it is mostly a consequence of the under-estimate of wind forcing. The CTR (black dashed) distribution gives higher values than the CO_2 (light grey) distribution for short return periods. For periods larger than 10 years the situation reverses and the CO_2 scenario presents higher extremes. However, although this might indicate that the CO_2 scenario would show higher extreme events for long return periods, the difference is not significant according to the error of the estimate (see Table 8.1).

CONCLUSIONS

The synoptic patterns determining surge in the Gulf of Venice are associated with a low-pressure system with a minimum above central Europe or Northern Italy. Although this short-term dynamic is well known, the mechanisms responsible for its frequency and intensity need further investigation. It appears that periods with extreme surge events are characterized by a general circulation anomaly, represented by a pattern with a negative centre of action above the Eastern Atlantic, according to which cyclones deviate southeastwards and penetrate into the Mediterranean Sea from the North-West. This pattern is different from the NAO dipole whose time behaviour is not correlated with that of the highest surges in the Gulf of Venice. It appears therefore that the storminess associated with the floods of Venice is not related to the NAO but to a structure with a centre of action located nearer to Europe. This situation is common to other aspects of the storm track over Europe (Rogers, 1997).

The analysis of the observed levels does not show a trend in extreme events but is associated with the local loss of ground level; this was particularly large in the years from the 1950s to the 1970s. There is no evidence of a correlation between extreme surge levels and the ongoing hemispheric warming trend. If the effect of regional sea-level rise is subtracted from the data, the record of extremes is dominated by a large inter-decadal variability, with respect to which an (eventually present) residual trend has a minor importance. The understanding of this variability, and of its past and future evolution, is a most inter-

esting scientific and practical issue. It will be important, on one hand, to investigate the variability of large scale patterns associated with it, and, on the other hand, to find the explanation for its correlation to the periodicity of the sun spot number.

The meso-scale features involved in the production of surge at Venice cannot be resolved in currently available global climate models and a downscaling procedure is needed. Since the procedure cannot completely compensate for lack of resolution and for other regional processes not adequately represented, an underestimation remains present. These results should be considered carefully because of this systematic error. However, they do not show any large systematic trend, although an intensification of extreme storms and surge levels cannot be ruled out. The difficulty of describing directly the impact of a climate change on floods in Venice suggests improvements and new approaches are needed. In particular, the understanding of the dynamics responsible for the frequency and intensity of the synoptic situations producing the surge is most important for the prediction of future climate scenarios.

ACKNOWLEDGEMENTS

The sea-level data at 'Punta della Salute' were kindly provided by 'Ufficio Idrografico del Magistrato alle Acque'. The author is indebted to A. Tomasin for his help on the hourly sea-level data and surge level computation. Yearly mean sea-level values were provided by the Permanent Service for Mean Sea Level, hosted at the Proudman Oceanographic Laboratory. The NAO data were downloaded from the CRU (Climate Research Unit, University of East Anglia) web page http://www.cru.uea.ac.uk. The geopotential height data of the NCEP reanalysis were downloaded from http://www.cru.uea.ac.uk/cru/data/ncep. Figure 8.3 is based on a simulation carried out by A. Nuhu. This paper partially includes the results of the project STOWASUS-2100 (STOrms, WAves, SUrges, Scenarios in the 21st century) of the Environment and Climate programme (contract ENV4-CT97-0498).

REFERENCES

Buzzi, A., and S. Tibaldi. 1978. Cyclogenesis on the lee of Alps: a case study. *Quarterly Journal of the Royal Meteorological Society* **104**, 171–287.

Camuffo, D. 1993. Analysis of the Sea Surges at Venice from A.D. 782 to 1990. *Theoretical and Applied Climatology* **47**, 1–14.

Carbognin, L., Gatto, P., Mozzi, G., Gambolati G., and G. Ricceri. 1976 New trend in the subsidence of Venice. *I.A.H.S.* **121**, 65–81.

Gleisner, H., and P. Thejll. 2003. Patterns of tropospheric response to solar variability. *Geophysical Research Letters* **30**, 1711.

Hurrell, J. W. 1995. Decadal trends in the North Atlantic Oscillation and relationships to regional temperature and precipitation. *Science* **269**, 676–9.

Kalnay, E. and co-authors. 1996. The NCEP/NCAR 40-year re-analysis Project. *Bulletin of the American Meteorological Society* **77**, 437–71.

Lionello, P., Dalan, F., and E. Elvini. 2002. Cyclones in the Mediterranean Region: the present and the doubled CO2 climate scenarios. *Climate Research* **22**, 147–59.

Lionello, P., Elvini, E., and A. Nizzero. 2003a. Ocean waves and storm surges in the Adriatic Sea: intercomparison between the present and doubled CO2 climate scenarios. *Climate Research* **23**, 217–31.

Lionello, P., Martucci, G., and M. Zampieri. 2003b. Implementation of a coupled atmosphere-wave-ocean model in the Mediterranean Sea: sensitivity of the short time scale evolution to the air-sea coupling mechanisms. *The Global Atmosphere and Ocean System* **9**, 65–95.

May, W. 1999. A time slice experiment with the ECHAM4 A-GCM model at high resolution: The experimental design and the assessment of the climate change as compared to a greenhouse gas experiment with ECHAM4/OPYC at low resolution. *DMI Scientific Report* **99–2**, 93 pp.

Pasaric, M., and M. Orlic 2001. Long-term meteorological preconditioning of the North Adriatic Coastal floods. *Continental Shelf Research* **21**,263–78.

Robinson, A. R., Tomasin, A., and A. Artegiani. 1973. Flooding of Venice, phenomenology and prediction of the Adriatic storm surge. *Quarterly Journal of the Royal Meteorological Society* **99**, 688–92.

Rogers, J. C. 1997. The North Atlantic storm track variability and its association to the North Atlantic Oscillation and climate variability of the Northern Europe. *Journal of Climate* **10**, 1635–47.

Smith, R. L. 1986. Extreme value theory based on the n-largest annual events. *Journal of Hydrology* **86**, 27–43.

Tawn, J. A. 1988. An extreme value theory model for dependent observations. *Journal of Hydrology* **101**, 227–50.

Tomasin, A. 2002. The frequency of Adriatic surges and solar activity. *ISDGM Technical Report* **194**, 1–8.

Trigo, I. F., and T. D. Davies. 2002 .Meteorological conditions associated with sea surges in Venice: a 40 year climatology. *International Journal of Climatology* **22**, 787–803.

9 • Forecasting the water level in Venice: physical background and perspectives

A. TOMASIN

INTRODUCTION

The exceptional flood of 4 November 1966 marked a milestone in the international concern for Venice, and this is also true for the efforts for the quantitative forecasting of the water level in the lagoon. Around 1970 the interest in this challenge involved many scientific centres all around the world, so that in 1972 the oceanographer W. Munk reported thirteen research institutes developing prediction techniques for Venice (Munk and Munk, 1972). The results were significant, even though the complexity of the problem does not allow, even today, the claim that a definitive solution has been found.

Over the years, these research interests have been paralleled by the development of questions from an operational point of view. The historical Hydrographic Office of the Magistrato alle Acque is complemented by a service of the city of Venice, the Centre for Tides, committed to alerting the city, initially by using sirens, and, step by step, by the various modern tools of high technology, like the internet and the Short Message System (SMS). Moreover, a deeper interest has grown from the perspective of the mobile barriers, for which a long-term forecast becomes important. It is not correct to say that the need for prediction is limited to flood events. For a town like Venice, for example, the very low tides are also dangerous; one thinks of their effect in each 'rio' (small, internal canal), on commercial traffic and much more on the rescue boats. Many considerations force a general request for the forecasting of all water levels, including the intermediate ones (Boato et al., 2001).

PHYSICAL BACKGROUND

A scientific approach to the predicting instrument requires a good understanding of the physical background. It is clear that the system to be investigated is primarily the Adriatic Sea; in the facts considered here the dynamics of the lagoon has definitely a minor relevance, although it should not be entirely forgotten.

In the Adriatic, like in all seas, the first element to be considered is the astronomical tide, or the ordinary cycle in sea level and currents which, in addition, supplies clean water to the Venice Lagoon. The North Adriatic area which is of interest here has a very active tide, compared to most of the Mediterranean, even though its range of about one metre is certainly much smaller than that observed in many oceanic harbours. An explanation is available for such a local increase. It turns out that the Adriatic is not far from a resonant condition: it has, like most physical systems, a proper frequency of vibration, like the tone emitted when a table (or a glass) is knocked. Since the 'tone' of the astronomical tide is rather similar to the proper one of the Adriatic, a forced oscillation takes place which approaches a resonant, and hence amplifying, condition. The characteristic frequency of the Adriatic is shown by its free oscillations, or 'seiches' (discussed below).

Another well-known feature of the Adriatic tide is related to the Earth's rotation, which gives rise to a kind of gyre (an 'amphidromy') for two dominant components of the tide (M2, the semidiurnal lunar oscillation, and S2, the solar one). Thinking of them as waves, we observe that they reach Venice coming from the north east, and not from the south, as would be expected since the tide is essentially stimulated by the Mediterranean. The same holds for other dynamic features of the Adriatic Sea.

Flooding and Environmental Challenges for Venice and its Lagoon: State of Knowledge, ed. C. A. Fletcher and T. Spencer.
Published by Cambridge University Press, © Cambridge University Press 2005.

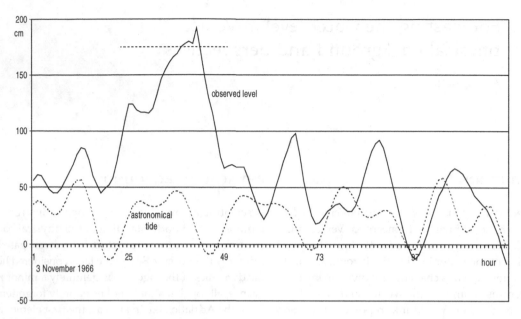

Fig. 9.1   The flood of 4 November 1966. The tide gauge could not measure above 175 cm, so the upper part of the curve is uncertain. Levels refer to the traditional local datum, mean sea level in 1897.

The astronomical tide is well known and predictable, thanks to the long record available for many stations: mathematical simulations, as will be seen, can also account for it. So it is definitely neither a problem for forecasters, nor for safety, since the tide by itself is not able to cause a flood, in whatever astronomical condition. It will become clear that the menace is elsewhere.

The atmospheric forcing on the sea surface is also important but unfortunately less easy to predict. Indeed, the conditions are relatively clear: wind pushing water and low atmospheric pressure can create an independent addition to the ordinary tide and give rise to flooding. The opposite also holds true but since the reverse causes low water conditions, this situation is of interest but less dramatic. The dominant wind for the present considerations is the *sirocco*, the SE flow, whose active area (the fetch) occupies frequently the whole Adriatic so that water piles up at the closed end. The *bora*, the NE wind, is frequently stronger, but the fetch against Venice is short and there is not any 'dead end' effect for it as water can escape sideways. Atmospheric pressure, instead, has little to do with the shape of the Adriatic,

since it is well-known that a local minimum causes the level to rise, attracting water from the areas where the pressure is high and it pushes down the sea. Now, all the above atmospheric effects are added to the ordinary tide, with a certain independence: in mathematical terms, it can be defined as an almost linear condition. In other places, like the North Sea, things happen differently, and the two terms interfere, due, roughly speaking, to a different ratio between the water depth and the range of the considered waves.

The Adriatic condition obviously makes things easier: given a tidal record, the astronomical part can be simply subtracted. The remainder is surge, a general name for the meteorological contribution (forgetting a negligible error due to non-linearity), and the research effort concerns this phenomenon. Now it is clear that when a flood occurs, a variety of background conditions are possible. Indeed, the duration of the 'acqua alta' is, in most cases, very limited, usually less than four hours, which means that the relative phase of the surge and the tide is critical. Severe surges have occurred that were not detected by the Venetians since at the time the astronomical tide was

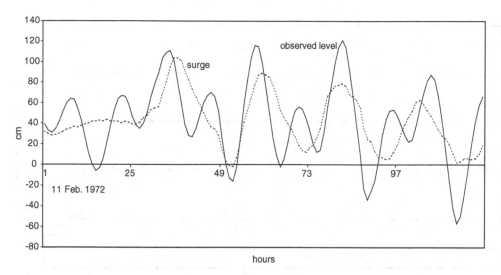

Fig. 9.2   Flooding in February 1972 showed the importance of seiches after a surge, with observed water levels higher in the days following the surge. Levels refer to the traditional local datum, mean sea level in 1897.

very low. By comparison, the famous case of 1966 is only attributed to surge since the astronomical factors gave a feeble oscillation of the neap tide. Under other conditions, its impact would have been much different (Fig. 9.1).

An important feature of Adriatic dynamics is given by 'seiches'. The free oscillations of the sea are observed for many days following a storm surge. Due to its shape, the Adriatic has a long-duration memory, since seiches do not quickly radiate their energy outside this area. Again, this is very important: to everybody's surprise, when a storm has passed, and the surge seems to be completed, over the following days, further floods can occur, sometimes worse than the first, surge-related one. This is due to the seiche return which in this case becomes synchronous with the ordinary tide (Fig. 9.2).

Seiches are really a general motion of the Adriatic, not restricted to the vicinity of Venice. The fundamental one, the most dangerous, is simply a kind of standing wave, slowly damped, with a node at the southern outlet of the sea (the Otranto Strait) and an anti-node (the place of maximum range) in the Gulf of Venice. The graph of the phenomenon at different stations in the Adriatic demonstrates its development (Fig. 9.3).

STATISTICAL MODELS

Munk and Munk (1972) observed that the research on prediction of water levels in Venice involved both correlation (black box) techniques and various analytical and numerical methods. This distinction still holds, and it is worth following it. Statistical methods appear as simple formulas, usually linear, that do not seem to explain what is going on, but their grounds are based on past data: 'should somebody, that day, have used these coefficients, he would have obtained a good prediction'. So, a large data base is required, from which, using proper numerical methods, the best coefficients can be obtained. This is the case both for the city of Venice and for the open sea that faces the lagoon, where a scientific platform is operating (Cavaleri, 1999), and the method will be adequate, in its 'outside' version, also in case of morphological changes of the inlets.

The simplest possible model of this kind is autoregressive, i.e. making use, to predict future water levels, only of observed levels from the last few hours. Since the sea has a long memory, with seiches that persist for many days, it sounds obvious that one can estimate the existing oscillation and predict its value. This is obtained with a filter, a simple numerical

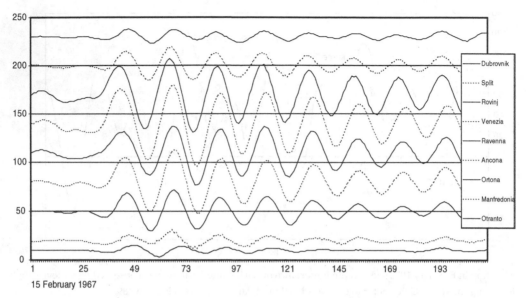

Fig. 9.3 Simultaneous plots of a seiche from February 1967 for nine stations in the Adriatic Sea.

scheme, called autoregressive, as stated above. Clearly, such prediction will be successful in the days that follow a surge (since the sea is in a free oscillation), and also on most occasions, due to the almost continuous presence of these vibrations. But the scheme will fail with the onset of a surge and on all those occasions when the oscillations are started or modified by weather conditions. So we must consider such an approach as a simple introductory example: it would have been successful for many days, for example, in the case of February 1967 (Fig. 9.3) due to the stability of the seiche.

Matters become more and more sophisticated when other observations are included in the predictive scheme, in addition to sea level. Wind data is the most obvious addition, but its reliability is frequently questionable. Atmospheric pressure is better for smoothness and significance. A simple trick is used for an indirect estimate of the wind forcing, since it strongly depends for its component along the Adriatic on the pressure difference across the sea: this is the well-known geostrophy. And since wind stress (what really matters) has a quadratic relationship to wind velocity, the difference is taken squared and used as a predictive parameter.

In conclusion, statistical models use simple formulas that add various terms (previously observed values of level, pressure, differences in pressure and so on), each with a coefficient obtained from the data from the past. Computational work can say in detail which parameters are useful, out of the large amount of possible measurements and possible combinations of them (like the cross-difference of pressure).

DETERMINISTIC MODELS AND COMPARISONS

Deterministic models, by comparison, presume a knowledge of physical processes and of the equations that describe them. Very few attempts have been made to solve them by analytical methods, and then by only trying to explain what is observed, and not to predict. Instead, the numerical solutions, favoured by the efficiency of modern computers, are the current tool. It means that the equations that describe the motion of a water particle, packed (i.e. integrated) for the full depth or for a certain number of layers, give the time evolution of the relevant quantities (level and currents) using reliable numerical methods for their solution. It is to be observed that this tool does not need, in principle, any large data base, except for the tuning of certain parameters.

Many possible choices are eligible, concerning the integration algorithm, or the 'grid' of depth data used, or the area covered by the numerical scheme, but other points of interest will be considered here. First, assimilation. Keeping the prediction purpose in mind, the model will be started, with its simulation of reality, a few days before the event, so it is reasonable that 'now' it reproduces the present conditions (of level, of course), since the values of wind and pressure, observed up to now, have been introduced. It is regrettable that the observed values of sea level up to the present time are not exactly satisfied and cannot simply be used to force the model. Assimilation means inserting in a proper way these observations, which is difficult, and which requires many additional calculations. This problem clearly does not exist for statistical schemes.

Then, meteorology: values must be available for both the historical record (for a correct simulation up to now) and the future, so meteorological forecasting is required. If weather is 'frozen' to the present condition, in most cases a decent prediction is obtained for three to four hours ahead, due to the inertia of the system. Otherwise, one depends upon the success of the meteorologists. This holds also for the statistical approach, since there is little hope for a statistical prediction of weather (e.g. Palmieri and Finizio, 1970).

Deterministic models describe rather easily the observed ordinary tide in the various stations, or more exactly this is a way to tune the scheme in a preliminary stage. This means that a model of this kind can simultaneously consider tide and surge, and so take into account the interaction between them (the non-linearity) that turns out to be pretty small.

For all kinds of models, as was anticipated, the problem remains of the connection between sea and lagoon. The statistical schemes seem, at first glance, unaffected, since the 'black box' used so far simply gives the prediction for the city of Venice (and independent algorithms could be set up for the external conditions). Indeed, all kinds of change in morphology (essentially, in the lagoon inlets) can menace the importance of the data bank used, if it refers to something different from the new situation. This occurred (with little effect) when the intermediate 'Malamocco mouth' and the corresponding canals in the lagoon were modified (1965-70): fortunately, most of the data used today for the statistical models were taken after that time. But what if new breakwaters are built in front of the inlets? For deterministic models one can remark that they typically refer to the Adriatic, and the propagation into the lagoon can only be given by some special model that describes it. Certainly, should any mobile defence for the lagoon be built, predictive models would refer, as a first step, to the sea level at the entrance to the lagoon. Then a management problem would arise, concerning the effect of the gates on the internal water masses. There would be inertial phenomena to be considered, rain contribution and local wind forcing, whose effect is relevant (see, for example, Ferla and Rusconi, 1994). Statistical models could be of very little help in this phase of transition.

OTHER PROBLEMS

A different scale of forecasting can be considered, with obvious caution, in the yearly order of time. Briefly, the question arises about years (or decades) in which surges were more or less frequent. An attempt has been proposed, and is shown here, to correlate surges and solar activity. A preliminary result is given, stimulating possible developments in this direction. For the sun, the simple parameter of the annual Wolf's number was taken, i.e. the yearly mean count of sunspots, since they measure the activity of our star (Hoyle, 1955). For surges, it was decided to assign to every year its mean sea level as the reference level, then to consider the yearly number of hourly measurements over the threshold of 40 centimetres. The reason for the temporary mean level to be subtracted lies in the large problem of the varying subsidence rate of Venice, and, in general, in the attempt to separate different geophysical phenomena. The threshold is a compromise between poor statistics (higher cut) and the smearing effect of seasonal variations (lower cut). The two sequences of yearly values thus obtained were smoothed by a three-year moving average and then compared. The shape of the graph encourages further study in this direction, since the linear correlation index is r = 0.69

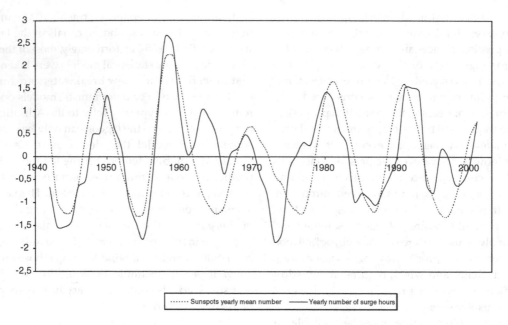

Fig. 9.4   A comparison between the mean number of sunspots each year with the annual number of surge hours (the variables are statistically normalized). The sequences have been smoothed with a three-year moving average.

(Fig. 9.4) with 60 degrees of freedom, subject to criticism, for example, for the correlation of residuals (Durbin and Watson, 1971). It is doubtful that the latter statistical refinements would properly apply to similar cases, and certainly different criteria could be used to improve the graph and the investigation, but the purpose here is only to suggest that some connection exists.

With all possible precaution, the suggestion is the following: even though nothing can be said about mean sea level (it was subtracted, as explained), the frequency of surges is in some way related to the solar activity. Obvious questions arise about the mechanism that connects solar activity and surges. If it really exists, there should be a connection, for example, between sunspots and some large-scale meteorological parameter. This is certainly not the place for climatological discussions to explain this fact, and the purpose is only provocative. If there is any truth, a glance at the expected trend of sunspot activity suggests a decrease of frequency of surges about three years ahead, with an increase afterwards, as it appears in Fig. 9.5, from the US NASA website

(NASA website, 2003). As observed above, surges are not floods, but a possibility of floods, so this is definitely not a prophecy about 'acqua alta'.

An unusual aspect of the forecast will be considered now, certainly of minor interest but in some way, again, provocative. It stems from the very recent observation, on 14 August 2003, of the arrival of a possible little 'tsunami' in Venice. Tsunami is the Japanese word for a special wave, sometimes very dangerous, produced by earthquakes or landslides (underwater or connected someway with the sea). Figure 9.6 shows the water level in Venice, that day, where the ordinary tide has been subtracted: the pulse has a range of only 15 cm and involves periods shorter than an hour. Since the transit from sea to lagoon normally damps the waves of high frequency, the record is surprising. It can be precisely connected to an earthquake of magnitude on the Richter scale of 6.3 that occurred about nine hours before in the Ionian Sea, about 200 kilometres from the Otranto Strait, in a seismically very active area. Data from other stations are being collected, and specific runs on numerical models show a decent agree-

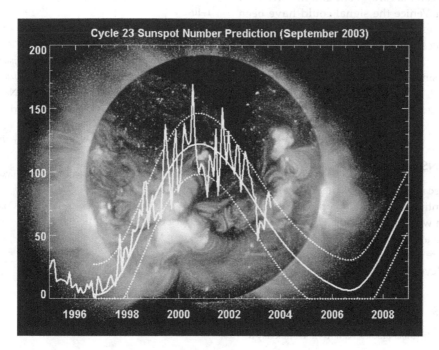

Fig. 9.5   Forecasting sunspot activity (courtesy: US NASA).

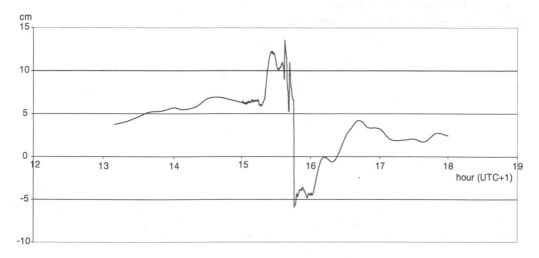

Fig. 9.6   A possible 'tsunami' recorded at Venice, 14 August 2003.

ment on timing. If this is confirmed, it is clear that in the hours between the earthquake and the arrival of the tsunami at Venice the signal could have been detected in the existing stations along the Adriatic. Then, a warning system like the one operating in the Pacific (Bascom, 1964) could be organized here with little difficulty, since the monitoring service for floods exists and will be further developed, and even though the danger for tsunamis is, at present, hypothetical.

CONCLUSIONS

In the last decades, big steps have been taken towards a quantitative prediction of the water level in Venice and a warning for floods. The effort for the future is aimed at better accuracy and a larger forecasting lag; there seem to be enough experience, and possibility of development, to give hope of success.

REFERENCES

Bascom, W. 1964. *Waves and Beaches*. New York: Doubleday & Co.

Boato, L., Canestrelli, P., and M. Mandich. 2001. Tidal information and alarm system in Venice. In *EOSS COST Action 40 Final Workshop Report, Sea Level in Europe: Observation, Interpretation and Exploitation*, Dubrovnik, Croatia, 11–8.

Cavaleri, L. 1999. The oceanographic tower 'Acqua Alta': more than a quarter of century of activity. *Il Nuovo Cimento* **22C**, 1–109.

Durbin, J., and G. S. Watson. 1971. Testing for serial correlation after least squares regression, III. *Biometrika* **58**, 1–19.

Ferla, M., and A. Rusconi. 1994. *L'evento di 'acqua alta' 921208 in Laguna di Venezia, indagine sui sopralzi differenziati*. Venice: Servizio Idrografico e Mareografico Nazionale.

Hoyle, F. 1955. *Frontiers of Astronomy*. New York: Harpers & Brothers.

Munk, J., and W. Munk. 1972. Venice hologram. *Proc. Amer. Phil. Soc.* **116**, 415–42.

NASA website, 2003. http://science.msfc.nasa.gov.

Palmieri, S., and C. Finizio. 1970. *The Prediction of Tides at Venice*. Special Technical Reports, 11. Rome: Consiglio Nazionale delle Ricerche – IFA.

10 · Meteo-climatic features of the Venice Lagoon

A. BARBI, R. MILLINI, M. MONAI AND S. SOFIA

CLIMATE AND CLIMATIC CHANGES

Climate summarizes the meteorological conditions that happen in a specific place over a restricted period of time. Climate is characterized by a typical variability, changing from region to region and from time to time. But when changes in variability are marked we can speak of 'climatic changes' and in particular of 'global changes', given the planetary scale of many of these phenomena. Climatic changes may be specifically related to human activities; these are mainly connected with the modification of the chemical composition of the atmosphere due to the increased concentrations of greenhouse gases. The 2001 IPCC report pointed out the ever higher concentration of greenhouse gases in the atmosphere, the slow but continued increase in the Earth's temperatures, and the necessity for a sensible reduction of emissions.

GENERAL TRENDS

The IPCC report shows increased warming, even though important parameters seem not to be affected and the warming is not completely widespread (for example in the southern oceanic and Antarctic zone). Forecasts estimate an increase in temperature of 0.6–2.5 °C over the next 50 years and of 1.4–5.8 °C for the next 100 years, with significant local peaks. An increase in temperature will mean a subsequent increase in global mean evaporation and precipitation. Local variability is more difficult to predict than global change and therefore assessing whether local climates will be more or less changeable is not an easy task (IPCC, 2001: 72).

OBSERVED TRENDS IN THE VENETO

The Veneto is a region of transition between continental central Europe and the Mediterranean Sea. Three different climatic regimes exist in this region, closely linked with topographic features and the presence of the Adriatic Sea: the alpine area, the plain area and the lagoon and coast area (Fig. 10.1).

To evaluate temperature and precipitation trends in this region, the daily data series of 18 stations, for precipitation, and 12 stations, for temperature, were analysed for the period 1961–90. The stations were selected from the series supplied by Servizio Idrografico e Mareografico Nazionale (SIMN) and Aeronautica Militare (AM), on the basis of meeting the necessary quality criteria.

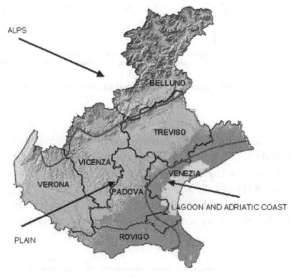

Fig. 10.1 Provinces and topographic characteristics of Veneto region.

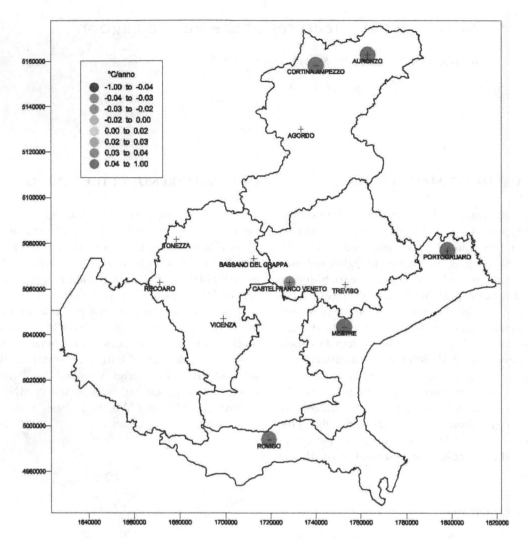

Fig. 10.2   Significant trends detected in mean annual values of maximum temperature (1961–90).

To detect monthly, seasonal and annual significant trends (at 95% and 90% significance levels) in the time series, the sequential Mann-Kendall test[1] was chosen (Lopez, 2000).

The principal results of this analysis show trends in this region in accordance with the more general scenarios, i.e., a slow but progressive increase of temperature. In fact, in the Veneto maximum temperatures have become higher and higher; 6 of the 12 stations demonstrate a positive trend in the mean annual values for maximum temperature, as shown in Fig. 10.2, more frequent during the winter season and the first part of the autumn. A reversal of the trend is registered in April at some stations.

There seem not to be substantial differences of increase between the two extreme values. In fact, also for minimum temperatures, Fig. 10.3 shows

---

[1] Description of this non-parametric test can be found in Sneyers (1975).

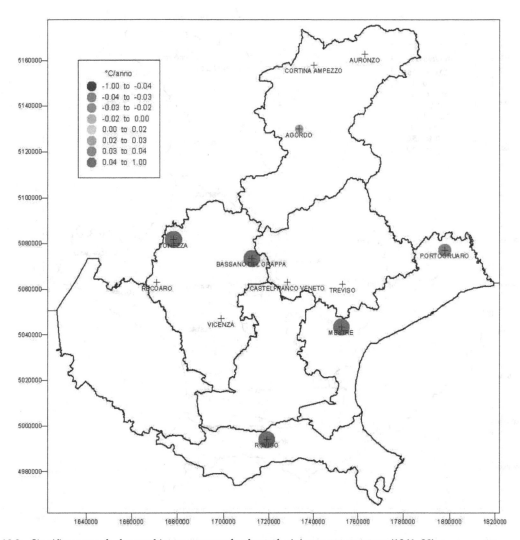

Fig. 10.3   Significant trends detected in mean annual values of minimum temperature (1961–90).

positive trends for mean annual values at various stations. A reversal of the trend is registered in June, in some mountain locations, and in November.

In the 1961–90 period, annual rainfall, as in most of the region south of the Alps, decreased; Fig. 10.4 show the significant trends detected in the Veneto. The monthly analysis highlights this reduction being more evident in November, while a reversal of the trend is registered in October.

EXTREME EVENTS

The trends highlighted so far are related to unstable and sometimes changeable climatological conditions; extreme phenomena may also be in conflict with these trends. Recent examples in the Veneto include:

- the 2000–1 winter. Very wet and snowy in the mountains (i.e. the station of Auronzo, in the Dolomites of Belluno, registered a total precipitation of 257 mm vs 165 mm 1961-1990 average). By comparison, the 2001–2 winter was very dry and

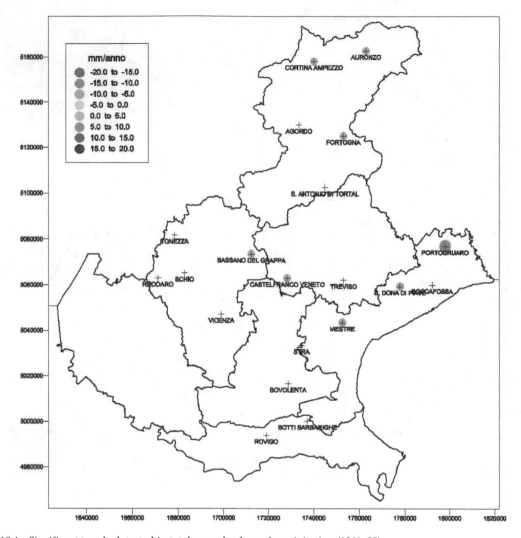

Fig. 10.4   Significant trends detected in total annual values of precipitation (1961–90).

in accordance with the last decade's trend (i.e. the same station registered a total precipitation of 56 mm vs 165 mm (1961–90 average);

- the hot and dry autumn of 2001. These conditions contrasted with the exceptionally rainy autumn of 2002, in particular in the mountains where rainfall amounts were higher than those experienced at the time of the 1966 flood (i.e. the station of Agordo registered in autumn 2001 a total of 208 mm vs 383 mm (long-term mean), while in autumn 2002 it registered a total of 1202 mm vs

the previous record registered during autumn 1966 of 844 mm);

- the cold 2001–2 winter. Contrary to general trends and to the previous winter when values were above the mean (i.e. the station of Legnaro in the plain near Padua registered a mean minimum temperature in December 2001 – January 2002 of –3.2 °C vs the long-term average of –1.3 °C).

CONSEQUENCES OF CLIMATE CHANGE FOR THE VENETO REGION

Some expected consequences for this region are:

- dry and warm winters: increase in forest fires, decrease in snow precipitation and consequent reduction of alpine glaciers, critical availability of water resources during the cold season, increase of atmospheric pollution due to persistence of atmospheric pollution;
- intense rainfall: more intense events with worsening of hydro-geological instability (this last aspect is, however, strictly determined by land management and topographic setting) and problems about optimization of the water regime in the urban environment;
- extension of summer season to the first part of the autumn: possible increase of short but intense precipitation; the autumn is typically characterized by consistent rainfall but alluvial, stormy events might become more frequent. One of the positive effects might be an extension of the bathing season;
- global warming: increase of respiratory allergies (due to persistence of pollen and atmospheric pollution).

SPECIFIC CONSIDERATIONS ABOUT THE VENICE LAGOON

Climatic characteristics of the lagoon around Venice are slightly different from the features of the conti-nental plain. Peculiarities of this area are determined by the proximity of the sea whose influence of humid winds and breezes affect the nearby zones. The mitigating action of the sea is limited by the shallow water of the lagoon and by its position, only influencing air masses blowing from the SE and E. Both minimum and maximum temperatures are affected, with slightly higher values for the former and slightly lower values for the latter in summertime.

The behaviour of the winds is quite similar to the coastal regime: during the hot period and mainly under anticyclonic situations, a weak pressure gradient from one side and a strong thermal difference between sea and land from the other set up local winds, known as breezes. During the day there is a sea breeze whose maximum intensity is reached in the afternoon and which comes from the SE. At night the breeze blows from the NE; it is not perpendicular but parallel to the coast due to the broader interaction of the Alps with the Adriatic Sea. With respect to rainfall, there are no particular differences with the rest of the plain.

Temperature and rainfall data for the last 40 years (1961–2000) for Venice Lagoon have been taken into consideration, with some evaluation of their annual fluctuations. In this case the station considered is Venezia Istituto Cavanis, located very close to the lagoon (dataset: Meteorological Observatory of Istituto Cavanis – Venezia). Figures 10.5 and 10.6 show the anomalies of the annual mean, minimum

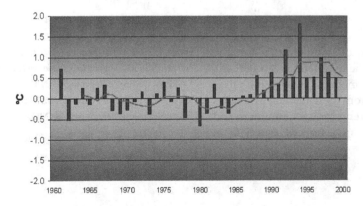

Fig. 10.5   Anomalies of annual mean, series of minimum temperature for 1961–2000 period, over the long-term mean 1961–90, with the corresponding smoothed series using a five-year moving average.

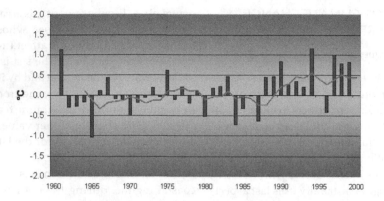

Fig. 10.6   Anomalies of annual mean, series of maximum temperature for 1961–2000 period, over the long-term mean 1961–90, with the corresponding smoothed series using a five-year moving average.

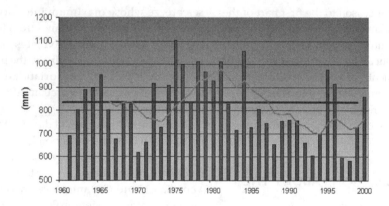

Fig. 10.7   The total annual rainfall for the 1961–2000 period and the long-term mean 1961–90, with the corresponding smoothed series using a five-year moving average. Rainfall values show a relatively rainy period in the 1970s, while the end of the 1960s and 1980s and the 1990s are characterized as dry periods.

and maximum series of temperature for the 1961–2000 period over the long-term mean 1961–90, with the corresponding smoothed series using a five-year moving average. From these thermal anomalies a generally increasing trend for both minimum and maximum temperature can be noticed, especially in the 1990s.

REFERENCES

IPCC. 2001 *Climate Change 2001. The Scientific Basis.* Cambridge: Cambridge University Press.

Lopez, P. 2000. *Variaciones en la precipitación de la Región del Veneto y su relación con los cambios en la temperatura.* Padua: ARPAV Centro Meteorologico di Teolo, Internal Report.

Sneyers, R. 1975. *Sur l'analyse statistique des series d'observation.* Technical Note 194. Geneva: WMO.

11 • Sea-level forecasting at the Centro Previsioni e Segnalazioni Maree (CPSM) of the Venice Municipality

P. CANESTRELLI AND L. ZAMPATO

INTRODUCTION

High water events constitute a very pressing problem for the city of Venice, because they influence both daily life and tourism. The CPSM-Centro Previsioni e Segnalazioni Maree of the Venice Municipality is responsible for the sea level forecast and early warnings to the city, in case of flooding events. In this chapter a brief description of the CPSM's activity is given, focussing on the forecasting activity. The different numerical models available at the Centre, both statistical and hydrodynamic deterministic, are described. The results obtained in the forecasting of some high water events of recent years, compared with observations, are shown. The experience gained in operational forecasting and the potential of recently developed modelling tools, such as data assimilation methods, and new datasets, such as wind fields from satellite scatterometry, allow us to outline some promising future perspectives for sea level forecasting in Venice.

The geographical position of Venice, in the centre of the Venice Lagoon at the closed end of the Adriatic Sea, and the peculiar urban structure of the city that lives in symbiosis with water, make it particularly exposed to flooding events. In fact, the morphology of the Adriatic Sea, deeper in its southern part and shallower in the north, and the basin's shape – elongated, semi-enclosed and surrounded by mountain chains – permits the occurrence of intense storm surge events: an anomalous sea level rise along the northern coasts of the Adriatic Sea, due to adverse meteorological conditions, such as the crossing of cyclones over the Mediterranean Sea during the autumn and winter. Typically these events are characterized by low atmospheric pressure in the Tyrrhenian area and strong *sirocco* winds over the Adriatic Sea, blowing along the basin axis and pushing the water towards the northern closed end. When the sea-level rise due to storm surge events occurs in phase with a maximum of the astronomical tide, the flooding of the coast which results is intensified and in Venice a flood event happens. Moreover, the Adriatic Sea geometry causes the setup of free oscillations, called seiches, that persist for several days after their generation by meteorological forcing. The seiches have to be taken into account in any attempt to forecast sea level in Venice.

An accurate and reliable forecast of sea level is indispensable to the city, enabling it to face any emergency due to the flooding. During exceptional high tide events, it is difficult to travel around Venice by foot or boat. Both ground floor flats and shops can be flooded; public services, such as water, gas and electricity, can also be at risk. To provide the city with an early flood warning system, the CPSM-Centro Previsioni e Segnalazioni Maree (Tidal Forecasting and Early Warning Centre) was founded in 1980. The primary goals of the municipal operations centre are to study and forecast the storm surge events and alert the city in case of flood events. In the second section of this chapter, a short description of the CPSM's activity is given. The third section explains in greater detail the forecasting of sea level: different numerical models available at the CPSM are presented, with examples of results obtained in the operational context. These results are compared with observed sea level data. Finally, in the fourth section, some future developments are outlined that should improve operational sea-level forecasting in Venice.

Flooding and Environmental Challenges for Venice and its Lagoon: State of Knowledge, ed. C. A. Fletcher and T. Spencer.
Published by Cambridge University Press, © Cambridge University Press 2005.

THE CPSM'S ACTIVITY

The activity of CPSM can be summarized as the following tasks:

* observation of sea level and some relevant meteorological parameters;
* provision of short-term forecasting of sea level in Venice; and
* issuing of tide information to the city and alerts in case of high water events.

The observation of sea level and meteorological parameters is carried out through a monitoring network that gives a real time view of marine and weather conditions in the Venice Lagoon and coastal zone on the Adriatic Sea (Canestrelli *et al.*, 2001).

The monitoring network (Fig. 11.1) is constituted by eleven tide gauge / weather stations: seven stations are located in the inner Venice Lagoon, one at each of the three tidal inlets communicating with the open sea and one at the CNR oceanographic platform *Acqua Alta*, located 15 km off the northern Adriatic coast. All stations measure sea level through a mechanical or ultrasonic gauge. Certain stations also collect meteorological parameters: air pressure, humidity, air temperature, wind (direction, average speed, gust), waves (maximum wave height, significant wave height). The reporting frequency is 5 minutes, except for wave related parameters which are transmitted every 15 minutes. The average wind speed and direction are calculated on data acquired every 2 seconds in the previous 10 minutes. Significant

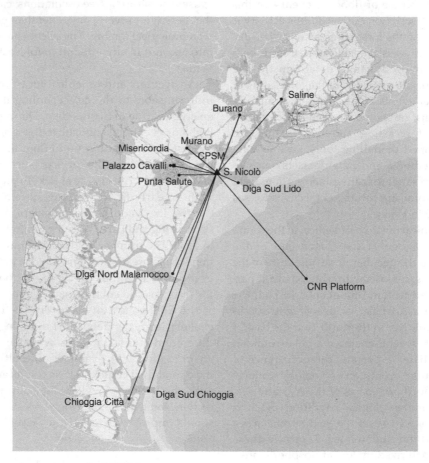

Fig. 11.1   The monitoring network of CPSM.

wave height and maximum wave height are computed on data acquired in the previous 15 minutes, from 4 measurements every second. The peripheral stations transmit, in real time, the measured data to a central station, located in the CPSM office, where they are validated and saved in a database. The validation procedure is carried out over a number of steps. First, the central station automatically checks the measured parameters to verify that they are in the correct range. Then missing data are replaced, when possible, by data from supplementary archives (local archives of the measurement stations or memory card). An accurate control is realized manually on sea level data that are graphically examined, to detect eventual spikes. A consistency check on sea level measurements is also possible, since the stations are close to one another: incongruent values are easily spotted. Validated data are distributed to interested users, who can obtain them on request. A book of hourly data, with some statistical and graphical elaborations, is published annually. Historical data since 1981 and real time non-validated data, collected by the monitoring network in the previous 3 months, are available on the CPSM's website (http://www.comune.venezia.it/maree/).

The sea-level forecast is performed through a set of numerical models, subsequently described in this chapter. They calculate the sea level component due to the atmospheric conditions, called the meteorological residual or surge. It is assumed that the astronomical component, related to the motion of astronomical bodies (moon and sun), accurately calculated through the harmonic constants, can be linearly superimposed on the surge, giving the total sea level. This assumption of decoupling between tide and surge is not verified in general: non linear effects, related principally to the quadratic bottom friction, play an important role in shallow waters, generating 'high order' harmonic, as M4 or similar, observed for example in the North Sea (see Bode and Hardy, 1997; Pugh, 1987). This is not the case for the Adriatic Sea, where the tidal amplitude, reaching at maximum one metre on the northern coast, is very small with respect to the total water depth and where 'high order' harmonics have never been observed: the amplitude of M4, computed from extended time series of sea level at the CNR platform, has been

shown to be less than 1 mm and negligible with respect to the measurement errors (A. Tomasin, personal communication 2004). These considerations have been confirmed also by numerical simulations performed through hydrodynamic deterministic models: in the northern Adriatic, the sea level computed taking into account tide and surge together does not show appreciable differences with respect to the value obtained by the sum of two terms separately computed (Lionello *et al.*, 2002).

The technical staff of the CPSM control and interpret the model results, integrate them with observed data from the monitoring network and other sources (METEOSAT images; wind and atmospheric pressure from the SYNOP network stations, according to WMO; weather parameters and sea-state from offshore oil platforms in the Adriatic Sea), formulate the sea level forecast for the next two days and circulate the information throughout the city. A high quality sea level forecast is needed at the CPSM to accomplish its third task: informing and warning residents. By simply telephoning the CPSM 'tide-line' number, an automatic answering machine gives the daily forecast for high and low tide values of the sea level, for the current and the next day. General and specific information about tidal phenomenon and data from the monitoring network can be gained through the website or upon request from CPSM technical personnel. Real time digital tide level displays, installed at strategic points around the city, are available. In case of severe flood tide events, the CPSM alerts the city. If the forecasted tide is greater than 110 cm above the local datum of Punta Salute, a network of sixteen sirens, placed over the whole municipality area and activated by CPSM staff, warns the population three to four hours before the event. An automated Call Manager system calls and alerts by telephone residents living on ground floors and shop owners. Additionally, a SMS (Short Message Service) service operates. The CPSM also contacts the companies responsible for the management of public services.

THE FORECASTING ACTIVITY

Different numerical models for the sea level forecast are available at the CPSM. Some of these are based on statistical methods: they have been operational since the CPSM's foundation and although simple, they have supplied satisfactory results to the present day. Other models follow a deterministic approach: they solve using a range of techniques the hydrodynamic equations for a basin, such as the Adriatic or the Mediterranean Sea.

The statistical models

A set of statistical models of increasing complexity is operational at the current time at the CPSM. They are based on a linear, autoregressive model (Tomasin and Frassetto, 1979) that relates the forecasted sea level in Venice to the observed sea level and pressure:

$$h_{(t+\tau)} = \sum_i (a_i^{\tau} h_{(t-i)} + b_i^{\tau} p_{(t-i)}) \qquad Eq.\ 11.1$$

The sea level h at time $(t+\tau)$, where the forecast is realized at time t with forecast lag τ, is given by a linear combination of predictors and coefficients. The predictors are the values of sea level $h_{(t-i)}$ and pressure $p_{(t-i)}$ observed in the previous hours t-1, t-2,... The coefficients a_i^{τ}, b_i^{τ} have been estimated by means of statistical methods based on the least squares theory: this calibration procedure has been realized by means of a 25-year (1966–90) database of sea-level data, observed by the CPSM monitoring network and by the Italian Hydrographic Service of Venice, and of pressure data, supplied by the Italian Air Force Meteorological Service. Because of the fundamental role played by the sea-level values observed in the past, this model is particularly reliable in reproducing the seiches in the Adriatic Sea.

Different choices of the predictors characterize different versions of the statistical model. The first statistical model, called MARCO, was set up operationally in 1981 (Canestrelli and Tomasin, 1981). It uses predictors of sea level and pressure in Venice. After this, in 1986, the *Semplificato* Model was developed (Canestrelli, Pastore and Tomasin, 1986) and is still operational today. It uses predictors of sea level in Venice and pressure in Venice, Genoa, Alghero and Bari and produces the sea level forecast for the next 36 hours. This model takes into account the effect of the cyclonic perturbations crossing the Mediterranean Sea, that are first detected over the Tyrrhenian Sea and which subsequently arrive over the Adriatic Sea.

An interesting development of the *Semplificato* Model is the *Completo* Model, operational since 1986. It is based on the same scheme as the earlier model, but it adds new predictors. These are the pressure differences between the two opposite coasts of the Adriatic Sea, measured at the following pairs of stations: Pola – Rimini; Zara – Ancona/Pescara; Split – Termoli; Dubrovnik – Bari. This pressure difference is representative of the strength of the wind field blowing along the Adriatic Sea axis. Unfortunately, even if it gives a more complete representation of the phenomenology, this model has not brought any substantial improvement in the forecast, with respect to the *Semplificato* Model.

The most recent statistical model is the *Esteso* Model, operational since 1996. Besides the observed values of sea level in Venice and pressure at the same stations as the previous models, new predictors have been added. These are the forecast pressures for the same set of stations, calculated by ECMWF, Reading (UK) and distributed in real time by the General Office for Meteorology (UGM) of the Italian Air Force. The computation of the statistical model is recursively repeated, taking the sea level computed in the previous step as the observed level for the next step. In this way it is possible to obtain a sea level forecast for the next 144 hours, the period for which the forecasted ECMWF pressures are available. The *Esteso* Model is the most accurate statistical model at forecast lags larger than 10 hours. It is currently used to formulate the official sea level forecast at the CPSM for the city of Venice.

Table 11.1 gives an accuracy index of the different statistical models, computed during the calibration procedure. The accuracy index is here defined as the average error ± 2 times the standard deviation, corresponding to an interval in which 95 per cent of the data are included. Until now the *Semplificato* Model has proven to be the most accurate, despite its simplicity, in the first 10 hours, because of its great sensitivity to the sea level values observed in the immediate previous hours.

In Figs. 11.2 and 11.3, two examples are reported of the *Esteso* Model's results in the operational phase. An example of accurate forecasting is shown in Fig. 11.2:

Table 11.1   *Errors of the three statistical models (in cm) in the calibration phase (Canestrelli and Pastore, 2000).*

	1 hour	3 hours	6 hours	12 hours	24 hours
Semplificato model	0 ± 4.0	0 ± 9.6	0 ± 13.0	0 ± 14.2	0 ± 17.0
Completo model	0 ± 4.0	0 ± 9.2	0 ± 12.2	0 ± 14.0	0 ± 17.6
Esteso model	0 ± 4.2	-0.1 ± 9.8	-0.1 ± 13.2	-0.2 ± 13.6	0 ± 14.6

it refers to the high water event of 21 November 1999 when the water level in the historical centre of Venice reached the value of +121 cm above the local datum of Punta Salute. The forecast was very good, with errors lower then 8 cm, until 24 hours before the event.

The forecast is not always so precise. Figure 11.3 refers to the high water event of 6 November 2000 when the tide level reached +144 cm above the local datum. The *Esteso* Model provided a sea-level forecast underestimated by more than 30 cm 3 hours

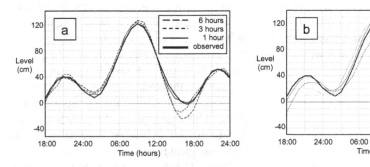

Fig. 11.2   The high water event of 21 November 1999. Results from the *Esteso* Model. (a) Observed level (solid line) at Punta Salute and model results at time lags of 1, 3 and 6 hours. (b) Observed level (solid line) at Punta Salute and model results at time lags of 12, 24 and 48 hours. Sea-level values are referred to Punta Salute datum (Canestrelli and Moretti, 2003).

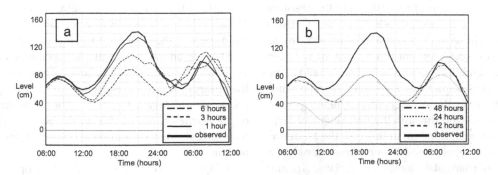

Fig. 11.3   The high water event of 6 November 2000. Results from the *Esteso* Model. (a) Observed level (solid line) at Punta Salute and model results at time lags of 1, 3 and 6 hours. (b) Observed level (solid line) at Punta Salute and model results at time lags of 12, 24 and 48 hours. Sea-level values are referred to Punta Salute datum (Canestrelli and Moretti, 2003).

before the event (Fig. 11.3a). The error was about 50 cm 6 hours before the event (Fig. 11.3a); a substantial contribution to this error was given by the meteorological input that was in this case very inaccurate (see below). These examples show the capability of the *Esteso* Model to extend the forecast to forecast lags of the order of days, as a consequence of the utilization of the forecasted pressures. On the other hand, the quality of the results is very dependent on the accuracy of the meteorological fields used as input.

The finite element hydrodynamic model SHYFEM

The hydrodynamic deterministic model SHYFEM (Shallow water HYdrodynamic Finite Element Model) has been developed at the ISMAR-CNR Institute of Venice (Umgiesser, 1986; Umgiesser and Bergamasco, 1993). It was operationally implemented at CPSM at the end of 2002 (Canestrelli *et al.*, 2003a). The model resolves the depth integrated shallow-water equations in their formulation with water level ζ and barotropic transports U, V:

$$\frac{\partial U}{\partial t} - fV + gH\frac{\partial \zeta}{\partial x} + RU + X = 0$$

$$\frac{\partial V}{\partial t} + fu + gH\frac{\partial \zeta}{\partial y} + RV + Y = 0$$

$$\frac{\partial \zeta}{\partial t} + \frac{\partial U}{\partial x} + \frac{\partial V}{\partial y} = 0 \qquad\qquad Eq.\ 11.2$$

where g is the gravitational acceleration, $H=h+\zeta$ the total depth, h the undisturbed depth, t the time, R the friction coefficient, proportional to the absolute value of the transport $(U^2+V^2)^{1/2}$, f the Coriolis parameter. The terms X, Y contains all other terms, as the wind stress and non-linear terms that do not need to be treated implicitly in the time discretization. The hydrodynamic equations are solved using a semi-implicit time resolution algorithm. The integration in space is realized through the finite element technique: the spatial domain, in this case the Mediterranean Sea, is represented by a computational grid made out of triangular elements of variable size and shape (see Fig. 11.4). The advantage of such a grid is its ability to accurately describe regions with complex bathymetry, such as the northern Adriatic

coasts, without needing too large a number of elements. The size of elements varies from about 35 km in the southern Mediterranean to about 1.5 km in the northern Adriatic Sea, in front of the Venice Lagoon. The operational model, implemented at CPSM, is driven by the meteorological fields, pressure and wind, computed by the atmospheric model of ECMWF.

Figure 11.5 refers to the hindcast simulation for the month of November 2000, realized with the SHYFEM model, during the preliminary phase of investigation and calibration. An example of the forcing wind field, from ECMWF, is reported in Fig. 11.5a; the time series of wind speed and of residual level simulated by the model and observed at the CNR oceanographic platform, located in the northern Adriatic Sea, can be seen in Fig. 11.5b and Fig. 11.5c respectively. It appears clearly that, in this month, the ECMWF wind over the northern Adriatic Sea is underestimated, in particular during the severe event of day 6 (see above). The simulated residual level at the CNR platform follows the low frequency behaviour of the observed data, but the peaks are generally underestimated, giving an error of more than 20 cm on day 6. Moreover, the signals are not always in phase. This inadequacy of the meteorological forcing stimulated an interest to looking for alternative meteorological fields.

Figure 11.6, from a recent study (Canestrelli *et al.*, 2003b), shows an example of a simulation of the same period, realized merging the ECMWF wind fields with wind fields measured by the QuikSCAT scatterometer on board the NASA satellite 'Seawind'. An example of the satellite wind field is shown in Fig. 11.6a. During the month in question, the satellite wind speed at the CNR platform is coherent with the wind observed using a traditional anemometer (Fig. 11.6b) and the simulated sea level is much closer to the observed sea level, especially in terms of phase and maximum levels (Fig. 11.6c). The overestimation of the seiches could be due to the bottom friction term that seems to require a further and finer calibration.

In Fig. 11.7 an example is given of the operational simulation of the event of 16 November 2002, when the water reached the level of +130 cm at the CNR Platform and +147 cm in the city, relative to the Punta

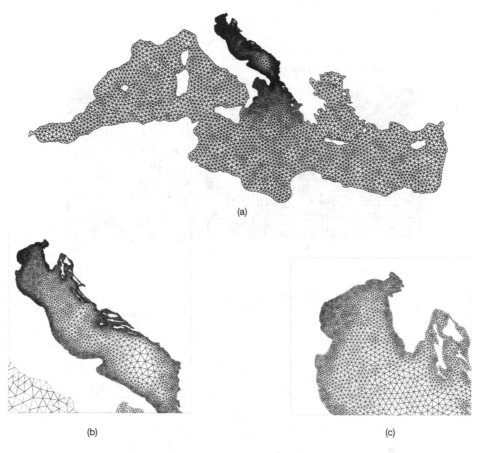

Fig. 11.4   The computational grid of the SHYFEM hydrodynamic model. (a) Mediterranean Sea. (b) Adriatic Sea.
(c) Northern Adriatic Sea.

Salute datum. In the figure, the solid line represents the observed sea level at the CNR Platform; dashed lines show the sea level modelled in the operational simulations realized, respectively, 24, 48, 72 and 96 hours before the event. Each operational simulation includes a spin-up time of 7 days, forced by ECMWF atmospheric fields of analysis and a forecast of 6 days, forced by ECMWF forecast fields: in this way, for example, the simulation labelled f+24 has been realized on 15 November, using ECMWF fields of the period from 8 November to 21 November. In this case, the SHYFEM model gives a quite good forecast at large forecast lags, with an underestimation of the peak of about 8 cm, 24 hours before the event; in

addition it reproduces very accurately the maximum of the previous day.

The finite differences hydrodynamic model HYPSE

The hydrodynamic model HYPSE (HYdrostatic Padua Sea Elevation model) has been developed at the University of Padua (Lionello, 1995; Lionello *et al.*, 1998) and its operational implementation at CPSM is in its realization phase (Lionello *et al.*, 2002). The primitive equations in the shallow-water approximation are solved on a finite difference C-Arakawa curvilinear grid of the Adriatic Sea. The

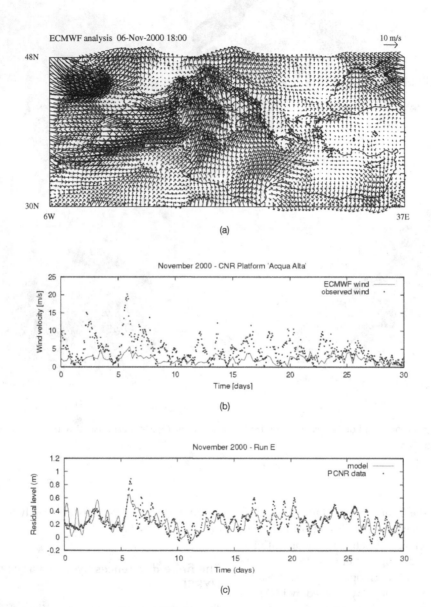

Fig. 11.5   Hindcast simulation of November 2000. SHYFEM model with ECMWF forcing. (a) The ECMWF wind fields over the Mediterranean Sea, 6 November 2000 at 18:00. (b) Observed and ECMWF wind speed at the CNR Platform. (c) Residual level observed and calculated by the SHYFEM model at the CNR Platform.

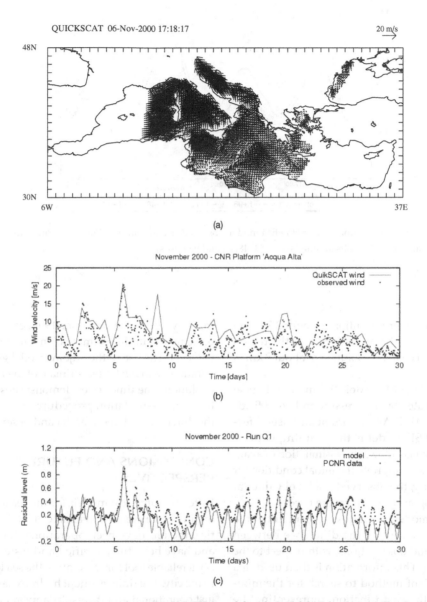

Fig. 11.6   Hindcast simulation of November 2000. SHYFEM model with QuikSCAT forcing. (a) The QuikSCAT wind fields over the Mediterranean Sea, 6 November 2000 at 17:18. (b) Observed and QuikSCAT wind speed at the CNR Platform. (c) Residual level observed and calculated by the SHYFEM model at the CNR Platform.

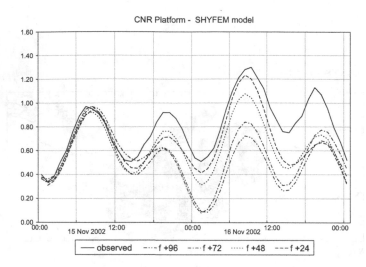

Fig. 11.7   An operational simulation of the SHYFEM model: observed sea level at the CNR platform in the event of 16 November 2002 and model results at time lags of 24, 48, 72 and 96 hours.

HYPSE grid has a finer resolution of about 2.7 km in the northern part of the Adriatic Sea, in the region near Venice, and a coarser resolution of about 8 km in the South.

As in the SHYFEM model, the meteorological forcing is constituted by the pressure and wind fields supplied by ECMWF. An important and useful feature of the HYPSE model is the coupling with an adjoint model, which allows the assimilation of available observations to improve the initial condition for HYPSE. Presently, the observed sea level values at the CNR oceanographic platform are used.

The adjoint model computes the gradient of the cost function representing the difference between HYPSE results and observations, with respect to the initial condition. This information is then used in a conjugate gradient method to search for the minimum value of the cost function, representing the optimal initial condition. In Fig. 11.8 the results of a hindcast simulation, for the month of November 1996 are reported. In Fig. 11.8a, 11.8b, 11.8c the meteorological residual at the CNR platform calculated by the basic version of the model is shown, together with the observed data. In Fig. 11.8d the results of the adjoint model, for a six-day period including the

main event of 18 November 1996, can be seen. The basic model underestimates, on average, the residual level. The use of the adjoint model gives, after 14 iterations, a much better estimate of the residual level evolution. The time series demonstrates the capability of the assimilation procedure to compensate for the difference between model and observations.

CONCLUSIONS AND FUTURE PERSPECTIVES

Different numerical models for the sea level forecasting in Venice are now available at CPSM. The statistical models have been operational for many years and have been fully verified and tested. They supply a reliable tool for calculating the sea level forecast to the city. The finite element hydrodynamic model is just operational and its results appear coherent with the results of the statistical models, showing a good capability of forecasting the arrival of a high water event. The finite difference model, even if not yet operational, has given very promising results: the data assimilation of water levels is hopefully a way to obtain a high quality forecast of the sea level in Venice. Hydrodynamic and statistical models appear

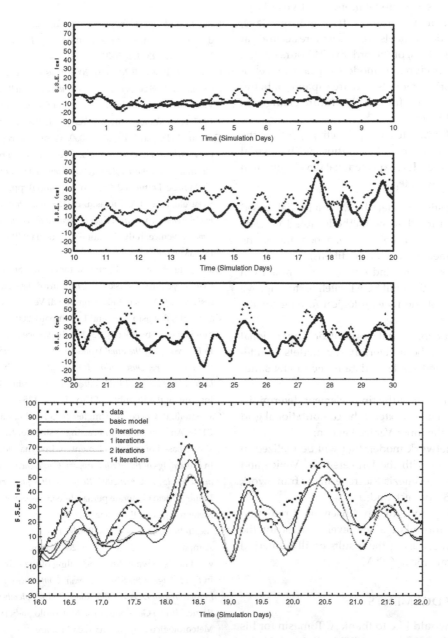

Fig. 11.8 Hindcast simulation of November 1996. (a), (b), (c) Residual level observed and calculated by HYPSE basic model. (d) Residual level observed and calculated by HYPSE after assimilation with adjoint method; after 14 iterations the model's results are very close to the observed data.

particularly efficient if used together. In fact hydro-dynamic models seem to be more sensitive at large forecast lags, of the order of two or three days, whereas statistical models are much more accurate at shorter forecast lags, of the order of 24 hours or less. Moreover, hydrodynamic models appear more elastic than statistical models and are capable of being modified to account for possible future changes in coastal morphology.

The CPSM promotes research activity, both in the centre itself and in collaboration with several research institutes. It is expected that development in the future will include:

- a new statistical model, the *Esperto* Model, based on artificial intelligence methods, able to decide which set of coefficients to use, depending on the observed meteorological conditions;
- the use of satellite wind fields, for example from 'SeaWind' on the QuikSCAT satellite, to improve the quality of the meteorological forcing for the deterministic models;
- the introduction of data assimilation operational procedures in both deterministic models, to integrate observed sea level data in the model simulations;
- an improvement of the finite element deterministic model that will extend the computational grid to include the inner Venice Lagoon;
- a neural network model, that will be realized in collaboration with the University of Venice and the Consorzio Venezia Ricerche, in the framework of a MURST project; and
- the development of an Ensemble Prediction Model, to formulate a sea level forecast for the city of Venice, using the results of all numerical models available at CPSM.

ACKNOWLEDGEMENTS

The authors would like to thank A. Tomasin for his constant collaboration to the CPSM activity and helpful suggestions in preparing this text; they are also indebted to Jane Frankenfield for her precious help in editing this chapter.

REFERENCES

Bode, L., and T. A. Hardy. 1997. Progress and recent developments in storm surge modeling. *Journal of Hydraulic Engineering* **123**, 4, 315–31.

Canestrelli, P., and F. Moretti. 2003. *Il sistema statistico del Comune di Venezia per la previsione del livello della marea in città. Valutazione comparativa dei risultati in fase di operatività 1997–2000.* Venice, Italy: Ist. Ven. Sc. Lettere ed Arti.

Canestrelli, P., and F. Pastore. 2000. Modelli stocastici per la previsione del livello di marea a Venezia. In *La ricerca scientifica per Venezia. Il Progetto Sistema Lagunare Veneziano.* Ist. Ven. Sc. Lettere ed Arti, vol. II, tomo II, pp. 635–63.

Canestrelli, P., and A. Tomasin. 1981. *Sull'attendibilità di un modello empirico-statistico per la previsione dell'acqua alta a Venezia.* Venice, Italy: Comune di Venezia (CPSM), ISDGM-CNR.

Canestrelli, P., Pastore, F., and A. Tomasin. 1986. *Sviluppi di un modello operativo previsionale delle maree di Venezia e revisione di casi rilevanti.* Venice, Italy: Comune di Venezia (CPSM).

Canestrelli, P., Pastore, F., and L. Zampato. 2001. The monitoring network of CPSM in the Venice Lagoon. In *Proceedings of the Final Workshop of COST Action 40. Sea Level in Europe: Observation, Interpretation and Exploitation, Dubrovnik, Croatia, 19–21 September 2001.* Split: Hydrographic Institute of the Republic of Croatia, pp. 125–28.

Canestrelli, P., Cucco, A., Umgiesser, G., and L. Zampato. 2003a. An operational forecasting system for the sea level in Venice based on a finite element hydrodynamic model. In *Proceedings of the Sixth European Conference on Applications in Meteorology ECAM 2003, Rome 15–19 September 2003.* Rome: Ufficio Generale per la Meteorologia-Servizio Meteorologico dell'Aeronautica Militare.

Canestrelli, P., Cucco, A., De Biasio, F., Umgiesser, G., Zampato, L., and S. Zecchetto. 2003b. The use of QuikSCAT wind fields in water level modelling of the Adriatic Sea. In *Proceedings of the Sixth European Conference on Applications in Meteorology ECAM 2003, Rome 15–19 September 2003.* Rome: Ufficio Generale per la Meteorologia-Servizio Meteorologico dell'Aeronautica Militare.

Lionello, P. 1995. Oceanographic prediction for the Venetian littoral. *Il Nuovo Cimento* **C18**, 245–68.

Lionello, P., Zampato, L., Malguzzi, P., Tomasin, A., and A. Bergamasco. 1998. On the correct surface stress for the prediction of the wind wave field and the storm surge in the Northern Adriatic Sea. *Il Nuovo Cimento* **C21**, 515–32.

Lionello, P., and co-workers. 2002. *Studio: sistema operativo di previsione del livello del Mare Adriatico basato su un modello numerico di tipo 'deterministico' della circolazione marina.* Report to CPSM, Venice (Phase I and II).

Pugh, D. T. 1987. *Tides, Surges and Mean Sea Level.* New York: John Wiley & Sons.

Umgiesser, G. 1986. A model for the Venice Lagoon. Unpublished Master's thesis, University of Hamburg.

Tomasin, A., and R. Frassetto. 1979. Cyclogenesis and forecast of dramatic water elevations in Venice. In J. C. J. Nihoul (ed.), *Marine Forecasting.* Amsterdam: Elsevier, pp. 427–38.

Umgiesser, G., and A. Bergamasco. 1993. A staggered grid finite element model of the Venice Lagoon. In J. Periaux, K. Morgan, E. Ofiate, and O. C. Zienkiewicz (eds.), *Finite Element in Fluids.* Swansea: Pineridge Press, pp. 659–68.

12 • APAT duties and techno-scientific activities regarding the Lagoon of Venice

M. FERLA

INTRODUCTION

In 2003, a new organization was set up in Italy in the form of the National Agency for Environmental Protection and Technical Services (APAT). The new Agency, which is supervised by the Ministry of Territorial and Environmental Protection, has reorganized and streamlined the responsibilities previously performed by the National Environmental Agency, the National Hydrographic and Oceanographic Service and the National Geological Service. In general, APAT duties involve technical and scientific activities regarding environmental protection, water resources management, water control and water pollution. In this case, the Agency has inherited the duties of the Hydrographic Office concerning the Lagoon of Venice. The Hydrographic Office has been a special technical and fact-finding body that has been operating in the lagoon since 1907 within the Magistrato alle Acque (the Water Authority responsible for safeguarding the Venice Lagoon). These responsibilities have now been attributed to the Department for Protection of the Inner and Marine Waters of the Agency and are carried out by the Venice Lagoon Service (VLS) a unit whose head offices are in Venice at Palazzo X Savi, formerly the Venetian Hydrographic Office. Its responsibilities include:

- systematic observation of water levels and parameters influencing meteo-marine phenomena in the lagoon and along the north Adriatic coast;
- management of the real time tidal gauge system of the Lagoon of Venice (RTLV);
- daily tidal forecasting service and forecasting and warning for exceptionally high tides (flooding or *acque alte*) for the benefit of all institutions associated with safeguarding the lagoon, civil defence, as well as coastal risk and hydrological risks at the large river mouths of north-east Italy, lagoon navigation and general population;
- elaboration, validation and distribution of data, especially analyses regarding observations from extreme tide events;
- analyses of historical time series for tide data, also in relation to subsidence and eustatism in the northern Adriatic region; and
- measurement of current parameters at the inlets in relation to lagoon-sea exchanges under different tide conditions.

This chapter aims to illustrate some of these activities and their scientific and technical and operating importance for the duties concerning protection of the Venice Lagoon.

THE REAL TIME TIDAL GAUGE SYSTEM OF THE LAGOON OF VENICE

The VLS manages a network of 52 tide gauge stations equipped for the systematic measurement of water level and other related parameters, such as wind direction, wind speed, atmospheric pressure, precipitation, and wave-height into the Lagoon of Venice and in the north-western Adriatic coastline (Fig. 12.1). The greater part of the 52 tide gauge stations have been operating for several decades; therefore, today VLS manages time series of tide data lasting more than 120 years.

Of these stations, 25 have been operating for more than 20 years with real time data transmission to the processing centre of the VLS. Correct functioning of this system is fundamental for warning and prediction of exceptional or atypical tides (storm surges)

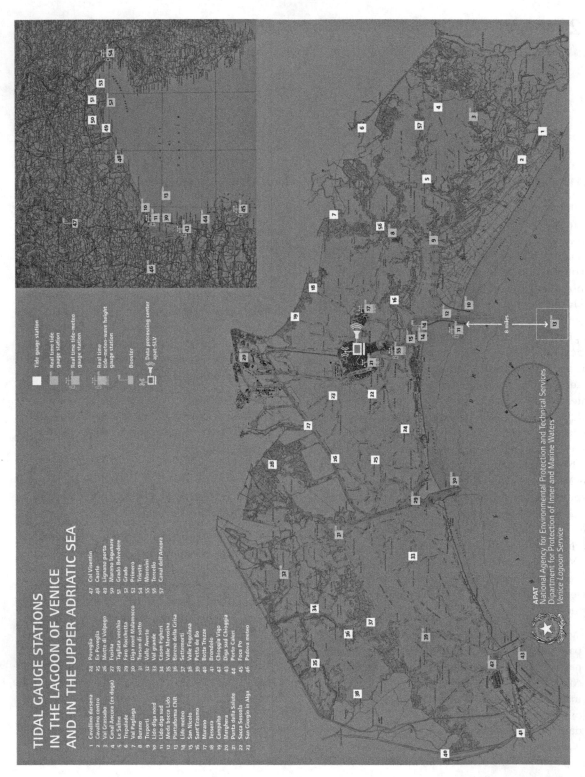

TIDAL GAUGE STATIONS
IN THE LAGOON OF VENICE
AND IN THE UPPER ADRIATIC SEA

1   Cavallino darsena
2   Cavallino centro
3   Val Grassabo
4   Canal Ancora (ex doga)
5   Le Saline
6   Trepalade
7   Val Pagliaga
8   Burano
9   Treporti
10 Lido diga nord
11 Lido diga sud
12 Meda bocca Lido
13 Piattaforma CNR
14 Lido meteo
15 San Nicolo
16 Sant'Erasmo
17 Murano
18 Tessera
19 Campalto
20 Marghera
21 Punta della Salute
22 Sacca Sessola
23 San Giorgio in Alga

24 Poveglia
25 Ex Poveglia
26 Motte di Volpego
27 Fusina
28 Tagliata vecchia
29 Faro Rocchetta
30 Diga nord Malamocco
31 Torson di sotto
32 Valle Averto
33 Val grande
34 Cason Figheri
35 Valle Morosina
36 Barene della Grisa
37 Settemorti
38 Valle Fogolana
39 Porta de Bo
40 Botte Trezze
41 Brondolo
42 Chioggia Vigo
43 Diga sud Chioggia
44 Porto Caleri
45 Foce Po
46 Padova meteo

47 Col Visentin
48 Caorle
49 Lignano porto
50 Maremo lagunare
51 Grado Belvedere
52 Grado
53 Primero
54 Trieste
55 Morosini
56 Torcello
57 Canal dell'Ancora

Tide gauge station

Real time tide
gauge station

Real time tide-meteo
gauge station

Real time
tide-meteo–wave height
gauge station

Booster

Data processing center
apat-SLV

APAT
National Agency for Environmental Protection and Technical Services
Dipartiment for Protection of Inner and Marine Waters
Venice Lagoon Service

Fig. 12.1   Tide guage system of the Lagoon of Venice.

and for the management of the lagoon hydraulic system. The 52 tide gauge stations, the data processing centres and the radio relay network constitute the real time tidal gauge system of the Lagoon of Venice (RTLV).

This system is a fundamental part of the weather and marine monitoring systems of the Italian seas, controlled by the APAT, including the national tide gauge network and national sea waves measuring network. In addition, the VLS has the capability to exchange data measured in real time between the networks of the environmental operating centres of the north-eastern regions of Italy.

THE DAILY TIDE BULLETIN AND THE TIDE GAUGE STATION LOCATED AT THE PUNTA DELLA SALUTE

Production of the daily tide bulletin (DTB) is one of the main activities of the VLS. The DTB is sent to the Water Authority, the Ministry for the Environment, Prefecture Offices, the Municipality of Venice, Steersmen of the Port, Police, the City Fire Department, and other bodies; it is also published through information broadcasting units and an internet website (www.apatvenezia.it). A telephone messaging service and is posted in the main city points of transit (St Mark's Square, the Rialto Bridge and Piazzale Roma). The DTB gives details of both the measured water level and the predicted level at the fundamental tide gauge at Punta della Salute, located in the historic centre of the city. The DTB is published every day at 10:00 a.m. and reports:

- the tide observed from midnight of the day before until 10:00 on the date of publication;
- the tide expected from 10:00 a.m. of the date of publication until midnight of the following day; and
- the astronomic tide.

In case of a storm surge warning, measurements of water level and the prediction of the level at that station (1) determines the closure of small sluice-gates to protect several of the smaller inhabited centres in the lagoon (e.g. Malamocco, Cavallino); (2) regulates the proper operation of land reclamation systems in the immediate lagoon hinterlands; and

(3) instigates the laying out of the pedestrian walkways along city streets, squares and piazzas.

The regular tidal observations made in Venice are among the oldest and most reliable in Italy and in the Mediterranean in general. In the 1960s, the Hydrographic Office of the Magistrato alle Acque established a correlation between the tide levels recorded at Punta della Salute, instituted in 1923, and levels recorded in Venice in the preceding decades. The result was to put together a continuous string of data that dates back to 1872. The reference benchmark, called *Zero Mareografico di Punta della Salute* (ZMPS) corresponds to this day to the mean sea level recorded in 1897 (Fig. 12.2). The tide levels across the lagoon are measured with respect to this benchmark. Therefore, if we can legitimately assume this reference to be integral with the soil level of Venice, the average sea level calculated today at Punta della Salute gives us an accurate indication of how far Venice has subsided with respect to the sea in just over a century. Such sinking is due to the combined effect of eustasy and subsidence. The most recent estimates give an average level of subsidence of about 24 cm (Bonato *et al.*, 2001a).

STORM SURGE EVENTS

It is well known that the occurrence of storm surge events in the northern Adriatic Sea is due to the passage of deep fields of low pressure moving from south of the Alpine crescent and the consequent persistence of south easterly winds blowing across the Adriatic Sea. The primary effect is that the mass of water is squeezed towards the northern Adriatic coast and therefore into the Venice Lagoon. This is followed by a seiche, a long wave with a period of about 22 hours in oscillation along the Adriatic in a south-east/north-west direction.

Table 12.1 charts the most severe cases of exceptionally high water level recorded over more than 120 years of tidal observations. Figures 12.3a and 12.3b show some recordings of the data relating to the event on the 16 November 2002, the seventh highest water level recorded: the level of the sea at Punta della Salute and the related rise of the storm surge above the astronomically predicted sea level.

When conditions leading to the possible occurrence of exceptionally high tides begin to appear, the

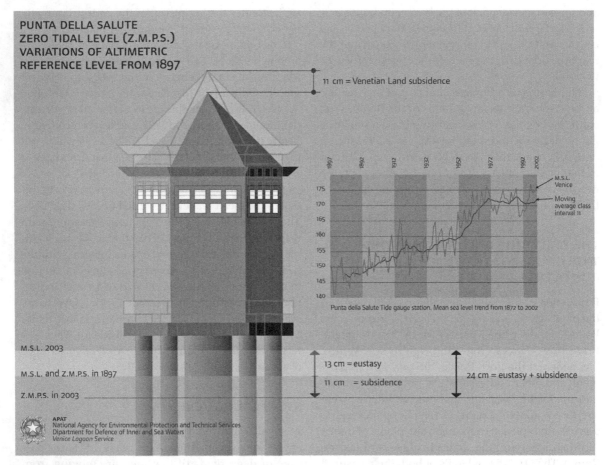

PUNTA DELLA SALUTE
ZERO TIDAL LEVEL (Z.M.P.S.)
VARIATIONS OF ALTIMETRIC
REFERENCE LEVEL FROM 1897

11 cm = Venetian Land subsidence

M.S.L. Venice

Moving average class interval 11

Punta della Salute Tide gauge station. Mean sea level trend from 1872 to 2002

M.S.L. 2003

M.S.L. and Z.M.P.S. in 1897

Z.M.P.S. in 2003

13 cm = eustasy

11 cm = subsidence

24 cm = eustasy + subsidence

APAT
National Agency for Environmental Protection and Technical Services
Dipartment for Defence of Inner and Sea Waters
Venice Lagoon Service

Fig. 12.2   *Zero Mareografico di Punta della Salute (ZMPS). Mean sea-level variation since 1897.*

VLS puts into action its own 24-hour warning and monitoring service (WMS) that continually monitors the water levels and makes adjustments to forecasts. Along with this, the WMS also provides a service of information broadcast, assistance and alerting for the lagoon protection activities, environmental services and navigation, transmitting dispatches of data updates and forecasts.

ANALYSIS OF THE DATA RELATING TO STORM SURGE EVENTS

The level at the Punta della Salute tide gauge is not on its own enough to represent the conditions of the lagoon basin during occurrences of storm surge. Other critical factors ensue, for example, from the winds that are localized in the Upper Adriatic. Tidal observations have shown that the forcing action of the wind along lagoon surface gives rise to considerable pressure on the water against the southern or northern boundaries of the lagoon, depending on the wind direction, either north-easterly (*bora*) or south easterly (*sirocco*) in direction. The maximum levels of the tide are significantly different compared with the levels recorded in Venice in the larger inhabited centres, such as Chioggia in the southern section of the lagoon (Plate 5), or Burano in the northern portion (Plate 6).

Table 12.1 *Worst cases of exceptionally high tides recorded in Venice at Punta della Salute.*

Date	Above ZMPS
4 November 1966	194 cm
22 December 1979	166 cm
1 February 1986	158 cm
15 January 1867	153 cm
12 November 1951	151 cm
16 April 1936	147 cm
16 November 2002	147 cm
15 October 1960	145 cm
3 November 1968	144 cm
6 November 2000	144 cm
8 December 1992	142 cm
17 February 1979	140 cm
5 November 1967	138 cm
26 November 1969	138 cm
22 December 1981	138 cm

In these varying weather conditions, sea level differences between the various parts of the lagoon and especially between the lagoon and the sea (or vice versa) can determine asymmetrical hydrodynamic conditions at the inlets. Such phenomena were observed during the exceptional high tide that occurred on the 8 December 1992, when, for a few hours, the local wind was blowing from a north easterly direction. The tide entered from the Lido inlet and, simultaneously, exited from the Chioggia inlet (Ferla *et al.*, 1999). The opposite effect occurred on the high tide event of the 6 November 2000 while the local wind was blowing from a south-easterly direction (Bonato *et al.*, 2001b).

These singularities have been also confirmed by the results of a finite element hydrodynamic model developed by CNR. – Institute for Study of the Dynamics of Large Masses in Venice, and calibrated with data collected by RTLV (Melaku Canu *et al.*, 2002). All these observations and studies have laid the basis for further investigation on the dynamics of the sea and lagoon exchanges by means of the direct measurements of current speed and discharge through the three inlets.

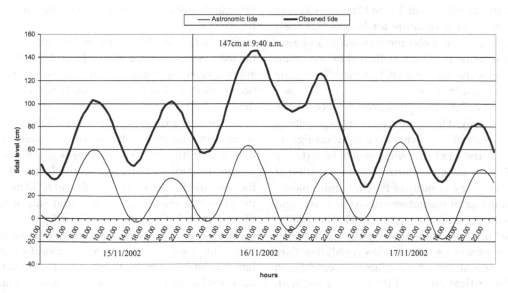

Fig. 12.3a Storm surge event of 16 November 2002. Recorded water level and astronomic predicted tide at Punta della Salute gauge station.

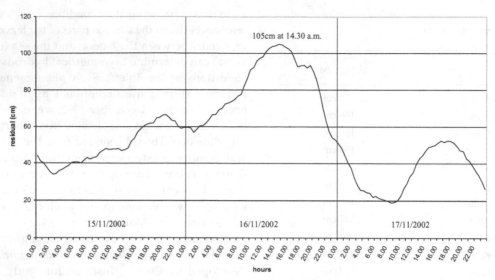

Fig. 12.3b   Storm surge event of 16 November 2002. Recorded water level fluctuations at Punta della Salute gauge station.

MEASUREMENTS OF CURRENTS AT THE LIDO INLET

In 2002, the Hydrographic Office of Venice and the National Research Council (CNR) – the Institute for Study of the Dynamics of Large Masses in Venice (now known as the Institute for Marine Sciences) started up a current meter measurement campaign with a view to developing a method of investigation that includes the use of ADCP devices (Acoustic Doppler Current Profiler).

The results of the first phase, which involved the Lido inlets, brought to light the excellent capability of the tool in describing the variations in the configuration of the current field, proving itself to be certainly appropriate for the characterization of the section of measurement. The rapidity of the procedure enables a large number of measurements to be taken and recorded during a normal cycle of tides. The profiler is installed on board a small, flat-bottomed boat that moves along the cross section at low speeds (less than 1 m s^{-1}). This installation has enabled researchers to record the vertical profile of the tidal current with constant scanning over time. The current speed values are given by the instrument for 'cells' whose

height is set between 0.50, 1.00 and 2.00 metres, depending on the water depths across the section. In the shallow water ends of the section the heights of the cells are set to values of 5.00, 10.00 and 25.00 cm. The width of the cells depends on the speed of the boat; in this case, of the order of a few tens of centimetres.

On 17, 18 and 19 September 2002, the boat conducted 52 surveys in the inlet of Lido. The tide was in approximately spring conditions. The weather conditions were good. Each transect took 15 minutes. A GPS device fitted with a cartographic plotter enabled researchers to precisely control the trajectory and the speed of the boat. In the phases of falling tide and with a tidal amplitude of 35 cm, the peak discharge was around 4000 m^3 s^{-1}. With greater amplitude (44 cm), the peak discharge reached a value of 6350 m^3 s^{-1}. In the phases of rising tide with a tidal amplitude of 64 cm, the peak discharge was 7350 m^3 s^{-1}.

These measurements were made with the intention of implementing a continuous monitoring system for the experimental determination of rates of flow exchanged at the inlet mouths, including exchanges under exceptionally high tide conditions. The results of these measurements have been also utilized, within a research project entitled '*Quantità e qualità degli scambi tra laguna e mare*' (CORILA – Venice) which involved studying the interaction with the shoreline current investigated by means of a CODAR device. A second phase of activity is currently in the process of inception between the APAT and the CNR where the parameters measured will be extended to include suspended sediment discharges. Further investigations are being developed through other activities, including an agreement recently signed between the APAT–SLV and the Magistrato alle Acque, to elaborate a calculation procedure for the variation in water volume in the lagoon basins, based on real time tide data.

ACKNOWLEDGEMENTS

Our thanks to M. Sbavaglia from APAT – Dipartimento per la Tutela delle Acque interne e marine – Servizio Laguna di Venezia for kindly supplying us with the translation of our contribution to the meeting in Cambridge in September 2003; to G. Baldin and G. Pavon from APAT – Dipartimento per la Tutela delle Acque interne e marine – Servizio Laguna di Venezia for furnishing us with tidal data; and to F. Iozzoli from APAT – Roma for poster graphics.

REFERENCES

Bonato, N., Egiatti, G., Ferla, M., and M. Filippi. 2001a. Tidal Observations in the Venetian Lagoon. Update on sea level change from 1872 to 2000. In *Proceedings of the final workshop of Cost Action 40. Sea level in Europe: Observation, Interpretation and Exploitation, Dubrovnik, Croatia, 19–21 September 2001.* Split: Hydrographic Institute of the Republic of Croatia.

Bonato, N., Ferla, M., Umgiesser, G., and G. Zen. 2001b. L'evento di acqua alta del 6/11/2000 in Laguna di Venezia. Approfondimenti e confronti sugli effetti di circolazione indotti dal vento 2001b. In *Atti dei Convegni Lincei 181–XIX Giornata dell'Ambiente. Convegno Il Dissesto Idrogeologico: Inventario e Prospettive.* Rome: Accademia Nazionale dei Lincei.

Ferla, M., Rusconi, A., and G. Zen. 1999. Indagine sperimentale sui cicli di invaso e di svaso nella laguna di Venezia in condizioni meteo avverse. In *Atti dei Convegni Lincei 161– XVII Giornata dell'Ambiente. Convegno Venezia: città a rischio.* Rome: Accademia Nazionale dei Lincei.

Melaku Canu, D., Umgiesser, G., Bonato, N., and M. Ferla. 2002. Analysis of the circulation of the lagoon of Venice under sirocco wind conditions. Scientific Research on safeguarding of Venice. In *CORILA Research Program 2001 Results.* Venice: Istituto Veneto di Scienze Lettere ed Arti, pp. 515–30.

13 • The third dimension in Venice

A. J. AMMERMAN

INTRODUCTION

In the literature on flooding in Venice, it is common to focus on natural aspects of the problem. There is accordingly a major interest in questions such as the trends observed in tide-gauge records, the meteorological factors that produce storm surges at the head of the Adriatic, the processes involved in the formation of marsh islands and so forth. In short, the way in which flooding is normally treated by the scientist and the engineer is as a problem that stems from natural forces that are, for the most part, independent of the life of the city. Venice is taken to be something that is simply given. There is, in other words, a static way of looking at the city as a two-dimensional place standing just above the water level. What is missing is an understanding of the third dimension. There is a failure to see the city as a dynamic, man-made environment that progressively took shape over the centuries in response to a long-standing test of endurance with *acqua alta*, the local name given to a tide that reaches an exceptional height. In fact, the build up of ground levels in the city and the history of flooding in Venice are not separate matters; they are closely related to one another. It is, however, only the historian of urbanism and the archaeologist who find it 'natural' to think about Venice in this way. Thus, my own perspective on the question of flooding is quite different from that found in most of the other contributions to this volume. Special attention will be paid in this chapter to the role of human action (both positive and negative) in producing the city's vertical dimension and to placing the question of flooding in historical context.[1]

The subject of the first part of this chapter is the third dimension in Venice – the elevation of the man-made surfaces (church pavements, house floors and street levels) that date to different times in the past. On the basis of recent work at archaeological sites in the city and the lagoon, the anthropic character of the build up of ground levels is now well established in Venice (Fig. 13.1). The earliest sites go back to Roman times. Thus, there is the opportunity to trace the evolution of the height of man-made surfaces over the last eighteen centuries. It is also possible to estimate over this span of time the average rate of rise per century in the elevation of such surfaces. This value, as we shall see below, turns out to be more than 13 cm per century. In addition, the same data on the height and the age of man-made surfaces can be used as a proxy in reconstructing a curve that shows the long-term trend in relative sea level (RSL) at Venice. In previous work, we put forward the first RSL curve of this kind for Venice (Ammerman *et al.*, 1999: Fig. 3). Here, based on a larger data set, a second approximation is presented (see Fig. 13.2) which yields basically the same result. As more excavations are done at archaeological sites in the years to come, this curve can be further refined.

In the second half of the chapter, an attempt is made to consider some of the implications of what has recently been learned about the third dimension. Emphasis is placed on the importance of taking the long view – both in looking back on what has hap-

---

[1] Support for the research was provided by grants from the Delmas Foundation, the Kress Foundation and the National Geographic Society. I would like to thank Maurizia De Min, who currently directs the Archaeological Superintendency of the Veneto region, for the opportunity of working at many different archaeological sites in Venice since 1989. For their collaboration on environmental studies in the Venetian lagoon, I would also like to thank Rupert Housley and Charles E. McClennen. Special thanks are extended to Ian Agnew and Joel Katz for drawing the two figures.

Flooding and Environmental Challenges for Venice and its Lagoon: State of Knowledge, ed. C. A. Fletcher and T. Spencer.
Published by Cambridge University Press, © Cambridge University Press 2005.

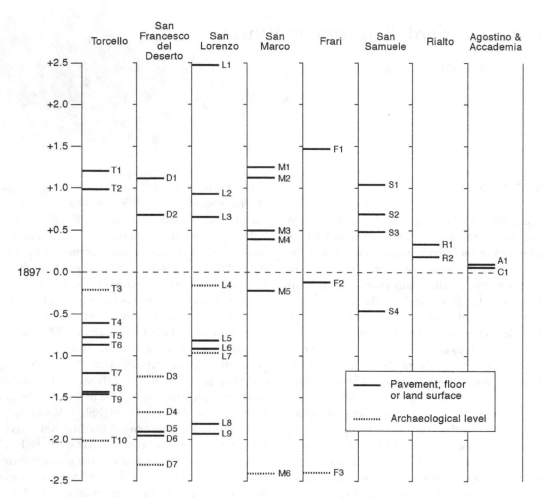

Fig. 13.1   Sequences at archaeological sites: the elevations of dated man-made surfaces and levels at nine sites. The elevations are measured with reference to the 1897 tide-gauge standard in Venice.

pened over the centuries and in looking forward when it comes to the future prospects for the city and the lagoon. Flooding is an old problem in Venice – one that goes back to the origins of the city. It will also be a serious challenge that Venice will have to face in the year 2100, in the year 2200, and beyond. In short, flooding is a problem that will not go away. Choices that are made at any one time will condition the options that are available in the future. One of the weaknesses of the current literature on flooding in Venice is the tendency to take the short view. This shortcoming applies both to the analysis of the problem and to the solutions

that are currently proposed. Even though such a short-term approach appears to be appropriate in the fast-paced world in which we live today, it may not stand up all that well at the bar of history. Indeed, in the long run, as seen for instance by the environmental historian writing at the end of the present century, much of what was done in Venice in the twentieth century is likely to be viewed in a negative light. We shall return to this subject below. There are, of course, many uncertainties – notably the question of global warming – that surround the problem of flooding in Venice and that make it difficult to foresee future conditions and

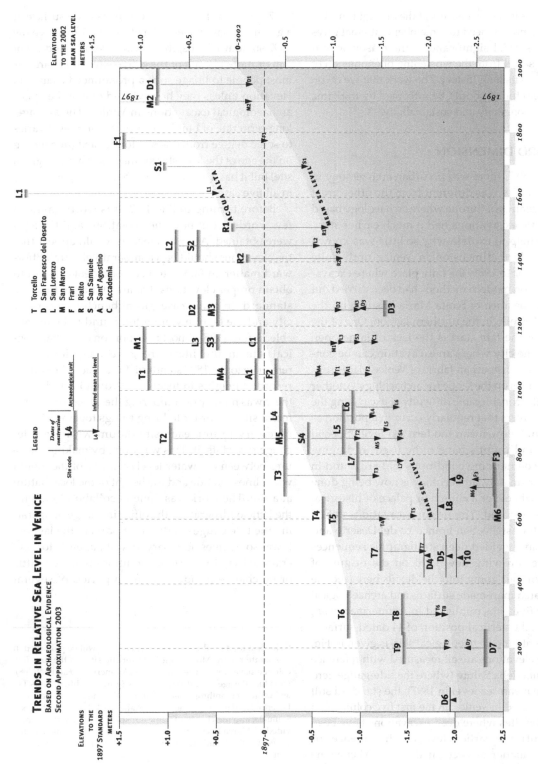

Fig. 13.2 Diagram showing the elevation and the age of man-made surfaces and archaeological level at nine sites in Venice. This is a second approximation, which makes use of such levels as a proxy for the long-term trend in relative sea level at Venice. For the first approximation and an explanation of the lines with arrows below the respective units, see the text and Ammerman *et al.* (1999).

to come up with clear answers at the present time. On the other hand, we need to remember that what gives Venice its special significance – the reason why it should be saved from the worsening phenomenon of *acqua alta* – is its long history. The work that we do on the problem today should be informed by the same concern for longevity and sustainability.

THE THIRD DIMENSION

As late as 1985, there was no urban archaeology in Venice. The city was different from the other great historical centres of Europe where archaeology based on scientific excavations had a much earlier start. This may help to explain why so little was known about the third dimension in Venice until quite recently. Prior to 1985, the only place where excavations based on modern methods had been carried out was at the Basilica of Santa Maria Assunta on the island of Torcello in the northern lagoon. One of the reasons for the late start is the lack of large open spaces in the city where an excavation can be conducted. The dense urban fabric of Venice places serious limitations on the scope for archaeology. Another reason is the considerable difficulty of excavating the wet, early layers that regularly occur in a position as low as 2 m below modern sea level. Special methods and pumps have to be used in digging under the demanding conditions that are found in Venice. Fortunately, archaeology is now being done regularly whenever a church or palace of historical interest is restored. The recent excavations at San Lorenzo di Castello, San Francesco del Deserto and Torcello have yielded deep stratigraphic sequences which are throwing new light on the origins of Venice. Without going into the details here (see the table of dated man-made surfaces and archaeological levels for five sites published in Ammerman *et al.*, 1999: 308), the vertical position of 44 dated surfaces and levels at nine archaeological sites is given in Fig. 13.1.[2] The elevations are all measured with reference to the Punta della Salute (where the tide-gauge zero equals the mean sea level in 1897), the standard still commonly used in Venice. In the first five columns on the left (the sites where deep excavations have been carried out), the earliest levels with evidence for human occupation all occur at around 2 m below the

1897 standard. The uppermost man-made surface at a given site has an elevation of at least 1 m above the 1897 standard. Thus, there is an overall build up of 3 m or more at each of these five sites. It is hard for most people to image such a pronounced change in elevation unless they have visited one of the major archaeological excavations in Venice. The average, long-term rate of build up in the ground level varies to some degree from one site to the next (depending on the age of the lowest man-made surface at a given site) but it has a value of more than 13 cm per century in all five cases.

Before turning to Fig. 13.2, it is worth saying a few words about how the elevations and the ages were obtained. At first glance, it would appear that the elevations should be a comparatively straightforward matter. In fact, it took a good deal of effort to obtain proper elevations that are all tied to the 1897 standard. In many cases, local benchmarks in the city and the lagoon were often found to be unreliable. Before 1995, almost no report on an archaeological site in the literature gave elevations with reference to the 1897 standard. Since each site simply recorded heights in terms of its own local datum, there was no way of comparing the elevations at different sites. In order to bring things together, it was necessary to link each site datum with the tide-gauge record itself. This was done by making measures between the water level on a day of the month with almost no tide and the height of the local datum at a site. The work was done in collaboration with the Ufficio Idrografico, the office in charge of operating the tide gauges in different parts of the lagoon. It was sometimes necessary to wait patiently for several weeks until the right conditions obtained (little or no change in tide level for a period of several

---

[2] The data used for drawing Fig. 13.1 and 13.2 will be published in a separate report. Most of the data for the sites in the first five columns have already been published (Ammerman *et al.* 1999: 308). Note that the age of unit T6 has been slightly modified; units T11 and D8 are new anthropic surfaces at Torcello and San Francesco del Deserto. The data for the sites of the Rialto market, the church of San Samuele, and the church of Sant' Agostino were generously provided by Marco Bortoletto. The twelfth-century pavement of the convent at the Accademia, unit C1, was excavated by Maurizia De Min.

hours and no wind in the lagoon) before taking the measurements at a given site.[3]

In terms of chronology, as well as using more traditional methods such as stratigraphy and the dating of individual levels by their pottery, there was the opportunity to work in close collaboration with the Radiocarbon Accelerator Unit at Oxford University. In all, more than 70 ^{14}C dates from archaeological sites in the city and the lagoon have been run at Oxford. The new method of radiocarbon dating based on accelerator mass spectrometry (AMS), which only became available to the archaeologist in the 1980s, has made a major contribution to the work on the third dimension. The AMS method permits the dating of samples that are much smaller in size. This makes it possible to be much more selective in the choice of the samples, which leads in turn to more reliable ^{14}C dates. By doing a series of AMS dates for each major site, a robust framework of absolute chronology is obtained for dating the individual man-made surfaces at the site. The ^{14}C dates have been calibrated using the method developed at Oxford, and are reported in the datelists published in *Archaeometry*.

As shown in Fig. 13.2, it is also possible to plot the elevations and the ages of the 44 surfaces and levels (presented in Fig. 13.1) on the same diagram. Here the elevations are given on the vertical axis (again with reference to the 1897 standard on the left side and to the mean sea level in 2002 on the right), while the ages are plotted on the horizontal axis. A vertical line with an arrow represents the relationship between a given anthropic surface and mean sea level at the time it was built or laid down (for an explanation of the five treatments that were used in the first approximation, see Ammerman *et al.*, 1999: 307). For example, the floor of a house is taken to stand at +100 cm above mean sea level at the time it was built. The Lagoon of Venice is a dynamic environment and tides have long conditioned building practices there. Habitation is keyed not to mean sea level but to the high water levels commonly reached during the course of monthly and annual cycles as well as storms. To avoid flooding, the Venetians have long known from practical experience that a house floor or a street pavement has to stand somewhat above the level of high tides and seasonal storms at the time it is built. As mentioned above, the man-made surfaces documented at archaeological sites can be used in this way as a proxy for the long-term trend in relative sea level. In a previous study (Ammerman *et al.*, 1999: Fig. 3), we put forward the first long-term curve for RSL in Venice. This curve, which placed emphasis on the time between AD 400 and AD 1400 (in line with our interests in the archaeology of early Venice), will now be called the first approximation. In the second approximation presented here, there are two new anthropic surface dating to the time before AD 300 (units D6 and T9, which give better definition to the early part of the curve) as well as seven new surfaces (from the archaeological sites of San Samuele, the Rialto Market, Sant' Agostino and the Accademia Gallery) that now help to fill in the part of the curve between AD 1000 and AD 1700. For the eighteenth century, there is now good evidence for values of RSL from the height of the algal front on palaces, as seen in paintings by Canaletto and Bellotto (Camuffo and Sturaro, 2003; Camuffo *et al.*, chapter 16). Starting in 1872, there is the record of regular tide-gauge monitoring in Venice.

In previous work, we were able to show that, below the archaeological sites (Torcello, San Francesco

---

[3] The potential complication that a man-made surface may find itself in a lower position in the ground today (relative to the other anthropic surfaces at a site and to the boundary at the base of the lagoon) than it once had at the time it was laid down (due to post-depositional compaction) has been taken into consideration and appears to be negligible. To begin with, few of the units shown in Fig. 13.1 involve freshly laid down layers of natural sediment. Most of the surfaces were subject to active human-induced compaction at the time they were originally laid down. In addition, only 3 m of natural lagoonal sediments (in composition usually 70 to 75 per cent silts) occur between the lowest archaeological level at a given site and the boundary at the base of the lagoon (see Ammerman *et al.* 1999: Fig. 13.2). Since these sediments are standing in water, they have a reduced weight (due to Archimedes principle). In effect, there is not much weight to compress the upper few metres in a given sequence. The main concept that is involved here is known as Terzaghi's principle of effective stress. On the other hand, compaction in the deep column of Pleistocene sediments that occurs below the lagoon boundary is a component of subsidence, which contributes to the change in RSL that one is trying to measure. I would like to thank Ahmet Çakmak of Princeton University and Roy Butterfield of the University of Southampton for their helpful discussions on this question.

del Deserto, San Lorenzo, San Marco and Frari), the boundary at the base of the lagoon commonly stands in a position at about 5 m below the 1897 standard (Ammerman *et al.*, 1999: Fig. 2). The 'high' position of the lagoon boundary is also documented by sub-bottom profiling and by off-site coring (McClennen *et al.*, 1997). In addition, there are four AMS dates on samples of plant fibre (recovered at the boundary), which date the formation of the lagoon to some 6000 years ago. Given that the earliest archaeological levels at sites such as Torcello and San Francesco occur in a position of about 2 m below the 1897 standard and that they date to late Roman times, one can obtain a value of 7 cm per century for the average rate of rise in RSL for the period between 4000 BC and AD 400. It is worth noting that this value is more or less the same as the long-term natural component of subsidence for Venice ($0.7-1.0$ mm yr^{-1}) estimated by Eugenio Carminati (see chapter 4). In contrast, the average rate of rise in RSL between AD 400 and 1897 is about 13 cm per century (both for the first approximation and the second one). The reason for this higher rate in an anthropic context (over a span of 15 centuries) is not well understood at the present time. The rate of change in RSL for the last 100 years, as revealed by tide-gauge records, has further increased to 23-25 cm. In the latter case, some 10 cm of the rise is attributed to the pumping of groundwater at the industrial complex of Porto Marghera in the years between 1930 and 1970 (Teatini *et al.*, 1995; Carbognin and Taroni, 1996; Pirazzoli and Tomasin, 2002). It is of interest that the long-term, historical trend in RSL (13 cm per century) in combination with the effect of pumping (10 cm) yields a value close to the observed rise in RSL since 1897. As more archaeological evidence becomes available in the future, the curve can be better documented and there may be a chance to recognize local variations (between different areas of the lagoon and between different centuries) in the long-term trend in RSL rise. The important thing is that a start has now been made. Only six years ago, Venice had no long-term curve for RSL of any kind. It is not without irony that the third dimension is finally beginning to come into clearer focus through the study of cultural artefacts of the city's history: archaeology and the paintings of Canaletto.

TAKING THE LONG VIEW

The history of human habitation in the Venice Lagoon, as mentioned before, goes back to Roman times. It was by means of sustained human action that ground levels in Venice were built up in a series of steps over the centuries. From the beginning, the Venetians have always lived with flooding as a rare event that happens whenever there is an exceptionally high tide. If one takes the long view, flooding is a very old story (e.g., the list of major historical floods in UNESCO, 1969). The traditional way that the Venetians responded to the threat of *acqua alta* was by raising the level of their city. Today this takes the form of the project called INSULA. Since the time of the fall of the Republic in 1797, insufficient attention was paid to maintaining the elevation of much of the city. Records for the twentieth century show that *acqua alta* is a worsening phenomenon. The decade with the highest number of such tidal events is the one that just ended. One of the worst seasons of flooding in the city's history was the winter of 2002–3. In short, the third dimension has not kept pace with the life of the city. Human action is, in part, responsible for this situation. As mentioned above, the city lost what amounts to 10 cm in its elevation because of the extraction of groundwater at Porto Marghera. In addition, because of the creation of deep navigation channels and the closing of large areas at the back of the lagoon for fish farms, Venice became much more vulnerable to storm surges in the twentieth century (e.g. Umgiesser, 1999). Thus, if one is really interested in the question of flooding, Venice seems to have made all of the wrong moves in the twentieth century. At the bar of history, much of what happened there in the last century will be viewed in a negative light. In terms of what is now known about the long-term trend in RSL (13 cm per century), the city lost, in effect, a century in its battle against the sea because of ill-advised human choices. In historical terms, INSULA, which began its work of cleaning out the canals, repairing the foundations of buildings, and raising street levels only in the 1990s, can be regarded as a last-minute effort to regain a lost century. It is an old-fashioned ray of sunshine in what is otherwise a rather dismal picture. At the same time, INSULA, as Bruno Dolcetta notes (see chapter 19), may represent the last time that the solution of building up the ground level can be employed without compromising the integrity of the city's architecture (as street levels and thresholds keep moving up in Venice, doorways become ever lower).

The situation has now reached the point, especially in light of the prospect of global warming (IPCC, 2001), where something more than raising the ground level is called for in Venice. The much-debated project known as MOSE seems to have received a green light from the current government, although there is still some question about whether or not the environmental impact assessment procedure required by an Italian law passed in 1995 has been properly completed. In retrospect, there are also serious questions about the quality of the scientific analysis in the impact studies (Ammerman and McClennen, 2000). For example, two out of the three projections for the rise in RSL by the year 2100 (the proposed life of the gates) used in the official studies (Magistrato alle Acque e Consorzio Venezia Nuova, 1997; Collegio di Esperti di Livello Internazionale, 1998; Ministry of the Environment, 1998) take no account of global warming. Today, only a few years later, this formulation of the problem (the three 'scenarios') would appear to make little sense, especially the low projection of a rise of only 4.4 cm in RSL by the year 2100. It will be recalled that the middle scenario was 16.8 cm. The only projection to include a consideration of global warming was the high one of 53.4 cm, which was taken to be an extreme or unlikely case in the impact studies.

In fact, the low projection provides a good example of what happens when one takes the short view. Because of the pumping of groundwater at Porto Marghera (1930-70), the tide-gauge record for Venice is unfortunately broken into three short pieces. If one looks at the short tidal record since 1970, it suggests that there has not been much change in RSL over the last 30 years. There is the paradox here that while various types of short-term measurements appear to show no change in the city's height (e.g. Tosi *et al.*, 2002), the worsening phenomenon of *acqua alta* implies just the opposite (e.g. Pirazzoli and Tomasin, 2002; Camuffo and Sturaro, 2003). Moreover, the value of 4.4 cm itself (used for the low projection in all three reports) can

be traced back to a regression analysis done by Gatto and Carbognin (1981: Fig. 10) based on a tide-gauge record that spans only 18 years (1908-25). Again, the weakness of their analysis, from the viewpoint of oceanography, is the shortness of the time frame. In the literature after 1998, there is a clear shift in ideas about projections for Venice by the year 2100. In a study undertaken by the consortium of universities and research centres in Venice and Padua (CORILA, 1999), the low scenario (SPP) is taken to be 16.4 cm and the high scenario (Spe) sees a rise of 31.4 cm in RSL by 2100. Our own attempt to estimate a low projection yielded a rise of 30 cm over the next 100 years (Ammerman and McClennen, 2000). This value (30 cm) is incidentally seven times larger than the low scenario used in the three official impact reports. And in the most recent literature (see, for example, Frassetto, chapter 5), a net rise in RSL on the order 50 cm is no longer considered to be an extreme value but a likely one.

The key question is the high frequency of closure in the later years of the gates' life. One of the weaknesses of the impact studies was not to take proper account of seasonality and the variability between good years and bad years (Ammerman and McClennen, 2000). It is well known that two-thirds of the *acqua alta* events in Venice commonly take place in the months from October through January (Venice City Council, 1995). If the gates are closed whenever a tidal event reaches a height of 100 cm or more above the 1897 standard, the average number of tidal events per year having such a height under a rise of 30 cm in RSL will be 94. If there is only closure for tides reaching 110 cm in height, then this same number will hold under a rise of 40 cm in RSL. However, if allowance is made for bad years and for false alarms, the number of closures will be much higher in some years. The implication is that under a 30 to 40 cm rise in RSL, gate closure may take place day-after-day and week-after-week in a month such as October during a bad year in the later years of the gates' life. It has been argued that the seasonal concentration of gate closures will not pose a serious problem, since what is involved are winter months (today the winter months have low levels of biological productivity). But can we be fully confident that October will still be a winter month (and not a summer one) when temperatures increase by 2-3 degrees centigrade as they may well do by 2100 under the most recent projections for global warming?[4] Under such a marked rise in temperature, it is difficult to predict what will happen to the ecology of the lagoon even without the gates. It will be recalled that the 1984 law authorizing the search for a solution to the problem of *acqua alta* states that the proposed project should have no negative impact on the activity of the port, on the quality of the water in the lagoon, and on the ecology and morphology of the lagoon. These are, of course, tough conditions to meet. In addition, the same law requires that the project should be reversible. This means (at least in theory) that the mobile floodgates will have to be removed when they become obsolete.

In taking long view, one of the important questions to ask concerns the effective use life of the floodgates (Pirazzoli, 2002). Once they are built, how many years will elapse before the gates become obsolete? This is, of course, an open question at the present time.[5] Much will depend on the rise in RSL due to global warming. The contribution of global warming should be somewhere in the range of 9 to 88 cm by the year 2100, according to the most recent report of the IPCC (2001).[6] There is still much uncertainty on this matter. The situation is likely to become clearer by the time of the next IPCC report, which is scheduled to appear in 2007. At that time (assuming that

the gate solution continues to move forward), the construction of the floodgates will be half way to completion. Thus, even before they are ready to be put into operation (after 2010), there will be a much better idea about whether or not they represent a sound, cost effective solution to the problem of flooding in Venice. If the net rise in RSL turns out to be a moderate one (say 50 cm by 2100) and the gates continue to function well through the year 2080 (without having a negative impact on the ecology of the lagoon), then the project will pass the bar of history. On the other hand, if the rise in RSL is higher (say 80 cm by 2100), then the gates may already become obsolete before the year 2060 (that is, they will have an effective use life of less than 50 years), and there will be serious questions about the wisdom of going ahead with the project. In the latter case, there is a good chance that the floodgates will never be used to protect the city against an extremely high tide of the kind that occurred in November of 1966 (the original motivation for the design of the gates). Attention will then have to turn to finding new solutions to the problem. And the gates themselves will probably not serve as a helpful building block in the search for a solution that takes a longer view.

One of the ironies in all of this is that a key factor in determining the intensity of global warming over the next 100 years is likely to be human action itself – the ability of the nations of the world to limit emissions linked with the greenhouse effect. Again, human behaviour, which was at the heart of the build up of ground levels in Venice over the centuries and which is also one of the main reasons for the problem of *acqua alta* today, will be at the centre of the city's future. Now the human choices that count will be those at the global level. What will happen to Venice in the twenty-second century? Only time, the fourth dimension, will tell.

---

[4] Note that the long-term steric change in the volume of the world's oceans produced by such a rise in temperature will put Venice in a very difficult position by the year 2200.

[5] One of the factors influencing the gates' life will be the success of INSULA in raising ground levels in Venice. If most street levels can be brought up to an elevation of 110 cm above the 1897 standard, this will allow the gates to be closed only when tidal events reach this height (and not 100 cm as originally planned). In turn, this will mean fewer gate closures for a given rise in RSL. In effect, INSULA (often seen as a competing project) may 'save' MOSE by slowing down the rate at which the gates become obsolete because of global warming.

[6] At Venice Workshop No: 5 held in 2002, Peter Stone of MIT suggested in his abstract, 'Global sea-level rise to 2100', that there is a 90 per cent probability that the value for the rise in RSL will fall in the range of 25 cm to 75 cm.

REFERENCES

Ammerman, A. J., and C. E. McClennen. 2000. Saving Venice. *Science* **289**, 1301–2.
Ammerman, A. J., McClennen, C. E., De Min, M., and R. Housley. 1999. Sea-level change and the archaeology of early Venice. *Antiquity* **73**, 303–12.

Camuffo, D., and G. Sturaro. 2003. Sixty-cm submersion of Venice discovered thanks to Canaletto's paintings. *Climate Change* **53**, 333–43.

Carbognin, L., and G. Taroni. 1996. Eustatismo a Venezia e Trieste nell'ultimo secolo. *Atti Istituto Veneto di Scienze Lettere ed Arti* **154**, 281–98.

Collegio di Esperti di Livello Internazionale. 1998. *Report of the Collegio di Esperti di Livello Internazionale on the Mobile Gates Project for the Tidal Regulation at the Venice Lagoon Inlets.* Venice.

CORILA. 1999. Scenari di crescità del livello del mare per la Laguna di Venezia. Venice: CORILA, pp. 1–40.

Gatto, P., and L. Carbognin. 1981. The Lagoon of Venice: natural environmental trend and man-induced modification. *Hydrological Sciences Bulletin* **26**, 379–91.

IPCC. 2001. *Climate Change 2001. The Scientific Basis.* Cambridge: Cambridge University Press.

McClennen, C. E., Ammerman, A. J., and S. Schock. 1997. Framework stratigraphy for the Lagoon of Venice, Italy, revealed by new seismic-reflection profiles and cores. *Journal of Coastal Research* **13**, 745–59.

Magistrato alle Acque e Consorzio Venezia Nuova. 1997. *Environmental Impact Study for the Preliminary Design of the Interventions at the Lagoon Inlets for Tidal Flow Control.* Venice: Magistrato alle Acque.

Ministry of the Environment. 1998. *Interventi alle bocche lagunari per la regolazione dei flussi di marea.* Rome: Ministry of the Environment.

Pirazzoli, P. A., 2002. Did the Italian government approve an obsolete project to save Venice? *EOS* (Transactions of the American Geophysical Union) **83**, 20, 217–18.

Pirazzoli, P. A., and A. Tomasin. 2002. Recent evolution of surge-related events in the northern Adriatic area. *Journal of Coastal Research* **18**, 537–54.

Teatini, P., Gambolati, G., and L. Tosi. 1995. A new three-dimensional nonlinear model of the subsidence at Venice. In F. B. J. Barends, *et al.* (eds.), *Land Subsidence.* IAHS 234. The Hague: IAHS, pp. 353–361.

Tosi, L., Carbognin, L., Teatini, P., Strozzi, T., and U. Wegmüller. 2002. Evidence of the present relative land stability of Venice, Italy, from land, sea, and space observations. *Geophysical Research Letters* **29**, 12, 3–1–3–4.

Umgiesser, G. 1999. Valutazione degli effetti degli interventi morbidi e diffusi sulla riduzione delle punte di marea a Venezia, Chioggia e Burano. *Atti dell'Istituto Veneto di Scienze, Lettere ed Arti* **155**, 231–86.

UNESCO. 1969. *Rapporto su Venezia.* Milan: Mondadori.

Venice City Council. 1996. *Definition of the Risks of Flood Tides to the City of Venice. Current Organizational State of the City.* Venice: Venice City Council.

14 • Geoarchaeology in the Lagoon of Venice: palaeoenvironmental changes, ancient sea-level oscillation and geophysical surveys by acoustic techniques

S. BUOGO, E. CANAL, G. B. CANNELLI, S. CAVAZZONI, S. DONNICI AND A. LEZZIERO

INTRODUCTION

A geoarchaeological approach to the reconstruction of the history of the Venice Lagoon is indispensable, giving a degree of reliability to the framework of the historical, environmental and socio-economic evolution of this territory. Over the centuries, the Lagoon of Venice has suffered continuous modification by natural processes, of which the principal ones are soil subsidence and oscillations in mean sea level. One of the more evident manifestations of these processes has been the alternation of appearances and disappearances of the *barene*, or salt-marshes.

The presence of buried archaeological sites in the lagoon floor is therefore correlated to particular marine environmental conditions. Their detection and unearthing thus helps to improve the knowledge of the geomorphological evolution of the area, as well as to yield new evidence on trends in sea-level change for the lagoon. Over the last 40 years, archaeological studies have revealed hundreds of ancient sites in the city of Venice and in the lagoon. Archaeological remains are buried in lagoon deposits at different depths but mainly between 2 and 3 m. In some cases, they lie on the lagoon floor or on the surface of the islands. The sediments found in these sites of interest have been studied by the methods of Quaternary geology. In particular, preliminary surveys have been performed in very shallow waters using high-resolution acoustic systems; this technology has been shown to be capable of detecting the distribution of structures buried in sediments and their size.

GEOARCHAEOLOGICAL RESEARCH

Archaeological research in the Lagoon of Venice began forty years ago with the activity of the Honorary Inspector Ernesto Canal, often in collaboration with W. Dorigo. Activities consisted of archive research, manual core sampling and underwater investigations (Canal, 1978, 1995, 1998; Canal *et al.*, 1989; Dorigo, 1983). In the past two decades, researchers of the Italian National Research Council (CNR) have extended Canal's investigations with geological and micropalaeontological analyses, in order to reconstruct palaeoenvironmental conditions at some buried sites (Favero and Serandrei Barbero, 1983). Geoarchaeological investigations are currently being carried out in every archaeological area by the Italian northern and middle Adriatic Sea Wet – Underwater Archaeological Group (NAUSICAA) (Lezziero, 2000; Serandrei Barbero *et al.*, 2001). Joint archaeological and geological data coming from stratigraphical excavations carried out in the city of Venice and in lagoon islands has made it possible to plot a sea level oscillation curve for the past 2000 years. The curve is constructed from both archaeological findings (from 70 BC) and medieval remains and from using the mean levels of the original floor of 80 Venetian palaces, grouped by century from thirteenth to the nineteenth centuries (Dorigo, 1983). These data have been corrected through the calculation of values for three important factors: subsidence, safety leeway and site geological conditions. The present curve (Fig. 14.1; Canal *et al.*, 2001) is a revision of the earlier study of Canal and Cavazzoni (1996) (Benardi *et al.*, 1999).

Of the other published sea level curves, one (Camuffo and Sturaro, 2003; see also Camuffo *et al.*, chapter 16) is based on water levels deduced from the paintings of Canaletto covering a time of only 31 years (1727-58), the other (Ammerman *et al.*, 1999; see also Ammerman, chapter 13) interpolates (often contradictory) data between the second and the fif-

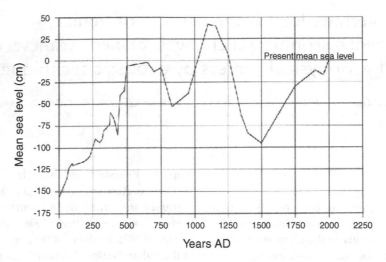

Fig. 14.1   Mean sea-level oscillations calculated from archaeological findings (Canal *et al.*, 2001), medieval remains and original ground floors of 80 Venetian palaces from the thirteenth to nineteenth centuries. The minimum level, at about AD 1450–1500, corresponds mainly to Venetian Gothic and Renaissance palace ground floor levels lower than Roman–Byzantine (thirteenth–fourteenth centuries) levels.

teenth centuries with modern values, with a lack of data for the sixteenth, seventeenth and eighteenth centuries.

OTHER STUDIES

Palaeoecological studies provide further information about the paleo-environmental conditions of the archaeological sites. The basin of the lagoon started to develop when the coastline of the Adriatic Sea had stopped not far from its present position, where some river mouths were located. Previously, during the last glaciation, the sea level was much lower than the present level and most of the Adriatic basin was occupied by an alluvial plain (Bortolami *et al.*, 1977). In the subbottom of the Venice Lagoon, alluvial deposits, hundreds of metres in thickness, are covered with a few metres of lagoonal sediments (Favero and Serandrei Barbero, 1978, 1980).

To discover former hydrological conditions in the lagoon, deposits have been described using sedimentological criteria and samples have been collected by core sampling and manual sampling at both submerged and emergent sites. Muddy layers without particular sedimentary structures, alternating layers of different grain size due to tidal phases (tidal bedding), massive sand deposits with 'wavy bedding' structures, horizons rich in organic matter and peat connected to salt marshes have all been distinguished. Site chronologies have been provided by archaeological data, dendrochronological and radiocarbon dating, coming from samples of ancient wooden anthropogenic structures or natural organic matter.

ACOUSTIC METHODS

Acoustic techniques applied to the deposits of the lagoon subsoil have been used to correlate levels intersected by individual drilling locations. Studying the images produced by acoustic surveys (Cannelli and D'Ottavi, 1995) permits the discrimination of the different acoustic characteristics of different layers which can then be defined more clearly by further spot drillings. Previous acoustic investigations in the Lagoon of Venice (McClennen *et al.*, 1997) used frequency ranges too low to work in very shallow water and to detect strata in the first 2–3 metres. However, in the last few years, teams from two CNR Institutes, ISMAR of Venice and IDAC of Rome, have conducted underwater acoustic surveys in the shallow waters (0.5 to 2 metres) of the Venice Lagoon, to search for, and map, subbottom remains (Buogo *et al.*, 2001). The aim has been to explore the first layers below the lagoon bottom down to a depth of 5-10 metres, where the base of lagoonal sediments has been found (Serandrei Barbero *et al.*, 2001).

Two experimental setups have been used. The first one is an echosounder ELAC LAZ 72 with 30 KHz transducer and equipped with a 'sediment card' that enhances the received echoes. The second approach is a high resolution echographic system, designed by IDAC, which has already given positive results elsewhere in landscape archaeology (Cannelli *et al.*, 2000). It has been improved for marine archaeology applications (Cannelli, 2002) within the framework of the Italian 'Special Project on Protecting and Safeguarding Cultural Heritage'. This system works in a broad frequency band from a few kHz to about 300 kHz and offers the possibility of choosing the optimal resolution frequency for various targets. The first field experiments demonstrated that the ELAC echosounder can be deployed in very shallow waters (up to 0.5 metres) and survey large areas with a resolution of the order of 10 cm, even if the boat speed is relatively fast (1-2 m s^{-1}).The echosounder was originally equipped for the internal recording of analogue signals on dry recording paper. The device has been now modified to allow the acquisition of received signals by means of a high-speed digitizing board with 12-bit resolution installed on a portable PC. The timing for acquisition is provided by a trigger pulse from the transmit module of the echosounder, synchronized with the electric pulse to the transducer. A typical pulse repetition rate is 20 pulses s^{-1}, with a sampling rate of 200 kHz. Each acquired trace is stored on disk for further processing and visualization and the produced data files can be directly checked against the standard dry recording paper record.

Many experiments have been carried out in areas where archaeological layers were previously discovered by means of discrete soundings made by E. Canal on the basis of ancient documents in the archives of Venice (Buogo *et al.*, 2001). Other surveys have been performed in areas of geomorphologic interest, such as those where ancient meandering rivers flow into the lagoon.

PROJECT ECHOS

The project ECHOS, financed by the Venice Water Authority, is now in progress. It has the task of reconstructing the natural environment and the distribution of inhabited areas over the past centuries in the Venice Lagoon. The methods being used are acoustic surveys and palaeoecological and radiochronological analysis on core samples from selected lagoon areas. On the basis of previous investigations, the selected areas are defined by whether the predominant interest is archaeological or geological. A submerged archaeological area is studied by very thick acoustic sections and controlled by geoarchaeological analyses on the sedimentary sequence. An area of geological interest, generally wider than an archaeological one, is investigated with the same methods, but using a larger grid of acoustic sections. The aim of the project is to draw lagoon maps for different centuries using the new geoarchaological data linked to archaeological and archival information coming from two other projects linked to ECHOS. As an example of the surveys already performed in the northern Venice Lagoon, Fig. 14.2 shows the geological structure underlying the subbottom: a buried palaeochannel filled with a thinning-upward sequence (note exaggerated vertical scale). The erosive base of the channel is clearly identified by the inner structures of the filling deposit. These point bar structures show the migrating direction of the active channel and meander concavity; they also provide detailed data to reconstruct the palaeoenvironmental changes in the area. It should be possible to understand the nature of the fill once sediment analysis of core samples has been made. Figure 14.3 shows the same geological structure seen in Fig. 14.2 but here the acoustic section follows the palaeochannel flow direction. In this way a marked horizon is very reflective and represents the palaeochannel base lag. It is probably composed of coarse sediments, but this hypothesis needs to be proved by further research. Other acoustic surveys are in progress and they will complete our present knowledge of the archaeological and geomorphological features of the two areas. A further purpose of the project is to create an interdisciplinary database, allowing access to geoarchaeologically-structured information to support decision-makers.

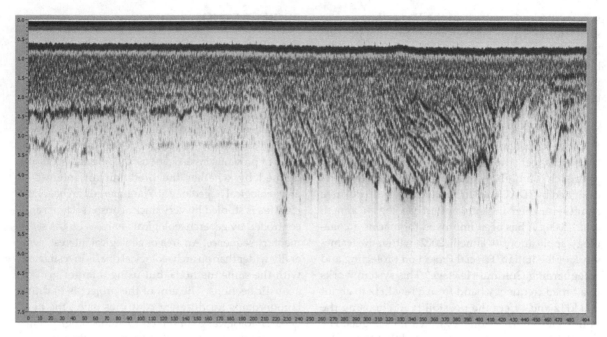

Fig. 14.2   An acoustic section showing a geological structure underlying the bottom of Venice Lagoon: a buried palaeochannel filled with point bar sediments. Scales are in metres.

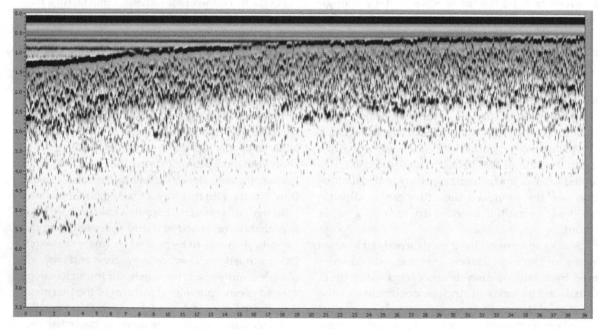

Fig. 14.3   An acoustic section following the palaeochannel flow direction: a marked horizon is very reflective and represents the palaeochannel base lag deposit. Scales are in metres.

REFERENCES

Ammerman, A. J., McClennen, C. E., De Min, M., and R. Housley. 1999. Sea-level change and the archaeology of early Venice. *Antiquity* **73**, 303–12.

Bonardi, M., Canal, E., Cavazzoni, S., Serandrei Barbero, R., Tosi, L., and S. Enzi. 1999. Impact of palaeoclimatic fluctuations on depositional environments and human habitats in the Lagoon of Venice (Italy). *World Resource Review* **11**, 2, 247–57.

Bortolami, G. C., Fontes, J. C., Markgraf, V., and J. F. Saliege. 1977. Land, sea and climate in the northern Adriatic Region during Late Pleistocene and Holocene. *Paleogeography, Paleoclimatology and Paleoecology* **21**, 139–56.

Buogo, S., Canal, E., Cannelli, G. B., and S. Cavazzoni. 2001. High-resolution acoustic prospecting for detecting the ancient Roman Port of Altino in the Lagoon of Venice. 3nd International Congress on Science and Technology for the Safeguard of Cultural Heritage in the Mediterranean Basin, Alcalà, Madrid, 9–14 July 2001.

Camuffo, D., and G. Sturaro. 2003. Sixty-cm submersion of Venice discovered thanks to Canaletto's paintings. *Climatic Change* **58**, 333–43.

Canal, E. 1978. Localizzazione nella laguna veneta dell'isola di San Marco in Bocca Lama e rilevamento di fondazioni di antichi edifici. *Archeologia Veneta* **10**.

Canal, E. 1995. Le Venezie sommerse: quarant'anni di archeologia lagunare. In UNESCO (ed.), *La Laguna di Venezia*. Verona: Cierre.

Canal, E. 1998. *Testimonianze Archeologiche nella Laguna di Venezia. L'età antica*. Venice: Ed. del Vento.

Canal, E., and S. Cavazzoni. 1996. Indagini geoarcheologiche. In M. Bonardi (ed.) *Studio dei processi evolutivi di alcune barene della Laguna di Venezia (Bacino Nord) in relazione alle variazioni del livello marino*. Rapporto 5, CNR-ISDGM, Venice.

Canal, E., Fersuoch, L., Spector, S., and G. Zambon. 1989. Indagini archeologiche a S. Lorenzo d'Ammiana (Venezia). *Archeologia Veneta* 12.

Canal, E., Fozzati, L., and A. Lezziero. 2001. Geoarchaeology in Venice Lagoon: palaeoenvironmental changes and ancient sea level oscillation. Paper presented at MEDCOAST 01, 5th International Conference on the Mediterranean Coastal Environment, 23–27 October 2001, Hammamet.

Cannelli, G. B. 2002. High-resolution and high-power ultrasound method and device for submarine exploration. *Intern. Applic. Publ. under the Patent Cooperation Treaty*, PCT IT03/007S1 (11 Nov. 2003) (Extension of Italian Patent no. RM 02A 000581 (19.11.2002)).

Cannelli, G. B., and E. D'Ottavi. 1995. High-resolution acoustic system for sea subbottom imaging of archaeologic remains. In *Proceedings 1st International Congress on Science and Technology for the Safeguard of Cultural Heritage in the Mediterranean Basin*, vol. I, pp. 373–78. (Catania, 26 Nov. – 2 Dec. 1995.)

Cannelli, G. B., D'Ottavi, E., and A. Alippi. 2000. Acoustic Methods in Archaeology. In *The Archaeology of the Mediterranean Landscape: Non Destructive Techniques Applied To Landscape Archaeology*, vol. 4. Oxford: Oxbow Books, pp. 155–65.

Dorigo, W. 1983. *Venezia, origini*. Milan: Electa.

Favero, V., and R. Serandrei Barbero. 1978. La sedimentazione olocenica nella piana costiera tra Brenta e Adige. *Memorie Soc. Geol. It.* 19, 337–43.

Favero, V., and R. Serandrei Barbero. 1980. Origine ed evoluzione della Laguna di Venezia, bacino meridionale. Lavori Soc. *Veneziana Scienze Naturali* **5**, 49–71.

Favero, V., and R. Serandrei Barbero. 1983. Oscillazioni del livello del mare ed evoluzione paleoambientale della Laguna di Venezia nell'area compresa tra Torcello e il margine lagunare. Lavori Soc. *Veneziana di Scienze Naturali* **8**.

Lezziero, A. 2000. Il sottosuolo di Venezia: sedimentologia e paleoambienti. In Proceedings of Congress, *Le Pianure: Conoscenza e Salvaguardia*, Ferrara: Emilia-Romagna Region.

McClennen, C. E., Ammerman, A. J., and S. G. Schock. 1997. Framework stratigraphy for the Lagoon of Venice, Italy: revealed in new seismic-reflection profiles and cores. *Journal of Coastal Research* **13**, 3, 745–59.

Serandrei Barbero, R., Lezziero, A., Albani, A., and U. Zoppi. 2001. Depositi tardo-pleistocenici ed olocenici nel sottosuolo veneziano: paleoambienti e cronologia. *Il Quaternario* **14**, 1, 9–22.

15 • A consistent interpretation of relative sea-level change in Venice

R. BUTTERFIELD

INTRODUCTION

The combined effect of ground subsidence and eustatic rise in sea level (i.e. the reduction in relative sea level, RSL) is reducing the freeboard of Venice annually by around 1.5 mm yr^{-1}. However, there are still conflicting views concerning the annual rate of reduction and its variation over the past two millennia. For example: (1) Pirazzoli (1986) quotes a reduction in RSL of 280 mm between 1872 and 1984 at a mean rate of 2.5 mm yr^{-1}; (2) Cavazzoni (1977) quotes a reduction in RSL of 235 mm between 1897 and 1977 at a mean rate of 2.92 mm yr^{-1}; and (3) Gatto and Carbognin (1981) a reduction in RSL of 224 mm between 1908 and 1980 at a mean rate of 3.1 mm yr^{-1}, results which imply that the RSL is reducing at a significantly accelerating rate. This chapter provides an alternative interpretation of

the data as an approximately constant, background rate of change of RSL over the past several hundred years, augmented by additional subsidence due to the excessive ground-water extraction that occurred between 1930 and 1970.

ARCHAEOLOGICAL DATA

Ammerman *et al.* (1999) retrieved artefacts from the lagoon spanning the period AD 200 to AD 1900. By measuring the depth from which they were recovered, relative to a sea level recorder (installed on the Salute in 1897), and making reasonable assumptions about their original height above mean sea level, together with either historical or carbon-dated assessments of their age, they were able to construct Fig. 15.1. This figure shows, as bold points, the deduced location, in

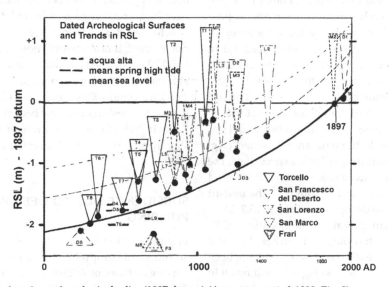

Fig. 15.1   Age – RSL data for archaeological relics (1897 datum) (Ammerman *et al.* 1999; Fig. 3).

Flooding and Environmental Challenges for Venice and its Lagoon: State of Knowledge, ed. C. A. Fletcher and T. Spencer.
Published by Cambridge University Press, © Cambridge University Press 2005.

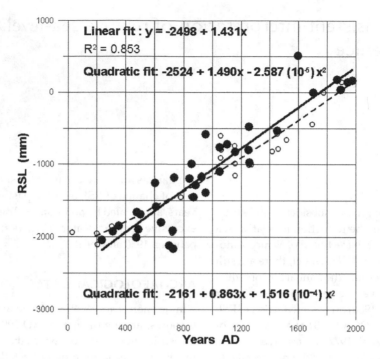

Fig. 15.2   Best-fit lines to relic levels in Fig. 15.1.

time and depth, of some 30 such relics. A line through these points provides a long-time-series estimate of the RSL (relative to the 1897 datum). Points T_1, T_2, M_1 (locating pavement relics in major churches) appear to be anomalously high, probably due to the assumption that such pavements were constructed 1.5 m above mean sea level. If this estimate were revised to be 2.0 m (about 6 Venetian feet) these points become embedded in the main body of the results. The full curve in the figure is their 'trend line' estimate of the change of RSL with time. This is misleading and if statistical best-fit lines to the data are calculated (Fig. 15.2), for either linear or quadratic expressions, the rate of change of RSL from (200 – 1900) AD is found to be virtually constant at 1.43 mm yr^{-1}. The best-fit equations are given in the upper part of Fig. 15.2. More recently, Ammerman (private communication, 2003; and see chapter 13) has obtained further data and adjusted some of his earlier points. These have been superimposed (open circles) on Fig. 15.2 and result in a modified quadratic-fit equation (shown at the bot-

tom of the figure and as a dashed line in it) that predicts a slow increase in the rate of reduction of RSL over this period. However, the mean rate of change over the past 200 years remains at 1.43 mm yr^{-1} and the conclusions drawn below from his original data are not affected. It is also worth noting that if the span of Fig. 15.2 is extended backwards in time to include reliable dating of the lagoon boundary (McClennen et al., 1997) at around 4000 years BC, and 5m below the 1897 datum, the best-fit quadratic becomes $-2220 + 1.011x + 7.953(10^{-5})x^2$. The predicted rate of increase of RSL is slightly reduced thereby, becoming 1.33 mm yr^{-1} in AD 2000.

SUBSIDENCE DUE TO DEEP WELL-PUMPING

Figure 15.3 (Ricceri and Butterfield, 1974) shows a summary of comprehensive data on mean ground-surface settlement within Venice from 1900 to 1972. The key element here is the rapid increase in subsi-

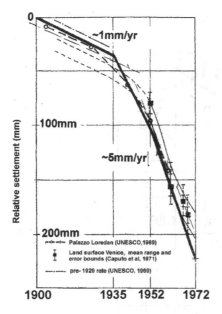

Fig. 15.3   Ground subsidence records within Venice, 1900 to 1972.

dence caused by deep-well water extraction, centred on Marghera, for industry between 1930 and 1970. Thereafter, deep-well pumping ceased, a rebound in ground level of about 20 mm occurred very rapidly and the subsidence regime would have reverted to its previous pattern. Fig. 15.4 shows the effect of this interlude superimposed on the 1.43 mm yr^{-1} RSL change rate established from the Ammerman et al. (1999) data. The additional subsidence cause by the pumping is seen to have potentially shortened the 'life' of Venice by about 84 years (Butterfield, 2004).

TIDAL RECORDS

Camuffo and Sturaro (2004) deduced the trend in RSL (1879–2000) from the Salute tidal records but, without making use of the anthropic deep-well pumping data, they were only able to propose rather unconvincing assessments of its rate of change. Figure 15.5, which shows their tidal data with Fig. 15.4 superimposed, does provide a much more satisfactory interpretation.

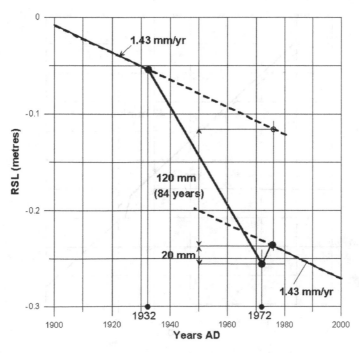

Fig. 15.4   Deduced RSL values, 1900 to 2000.

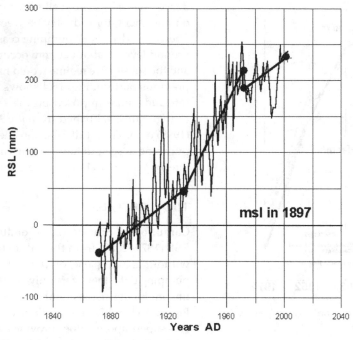

Fig. 15.5   RSL from Fig. 15.4 superimposed on sea-level data.

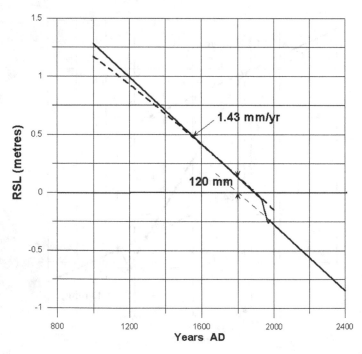

Fig. 15.6   Freeboard vs time for a station in Venice at MSL in 1897.

CONCLUDING REMARKS

Figure 15.6 is a synthesis of the data presented in this chapter. The conclusion is that the changes in RSL over the last few hundred years can be usefully interpreted as a continuous 1.43 mm yr^{-1} reduction augmented by a rapid net decrease, virtually a step-change, of about 120 mm due to anthropic activity from 1930 to 1970 (the dashed curve in this figure shows Ammerman's updated trend). This figure not only generates values for RSL change which agree almost exactly with those quoted in the introduction to this chapter from Pirazzoli (1996), Cavazzoni (1977) and Gatto and Carbognin (1981) but also provides a consistent interpretation of the rate at which RSL has been changing over the last 2000 years and, by extrapolation, a prediction of future changes in RSL. There is no obvious reason why this process should change significantly over the next few centuries other than through a calamitous increase in the rate of sea level rise due to global warming.

The data interpreted in this chapter incorporates the most reliable, long time-series records from recovered artefacts yet available and detailed land survey results throughout Venice itself. Elsewhere in and around the lagoon the anthropic water-extraction effects were different (for example, much higher subsidence around the major well heads in Marghera). Subsidence rates along the coastline at Jesolo and Chioggia are also higher and have been recently confirmed by satellite and radar measurements (Strozzi et al., 2002). There have also been conjectures that the 'regional eustatic' rate has decreased significantly since 1970 (Woodworth, 2003). However, Woodworth is at pains to point out that the evidence is inconclusive because 'the measured trend since 1970 has such a large standard error'. This data, shown in Fig. 15.5, is clearly insufficient to support any claim of a recent, significant change in the eustatic regime.

REFERENCES

Ammerman, A. J., McClennen, C. E., De Min, M., and N. R. Housley. 1999. Sea level change and the archaeology of early Venice. *Antiquity* **73**, 303–12.

Butterfield, R. 2004. On subsidence and eustacy in relation to Venice. *Advances in Geotechnical Engineering (Skempton Conference) ICE London*, 1231–42.

Camuffo, D., and G. Sturaro. 2004. Use of proxy documentary and instrumental data to assess the risk factors leading to sea flooding of Venice. *Global and Planetary Change* **40**, 93–103.

Cavazzoni, S. 1977. Variazioni batimetriche e idrografiche nella Laguna di Venezia intercorse tra il 1933 e il 1971. *Istituto Veneto di Scienze, Lettere ed Arti, Rapporti e Studi*, 7, 1–17.

Gatto, P., and L. Carbognin. 1981. The lagoon of Venice: natural environmental trend and man-induced modification. *Hydrological Sciences Bulletin des Sciences Hydrologiques* **26**, 379–91.

McClennen, C. E., Ammerman, A. J., and S. Schock. 1997. Framework stratigraphy for the Lagoon of Venice, Italy, revealed in new seismic-reflection profiles and cores. *Journal of Coastal Research* **13**, 745–59.

Pirazzoli, P. A. 1986. Secular trends of relative sea level (RSL) changes indicated by tide-gauge records. *Journal of Coastal Research* **1**, 1–26.

Ricceri, G., and R. Butterfield. 1974. An analysis of compressibility data from a deep borehole in Venice. *Géotechnique* **24**, 2, 175–92.

Strozzi, T., Tosi, L., Wegmuller, U., Teatini, P., Carbognin, L., and R. Rosselli. 2002. Thematic and land subsidence maps of the lagoon of Venice from ERS SAR interferometry. *Geophysical Research Letters* **29**, 13, 345–7.

Woodworth, P. L. 2003. Some comments on the long sea level records from the Northern Mediterranean. *Journal of Coastal Research* **19**, 1, 212–19.

16 • The extraction of Venetian sea-level change from paintings by Canaletto and Bellotto

D. CAMUFFO, E. PAGAN AND G. STURARO

SUMMARY

The high water in Venice over the last three centuries has been investigated by using a proxy indicator of mean sea level: the height of the green belt of algae which live in the intertidal zone and whose upper front indicates the average high tide level. In the first half of the eighteenth century, this indicator was accurately drawn by Antonio Canaletto (1697-1768) and his pupils, mainly Bernardo Bellotto (1722-80), in their 'photographic' paintings made with an optical *camera obscura*. Canaletto and Bellotto paintings extend our knowledge about Venice submersion back in time for almost three centuries, which includes the end of the Little Ice Age and the recent warming. The everyday wetting has increased by 69 ± 11 cm and high tides and flooding waters reach brick and plaster which are being rapidly destroyed by salt crystallisation cycles. The rise of the algae belt, however, has also increased by some 8 cm, as a sum of two contributions: the increased height of waves generated by motor boats and the dynamic increase of the tidal wave after the excavation of some canals. With the above corrections, the bulk submersion of Venice estimated from all the analysed paintings is 61 ± 11 cm, with an average yearly trend of 1.9 mm yr^{-1}. The city's submergence is due in part to land subsidence and in part to sea-level rise in the Mediterranean.

INTRODUCTION

The *camera obscura* (Latin for dark room) is an optical instrument, consisting of a box with a lens, similar to a modern camera, or a mobile tent model, which projects onto a sheet of paper the image of the view, with the artist carefully following all the contours thus revealed. The light beam with the image to be reproduced passes through an objective lens, is reflected by a mirror and is projected onto a glass surface. Here the artist places the sheet of paper, obtaining a 'photographic' painting.

The *camera obscura* was initially a dark room with a hole in one side, useful for astronomical observations. It has been known since the time of Aristotle (*c.* 300 BC). The Arabian scholar Hassan ibn Hassan (also known as Ibn al Haitam), in the tenth century, and Roger Bacon around AD 1300, may have used such a device to observe solar eclipses. The progress towards a *camera obscura* as an aid to drawing and perspective is due to Leonardo da Vinci (1452-1519), who discussed the physical operating principle, and to Cardano (1550), who applied a lens to the hole. The Venetian Daniel Barbaro (1568) wrote:

> Close all shutters and doors up to light so that it does not enter the camera except through the lens. Opposite to the lens, place a piece of paper, which you move forward and backward until the view appears in the sharpest detail. The whole view will appear on the paper as it really is, with distances, colours, shadows and motion, clouds, water twinkling, birds flying. If you keep the paper steady, you can trace the whole perspective with a pen, you can shade it and delicately colour as it is in the nature.

Some decades later, the *camera obscura* was a manageable device with its size reduced to a portable box and by the 1600s was a fully portable item. In 1620 Johan Kepler used a tent model for scientific observations. Giovanni Battista della Porta (1589) and then Athanasius Kircher (1646) published that the camera was a precious aid to drawing in a period in which artists (e.g. Van Eyck and the other Flemish masters)

Flooding and Environmental Challenges for Venice and its Lagoon: State of Knowledge, ed. C. A. Fletcher and T. Spencer.
Published by Cambridge University Press, © Cambridge University Press 2005.

sought to be as precise as possible in reproducing their indoor or outdoor views. The first painter to use the *camera obscura* was Viviano Codazzi in the seventeenth century. This methodology, based on an 'artificial eye' was theoretically and practically explained by a number of authors in various treatises. Johan Zahn wrote the paper 'Oculus artificialis teledioptricus' (1702) and the Venetian Francesco Algarotti, in his treatise: *Saggio sopra la pittura* (1756), wrote that the *camera obscura* was considered as necessary a tool for painters as the telescope was for astronomers or the microscope for physicists. William Hyde Wollaston (1807) published and patented another version of the camera, without a box: a prism on the top of an adjustable stand. This was soon abandoned.

The most famous painter to use this method was Antonio Canal (1697-1768), nicknamed Canaletto, and his pupils: Bernardo Bellotto (1720-80); Francesco Guardi (1712-93), Michele Marieschi (1696-1743), Gabriele Bella (1730-99) and Giuseppe Bernardino Bison (1762-1844). Canaletto derived this technique from his father, Bernardo, who was a scene painter. Also Luca Carlevarijs (1665-1731) used it in his Venice views. Two original cameras used by Canaletto are still preserved at the Correr Museum, Venice (Fig. 16.1). Before Canaletto, artists painting '*vedutas*' generally made a quick draft on the site and

then completed paintings in their workshops, including imagined details. At the beginning of the eighteenth century, the *camera obscura* was considered a useful tool for artists to exactly reproduce views. Either for this reason, or from the need for money to keep up his standard of living, Canaletto used a quick way to produce his paintings with a *camera obscura* on site (Pignatti, 2001). For this reason his paintings constitute an objective view of the buildings, similar to a modern photograph. From this standpoint, sound observations can be made and also quantitative evaluations are possible (Camuffo, 2001). In particular, Canaletto was very accurate in reproducing both buildings and details (Fig. 16.2) He had exceptional eyesight; he wrote at the end of a drawing made with low light in the interior of the Basilica of San Marco that he drew without glasses at the age of 68.

The paintings made in Venice have a detail which is highly relevant to determining the local sea-level rise that has occurred there over the last three centuries. Knowledge of the local sea-level rise in recent centuries is relevant to safeguarding the city; the IPCC (2001) has suggested that by the end of this century an increase in sea level of between 10 to 80 cm is expected, with the highest probability of a rise between 30 and 50 cm. In a number of paintings by Canaletto and his pupils the brown-green front left by

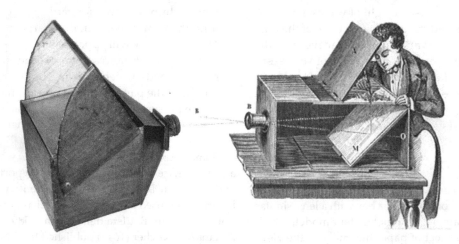

Fig. 16.1   Canaletto's original camera (by courtesy of the Correr Museum, Venice). Principle of operation of the *camera obscura* in a compact box model (after Ganot, 1860).

Fig. 16.2   Testing painting accuracy: cross comparison between Grimani Palace reproduced by Canaletto (1735, left) and by a photographic camera (1880-90, right). Heights of floors (evidenced with lines) and other details are carefully reproduced.

algae is visible on the buildings facing the canals. This front is a valuable biological indicator of the average high tide level (HTL) made by some algae which live in the belt between high and low tide. Most of the paintings by Canaletto and his pupils, who used the same methodology, are like real photographs but taken one century before both the invention of the photograph by Daguerre (1839) and the beginning of instrumental observations. It should be noted that the views in paintings are not always comparable with a photograph. Artists used to reproduce some parts of the view, e.g. a building, and then continue with other buildings, or the background, by moving the *camera obscura* to a more appropriate or comfortable position. A view, therefore, may be the combination of more than one partial views (Pedrocco, 2001). However, this is not a problem in our case, because we are interested only in a portion of a building and not in the entire view.

THE ALGAL BELT

Relative sea-level rise (RSLR) in Venice is the combination of two factors: land subsidence and actual sea-level rise (eustatism) in the Mediterranean. The relative sea level corresponds to the height reached by the sea level in Venice with reference to the local soil and the buildings, i.e. the tide gauge observation. Sea level leaves a characteristic mark on walls. Some algae live in the intertidal zone and the upper front of the brown-green belt that they form on walls indicates the upper limit to tidal influence and the growth limit to coastal algae which need water, air and sunshine. In Venice, the brown-green algal front, called the *Commune Marino* (CM), has been considered as an official reference for the average sea level over which to establish the height of bridges, floors and buildings. It was measured to be 31 cm above mean sea level (MSL) and practically coincident with the average high tide level (HTL) (Rusconi, 1983). In some cases the CM was also engraved on buildings with the initials CM although this information is not usable for the earliest times because labels or dates have in many cases been eroded and lost. An analysis of the displacement of the algal belt before the instrumental monitoring of tide level (started in 1872) may show the long-term trend of the apparent sea-level rise. From this information it may be possible to obtain information concerning natural and recent anthropogenic contributions. This is the aim of this chapter.

The algal growing season varies from species to species but it generally ranges between late winter to early autumn (La Rocca, 2002). The tide level however, may change a bit from one year to another and the algae are conditioned by this slightly variable height. The standard deviation, s, of the yearly average sea level around the least squares best-fit regression line representing the long-term trend of the RSLR is $\sigma = 3.8$ cm i.e. the yearly average sea level varies within $2\sigma = \pm 7.6$ cm (95% confidence). However, the algal variability is smaller. Should the HTL rise for one year, the algal belt also rises; should it decrease a bit the next year, the living algae tend to be lower, in any case being partially protected by the old colony which tends to survive, being kept soaked by waves. The algae colony resists better when it is thicker as a result of several years growth, as a thicker layer is better soaked and the internal individuals are protected by the external ones. Exceptional peaks of yearly HTL may lead to a temporary colonization which is easily destroyed by dryness and sunshine in the following years. Only in the case where the sea level is substantially lowered for a number of consecutive years does the upper, dry part of the colony die under the sunshine, forming a thin black strip. This may sometimes happen, and in this case we have a double reference level: a green line with fresh algae that refers to the present year and a black line which refers to the highest yearly level reached in the past five or six years. Such a detail is useful only when assessing the present day level as it is not distinguishable in the paintings. In conclusion, the algae level may fluctuate around the mean yearly trend of sea level but to an extent less than the yearly sea level, like a running average. No specific studies exist about this topic, and we could evaluate the uncertainty looking at the yearly average sea level variability and disregarding the extreme fluctuations i.e. assuming as error the above s. In reality, this assumption is severe when compared with field observations which generally give no more than a 2 cm fluctuation in the height of the algal front. Algae may show slightly different growth on sunny or shaded walls, and on porous bricks or less porous stones, with the result that their level may slightly differ in various parts of the city. However, this is not a problem in this analysis because we make reference to the algal belt displacement that has occurred on the same wall. Venice hosts several species of algae, known for a long time (Correr *et al.*, 1847; Curiel, 1999; La Rocca, 2002). For our aim, it is fundamental to know that in the past, and today, algae colonizing the marine wetting front provide us with a precious biological marker; the specific names of the species are less relevant.

METHODOLOGY

The methodology simply consists in measuring how far the front of the algal belt has been displaced since the time when Canaletto painted the scene. Not all the paintings are equally accurate and the use of a painting requires some preliminary tests to assess their reliability. The building too should be untouched by restorations or other works that may have changed its basement, and this is another condition that has to be verified. Once a painting and a building have been recognized to be reliable, then the measurements start on the site and continue in the laboratory. On the site, during a low tide, the basement of the actual building is cross compared with the corresponding expanded view from a painting. It is also possible to recognize where the algae belt was, making reference to some decorative or architectonic features. Its displacement is measured. The next logical step is to connect the algal displacement with the sea level, and this is related to the average high tide level. After the algal front displacement has been assessed, this finding is reduced correspondingly to allow for the over-evaluation for the higher height of modern waves generated by motorboats (although their speed is low and strictly controlled) and other minor factors, as we will see later. The confidence limit is assessed, following the theory of errors currently used in physics for experimental observations (Barford, 1990). By summing up all of the errors, the uncertainty range is around ±10–12 cm. This range constitutes the main limit of this methodology. This limit is fully acceptable for the long-term RSLR because the algal belt has been displaced by a relevant height. When a case study presents elements of larger uncertainty, it is discarded.

TESTING THE PAINTING ACCURACY AND CROSS COMPARISON IN THE FIELD

The first step in our analysis is to check the painting accuracy in reproducing all details. Two kinds of details can be controlled: the architectural and decorative features of buildings and the pattern of the black crusts due to pollutants. The former can be made by cross comparison with the present day situation and is an easy test where the building has not been restored and altered in the meantime. The latter can be controlled by carefully inspecting whether the black crusts are found exactly where the building surface prevents run-off due to protrusions i.e. where black smoke particles are deposited but are not subsequently removed by run-off. Similarly, where running water converges and forms small rivulets, or drops, the surface is continually washed out and appears white. All of these details show the care used in drawing the view.

When a painting shows the algae belt on buildings that have not been transformed in the meantime, the image is recorded using a camera or a scanner. A boat field survey, performed during a low tide, permits a careful analysis of all possible building details (e.g. a specific decoration, the rows of stones, a window) to precisely establish the position of the algae as they appear in the painting and as it is today. Measurements are carried out to establish lengths, heights, and distances between recognizable decorative or architectonic elements which may be useful when the evaluation is repeated later in the laboratory. All the potentially useful details are photographed on the site with a rule in the images for scale. Later, in the laboratory, all the corresponding details from the painting and the survey are projected on a white screen and expanded, making possible an easier and more accurate quantitative evaluation of the position of the algae belt then and today, in order to evaluate their displacement. Should some doubt remain, survey and *in situ* measurements are repeated. In the case of more than one possible reference, the case leading to less uncertainty is preferred e.g. a direct measurement of algae appearing at the base and the top of a clearly recognizable row of decorative stones 70 cm thick is preferable to making reference to the height of a door, which may be five times taller, the uncertainty increasing accordingly. When a building has been restored or transformed in such a way as to change the features useful to this aim, it is discarded.

When a painting is reliable, with the lower part of the building remaining untouched and the algal belt occurring alongside recognizable detail, it is possible to measure on the site how much the present day level is above that detail. This kind of measurement is of course independent from dimensional distortion of the image. If the algae belt falls in the middle of a recognizable stone in the painting, the limit of uncertainty is connected with the precise evaluation of the proportion of stone covered with algae in comparison with the part free of algal cover. Once this has been established, it is possible to measure the algae displacement on the site. In this case the uncertainty is a fraction of the brick or stone size, and depends on the size of the detail in the painting. For instance, in the case of algae apparently covering half of a typical brick row 8 cm in width, the maximum error is ± 4 cm.

Another test is based on the cross-comparison between the size of the architectonic features as they appear in the painting and in a photograph of the actual building today. This test is useful in two cases. The first, when in the painting the algae belt does not relation to any architectural detail, so that its position is determined by making a quantitative proportion with other recognizable items, e.g. the distance between the algae belt and the first row of windows corresponds to a given fraction of the windows. In this case we should know accurately if the proportions in the painting are accurate and then the actual value of the reference item can be used to establish a quantitative scale in known units. The second case is when we have more than one painting of the same view and an objective criterion is needed to establish which painting should be analysed. It might be expected that the first painting made on the site is the most accurate one, and in the case of reproductions made in studio the precision is diminished. We will see later that this is not necessarily true, as it seems that Canaletto and Bellotto used to take photographic views on paper sheets on the site and then transformed these views into paintings in the workshop (Gioseffi, 1959; Pignatti, 2001). In these cases the accuracy is independent from the order of production. However, when more than one dated painting is available for the same view, we followed the most severe hypothesis that the first is likely to have been made on the site, and therefore preferred, the later ones possibly being studio replicas made on the ground of early drawings by the artist.

However, an even better method can be used to establish the general accuracy of a painting. It can be explained by taking the same view as represented by paintings at different levels of accuracy. With this method, we measure, in each painting, the size of

some architectonic items $A_{1p}, A_{2p}, A_{3p}\ldots$ e.g. height of doors, windows, distance between balconies, and we normalize each measurement by dividing it by the height, H_p, of the building in the painting i.e. $R_{1p} = A_{1p}/H_p$, $R_{2p} = A_{2p}/H_p\ldots$ All of these normalized measurements are now independent from the size of the painting and can be compared with the similar normalized distances of the corresponding elements of the actual building, measured from a modern picture taken from the same position (i.e. $R_{1ab} = A_{1ab}/H_{ab}$, $R_{2ab} = A_{2ab}/H_{ab}\ldots$). The cross comparison is made by calculating the ratios between the corresponding distances on the paintings and on the modern image (i.e. R_{1p}/R_{1ab}, $R_{2p}/R_{2ab}\ldots$). All of these ratios are close to 1, some distances being overestimated (>1) and some underestimated (<1). Replicas of paintings derived from the same original painting should repeat the minor departures, keeping for each of them the same direction (e.g. either >1 or <1). Independent (or badly reproduced) paintings have random departures. The accuracy of a painting can

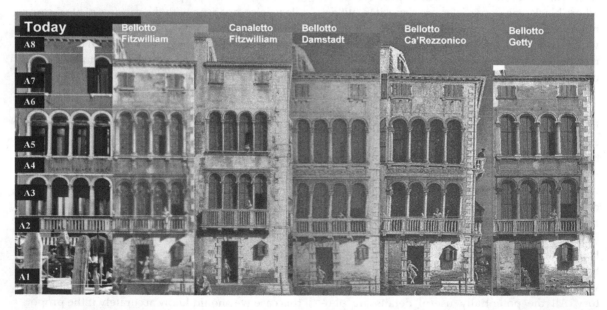

Fig. 16.3   Example of the Standard Deviation methodology to assess the accuracy of the architectural features of the Pisani-Gritti Palace in Canaletto and Bellotto paintings. For each painting, the building is divided into a number of strips corresponding to natural partitions of the façade, and the width of each strip is normalized expressing it as a fraction of the total building height. Then the ratio is computed between each normalized size in the painting and the corresponding size in the picture of the building today, obtaining values close to 1.

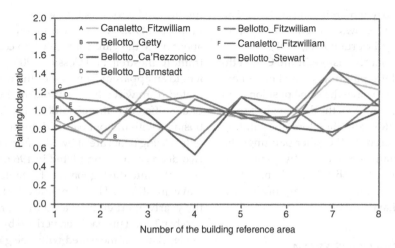

Fig. 16.4   Standard Deviation of five paintings with the same view: paintings with the Grand Canal and the Salute viewed from Campo Zobenigo.

be established with a statistical parameter that evaluates departures,the standard deviation. The smaller the standard deviation, the better the accuracy.

An example of this method is provided by the famous view of the Grand Canal and the Salute viewed from Campo Zobenigo. Two paintings of this view, one by Canaletto and one by Bellotto, are at the Fitzwilliam Museum, Cambridge. Four other paintings of the same view by Bellotto are at the Getty Museum, Malibu (Los Angeles); Ca' Rezzonico, Venice (anonymous, possibly Bellotto); private collection, Darmstadt; and the Stewart Collection, New York. In this view, only the Pisani-Gritti Palace, painted on the left side, is helpful for the study of sea-level rise in Venice because the buildings on the right side of the Canal have been changed. The Pisani-Gritti Palace was, however, transformed into a hotel in the nineteenth century but, fortunately, the work did not affect the basement. For each painting, the building is divided into a number of strips corresponding to natural partitions of the facade, and the width of each strip is normalized, expressing it as a fraction of the total building height. The same is made with a picture of the building today (Fig. 16.3). The most accurate painting is Bellotto's at the Fitzwilliam (SD = 0.13), followed by Canaletto's (SD = 0.23), and others (Fig. 16.4 and Table 16.1). This analysis suggests that some decades after the first

painting by Canaletto in the 1730s, Bellotto made on the same site, and more accurately, another painting of this famous view which is not a replica of the Canaletto. Then Bellotto made other paintings with the same view, but with lower accuracy (double the standard deviation). The departures are random but this does not exclude the possibility that they are replicas, as the randomness might be a consequence of the lower accuracy. This discussion has been reported to clarify the methodology, not to claim

Table 16.1   *Paintings with the Grand Canal and the Salute viewed from Campo Zobenigo, and accuracy (Standard Deviation) in reporting the architectural features of Pisani-Gritti Palace.*

Painting	Std Deviation
Canaletto, Fitzwilliam (1730s)	0.23
Bellotto, Fitzwilliam (>1738)	0.13
Bellotto, Darmstadt (1740-1)	0.19
Bellotto, Stewart, N.Y. (1743-4)	0.22
Anonymous - Bellotto? Ca'Rezzonico (>1738)	0.26
Bellotto? Getty, Malibu (1743-4)	0.27

advances in art history which is not our field. The dating here reported follows a recent exhibition in Venice (Kowalczyk and Cortà Fumei, 2001).

The algal belt is clearly visible in the Canaletto painting. In the Bellotto painting, at the Fitzwilliam Museum, algae seem to have the same position as in Canaletto's painting; however, the dark might be also interpreted as a shadow and therefore this painting has not been considered. In the other paintings the algal belt is uncertain, or masked by planks, or absent. The observed algal shift in the Canaletto is 73 ± 10 cm i.e. 4 cm more than the average of all of the other paintings we have analysed.

RESULTS: OBSERVATIONS FROM PAINTINGS AND CORRECTIONS

The results of the observations of the algae displacements in reliable paintings are reported in Table 16.2, together with the corrections for reporting the disturbing factors, as explained below, to the extent they had when Canaletto and Bellotto painted. The observed shift of the algae belt is primarily determined by relative sea-level rise and, secondarily, by disturbing factors, as follows. Waves generated by motor boats have a typical height of some 10 cm, about twice the value of eighteenth century row-boat traffic (Canestrelli and Cossutta, 2000). This is equivalent to an apparent 5 cm rise of the wet front, as estimated from wave observations in the Grand Canal under different traffic conditions, i.e. during everyday business with motor boats and during a regatta when only rowing boats are allowed. After the excavation of two deep canals, completed in 1969, penetration of sea water into the lagoon was facilitated and the tidal wave amplified. This dynamic effect contributes to the yearly average tidal amplitude, raising the CM another 3 cm. This has been verified by comparing the sea level height measured with tide gauges in Venice and 16 km offshore where a permanent CNR platform measures both marine and atmospheric variables. This result is in accordance with a different evaluation by Rusconi (1983) who found that in 1979, after the end of the hydraulic works, the tidal range in Venice had increased by 3.2 cm compared with the beginning of the series (1872).

Essentially, the observed shift should be corrected by subtracting 8 cm from the figures obtained from

Table 16.2 *Observations concerning the algal displacements in reliable paintings: the site and date of painting, the observed algal displacement and the corrected values, i.e. the RSLR.*

Site	Date of painting	Observed algae displacement – ΔCM_{obs} (cm)	RSLR ΔCM_{cor} (cm)
1. Punta Dogana	1727	69 ± 16	61 ± 17
2. Fontego Tedeschi	1727	68 ± 8	60 ± 9
3. Balbi Palace	1730/31	67 ± 6	59 ± 7
4. Ca' Grande Palace	1732	70 ± 9	62 ± 10
5. Emo Palace – Cannaregio	1735	71 ± 10	63 ± 11
6. Giustinian-Lolin Palace	1735 (?)	66 ± 10	58 ± 11
7. Grimani Palace	1735	66 ± 10	58 ± 11
8. S. Stae Church	1740	77 ± 18	69 ± 19
9. S. Giovanni e Paolo	1741	77 ± 10	69 ± 11
10. Flangini Palace	1741	71 ± 12	62 ± 13
11. S. Sofia Church	1758	65 ± 7	57 ± 8
12. Pisani-Gritti Palace	1730s	73 ± 10	61 ± 10

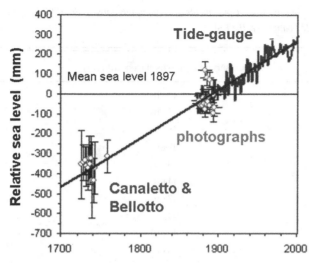

Fig. 16.5   Apparent sea-level rise (RSLR) in Venice from tide gauges (continuous grey line, period 1872-2002), from Canaletto and Bellotto's paintings (white dots with error bars, period 1727-58) and from early photographs (small dots, period 1880-90). RSLR from paintings (and photographs) was estimated from the difference in level of the algal belt as it was in the paintings and as it is today. The best-fit trendline has been calculated after the tide gauge data with the least squares method (R^2= 0.85, trend: 2.4±0.1 mm yr^{-1}).

the Canaletto and Bellotto paintings. It might also be considered, however, that today the wave height is slightly more increased by the impact on buildings and that in the eighteenth century the width of the tidal wave was attenuated by the lagoon marshes. These two effects are thought to be relatively modest and tend to oppose each other. As the bulk correction is small (of the order of a few cm) with the uncertainty of the same order of magnitude as the correction, it seemed preferable to omit it and increase correspondingly the final bulk uncertainty. The data obtained from the paintings are reported in Fig. 16.5, together with the tide gauge instrumental observations available since 1872. The data are self-consistent, being close to each other, and fall along the line of best fit obtained from the tide gauge observations. This means that the RSLR has been substantially unchanged over the last three centuries. The average yearly trend for the period 1727-2000 is 2.3 ± 0.4 mm yr^{-1}, close to that in the instrumental period (2.4 ± 0.4 mm yr^{-1}) and becomes 1.9 ± 0.4 mm yr^{-1} if the recent man-made subsidence for water pumping i.e. 10-12 cm (Teatini *et al.*, 1995; Carbognin and

Taroni, 1996; Pirazzoli and Tomasin, 1999) is removed (Camuffo and Sturaro, 2003, 2004).

OBSERVATIONS FROM EARLY PHOTOGRAPHIC PICTURES

The same methodology has been applied to early photographic pictures as a further test on the reliability of the method. Some archives of historical photographs exist: loose pictures, Foto Alinari (Florence), Foto Blitz (Venice) as well as some publications (e.g. Ongania (1890-1)) with pictures without indication of the date. The problem is the exact dating as sometimes the same pictures appear in different collections, differently dated. The method has been tested on early photographs after Ongania (1890-1) (Plate 7). The date of publication of the book (1890-1) evidently shows the upper age limit and the specific type of pictures (i.e. heliotypography) the lower limit, i.e. 1880. Some details can sometimes help to provide a more precise date for an individual image: thus, for example, a picture with snow and ice slabs in the canals can be dated 1880 or 1881 as we know of

Table 16.3   *Observations concerning the algal displacements in photographs dated 1880-1890 (Ongania, 1890-1891): the observed algal displacement and the corrected values, i.e. the RSLR.*

Building or site	Observed algae displacement (cm)	RSLR (cm)
Ca' D'Oro Palace	45±10	37±10
Rio Priuli	47±10	39±10
S. Barnaba	48±10	40±10
Ca' Rezzonico Palace	47±10	39±10
Fondamenta della Misericordia	48±10	40±10
Ca' Papadopoli Palace	44±10	36±10
Rio Angelo	47±10	39±10
Corner-Spinelli Palace	49±10	41±10
Rio S.Antonio	45±10	37±10
Rio S.Apollinare	45±10	37±10
Rialto Bridge	47±10	39±10
Ca' Dario Palace	46±10	38±10
Traghetto S.Tomà	44±10	36±10
Rio dei Greci	47±10	39±10
Grimani Palace	48±10	40±10
Average	46±10	38±10

such events from local meteorological records. The test performed with early photographs (Table 16.3, Fig. 16.5) gave results coherent between them. The findings are in accordance (within the confidence range ± 10 cm) with the tide gauge observations (30 ± 1 cm), thus confirming the reliability of this approach. The SLR average from photographs seems to slightly overestimate the RSLR; this might be referred to an underestimate of the 8 cm correction factor. However, in the short term (110-20 years) the RSLR is relatively modest and the sum of errors and uncertainties is too large compared with the RSLR, i.e., 25% of the signal. The above results are consistent with the city sinking found by Rusconi (1983) who measured the RSLR from the displacement of two reliable and complete CM marks, one on the railway bridge where the algae were displaced by 48 cm in the period from 1841 to 1980 and one on the Loredan Palace where the algae were displaced by 44 cm in the period from 1871 to 1980. Since 1980 rela-

tive sea level has risen by 5 cm. The displacements quoted by Rusconi were accurately determined with topographic levelling to ± 0.2 cm. The problem remains on the precision of the determination of the actual algal level when it was engraved on the label, two centuries ago and today; that is the same problem we encounter in our surveys.

CONCLUSIONS

Paintings made by accurate artists, with the help of a *camera obscura*, have been shown to furnish sound information about RSLR in Venice. The methodology gives reliable findings and has been tested with early pictures. An example of the application of this methodology to a series of paintings with the same view shows that the same subject can be represented with different levels of accuracy. In addition, it allows us to establish, on purely mathematical grounds, that the Bellotto painting at the Fitzwilliam Museum is not an updated replica of the earlier Canaletto painting but an original view, made later on the site. Surprisingly, it might help not only climatologists but art historians also.

Canaletto and Bellotto paintings extend our knowledge about Venice's submergence back in time for almost three centuries, a period which includes both the end of the Little Ice Age and the recent warming. The bulk city submersion estimated from all the analysed paintings is 61 ± 11 cm with an average yearly trend of around 2 mm yr^{-1} (Camuffo and Sturaro, 2003, 2004). This finding is in agreement with the results obtained from archaeological evidence (Ammerman et al., 1999, and see chapter 13) from which 60 cm are obtained over the same period.

The city's submergence has been partly due to land subsidence and partly due to sea-level rise in the Mediterranean. In the last three centuries, these two contributions have been of the same order of magnitude, as we can deduce from independent evaluations of land subsidence (Teatini et al., 1995; Bondesan et al., 2001; Carminati and Doglioni, 2003) and Mediterranean sea-level rise (Tsimplis and Baker, 2000). The Mediterranean sea-level rise has been smaller than the known rate on the global scale for the following reasons: The Mediterranean is a nearly closed sea, with a small gate to the Atlantic Ocean, i.e. the strait of Gibraltar; in the last decades the precipitation in the catchment area has decreased and the evaporation increased (Piervitali et al., 1998), the balance with Black Sea has changed after the water of its tributaries has been extracted for irrigation purposes; and finally the atmospheric pressure has increased by 1-2 hPa which is equivalent to a 1-2 cm sea level depression. It might be useful to specify that in the expected scenario, the global sea-level rise in the oceans will not necessarily be followed by the same rise in the Mediterranean Sea, and that further studies are necessary to establish this relationship.

The results obtained from this study are useful in view of deciding measures to save Venice and its historical buildings because they constitute a long-term observation of the local sea-level rise. Another concern is the sustainability of apparent sea-level rise and high waters. The Venice palaces were originally protected against groundwater rise by a belt of non-permeable Istria stone, but now the protective belt has sunk along with the city, under the combined effect of natural and anthropic factors. The everyday wetting has increased by 69 ± 11 cm and high tides and flooding waters reach brick and plaster which are rapidly being destroyed by salt crystallisation cycles. In the long run, with the cumulative effect of the crystallisation-dissolution cycles of the NaCl adsorbed into the masonry, all of these buildings are at risk of collapse.

ACKNOWLEDGEMENTS

The idea for this work arose under researches funded by the European Commission, DGXII (Environment and Climate Programme) and was developed with funding by CORILA, Venice. Special thanks are due to L. Alberotanza, Director of CNR-ISDGM, Venice, G. Cecconi, Consorzio Venezia Nuova and C. Fletcher, Churchill College, Cambridge for their precious help with this research.

BIBLIOGRAPHY

Algarotti, F. 1756. *Saggio sopra la pittura.* Venice: Graziosi (printed 1784).

Ammerman, A. J., McClennen, C. E., De Min, M., and R. Housley. 1999. Sea-level change and the archaeology of early Venice. *Antiquity* **73**, 303.

Bacon, R. (C. 1219–C. 1292). 1603. *Roger Baconis Angli De arte chymiae scripta cvi accesservnt opuscula alia eiusdem authoris.* Frankfurt: Schönvvetteri, Saurij, 408 pp.

Barbaro, D. 1569. *La pratica della perspettiva di monsignor Daniel Barbaro ... : opera molto vtile a pittori, a scultori & ad architetti: con due tauole, una de' capitoli principali, l'altra delle cose più notabili contenute nella presente opera.* Venice: Borgominieri, 195 pp.

Barford, N. C. 1990. *Experimental Measurements: Precision, Errors and Truth.* New York: John Wiley & Sons, 159 pp.

Bondesan, M., Gatti, M., and P. Russo. 2001. Vertical ground movements obtained from I.G.M. levelling surveys. In G. B. Castiglioni, and G. B. Pellegrini (eds.), *Illustrative Notes of the Geomorphological Map of the Po Plain, Suppl. Geogr. Fis. Dinam. Quat.*, vol. **4**.

Camuffo, D. 2001. Canaletto's paintings open a new window on the relative sea-level rise in Venice. *Journal of Cultural Heritage* **4**, 277–81.

Camuffo, D., and G. Sturaro. 2003. Sixty-cm submersion of Venice discovered thanks to Canaletto's paintings. *Climatic Change* **58**, 333–43.

Camuffo, D., and G. Sturaro. 2004. Use of proxy-documentary and instrumental data to assess the risk factors leading to sea flooding in Venice. *Global and Planetary Change* **40**, 1–2, 93–103.

Canestrelli, P., and L. Cossutta. 2000. *Moto ondoso a Venezia.* Venice: Centro Previsioni e Segnalazioni Maree.

Carbognin, L., and G. Taroni. 1996. Eustatismo a Venezia e Trieste nell'ultimo secolo. *Atti Istituto Veneto di Scienze, Lettere ed Arti,* **154**.

Cardano, G., 1550. *Hieronymi Cardani ... De subtilitate libri XXI.* Nuremberg: Petreium, 371 pp.

Carminati, E., Doglioni, C., and D. Scrocca. 2003. Apennines subduction-related subsidence of Venice (Italy). *Geophysical Research Letters* **30**, 13, 1717, doi:10.1029/GL017001.

Correr, G., Sagrado, A., Priuli, N., Pasini, L., and L. Carter. 1847. *Venezia e le sue Lagune.* Venice: Antonelli.

Curiel, D. 1999. Aggiornamento di nuove specie algali per la laguna di Venezia. *Lav. Soc. Ven. Sci. Nat.* **24**, 55–66.

Della Porta, G. B. 1589. *Io. Bapt. Portae Magiae natvralis libri XX / ab ipso authore expurgati, & superaucti, in quibius scientiarum naturalium diuitiae, & delitiae demonstrantur.* Naples: Horatium Saluianum, 303 pp.

Ganot, A. 1860. *Traité de physique expérimentale et appliquée, et de météorologie.* Paris: chez l'auteur-éditeur.

Gioseffi, D. 1959. *Canaletto, il Quaderno delle Gallerie Veneziane e l'impiego della Camera Ottica.* Trieste: Università degli Studi di Trieste. Facolta di Lettere e Filosofia Istitute di Stonà dell'Arte Antica e Moderna No. 9.

IPCC. 2001. *Climate Change 2001: The Scientific Basis.* Geneva: WMO–UNEP.

Kircher, A. 1646. *Athanasii Kircheri Fvldensis Ars magna lvcis et vmbrae [Microform]: in decem libros digesta: qvibvs admirandae lvcis et vmbrae in mundo, atque adeò vniuersa natura, vires effectus[que] vti noua, ita varia nouorum reconditiorum[que] speciminum exhibitione, ad varios mortalium vsus, panduntur.* Rome: Ludouici Grignani, 935 pp.

Kowalczyk, B., and M. da Cortà Fumei. 2001. *Bernardo Bellotto, 1722–1780.* Milano: Electa, 280 pp.

La Rocca, B. 2002. *Le alghe della laguna di Venezia.* Venice: Arti Grafiche Venete, 120 pp.

Ongania, F. 1890–1. *Calli e canali in Venezia.* Venice: Ongania

Pedrocco, F. 2001. *Il Settecento a Venezia i Vedutisti,* Milan: RCS Libri S.p.A., 239 pp.

Piervitali E., Colacino, M., and M. Conte. 1998. Rainfall over the Central-Western Mediterranean basin in the period 1951–1995. Part I: precipitation trends. *Il Nuovo Cimento* **21C**, 3, 331–44.

Pignatti, T. 2001. *Canaletto.* Florence: Giunti, 208 pp.

Pirazzoli, P. A., and A. Tomasin. 1999. L'evoluzione recente delle cause meteorologiche dell' 'Acqua Alta'. *Atti Istituto Veneto di Scienze, Lettere ed Arti* **157**.

Rusconi, A. 1983. *Il Comune Marino a Venezia: ricerche e ipotesi sulle sue variazioni altimetriche e sui fenomeni naturali che le determinano.* Pubbl. **157**. Venice: Uff. Idrogr. Magistrato Acque.

Teatini, P., Gambolati, G., and L. Tosi. 1995. A new three-dimensional nonlinear model of the subsidence at Venice. In F. B. J. Barends *et al.* (eds.) *Land Subsidence.* The Hague, *IAHS* **234**, 353–61.

Tsimplis, M. N., and T. F. Baker. 2000. Sea level drop in the Mediterranean: an indicator of deep water salinity and temperature changes? *Geophysical Research Letters* **27**, 12, 1731–4.

Wollaston, W. H. 1807. Description of the Camera Lucida. *Nicholson's Journal* **17**.

Zahn, J. 1702. *Oculus artificialis teledioptricus sive telescopium: ex abditis rerum naturalium & artificialium principiis protractum novâ methodo, eâque solidâ explicatum ac comprimis è triplici fundamento physico seu naturali, mathematico dioptrico et mechanico, seu practico stabilitum.* Nuremberg: Johannis Ernesti Adelbulneri.

Part III · Urban flooding and the urban system

17 • Introduction: urban flooding and the urban system

T. SPENCER, R. J. S. SPENCE, J. DA MOSTO AND C. A. FLETCHER

The intimacy between water and buildings, pavements, passages and bridges provides that 'dense network of internal relations and uses of the city' (Spinelli and Folin, chapter 18) that is peculiarly Venetian. Over the centuries these relations have been repeatedly renewed and redefined by the replacement and alteration of structures throughout the city, set against a backdrop of repeated urban flooding events, themselves superimposed on the long-term rise in water levels. An 1800-year record of archaeological, historical and recent tidegauge measurements shows a long term sea level rise for the city which averages 13–19 cm per century (Spencer *et al.*, chapter 3). As early as 1725, a census of Venetian wells was required in order to identify those structures that needed to be modified to protect the city's drinking water cisterns and in 1786 Proto Mazzoni wrote of the need to elevate certain walkways (Spinelli and Folin, chapter 18). Today Venetians continue to try to protect properties and livelihoods from flooding. A recent survey of businesses in the city with operations at ground floor level (retail outlets, bars and restaurants, offices and hotels) showed that the raising of floor levels and the use of temporary barriers across doorways (*paratia*) are widespread practices, with some 'tanking' (*vasca*) of ground floor rooms and use of pumps to remove flood waters. Short-period, on-site damage due to high water events is estimated to cost c. 3.4 million Euros per year at the present time (Breil *et al.*, chapter 26).

The long-term impacts of centuries of flood events and progressive city submergence have been dramatic. Thus, for example, the crypt of St Mark's Basilica now lies 170 cm lower than when it was built, and the floor of the atrium is now in a perilous position in relation to present sea level (Vio, chapter 21).

Water invades the atrium between 150 and 180 times each year, often every day during the winter months. As well as sea level rise, average water levels have also been raised, by an estimated 5 cm (Camuffo *et al.*, chapter 16), due to higher waves generated by motorized boat traffic. Associated turbulence from propeller wash has led to weakening of underwater canal sides and the re-suspension of canal sediments blocks drains and antiquated sewage systems. For large sections of the canal network, a high fine sediment content has been a consequence of the lack of regular maintenance dredging between the 1960s and late 1990s (partly as a result of the problems of finding a disposal solution for dredged material classified as too contaminated under a change in environmental legislation).

In a periodic wetting and drying environment, from tidal fluctuations and wave splash, buildings show evidence of salt efflorescence, freeze/thaw effects, chemical and biological attack and surface erosion (Sandrolini *et al.*, chapter 22). These processes have been exacerbated in recent times. In the great period of city building between the tenth and fourteenth centuries, a key element in the construction of the masonry outer walls was the use of impermeable Istrian stone within the intertidal and immediate supratidal zone above a foundation of wooden piles driven to resistance into the basement clays of the lagoon. This acted as a damp-proof course, preventing (or substantially reducing) the vertical migration of saline waters into overlying brick, plaster (see Danzi *et al.*, chapter 24, for a review of Venetian plasterwork) and wooden materials. However, and with the added effects of wave action, water levels are now reaching above the Istrian stone layer increasingly frequently and saltwater penetrates the overlying porous brickwork. Rising through capillary

Flooding and Environmental Challenges for Venice and its Lagoon: State of Knowledge, ed. C. A. Fletcher and T. Spencer.
Published by Cambridge University Press, © Cambridge University Press 2005.

action, typically to heights of 2–3 m (Biscontin *et al.*, chapter 25) but exceptionally to +6 m in St Mark's Basilica (Vio, chapter 21), saline interstitial waters go through cycles of dissolution (dissolving and recrystallizing). Solution and migration to the air/wall interface occurs at times of high (up to 98–9%) relative humidity in summer and when precipitation occurs under the lower (32–5%) relative humidities of winter. As a result, salt crystallization in pore spaces leads to brickwork disintegration and the eventual collapse of sections of wall. The maintenance of walls is thus a great expense for those trying to live and work in the face of periodic flooding (Breil *et al.*, chapter 26), and affects all types of structures from palaces and churches to modern shops and offices. Details of some of the chemical and physical treatments being used to extract salts from brickwork and to protect bricks and plaster surfaces from future salt invasion are given by Dolcetta (chapter 19), Turlon (chapter 20), Vio (chapter 21) and Sandrolini *et al.* (chapter 22); the latter considers innovative electrokinetic methods to control rising damp. There is some concern, however, that the long-term impact of some modern techniques of reconstruction and rebuilding are not known and it has been argued (Spinelli and Folin, chapter 18) that, in spite of their high cost, interventions tied to traditional techniques and using materials that are physically and chemically compatible with the ancient building stock should be adopted wherever possible.

Such studies must sit, however, within the wider framework of protecting the urban fabric from flooding. The integrated system of measures to protect against flooding in Venice covers three groups of intervention: 'local measures' to raise the margins of islands and canals; the use of 'mobile barriers' to temporarily close the lagoon inlets; and morphological interventions at the inlets, within the lagoon and at the landward margins of the lagoon. The first of these groups is considered in this section (note: chapter 29 introduces the mobile barriers and interventions at the inlets and chapter 42 the morphological activities in the lagoon).

Local defence measures involve the raising of pavements, walkways, embankments and bridges and 'flood-proofing' the ground floor levels of buildings in the lowest parts of the city. Associated with

this work has been the renewed programme for dredging canals, the repair of canal side walls, culverts and sewage outlets, the renovation of bridges (Turlon, chapter 20) and the installation of cabling for modernized services. All of this work is being carried out by Insula SpA, an urban maintenance company owned by the Venice City Council and the four main utility companies (electricity, gas, water and telecommunications) (Dolcetta, chapter 19). The aim of raising urban surfaces is to exclude medium-high flood events (water levels up to +110 cm above the reference level) and where possible to attain a protection level at +120 cm. Insula has identified 100,000 m^2 of the city for defence level raising and a further 15,000 m^2 of walkways where the protection can be raised to +120 cm. In addition a further 30,000 m^2 of embankments has been assigned to the Venice Water Authority – Consorzio Venezia Nuova (MAV-CVN) for similar modification. This work is, however, not without controversy. New structures can threaten the architectural integrity of buildings, both directly, by obliterating the bases of columns and changing the proportions of doorways and windows, and indirectly, by changing the appearance of facades by fronting defences. Different studies, employing different criteria for the acceptability of such modifications, have arrived at different conclusions as to the degree to which these 'local measures' can be applied throughout the city. What is clear, however, is that each proposed modification must be evaluated on a case by case basis, with sympathy for the individual setting (Spinelli and Folin, chapter 18). Such an approach requires the detailed and reliable cataloguing of the locations and characteristics of all public and private buildings and their external and internal artefacts. Such activity can be seen as part of a much wider information gathering exercise on the whole urban infrastructure of Venice. Several chapters (Spinelli and Folin, chapter 18; Carrera, chapter 27; Mancuso and Pypaert, chapter 28) show the power and potential of modern databases and GIS systems to support municipal operations, from building upkeep schedules to more sophisticated urban planning which might help meet environmental goals. A prime example of the latter is the design of a cargo delivery system by destination in the city rather than by product; this offers the possibility of reducing cargo boat movements, and by implication their negative effects, by 90 per cent, with no loss of full time jobs (Carrera, chapter 27).

In some instances whole islets (*insulae*) within the city have been the subject of more extensive and wider-ranging interventions, involving not only the raising of the islet perimeter but also the replacement of surface paving and the renovation of underground drainage and sewage systems. Brotto (chapter 23) outlines this approach as it applies to St Mark's Square. More radical solutions have also been proposed. One of these involves reversing the effects of subsidence and sea level rise through the raising of ground surfaces by deep injections, of materials or fluids. These ideas were first tested in a trial at Poveglia, one of the lesser islands of the lagoon, in the 1970s, with results which were not encouraging. More recently, Comerlati *et al.* (2004) have proposed the injection of seawater into a brackish water aquifer at depths of 600–800 m beneath the Venice Lagoon, via a necklace of 12 wells at a 5 km radius from the Rialto Bridge. The expected surface uplift after 10 years of seawater injection under the city is 11–40 cm, with a most likely value of 25 cm. However it is known that the subsurface geology of the Venice Lagoon is highly heterogeneous (see Spencer *et al.*, chapter 3), raising the problem of differential responses to injection within islets or even possibly within individual buildings. A full investigation of the spatial variability of the geologic, hydraulic and geomechanical characteristics of the lagoon subsurface, particularly the lithostratigraphy of the unit to be injected, as well as further pilot scale experiments, will be required before this particular project can be taken further forward.

REFERENCE

Comerlati, A., Ferronato, M., Gambolati, G., Putti, M., and P. Teatini. 2004. Saving Venice by seawater. *Journal of Geophysical Research* **109**, FO3006 doi:10.1029/2004JF000119.

18 • Local flood protection measures in Venice

M. SPINELLI AND M. FOLIN

Venetorum Urbs divina disponente Providentia in Aquis Fundata aquarum ambitu circumsepta aquis pro muro munitur. Quisquis agitur quoquomodo detrimentum publicis aquis inferre ausus fuerit et hostis Patriae judicetur nec minore plectatur poena quam qui sanctos muros patriae violasset. Huius edicti jus ratum perpetuumque esto.[1]

The city of Venetians according to divine providence was founded on water and by water surrounded and by water protected as by a wall. Whoever acts in whichever damaging way toward the public waters will be judged an enemy of the nation and will not have a punishment inferior to those who violate the sacred walls of the nation. May this edict have eternal validity.

INTRODUCTION

The problem of defending the historical centre of the city of Venice from tidal waters is certainly a particular and well-known phenomenon. However, the fact that the Lagoon of Venice stretches over an expansive area, and that it is composed and characterised by ecosystems, settlements and different ways of using the territory, is not always so clear to everyone. To locally intervene on the effects caused by the Venice Lagoon's tidal waters means to intervene, often with profound impact, within the most delicate sections of the entire lagoon's ecosystem: the limits or borders between land and water. These embankments, of an

often indeterminate nature, are most definitely vital for the conservation of the ecosystem itself. The fragility of the environment and its importance therefore demands an intimate understanding not only of the limit itself but also of the area in which it is situated and the reciprocal relations that have developed therein. What remains is the concrete risk of effecting intervention operations that are indifferent to the surrounding context, or to the genius loci of the area. Consequently, to set forth interventions that repeat themselves in the same manner regardless of the context, be it urban, historical, natural, or agricultural, can further jeopardise an already unstable and precarious environmental condition. The risks of local flood protection measures, which are poorly designed and lacking in true knowledge of the subject at hand, are particularly high in the case of historically significant water embankments (*marginamenti*), where the structure should be preserved and considered as part of the lagoon's architectural heritage. Paradoxically, the consciousness that a historically significant water embankment can and should be considered an architectural monument is not as widespread as one would imagine it to be in a city like Venice. There is still much to be done on a cultural level with respect to this subject. It is precisely on this subject that, in recent years, the IUAV (University Institute of Architecture in Venice), through its research and design studio ISP (IUAV Studi e Progetti), has conducted a series of investigations focussed on possible research methods for the study of the Venetian water embankments and their surroundings. These research studies have also focussed on confronting the problem at various levels and scales, from a general environmental perspective (at urban, regional and natural scales) to a more detailed analysis (including morphologies, stratification and deterioration).

---

[1] Archivio di Stato di Venezia, *Savi et esecutori alle Acque*, b. 559. *Archivio proprio Poleni*, vol. 27.

Flooding and Environmental Challenges for Venice and its Lagoon: State of Knowledge, ed. C. A. Fletcher and T. Spencer.
Published by Cambridge University Press, © Cambridge University Press 2005.

A HISTORICAL PERSPECTIVE OF FLOODING AND LOCAL DEFENCE MEASURES

The phenomenon of high waters and the resulting 'struggle' for the protection of the lagoon's built environment is an age-old problem, perhaps intrinsic to the city itself. Recent decades have witnessed a rapid increase in this problem, for reasons tied to natural and man-made causes. In historical descriptions of the Veneto region's coastal lands, Roman and Byzantine geographers made considerable reference to the effects that the continuous and conspicuous tidal flows of the Adriatic Sea had in these areas. This geographic region was inhabited in great part by rural populations in mostly agrarian settlements and the attention of those studying the area was mostly focussed on the 'picturesque' aspects of the territory's dynamic: a landscape that repeatedly during the course of the day witnessed the transformation of its plains into estuaries and vice-versa, according to the tidal flux of the Adriatic waters that flowed into the lagoon, receding once again six hours later.

More detailed records start to come forth during the eighth century with archival testimonies of exceptionally high water levels of the Adriatic's tides that 'nearly submerged almost all of the lagoon's smaller islands'. It is with an increase of urbanised settlements within the lagoon that exceptionally high tide levels are registered as '*Sovracomun*' (exceeding common tidal levels) by the chroniclers of that period. They started to assess the high tides' devastating consequences and damage to agricultural cultivation as well as to buildings and, above all, to the merchandise and goods that were stocked in the lagoon's *fonteghi* (covered markets) and *magazeni* (storehouses). Herein after, the first documents that make explicit reference to the flooding of buildings and churches in Venice date back to the mid ninth century (Zucchetta, 2000). It is interesting to note how historic documentation describes a discontinuous progression in the tidal patterns of the lagoon. For example, in the fifteenth century, there was a tidal regression that brought the average sea level to about one metre lower than it is today. Subsequently, for that period, there is no record of flooding.

Much debate surrounded the flooding even back in the sixteenth century. An obvious consequence of

the events of flooding and urban inundation, capable of bringing about serious economic losses and hardships to the local population (historic chronicles refer even to cases of drowning and hypothermia in periods of high water during the winter), is the heated debate on the causes for these tidal cycles and the possible ways to contrast their effects. One of the more heated debates regarding the causes for the high tides that flooded Venice and the lagoon's islands (at times flooding even the castle of Mestre along with other inland areas) and how to face the problems of conservation for the city and its lagoon was held during the course of the sixteenth century with the active participation of Alvise Cornaro (1470-1566) and Cristoforo Sabbadino (1487-1560). In setting aside the debate on the historic interventions of hydraulic engineering (the re-routing of rivers, the closing of portal outlets, the widening of the lagoon) as an entirely different area of research, it becomes evident how the interventions and procedures, which are today defined as *difese locali* (local flood protection measures), were already being used in those times. As it was quite clearly stated in an early Venetian edict: 'Due in part to the excessive number of pedestrians, the ground level is sinking, hence there is nothing surprising about the fact that the older buildings are at a lower level and that it will be necessary to continually raise both the city's walkways and its squares.'[2]

LOCAL FLOOD PROTECTION MEASURES

As in many other historic European cities, the urban development of Venice progressed through a process of subsequent stratification that led to the continuous raising of its ground level surfaces. This was not always the desired outcome of man's interventions in the lagoon-city, although, for many centuries, the Venetian Republic's magistrates and their technical experts were well aware of the problems they would

---

[2] 'La tera se è arbasciata, anche per tropa gente che cammina' e perciò non è da meravigliarsi se 'le fabriche vechie sono al basso e se le strade e piazze si convengono alciar': State Archives of Venice, *Savi et esecutori alle Acque*, b. 231, *N.5 libri dell'ingegnere Sabbadini e Cornaro circa la Laguna*.

face if they abstained from dealing with continuously shifting geological balances under the pressures of hydraulic dynamics. There was henceforth a raising of the pavement level surfaces in order to *continually* adapt the height of the city's walkways and squares according to the periodic variations of normal tide levels, changing tide level ranges and to the ever more frequent and exceptional *sovra comun* high tides. By way of illustration, in 1786, Proto Mazzoni[3] had proposed the elevation of certain Venetian walkways, in that the 'the ground level of the walkways … is quite low and, at every slight swelling of the lagoon, the city's inhabitants are blocked from freely entering and exiting their very own homes'. Moreover, following the elevation of the normal '*commune*' tide levels, there were periodic interventions to raise the height of the city's bridges in order to facilitate boat transportation. Today however, interventions of this type are rarely carried out in project plans for the renewal of Venice's public walkways.

As time went on, the problem of the high water phenomenon gradually became less of an exceptional event and started to affect a larger part of the city. As a matter of fact, during the second half of the nineteenth century, Giuseppe Bianco, Venice's head municipal engineer,[4] stated that: 'since the public pavement levels of Venice are at an average of only 80 cm above common high tides, and the exceptionally high tides have reached levels of 120 cm more than twice in 10 years, it would benefit the city to raise its public ground level surfaces by 50 cm. Moreover, the private citizens will be required to raise their ground floor levels, along with the doors and windows, of their homes'. The tide level ranges that were flooding the city did not only represent a problem regarding the accessibility of the city, the use of the ground floors in homes, or the maintenance of merchandise stocked in the city's store-houses (*magazzini*). Historically, the water supply for the city was obtained through a well system that was filled by rainwater collected in the city's small squares where the wells were located. When it rained, the waters that accumulated at the pavement-level of many squares and courtyards (both public and private) poured down through a selected series of white stone blocks (*pillelle*) equipped with openings (*gatoli*), and were then filtered through layers of sand and gravel that purified the water before ultimately leading it to the wells' cisterns. One can thus understand how vitally important it was to avoid any salt-water flooding that could invade the walkways and squares, which could then reach the wells' *gatoli* and contaminate the city's drinking water. Therefore, the only possible effort to correct the serious damage caused mainly by the infiltration of brackish-waters into the cisterns of drinking water, during the periods of exceptionally high tides, was to continue raising the city's pavement levels (Zucchetta, 2000). In regard to this endeavour, there are the documents of the city's superintendents, dating back to 1725, in which a census of the wells was requisitioned in order to indicate which ones had been restored so as to 'preserve them against future high-tides…' (Zucchetta, 2000: 64). The interventions for raising pavement levels were often quite conspicuous with quota level increases that could exceed 50 cm. For example, in 1773, to preserve the well of Santa Maria Formosa, there was a 4-foot pavement level increment planned to defend it from seawater in the case of exceptionally high tides, without compromising access to the area's shops and homes.[5]

Even today, after decades of inactivity, local interventions of raising the city's surface levels for various urban routeways has returned to being one of the most effective instruments for salvaging the practicability and access of Venetian public spaces, as well as for the protection of buildings at their ground floor levels. Although effective, these intervention operations represent a potential risk for the dependent relationship that the city has with its historical heritage.

---

[3] '*[I]l piano della calle … è assai basso e ad ogni piccola escrescenza resta impedito agli abitanti il libero ingresso e regresso alle proprie case*': State Archives of Venice, Senato, Terra, filza 2837, Febbraio 1785, seconda.

[4] Archivie Municipale di Venezia (AMV), 1860–90, IX-6/22.

[5] '[S]opra al Comun dell'Acqua … per difenderlo dalle salse nelli casi di estraordinarie escrescenze … a norma degli altri che furono rinnovati e che riuscirono con ottimo effetto, senza pregiudicare l'accesso alle case ed alle botteghe': State Archives of Venice, Senato, *Terra, reg. 384, March 4 – August 5 1773, c.209v; Decree 9 June 1773*.

Presently, as in the past, such interventions must pay careful attention not to impede free access to the city's shops and homes; moreover, in a city that is considered a heritage of mankind, they must also recognise the importance of reducing impact on its priceless and delicate historic architectural patrimony to a minimum. Therefore, more than ever, it becomes of primary importance for project designers, planners and decision makers to practice and adopt an attentive evaluation of every possible social and environmental repercussion for both preliminary and in-progress intervention operations. It is precisely with such a preventive assessment of architectural and environmental impact that, over the past few years, the research for 'Local Flood Protection Measures' has been conducted by the offices of ISP (IUAV Studi e Progetti), under request from the City Department for Water Management (Magistrato alle Acque), in their evaluation of a prospective elevation to 120 cm (above the Punta della Salute datum) of canal embankments and public pavement surfaces in certain areas of Venice.

STUDY OBJECTIVES

High water has been a problem for Venice and continues to heavily influence its economic development as well as the quality of life of its residents. It hence remains quite apparent how every possible solution to contrast such a phenomenon will have significant repercussions on the inhabitants and their activities within the lagoon.

In light of this fact, in March 1999, the national committee for policy co-ordination and control, working to implement the 1984 law no. 798 regarding new interventions for the recovery and protection of Venice, summoned the Municipality of Venice and the Department of Water Management to verify the possibility of elevating the height level quota for local flood protection from a 100 cm to a 120 cm level as an operation of architectural tutelage.

The study was conducted in two phases: the first of which, completed in October 1999, consisted in an investigation on the two distinct island districts of the Dorsoduro quarter and the island of Rialto in the San Polo quarter of the city; in the second phase of investigation, completed in March 2000, the San Marco area and its prisons were examined. Overall, the study area focussed on 18 per cent of the area of the historic centre of Venice that is subject to flooding. The research project had the objective of collecting and elaborating all available data on the architectural and environmental nature connected with the areas of study at the ground levels of the principal buildings. The investigation was concluded with the deliberation of all relevant elements or factors of criticism to the project, which regarded the complexity of the structure of the city of Venice and served to determine the feasibility of the planned intervention.

Through the building of a multimedia data bank on a GIS platform, it was possible to simulate and make a detailed analysis of the effects that the intervention operations for raising the pavement surfaces would have on the existing architectural and urban heritage. 41 km of building front façades were analysed, in their altimetrical and morphological detailed aspects, by means of 9300 survey cards, which helped to define the possible impacts that the intervention operations would have on entire urban sectors of the city, as well as on its architectural ornaments and elements of street décor, including doorways, lintels, fountain and well curbs, bollards and pavement types. The GIS instrument proved to be of great use as it allowed for both the precise evaluation of the consistency of architectural patrimony in the intervention zones, as well as the subsequent assessment of 'expenditures' that such interventions would represent for the city, for its artistic-archival value and for its use. Therefore the challenge was to define an approach, in a very short time period, which could allow for the accurate monitoring of the city's detailed building texture, highlighting the manifold relations that exist between the architectural elements and, concurrently, guarantee a much larger-scaled perspective of the urban entity.

CONCLUSIONS

The research confirmed that the raising of the city ground level historically was a standard practice. At the same time, however, it's equally clear how this practice cannot be implemented for much longer, for two reasons. The first, being strictly cultural, derives

Fig. 18.1   Graphic simulation in the Fondamenta delle Romite.

from the fact that a common understanding has developed in almost every political and social sector regarding the necessity of safeguarding and conserving Venice's architectural heritage as a historical testimony of the city's history. The second group of reasons derives in part from the first. It regards the limited possibilities and opportunities for continuing to practise such interventions. It is quite evident that any further increase in the city's ground-level paving would make certain doorways unserviceable and certain ground floor levels impractical; furthermore, such operations could cancel out the many architectural elements that are either isolated or attached to the building front façades.

The historic customs of elevation interventions for the urban walkway levels are rightfully correlated with today's extensive work-sites for urban maintenance and in particular with the restoration of the embankments. It is however the knowledge acquired of the tight knit relations existing among the city's urban elements (its buildings, access points, routeways, components of urban décor, the embankments

etc.) that highlights the gap to the historical Venetian tradition of maintenance. For these are the relations that reveal the constant and natural toil on and of the city, the daily activities that occur therein, the culture of craftsmanship and of the artisans that are true masters of their trade, and ultimately the modern culture of emergency with its grand intervention operations that are often incapable of interpreting and respecting this dense network of internal relations and uses of the city. This problem, and the problem of having lost so many experienced artisans and technical experts, becomes particularly apparent in the cases of urban interventions, especially for the maintenance and consolidation of the embankments for secondary canals. The character of impending emergency given to such interventions often has the effect of depriving the embankments of all of their characteristic features, except for the strictly practical ones connected to their function of both containing water and confining terrain.

In our contemporary Venetian experience, when intervening on the city's water embankments, we should always remember that we are faced with 'an ensemble of distinct artefacts of a quite assorted constitution'. This assortment of embankments can vary from very simple and rudimentary models to much more complex and monumental ones, yet they are all full of formal and symbolic values. Moreover, as architectural elements, they represent a great range of constructive and technical solutions that are always of relevant meaning for their historical testimony. These solutions deserve – or should deserve – the same attention in terms of protection, conservation and restoration that a historic building does (Piana, 2001). In remaining convinced of the absolute necessity for intervention operations on the water embankments to remedy the serious state in which they find themselves and provide for their maintenance, reparation, reinforcement and restoration, we must also be aware during every phase of intervention, from its conception to the building site and to the following verifications, that any intervention will inevitably bring about a tampering with the artefact, a loss of its original materials, and even an extended and definitive loss of the history of the artefact. Therefore, the problem today remains not *if* but *how* to take action.

THE WATER EMBANKMENTS OF THE VENICE *ARSENALE*

As it happens with the approaches adopted in the restoration of historical buildings (and water embankments are, in most cases, considered historical artefacts), it becomes ever more necessary to appropriate an intimate understanding of the building's dynamic, the possible modifications and transformations that it undergoes in time, its design and structural arrangements, as well as the devices or artifices adopted for the building of these embankments. It is therefore necessary to base the intervention operation on a sound apparatus of analysis and knowledge, the result of investigations that were completed prior to design planning and that are continuously verified and up-dated throughout the building process. This apparatus should also help to guide the project designer throughout the phases of project conception, which must remain open and attentive to verifications and new understandings that can be acquired after the opening of the building site. The concrete risk, experienced daily, is otherwise that of *standard* interventions, which are often useless or disproportionate in relation to real necessities. Such interventions can also be indifferent to the specificity of each single artefact, being based on techniques and materials that are already questionable in regard to ordinary building procedures. In the case of Venetian water embankments, these interventions become devaluating, in that they can cancel out entire portions of the history of traditional Venetian building techniques.

The awareness that the water embankment system in Venice is often an artefact of extraordinary historical and architectural value and that it should be treated as such, has dictated the methods of analysis of the campaign entitled: 'Investigations for the technical-historical characterization of the embankments of the Venice Arsenale', which was completed in 2001 for the Magistrato alle Acque. There are five kilometres of detailed surveys on the city water walls of this historical Venetian complex that pay special attention to the intertidal tracts and to the areas immediately above them. This object of study was investigated indirectly through bibliographic, iconographic and archival research, but also directly with on-location

Fig. 18.2   Example of mapping of units of transformation (UTR).

survey examinations. The campaign for a geometric and photographic survey (output scale 1:20) has constituted the basis for the providing of a morphological relief, which clearly portrays both present transformations and stratigraphic descriptions. The mapping of the transformations by photography singles out the homogeneous areas of deterioration on the wall surfaces that are defined as 'units of transformation' (UTR) (Fig. 18.2).

A single UTR can be of variable dimension, at times reduced to the size of a single brick. Each UTR has been catalogued in a digital data bank that describes its geometric characteristics (geographic orientation, sea-level position, and size) and also its morphological and statigraphic characteristics (according to the materials that constitute it). The stratigraphic units instead depict the levels of deterioration, pathology, chronological order, and structural hazards. Only such a complex databank can allow for an integrated reading of the transformational phenomena that are occurring along the water embankments. This can also enable us to trace the first causes of the same phenomena and, consequently, to determine the most appropriate forms of intervention, thus reducing the impact of interventions on the artefact to a minimal level. The databank has allowed for an accurate real-

Fig. 18.3   The mapping of the tracts exposed to major structural risks: the shaded areas depict the tracts of higher risk.

ization of an integrated cartography that delineates possible structural hazards by representing the UTRs in five incremental danger levels that can be easily interpreted (Fig. 18.3).

Similarly, a stratigraphic survey, through a historical-archival analysis, has been completed in order to define the phases of development and maintenance for the water embankments (Fig. 18.4). The amazing overlapping of constant, almost daily, maintenance interventions that have marked the artefacts' existence – allowing them to endure up to now and clearly showing the close relationship between the materials used for maintenance interventions and the degradation upon them that was later generated – has been an element of great interest.

We should therefore reflect upon the necessity of using the most experienced and verified technologies in order to make maintenance a planned and steady activity with known and foreseeable effects (which is often not the case in the Venetian site locations for building renovation). Particular attention should henceforth be given to the very common use that is made of mortar and lime-based colloidal gel injections. They can radically modify the structural

Fig. 18.4   Detail of the representation of the developing phases of the artefact.

behaviour of the artefact as the final results of these interventions are often difficult to predict and verify. As a matter of fact, it is nearly impossible to establish exactly how and where the injection will expand within a wall's mass. It is also quite difficult to fully comprehend all of the implied risks that come with this type of intervention, along with the possible irreversible processes of structural transformation (or even damage) that can be triggered.

In general, 'light' interventions that are tied to traditional techniques (and thus compatible with the pre-existing water embankments) are especially suitable to the typology of the lining containment walls of the Venetian canals. The experience of direct analysis and, in particular, the completed stratigraphic and degradation surveys demonstrate how the first answer to the problem of urban preservation in its entirety must inevitably return to a common sense logic of pure maintenance. In other words, a constant and steadfast attention placed on the degradation of the artefact and a quick enactment of repairs will normally prevent an aggravation of those damages and render superfluous every need to start grand and committing works that can be economically much more costly and even risky for the preservation of the artefact itself (Piana, 2001).

Translated into operational terms, the approach to project design that views the water embankment as an artefact to be studied, understood and safeguarded, is an approach that aims at *improvement*. It clearly requires neatly arranged decisions in order to reduce interventions that plan for 'heavy' structural modifications and/or the demolition and rebuilding of existing structures with modern techniques and materials. All this should thus give preference to traditional techniques, such as the *'scuci-cuci'* wall rebuilding methods and the re-working of mortar amalgamations (Modena, 2001), and include a careful selection of materials that are chemically and physically compatible with the object for which intervention has been planned.

THE VENETIAN LAGOON'S WATER EMBANKMENTS

The central role that the analytical aspect takes in every intervention of local flood defence requires methods and instruments that can answer with simple clarity to the complexity of the issue at hand. The analytical approach must comprehend not only the artefact itself, its evolution in time, its morphologic and generating aspects, its state of conservation and dynamics in progress, but also (and perhaps above all) the complex relations that the artefact has with the environment in which it is situated and the ensuing priorities for intervention.

The definition of the state of conservation itself, in particular, shall be the result of the use that has been made of the water embankments, of their context and of the true expectations that regard them. The same water embankment inserted in different contexts will clearly express different levels of functional adequacy. Today, to intervene on Venice's water embankments often means to adopt a complete indifference, regarding their materials, restoration techniques and their morphology. It becomes a typological flattening that risks creating uniformity for the entire lagoon system with a permanent loss of the numerous intervention methods that have characterized the history of the Venetian Republic. This is one of the main considerations that came forth during the investigations that aimed at realizing an 'abacus on the true state of the lagoon's water embankments'. If the investigation on the Arsenale's water embankments dealt mainly at a level of architectural detail, the large-territorial scale was ultimately the principal object of this study, completed between 2002 and 2003, upon request of the Magistrato alle Acque.

What we assumed in starting this research is that every water embankment is tied to the complex relations with its environment, so the analyses needed to be flexible and capable of adapting themselves to the different characteristics surrounding each embankment. It is only by means of analysing the morphological, structural and functional profile, as well as the profile of degradation of the water embankments seen from regulating, historical, archival (Fig. 18.6), urban planning, naturalistic (Fig. 18.7) and perceptive points of view, that we can investigate the relations between water embankments in the most diverse contexts: agricultural, natural, urban and industrial.

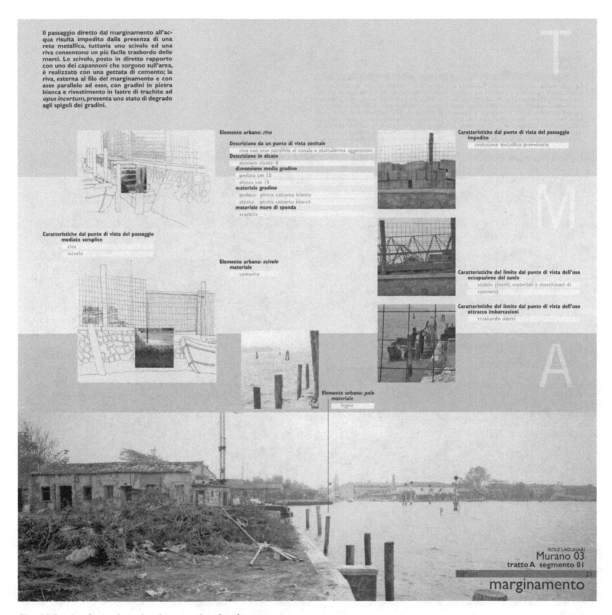

Fig. 18.5   Analysis chart for the use of embankments.

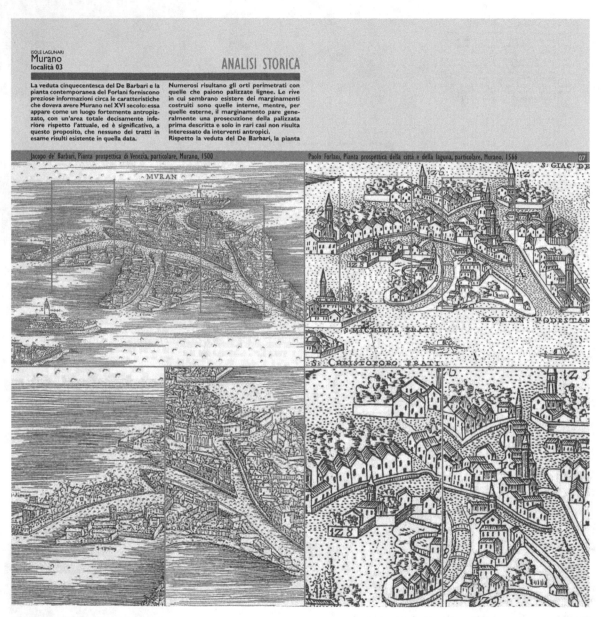

Fig. 18.6   Historical – archival analysis.

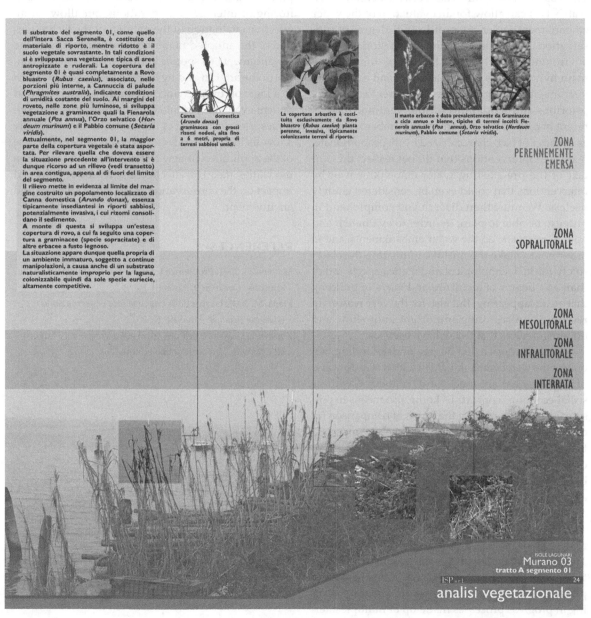

Il substrato del segmento 01, come quello dell'intera Sacca Serenella, è costituito da materiale di riporto, mentre ridotto è il suolo vegetale sovrastante. In tali condizioni si è sviluppata una vegetazione tipica di aree antropizzate e ruderali. La copertura del segmento 01 è quasi completamente a Rovo bluastro (*Rubus caesius*), associato, nelle porzioni più interne, a Cannuccia di palude (*Phragmites australis*), indicante condizioni di umidità costante del suolo. Ai margini del roveto, nelle zone più luminose, si sviluppa vegetazione a graminacee quali la Fienarola annuale (*Poa annua*), l'Orzo selvatico (*Hordeum murinum*) e il Pabbio comune (*Setaria viridis*).

Attualmente, nel segmento 01, la maggior parte della copertura vegetale è stata asportata. Per rilevare quella che doveva essere la situazione precedente all'intervento si è dunque ricorso ad un rilievo (vedi transetto) in area contigua, appena al di fuori del limite del segmento.

Il rilievo mette in evidenza al limite del margine costruito un popolamento localizzato di Canna domestica (*Arundo donax*), essenza tipicamente insediantesi in riporti sabbiosi, potenzialmente invasiva, i cui rizomi consolidano il sedimento.

A monte di questa si sviluppa un'estesa copertura di rovo, a cui fa seguito una copertura a graminacee (specie sopracitate) e di altre erbacee a fusto legnoso.

La situazione appare dunque quella propria di un ambiente immaturo, soggetto a continue manipolazioni, a causa anche di un substrato naturalisticamente improprio per la laguna, colonizzabile quindi da sole specie euriecie, altamente competitive.

Canna domestica (*Arundo donax*) graminacea con grossi rizomi nodosi, alta fino a 6 metri, propria di terreni sabbiosi umidi.

La copertura arbustiva è costituita esclusivamente da Rovo bluastro (*Rubus caesius*) pianta perenne, invasiva, tipicamente colonizzante terreni di riporto.

Il manto erbaceo è dato prevalentemente da Graminacee a ciclo annuo o bienne, tipiche di terreni incolti: Fienarola annuale (*Poa annua*), Orzo selvatico (*Hordeum murinum*), Pabbio comune (*Setaria viridis*).

ZONA PERENNEMENTE EMERSA

ZONA SOPRALITORALE

ZONA MESOLITORALE

ZONA INFRALITORALE

ZONA INTERRATA

(ISOLE LAGUNARI)
Murano 03
tratto A segmento 01

ISP s.r.l.   24

analisi vegetazionale

Fig. 18.7   Analysis of vegetation.

RESULTS AND CONCLUSIONS

The result of these studies has been the definition of a matrix that can allow for an evaluation of the water embankment under the profile of its degree of conservation, contextual functionality, historical value, environmental value and landscape impact. The main hypothesis is that an analytic and evaluating instrument can help the project designer to define the priorities and modalities of intervention. Today's recent history of the lagoon's water embankments is, as a matter of fact, characterized by disproportionate, oversized interventions that do not respect the context of the pre-existing water embankment or by interventions that could even be considered entirely useless. We must then discern and comprehend in order to be able to plan, in order to evaluate if and how to intervene on the water embankments, and in order not to make the mistake of treating the object solely for its technological and static aspects rather than as a bearer of a culture and ways of building that are disappearing. But also for the very reasons of rescuing and restoring important know-how and building devices that today have been lost.

We must adopt a 360-degree understanding, not only of the water embankment in a strict sense but also of the uses that are made of it, connected with its social-economic context. To know also means to possess the instruments and the terms of comparison for conscientiously evaluating results that the intervention, be it of maintenance or of a new construction, has on the environment, understood as an entirety of biotic, a-biotic and man-made components. This approach is indispensably a multi-disciplinary one and it must view the water embankments as part of a complex environment, thus adopting the logic that is at the basis of a strategic environmental evaluation. The fragility of the situation – the reason for this chapter – and of the Venice Lagoon's environment requires project planning procedures that can foresee evaluating in progress processes made up of multiple alternative scenarios among which we can choose in order to arrive at an intervention approach that will be judged 'appropriate' in regard to the real needs of urban preservation. Further requirements regard project procedures that can make use of the participation and interaction of those who live the territory as an essential and unyielding moment for the acquisition of indications, project requests, and verification of the proposed intervention hypotheses. A continuous monitoring of intervention operations can allow for an effective evaluation of the validity of the techniques and of the adopted materials by contributing to the advancement of a specific knowledge for a steady control of the artefact and by helping the project to return to a logic of 'maintenance'. A gradual, extended and patient evaluation of intervention processes that continues after the closing of the worksites can ultimately allow us to better understand, especially in the case of the application of new techniques and intervention modalities, their real effectiveness, reproducibility and impact on the conservation of the city's artefacts and environment.

REFERENCES

Modena, C. 2001. Problemi e tecniche di consolidamento delle strutture. *Quaderni* **9**, Anno III, 89–96.

Piana, M. 2001. Le rive della città e la loro conservazione. *Quaderni Insula* **9**, Anno III, 15–20.

Zucchetta, G. 2000. *Storia dell'acqua alta a Venezia dal Medioevo all'Ottocento*. Venice: Marsilio editore.

19 • Urban maintenance in Venice

B. DOLCETTA

HOW THE CITY WAS BUILT

In the early Middle Ages, from the fifth century onwards, the people in this area, under the pressure of the devastation wrought in their territory first by Attila and then by the Lombard conquerors, began to abandon their homes on the mainland to settle in the area of the lagoon, where the stable islands that had emerged from the waters were scattered and still quite small. From the sixth century onwards, this population movement increased and the bishop's seat was transferred from Altino to Torcello. The island had previously been inhabited under the Romans, as borne out by the succession of archaeological excavation campaigns on this site. The strategic importance of the higher elevations (*tumbae*) concentrated in the centre of the lagoon further south, however, gradually emerged even during the centuries in which Torcello performed its significant function as an important centre, being referred to at least until the ninth century as *emporium mega*. These other islands lay along the sides of a riverbed that pursued the winding course of its natural banks through the low surrounding waters. A wide watercourse (now the Grand Canal) curved through them like a spine, they were the nearest islands to the channels left open through the coastal dunes for contact with the sea and were served by canals that were deeper and safer for maritime traffic, which constituted the vocation and the economic and political *raison d'être* for the settlements on the lagoon. Building began on the *rivus altus* and on the neighbouring islands and the outline of the port-city, the *miraculusissima civitas*, the Venice we know today, began to take shape.

Attention is limited here to the material formation of the city. This is a theme in which hugely important historical, civil and cultural questions interlock. The process of physical construction witnessed the convergence of technical and organizational skills, the roles of public and private domains in society, financial investments, procedures and finally institutions and magistracies dedicated to the various aspects of the creation of the city structure. These are all constituent elements of Venetian civilization and if we know nothing about them we cannot hope to understand much about the spaces or the forms in this city. In the long process of building Venice and all the settlements in the lagoon, a process that has renewed itself uninterruptedly over the centuries and continues to this day, the basic preliminary actions were to construct the soil on which to build and then to define the boundary between land and water. The land was reclaimed as the growth of the city demanded; *insulae* were formed by accumulating silt and mud inside a border marked by piles until a height was reached that was thought sufficient to keep the new island consistently above the variations in tidal levels. The limits of land and water were defined in many different ways in past centuries, linked to the use of the land spaces on the water's edge, first for the more densely built-up areas and later for the outlying islands. These operations allowed the beginning of the *aedificatio*, as it is described throughout the early Middle Ages. The chronicles speak of an endless building site where work proceeded on 'hedificare, fabricare or construere castles and cities – castra civitesque – palacium, domum episcopalem, pontem – a palace, a bishop's residence, a bridge' (Concina, 2000).

The definition of the city's layout and the refinement of its functions took many centuries, from the ninth to the sixteenth century, when Venice might be said to have completed its fundamental structures

Flooding and Environmental Challenges for Venice and its Lagoon: State of Knowledge, ed. C. A. Fletcher and T. Spencer.
Published by Cambridge University Press, © Cambridge University Press 2005.

and consolidated the extension it would reach in our time. There would be only one further season of development and innovation in the physical structure of the lagoon in the nineteenth century, with the advent of the industrial revolution. The city acquired form beyond its centre on Rivoalto during the period from the tenth to the fourteenth centuries. The expansion was generated predominantly by private interests while government powers gradually worked out rules in the interest of the entire community, in response to the disputes and day-to-day problems that arose in the city from having to deal with the organization of life in a restricted space. The great landowning families defined the functions of most of the land area: accessory buildings placed beside the churches and the noble palaces, the *domus magnae*, on which land possession and its organization were based: warehouses, factories and rental properties. Building was discontinuous, and for a long time, large areas were dedicated to growing fruit and vegetables and other crops, even along the Grand Canal. This only appeared in its present completely built state in relatively recent times. In colonising the *insulae*, therefore, the landowners laid out the boundaries with *fossae* or *rivi* which they dug to provide their homes with landing stages and to keep sewage and sludge away (*meati ad rigandum*, the *cloacae*). When they carried out works on a larger scale, they reclaimed vast internal lakes and marshy areas, claiming new land from the water to build on.

Over the same time period, the second system of communications also began to take shape: the streets. This network was private at first, confined to the *insulae*, forming irregular pathways that were strictly functional to the properties. The various pathways were later connected into a more unified system and became public thoroughfares, especially when the inhabitants began to reach beyond the banks of their *insulae* by constructing bridges, beginning with the Rialto Bridge. This proved to be a real innovation and one that had a decisive influence on the increasing significance of the pedestrian system.

Both the land and water circulation systems, in terms of geometry, section and density in the different areas of the city, were affected by their piecemeal fabrication and by the polycentric and discontinuous fabric they originated from. Only later would

they be brought to their final union, when private rationale gave way to public administration (Dorigo, 2003). The care and maintenance of banks, bridges, land and public works were supervised and conducted with unremitting zeal by specific magistracies that not only took action to strengthen the coastal defences in the lagoon against erosion wherever it was necessary and to prevent sludge from pouring into the canals, but also to settle private disputes and set up accurate public land registers (Calabi, 2000, 2003). Starting in the fifteenth century, the boundary between land and water was consolidated by building public embankments, using the same techniques for the entire city, employing walls of brick and white stone from the quarries of Verona, or more frequently from Istria. These two materials, as well as trachyte from the Euganean Hills, were used for the pavements of the city and thus all the public spaces in Venice acquired their present appearance.

TWO QUESTIONS FOR THE SAFEGUARDING OF VENICE

The extraordinary thing about the city is that it is *suspended between two elements*, as if it were *an image of the earth* – according to the idea the ancients had– *in the middle of the ocean*.

(Ammerman, 2001)

Suspended: the term supplies an excellent definition of the nature of Venice's invention because it conjures up the idea of a site in which man embarks on a difficult and never-ending contest with nature. This contest engages us every day, because the space in which Venice exists is constantly evolving, whereas the city was conceived for the long term and for a future which we would like to believe will never end. With regard to these two elements the city is now subject to two kinds of problems, both of them critical:

- on the one hand, its very inhabitability is threatened, as a result of the increasing frequency with which tidal water invades large parts of the public spaces and the private homes themselves; and
- on the other hand, its vital structures are in a precarious state of repair and stability: the embankments, the public quaysides along them, the banks

and the wing walls of the houses and palaces that border them, the bridges; the city is in a state of functional deterioration and there are seriously dangerous conditions throughout the fabric of the city that affect its very physical stability.

There are many questions to consider involving the first point, above all the inexorable progressive subsidence of the soil (about 10 cm a century according to reliable and up-to-date studies). The phenomenon was aggravated during the past century by imprudent behaviour on our own part, misled by the promise of industrial progress whose cost in this sense was not taken into consideration. This is also confirmed by archaeological excavations that have found sterile layers of original soil on which the foundations of Roman buildings rest, but also those of Late Medieval buildings, well below present sea level (Concina, 2000). The evolution of this phenomenon is completely predictable, even in its quantitative aspects, and the tendency is for the invasion of the city by seawater to be irreversible. The solution lies in regulating the relationship between sea and lagoon, and as far as timing and urgency are concerned, this should be considered from the point of view of the evolution of climate and its consequences on mean sea level. A large proportion of urban studies in Venice are devoted to this subject.

At this point it is necessary to deal with the second group of problems, strongly emphasizing the fact that the integrity of the city is a priority and a primary responsibility, and is universally perceived as such, judging from the number and the importance of the private international Committees for the Safeguarding of Venice. The protection and maintenance work must be carried on with continuity, great determination and with the degree of organizational and technical skill required by the prestige and complexity of these Venetian problems.

TASKS INVOLVED IN THE MAINTENANCE OF THE CITY

We have seen, very briefly, that the construction of the physical city engaged the entire community for centuries. We find ourselves, therefore, in the presence of a palimpsest that turned into a consolidated struc-

ture only after an infinite series of adjustments and changes, one that preserves in all its parts the stratification of the decisions and the works that men have produced, a living document of the art of building a city. The technical solutions adopted are unquestionably appropriate. They are practices originating in the very art of building in the Late Classical age (Concina, 2000) which were developed exploiting skills accumulated and refined within the great building site constituted by Venice itself, and in the countless constructions it disseminated throughout its Mediterranean possessions and beyond. Trials and errors were later checked with careful monitoring over time, during which the solidity of the buildings was examined to assess their resistance to attacks by nature, by water in particular.

In past centuries Venice never failed to provide assiduous and expert maintenance of the entire complex of maritime works inside and outside the city, well aware that the slightest neglect could lead to a substantial loss of stability and efficiency and imperil the delicate balance of the system, and that delays in repair work would entail much higher costs. The subject dealt with here, therefore, concerns the complex of works that Venice has planned, and has actually started carrying out in the context of an investment programme stretching over many years. The task facing this complex plan is to raise the entire structure of the city to a high level of efficiency, and to create a modern system of planned maintenance aimed at preventing the loss of efficiency, safety, and public image in future years.

THE NATURE OF THE PROBLEMS

Navigability in the internal canals

First of all, it must be remembered that the perception of the seriousness of the problem of maintenance in Venice emerged at the end of the 1980s when it became evident that an emergency of a very particular sort, though undoubtedly obvious to everyone, had become acute: the gradual loss of navigability in the internal canals due to the sedimentation of sludge that had accumulated for too long without dredging. The reason was that the well-being of the canals inside the city is related to the volume of water they carry, and the reduction in their navigability (one must bear in mind that public emergency services such as ambulances and fire brigade teams travel by water, as do goods and passengers) is a source of very real danger to individuals and to the community, causing an unsustainable increase in the day-to-day cost of running the city.

Improvement of the pedestrian system

Against the 'high waters' that afflict the city many times a year the ancient practice of raising the heights of pedestrian thoroughfares by a few centimetres may be reinstated, thus reducing the frequency of flooding for several decades to come.

Stability of constructions

Every work of man requires care and maintenance. But the walls of Venice are continuously immersed in an aggressive saltwater environment. In recent decades they have also suffered the impact of the wave motion caused by the introduction of the motors that now propel all watercraft, above all the heavy transportation barges, and this provokes destructive turbulence in the water spaces within the city.

Renewal of underground services

The underground utility systems (electricity, water, telephone) are obsolete and must be completely replaced and reinstalled. The sewer system also has to be integrated, partly by restoring the efficiency of the historic trunk lines and partly by integrating the system with modern technology.

These are the problems that the city's maintenance programme aims to confront with a complex, integrated project.

THE MAINTENANCE PROGRAMME

After the fall of the Most Serene Republic in 1797, the patient work of canal maintenance was carried out without interruption during the first period of Austrian domination, from 1799 to 1805. Under French rule on the other hand, from 1805 to 1815, the

urban infrastructure efficiency maintenance sector remained rather inactive. The return of the Austrians coincided with the launching of a vast programme of works from 1818 to 1821, which it was felt could not be postponed in the interests of hygiene and sanitation. The initial programme was followed by substantial maintenance campaigns from 1835 to 1849. When Venice was annexed to the Kingdom of Italy, maintenance works (1869–75) began once again to continue 'until the completion of the rational plan, which should restore all the city's embankments to a normal state that is to be permanently maintained', in the words of a Council resolution. In 1892, large-scale dredging was resumed and about a metre of sludge was removed from the canals. The years following the world wars again witnessed severe cases of deterioration in the canals of Venice. After the First World War, conditions were so serious as to require an extraordinary intervention by the State (1922–24) through the Magistrato alle Acque, the local arm of the Italian Ministry of Works. This situation presented itself again after the mid-1940s, when maintenance work resumed with greater intensity, in part because of the resources made available by the budget allocated by the special law for Venice. Work slackened again in the 1970s and 1980s. This was caused by various concomitant factors: scarcity of resources, priority given to questions that were considered more urgent and the problem, which involved new considerations, of the difficulty in disposing of the sludge removed from the canals as a result of the evolution of environmental protection legislation.

The reasons underlying the accumulation of delays were overcome between February 1992 and the summer of 1993, a period in which three important provisions saw the light: the Special Law for Venice 139/92, the Protocol of Understanding on the classification and isolation of sludge and the Preliminary Agreement signed on 3 August 1993. Law 139/92 classified maintenance work in the city as structural, and stated that its implementation should proceed by means of an integrated project. The works included in the context of the Integrated Canal Project, as specified by the law, must be carried out in unified form to ensure technical design homogeneity, coordination during execution and the integration of financial resources. The Protocol, stipulated with the Ministry for the Environment on 8 April 1993, solved the hoary problem of the disposal of the sludge excavated from the canals: it laid down the provision that some of the sludge containing pollutants required specific procedures and sites for their removal and storage. The Preliminary Agreement, signed on 3 August 1993 by the State through the Magistrato alle Acque, the Veneto Regional Council and the Venice City Council, addressed the need for bodies with different administrative functions to operate in a coordinated manner; this document appointed the City Council as the party responsible for implementing the Integrated Canal Project. On the basis of these instruments, the Venice City Council drew up the Integrated Intervention Programme Plan for the improvement of the sanitary sewer system and the buildings in the city of Venice, a project that had been provided for by the programme agreement and promoted by Article 5 of Law 139/92.

Based on a programme of urban maintenance works more succinctly entitled 'Integrated Canal Project' (because the most urgent action was dredging the city's canals and repairing the embankments), the 1994 Programme Plan, in short, involved the completion of works and actions costing about 723 million euros over a period of 23 years. This time period was calculated in consideration of the capability of the city of Venice to sustain the works. As early as 1999 the mass of design data and works accounting that had been collected led to an update of the Programme Plan to provide a more accurate and detailed estimate of the annual and total flow of financial resources necessary to complete the cycle of extraordinary urban maintenance set up by Law 139/92. The completion of this cycle for all the works included in the plan extends the schedule up to the year 2025, with a budget of about 985 million euros (1213 in all, less 228 already spent by the end of 2003), and an average annual budget allocation in the range of 45 to 50 million euros.

THE GEOGRAPHICAL AREA

The urban structures involved in the plan consist of the historic centre of Venice, the built areas on the islands of the lagoon (Murano, Burano and Pellestrina) and several urban areas on the Lido. The geographical area involved in the Integrated Canal Project corresponds to the allocation of responsibility for the city's large and small canals as defined in Royal Decree 721 of 1904 which made the City Council responsible for all the canals within the perimeters of the islands, while the State was given responsibility for all the canals around the perimeters of the islands and their embankments.

Now, as a result of subsequent agreements, the area is divided into three different jurisdictions: the Port Authority, the Magistrato alle Acque and the City Council. The maps regulating this breakdown of responsibilities are not detailed enough to define responsibility unequivocally, but as a general rule all the canals inside the city and the islands are the responsibility of the Council, while the banks on the lagoon, the Giudecca Canal, the Bacino di San Marco, all the embankments around the perimeter of the historic centre, the Giudecca and the islands are the responsibility of the Magistrato alle Acque.

THE INTEGRATED CANAL PROJECT AND INSULA S.P.A.

The Integrated Canal Project is divided into a first phase involving works to restore the full hydraulic, structural and sanitary efficiency of the canals and their embankments (water sites) and will end in 2014, gradually slowing down from 2005 onwards. After this date the second phase will gradually expand, involving radical renewal of the underground service grids in all parts of the city not in direct contact with internal embankments, improvement to the system of collecting and disposing of sewage and the maintenance of the pedestrian areas concerned in these works (land sites).

Such complex works in a fabric as particular as that of Venice called for a dedicated structure to be created, in which the City Council was to be represented, as were the underground utility companies. Apart from the need to provide an efficient business management, this was the reason for the creation of Insula S.p.A. on 10 July 1997, an urban maintenance company consisting of the Venice City Council (with a 52 per cent share of the capital) and, in equal proportions, Aspiv (now Vesta), Ismes (now Enel Hydra), Italgas and Telecom Italia. The utility companies are involved so that the work they do on underground services can be coordinated, with reciprocal benefits in terms of economies of scale, quicker implementation of plans for resetting and improvement, but above all in terms of reducing inconvenience to the inhabitants. The criteria for the planning and implementation of maintenance works were defined in 1998 by the Regulations for the Coordination of Joint Works Underground, a document that lays down rigorous procedures for coordination and collaboration, for which a special working group has been formed to supervise application.

Insula's tasks involve the design and execution of the works contemplated in Law 139/92, the Preliminary Agreement of 3 August 1993 and the Programme Plan, and particularly the acceleration of programmes for canal dredging, structural repairs, renovation of sewer systems, resetting underground utility grids, protection from medium to high tides and a series of works aimed at safeguarding the physical integrity of building foundations. This is a wide-ranging and systematic programme of works which also includes point repairs that are urgent or constitute a priority (single bridges or embankments) and are not integrated with other projects, or that involve works on a larger scale (wet dredging of canals). These activities may now be observed in various parts of the city for the first time in more than thirty years.

URBAN PATHOLOGIES AND THE MAINTENANCE PROJECT

In carrying out the maintenance works, as we have already mentioned, problems arising from serious urban pathologies are immediately dealt with and resolved.

The categories of inconvenience

Canal dredging

Periodical dredging is necessary to ensure routine navigation and proper sanitary conditions for the inhabitants. The excavation work can be done either on the dry bed of the canal or in the presence of water. The digging proceeds on the basis of a section that does not affect the safety of the building structures and foundations at the sides, while the turbid state of the water is contained by shutting off the section involved with drainage cloth barriers. Logistics and costs are affected by the presence of underground utility systems (surveys are carried out to detect these systems) and legislation governing sludge disposal (the method of disposal and the destination of sludge are determined by chemical and physical tests). The decision whether or not to dry out a canal depends on whether action has to be taken on the entire system constituted by the canal and its banks; in this case sludge removal is one of several operations carried out, but not the decisive one.

Bridge restoration

Bridges are indispensable connecting features within the fabric of the city, and constitute an integral part of the pedestrian system. In the Middle Ages, they were constructed mainly out of wood and presented an elementary structure, as noted by W. Dorigo, who examined their early sixteenth-century appearance in De Barberi's perspective map of Venice They are now built using different techniques, with a prevalence of stone arched bridges, followed by iron or cast iron structures and finally a small number of wooden bridges. These artifacts need continual repairs to keep them in working order. Perceptible damage generally includes deteriorating masonry, detached plaster, Istria stone arch segments and trusses, cracks formed along the intrados of the vault (a sign that there are utility pipes within the fabric of the arch) and the bridge back, and distortion of the vault. Consolidation of the foundations and the piers is followed by restoration of the stone arch segments and repair of the brick vault (injecting binders to fill the gaps in the masonry), and reconstruction of the parts that are missing as a result of the introduction of services ducts. Appropriate restoration methods are adopted for bridges made of other materials and with other techniques. Finally new utility pipes are placed inside the superstructure and the bridge is paved.

Alleviating the inconvenience caused by medium to high tides

A second very important source of inconvenience to movement and city life is represented by medium to high tides. The remedy is to raise the pavements wherever this is feasible, up to about 120 cm above mean sea level (as defined by the Punta della Salute datum). Before deciding on this kind of work the impact on architectural structures is assessed and a cost/benefit analysis is made. One basic criterion is to measure the effectiveness of raising the pavement, in terms of the reduction in the number of high water occurrences in the section in question. The works involved consist of removing the pavement and raising its edge by adding more of the same material the wall is made of (brick or stone), placing it under the coping until the intended level is reached, without altering the original appearance of the surface. The pavement is then put back into place introducing a slope so that the level corresponds to the previously existing level of the houses at the side.

The categories of instability

The most important category is the stability of the buildings in contact with the water inside the city. The reasons are those already mentioned, but they reveal their gravity and magnitude when the sides of the canals are exposed to do a thorough job, and the operators work directly on the walls down to the level of the base of the foundation using all available techniques. The main pathologies that emerge are:

- absence of soil at the base of the foundation;
- loss of binding mortar;
- disgregation with a physical loss of materials;
- washing away and loss of the finer part of the soil; and
- deterioration as a result of biological action.

The work aims to respect the historic artifact, acknowledging its original construction consistency, and applying restoration techniques. In some cases corrections are made using modern technology and materials. The essential action taken to secure the stability of embankments, canal walls and buildings includes:

- fixing wooden piles along the edge of the foundation to support it and prevent future alterations to the verticality of the wall; alternatively, in specific cases assessed on site, a metal barrier is installed;
- injections of mortar to fill in gaps and reconstruct the physical continuity of the wall;
- regeneration of the mortar joints, surface injections, replacement of or additions to the brick or stone wall adopting restoration techniques to reconstruct the missing surface areas;
- reclaiming the soil behind the masonry and reconstructing sewage drains if any; and
- construction correctives or partial reconstruction to rectify any bulging or shifting that may have changed the original geometry of the load-bearing masonry of embankments or buildings.

Categories of inefficiency: underground services and sewers

The underground ducts supplying primary services have always demanded continual works of maintenance and adjustment. We must remember, however, that public land in Venice, whose historical origins were described earlier, is scarce with many critical points. The utility grids cross from one bank to another under the bridges. Furthermore, work on these services has been going on for more than a century, so that there are critical points in the systems as regards both materials and layout, the different services get in each others' way and there is only very approximate understanding of their characteristics. In such an environment, maintenance is difficult and expensive. The decision was made to renovate them completely. When work is done on quaysides, pavements and bridges the utility grids are laid bare and the service companies can act, replacing or integrating the systems as required. At the same time the sanitary and environmental conditions are restored, including renovation of the sewage disposal system.

Resetting utility grids

The work of the utility companies (Enel, Italgas, Telecom Italia and Vesta) is coordinated by Insula through a complex operational plan whose purpose is to harmonize the plans, investments and working procedures of the single entities. These jobs are entrusted to the same firms that are engaged on the specific project in order to ensure the homogeneity of the various work sites and reduce completion times. In collaboration with Insula, Vesta is also setting up a fire hydrant water supply network based on use of the normal water system, but with the water distributed at a pressure high enough to reach the upper floors.

Improvement of the sewer system

The Programme Plan divides the city into two areas that correspond to specific intervention criteria. The historic areas will be subjected to restoration and optimization of the existing system, consisting of the combination of house drains (septic tanks, small purification plants, etc.), *gatoli* (the traditional Venetian sewers) and canals. During restoration work on the canal walls all drain outlets will be brought to 75 cm below sea level (with reference to the Punta della Salute datum), as required by city sanitation regulations. For the peripheral areas of the island of Venice, where environmental conditions are not as critical, the most suitable technical solutions are being considered for each different context. In these areas we can work towards creating a modern sewage system, which will separate rainwater from sewage, delivering the latter to centralized treatment plants located outside the historic centre.

A COMMITMENT IN PROGRESS

The plan for urban maintenance in Venice and the other built areas in the lagoon is proceeding in strictest compliance with objectives and scheduled completion dates. The investment capacity has been established at around 50 million euro a year, which is considered the maximum compatible with the impact of the works on the city, from the point of view of the inconvenience the works cause to normal business and city life. The public funds set aside have proved adequate and have until now sustained Insula's planning, within the framework of a rolling three-year plan approved annually by the City Council.

Collaboration with the offices and organizations that must examine and approve the projects is effi-cient and has not led to any disputes that might delay the opening of building sites or the proper execution of the works. In particular, there is continuous collaboration with the design offices of the utility companies, with the City Council and the Superintendency for Architectural Heritage, which exercises general supervision over all works in the Venice area. Experts from the Archaeological Office are constantly present on site when finds of different types are uncovered during works (building structures, ancient canal banks, pottery or other material).

In recent years Insula has acquired technical and organizational skills in maritime works and water sites that can be well defined as rare and extremely precious. In addition to the work in design, site management, accounting and inspection it carries out on the building site, Insula has put together a complete territorial information system regarding the city and the works that have been completed, offering a wide range of services to all, institutions and citizens, using direct web technology or more technical and sophisticated forms. Finally, work has started on the protection and integrated maintenance of the island and residential area of Burano, where advanced technical solutions in various spheres will be experimented.

The organizational, financial and technical solution adopted by the Venetian community to provide a prompt remedy to the serious situation of deterioration and inefficiency of the historic urban system seems, in the light of experience, to be very effective. It can therefore be hoped that during the next ten years the inherited delay will have been entirely made up for; moreover, planned maintenance, with measures of a different nature, but complementary and necessary to protect the physical integrity of the city (such as strict regulation of wave motion and technical and organizational innovations in water transport vessels and management) should be able to keep the entire system under control in the future. We proceed confidently, therefore, committed to ensure for our city the boundless future that inspired the builders of Venice and the tireless magistrates that conserved its form and substance so that present generations might experience and admire it. From this point of view, urban maintenance is an unavoidable and extremely challenging daily responsibility.

REFERENCES

Ammerman, A. J. 2001. *Venice before San Marco, Recent Studies on the Origins of the City*. Exhibition and Conference, Colgate University, Hamilton, New York.

AAVV. 2002. *Insula Quaderni* **13**.

Calabi, D. 2000. Acqua e suolo. In *Tra due elementi sospesa. Venezia, costruzione di un paesaggio urbano*. Venice: Marsilio, pp. 53–97.

Calabi, D. 2003. Definire il limite a Venezia in età moderna. *Insula Quaderni* **17**, 7–12.

Concina, E. 2000. *Tra due elementi sospesa. Venezia, costruzione di un paesaggio urbano*. Venice: Marsilio, 15, 17.

Dorigo, W. 1983. *Venezia origini: fondamenti, ipotesi, metodi*. Milan: Electa. (For hypotheses and interpretations regarding the birth of the lagoon settlements.)

Dorigo, W. 2003. *Venezia romanica*. Sommacampagna: Cierre, pp. 117–65.

Fig. 19.1   Lack of maintenance can lead to serious instability phenomena, even to the collapse of canal walls. This is what happened in the Rio di San Barnaba in the summer of 2003.

Fig. 19.2   An unknown city, with its submerged architecture, is revealed when a canal is dried out: here repairs are made by injecting mortar to restore the waterproof qualities of the canal wall (Rio della Pergola, May 2003).

Fig. 19.3   Working in Venice still requires a man's hand and the experience of an artisan, assisted when it is possible by modern technologies: in the two illustrations, the reconstruction of a crumbling canal wall and the insertion of a block of Istrian stone in the embankment (Rio del Mondo Novo, November 2000 and May 2001).

Fig. 19.5 (*above*)   Rehabilitation also includes improving and resetting underground grids, which are not easy to lay out in a city dominated by water, where land is precious (Rio del Gaffaro, Fondamenta Minotto, June 2002).

Fig. 19.4 (*left*)   Restoration work on embankments is the opportunity to intervene in the sewer systems in the areas on the canals: here work is being done to renew a canal drain (Rio del Malcanton, December 2001).

Fig. 19.6 Bridge restoration sometimes even includes reconstruction of the vault when the structure has serious static defects (Rio di San Provolo, Ponte dell'Osmarin and Ponte dei Greci, June 2003).

Fig 19.7   Bridges allow pedestrian passage from one Venetian islet to another, and also accommodate utility lines: here is an unusual view of infrastructure that is normally unseen. Placing such infrastructure requires careful design and planning with the combined efforts of all the utility companies (Rio and Ponte della Croce, October 2001).

20 • Methodologies for the functional restoration of a historic urban system

I. TURLON

INTRODUCTION

The overall programme of repairs to the embankments of the canals of Venice, given the peculiarity of their construction and the historic value of the way they were built, has demanded a design approach focussed simultaneously on conserving the artefacts and restoring their proper function. It is vitally important to know the object that we are working on: we analyse its history, the state of the material of which it is made, its geometry and its function. We interpret the differences attributable to the alterations in these elements, characteristic of the original construction, as 'pathologies', to which we respond with corrective work, providing an 'improvement' in the structure's condition sufficient to satisfy its contemporary function and conserve its historic value. We examine the geometry of the structures (translation, rotation, flexure), the state of conservation of the materials that compose it, the physical and mechanical functions for which it was built and those for which it is still efficient. The work to be done is defined as a response to the level of planned maintenance. The techniques and technologies adopted for the repairs spring from the ancient art of building, involving manual work and the use of materials that are compatible with each other chemically, mechanically and aesthetically. Corrective work may involve both geometry and matter or function, individually or in combination.

THE BRIDGES

Specific problems arose in the structural and functional restoration of some arched stone or grey cast iron bridges. In the cases presented here, the necessity of keeping the artefact in the same spot, maintaining the essence of its structural appearance, in addition to the necessity of restoring the function of the arch, badly affected by the traumatic introduction of the old water and gas distribution system, required the use of advanced technologies such as composite fabric fibres.

Flooding and Environmental Challenges for Venice and its Lagoon: State of Knowledge, ed. C. A. Fletcher and T. Spencer.
Published by Cambridge University Press, © Cambridge University Press 2005.

CORONA CAST IRON BRIDGE CONSTRUCTED IN 1851

The bridge (in its deteriorated state in Fig. 20.1) was restored applying AFRP (Aramidic Fibre Reinforced Polymers). After a mechanical cleaning of the bridge, an anti-corrosion primer was laid (Fig. 20.2), then the fabrics were cut, impregnated and stuck to the walls of the arches (Figs. 20.3 to 20.5). The resin was polymerized and some pieces were clipped and shaped, and then the pigmented resin was laid in anthracite grey.

RUGA BELLA STONE BRIDGE CONSTRUCTED IN 1800

The bridge (damage to the vault due to insertion of utility lines (Fig. 20.6)) was restored using unidirectional technique carbon fibre bands. The following figures show the different stages in the work: laying epoxy mortar (Figs. 20.7, 20.8) to form a level surface, a fundamentally important operation for the correct functioning of the carbon fibres (Figs. 20.9–20.12).

21 • St Mark's Basilica as a case study in flooding issues for historical Venice

E. VIO

ST MARK'S BASILICA AND THE *ACQUA ALTA*

St Mark's Basilica is the oldest building in Venice. Its structure bears evidence of the decay produced by flooding waters. The flooding of historical Venice is an event that has been occurring for centuries, connected to the high tides of the lagoon during the equinoctial seasons, especially in the months of February and November. The earliest documentation on floods are found in the thirteenth century, during the years 1282, 1284 and 1297. For the fourteenth century, the State Archives record the floods of 1314 and of 9 December 1386, the last being described as '...very high water, eight feet higher than usual'. With regard to the fifteenth century, there are records of floods in 1410 that submerged the whole city; very high water in 1423, very high water on 2 March 1429, causing pollution to the wells; floods on 10 December 1430; and in December 1464. In the sixteenth century, floods were recorded 3 October 1535; on 21 November 1551; and on 12 October 1574, during which it was possible to use boats in St Mark's Square and the Calle dei Fabbri. In the seventeenth century, during the floods recorded on 21 November 1609, 7 December 1660 and 5 November 1686, it was possible to navigate all over the city and use gondolas between the Square and the area of the Mercerie. During the eighteenth century, there is a record of a very important flood on the 21 December 1727, with a level of water that reached the steps of the main altar of the church of St Antonin (about 2 m above mean sea level). Other floods of the eighteenth century were recorded on 31 October 1746, 9 November 1750, 14 November 1782 and 8 March 1783, the last being regarded as exceptionally high. In the twentieth century, the flood of 4 November 1966 inundated most parts of the entire city and the interior of the Basilica, the latter for the first time in history, reaching a level 1.92 m above mean sea level.

These high water events have been superimposed over the long-term subsidence of the ground level of the city. This has been a continuous phenomenon which has been witnessed since Roman times, with an average lowering of 14 cm per century. This quantity has increased or decreased over time, according

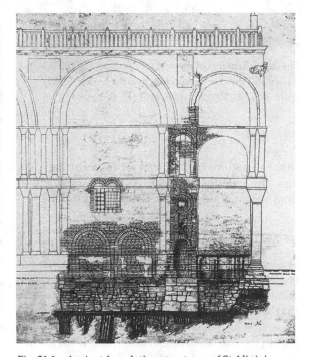

Fig. 21.1 Ancient foundation structures of St Alipio's corner of the Basilica: the brick floors (spicatum) and their levels are shown (PSM Archives).

Flooding and Environmental Challenges for Venice and its Lagoon: State of Knowledge, ed. C. A. Fletcher and T. Spencer.
Published by Cambridge University Press, © Cambridge University Press 2005.

Fig. 21.2   Remains of the first establishment of the Ducal Palace and the chapel built to house the remains of the body of St Mark (PSM Archives).

to the variations of the mean sea level during each age. The levels of the Basilica's floors and foundations show the subsidence of the ground surface.

ST MARK'S BASILICA

St Mark's Basilica, as chapel of the Doge of Venice, is an important part of the history of the city. It's architectural form is derived from eastern models that were adapted to pre-existing western ones. Thus St Mark's Basilica is one of the very few buildings in the world in which eastern and western cultures are expressed and live together. The Basilica was built in AD 829-32, for housing the body of St Mark. Dorigo proposes that evidence of the first

chapel can be found in St Mark's crypt from which the original plan form is derived. The comparison of the levels surveyed of the Ospizio Orseolo and the St Mark's bell tower with the levels of the crypt confirms the date of the construction of the first church. In AD 976 the Basilica burnt down; it was restored in 978 and then rebuilt in 1063–94. Between c. 1205 and 1265 the building was clad with stone and marbles brought to Venice from Byzantium. In 1563 the crypt floor was raised. However, it stopped being used in 1580 because of high tides.

The Republic of Venice came to an end in 1797 and in 1807 Napoleon separated the Basilica from the Doge's Palace. In 1866 Venice was annexed to Italy and in 1869-71 the Crypt was restored and re opened

Fig. 21.3   High tide in St Mark's Square, about 1 m above sea level. The narthex is flooded with 30–32 cm of water (photograph by the author, 1983).

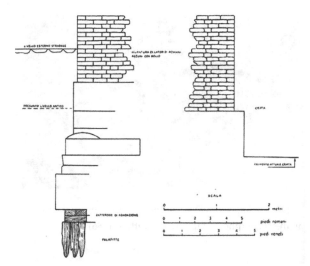

Fig. 21.4   Section though the present crypt and its foundations in which a stone from the Roman period, sculpted with the disc of the sun, was found (PSM Archives, 1974).

to use after an interval of almost three centuries. The 4 November 1966 floods inundated the whole Basilica, except for the Presbytery, for the first time in its history. In 1986–94 the Crypt was totally restored.

ST MARK'S BASILICA AND THE TIDES OF THE VENICE LAGOON

The ground subsidence in Venice has been calculated at 14 cm per century. The phenomenon is clearly visibile from the levels of the ancient brick paving floors in St Mark's Square. The subsidence is mainly shown in the lowering of the level of the most ancient buildings of the city, first amongst all in the Basilica of St Mark. Twelve centuries after its construction, the floor of the crypt of the Basilica is now 168 cm lower than it was in AD 829. During the restoration of the NW corner of the facade of the Basilica, the old foundations of the third church and the brick floor were revealed. The lowest point of the narthex of the Basilica today is 67 cm above mean sea level. This very low level causes the narthex to flood, either in whole or in part, between 150 to 180 times every year. Recently the number of floodings per year has reached 200–230. The levels of the interior floors of the Basilica vary between 178 and 188 cm above mean sea level; the Presbytery floor level is 80 cm higher. The crypt shows the first church's remains in St Mark's tomb, in the floor's ancient parts and in some of the walls. The level of the crypt floor is 20 cm below mean sea level and the sub-crypt floor is at -57 cm.

Fig. 21.5   Survey of the Basilica's EW section showing the crypt's levels and shape (PSM Archives, 1992).

THE RISING DAMP IN THE BASILICA'S BRICK WALLS

The structure of the Basilica is made of clay bricks, its columns of Greek marble and its internal and external walls clad with stone and marble panels. Rising damp in the walls reaches a height of several metres. The presence of salts in the crypt walls is very high, reaching 2.5 per cent.

The most common of these salts is sodium chloride. The relative humidity of the air reaches minimal levels of 32 to 35 per cent in winter and maximum levels of 98 to 99 per cent in summer. These variations cause the deterioration, flaking and powdering of bricks and mortar. The salts solidify at a relative humidity level below 72 per cent. This phenomenon is particularly critical for the tiled surfaces as it affects the mortars of the mosaics as well as the stone tiles themselves, causing the materials to fall.

INTERVENTIONS ON THE WALLS

The ancient Roman bricks of the type Sesquipedale (dimensions: 30 × 40 × 8 cm) are bound with mortar made with slaked lime and ground bricks of different grains (coccio pesto).

Decayed mortars are consolidated with injections of water-based solutions of resins with the addition of lime. The elimination of salts and rising damp are major problems. One of the difficulties is the taking

Fig. 21.7   Lagoon water coming out from the brick seats in the crypt (photograph by the author, 1986).

down of the stone cladding of the walls. In the past, the habit was to demolish the old cladding and build a new one but the last intervention of this type was carried out in 1865–75 in the Baptistry.

INTERVENTIONS ON THE MOSAICS

In the past, restoration meant substitutions and modifications to the building. But due to John Ruskin's strong opposition to this behaviour, with regard to the restoration of the Baptistry, nowadays the Basilica

Fig. 21.6   Water present in the crypt walls (photograph by the author, 1986).

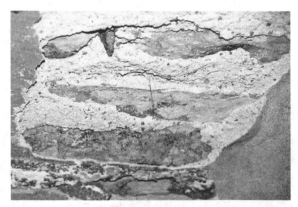

Fig. 21.8   Crypt mortar, named 'coccio pesto', made with slaked lime and ground bricks of different grain types (photograph by the author, 1992).

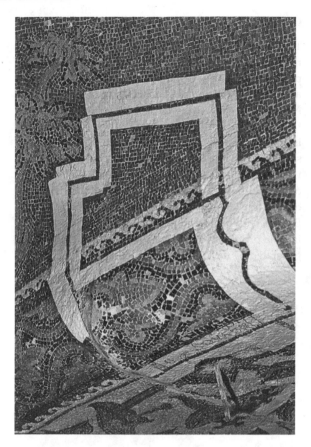

Fig. 21.9   Identification of a mosaic's deteriorated area requiring restoration (photograph by the author, 1989).

is regarded as a museum of architecture itself and every piece of the building is considered as worth respecting and preserving for the future.

As regards deteriorated mosaics, in the past they used to be remade. From the second half of the eighteenth century, however, they were removed from walls in patches of 40 × 40 cm, restored in the laboratory working from the back, then re-attached to the walls after the restoration of the mortar and the bricks. This procedure was applied to about a quarter of all the tiled surfaces of the Basilica. From the end of the eighteenth century, restoration of the mosaics was preceded by making impressions on paper or gypsum moulds of the deteriorated areas; the paper impressions shown during the international expo in Paris in 1888 are well-known examples of this process.

Now, except in the case of the total decay of the mortar, the removal of the mosaic layer is not undertaken anymore; rather they are consolidated with injections of water-based solutions of lime, acrilic resins and marble powder. Since the 1930s, deteriorated wall bricks have been substituted while keeping the mosaics in place, by means of special scaffolding.

Still not resolved are the problems connected to the rising damp in the narthex walls, reaching up to 6 metres in height, that cause the powdering of the mosaics.

THE CRYPT: ITS STRUCTURE, THE DETERIORATION, THE RESTORATION

The Crypt is the most ancient part of the Basilica and contains the remains of the first church. In the crypt evident damage has been produced not only by flooding water but also by the solutions that have been attempted in order to reclaim the crypt. Recent interventions have been carried out with the aim of eliminating all the cavities and the hollows existing in the crypt foundations.

Waterproofing resins were injected into the foundations. Once finished on the foundations, the works moved to the walls. The parts of the mortar bonds deteriorated by the salts were removed and substituted by a resin-based mortar capable of blocking the outward migration and crystallization of the salts contained in the wall, and the consequent damage to its surface. The porosity of bricks maintains the permeability of the walls. Venice is characterized by a wide variation of relative humidity between summer and winter and consequently the crypt is very dry in winter and very humid in summer. The main problem in the reclamation of the crypt has been to attain a balance between the water content of the air and that of the walls. In a situation of wide imbalance the walls take in humidity when the air contains a lot of water. When relative humidity is greater than 72 per cent, the salts in the walls dissolve and migrate to inside whereas when the air is dry, the humidity of the walls transfers to the air, carrying with it the salts which consolidate and damage the bricks.

Fig. 21.10   Salt efflorescence on brick surfaces
(photograph by the author, 1988).

CONCLUSIONS

The subsidence of the city in the St Mark's area has
reached a level which precludes further raising of the
paving level as this would alter the outward appearance of the monuments, particularly of the Basilica.
New paving also creates damage to those parts of the
facades which would be covered. Nevertheless, there
is an urgent need to find solutions to avoid the construction of new layers and to block the rising of sea
water in the walls. This will undermine the stability
of the building itself and threaten the condition of its
decorations and mosaics. Finding a solution for the
Basilica requires learning ways of effectively addressing the problems of Venice's historical architecture
generated by the tides of the lagoon.

22 • Challenging transient flooding effects on dampness in brick masonry in Venice by a new technique: the narthex in St Mark's Basilica

F. SANDROLINI, E. FRANZONI, E. VIO AND S. LONARDONI

INTRODUCTION

St Mark's Basilica has always shown peculiar dampness problems, particularly caused by water seepage and rising moisture. This leads to continuous decay of the ancient brick masonries due to effects which include soluble salt migration and crystallization in the pores of the material, efflorescence formation, freeze and thaw effects, chemical and biological attack and exterior erosion (Camuffo, 1995; La Iglesia *et al.*, 1997; Mulvin and Lewis, 1994; Saiz-Jimenez, 1993; Seaward, 1997; Sandrolini *et al.*, 2000). Considering the uniqueness of the internal wall and ceiling finishings (frescoes, mosaics, etc.), only non-damaging and non-disfiguring reclamation methods can be applied, leading to the search for newer and newer anti-dampness materials and techniques. Furthermore, any anti-dampness system is required to be reversible i. e. it can be removed when desired.

Unfortunately, most dehumidification techniques currently used in historical buildings seem unsuitable for application in the Basilica. Chemical cutting by resin injection gives a poor guarantee of effectiveness, due to the large masonry thickness and in any case it cannot be considered to be fully reversible. For the same reasons, mechanical wall cutting techniques are almost impossible (both sides of the brick masonry are seldom accessible due to the presence of marble structural elements and/or ornaments) and in any case it increases the seismic vulnerability of the structure. Active electro-osmotic systems are very difficult to place underground and it is very difficult to distribute and hide the governor apparatus in the architectural bodies of the Basilica.

Since 1998, some brick masonry of the narthex of St Marks in Venice has been subject to experimentation to reduce or eliminate rising dampness. This has included a new, simple and fully reversible technique based on spontaneous and automatic governing of the electrokinetic phenomena arising from water migration in porous building brickworks (Sandrolini, 1999; Sandrolini and Franzoni, 1999). This method does not require any external energy supply or controlling apparatus and has been successfully tested in several ancient buildings against rising dampness (e.g. baroque porticos of San Luca, Bologna; nineteenth-century church of the Blessed Crucifix of the monumental Cemetery, Ravenna; baroque church of St Caterina of Italy in La Valletta (Malta); and others (Sandrolini *et al.*, 1999; Sandrolini and Franzoni, 1999)).

PRINCIPLE OF THE RECLAMATION METHOD

Capillary rise height

Dampness in brick masonry rises through material capillaries primarily through adhesion forces between water and capillary surfaces that exceed the cohesion forces in the water itself. This effect leads to a steady-state rise of water in capillaries and yields concave water surfaces at the water-air interface. In ideal, cylindrical capillaries of radius r, after the Laplace treatment of the water surface meniscus formation (Dullien and Batara, 1970; Rostagni, 1951), the theoretical water-rise height, h, is found to depend on the contact angle (θ) between water and capillary surfaces, water density (ρ) and surface tension (τ) according to the Jurin law:

$$h = \frac{2\tau \cos \theta}{\rho g r} \qquad \qquad Eq.\ 22.1$$

Flooding and Environmental Challenges for Venice and its Lagoon: State of Knowledge, ed. C. A. Fletcher and T. Spencer. Published by Cambridge University Press, © Cambridge University Press 2005.

where g is the gravity acceleration (Amoroso and Fassina, 1983; Rostagni, 1951).

In the case of most building material/water interactions, substitution of τ, θ and ρ values for water leads to the simple equation:

$$h_{max} \approx \frac{15}{r} \qquad \qquad Eq. \ 22.2$$

where h_{max} is in m if r is given in μm. Thus if r is 1 to 10 μm, h_{max} spans 15 to 1.5 m respectively. Of course, in actual masonry an equilibrium between capillary water suction and evaporation at the water meniscus is attained and the water rise stops at an 'equilibrium line' which can be easily observed in any brick masonry subject to water capillary rise. Similarly, this means that a steady-state water migration occurs in brick masonry proportional to $t^{1/2}$ (where t is the time), depending on the sorption coefficient (or sorptivity) S of the building masonry, which, in turn, depends on the pore volume fraction (both local and average) and material permeability according to Darcy's law and the water dynamic viscosity η, besides the other parameters of the Jurin law (Amoroso and Fassina, 1983; Atzeni *et al.*, 1991). In this way the microstructural complexity of the brick masonry may be taken in account to some extent.

Electrokinetic effects

Stable electrical double layer phenomena arise at any aqueous electrolyte/solid interface and hence electrostatic phenomena involving both solid and liquid phases i. e. stable electrical polarization in both phases, related also to the dielectric constant (Creighton, 1948; Glasstone, 1956; Wulff, 1964). The application of an electric field to such a system may result, therefore, in a relative motion of the two phases, such as the fluid in the pores of a porous solid (electro-osmosis) or solid particles in an aqueous solid suspension (electrophoresis). These effects are exploited in specific industrial operations (Pisarenko *et al.*, 1996; Thony *et al.*, 1997). As rising dampness typically comes from several contacts with rocks, soils and polluted environments, it usually contains several salts, even if one neglects the water dissociation itself. Therefore, at any water/capillary interface

in porous solids, such as brick masonry, a more or less noticeable, but spontaneous polarization usually develops at the phase interfaces as a function of type, concentration and mass diffusivity of the ions in the aqueous solution, thus resulting in spontaneous polarization along the capillary path.

Principle of the method

The method (DLK, 2003) provides for the insertion of conducting bars (usually cylindrical bars of steel or stainless steel, electrically insulated by a polymeric sheath) in the water-flooded zone of the brick masonry (Fig. 22.1). The conducting bars immediately undergo an electrostatic induction from the steady-state, self-polarised charges in the moist masonry, which in turn provides an electrical field opposing the self-polarization in capillaries. Then, the electrostatically induced field opposes, according to basic chemical, physical and electrical laws (Le Chatelier, Lenz), the self-polarization in the water solution, thus contrasting the spontaneous capillary water rising effect. Length, slope and relevant bar distances are arranged to obtain the maximum counterbalancing effects in the brick masonry to be treated. No external energy supply or controlling apparatus is needed and the system is clearly reversible. It may also reinforce the brick masonry against lateral bending stresses by calculating the proper bar diameter and configuration (Sandrolini and Diotallevi, in progress): this possible reinforcement effect is very important when the brick masonry is not properly constrained at the summit. Finally, the system exhibits a clear self-regulating behaviour, mainly depending on the water-rise itself according to the above quoted basic laws.

ST MARK'S NARTHEX EXPERIMENTATION

The reclamation system and on site dampness measurements

St Mark's Basilica has undergone several restoration works since its construction; however, they are not entirely known today, in spite of the large corpus of archive information (Forlati, 1975; Proceedings, 1995; Vio, 1995, 1999). The experimentation of self-polar-

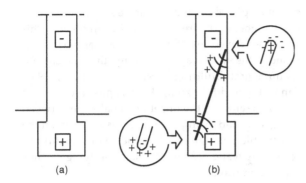

Fig. 22.1   Self-polarization in the masonry (a) and conducting bar electrostatic effect (b).

ization method in the right front narthex of St Mark's began in 1998, with reference to the placing scheme of Fig. 22.1 and plan of Fig. 22.2.

The experimentation zone (Fig. 22.2) was chosen as it bordered (on the right) onto the Zen Chapel, where serious damages recurrently occurred on walls, thus spoiling the existing ancient mosaics. In particular, zone 1 was left exactly in its former state as a control and zones 2 and 3 underwent the placing of the anti-dampness system, in both cases by the insertion of 2 m long AISI 316 stainless steel bars, coated by a vulcanised rubber sheath. The former (zone 2) was equipped with a double, and the latter (zone 3) with a single, order of bars. The placing procedure involves a double order of steel bars when the wall thickness is very large and its structure is not well known, to ensure electrokinetic effects as large and extended as possible.

It is important to note that dampness evolution was previously investigated, by different personnel, via the periodic withdrawal of samples of brick powder left in holes, made with a low rpm electrical drill at several wall heights. The samples were then dried in an air oven to constant weight to obtain the moisture content. After careful laboratory experimentation, since 2002 a new sampling and measurement protocol, purposely designed to avoid any systematic measurement error, was applied to new holes by the present authors. It was noticed that the moisture content of powder samples was much higher than that of the same solid brick in the same thermal/hygromet-

ric conditions, after the first measurement. This effect is connected to the actual masonry brick microstructure, which really controls the so called 'equilibrium' moisture of the material. The brick powder withdrawn from the holes exhibits an overall surface area much larger than the original brick sample; this leads to larger water absorption on the brick powder grains and hence to a water content that is systematically higher than that of the original brick (Franzoni and Sandrolini, 2003; Sandrolini and Franzoni, in press) after the first measurement.

In the new experimentation, samples of the 'same' solid brick where the new holes were made were introduced and sealed within the holes themselves, to bring them to the same equilibrium moisture content as the surrounding solid brick. Figures 22.3 and 22.4 show the heights of the new sampling holes: the reported brick masonry structure is only indicative, due to the lack of information about it.

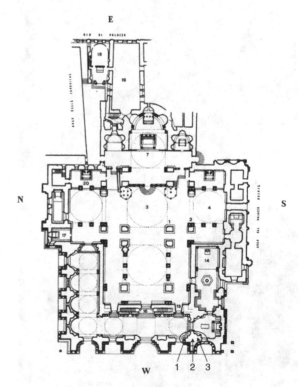

Fig. 22.2   Self-polarization system experimentation zones (1–3) at the St Clemente gate.

The results so far obtained

The anti-dampness technique exhibited a good control of the rising dampness, as can be seen from Table 22.1. In the first measurements only surface fragments, of course, could be sampled, together with hole powder. The results indeed point to a large and meaningful difference between the moisture contained in the samples withdrawn from sampling points 1 (treated) and 4 (not treated). However, in the narthex investigations a large moisture content at sampling points 2-3 and 5-6 can be seen in Table 22.1, much higher in the treated zone than in point 1. This is an unexpected result, which so far has never been found in the system application in other ancient and monumental buildings (Sandrolini *et al.*, 1999; Sandrolini and Franzoni, 1999). A possible cause of such behaviour is the onset of a challenge between rising and condensing dampness, as a survey on the practicable passage in the upper baluster did not reveal any water seepage in the underlying narthex ceiling and wall.

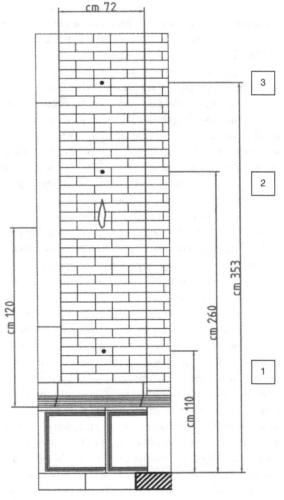

Fig. 22.3   Height of sampling holes for the treated area (zone 2) (schematic).

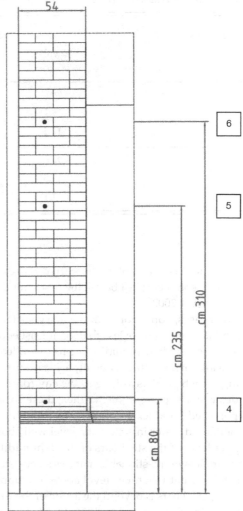

Fig. 22.4   Height of sampling holes for the non-treated area (zone 1)

Flooding, condensing moisture and electrokinetic effects

The problem of rising dampness in Venice appears to be complicated by the tidal flooding effects (*acqua alta*) on the pavements of the town. The lowest point in Venice is in St Mark's Square which is only 0.64 m above sea-level. The investigated narthex offers a shelter to flooding, raising the level of the St Clemente gate to *c.* 0.74 m, but its atrium level is well below the St Clemente threshold: therefore, during high tides it is completely flooded.

This may reasonably be the cause of the formation of the condensation humidity at high level in the narthex, where the now placed self-polarization system cannot act, simply due to its insufficient height and/or not suitable configuration, being designed and placed to address only the rising humidity. Indeed, the seawater frequently remains for long periods of time within the investigated area of the

Table 22.1   *Moisture content (% weight) of samples in different settings.*

| Sampling points | 30/09/2002 | | 18/02/2003 |
	Powders	Brick surface fragments	Hole cavity brick fragments
1	2.0	2.4	3.4
2	18.1	13.5	19.6
3	14.5	8.5	17.4
4	24.0	18.3	19.2
5	14.5	10.1	13.9
6	11.1	11.3	13.4

narthex, almost as in a pool; this occurred for seven days (21–27 September) just before the first sampling on 30 September 2002.

In such cases, some zones in masonry surfaces may exhibit a temperature less than or at least equal to the dew point, where humidity begins to condense on the wall surface, thus causing a local surface flooding, which persists as long as the environment provides water vapour. In the narthex, these conditions persisted for at least a week, thus providing a true condensed water injection at a level well above the upper level of the steel bars of the dehumidifying system. It is unquestionable that this steady state water injection at whatever level above the upper level of the steel bars is beyond the action of the electrostatic field of the system as it is placed, and can act exactly as a dampness feed from the top of the wall, while the rising dampness is fed from the bottom of the wall. Therefore, it cannot be contrasted nor mitigated by the conventional configuration of the dehumidifying system, placed only according to the specifications against rising dampness.

An electrokinetic system of the kind described here should surely operate in cases of large condensation surfaces like those found in Venice, due to the suction effect of pores of the smallest radius, according to the Jurin law; while the largest radius pores should only allow drop of the condensed water under gravity action, with a possible connection with the rising one. Of course, to properly configure and place the conducting bars to hinder the condensing humidity and, at the same time, the rising dampness, requires, first, further studies on the characteristics of the onset of the condensing humidity, and, secondly, finding new design criteria, as well as experimentation on specific cases.

CONCLUSIONS

From the investigation so far performed, the following conclusions can be drawn:

• Ancient and monumental buildings in Venice can suffer not only from current and distributed rising dampness and water seepage but also from condensing dampness problems caused by periodical tidal flooding effects (*acqua alta*).

• At present, no short-term and permanent provisions (such as the 'MOSE' protection barriers and so on) against tidal flooding effects (*acqua alta*) are foreseeable and hence against condensing moisture effects: local and distributed electrokinetic systems against both rising and condensing humidity should be welcome, to prevent permanent, heavy damage of the architectural heritage of Venice.

• Experiments with the self-polarizing method based upon the insertion of conducting (steel) bars (if properly configured and placed) will be very helpful, ensuring at the same time reversibility, self-regulation operation without any external power supply, possible structural reinforcement of the brick masonry and absence of the changes and/or losses in the architectural image of the ancient buildings. All the controls of the effectiveness of the new proposed technique should be made by the new sampling and measurement protocol quoted above (Franzoni and Sandrolini, 2003; Sandrolini and Franzoni, in press), with particular reference to the onset of condensing moisture at high levels above sea level.

• A fully multidisciplinary approach (chemical–physical, environmental, structural, architectural, restoration and conservation) to the Venice dampness problems is necessary to help and safeguard the architectural heritage of Venice, particularly while the city awaits a final and global solution to its dampness problems.

This goal can be attained only by new and multidisciplinary experimentation on both aspects of the whole problem, case studies of specific environmental problems and designing criteria for this new self-polarizing electrokinetic technique.

REFERENCES

Amoroso, G. G., and V. Fassina. 1983. Stone decay and conservation. *Materials Science Monographs* **11**, 18. 34 fols.

Atzeni, G., Massida, L., and U. Sanna. 1991. Considerazioni sull'applicazione della legge di Darcy nello studio dell'assorbimento capillare in rocce porose. In *Proc. Congr. 'Scienza e Beni Culturali Le pietre nell'architettura: struttura e superfici', Bressanone (Italy) June 25th – 28th 1991*. Palma: Liberia Progetto Editore, pp. 203–13.

Camuffo, D. 1995. Physical weathering of stones. *The Science of the Total Environment* **167**, 1–14.

Creighton, H. J. 1948. *Principles and applications of electrochemistry, Vol. I:Principles*. New York: John Wiley and Sons Ltd, pp. 143–65.

DLK. 2003. Verona (Italy), www.dlk2002.it.

Dullien, F. A. L., and V. K. Batara. 1970. Determination of the structure of porous media. In *Flow Through Porous Media*. Am. Chem. Soc. Publ. 17. Washington, DC: Am. Chem. Soc.

Forlati, F. 1975. *La Basilica di San Marco attraverso i suoi restauri*. Trieste: Ed. LINT.

Franzoni, E., and F. Sandrolini. 2003. Problematiche relative all'umidità nelle murature storiche ed alla sua misura. *INARCOS* **637**, 153–7.

Glasstone, S. 1956. *An Introduction to Electrochemistry*. Princeton: Van Nostrand, pp. 521–45.

La Iglesia, A., Gonzales, V., Lopez-Acevedo, V., and C. Viedma. 1997. Salt crystallization in porous construction materials I. Estimation of crystallization pressure. *Journal of Crystal Growth* **177**, 111–18.

Mulvin, L., and J. O. Lewis. 1994. Architectural detailing, weathering and stone decay. *Building and Environment* **1**, 113–18.

Pisarenko, D., Morat, P., and J.-L. Le Mouel. 1996. On a possible mechanism of sandstone alteration: evidence from electric potential measurements. *Compt. Rend. Acad. Sci. Paris. Ser. IIa* **322**, 17–24.

Proceedings. 1995. *Proc. Conv. Int. di Studi. 'Scienza e Tecnica del Restauro della Basilica di San Marco' Venezia, May 16–19th 1995*. Venice: Ist. Veneto di Scienze, Lettere ed Arti.

Rostagni, A. 1951. *Meccanica e termodinamica*. Padua: Libreria Universitaria G. Randi, 480 fols.

Saiz-Jiminez, C. 1993. Deposition of airborne organic pollutants on historic buildings. *Building and Environment* **1**, 77–85.

Sandrolini, F. 1999. Come invertire la polarità per prosciugare le murature. *Bioarchitettura* **15** (Oct.), 50–1.

Sandrolini, F., and E. Franzoni. 1999. Un nuovo sistema di deumidificazione muraria reversibile e non invasivo: sperimentazione nel Cimitero Monumentale di Ravenna. *Tema (Tempo Materia Architettura)* **2**, 39–46.

Sandrolini, F., and E. Franzoni. In press. A new operating protocol for reliable measurements of moisture in porous building materials.

Sandrolini, F., Cuppini, G., and E. Franzoni. 1999. La diagnostica come fondamento dei protocolli (manuali) di recupero: il caso di S. Caterina degli italiani a La Valletta. *Proc. II Congr. 'Materiali e tecniche per il restauro'. Cassino, Oct. 1st –2nd 1999*. Cassino: Idea Stampa Ed., pp. 243–57.

Sandrolini, F., Franzoni, E., and G. Cuppini. 2000. Predictive diagnostics for decayed ashlars substitution in architectural restoration in Malta. *Materials Engineering* **11**, 3, 323–37.

Seaward, M. R. D. 1997. Major impacts made by lichens in biodeterioration processes. *International Biodeterioration & Biodegradation* **40**, 2, 269–73.

Thony, J.-L., Morat, P., Vachaud, G., and J.-L. Le Mouel. 1997. Field characterization of the relationship between electrical potential gradients and soil water flux. *Compt. Rend. Acad. Sci. Paris. Ser. IIa* **325**, 317–21.

Vio, E. 1995. Il cantiere marciano: tradizioni e tecniche. *Proc. Conv. Int. di Studi. 'Scienza e Tecnica del Restauro della Basilica di San Marco' Venezia, May 16–19th 1995*. Venice: Ist. Veneto di Scienze, Lettere ed Arti, 79–141.

Vio, E. 1999. *La Basilica di San Marco a Venezia*. Florence: SCALA Group S.p.A.

Wulff, J. (ed.). 1964. *The Structure and Properties of Materials, Vol. IV*. New York: John Wiley and Sons Ltd, pp. 251–62.

23 • Venice high water protection measures in St Mark's Square

M. T. BROTTO

SAFEGUARDING VENICE AND ITS LAGOON

The Italian Republic has defined the problem of safeguarding Venice and its lagoon as of 'primary national interest'. Contributing to safeguarding the lagoon are: the State, responsible for defending Venice, Chioggia and other urban areas from high waters, protecting coastal areas from sea storms and restoring the environmental balance of the ecosystem; the Veneto Region, responsible for pollution abatement; and the local authorities, responsible for socio-economic development, maintenance and restoration of the architectural and built fabric. The system of safeguarding measures is directed, coordinated and controlled by a committee, chaired by the President of the Council of Ministers, consisting of representatives of local and national authorities and institutions. The plan to protect the St Mark's area from the most frequent flooding is described in this chapter. This plan was drawn up by the Ministry of Infrastructure and Transport - Venice Water Authority through the Consorzio Venezia Nuova[1] and is part of the General Plan of Interventions delegated to the State. This is the main instrument for planning and financing the safeguarding work. The local defence of St Mark's Square (Fig. 23.1) is an integral part of the system of measures to protect the city and lagoon from flooding which includes the MOSE[2] system to defend the entire lagoon area from all high waters, including extreme events.

LOCATION

The lowest-lying areas of Venice, usually the oldest, are subject to the most frequent floods. Particularly liable to high waters is the 'insula' of San Marco, once the political, religious and administrative heart of the Serenissima, today symbolic of the city itself in the eyes of the world. This is an exceptionally evocative architectural and monumental area, containing the city's most precious and representative buildings. These include the Basilica, constructed at the beginning of the eleventh century following Byzantine models and partly modified during subsequent centuries with the addition of significant decoration and ornamentation; the Ducal Palace, residence of the Doge and seat of the government and magistracy, built as a fortress with a square plan and transformed and extended between the fourteenth and fifteenth centuries to become one of the most important examples of Gothic art; the Procuratie Vecchie and Procuratie Nuove, once the seat of the procuratori of San Marco, the highest functionaries in the Republic after the Doge and the Biblioteca Marciana, an important example of Renaissance architecture.

HOW THE SQUARE IS FLOODED TODAY

The problem of high waters has worsened considerably over the last century as a result of a rise in sea level (eustatism) and a lowering in land level (subsidence). Together the two phenomena have caused a loss of *c.* 23 cm in the land level in the Venice Lagoon. In recent decades, high water in St Mark's Square has become an almost daily event. About 250 times a

---

[1] Consorzio Venezia Nuova, a consortium between many Italian engineering companies, is the concessionary of Ministry of Public Works in Italy for physical and environmental protection of Venice and its lagoon.

[2] You can find a detailed description in www.salve.com.

Flooding and Environmental Challenges for Venice and its Lagoon: State of Knowledge, ed. C. A. Fletcher and T. Spencer.
Published by Cambridge University Press, © Cambridge University Press 2005.

Fig. 23.1   St Mark's Square.

year, when the tide reaches a height of 60 cm, the water starts to invade the narthex of the Basilica and the paving in front of the entrance. The flooded areas increase as the tide gradually rises. With a tide of 80 cm, the water fills large areas of the square and with 90 cm, almost two thirds of the surface is flooded. With 100 cm (an average of seven times a year), the square and surrounding areas are almost completely submerged (Plate 8). As well as causing problems for the inhabitants of Venice and for socio-economic activities, the repeated high waters damage the paving and subsoil of the square. Deterioration of the stone blocks, the presence of generalized surface damage and the collapse of the underground rainwater drainage conduits are the main symptoms of a widespread degeneration which has accelerated during recent decades. The water invades the square in three ways: flowing over the lagoon bank, rising up through the drains and seeping through the subsoil.

HOW THE PROBLEM IS BEING TACKLED

The general plan

To combat flooding caused by water flowing over the lagoon bank, only the bank itself and adjoining paving will be raised to at least 110 cm, while the level of paving in the square will not be modified. The raising operation will be accompanied by consolidation of the bank to oppose the deterioration caused by wave motion resulting from the intense motorized water traffic. To avoid flooding caused both by flowback through the drains and by seepage through the subsoil, the old system of underground conduits will be isolated from water coming from the small canals surrounding the area. The structural deterioration of the conduits, resulting in the collapse of the paving in many points, will necessitate their complete restoration. At the same time, a new rainwater collection and conveyance system will be constructed, linked to a pumping station to be installed in the Giardini Reali, to enable the water to flow out to the lagoon during high tides. Consideration will also be given to the possibility of locating a waterproof bentonite membrane under the paving in the square to counteract flooding caused by seepage through the subsoil. As part of this work, the network of underground infrastructure will also be systematized, unifying the course of existing lines along established routes. This will simplify future maintenance and avoid the generalized tampering with the paving which occurs today.

Bearing in mind the importance of the St Mark's area, its strategic position and its symbolic value, the entire programme of work will be carried out in such a way as to avoid closing the square, involving a single limited area at a time.

The first phase

The first phase, begun in March 2003, involves a total of 150 m of quayside along the basin. The work includes raising, restoration and reinforcement of the quayside, also with a view to protecting it from wave motion; interception of discharges by construction of a new rainwater drainage system; and restoration of paving. The criteria and procedures for implementation of the plan have been agreed with the Venice Local Authority and the Superintendency for Architecture and the Landscape. The sequence of work has been established in agreement with those involved. The site will proceed with two successive 75 m sections and will include provision of provisional landing stages and protected access for craft.

OBJECTIVES AND CRITERIA

Objectives

To defend the San Marco insula from the most frequent high waters and ensure its accessibility up to +110 cm (with respect to the tidal datum), eliminating the destructive effects of the flooding; to restore the paving in the square and improve the condition of the subsoil.

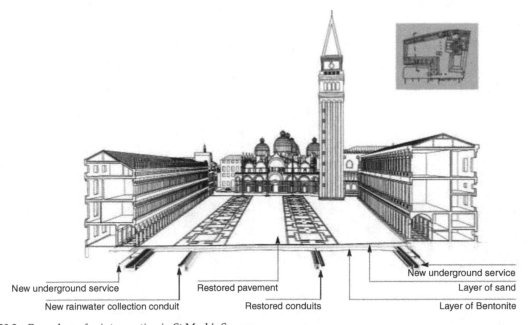

New underground service

New rainwater collection conduit

Restored pavement

Restored conduits

New underground service

Layer of sand

Layer of Bentonite

Fig. 23.2   Procedures for intervention in St Mark's Square.

Constraints and criteria

To draw up, partly through specific study and survey campaigns, a complete, detailed and up-to-date picture of the area, concerned including all elements defining the 'current situation'. To avoid modifying the height of the paving in order to avoid altering the architectural and compositional relationship between the buildings and between the buildings and the paving itself. To avoid compromising the static condition of buildings in the square, avoiding alteration of the relationships established between the foundation land, the presence of water in the subsoil and the stresses produced by the weight of the buildings. To employ conservative restoration techniques.

Studies and surveys

The choice of the general solution and the design of specific interventions has been based on studies, surveys, monitoring and historical-archaeological research involving every aspect of the St Mark's area affected by the project. Activities have involved general planimetric and topographic surveys to define the height of public surfaces (to draw up a map of areas flooded at various tidal levels); specific surveys of paving in the square, lagoon banks and ground floors of buildings affected by flooding to analyse the architectural, constructional and functional characteristics; studies and verification of the subsoil, both to determine the nature and characteristics of the land (composition, permeability, chemical-physical properties) and to assess the condition of areas underlying the square; surveying archaeological elements requiring protection; and identifying the network of old rainwater drainage tunnels and the system of underground infrastructure and verifying their state of preservation. Further specific analyses have involved data on rainfall in order to size the new rainwater collection and drainage system and on water depth near the lagoon bank in relation to consolidation work.

24 • Research for conservation of the lagoon building culture: catalogue of the external plasterwork in Venetian buildings

E. DANZI, A. FERRIGHI, M. PIANA, P. CAMPOSTRINI, S. DE ZORZI AND E. RINALDI

INTRODUCTION

In the context of a broader research project, supported by CORILA and co-ordinated by the Department of Architectural History at the University IUAV of Venice regarding lagoon constructions, accurate historical knowledge of the external plasterwork of Venetian buildings is considered the basis for their successful preservation. The general aim of the research presented in this chapter is to improve knowledge of the building systems in the lagoon culture, in particular to provide methodological criteria for future interventions by evaluating the efficacy of the various measures used to restore plasterwork. In order to supply quantitative and qualitative knowledge a complete survey of the historical centre of Venice was made. The resulting system – still being implemented and tested – is a GIS, at local scale, of the outer coatings of the city buildings, to facilitate rapid census-related activities. In the specific case of the plaster catalogue, 'knowledge' means the systematic collection and management of information, in terms of searchability, accessability and utility.

RESEARCH OBJECTIVES

The broader research on 'protection from high waters and architectural conservation' deals with the following themes:

- history of ancient and contemporary precautions devised against the dangers originated from the dynamics of environmental processes; comparison and evaluation of old and modern building techniques aimed at avoiding or neutralizing decay produced by subsidence, erosion, alterations to foundation systems, structures and materials;

- documents stored in the city's archives, the 'Archivio di Stato di Venezia', report the role of engineers, architects, builders and skilled workers who were involved either in the safeguarding or transformation of the city. They have been previously studied and organized into a database (Foschi and Morresi, 2002);

- identification of the techniques particular to the lagoon, precautions and building methods conditioned by the presence of water; catalogue of deterioration phenomena and identification of the degree of construction sensitivity towards the aggressive phenomena caused by water; study of structural elements which are most damaged by the destruction of the binding elements (Faccio, 2002);

- analysis of the foundation systems of buildings through comparison with information derived from archaeological excavations;

- analysis of below-ground sediments, with granulometric, mineralogical and micropaleontological surveys aimed at a more definite knowledge of the 'caranto';

- survey and systematic comparison of the methods used to solve the problems caused by re-ascending capillarity and high tide: reconstruction of exterior walls, replacement with the 'scuci-cuci' method, mechanical and chemical barriers to cut off the re-ascending capillarity, methods to remove salt from walls, underground catchments and drainage systems;

- evaluation of the work undertaken – not only from the point of view of technical efficacy but also of the compatibility with building materials and architecture – weighing the pros and cons, in order to proceed with testing methodologies, systems, materials and procedures;

Flooding and Environmental Challenges for Venice and its Lagoon: State of Knowledge, ed. C. A. Fletcher and T. Spencer.
Published by Cambridge University Press, © Cambridge University Press 2005.

- checking the efficacy and endurance of consolidation materials for stones exposed to the open air; and
- definition of specific procedures for structural consolidation, with suggested methodologies for analysis and project execution.

Another aim is to thoroughly catalogue all Venetian plasterwork and to forward this information to the relevant public bodies as a precious repository of knowledge, immediately employable in the safeguarding of the city and the control of specific restoration projects (Danzi and Piana, 2002; Ferrighi, 2002). A huge amount of ancient plasterwork still exists on the exteriors of the city's buildings: for example, there are several hundred examples of plasters, generally frescoed in the fourteenth and fifteenth centuries, sometimes very fragmentary but in other cases almost entirely preserved. In the absence of any kind of information concerning their nature and value, there is the risk of these plasters vanishing because of careless maintenance or restoration projects. A particular section of this research has thus been dedicated to a systematic survey of the historical exterior plasterwork of the city's buildings. It aims to gather thorough information about the nature of the plasterwork, its types and the causes of its decay. This catalogue has involved c. 15 000 buildings, from almost all the historical areas of Venice. Along with a precise description of the technical and formal evolution of the plasterwork, this work has permitted a reconstruction of the city's aspect over the centuries, giving an essential baseline for any further investigation in this field.

The building traditions of the lagoon have always assigned a protective role to exterior plasterwork. The comparative thinness of walls in Venetian buildings, due to the need for maximizing lightness, requires careful protection. Even a small decrease in their cross-section could trigger serious problems of instability, with consequent dangerous structural disruptions. Thus, besides their undoubted aesthetic and formal qualities, finishes and plasterwork have always been applied to the exteriors of Venetian buildings in order to prevent masonry decay. This is a sort of 'sacrificial surface' whose function is to take up the decay caused by the combination of weathering, saline aerosols and, towards the bottom of the building, capillary rise with the consequent cycles of salt crystallization. Studies concerning historical Venetian plasterwork, mostly carried out over the last twenty years, have only partially explained the technical and formal evolution of external plasterwork over the centuries (Armani and Piana, 1982, 1984, 1985; Piana, 1988), and partially addressed their composition and decay (Biscontin *et al.*, 1982; Charola *et al.*, 1985; Biscontin *et al.*, 1986). However, before this study, quantitative information was still lacking on the type, stratification, location in the city and present condition of Venetian plasterwork.

VENETIAN PLASTERWORK

There are several categories of historical plasters which have been classified according to age and type:

A) *stabiliture* (for example the *regalzieri*, plasterwork with monochrome painting, or decorated bands, or surrounding door and window openings, or painted layers applied to stone or masonry);
B) frescoes (figurative or otherwise);
C) *marmorini* (either single-layered or on a layer of *cocciopesto* – a sort of plaster made of lime and ground tiles – or sand, with white, pink or coloured body, provided with relief, coloured, engraved bands and with heavily defined edges); and
D) common plasterwork (made of lime and sand, rough *cocciopesto* or decorated). Also the plasters of twentieth and twenty-first centuries are classified here, even if in a simplified way, together with other categories of covering: stone, exposed brickwork, removed coverings, and so on (Armani and Piana, 1984, 1985).

The *regalzieri*, the figurative frescoes and the *marmorini* are among the most interesting categories of plasters on account of their technological and formal qualities, their importance and widespread occurrence. The *regalzier* is the most common kind of plasterwork on medieval buildings. It vanished from Venice at the end of the fifteenth century but remained popular in vernacular building throughout the Cinquecento, spreading to other parts of the peninsula and Europe (Valcanover, 1991). It consists of imitation brickwork, frescoed on a single thin layer

Fig. 24.1   Fragment of red and white *regalzier* with band showing vegetal decoration.

of plaster (Piana, 2000; Schuller, 2000; Wolters, 2000). The white pointing design is applied over a reddish background painted with large brushstrokes; this white-painted pointing often follows horizontal striations, traced out with nails on the fresh plaster in correspondence with the real underlying pointing. The seeming incongruity of an operation which methodically hides true brickwork with an imitation one was justified by the desire to standardize the colour and design of building surfaces. This was necessitated by the marked variability of Venetian medieval brickwork, built with bricks of different tonality and/or size. The difference of size was due to the practice of re-using material from demolished buildings or, the result of the long time between the commencement and completion of a building. Occasionally, in those parts of the brickwork processed by smoothing of the surface, the *regalzier* was painted with an oily binder paint directly onto the brick surface or stone, as on the external stone surfaces of the Cà d'Oro or the two-colour covering of the Doge's Palace (Cecchetti, 1886; Boni, 1883, 1887; Piana, 2000; Schuller, 2000). The imitation brickwork, widespread in both civil and in religious buildings, was often combined with painted bands with vegetative (Fig. 24.1), geometrical or marble-like decoration (Bristot, 2000; Piana, 2000).

Figurative frescoes are less common but they are very important for the history of Venetian art (Zanetti, 1760; Foscari, 1936; Wolters, 2000). Decorations appeared in the Renaissance (mainly in the sixteenth century) when nearly all Venetian painters of the time frescoed facades at one time or another: from

Fig. 24.2   Detail of *marmorino* decorated with a relief of an imitation band of Istrian stone like the windowsill shown on the right.

Giorgione and Titian to Veronese, Pordenone and Tintoretto (Plate 9). However, from the second half of the sixteenth century onwards, the taste for figurative decoration on facades declined, due more to a general cultural change than to reasons of decay due to weathering, as pointed out by Vasari (Vasari, 1550). The spirit of the Counter-Reformation tended to restrict scenes with human images to edifying duties, confining them to the churches and oratories.

Finally, the *marmorino*, also referred to as *stucco*, *terrazzo* or *terrazzetto* in the documents of the sixteenth to eighteenth centuries, is the most common type of plasterwork that is still prevalent in Venice. The *marmorino*, composed of lime and fragments of Istrian stone (Fig. 24.2), was applied directly to the brick-

work until the first decades of the seventeenth century, when it was applied onto a preparation layer composed of *cocciopesto*. These coverings are characterized by rigorous technical standards and specific procedures, including a final treatment with linseed oil, soap or wax (Viola Zanini, 1629; Biscontin *et al.*, 1982; Piana, 1984; Biscontin *et al.*, 1986). The *marmorini* appeared in the city between the Quattrocento and the Cinquecento, together with the dissemination of the new architectural language. The reason for their quick diffusion and huge success in the following centuries must be associated with the capacity of the *terrazzetti* to evoke, sometimes to imitate strictly, the aspect and consistency of stone materials in Venice, especially Istrian stone (grey-white). Coloured *mar-*

morini – pinkish, yellowish and grey-cerulean – were also used (Piana, 1988, 1989). A great proportion of the poor-grade plasterwork made with sand and *cocciopesto* used in the city during the sixteenth to nineteenth centuries tried to imitate the appearance of better-quality plasterwork, albeit poorly due to a simplification of technique.

STAGES OF THE RESEARCH

The catalogue includes sampling of the most important plasterwork by using chemical-physical analyses to determine the specifications of the material composition and techniques used in the various categories of covering (*regalzieri* and medieval decorations, *marmorini* in one layer or with *cocciopesto* background layer, etc.). Using the results from chemical-physical analyses, applied to representative samples of each category that are also datable with certainty, a database has been established which is set to become a reference for otherwise undatable plasters.

In the future, chemical-physical analyses will also be undertaken on the most significant works of consolidation, restoration and integration of plasterwork undertaken in the city over the last twenty years. These interventions, performed with different methods and materials, will be analysed in the final phase of the research, to assess the efficacy of the various different systems and the acceptability of the results from an architectonic and formal point of view.

The final aim of the research has been to provide methodological criteria for guiding future projects to conserve and restore the exterior plasterwork of buildings in Venice. The field survey has been completed. Out of 14 451 buildings, 14 260 UA – Architectonic Units – have been catalogued (three islands of the historic centre of Venice, such as Tronchetto, have been omitted). 1390 UA (*c.* 10%) have not been surveyed due to difficulties in reaching them or because at the time they were covered for restoration works. 1522 UA (*c.* 12% of the total) display the presence of historical plasters in categories A, B, C and D. For instance, the district of San Marco is composed of 1874 buildings: 1787 of them have been surveyed and 375 samples of historic plasters have been found on the exterior of the buildings (87 *stabiliture*, 9 frescoes, 160 *marmorini*

and 119 common plasters). The category C – *marmorino* – represents the most widespread and the best preserved among the different categories of historic plasterwork. At present a web platform, compatible with the CORILA system, is being realized (Piana *et al.*, 2003). The catalogue of Venetian plastered facades will be made available for the local government (Municipality and Monuments and Fine Arts Superintendency) and is destined to become an instrument of great value, capable of being updated and immediately usable for the control, protection management and programming of the restoration of Venice's architectural heritage.

REFERENCES

Armani, E., and M. Piana. 1982. A research programme on the plaster of historical buildings in Venice. *Mortars, Cements and Grouts used in the Conservation of Historic Buildings. Atti del Symposium ICCROM*. Rome: Sage, pp. 385–400.

Armani, E., and M. Piana. 1984. Primo inventario degli intonaci e delle decorazioni esterne dell'architettura veneziana; indagine e classificazione degli intonaci colorati di una città che fu policroma. *Ricerche di Storia dell'Arte* **24**, 44–54.

Armani, E., and M. Piana. 1985. Le superfici esterne dell' architettura veneziana. In G. R. Terminiello (ed.), *Facciate dipinte, conservazione e restauro, Atti del Convegno*. Genoa, 75–8.

Biscontin, G., Piana, M., and G. Riva. 1982. Research on limes and intonacoes of the historical Venetian architecture: characterization on some marmorino intonacoes from the 16th to the 17th century. *Mortars, Cements and Grouts used in the Conservation of Historic Buildings. Atti del Symposium ICCROM*. Rome: Sage, pp. 359–71.

Biscontin, G., Piana, M., and G. Riva. 1986. Aspetti e durabilità degli intonaci 'marmorino' veneziani. *Restauro e Città* **3/4**, 117–26.

Boni, G. 1883. Il colore sui monumenti. *Archivio Veneto XIII*, t.XXV, 26–43.

Boni, G. 1887. La Cà d'Oro e le sue decorazioni policrome. *Archivio Veneto XXXIV*, 115–32.

Bristot, A. 2000. Gli affreschi esterni di Santa Maria Gloriosa dei Frari. in F. Valcanover and W. Wolters (eds.), *L'architettura gotica veneziana. Atti del Convegno*. Venice: Istituto di Scienze, Lettere ed Arti, pp. 189–94.

Cecchetti, B. 1886. La facciata della Cà d'Oro dello scalpello di Giovanni e Bartolomeo Buono. *Archivio Veneto* **31**, 26–43.

Charola, A., Laurenzi Tabasso, M., and L. Lazzarini. 1985. Caratteristiche chimico-petrografiche di intonaci veneziani del XIV–XX secolo. In *L'intonaco: storia, cultura e tecnologia, Atti del Convegno di Bressanone*, Padua: Libraria Progetto Editere, pp. 211–21.

Danzi, E., and M. Piana. 2002. The catalogue of the external plasters in Venetian building. In P. Campostrini (ed.), *Scientific Research and Safeguarding of Venice, Corila Research Program 2001 Results*, Proceedings of the Annual General Meeting of Corila, 4–5 April 2002, San Servolo. Venice: Istituto Veneto di Scienze, Lettere ed Arti, pp. 87–95.

Faccio, P. 2002. The lagoon building and water. In P. Campostrini (ed.), *Scientific Research and Safeguarding of Venice, Corila Research Program 2001 Results*, Proceedings of the Annual General Meeting of Corila, 4–5 April 2002, San Servolo. Venice: Istituto Veneto di Scienze, Lettere ed Arti, pp. 97–108.

Ferrighi, A. 2002. The plasters data base. In P. Campostrini (ed.), *Scientific Research and Safeguarding of Venice, Corila Research Program 2001 Results*, Proceedings of the Annual General Meeting of Corila, 4–5 April 2002 San Servolo. Venice: Istituto Veneto di Scienze, Lettere ed Arti, pp. 97–108.

Foschi, S., and M. Morresi. 2002. Data-base 'Venetian skilled workers' (Maestranze veneziane) XVth–XVIIIth centuries. In P. Campostrini (ed.), *Scientific Research and Safeguarding of Venice, Corila Research Program 2001 Results*, Proceedings of the Annual General Meeting of Corila, 4–5 April 2002, San Servolo. Venice: Istituto Veneto di Scienze, Lettere ed Arti, pp. 109–23.

Foscari, L. 1936. *Affreschi esterni a Venezia*, Milan: Hoepli.

Piana, M. 1984. Una esperienza di restauro sugli intonaci veneziani. *Bollettino d'Arte* 6, Ministero per i beni Culturali e Ambientali, 103–6.

Piana, M. 1988. Gli intonaci veneziani. In *Primo corso di perfezionamento in restauro architettonico dell'Istituto Universitario di Architettura di Venezia*. Venice: Cluva, pp. 183–91.

Piana, M. 1989. Tecniche edificatorie cinquecentesche: tradizione e novità in Laguna. In *D'une ville a l'autre: structures matérielles et organisation de l'espace dans les villes européennes (XIII–XVI siècle). Atti del Convegno Ecole Française de Rome*. Rome, pp. 631–9.

Piana, M. 2000. Note sulle tecniche murarie dei primi secoli dell'architettura lagunare. In F. Valcanover and W. Wolters (eds.), *L'architettura gotica veneziana. Atti del Convegno*. Venice: Istituto di Scienze, Lettere ed Arti, pp. 61–70.

Piana, M., Danzi, E., De Zorzi, S., Ferrighi, A., and E. Rinaldi. 2003. Un GIS-WEB per la catalogazione degli intonaci esterni veneziani. Presented at Conferenza Nazionale ASITA, L'informazione territoriale e la dimensione tempo, Paper 7A, 28–31 October 2003, Palazzo della Gran Guardia, Verona.

Schuller, M. 2000. Le facciate dei palazzi medioevali di Venezia. Ricerche su singoli esempi architettonici. in F. Valcanover and W. Wolters (eds.), *L'architettura gotica veneziana. Atti del Convegno*. Venice: Istituto di Scienze, Lettere ed Arti, pp. 281–345.

Valcanover, F. 1991. Venezia e provincia. In *Pittura murale esterna nel Veneto*. Bassano del Grappa: Ghedina and Bassotti.

Vasari, G. 1550. *Vite de' più eccellenti pittori, scultori e architetti*. Florence.

Viola Zanini, G. 1629. *Della Architettura, libri due, I, XVI, Della prattica delle malte*. Padua.

Wolters, W. 2000. *Architektur und Ornament*. Venezianischer Bauschmuck der Renaissance. Munich: Beck, pp. 73–97.

Zanetti, A. M. 1760. *Varie pitture a fresco de' principali maestri veneziani*. Venice.

25 • Alteration of brickwork exposed to sea tides in Venice

G. BISCONTIN, E. ZENDRI AND A. BAKOLAS

INTRODUCTION

The walls in Venice can be exposed to water either directly, in the buildings along the canals, or indirectly. In the former case, the part of the wall underwater exhibits rather limited decay processes in the presence of a stable environment. Conversely, the part which is alternatively in contact with air or water is more deeply attacked, both by the action of tides and by the significant effect of waves. In the second case, capillarity phenomena can be observed typically for 2-3 metres above sea level, whereas the overlying brick is exposed only to the normal atmospheric environment (Biscontin and Driussi, 1988; Elert *et al.*, 2003; Charola and Koestler, 1982). Only a part of the walls is involved in cyclic contact with water as a consequence of tides, sea waves and the transit of motorboats. It is this part which is the subject of this investigation. Mechanical, physical, chemical and biological factors can all be involved. The Venetian walls in contact with water usually are covered by compact, carbonate stone (Istrian stone). This was intended to serve as protection, due to its low porosity (0.1–0.5%) which reduces rising damp and preserves the wall above it (Maretto, 1988).

However, in recent years sea level has often reached higher than the protected part of the walls, establishing direct sea contact with the bricks. The increasing presence of high water phenomena has lead to unexpected effects which have proved difficult to quantify. The passage of motorboats causes waves, and in this way an extra 40–50 cm of the walls are periodically forced into contact with the water. Such wave effects are surely responsible for a number of physical consequences but these are not yet clearly defined. One evident, direct consequence is a decrease in the cohesion of mortars and in their adhesion to the bricks, leading to bricks sloughing and falling.

This research is part of a study about the effects of marine water on traditional walls and the methodologies for their conservation. Here we report a preliminary study on chemical and physical changes on brickwork periodically in contact with seawater. Chemical analyses and knowledge of the ion composition profile can provide a useful indication of chemical processes due to sea tides on bricks. Porosity has been chosen as a significant physical parameter (Sandrolini *et al.*, 1983; Scherer, 1999). It can be related to the depth of the sample in the brick, in order to obtain indications about the erosion process.

EXPERIMENTAL DESIGN

Sampling

In order to maximize information, samples were collected according to the age of the wall. The surface of the wall involved in periodic contact with water was determined by taking into account mean sea level and tide level data, obtained from the Venice Municipality Office. Sampling was carried out by core boring into different walls in the mid-intertidal zone during low tide periods (Table 25.1). Exposure to the environment, and location in relation to the movement of motorboats, was the same for all walls. From each sample sections were cut at 0–0.5, 1.5–2.0, 3.0–3.5 and 5.0–5.5 cm from the external surface.

Analysis

Samples were dried and stored in sealed glass containers. Subsequently sections were cut at 0–0.5,

Flooding and Environmental Challenges for Venice and its Lagoon: State of Knowledge, ed. C. A. Fletcher and T. Spencer.
Published by Cambridge University Press, © Cambridge University Press 2005.

Table 25.1 *Sample locations and wall ages.*

Sample	Age of the wall	Provenance
1	13th century	Building along the Grand Canal
2	13th century	Building along the Grand Canal
3	17th century	Building along Rio di Noale
4	18th century	Building along Rio del Tenor

1.5–2.0, 3.0–3.5, 5.0–5.5 cm from the external surface. In order to obtain information about the chemical and physical brick transformations due to sea tides, analyses according to the following analytical procedure were carried out:

- extraction of soluble salts with deionized water and determination of the species Na^+, K^+, Ca^{2+} and Mg^{2+} by atomic absorption spectrophotometry (Perkin-Elmer 4000);
- treatment with concentrated hydrochloric acid and determination of the species Na^+, K^+, Ca^{2+} and Mg^{2+} (total ions) by atomic absorption spectrophotometer (Perkin-Elmer 4000) and determination of silica SiO_2 as total silica (silica + silicates not attacked by acid); and
- measurement of the cumulative volume ($mm^3\,g^{-1}$) of the samples with a mercury porosimeter (Carlo Erba 4000) in the range of pressure between 0.01 and 2000 atm.

The data reported are the mean of three measures for each sample and each depth, according to the Italian standard (Normal 4/80).

RESULTS AND DISCUSSION

The results of the chemical analysis for the total and water soluble Na^+, K^+, Ca^{2+} and Mg^{2+} at various depths are reported in Tables 25.2 and 25.3. Total ions (Table 25.2) show a variable trend, probably due to the different original composition of the bricks. However the trend of mean total ions in relation to depth (see Fig. 25.1) indicates an increase from the external to the internal part of the walls. This increase corresponds to a decrease in the mean values of the cumulative volume. The trend is seen better in Fig. 25.2 which reports the average variation in total ions against the cumulative volume, taking as a reference point the value of porosity corresponding to a depth of 5 cm. An increase in total ions can be explained by a decrease in the action of seawater on the bricks. The reduction of insoluble ions, which corresponds to the original composition of the bricks, provokes an increase in the cumulative volume. The soluble salts, coming from the seawater, cannot be responsible for the variations in porosity since they are always in solution.

The level of soluble ions (Table 25.3) shows small variations with depth. It is important to note that the walls exposed to the sea tide, every six hours, do not dry out at low tide and are characterized by the presence of algae. The presence of water prevents the accumulation of salts in the brick. There is no obvious relationship between the trend in the soluble ions with the age of the wall, the brick composition or microstructure. This means that in all likelihood the level of soluble ions is not a good indicator of the chemical decay of brickwork exposed to the sea tide.

The same trend results for the levels of total silica (SiO_2) (Table 25.4). The amount of silica in relation to the corresponding values of cumulative volume suggests a chemical attack occurring, on the average, in the first 3–4 cm of the material. The loss of silica and other ions, above all calcium and magnesium, certainly derive from a chemical transformation of the original matrix, which causes the formation of soluble products (Iler, 1979; Carlsson, 1988). These new salts are washed away by seawater and consequently the cumulative volume of the bricks increases. This is true for all the samples, although to varying degrees.

Table 25.2   *Depth profile of total sodium, potassium, calcium and magnesium.*

Sample	Depth cm	K⁺ %	Na⁺ %	Ca²⁺ %	Mg²⁺ %
1	0-0.5	0.20	0.48	1.49	0.58
	1.5-2	0.24	0.83	1.72	0.71
	3-3.5	0.28	1.22	2.66	1.02
	5-5.5	0.32	1.28	2.78	1.12
2	0-0.5	0.21	1.11	2.72	1.81
	1.5-2	0.21	1.01	1.70	2.21
	3-3.5	0.18	1.00	1.72	2.21
	5-5.5	0.19	1.01	2.87	2.02
3	0-0.5	0.35	0.71	0.83	0.79
	1.5-2	0.36	0.61	0.49	0.61
	3-3.5	0.32	0.62	0.44	0.49
	5-5.5	0.38	0.61	0.70	0.82
4	0-0.5	0.20	1.07	1.95	1.53
	1.5-2	0.19	1.11	1.07	2.61
	3-3.5	0.19	1.01	0.86	2.62
	5-5.5	0.18	1.01	0.95	2.51

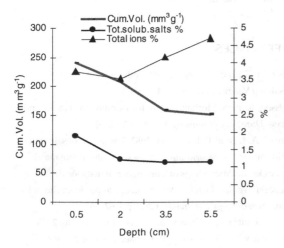
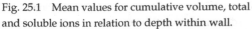

Fig. 25.1   Mean values for cumulative volume, total and soluble ions in relation to depth within wall.

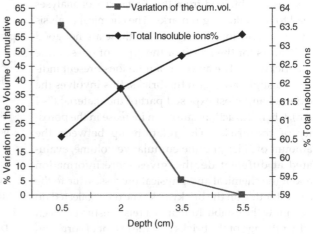

Fig. 25.2   Variations in cumulative volume and total insoluble ions in relation to depth (mean values).

Table 25.3   *Depth profile of soluble sodium, potassium, calcium and magnesium.*

Sample	Depth cm	K⁺ %	Na⁺ %	Ca²⁺ %	Mg²⁺ %
1	0-0.5	0.13	0.35	0.58	0.13
	1.5-2	0.18	0.57	0.42	0.16
	3-3.5	0.15	0.57	0.42	0.13
	5-5.5	0.15	0.58	0.41	0.13
2	0-0.5	0.14	0.79	0.52	0.34
	1.5-2	0.13	0.66	0.37	0.20
	3-3.5	0.16	0.65	0.17	0.20
	5-5.5	0.15	0.67	0.18	0.20
3	0-0.5	017	0.42	0.42	0.25
	1.5-2	0.17	0.38	0.28	0.19
	3-3.5	0.18	0.37	0.27	0.16
	5-5.5	0.17	0.38	0.24	0.16
4	0-0.5	0.11	0.57	0.59	0.31
	1.5-2	0.10	0.50	0.32	0.32
	3-3.5	0.10	0.52	0.26	0.31
	5-5.5	0.08	0.50	0.27	0.30

CONCLUSIONS

The results obtained from this first series of analyses lead to the following remarks. The chemical analysis of the soluble ions indicates that these are not good indicators for the study of the effect of the sea tide on brickwork. The amount of total ions present indicates a slight attack on the bricks; this involves the external and most exposed part of the material (3–4 cm). Chemical action causes an increase in the porosity of the bricks. The relationship between the amount of silica and the cumulative volume, evaluated at different depths, gives more information about the chemical and physical processes due to the seawater tide on the brickwork. The data collected do not allow the establishment of a relationship between either the age of the bricks and the extent of corrosion or between corrosion and the position of the bricks. Further studies are required to establish clearly a relationship between pore size and physical and/or chemical action.

REFERENCES

Biscontin, G., and G. Driussi. 1988. Indagini sull' umidità di risalita a Venezia. *Recuperare* 33, 78–82.

Carlsson, R. 1988. Mechanisms of deterioration in ceramic and glass. *Durability of Building Materials* 5, 421–7.

Charola, A. E., and R. J. Koestler. 1982. The action of salt water solutions in the deterioration of the silico-aluminate matrix of bricks. In Proceedings of Conference 'Il mattone di Venezia'. Venice, October. Venice: Instituto per lo studio della dinamica delle grand: masse, CNR, pp. 67–76.

Elert, K., Cultrone, G., Navarro, C. R., and E. S. Pardo. 2003. Durability of bricks used in the conservation of historic buildings – influence of composition and microstructure. *Journal of Cultural Heritage* 4, 91–9.

Table 25.4   *Depth profile of total silica and cumulative volume.*

Sample	Depth cm	SiO$_2$ %	Cum. Vol. mm^3 g^{-1}
1	0-0.5	55.55	315
	1.5-2	57.23	275
	3-3.5	57.15	260
	5-5.5	57.20	250
2	0-0.5	61.24	180
	1.5-2	61.83	155
	3-3.5	61.98	110
	5-5.5	62.03	105
3	0-0.5	56.32	21
	1.5-2	56.75	195
	3-3.5	57.44	105
	5-5.5	57.43	100
4	0-0.5	61.74	230
	1.5-2	62.33	190
	3-3.5	62.28	155
	5-5.5	62.31	130

Iler, R. K. 1979. *The Chemistry of Silica*. New York: John Wiley & Sons Ltd.

Maretto, B. P. 1988. *La casa Veneziana*. Marsilio Editore.

Normal 4/80 [Italian standard for cultural heriatge], Rome: CNR-ICR, 1980.

Sandrolini, F., Moriconi, G., Veniali, F., and G. Zappia. 1983. Mechanical properties and porosity of ceramic building materials. In J. M. Haynes and P. Rossi-Doria (eds.), *Proceedings of the RILEM/CNR International Symposium 'Principles and Applications of Pore Structural Characterization'*. Bristol: J. W. Arrowsmith Ltd, pp. 291–7.

Scherer, G. W. 1999. Crystallization in pores. *Cement and Concrete Research* **29**, 1347–58.

26 • Economic valuation of on-site material damages of high water on economic activities based in the city of Venice: results from a dose-response-expert-based valuation approach

M. BREIL, G. GAMBARELLI and P. A. L. D. NUNES

INTRODUCTION

The chapter focuses on the economic assessment of damages caused by high water in the city of Venice. In particular, we focus our attention on a valuation exercise that addresses the estimation of monetary, short period, on-site damages due to high water events on the different business activities located in Venice. We chose working with business activities, including *inter alia* hotels, bars, gift shop and restaurants, since these are very sensitive to high water events; the great majority are located at ground floor level. In addition, business activities constitute an important stakeholder in the Venetian society – keeping the ancestral trading tradition of the Republica Serenissima – and for this reason play a crucial role in the assessment of any high water management policy. Furthermore, the present study strictly refers to on-site damages since the valuation exercise focuses only on the monetary estimation of impacts due to high water on the structure and construction materials, including mitigation expenditures necessary to prevent and lessen these impacts. Finally, the valuation exercise concentrates exclusively on short-period damages given that it is anchored to the present number and distribution of business activities located in Venice, excluding any supplementary impact(s) that the high water can cause on the number and distribution of Venetian business activities.

In previous valuation studies, the economic impacts of high water have been frequently analysed in relation to the environmental assessment with respect to the much-debated mobile-gates project – called MOSE (*MOdulo Sperimentale Elettromeccanico*). However, structural damages have been often tackled and quantified in physical, non-monetary approaches, including the estimation of the number of affected agents. Recently, Cellerino (1998) developed and applied a valuation study that focussed on the monetary estimation of high water impacts, including the damage that these create on the tourism demand, as well as on additional expenditures for the maintenance of the walls of the buildings located in Venice. We propose to extend this pioneering attempt to quantify, from an economic point of view, on-site damages, such as inner and front door maintenance and cleaning of floors, that high water causes to all registered business activities today located in Venice. The economic valuation exercise is characterized by the use of an integrated dose-response modeling and an expert-based valuation approach. The ultimate aim is to combine dose-response relationships, which are based on the judgement of experts and describe the physical effects of high water on the different on-site damage categories, with specific market information on price and effectiveness of alternative remediation measures.

The organization of the paper is as follows. The first section introduces and the second discusses the phenomenon of high water in Venice and its main damage implications to business activities. The third section presents and analyses the economic valuation approach for on-site damages. The fourth section presents main valuation results and respective policy implications. The fifth section concludes.

BACKGROUND

The threat of *acqua alta* in the city of Venice

Acqua alta – the Italian name for high water events – has gone along with Venetians' lives from the very beginning of the history of the city. So, during the centuries Venetians have learnt to coexist and deal with

Flooding and Environmental Challenges for Venice and its Lagoon: State of Knowledge, ed. C. A. Fletcher and T. Spencer.
Published by Cambridge University Press, © Cambridge University Press 2005.

this periodical phenomenon by adapting their own behaviour and preparing the city to tackle with this sort of event. *Acqua alta* events, nevertheless, are of increasing concern to Venetian inhabitants and policy makers, as well as the Italian national and international community, because of their impacts on the architectural, cultural and artistic heritage, on the structural security of Venetian buildings, and the future development of the current number and distribution of socio-business activities. All these impacts, and respective economic damages, have witnessed an increased attention in the policy agenda. The last decades, indeed, have been characterized by a systematic increase in the intensity of the high water events, which is expressed in the upwards trend of mean tidal excursions in Venice, measured by the tidal observatory of the Municipality in terms of additional centimetres in the average water level with respect to the Venetian main reference tide gauge, the Punta della Salute tidal datum[1] (see Fig. 26.1).

---

[1] Since 1897, mean sea level in Venice has been measured with respect to a level of mark on land, the Punta della Salute, just across St Mark's Square. At the present time, this benchmark is approximately 23 cm below the current mean sea level.

There are many factors influencing such a pattern which include, *inter alia*, the increase of the Adriatic's relative sea level of about 23 cm during the twentieth century. The latter is due to both natural and human causes, including the subsidence of the island of Venice and more importantly due to global climate change events. According to the international scientific community, climate change is expected to continue in the following decades, along with an increase of the incidence of high water in Venice. These, in turn, cause important damage to the city in general, and to its population, in particular. High water economic damage can be defined in term of three main categories: on site-damages; off-site damages; and social damages. On-site damages are described by the negative impacts that high water causes on the structures and materials, including private mitigation costs necessary to prevent and lessen these impacts. Off-site damages are due to the limited usability of the city during high water events, including difficulties for supply and provisioning, reduced access to the work place by the personnel, and decreased flow of clients. Finally, social damages of high water are linked to a public protection perspective and to the impact of high water on the total economic value of the set of business activities located in Venice (e.g. persistent high water events can

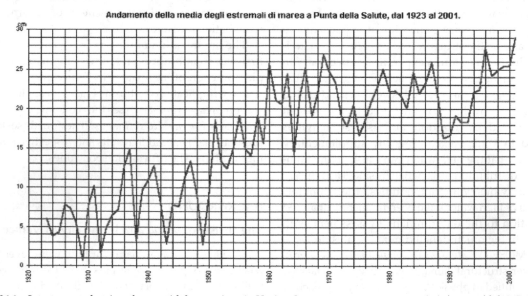

Fig. 26.1   Intertemporal series of mean tidal excursions in Venice. Source: www.comune.venezia.it/maree/dal1867.asp.

create uncertainty on the turnover of the business activities located in Venice that, in turn, can affect the number, type and diversity of activities.)[2] The present paper focuses on the first type of damages that are incurred by an important Venetian stakeholder: the business activities located and operating in Venice.

CHARACTERIZATION OF THE STAKEHOLDER UNDER ANALYSIS: THE BUSINESS ACTIVITIES BASED IN VENICE

As mentioned, this paper deals with on-site damages caused to business activities located and operating in Venice due to the high water. The estimation of on-site damages is based on data provided by a local governmental agency: COSES. Between 1999 and 2001, COSES conducted an on-site survey to all business activities located in ground floor units in the city of Venice. The survey offers information on the different type of activities, the nature and conditions of building materials, the presence and kind of mitigation measures for high water, as well as some architectural measurements. The present study works with a sample of 2598 respondents, on a total population of 5097 business activities registered in the city of Venice. Statistical elaborations of the survey's results lead to a better understanding of the number and

diversity of the socio-economic actives and highlight how they are affected by high water. The most frequent business activity in Venice is commercial (retailers of food, beverages, clothing, articles for tourists, etc.), followed by restaurants and bar and then by industrial activities, including glass manufacturers and local shipyards (Table 26.1).

A very important piece of information concerns the declared flooding level for each single business activity unit, i.e. the minimum level of high water that causes the flooding. According to Table 26.2, more than 80 per cent of the business activities that are located at the ground floor are affected by high water events when these are in the range from 111 cm to 130 cm.

In order to mitigate damages to structures and materials due to high water, as well as to limit the time and expenditure in cleaning floors, according to the COSES survey only 37 per cent of economic operators have adopted protection measures of different types (Table 26.3).

According to Table 26.3, the most widespread measures of protection are *paratia*[3] and raising of floor levels, while pumps and *vasca*[3] are not very common, which is probably due to their relatively

---

[2] See Nunes *et al.* (2004) for a detailed computation of the monetary value of social damages of high water events.

[3] Doors that give access to street are the most important mean of water infiltration to the business activities - for this reason they are often protected by mobile barriers, i.e. *paratia*, to keep water out. With the protection system called *vasca*, external walls and the floors of ground floor units are sealed with waterproof materials forming a basin to protect the inner parts of the building against high water.

Table 26.1   *Distribution of types of activity located at the ground floor.*

Type of activity	Frequency	Percentage
Industry	297	11.4
Shops	1518	58.4
Bars and restaurants	422	16.2
Offices	154	5.9
Hotels	116	4.5
Other	91	3.5
Total	2598	100.0

Table 26.2   *Distribution of Venetian business activities for flooding level.*

Flooding level	Frequency	Percentage	Cumulative %
< 90 cm	27	1.0	1.0
91–110 cm	472	18.2	19.2
111–130 cm	1628	62.7	81.9
131–145 cm	397	15.3	97.2
> 145 cm	74	2.8	100.0
Total	2598	100.0	

Table 26.3   *Distribution of Venetian business activities per type of mitigation measure.*

Protection measure	Percentage[a]
Yes	36.7
Raising floor levels	18.1
Paratia	19.3
Water pump	3.8
Vasca	7.4
No	63.3
Total	100.0

[a] The sum of the percentages of specific mitigation measures is higher than the percentage for 'yes' because a business activity can have more than one type of protection.

high financial costs. The distribution of the different protection measures among the business activities is an important element when assessing the economic cost of high water since different mitigation measures offer different degrees of protection. This effect is discussed in detail in the following section.

ECONOMIC VALUATION OF HIGH WATER DAMAGES

On-site material damages

Among on site damage components, the present study considers two types of costs connected to the impact of high water, i.e. mitigation and remediation costs. Mitigation costs are related to the adoption of high water protection measures and equipment like hydraulic pumps, *paratia* and rising of floors. These costs are essentially independent from the frequency of flooding experienced. Therefore, the category is referred to as fixed costs.

Remediation costs, on the other hand, are related to damages caused by flooding to building elements, including damages that high water and salinity causes to the walls. This category of costs is directly or indirectly connected to the number of flooding events. For this reason, these will depend on the unit's specific location in relation to the mean sea level. These costs are thus referred to as variable costs. The criteria for the selection of the types of damages to be considered looked at the availability of data and the existence of a direct relationship between the economic damage and the fundamental variable, which is the unit's specific level above sea and/or the frequency of flooding it has to face. The first criterion led to the exclusion of costs that require information that is not reported within the COSES data set. The second criterion implies that single categories of cost were included only if the connection between frequency of flooding or level above the sea and damage was identified in a sufficiently precise manner during interviews with experts.

In the present valuation study, the items considered for valuation are front doors and inner doors, floors, walls and the cost of the time employed in the cleaning of the floors and for re-establishing the functionality of the economic unit. Furniture had to be excluded from the estimation, as in the COSES database there was no sufficient information available on the characteristics of the furniture in each business activity. This may lead to an underestimation of variable costs, although in Venetian business activities furniture is frequently renewed, independently from the effects of high water.

Valuation approach

The economic valuation of on-site damages on structures and materials due to high water events is not straightforward. A literature review and a comparative analysis of previous studies about economic impacts of high water events highlight some important aspects which need to be considered in the assessment.[5] While some suggestions came from the methodological choices of previous studies, some other ideas derived from the attempt to overcome the limits of past assessments. In this context, we propose to work with dose-response valuation methods – see, for example, Adams and Crocker (1991). The dose response methods have in common that they put a value on environmental commodities without

retrieving people's preferences for these commodities. In this particular study, dose-response valuation technique is operated by (1) assessing the physical damage of high water on structures and materials and (2) estimating the economic value of the damage. For this reason we integrate the use of a response valuation technique with expert based information regarding both the increased maintenance and repair activities due to the high water and related market prices. This type of techniques have already been considered by Cellerino et al. (1997), Cellerino (1998) and Bertoldo et al. (1997), but under more restricted conditions.[5]

In particular, the following indications have been applied in the present study. First, damages have been assessed with respect to the specific characteristics of different business activities. In this study, and differently from previous ones, damages are referred to, and assessed, at the individual level, considering, among others, the type of activity, the location of the business activity with respect to the sea level and the use (or not) of mitigation measures (and which ones). Overall cost has been calculated by aggregating individual costs. Second, protection (or mitigation) measures have been included in the value assessment and modelled as follows: (1) costs sustained by economic operators for mitigation have been considered among on-site economic damages as fixed costs; (2) the effectiveness of protective measures in the mitigation of the high water damages has been attentively investigated and brought in the computations, and (3) high water damages have been disentangled from costs related to ordinary maintenance and substitution of structures and materials. A similar attempt was also tackled by Cellerino (1998). Finally, different scenarios with respect to the frequency of high water have been considered. The investigation of the consequences of different future scenarios is very useful for deriving

---

[4] For a complete overview of the literature see CORILA (2004).

[5] Dose-response-expert-based valuation method has also been applied by economists in the analysis and valuation of social and economic dimensions of climate change impacts and adaptation in Italy, including a case study that deals with an expected sea-level rise in the Fondi plane area located at the south-west of Rome (see Gambarelli and Goria, 2004).

Table 26.4   *Level of protection of mitigation measures across different building elements.*

	Pavements	Walls	Inner doors	Front doors
Raising of floors	+++	+	+++	+++
Vasca	+++	+++	+++	+++
Paratia	+	0	++	+++
Water pump	+	0	++	0
Water pump and *paratia*	++	0	+++	+++
Waterproof barriers for walls[6]	0	+++	0	0

Legend: +++ strong, ++ medium, + small, 0 no mitigation effect.

policy implications and it is a consolidated procedure in most economic assessment studies (see Magistrato alle Acque e Consorzio Venezia Nuova, 1997).

As mentioned, mitigation costs refer to all types of financial expenditure undergone to avert damages caused by flooding. Combining expert information on mitigation costs with the COSES information on the type of mitigation measures adopted, we were able to estimate the total cost with respect to mitigation incurred by each single unit. As we can see from Table 26.4, mitigation may consist in one or more interventions – e.g. *vasca* or water pumps and *paratia*.

The valuation of variable costs considers the infiltration of water causing damages to different building-elements, including floors, walls, inner and front doors. In addition, the presence of mitigation measures needs to be taken into account when calculating every type of variable cost. In practice, the estimation of economic damages is based on information produced by 'expert judgements' about the effects that infiltration of water has on the building materials, and the level of protection offered by the different measures (or a combination of measures) with respect to the specific elements of the unit (doors, floors, walls). Table 26.4 shows the relation existing

between different mitigation measures and the effects of high water on building-elements. It can be noted, for example, that the protection offered by a *vasca* is much higher than the protection offered by a *paratia*.

As an illustration, the cost of substitution of inner doors (IDSC) damaged by high water is estimated as follows:

$$IDSC = \frac{UCDI}{n}\lambda(e_j,n)\sum_{r^i}\sum_l e_l^{i,r}(p) \qquad Eq.\ 26.1$$

where *UDCI* denotes the unit cost of a inner double skin door in wood (expert based information) and n the number of tidal events after which a wooden double skin door has to be substituted (also expert based information). Therefore the first term, captures a monetary magnitude that reflects the physical *resistance* of inner wooden doors with respect to the tidal events, *ceteris paribus*. This measure is combined with two additional terms, λ and $e_l^{i,r}$, respectively. The latter term reflects the fact that doors are located in different rooms within each i business activity unit and each r room is associated to a given quota level l, which is a measurement above the Punta della Salute threshold sea level as indicated by the tidal observatory of the Venetian Municipality. Therefore, $e_l^{i,r}$ denotes the annual frequency of high water events that the room r at level j of the unit i is subject to. This term is function of p, which denotes the type of pro-

---

[6] Waterproof barriers are architectonical devices that prevent bricks from being soaked by tidal waters and thus protect the walls from damages caused by salty water.

tection measure adopted by the business activity unit, since we consider that the existence and type of mitigation measures influence the protection level and thus the number of events that each room is subject to. Finally, λ denotes a technical term that limits the maximum number of substitutions of inner doors at one per year. For this reason we represent λ as a function of e_j and n with $\lambda = 1$ if $e_j \leq n$ and $\lambda = n/e$ if $e_j > n$. As we can see from this illustration, the information contained in the COSES database on size, number of rooms, distance of the floor level from sea level together with expert based information allows us to calculate in an efficient and robust way the economic damage caused by flooding to inner doors. This methodology was also applied to the other remaining cost components.[7]

On-site damages can then be computed. In order to explore possible policy implications in reference to the protection, or absence, of high water we intro-

duce possible scenarios that describe alternative high water situations for the city of Venice. These will be discussed in the next section.

The formulation of *acqua alta* scenarios

In the present analysis, we consider four scenarios. The first refers to the current situation, which is denoted by the 'business as usual' (BAU) scenario. This scenario assumes the annual average frequencies of tidal events as registered by the Tidal Observatory of the Venetian Municipality during the period 1966–2001. In other words, the annual frequency of high water in BAU coincides with the historical frequencies (Table 26.5).

A second scenario, 'Defence 110', refers to the situation that is characterized by the use of collective, defensive measures against all tidal events above 110 cm with respect to mean sea level as measured by the Punta della Salute datum. For this reason the frequency above this threshold is zero. Below 110 the frequency is the same as BAU and coincides with the historical high water frequencies. A third scenario

---

[7] A more detailed discussion of these is presented in CORILA (2004).

Table 26.5   *Average annual frequencies of high water per scenario.*

Tidal level	Average annual frequency			
	BAU	Defence 110	Climate change	Climate change & Defence 110
+ 190 cm	0.0286	0.0000	0.0286	0.0000
180 – 190 cm	0.0286	0.0000	0.0286	0.0000
170 – 180 cm	0.0286	0.0000	0.0571	0.0000
160 – 170 cm	0.0571	0.0000	0.0862	0.0000
150 – 160 cm	0.0862	0.0000	0.2000	0.0000
140 – 150 cm	0.2000	0.0000	0.6250	0.0000
130 – 140 cm	0.6250	0.0000	1.5000	0.0000
120 – 130 cm	1.5000	0.0000	3.5000	0.0000
110 – 120 cm	3.5000	0.0000	8.5000	0.0000
100 – 110 cm	8.5000	8.5000	19.700	19.700
90 – 100 cm	19.700	19.700	49.200	49.200
80 – 90 cm	49.200	49.200	117.60	117.60
70 – 80 cm	117.60	117.60	241.30	241.30

considers an average sea-level rise of 10 cm due to climate change events,[8] and for this same reason this scenario is denoted as 'Climate change'. In this scenario, the number of high water events for a given sea level equals to the frequency registered in the BAU scenario and corresponding sea level immediately below. For example, the frequency of tidal events at 120 cm for climate change corresponds to the frequency reached at the level of 110 cm for BAU. Finally, we consider a mixed scenario that results from the combination of both 'climate change' and 'Defence 110'. In short, this scenario considers the defence of the whole lagoon against flooding events above 110 cm, however the frequency of the high water events below that threshold will increase and will remain equal to the rates of the 'climate change' scenario.

VALUATION RESULTS

Estimates of economic damages related to high water depend crucially on the floor level of the unit with respect to the mean sea level (Punta della Salute). In fact, the level above the sea is directly related to the frequency of high water events which affect a particular unit, and thus to the number of times a building element comes into contact with water. The latter has a direct effect on the frequency of restoration, implying either maintenance or substitution of the affected element, having in any case an economic impact. Even in the case of walls, where damages are more linked to action of saltwater in bricks than to the frequency of flooding, the unit's location with respect to the sea level is reflected in the proximity of walls to the underlying water, and this way capture the effect of saltwater damages in the building structure of the business activity. As a consequence, the more precise the knowledge of the unit's level above the sea is, the

more reliable the estimates of the economic damages will be. Unfortunately, in Venice a full GIS cover and measurement of each business units' front-door location with respect to the sea level is not yet available – even if there are a series of measurements either covering only part of the city, or different points of relevance (e.g. street level, bridges levels, etc). For this reason, the present study anchors the value estimates to the flooding level as declared by respondents on the COSES questionnaire. The underlying assumption is that economic agents who have been operating in Venice are familiar with the impacts that different levels of high water have on their units and thus are able to link their individual flooding level to the announced tidal levels.

In addition, we consider three alternative approaches with respect to the calculation of the flooded area for each business activity. In concrete terms, the three approaches denote three assumptions with respect to the percentage of the unit flooded: one corresponds to the percentage of the flooded surface as reported by the respondent in the survey (approach A); a second corresponds to the percentage of the flooded surface bearing in mind the different size and location with respect to the sea level of different rooms (approach B); and finally we consider the totality of the surface flooded (approach C). The combination of these three approaches can be interpreted as providing a confidence interval – when compared to the reported answers on the flooded surface. Approach C provides an upper estimate since it considers a flooded surface at 100%. In addition, approach B combines the use of survey information so as to get a more precise value estimate on the overall flooded surface.

Furthermore, the present valuation exercise is also flexible in accommodating the impact of key parameters – such as the number of high water events that cause the need of restoration for each considered building element – on the valuation results. In order to deal with impact, the valuation exercise considers two parameters with respect to the restoration and maintenance for each building element, reflecting two different rates of intervention with respect to inner doors, outer doors and floors. In the case of walls, such analysis takes account of the maintenance technique to be used for restoration, given the large

---

[8] The scenario is limited to a rather low rate of sea level increase – many climatologists support the idea that sea level will rise by as much as 61 cm over the next century (see United Nations Environmental Program, 2001; Ministero dell'Ambiente e della Tutela del Territorio, 2003). However, such a high rate would imply important changes in the number and distribution of business activities that would not be consistent with the short-term perspective embodied in the present study.

Table 26.6   *On-site economic damages – BAU Scenario.*

BAU	Approach		
	A	B	C
COST CATEGORY			
SUBSTITUTION OF INNER DOORS			
Higher Bound	1 253 745	1 441 144	1 707 537
Lower Bound	804 298	893 049	1 086 249
MAINTENANCE OF FRONT DOORS			
Higher Bound	n.a.	135 281	135 281
Lower Bound	n.a.	63 598	63 598
MAINTENANCE AND CLEANING OF FLOORS			
Higher Bound	73 833	78 797	112 695
Lower Bound	32 725	68 519	97 995
MAINTENANCE OF WALLS			
Higher Bound	n.a.	2 920 301	3 023 313
Lower Bound	n.a.	1 240 247	1 346 235
TOTAL ON-SITE ECONOMIC DAMAGES			
Higher Bound	n.a.	4 575 522	4 978 825
Lower Bound	n.a.	2 265 413	2 594 078

range of available interventions. The respective impacts on the estimation exercise are interpreted in terms of a sensitivity analysis, providing a lower and upper value estimate for each building element. Tables 26.6 to 26.9 report the estimation results for all the scenarios under consideration.

Table 26.6 shows that total on-site damages due to high water on the Venetian business activities, given the actual frequency of high water events, ranges from 2.26 to 4.97 million euros per year, with a best point estimate of 3.41 million euros. The later corresponds to the average estimate for the total economic value of on-site economic damages according to the valuation approach B, which is retained as giving a more precise calculation on the overall flooded area.

If the sea level rises by ca. 10 cm and no policy protection measures are adopted, then on-site damages will increase, giving costs estimated to range

between 3.53 and 6.47 million euros per year, with a best point estimate of 4.73 million euros (Table 26.7).

In addition, a policy of total protection against tides higher than 110 cm implies a reduction in mitigation costs, estimated to range between 1.61 and 3.65 million euros per year, with a best point estimate of 2.49 million euros (Table 26.8).

Finally, if a high water protection policy were to be adopted against tides higher than 110 cm in combination with the presence of climate change, the economic value of on-site damage estimates range from 2.05 to 4.05 million euros per year, with a best point estimate of 2.87 million euros (Table 26.9).

When we compare the alternative scenarios, two important outcomes emerge. First, estimation results show that the most relevant sources of mitigation costs are related to the maintenance of the walls and the restoration of inner doors. External doors and floors seem not to be highly affected, and this is

Table 26.7   *On-site economic damages – Climate Change Scenario.*

Climate Change	Approach		
	A	B	C
COST CATEGORY			
SUBSTITUTION OF INNER DOORS			
Higher Bound	1 862 771	2 233 543	2 557 842
Lower Bound	1 412 111	1 638 794	1 927 454
MAINTENANCE OF FRONT DOORS			
Higher Bound	n.a.	284 389	284 389
Lower Bound	n.a.	164 415	164 415
MAINTENANCE AND CLEANING OF FLOORS			
Higher Bound	176 313	188 408	269 292
Lower Bound	153 316	163 833	234 167
MAINTENANCE OF WALLS			
Higher Bound	n.a.	3 232 043	3 358 486
Lower Bound	n.a.	1 568 812	1 692 679
TOTAL ON-SITE ECONOMIC DAMAGES			
Higher Bound	n.a.	5 938 383	6 470 009
Lower Bound	n.a.	3 535 854	4 018 715

mostly due to the separation of ordinary maintenance costs from total restoration costs. Second, the comparison of the valuation results across the different high water scenarios brings along with it significant welfare changes, measured in terms of the variation of the on-site estimation results. Welfare changes, in turn, can be interpreted in terms of policy guidance. Therefore, any policy measures that do not involve a public protection mechanism that defends the city of Venice from high water events would always involve heavy effects on the total on-site damages and will thus be associated with a negative impact on the welfare. In particular, estimation results show that the welfare loss due to on-site, short-term damages supported by the business activities ranges from 3.41 to 4.73 million euros per year, depending respectively on whether the frequency of high water remains unchanged or increases as a result of climate change. Furthermore, the introduc-

tion of a public protection mechanism that defends the city of Venice from high water events superior to 110 cm will reduce the on-site damages supported by the business activities. In fact, in a scenario where climate change is unavoidable, the introduction of a protection mechanism would reduce on-site, short-term damages supported by the business activities by up to 2.87 million euros per year.

CONCLUSIONS

In order to estimate the economic value of on-site damages of high water on Venetian business activities, we took into account the combination of dose-response methods, which are able to describe and account for the physical effects of high water on the different on-site damage categories, together with the judgement of experts. Furthermore, the dose-response methodology is sufficiently flexible so that economic

Table 26.8   *On-site economic damages – Defence + 110 cm Scenario.*

Defence + 110 cm	Approach		
	A	B	C
COST CATEGORY			
SUBSTITUTION OF INNER DOORS			
Higher Bound	465 833	460 253	611 513
Lower Bound	410 342	402 604	538 237
MAINTENANCE OF FRONT DOORS			
Higher Bound	n.a.	108 467	108 467
Lower Bound	n.a.	63 598	63 598
MAINTENANCE AND CLEANING OF FLOORS			
Higher Bound	44 638	42 151	67 954
Lower Bound	38 816	36 653	59 090
MAINTENANCE OF WALLS			
Higher Bound	n.a.	2 773 044	2 870 125
Lower Bound	n.a.	1 106 891	1 208 347
TOTAL ON-SITE ECONOMIC DAMAGES			
Higher Bound	n.a.	3 383 916	3 658 059
Lower Bound	n.a.	1 609 746	1 869 273

valuation results are able to accommodate both the impact of key *input* parameters provided by the experts, such as the number of high water events that cause the need of restoration for each building element, as well as alternative approaches regarding the computation of the flooded surface.

In such a valuation framework, we started by conducting interviews with different experts, mostly architects working in Venice, retailers of doors, floors and other materials, and firms installing mitigation measures. Then we proceeded with the estimation of on-site damages not only for the business as usual situation, but also for three additional high water scenarios. Estimation results show that the welfare loss due to on-site, short-term damages supported by the business activities ranges from 3.41 to 4.73 million euros per year, respectively for the business as usual and climate change scenarios. Furthermore, the introduction of a public protection mechanism that

defends the city of Venice from high water events superior to 110 cm will reduce the on-site damages supported by the business activities up to 2.87 million euros per year.

All in all, the present economic valuation estimates should be considered at best as an underestimate of the unknown total value of the impact of high water events since this study only considers a single stakeholder – the business activities; only tackles on-site damages – including walls, doors and floors; does not include any architectural value considerations; and it is always contingent upon the current number, distribution and diversity of business activities, what is a short term perspective.

ACKNOWLEDGEMENTS

This research was funded by the research programme 'An economic evaluation of interventions

Table 26.9   *On-site economic damages – Defence + 110 cm and Climate Change Scenario.*

Defence + 110 cm and Climate Change	Approach		
	A	B	C
COST CATEGORY			
SUBSTITUTION OF INNER DOORS			
Higher Bound	465 833	460 253	611 513
Lower Bound	465 833	460 253	611 513
MAINTENANCE OF FRONT DOORS			
Higher Bound	n.a.	124 133	124 133
Lower Bound	n.a.	122 566	122 566
MAINTENANCE AND CLEANING OF FLOORS			
Higher Bound	106 235	100 401	162 029
Lower Bound	92 378	87 305	140 894
MAINTENANCE OF WALLS			
Higher Bound	n.a.	3 019 921	3 151 392
Lower Bound	n.a.	1 380 724	1 508 268
TOTAL ON-SITE ECONOMIC DAMAGES			
Higher Bound	n.a.	3 704 709	4 049 066
Lower Bound	n.a.	2 050 849	2 383 242

for the safeguarding and environmental protection in the Venice Lagoon' of CORILA, Consortium for Coordination of Research Activities concerning the Venice Lagoon System. Two anonymous referees provided valuable comments on an earlier draft of this chapter.

REFERENCES

Adams, R. M., and T. D. Crocker. 1991. Material Damages. In J. B., Braden and C. D. Kolstad (eds.), *Measuring the Demand for Environmental Quality.* Amsterdam, The Netherlands: North-Holland, pp. 271–302.

Bertoldo, M., Maggioni, E., and F. Sbetti. 1997. *Le Condizioni Abitative nei Piani Terra de Centro Storico e la Propensione al Risanamento.* Comune di Venezia, Coordinamento Legge Speciale, Osservatorio Casa.

Cellerino, R. 1998. *Venezia Atlantide: L'impatto Economico delle Acque Alte.* Milan: Franco Angeli.

Cellerino, R., Anghinelli, S., Brasili, A., Gianicola, L., Mucci, M.N., Negri, A. M., and D. Priolo. 1997. Effetti delle opere mobili contro i danni dell'acqua alta: strutture fisiche, turismo, immobili notificati e mercato immobiliare. Unpublished paper.

CORILA – Consortium for Coordination of Research Activities concerning the Venice Lagoon System. 2004. Valutazione Economica Degli Impatti Di Breve Periodo Del Fenomeno Delle Acque Alte Sulle Attività Economiche Della Città Di Venezia. Venice: unpublished scientific report, vii+220 pp.

Gambarelli, G., and A. Goria. 2004. *Economic Evaluation of Climate Change Impacts and Adaptation in Italy.* FEEM Working Papers, 103.04. Milan: Fondazione Eni Enrico Mattei.

Magistrato alle Acque e Consorzio Venezia Nuova. 1997. *Environmental Impact Assessment of the MO.S.E. Project.* SIA del Progetto di Massima della Chiusura delle Bocche di Porto, Allegato 3B, Allegato 6, Tema 7.

Ministero dell'Ambiente e della Tutela del Territorio (MATT). 2003. *La Risposta Al Cambiamento Climatico in Italia.* Rome: ENEA and Fondazione Eni Enrico Mattei.

Nunes, P. A. L. D., Galvan, A., and A. Chiabai. 2004.
  Combining Stated Choice Methods and Hedonic Valuation:
  Measuring the Economic Benefits from Protecting the
  Business Activities from High Water in Venice. Paper
  presented at 13th Annual EAERE Conference, Budapest,
  Hungary.
United Nations Environmental Program (UNEP). 2001.
  Climate Change 2001: Impacts, Adaptation, and Vulnerability.
  Contributions of Working Group II to the Third Assessment
  Report of the Intergovernmental Panel on Climate Change.
  Cambridge: Cambridge University Press.

27 • City knowledge as key to understanding the relation between waters and stones in Venice

F. CARRERA

INTRODUCTION

Flooding affects the physical infrastructure in Venice in a variety of ways, most of which have been recognized since the early days of the city's existence. Manuscripts in the Venice archives chronicle the frequent requests for maintenance along the inner canals of the city ever since records were kept (Caniato, 1999; Dorigo, 1999; Piasentini, 1999). Natural erosion due to fluctuating tidal water levels is a fact of life in a place like Venice. Construction materials are gradually weakened by the constant wet-dry cycles and by the natural salts and unnatural pollutants contained in the tidal waters. A simplified taxonomy of the primary elements of the city that are negatively affected by exposure to the salt water of the lagoon and canals includes:

- Above-ground Infrastructure
  - Streets and Pavements
  - Bridges
  - Docks
- Underground Infrastructure
  - Utilities
  - Sewers
- Private and Public Buildings
- Artistic and Architectural Heritage
  - Churches
  - Palaces
  - Convents
  - Public Art
  - Street Furniture
- Canal Walls

While there is no doubt that these elements of the built environment are to some degree harmfully impacted by the gradual, but incessant, assault of the waters of the lagoon, the precise extent of the overall impact of flooding on the state of conservation of today's architectural and urban structures is hard to measure and the costs nearly impossible to quantify.

What is clear is that there are several concurrent factors at play in the undermining of Venice's built environment. Perhaps flooding is not the most destructive of all of the forces participating in the constant interplay between the liquid and solid components of the city. Many see 'moto ondoso' (motor boat wake) as a major player in this arena (Pulliero, 1987; Mencini, 2000; Carrera, 2001). Another potential – though perhaps unexpected – culprit in this milieu is sedimentation. This is accused of engendering damage along canal walls through the clogging of underwater sewer outlets, leading to underground ruptures and thus to seepage and weakening of the mortars that bond together the bricks and stones of the canal walls (Carrera, 1994).

THE CASE FOR CITY KNOWLEDGE

Without aspiring to quantify the damage done to these elements of the urban landscape, this chapter will instead use these pressing issues to delineate a path that will make these and other analyses much more feasible in the long run. In this manner, this chapter will demonstrate that Venice, like any other city in the world, would immensely benefit from espousing a systematic approach called 'city knowledge' (Carrera, 2004).

The research introduces a 'middle-out' approach to the accumulation and sustainable upkeep of city knowledge, through a spatially-based, self-organising framework produced semi-independently by individual municipal departments and by other information producers (Carrera, 2004). The envisioned emergent knowledge infrastructure could permit the efficient

Flooding and Environmental Challenges for Venice and its Lagoon: State of Knowledge, ed. C. A. Fletcher and T. Spencer.
Published by Cambridge University Press, © Cambridge University Press 2005.

collection, organization, integration, distribution, use and re-use of urban facts by government agencies as well as by scholars and scientists. It could therefore guarantee the constant availability of information for the scientific analysis of environmental, social, economic and cultural issues (including the topic at hand), as well as for the development of concrete actions related to city maintenance, management and planning. Recent work on 'city knowledge' explores the feasibility and desirability of the creation of a reliable, permanent, updatable, maintainable, reusable and sharable knowledge infrastructure to support municipal operations at all levels, from the ordinary upkeep of basic city properties, to the complex envisioning of alternative futures through sustainable collaborative planning paradigms. The focus of city knowledge research is specifically on the mechanisms that can be employed to produce richly encoded information that can allow a shift from the current 'plan-demanded' mode of data collection, to a more 'plan-ready' approach to information accrual. This in turn can lead to 'plan-demanding' situations in which the mere existence of 'middle-out' urban knowledge could engender plans which may have never been envisioned otherwise. The complexity of the relationship between the stones of Venice and the waters of the lagoon is a great demonstration of the types of issues that could be better understood once a plan-ready city knowledge infrastructure becomes available. The rest of this chapter briefly explores the nature of the interactions between natural and human-caused phenomena vis-à-vis the maintenance of the aforementioned elements of the Venetian built environment. The entire discussion will then focus on the indispensable foundation of information that is needed whenever we try to explore complicated relationships such as the one between stones and water in Venice. This chapter is therefore designed to highlight the knowledge infrastructure that can support such complex analyses.

Since 1988, the Venice Project Center of the Worcester Polytechnic Institute (WPI; www.wpi.edu) has been at the forefront of the exploration of the causal relations that cumulatively produce the physical damage that is visible everywhere in Venice. In collaboration with UNESCO, the Center has systematically collected a wealth of information about the phenomena connected with architectural damage and decay, both along the canals and elsewhere. Though none of these data connect flooding to structural damage *per se*, numerous correlations were tested out and verified, by relating a variety of independent and dependent variables that link the 'waters' with the 'stones'. These include: traffic levels, boat wake-loading, sedimentation, hydrodynamics, construction materials and maintenance.

The numerous databases and Geographic Information System layers that have been created since 1988 make it possible to test many assertions and draw useful conclusions, but, despite their sophistication, they only hint at what could be possible if a true city knowledge infrastructure were created and maintained in Venice. This chapter now sketches out the type of information that one would need to have accrued in order to measure the 'before and after' of a situation to begin to hypothesise about the 'cause and effect' of phenomena that can – at best – be treated as 'natural experiments' over which there is very little design control.

DAMAGE TO PUBLIC INFRASTRUCTURE

If one considers the 'public domain' only, then the above-ground infrastructure in Venice can be succinctly defined as: streets, bridges and docks. Flooding certainly damages all of these three elements of the public realm. Moreover, 'moto ondoso' (motorboat wakes) also damage bridges and docks. The causal nexus between flooding, 'moto ondoso', and the corresponding damage to these artefacts is extremely complex to isolate, but one can start by simply knowing as much as possible about these three elements of the public realm that are subjected to the destructive force of water. The Venice Project Center has teamed up with a Venetian company called Forma Urbis to complete: the mapping of over 1 million square metres of Venetian street pavements for Insula (Fig. 27.1); the inventory of all of the 472 bridges in the city also for Insula (Bahn et al., 1998; Fig. 27.2); and the catalogue of all 1321 public docks in the city for the city's department of public services (Felices et al., 1994; Doherty et al., 1995; Fig. 27.3).

These three major physical inventories create the backbone of a possible study of the effects that tides

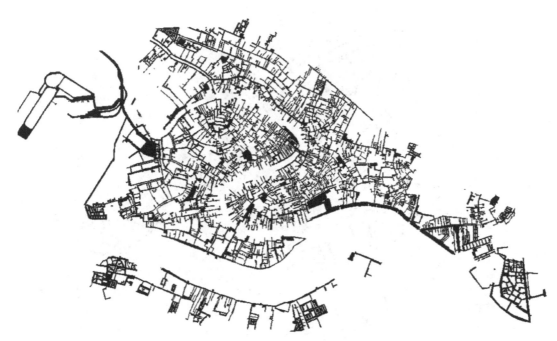

Fig. 27.1   The streets of Venice.

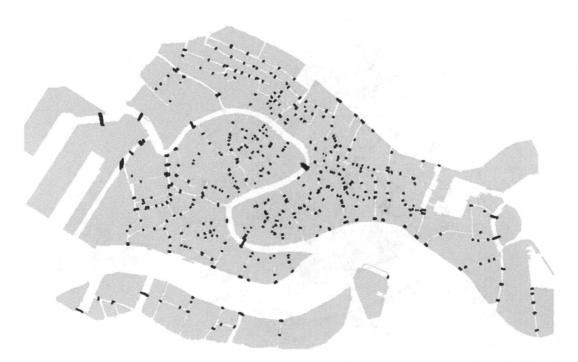

Fig. 27.2   The 472 bridges of Venice.

Fig. 27.3   The 1321 docks of Venice.

Fig. 27.4   Flooding in Venice up to 130 cm tides (dark gray).

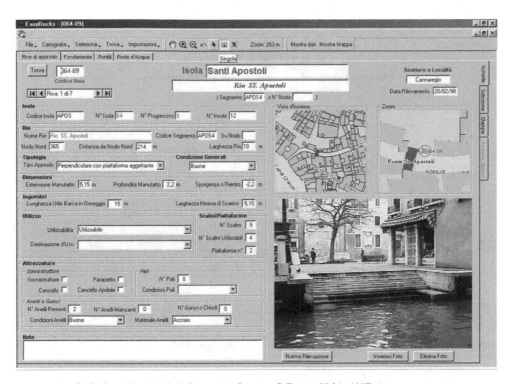

Fig. 27.5   Permanent dock data (*EasyDocks* Information System, © Forma Urbis 1997).

and moto ondoso have on them. By adding the digital elevation maps[1] that the Magistrato alle Acque has developed, through the Consorzio Venezia Nuova (1988), it is possible to produce maps of the flooding of the streets of Venice at a succession of tide levels (Fig. 27.4). Knowing how frequently these various tides occur leads to the identification of what portions of the city's streets, bridges and docks get wet and how often. So, with only a modicum of approximation and extrapolation, it is possible to arrive at reasonable estimates of the damage that can be caused by mere flooding.

It must be noted that these data-gathering efforts have not been aimed at simply producing these maps, but also entailed painstaking surveys of the artefacts inventoried. Each of the Forma Urbis catalogues has resulted in an information system application that bings together the permanent (Fig. 27.5) and ephemeral (Fig. 27.6) characteristics of both the bridges and the docks. In fact, whereas the state of conservation of an artefact (Fig. 27.6) will change and thus will need to be collected repeatedly over time (allowing the measurement of 'damage'), the permanent aspects (Fig. 27.5) will never have to be recorded again, which is one of the advantages of a city knowledge system.

These user-friendly, multimedia information systems make it possible for the dynamic data to be maintained up-to-date through the creation of a new, time-stamped, state-of-conservation assessment every time an intervention (or mere inspection) takes place. One can begin to see how useful such 'plan-ready' information would be if it were accessible to scientists and decision makers as well as to the front-line users who are in charge of the upkeep of the physical artefacts.

---

[1] All elevations in Venice are from the 'mareographic zero' of 1897, also known as the 'Punta Salute' datum.

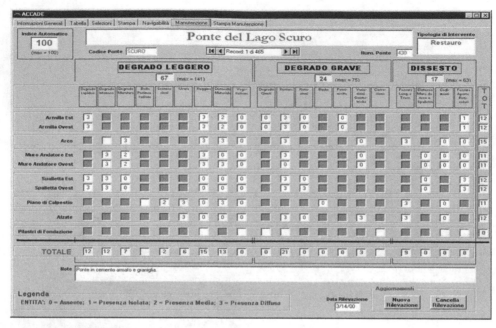

Fig. 27.6   Bridge conditions (*EasyBridge* Information System, © Forma Urbis 1998).

DAMAGE TO UNDERGROUND INFRASTRUCTURE

The next category of items that can be damaged by water includes objects under ground, namely the infrastructure for electrical, water, gas, telephone, street lighting and sewage services, composed of pipes, ducts, valves, manholes, inspection wells, cables, etc. Insula S.p.A. (see Dolcetta (chapter 19) and Turlon (chapter 20)) has worked on these sub-systems since it was originally chartered by the four main utility companies (electricity, phone, gas and water) who own 48% of the shares – the remaining 52% being owned by the City. One of the goals of Insula is to eventually produce a maintainable map of the underground infrastructure in Venice. The effects of flooding on subterranean utilities could be much more predictable with this sort of city knowledge at one's fingertips, avoiding waste and redundancy. Once in place, these systems would represent practical applications of city knowledge principles. The first major system to showcase these principles was *SmartInsula*, the pioneering and

award-winning application which formed the backbone of Insula's sophisticated information system that was developed by a UNESCO team in 1997 (Cozzutto, 1997) and has since evolved way beyond that initial application.

The challenge that Insula is now experiencing lies in keeping such rich systems up to date. Insula needs to constantly update bathymetric measurements, bridge conditions, work progress and much much more. Truly emergent, self-sustaining, city knowledge systems delegate data entry to the most appropriate external user, or to beneficiaries of the work – in other words to the end users of the system. In fact, the major technical difficulty for Insula so far has been in decentralizing the data entry, while reconciling accounting systems with technical or engineering systems (frequently CAD-based), and integrating them with geographical information systems (now web-GIS). Many difficulties are being eliminated by forcing compliance with a desired file format as part of a contractual stipulation with outside contractors. Internal adherence to this standard tends to be harder to enforce. As the entire GIS system is ported to the

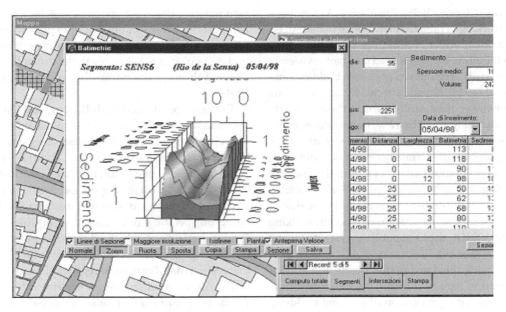

Fig. 27.7   The original canal maintenance system (*SmartInsula*, 1998).

internet, web-based applications to assist contractors in submitting the appropriate digital documents and files are beginning to relieve internal staff of data entry tasks (Todaro, pers. comm. 2003).

Another web-GIS that deals with the underground and embodies some of these principles has been developed by Dr Enrico De Polignol and Dr Lapo Cozzutto for the Environment Department of the City of Venice (www.comune.venezia.it/ambiente/sia). The Sistema Informativo Ambientale (SIA) was initially dedicated to the self-reporting of core-sample analyses about contaminated sites in Porto Marghera. Private companies are entering all of the data into this system, through a password-protected internet browser and the data are later analysed and mapped semi-automatically by the system. The system has recently been incorporated into a more ambitious 'Atlas of the Venetian Environment' which will contain several layers of environmental information for the benefit of Venetian citizens and for the delight of Venetian scholars and scientists.

What is still missing, to allow a sustainable use and re-use of these data repositories for more complex, higher-order analytical studies, is a clear definition of 'informational jurisdictions' and a mutually beneficial agreement to share the information between different agencies – two basic tenets of the city knowledge concept (Carrera, 2004).

DAMAGE TO PRIVATE AND PUBLIC BUILDINGS

The next big category of physical objects that could be impacted by frequent floods includes all buildings: public and private. By implication, this category also includes all of the stores, shops, restaurants and all other businesses housed in these buildings. Flooding like that shown in Fig. 27.4, affects all buildings (and businesses) in its path. *Moto ondoso*, on the other hand, only affects buildings along the canals. Using the information systems developed, it is possible to know how many buildings are affected by floods at any tide level. For example, during a tide that reaches 130 cm., 9124 buildings come into direct contact with the *acqua alta* out of a total of 15 486 buildings in the entire city (including the Giudecca). The system does not allow the prediction of whether or not the *interior* of each individual building actually gets inundated with tide water, although a specific inventory on the '*piani terra*' was conducted by the city in 1999 to answer just such a question (COSES, 1999). Insula S.p.A. has been actively working to raise the elevation of public streets to reduce the number of buildings reached by high tides, so the altimetric maps need constant revision to reflect the 'rialzi'. The owners of private buildings and commercial establishments affected by 'acqua alta' are doing whatever they can to protect their property from floods: using small barriers, impermeable membranes and 'vasche' to seal out the water, and even installing sophisticated drainage systems to direct the water to sump pumps that expel it from the inside of the building. Public buildings are similarly protected and access to many of them is guaranteed even during high floods, by the installation of wooden walking planks (by VESTA, the local public–private Water and Sanitation authority).

Quantifying the damage that floods do to buildings and stores may be difficult, but the expenditure related to the local prevention of flooding building-by-building and business-by-business should be somewhat easier to calculate, by inventorying and estimating the cost of all the measures that have been put in place to either protect private and public property from floods or to make them accessible during floods. Moreover, in addition to tallying the cost of preventive measures, one could also account for all of the restorations and repairs that were caused by particularly severe floods. It may be rather difficult to do so, however, unless the government was involved in disbursing emergency relief funds for such activities and thus records were kept of the repair costs incurred. City knowledge would help us with such difficult estimates, by telling us where all the buildings are with respect to the flood lines. A map of all stores that was produced in 2001 (Duffy et al., 2001) shows that 2862 shops (out of 4263) would get flooded by a 130 cm tide. Together, these two figures – the number of flooded buildings and number of flooded shops – represent a necessary place to start in an estimate of flood damage to private and public property. Since permits are necessary to install local barriers or to raise the ground floors by adding a step or two to the entrance, an estimate of the overall city-wide cost of localized flood prevention measures would be possible if a system was put in place to geographically archive permits that affect the external built environment. The city knowledge framework

recommends that such mechanisms for capturing administrative transactions be put in place to guarantee that the information systems are maintained up-to-date as piecemeal changes to the urban fabric are allowed through the permitting process (Carrera, 2004).

Research conducted with P. Pypaert (UNESCO; see Mancuso and Pypaert, this volume) has been actively promoting such a self-sustaining system, by bringing together data from (i) the former Assessorato all'Urbanistica, that keeps track of zoning and land use; (ii) the Legge Speciale department, that is in charge of disbursing restoration funds, based on specific work estimates; (iii) the Edilizia Privata department, which administers permits; and (iv) the Soprintendenza, which updates the 'vincoli' that restrict modifications to registered historic properties. All these organizations provide information for the benefit of the Commissione di Salvaguardia. This institution has the final word on all major modifications to buildings in the historic city and would greatly benefit from such contextual knowledge at its fingertips when making important decisions (Halloran et al., 2002). Being one of the institutions with a representative in the Salvaguardia commission, UNESCO has a vested interest in bringing about such a confluence of information from all of these sources, but so do all Venetians, who are in the end affected by the permanent changes that are approved by this regulatory body. Some of the buildings touched by high tides are more important than others and have therefore received more attention from public authorities and from philanthropic organizations such as the so-called 'private committees'. Palaces, churches and convents, which are part of Venice's more prestigious architectural heritage, are discussed separately in the section that follows.

DAMAGE TO PALACES, CHURCHES AND CONVENTS

Venice owes much of its fame to its aqueous 'forma urbis' and to the art and architecture it contains. In 1987, the whole city was inducted by UNESCO into the list of World Heritage Sites as a treasure that belongs to all humanity; the first city to receive such an honour in its entirety. When it comes to damage due to floods, the parts of Venice's heritage that stand to suffer the most are palaces and churches, which tend to have ground floors containing more elaborate craftsmanship and more precious materials.

Right after the 1966 flood, UNESCO funded the creation of catalogues of Venetian palaces, churches and convents. In the three decades since then, these catalogues have proved invaluable as a knowledge base, supporting the relentless efforts for the restoration of the priceless treasures of art and architecture first inventoried in the late 1960s. Starting in 1999, teams of WPI students began the task of revisiting and computerising the information known for all these historic structures (Donnelly *et al.*, 1999; Halloran *et al.*, 2002; Marchetti *et al.*, 2003). These efforts allow us to say, for instance, that out of the 383 palaces in Venice, 308 get wet with a 130 cm tide (Fig. 27.8), as do 46 out of 59 convents (Fig. 27.9). Similarly, we can identify all of the churches that would get flooded with the same 130 cm tide. Out of a total of 113 churches in Venice (including Giudecca), 86 are affected by this *acqua alta* (Fig. 27.10).

Nevertheless, in order to convert the knowledge of what gets wet at the various tide levels into a more useful estimate of the damage incurred by these artefacts when they get flooded, it is necessary to know a lot more about what's inside these historic buildings.

Starting in 2002, recording the tombstones, inscriptions and artefacts that are embedded in church floors has been initiated (Delaive *et al.*, 2002; Hayes *et al.*, 2003). In addition to being frequently at lower elevations due to the age of the original foundation, churches have the added handicap of being vulnerable to flooding through their floors, which are riddled with tombstones. The underground cavities where the entombments took place are a conduit through which high tides can quickly reach the artefacts on the floor's surface. This process is abetted by the high permeability of the bottoms of the tombs, which were purposely constructed in such a way as to allow tide waters into them, so that the mortal remains could be rapidly washed away and the tomb could therefore be recycled and reused rather promptly. Thus, for example, recent excavations have shown huge gaps purposely

Fig. 27.8   The 308 palaces flooded with 130 cm tide.

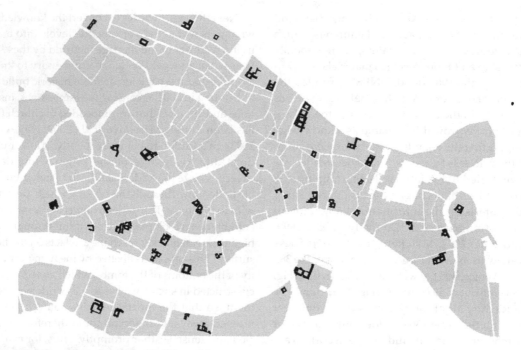

Fig. 27.9   The 46 convents flooded with 130 cm tide.

Fig. 27.10   The 86 churches flooded with 130 cm tide.

left between the planks laid at the bottom of a tomb under the church of San Samuele.[2]

Once this additional city knowledge catalogue of church floor artefacts is finished (over 80 per cent of the surveys have been completed), a more accurate assessment of the potential damage inflicted upon church floors by frequent floods will be possible. Arriving at a similar inventory for the ground floors of palaces and convents would also be useful in this regard.

The overall impact of flooding on churches, convents and palaces can thus include a better estimate of the damage to their floors, but should also include the deleterious effects of salt water on any other artefact that may be touched by tidal waters in the interior of these historic structures. Appropriate monitoring of the decay by the Curia and Soprintendenza could help maintain these catalogues up-to-date and thus prevent catastrophic damage to these important artefacts.

DAMAGE TO ARTISTIC AND ARCHITECTURAL HERITAGE

After the flood of 1966, most, if not all, precious paintings in Venice have been moved up and out of the reach of even the highest high tides. Practically all damage to heritage would now be limited to fixed and immovable structures, such as floors, bases, pedestals, columns, steps and other artfacts within a 2-metre band from ground level (which translates to over 3 metres above sea level).[3] With the exception of the work on church floors (Delaive et al., 2002; Hayes et al., 2003), there are no systematic assessments of the artistic or historic heritage contained in this 'danger zone' in the entire city. Common sense suggests that everything that could be moved away

from this perilous band should have been already moved, though it is quite possible that some artwork might still be in a vulnerable location to this day. Estimating just how many works of art still remain within the 'danger band' is arduous at best, whereas a fully developed (utopian?) city knowledge system could provide the answer to this enigma in a few seconds. Under such a system, the various authorities in charge of heritage collections (like the municipal, provincial and regional governments, the Curia, the Soprintendenze, the Archivio di Stato and the two main libraries – Querini and Marciana – to name a few) would have already catalogued all of the objects that they are respectively in charge of, namely the buildings, properties, church floor artefacts, paintings, mosaics, manuscripts, parchments, books and so forth. The information contained in the Venice archives and in the historic libraries would be even better protected if electronic transcriptions of the manuscripts were produced using the 'Emergent Transcription Assistant System' and the ultraviolet scanner being developed at WPI.[4]

With some foresight, these computerized catalogues could have included a field for the height of the artefact from the floor, which in turn would allow one to simply select all objects whose distance from the floor was less than two metres.

As partial demonstration of the benefits of having city knowledge systems in place, it has been possible to select, from the 2930 pieces of outdoor sculpture that have been catalogued over the last decade (Fig. 27.11), the ones within 2 m of ground level. This search instantly reports that 69 artworks are on public display at a height of 2m or less. Similarly, one could also include in the 2-metre band all 232 wellheads from the wellhead catalogue (Fig. 27.12), since they all sit at ground level, as do all 22 historic flagpole holders that dot the city. More specifically, with the same 130 cm flood used as an example, 122 of the 232 wellheads would get wet (Fig. 27.13). Fortunately, though, most of the 4500+ pieces of public art and street furniture that have been inventoried in the calli and campi of the city are

[2] Courtesy of Dr Luigi Fozzati, Soprintendenza Archeologica, and Dr Marco Bortoletto, archeologist.

[3] This measure has been picked somewhat arbitrarily to reflect the approximate height of the theoretical maximum tidal surge. A one metre threshold would reflect the actual street flooding during the historical maximum level of 1.92, recorded in 1966. The exact dimensions of this 'danger zone' are irrelevant to this discussion.

[4] More information at www.wpi.edu/~carrera, under WPI research.

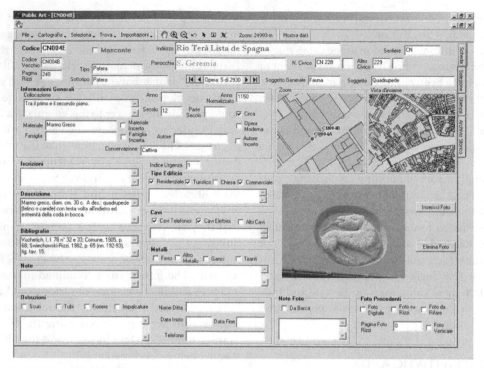

Fig. 27.11   Multimedia GIS application to manage the 2930 erratic sculptures of Venice (*Arte Pubblica* Information System, © Forma Urbis 1997).

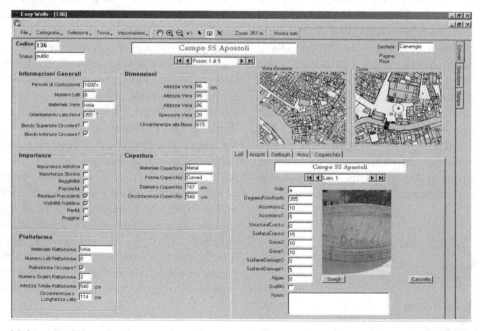

Fig. 27.12   Multimedia GIS application to manage the 232 wellheads of Venice (*Easywells* Information System, © Forma Urbis 1998).

Fig. 27.13   Some of the 122 wellheads that get flooded by a 130 cm tide.

outside the 2-metre danger-zone. Once again, the availability of a city knowledge system would make many of these preliminary assessments instantly possible, as long as each institution in charge of artistic treasures had developed its own catalogue according to the aforementioned informational jurisdictions. It is also obvious that a disaster relief agency, such as the Italian Protezione Civile, would greatly benefit from a distributed, yet interconnected city knowledge system that was able to direct emergency crews to the exact locations where works of art were in danger of being flooded during an *acqua alta*. The key here is not to focus on creating a centralized know-it-all system, but to foster the emergence of a distributed network of smaller (and more manageable and maintainable) systems, through a process possibly termed 'middle-out', to differentiate it from 'top-down' or 'bottom-up', both of which have demonstrated severe shortcomings.

DAMAGE TO CANAL WALLS

In concluding this brief *excursus* into the various points of contact between waters and stones in Venice, this section discusses what is perhaps the most critical interface between these two fundamental elements: canal walls. This section illustrates how a systematic and methodical approach to the accumulation of urban information can yield 'deep knowledge' about the causal nexus between phenomena. In 1998, Babic *et al.*, using the accumulated storehouse of knowledge on traffic, wall damage and bathymetries, concluded that the root cause of wall damage is lack of dredging, which is only later compounded and exponentially accelerated by traffic and wakes (Babic *et al.*,1998; Fig. 27.14). Based on this knowledge, a further study was conducted in 1999 (Borrelli *et al.*, 1999) to quantify the relative and absolute contributions for a variety of possible sediment sources, including the debris produced by crumbling masonry, to the accumulation of sediment at the bottom of canals.

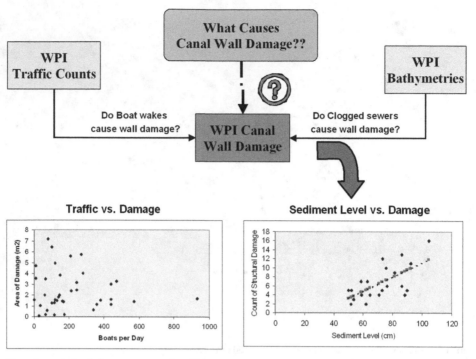

Fig. 27.14   'Deep knowledge': the root cause of wall damage.

This study led to the proposal for a sedimentation model to predict the rate and location of mud accumulation at the bottom of canals, which would allow a more effective and efficient scheduled maintenance before more wall damage was generated by the clogging of sewer outlets. In 2002, the concept of a 'moto ondoso index' was developed, that translates levels of boat traffic (i.e. number of boats) to levels of 'wake-loading' (how much wake energy is discharged in the canal), which helps to better correlate traffic to damage (Chiu *et al.*, 2002).

More recently, WPI students (Chiu *et al.*, 2004) designed and successfully tested an instrument that maps the locations where motorboats discharge energy into the canals when manoeuvring to make turns, or otherwise stopping abruptly by shifting into reverse when an approaching boat threatens a collision, or even when moving back and forth near a dock to tie up the boat and unload people or cargo. This custom device is equipped with a differential GPS, a triaxial accelerometer and an RPM meter and

will produce the first ever map of 'turbulent discharges' in the inner canal network, further facilitating the prediction of future damage along canal walls.

The application of city knowledge principles has paid off dramatically in another project related to moto ondoso entitled 'Re-Engineering the City of Venice's Cargo System' (Duffy *et al.*, 2001). The project has demonstrated the plausibility of 'plan-demanding' knowledge as a consequence of 'plan-ready' information, in opposition to the traditional modus operandi of 'plan-demanded' data collection. Here, work on the optimization of canal closures (Amlaw *et al.*, 1997), which produced plan-ready information on the amount of cargo delivered to each Venetian island, has led directly to the spontaneous emergence of the need to develop a plan (hence the term 'plan-demanding') to restructure the way deliveries are conducted in Venice. The 2001 award-winning project was conducted with and for the former Consorzio Trasportatori Veneziani Riuniti

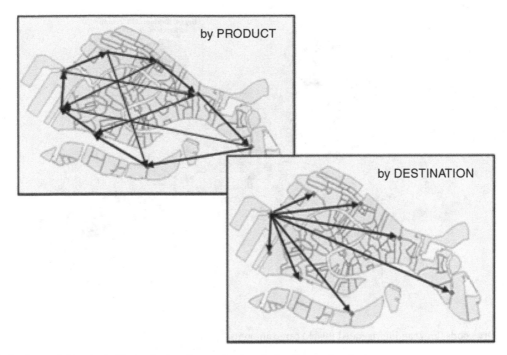

Fig. 27.15   Cargo deliveries in Venice: before (by product) and after (by destination).

(a group representing about 70 per cent of all cargo boat drivers in 2001) and resulted in a revolutionary proposal to redistribute merchandise 'by destination' instead of 'by product' (Fig. 27.15). Cargo demand has been estimated by inventorying all shops and stores in Venice (Plate 10) and by surveying representative samples of each typology of commercial establishment to quantify the number of parcels delivered to each store of that type everywhere in the city.

The map in Fig. 27.16 shows the result of this effort to estimate the demand for cargo deliveries for each island in the city. This research also proposed the creation of a cargo warehouse in the Tronchetto area to allow for the sorting of the merchandise arriving by truck. The study envisioned that the city would be divided into 16 zones with commensurate delivery demands. The warehouse would mirror such a division, reserving a loading bay for each zone (Plate 11).

In the end, it was possible to estimate the total number of boats (for dry and refrigerated goods) that would be necessary to deliver all necessary merchan-dise to all the stores in each zone (Plate 12). Boat drivers were repeatedly engaged in public meetings as well as surveyed and interviewed during the development of this plan. This participatory process created consensus over a plan that would make all licensed drivers shareholders of a single, large logistics and distribution company delivering most of the goods in the city. Boat drivers would be able to deliver to their destination dock at a set time and without interference from other cargo boats. Vacations and sick days would now be possible without the fear of losing clients. The great majority of boat drivers that participated in the public meetings found the benefits of such a system advantageous. The new system would reduce cargo traffic by about 90% while preserving all of the non-seasonal jobs. The entire city would benefit from a proportional decrease in 'moto ondoso' (Carrera, 2002). In the summer and autumn of 2001, the proposal was presented to the Mayor and the Vice Mayor who espoused it. The project, which is currently well on its way to being realized, was a triumph of city knowledge

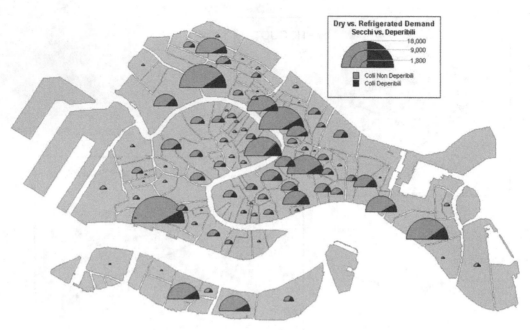

Fig. 27.16   Dry goods (grey) and refrigerated (black) cargo amounts.

principles, demonstrating the 'plan-demanding' potential that such an approach entails. A plan-demanded study dedicated to minimizing disruption to deliveries during canal closings for maintenance and dredging operations led to the accumulation of enough plan-ready information about the inefficiencies of the system of deliveries 'by product' to spur the inception of a follow-up study to explore the boat-drivers' perspective on the revolutionary approach of deliveries 'by destination' that was proposed. The project went full-circle, from plan-demanded to plan-demanding. Presently, the municipality of Venice is moving forward with the actual implementation of this plan. This proposal will reduce cargo-generated *'moto ondoso'* by a substantial amount. Point-to-point cargo journeys will be reduced to one or two per boat, instead of the dozen or more segments travelled each day under the current system. A reduction of the order of 90 per cent of the wall damage induced by cargo boats is therefore not implausible.

THE PROMISE OF CITY KNOWLEDGE

This chapter argues that not only should Venice entertain the notion of a 'central bank' guaranteeing the free flow of data, to be open to all and transparent, but that the city should more importantly employ a sustainable methodology to allow such a bank to emerge from the middle out (not from the top down or from the bottom up) as the sum total of the data produced by a whole variety of contributors distributed in the territory. For such an endeavour to really have staying power and to take on a life of its own, it will be necessary to forego the old-fashioned notion of a 'central bank' and replace it with the principles of city knowledge introduced in this chapter. The basic tenets of city knowledge which are aligned with some of the most recent trends in Geographic Information Systems (Carrera, 2004), suggest the adoption of a middle-out approach based on clear 'informational jurisdictions' assigned to the producers of urban information, starting with the plethora of municipal offices which, through the approval of permits or the assignment of licenses or other administrative acts actually cause – or more frequently allow – the city to change ever so slowly. Intercepting these administrative transactions will enable the city knowledge systems to maintain the information up-to-date more or less automatically. Self-interest is the fuel that will make the city knowledge system thrive since all offices will have self-serving incentives for making their operations smoother and more citizen-friendly. Once a number of departments in the city have embarked in the full life-cycle analysis of the information flows that guide their actions, leading to the identification of their specific jurisdictions, and once the backlog of existing information is captured in databases and GIS, scholars and scientists, planners and decision-makers, as well as citizens at large will be able to enjoy the benefits of the synergic, emergent properties of a connected and shared city knowledge system (as partially demonstrated by Hart *et al.*, 2004). Although the path to a full city knowledge system will suffer from typical implementation woes (Carrera, 2004), such an approach would create a virtual 'central bank' on everyone's desktop and would make Venice a model for sustainable municipal information systems all around the world.

REFERENCES

Amlaw, K., Kervin, C. L., Mondine, I., and C. Vepari. 1997. *Optimization of Cargo Boat Deliveries Through the Inner Canals of Venice*. WPI Interactive Qualifying Project Report. Worcester Polytechic Institute: Worcester, MA.

Babic, K., Leeds G., Sidiroglou, S., and M. Borek. 1998. *Analysis of Sewer Holes and Canal Wall Damage in Venice, Italy*. WPI Interactive Qualifying Project Report. Worcester Polytechic Institute: Worcester, MA.

Bahn, R., Deliso, A., and S. Hubbard. 1998. *The Inventory and Analysis of the Bridges and Pedestrian Traffic in Dorsoduro, San Polo, and Santa Croce Sestieri of Venice*. WPI Interactive Qualifying Project Report. Worcester Polytechic Institute: Worcester, MA.

Borrelli, A., Crawford, M., Horstick, J., and H. Ozbas. 1999. *Quantification of Sediment Sources in the Canals of the City of Venice, Italy*. WPI Interactive Qualifying Project Report. Worcester Polytechic Institute: Worcester, MA.

Caniato, G. 1999. La manutenzione dei rii in epoca moderna. Politiche e modalità d'intervento. In *Venezia la Città dei Rii*. Verona: Cierre Edizioni.

Carrera, F. 1994. *I Rii e la Qualità della Vita a Venezia*. Venice: Coses Informazioni.

Carrera, F. 2001. *City Under Siege*. National Geographic Video. Aired around the globe on the National Geographic Channel.

Carrera, F. 2002. *Trasporto e distribuzione di merci nel centro storico di Venezia: spontaneismo e riorganizzazione. Quaderni di Insula*, September 2002, 29–33.

Carrera, F. 2004. *City knowledge: an emergent information infrastructure for sustainable urban maintenance, management and planning*. Cambridge, MA: Massachusetts Institute of Technology Ph.D. dissertation. See www.wpi.edu/~carrera/dissertation.html.

Chiu, D., Jagganath, A., and E. Nodine. 2002. *The Moto Ondoso Index: Assessing the Effects of Boat Traffic in the Canals of Venice*. WPI Interactive Qualifying Project Report. Worcester Polytechic Institute: Worcester, MA.

Chiu, D., Lacasse, C., and K. Menard. 2004. *Mapping Turbulent Motorboat Discharges in the Canals of Venice, Italy*. WPI Major Qualifying Project.

Consorzio Venezia Nuova. 1988. *Livellazione di dettaglio del Centro Storico e delle isole*. Venice.

COSES. 1999. *Indagine sui danni provocati dagli eventi di marea sulle Unità Immobiliari pubbliche e Private ai piani terra del Centro Storico di Venezia*. Rapporto COSES no. 46/1999.

Cozzutto, Lapo. 1997. *Rapporto sull'Applicativo 'SmartInsula'*. UNESCO Report in the framework of the 'Venice Inner Canals' project.

Delaive, A., Kristant, E., Petrowski, C., and L. Santos. 2002. *The Church Floors in Venice, Italy: An Archeological Study and Analysis*. WPI Interactive Qualifying Project Report. Worcester Polytechic Institute: Worcester, MA.

Doherty, K., Maraia, J., Parodi, C., and F. Souto. 1995. *A Documentation and Analysis of the Traffic, Cargo Deliveries, and Docks within the Insulae of Santa Maria Formosa and Frari*. WPI Interactive Qualifying Project Report. Worcester Polytechic Institute: Worcester, MA.

Donnelly, B., Hart, B., Pilotte, M., and T. Scherpa. 1999. *Safeguarding the Churches of Venice, Italy: a Computerized Catalogue and Restoration Analysis*. WPI Interactive Qualifying Project Report. Worcester Polytechic Institute: Worcester, MA.

Dorigo, W. 1999. I rii di Venezia nei secoli IX–XIV: un profilo storico. In *Venezia la Città dei Rii*. Verona: Cierre Edizioni.

Duffy, J., Gagliardi, J., Mirtle, K., and A. Tucker. 2001. *Re-Engineering the City of Venice's Cargo System for the Consorzio Trasportatori Veneziani Riuniti*. WPI Interactive Qualifying Project Report. Worcester Polytechic Institute: Worcester, MA.

Felices, D., Moreno, C., Munoz, A., and B. Smith. 1994. *A Documentation and Analysis of the Docks, Cargo Deliveries, and Boat Traffic within the Santa Maria Zobenigo Insula*. WPI Interactive Qualifying Project Report. Worcester Polytechic Institute: Worcester, MA.

Halloran, A., Rohleder, K., Malik, R., and K. Fletcher. 2002. *An Integrated, Multi-Agency Approach to the Preservation of Venetian Palaces*. WPI Interactive Qualifying Project Report. Worcester Polytechic Institute: Worcester, MA.

Hart, A., Hetrick, T., LeRay, D., and E. LoPresti. 2004. *Boston City Knowledge*. WPI Interactive Qualifying Project Report. Worcester Polytechic Institute: Worcester, MA.

Hayes, H., Liu, J., Salini, C., and A. Steinhart. 2003. *Church Floors and Archeology*. WPI Interactive Qualifying Project Report. Worcester Polytechic Institute: Worcester, MA.

Marchetti, P., Mazza, K., Hoey, S., and M. Kahan. 2003. *Adaptive Reuse of Venetian Palaces and Convents*. WPI Interactive Qualifying Project.

Mencini, G. 2000. *Sull'Onda Viva del Mare*. Venice.

Piasentini, S. 1999. Aspetti della Venezia d'acqua dalla fine del XIV alla fine del XV secolo. In *Venezia la Città dei Rii*. Verona: Cierre Edizioni.

Pulliero, A. 1987. *Canal Grande Mare Forza Tre*. Venice.

28 • UNESCO contribution to a better understanding of the Venice urban system

A. MANCUSO AND P. PYPAERT

INTRODUCTION

Since the beginning of the 1990s, the UNESCO Regional Bureau for Science in Europe (UNESCO ROSTE) has implemented various research projects in Venice which are concrete examples of the organization's overall commitment to the integrated conservation of the historic city of Venice and its lagoon.

As a direct follow-up to the 'Venice Inner Canals' Project, which aimed to develop, calibrate and validate a first water quality model for the Venice inner canals, and during which very extended databases and a related Geographical Information System (GIS) were developed on subjects such as canal bathymetries, hydrodynamics, boat traffic, sewer outlets and damage to foundations (see www.unesco.ve.it), UNESCO-ROSTE has developed a web-based GIS application intended as a decision support system for the urban planning and management of the city of Venice.

The starting point of this work has been the study and the analysis of the methods of town planning used in Italy today. The main aim of this project has been the realization of a system which could actually be used in the management of the city and not just a piece of speculative research. For this reason, the system was conceived with the active participation of several organizations working on city planning and safeguarding measures, both to collect information, documentation and records on the territory of Venice, and to check the technical support needed in such a system. In particular, the organizations involved were:

- UNESCO Venice Office;
- Ufficio Urbanistica-settore sviluppo del territorio e mobilità (Town Planning Office, Department of territory development and mobility) of Venice municipality;
- Soprintendenza ai beni Artistici e Architettonici (Monuments and Fine Arts Office) of Venice;
- Commissione Salvaguardia (Venice Safeguarding Commission); and
- Worcester Polytechnic Institute, USA.

The analysis of the different responsibilities of these organizations, the study of the information management processes which these organizations employ today and their reciprocal information needs have been decisive in choosing the right paths to follow; in fact, to have a really effective and working system it must answer, to its best ability, to users' needs and it must offer a tangible answer to problems, interacting with pre-existing systems without substituting for them.

For this reason, the base of this work has been the specific information demands and the true issues faced by users. Moreover, even if the system may have specific configurations, it must have an information platform common to a number of different users. Consequently, it has to be configured as a 'public system'; that is to say, it has to be able to share information. To achieve this outcome, the system has to be sufficiently flexible to adapt itself to different needs and it has to be able to communicate with the information systems already in use. Furthermore, it has to be easily implemented and updated, through the introduction of new information and functions to answer to needs generated by the dynamics of social and territorial changes.

The analyses that have been undertaken over the course of this project underline, on the one hand, that the different organizations working on city planning and safeguarding measures have specific internal

Flooding and Environmental Challenges for Venice and its Lagoon: State of Knowledge, ed. C. A. Fletcher and T. Spencer.
Published by Cambridge University Press, © Cambridge University Press 2005.

demands but that, on the other hand, they also need to share information and interact with one another. The information system developed by UNESCO is not a substitute for old individual systems; on the contrary, it can communicate with these systems and it makes their data reciprocally available to them. In addition, the information collected by every organization has been made available. Interrogating information systems at different levels produces the flexibility necessary to best address any problem of interest. The user needs described above have led to the development of a GIS application based on a web platform.

A WEB-GIS FOR THE VENICE URBAN SYSTEM

The main functions of the developed web-GIS system are described below in detail. The application can be directly consulted on the following website: www.intelligencesoftware.it/unesco/venezia (username: guest, password: guest).

The web-GIS system allows the user to interactively consult the cartography of the historic centre of Venice. It is also possible to locate a specific area, either from the street number or by a toponymical research mode. The different information levels available through the system allow the user to develop an interactive and customized consultation, according to his/her individual demands. At present, reference to the following information levels is possible: (basic cartography) islands, *rii terà* (filled-in canals), architectonic details, buildings, green spaces, street numbers, toponyms and flooding levels. All these information levels can be contextualized by superimposing them onto the orthopicture. This allows a clear representation up to a scale ratio of 1:500.

Some of the details from the map can be selected to give a wide range of data. Thus, for example, in selecting a building the system provides basic information such as the architectonic typology, the height of the ground floor above sea level, the height of the eaves, the size of the surface and all the provisions of the law stated by the master plan for that kind of building. Furthermore, interactive consultation of the cartography allows the user to find information on

several objects. Thus, for example, it is possible to find out if there are works of art in a particular area. The query named 'works of art in public overlook' provides access to a form which shows a picture of the works of art and some historical and artistic information. (The use of this data bank has been kindly granted by the Venice Project Center of the Worcester Polytechnic Institute.)

The system also contains information on bridges, bell towers, canals and buildings. In addition, the system allows for the consultation and management of the data bank relating to building legislation of the Venice Monuments and Fine Arts Office. In this way, the Monuments and Fine Arts Office can transmit information on laws in real time to all the organizations needing it. Similarly, the system allows for the consultation and management on a cartographic basis of all the files of the Venice Safeguarding Commission concerning the transformation of buildings. This function is a robust instrument to support the decision-making process because it allows the valuation of interventions in building transformation in their urban context, interacting with different kinds of information. In this way, it is possible to make specific urban analyses to determine long-term political strategies or undertake decisions on building transformations.

At the present time, the system is being tested by the organizations involved in the project but it is hoped that in a short while it will be used as a supporting instrument for the safeguarding of the artistic and architectonical heritage of Venice. For this reason, UNESCO is now developing, in a collaboration with the Palazzo Cappello Venexia scientific institution, the integration of this application with another system, aimed at monitoring and managing conservation and restoration interventions on historical buildings.

Part IV • Large-scale engineering solutions to storm surge flooding

29 • Introduction: large-scale engineering solutions to storm surge flooding

T. SPENCER, P. M. GUTHRIE, J. DA MOSTO AND C. A. FLETCHER

There is no doubt that the most intensely debated element in the range of flood protection and environmental restoration measures being implemented and proposed for Venice (Scotti, chapter 30, identifies six types of intervention) is the plan to temporarily close off the inlets to the Venice Lagoon at times of high water levels by what have become known as 'mobile barriers'. These are retractable barriers which can be deployed rapidly. Proposals for an intervention of this kind have a long history; notably the 'competition of ideas' in 1971 which followed the 1966 flood; through the establishment of the Consorzio Venezia Nuova in 1984, charged with the sole responsibility for implementing the barrier solution; the design, construction and testing of the first experimental prototype, the MOSE ('MOdule Sperimentale Elecctromeccanico') between 1988 and 1992 near the Lido inlet; the construction of the physical scale model at Voltabarozzo (Padua); and other specialized and detailed design and modelling initiatives in collaboration with international experts. Throughout, the choice of construction has been driven by a diverse set of requirements: barrier performance (including speed of deployment, reliability and maintenance); aesthetics (in terms of the lack of visibility of the structures from the city); avoidance of interference with expanding port activities (Razzini, chapter 32); and the need to meet environmental concerns with regard to impacts on lagoon morphology, ecology and water chemistry during both barrier construction and operation, the last condition still not satisfactorily met for some parties.

The current design envisages a total of 78 buoyant steel gates pin jointed at their base on the seabed, and associated jetties and breakwaters, deployed across the Chioggia (18 gates), Malamocco (19 gates) and Lido (41 gates, in two sections around an artificial island) inlets. The Malamocco and Chioggia inlets will also incorporate a navigation lock (Scotti, chapter 30). The individual gates are 18–30 m high (although they will never protrude more than a few metres above the water surface), 3.5–5 m thick and 20 m wide (Eprim, chapter 31); it is clear that in order to survive in a corrosive marine environment, with the attendant risk of biofouling and to meet necessary performance levels – primarily the minimization of water leakage between individual gates – construction and installation protocols will need to be to high levels of accuracy found only in a few large scale civil engineering works. Under non-storm conditions, the gates will sit underwater within caissons of reinforced concrete on the inlet floors. When storm tides are forecast, compressed air will be forced into the gates. The gates will then rise, rotating on a hinged axis with the caisson up to an angle of 45°, to block the flow. The inlets will remain closed only for the duration of the high water event and the time required to manoeuvre the gates between their 'down' and 'up' positions. The average gate closure operation has been estimated at five hours but there are likely to be times (as in November 1966; see Tomasin, chapter 9) when the occurrence of seiches will prolong the period of closure considerably. Assessments of the operation of sophisticated storm surge barriers elsewhere identify the likely points of failure as being electrical and electronic control systems (Wilkes and Lavery, chapter 35) and human factors, particularly where control systems are highly automated (Bol, chapter 38). One particular problem, seen in Cardiff Bay, UK (Faganello and Dunthorne, chapter 37) is 'gate hunting' where the system tries to follow long period water level oscillations, themselves induced by sudden gate movements.

Flooding and Environmental Challenges for Venice and its Lagoon: State of Knowledge, ed. C. A. Fletcher and T. Spencer.
Published by Cambridge University Press, © Cambridge University Press 2005.

The nature of the Venice flood defence system clearly calls for the accurate identification of those flood events for which gate closure will be required and a minimization of the number of 'false alarm' closures. These rules necessitate development of forecasting models, data assimilation schemes and the reanalysis of databases of both historic environmental information and previous forecasting model performance (see Spencer *et al.*, chapter 3 for current forecasting approaches). These are complex issues as water level predictions, and hence closure decisions, are not only dependent upon flows through the lagoon inlets but are also affected by rain falling on the lagoon, runoff from the large surrounding watershed and wind set-up and set-down effects over the lagoon (Eprim *et al.*, chapter 33). And the issues of closure frequency and duration highlight the delicate balance that is required in such schemes between effective flood protection on the one hand and, on the other hand, the maintenance of the ecologically valuable yet fragile transitional environments of estuaries and coastal lagoons.

It is useful to consider these issues for Venice alongside experience gained from the construction and implementation of large scale flood protection structures elsewhere in Europe. In particular, it is instructive to gain feedback from schemes at different stages in the storm surge barrier life cycle: from the immense 'Delta Plan' of The Netherlands, developed and implemented progressively between the late 1950s and the late 1980s; from the Thames Barrier, first used operationally in 1983; from the Cardiff Bay barrage completed in 1999; and from the as-yet-uncompleted St Petersburg Flood Protection Barrier. Insights can be gained from the design, construction and operational phases and, in the case of long-completed structures, how the enclosed ecosystem has adapted to changes in tidal regime, water residence times and water body type, sediment supply and distribution, and water quality.

From a civil engineering viewpoint, it is clear that the challenges have not been exclusively technical. Hunter (chapter 36), in reviewing the construction of the Cardiff Bay, UK barrage, outlines the good practice that was required at the planning and design, procurement and construction phases, the importance of clear and fixed goals, and the need for maintenance of consultation with, and communication between, all project stakeholders. All these challenges will need to be faced if the Venice intervention is to reach successful completion. Public information and involvement is normally vital. In St Petersburg, public concern over the environmental impacts of the Flood Protection Barrier, combined with a loss of faith in the government, effectively led to a halt in construction in 1987, when the barrier was two thirds complete. Gerritsen and co-workers (chapter 41) argue that public participation in the evaluation of a new environmental impact assessment for the project has been a key element in generating support from both the public and European financial institutions to now see the project through to completion. For long-completed storm barriers, it appears that environmental impacts cascade through the ecosystem over time. Thus as sediments in the now sediment-starved Eastern Scheldt estuary have been transferred to lower tidal levels, the loss of higher mudflats has reduced the forage time for intertidal wading birds, suggesting lower wader population carrying capacities in the future. At the same time, the change in water residence times has resulted in more rapid spring warming of waters which may have triggered the expansion of the exotic Pacific oyster (Saeijs and Guerts van Kessel, chapter 39). The time lag of such changes, from initial political decisions on implementation and construction, highlights the critical importance of on-going environmental monitoring and integrated evaluation of monitoring results in all such large scale engineering interventions.

Case studies show that it is possible in the design stage to increase the eventual flexibility of use. Increasingly, this suggests that structures originally constructed solely for flood control purposes might be used for more sophisticated water management which could help offset some of the ecological difficulties described above. Such strategies have already been used in parts of the deltaic plain of The Netherlands and are also evident in the changing closures strategy of the Thames barrier, which now incorporates river flow management as well as tidal flow control. Experimental studies also suggest that sectional opening and closing of the St Petersburg Flood Protection Barrier could be similarly used to improve ecological conditions in Neva Bay

(Mikhailenko *et al.*, chapter 40). Such manipulation of a defence system has been suggested for the Venice mobile barriers too (Harleman, chapter 34). Of course, all of these issues are a moving target in the context of global environmental change. These processes shift the baseline conditions against which storm surge events take place. They incorporate near future sea level rise (where the envelope of potential rates of rise is large), changes in storm incidence over the next century which are similarly highly uncertain, and where the possibility exists of the drainage network into the lagoon being overwhelmed by intense rainstorms (Spencer *et al.*, chapter 3).

Finally it is clear from much of the foregoing discussion that barrier structures and their operation sit within wider changes in social, economic and political forces. Thus Saeijs and Guerts van Kessel (chapter 39) demonstrate how the growing public concern over ecological issues in the 1980s in The Netherlands – how, as they put it, 'safety or environment' became 'safety and environment' – saw the plan for total closure of the Eastern Scheldt estuary and the creation of a freshwater lake in the 1960s ultimately replaced with a much more flexible storm surge barrier solution and, through compartment dams, the preservation of a gradient from fully saline to freshwater environments. As Wilkes and Lavery (chapter 35) note, when the Thames Barrier was designed and built in the 1970s and 1980s 'the vocabulary of sustainability was not in use' whereas discussions as to the second generation of Thames estuary defences (the present structure reaches the end of its design life by 2030) must now be considered within a framework of sustainable long-term flood management. Such a framework may generate, *inter alia*, quite different, and perhaps more radical, flood aversion strategies. Furthermore, it is certain that future strategies will be more closely constrained by the need to comply with European Union-driven environmental legislation, such as the successors to the Habitats and Birds Directive and the Water Framework Directive.

Coastal barriers only buy time in the search for more sustainable flood management. They are unlikely to be single-generation features of the coastal landscape. As flood risk, and, just as importantly, perception of flood risk, changes so existing structures are likely to need to be modified and/or replaced under new strategies (see, for example, Wilkes and Lavery, chapter 35, for such a debate on the future of the Thames Barrier). Venice is still just at the start of this process, although the commitment to the MOSE seems irrevocable and the extent to which flexibility can be incorporated into use of the system needs to be further examined.

30 • Engineering interventions in Venice and in the Venice Lagoon

A. SCOTTI

INTRODUCTION

The actions decided to recover and to protect the city of Venice and its lagoon are frequently named as 'engineering interventions' due to the presence of the mobile barrier, which is certainly a sophisticated engineering challenge. In reality, the actions decided upon are much more wide ranging and can be described as a huge and unique Environmental Plan. The engineering component of the Plan is certainly important, but, alone, can not solve the long lasting problems of the lagoon. Other components, that are not less important, have to be considered such as the morphology, the biology, the chemistry, the landscape, the culture and the history of the lagoon. This chapter outlines the different problems of the lagoon and gives a short description of the interventions needed and decided upon to cope with these problems

THE LAGOON: A VERY SPECIAL ECOSYSTEM

We all know that an ecosystem is a system of relationships where stability is maintained if the relationships are not modified. But if something changes (e.g. the temperature, the wind, the human intervention) the equilibrium is lost and another level of equilibrium is reached. In general only specialists can understand those changes and their causes since they occur over long timescales while the information is commonly only available for shorter periods of time.

The Lagoon of Venice is a great example of this. In the lagoon numerous factors interact, each linked to many others by intimate relationships of cause and effect involving chemical, biological, physical and morphological processes, natural phenomena typical of alluvial areas, complicated by other dynamic processes on a planetary scale (such as eustasy) and further and highly complicated by the massive presence of man with his economic and social activities.

What is special in the Lagoon of Venice, in comparison with any other ecosystem, is not only the historical and monumental value of Venice but also the amount of available information, going back centuries. It should be noted that the existence of Venice, and its military security and health conditions, have always been dependent on the environment, which has both encouraged and pushed the study and the understanding of the environment.

It is very special also for the speed of the environmental changes: many of the changes can be even detected by the direct experience of those who live and work in the lagoon. The capability of observing the changes should not be confused, however, with the capability to understand the causes of such changes. The interactions between different processes remain in fact very complex and the timescales that the changes occur on does not readily facilitate the interpretation of the environment. Rather, it often increases the number of possible factors to be studied and compared before reaching a conclusion.

When we have studied the lagoon and interventions (engineering or not) have been proposed, we have always kept in mind how special the lagoon is and made sure that any interpretation of the complex relationships, even if very reasonable and considered valid in other cases, could only be accepted if verified in depth with relevant data and state of the art mathematical models. Mathematical models allow a comparison between the importance of each of the possible factors identified.

Flooding and Environmental Challenges for Venice and its Lagoon: State of Knowledge, ed. C. A. Fletcher and T. Spencer.
Published by Cambridge University Press, © Cambridge University Press 2005.

THE PROBLEMS THAT NEED INTERVENTIONS

As mentioned above, the lagoon is an unstable place, continuously in conflict between the land and the sea. Its survival is thus under perpetual threat. With immense work, Venetians have delivered the lagoon from the dominion of the rivers and defended it from the sea. For centuries, they have sought and found, lost then re-conquered, an 'impossible' equilibrium. With unflagging and complex tenacity, they have 'artificially' maintained a natural place.

But the twentieth century gave us a city of Venice and its lagoon, definitively at risk. As time goes by, the lagoon is becoming deeper (by more than 60 centimetres in one century), and 50 per cent of the marshland existing at the beginning of 1900 has been lost, as can be easily detected comparing the lagoon bathymetry measured by the Magistrato alle Acque in 1900, 1930 and 1970 and the bathymetry measured by the Magistrato alle Acque in 1989 and 2000. Furthermore, in the same period, the volume of water exchanged twice a day between the sea and the lagoon has increased, the volume of pollutants discharged by industry, agriculture and the cities has also increased, the protection from the sea built in the past centuries has gradually lost its strength and the city is flooded ever more frequently and with ever greater intensity. Meanwhile, the use of the lagoon has rapidly increased. The port of Venice has grown, becoming the fifth largest in Italy, fishing in the lagoon is today a real industry and each year 6 million tourists visit the city of Venice.

It should also be noted that the relative level of the land dropped by 23 cm (as demonstrated by CNR – Gatto Carbognin) with respect to the sea and the tide rose by a further 8 cm as a result of major morphological modifications occurring in the lagoon basin up until the 1970s (from an analysis performed by the Magistrato alle Acque, confirming the results presented by Pirazzoli). At the beginning of the twentieth century, the lowest parts of the city flooded an average of 90 times (calculated over 100 years). Today this average has risen to 3,900. The resultant damage is insidious and not yet visible. Until now, attention has been focussed on damage caused in the past. But it is the cumulative effect which escapes predictions, although its symptoms are obvious: physical structures degrade slowly at first, then once certain thresholds are passed, tumultuously. At the beginning of the twentieth century, the return period of a tragic event such as the one that occurred in 1966, flooding the whole city under a metre of water and causing devastating damage, was once every 1,000 years (in other words, it was almost impossible). Today, the same event could take place every 140 years and, with a 20 cm rise in sea level, every 40 years. These evaluations are possible as more than 100 years of water level measurements are available for Venice. The past loss in the relative level of 23 cm and the future eventual further loss due to subsidence, eustatic rise and perhaps greenhouse effects, can be estimated by translating the reference level by the same amount. So extreme events are no longer a remote possibility. The only indefinite variable is when and how they will occur. Venice is now in the 'inter-tidal band', no longer above it as it was when it was built. Massive documentation on the water levels proves this scientifically.

THE SPECIAL LEGISLATION AND THE PLANS TO SAFEGUARD VENICE

During the last thirty years, the uniqueness of the Venice case, regulation of the lagoon and the physical defence of a city considered priceless to the world, has produced a body of ordinary and special legislation in every way exceptional (Special Laws no. 171/73, 798/84 and 139/92), making the safeguarding of Venice a matter of 'priority and national interest' and linking the problem of protecting it from high waters to all the solutions for the restoration of environmental balance and socio-economic development hypothesised for the city. Up until now, the funding provided for by Special Law no. 798 amounts to more than 5 billion Euro distributed among the implementing bodies. With its share of the available funding, each body carries out the activities for which it is legally responsible. The *State* works towards physical safeguarding and restoration of environmental balance in the lagoon basin; the *Veneto Region* is responsible for pollution abatement in the waters of the drainage basin and *Venice* and *Chioggia Municipalities* are responsible for economic and social development and the restoration and conservative improvement of the urban structure.

The special legislation state that all the activities for the safeguarding of the lagoon are executed under the political control of an interministerial committee composed by the Ministry of Infrastructures and Transportation, the Ministry of Environment, the Ministry of Culture and Heritage, the Veneto Region and the Municipalities of the cities of the lagoon, under the technical control of the highest technical committee existing in Italy at the institutional level. These committees have operated since 1994, governing every decision taken since then. They have approved the plan of intervention and the design of each specific project. In particular, following the approval of a General Plan of interventions to safeguard the Lagoon of Venice in 1994, the following detailed plans have also been approved:

- the Plan to reduce the pollution from the drainage basin;
- the Plan to implement a sewerage system of the historical city;
- the Plan for coastal reinforcement;
- the Plan to restructure the jetties;
- the Plan for morphological recovery;
- the Plan to stop environmental decay and to promote its improvement;
- the Plan to assess the feasibility of re-opening fish farms to tidal expansion;
- the Plan to reduce the risk induced by oil tanker traffic in the Venice Lagoon; and
- the Plan to protect all living areas in the lagoon from flooding, considering both, the frequent and the exceptional floods.

The first Plan to reduce the pollution from the drainage basin has been developed under the responsibility of the Regione Veneto. The responsibility of the second Plan (the sewerage system of the city) belongs to the municipalities. All other Plans must be developed by the State through its concessionary, the Consorzio Venezia Nuova. Up until now, the Consorzio Venezia Nuova has executed more than 100 projects. In January 2004 the construction of the interventions to be built at the lagoon inlets was also started.

THE STUDIES AND EXPERIMENTATION SUPPORTING THE PLANS

The need to solve many different problems, to understand their causes, and to identify the remedial actions, have made it necessary to develop a better knowledge of the dynamics of each component of the ecosystem and, therefore, to carry out, for each of these components, studies on the current situation and on the past trend. Experiments in the laboratory and on site have been undertaken to indicate the possible future trend, with and without remedial interventions. It has also been necessary to use the best available mathematical models, which have been calibrated and validated with reference to data on past trends, and to develop new ones, specific for the lagoon.

Numerous scientific processes needed to be investigated in depth and their degree of importance balanced alongside each other. Among these it is worthwhile to mention:

- the natural forcing components, such as the tide, the currents, the wind, the waves, the subsidence, the eustasy, the temperature and in general the local climate, salinity, and river inputs;
- the anthropic forcing components, such as navigation, dredging for navigation, fishing, fish farming, the activities of the port, industry and cities present in the lagoon and the use of the drainage basin;
- the processes that explain and quantify the lagoon changes, such as the currents induced by waves, wind and tide, erosion, sedimentation, biological growth of the flora (submerged and terrestrial) and of the fauna (aquatic and terrestrial) and chemical and biochemical reactions; and
- the processes that explain and quantify the relationships existing between human activities and the changes in the lagoon.

This has been a massive commitment. From the start of activities until today, the Water Authority and Consorzio Venezia Nuova have completed a significant inter-disciplinary programme to study, survey and monitor the interventions (before and after implementation). This programme was started more than 15 years ago and has been up-dated, integrated and certified by experts down the years. The work has been carried out with the collaboration of universities, public and private national and international research centres, leading Italian and foreign laboratories, engineering companies, professional studios and top experts in the individual sectors.

An Information Service has been organized, a massive data bank containing the historic memory of the lagoon and one of Italy's most advanced structures in terms of information technologies applied to environmental management. All this information is available, upon request to the Magistrato alle Acque, depending on the use of the information and the organization requiring them. Finally, at the Ministry of Public Works – Venice Water Authority Voltabarozzo Experimental Centre for Hydraulic Models, near Padua, studies and tests relating to implementation of the safeguarding measures have been carried out using physical and mathematical models. As a result of this work, the Voltabarozzo Centre now has the equipment, facilities and experience which makes it the most important Italian centre for research in this field.

INTRODUCTION TO THE INTERVENTIONS

When describing the interventions that can solve the lagoon problems, it is always difficult to explain that each of them is part of an overall plan. It is possible to mention it, but the understanding is not easy, since it implies a deep knowledge of many natural processes at a scientific level and also a deep integrated knowledge of the information available. Furthermore, the demonstration can be found only in the plans prepared for each problem while, often, the impact of a group or of a single intervention seems in conflict with experience. Each of the several objectives can in fact be reached by acting on many components of the lagoon ecosystem and, for each component, by many types of interventions located in the different parts of the lagoon. Furthermore, it is possible that one intervention can achieve different objectives.

Therefore, in reading or listening to the description of each intervention one cannot expect to find the solution to all the problems of the lagoon. When this has been attempted, expectations have been dis-

appointed. For example: the expectation that the interventions that eliminate floods in Venice should also solve the erosion of the lagoon. Or the opposite: that the interventions against erosion can also eliminate the floods. Those statements were certainly induced by the wish to find a short cut, a solution that solves all the problems. That this approach is wrong can be easily demonstrated by the fact that Venice has always periodically flooded even when the lagoon was filling up with sand or when it was eroding heavily. In reality, the safeguarding of the lagoon can be obtained by acting in many directions, as already mentioned, and also by taking care of the environment through appropriate maintenance and with new regulations for the use of the lagoon. The description presented hereafter is certainly very simplified since it is limited to a few pages and these pages are mostly used to describe innovative interventions. Furthermore, no reference is made to the new regulations introduced for the use of the lagoon. The interventions that are part of the overall master plans are the following:

- the intervention for the protection of the coastline from storms in the Adriatic Sea;
- the intervention for the restoration of the environment; and
- the intervention for the protection against flooding.

INTERVENTIONS FOR THE PROTECTION OF THE COASTLINE FROM THE SEA STORMS IN THE ADRIATIC SEA

Coastal protection

From north to south, the watershed between the Adriatic Sea and the lagoon is defined by the coastlines of Jesolo, Cavallino, Lido, Pellestrina, Sottomarina and Isola Verde, a 60-kilometre-long strip of land interrupted by the mouths of the rivers Sile (to the north) and Brenta (to the south) and the three inlets of Lido, Malamocco and Chioggia. As the first line of defence against sea storms, down the centuries, the coast has literally lost its function. Waves and currents together with a series of other factors such as land subsidence have progressively eroded the beaches, reducing their width or even completely eliminating them. The built-up areas near the sea have found themselves ever more exposed to risk.

These problems have been tackled with a complex intervention programme including, depending on local conditions, the creation of new beaches or widening of existing ones; construction or reinforcement of 'breakwater' shore structures; restoration of the ancient seawall defences (*murazzi*) and, where possible, restoration of the band of dunes. This latter type of measure, as well as providing protection against sea storms, is also aimed at restoring elements once typical of the environment and coastal landscape and favoured habitats for characteristic species of animals and plants. The system of interventions on the seaward side of these barriers is integrated with the programme of interventions being implemented on the lagoon side for the local defence of built-up areas on the coastal strip from high tides (see Fig. 30.1).

Fig. 30.2   Malamocco north jetty during interventions.

Achievements to date

To date, work has been completed on five of the six lengths of coastline for a total of almost 40 km out of the 50 km specified in the General Plan of Interventions (80%).

Revamping of the jetties

The aim is to consolidate and restore the stability of the jetties, definitively opposing the deterioration which previous ordinary maintenance work had tackled by remedying the most immediate damage only, without eliminating the causes.

Achievements to date

All the main work specified has been completed for a total length of 10 km out of the 14 km specified in the General Plan of Interventions (71%). The remainder is accounted for by a series of complementary measures currently being implemented (see Fig. 30.2).

INTERVENTIONS FOR THE RESTORATION OF THE ENVIRONMENT

Recovery of the lagoon morphology

The morphological wealth and complexity of the lagoon ecosystem are gradually disappearing due to the escalating action of a series of factors linked in a vicious circle. Erosion, high waters and wave motion are gradually transforming the lagoon which is thus

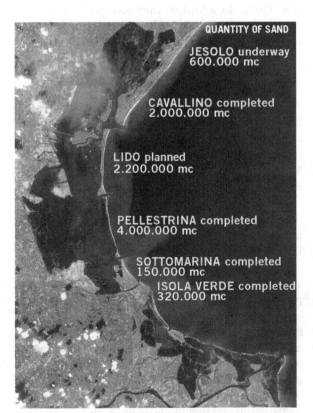

QUANTITY OF SAND

JESOLO underway
600.000 mc

CAVALLINO completed
2.000.000 mc

LIDO planned
2.200.000 mc

PELLESTRINA completed
4.000.000 mc

SOTTOMARINA completed
150.000 mc

ISOLA VERDE completed
320.000 mc

Fig. 30.1   Coastal reinforcement: interventions.

losing the physical characteristics of a wetland and gaining the more simplified and uniform characteristics of a marine environment.

The first objective of the morphological recovery measures is to oppose erosion by retaining part of the sediment, which would otherwise be dispersed at sea. The objective is then to restore the environment, hydrodynamics and natural habitat associated with the individual elements of the lagoon environment, thus contributing to restoring the balance of the ecosystem as a whole. Currently, of the 1,100,000 m³ of sediment previously lost at sea, 400,000 m³ is retained in the lagoon as a result of work carried out by the Water Authority – Consorzio Venezia Nuova. After years of study, experimentation and monitoring of the outcome of the work, the plans and measures now have a more natural and environmental orientation. The programme of measures ranges from the dredging of lagoon channels and use of the dredged sediment to construct mudflats and salt marshes to the protection of existing salt marshes using the most advanced natural engineering techniques and, to reduce wave motion, the raising of the lagoon bed and its consolidation by planting eelgrass. This programme also incorporates work to restore the shores of minor islands (see Figs. 30.3 and 30.4). Contributing to identification of the intervention areas are a series of specific studies enabling the lagoon to be divided into homogeneous areas defined on the basis of local hydrodynamic, morphological, environmental and functional conditions.

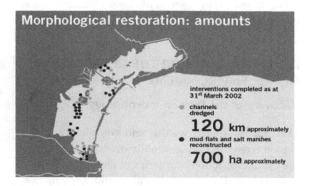

Fig. 30.3   Morphological restoration: amounts.

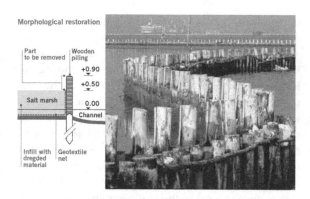

Fig. 30.4   Reconstruction of salt marshes.

Achievements to date

To date, about 400 ha of mudflats and salt marshes have been reconstructed out of the 800 specified in the General Plan of Interventions (50%) and 70 km of channel have been dredged out of the 100 km specified (70%). In addition, the shores of 7 out of 14 minor islands (50%) have been reconstructed for a total of 5 km.

Arrest and reversal of environmental decay

Abatement of pollution in the waters flowing into the lagoon basin from the hinterland is largely the responsibility of the Veneto Region. The scope of those measures, falling within the competence of the Water Authority – Consorzio Venezia Nuova, is to improve the quality of the lagoon with intervention in the lagoon. This means, above all, preventing the dispersion of contaminants from dumps used in the past to dispose of waste of various origins (including industrial waste) and to reduce pollution in the channels and canals of Porto Marghera (bottom and sides) by removing the residues of industrial processes which have accumulated in the sediments over the years. The full gravity of this problem has become evident only in recent years and it is being tackled systematically. The proposed work is covered by a Programme Agreement for the Porto Marghera chemical industry signed in October 1999 and includes numerous and specific studies (partly completed) designed to provide a precise and up-to-date picture of the actual conditions and danger to the

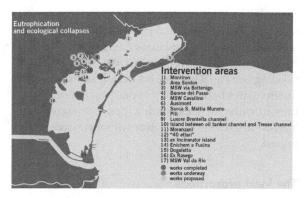

Fig. 30.5   Abandoned dumps.

environment (see Fig. 30.5). New bed cleaning-up techniques are being tried out in other lagoon areas to block the release of the pollutants which have accumulated in the sediment over the years. In order to prevent the risk of environmental crisis due to the proliferation of macroalgae, periodic 'harvesting' campaigns are also carried out.

Achievements to date

To date, four of the six major lagoon dumps specified in the General Plan of Interventions (67%) have been sealed. At Porto Marghera, 7 km of industrial canal banks out of the 28 mentioned (25%) are being reconstructed and 90,000 m³ of polluted sludge is being dredged out of the 400,000 specified (23%). In the annual macroalgae harvesting campaigns carried out to date, 250,000 m³ of algae have been removed from critical areas of the lagoon.

Re-opening of fish farms to tidal expansion

The re-opening of fish farms, all situated along the edge of the lagoon, could improve the hydrodynamics and quality in peripheral areas of the lagoon with poor water exchange. This is the objective of the re-opening activities included in the list of works falling within the competence of the Venice Water Authority and has been initiated with a preliminary pilot project in a fish farm in the south lagoon. However, it has been demonstrated many times by different mathematical models that this type of intervention has an extremely modest effect on the reduction of tidal levels in the lagoon and, as a consequence, on the frequency of flooding. Re-opening of the fish farms would produce an average reduction in tidal levels in Venice of just over 1 cm. After many years of discussions on this issue, finally this result has been shared also by the opponents of the mobile barriers.

Achievements to date

The first of the four pilot projects specified in the General Plan of Interventions (25%) has been implemented in the Valle Figheri.

Exclusion of oil tanker traffic and aids to navigation

An Operational Plan has been drawn up to eliminate oil tanker traffic from the lagoon with the aim of reducing the extremely serious risk of pollution or even ecological catastrophes associated with the transit of large petrochemical tankers. The project, has been developed in a number of phases and involved consultation with relevant stakeholders. The latest plan, involves closing the Marghera terminal of the oil pipeline supplying the Mantua refinery and using the Genoa-Cremona pipeline in its place, which will run as far as Mantua. The corresponding traffic to Genoa will be diverted and the production and transport of petrochemical products in the Adriatic will be reorganized in order to reduce the negative impact on Marghera as far as possible. In addition, measures to prevent ships which do not conform fully to international safety standards from entering the lagoon will be implemented.

Achievements to date

To make the transit of large ships heading for Porto Marghera through the lagoon safer, the Water Authority – Consorzio Venezia Nuova have installed a system of lights and a series of auxiliary electronic aids along the so-called Oil tanker-Canal linking the Malamocco inlet with the industrial zone. The new illuminated route (*sentiero luminoso*) is 15 km long and increases the safety of shipping even at night or when fog causes poor visibility.

Interventions for the protection against flooding

An integrated system of interventions

Current legislation establishes the objective of guaranteeing the total defence of all built-up areas of the lagoon from all levels of high water, including extreme events. In selecting the intervention which best fulfils this objective, the designers assessed the effectiveness and technical and economic feasibility of a number of alternative design hypotheses grouped into the following categories:

- interventions acting on the physical structure of the lagoon (morphological);
- intervention for the opening of the fish farms;
- interventions for the reduction of the volume of water exchanged with the sea;
- interventions aimed at defending built-up areas of the lagoon by raising shores and banks
- interventions to temporarily close one or two lagoon inlets
- interventions to temporary close the three inlets.

Numerous possible combinations of these categories of solution were also studied as illustrated in the General Plan and in the Environmental Impact Study.

The solution chosen to provide a full response to the complex problem of high waters involves a combined system of interventions including: temporary closure of all three lagoon inlets by means of a row of mobile gates; local measures to raise shores and banks, compatible with the architectural and socio-economic structure of the individual built-up areas; and wide area morphological measures compatible with the environment to protect against the most frequent flooding (see Fig. 30.6). This is a flexible strategy which differentiates between the various prevention measures in order to cope with a possible significant increase in sea level in the future without unacceptable impact.

The local protections

Many alternative solutions were studied such as: raising of pavements; raising of perimeters of the islands; raising of each building; raising of large areas by injecting the soil. The first two solutions were finally adopted, while the other two are presently not practical or feasible. Local defences involve more than raising the surface level of shores, banks and public paved areas: through complex operations designed to avoid filtration of water from the subsoil and flow back from drains; protection of ground floor areas of buildings from flooding; modernisation of the waste water drainage system and reorganization of the network of underground service infrastructure. This type of measure has precise limits dictated by the urban, architectural and monumental contexts of the individual areas. Local defences need to therefore be integrated with the mobile barriers so that the level of flood protection needed in the built up areas can be achieved.

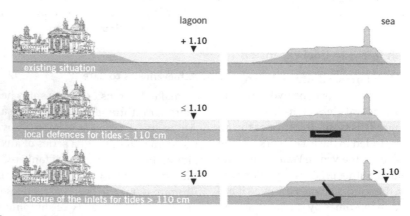

Fig. 30.6   High tide protection system.

The different urban situations require different types of intervention, with the dual aim of eliminating the problems caused by the most frequent flooding as far as possible, while at the same time avoiding invasive measures affecting the built fabric by forcing the raising of surface levels. In the towns and villages along the barrier islands, for example, the sparser and less fragile nature of the built-up areas enables the raising to be carried out to a greater height, providing a relatively high level of protection (between +130 and +180 cm). In the historic cities, towns and villages within the lagoon, in other words the lowest lying areas of Venice and Chioggia together with the islands of Murano and Burano, the operation is much more delicate and complex and the level of defence possible is considerably lower than in the built-up areas on the coastal strip. The projects approved and measures implemented to date demonstrate that without unacceptable alterations to the architectural elements and the relationship between the level of public areas and the level of ground floors, Venice can be protected up to a generalized and homogeneous level (therefore functional in terms of management of the mobile barriers) of +100 to +110 cm. These heights have, for example, been achieved along certain sections of Tolentini and are specified in the final design for the defence of the San Marco 'insula', approved (among others) by the City Council. This confirms that the oldest and most valuable areas cannot be raised or otherwise protected above these levels.

Achievements to date

To date, measures have been implemented along more than 50 km of the 90 km of shores and banks specified in the General Plan of Interventions (58%). In total, an area of 960 ha has been secured out of the 1,200 ha mentioned (80%).

The mobile barriers

Objectives

The objective attributed by legislation to the Design for Mobile Barriers is of the utmost importance. It requires defence of the cities of Venice and Chioggia and other historic towns and villages in the lagoon together with the entire lagoon basin itself, from the damaging effects of medium high tides and the devastating effects of exceptional tides. The plan to provide protection against high waters has been given immediate priority in the special legislation. A selection was made from a number of alternative designs and the design adopted was considered the constraints inherent in protecting and restoring the balance of the lagoon ecosystem and in accordance with the precise instructions of the Higher Council of Public Works, namely, to avoid interfering with port activities and the landscape, modifying water exchange between the sea and lagoon or damaging lagoon morphology and water quality. The design for the Mobile Barriers has an extremely precise role but is nevertheless just one element in a much vaster Plan of Interventions.

The solution involving mobile barriers separating the lagoon from the sea during exceptional high tide events has been elaborated at such length that few projects in the world have been studied and experimented in such depth. The hypothesis (advanced more than once) that a storm surge event could be opposed simply by acting on the physical structure of the lagoon by means of wide area morphological measures has, on the other hand, been shown to be completely ineffective in reducing water levels in Venice. The wide area morphological measures, compatible with the environment, are nevertheless indispensable to achieve various general objectives (and with this scope in mind are currently being implemented) such as the safeguarding and restoration of the lagoon environment and improvement of water quality. Local morphological measures affecting the cross section of the lagoon inlets, with compatible modification of volumes exchanged between the sea and the lagoon are, on the other hand, able to reduce high water levels in Venice only by a few centimetres, depending on the characteristics of the tidal events. On its own, this type of measure, although it may partly limit the number of floods, cannot therefore resolve the problem definitively and, in particular, is not able to provide any form of protection from exceptional events.

Nature and operation of the mobile barriers

The mobile barriers consist of rows of gates installed on the seabed at the lagoon inlets (Figs. 30.7–30.9).

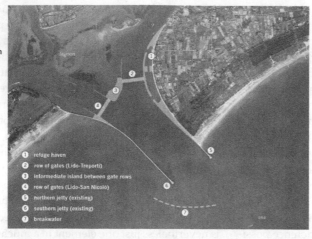

The new
arrangement
of the inlet after
the realisation
of the MOSE system
for the defence
from high tides

① refuge haven
② row of gates (Lido-Treporti)
③ intermediate island between gate rows
④ row of gates (Lido-San Nicolò)
⑤ northern jetty (existing)
⑥ southern jetty (existing)
⑦ breakwater

Fig. 30.7   Lido inlet.

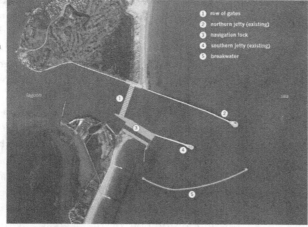

The new
arrangement
of the inlet after
the realisation
of the MOSE system
for the defence
from high tides

① row of gates
② northern jetty (existing)
③ navigation lock
④ southern jetty (existing)
⑤ breakwater

Fig. 30.8   Malamocco inlet.

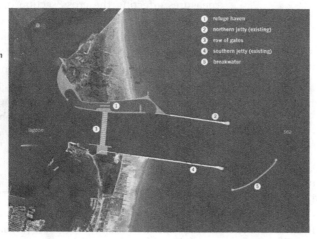

The new
arrangement
of the inlet after
the realisation
of the MOSE system
for the defence
from high tides

① refuge haven
② northern jetty (existing)
③ row of gates
④ southern jetty (existing)
⑤ breakwater

Fig. 30.9   Chioggia.

They are defined as 'mobile' as under normal tidal conditions they are full of water and rest in recessed caissons on the seabed. Each gate is attached to the caisson via hinges. When tides above the established height are forecast (if the height is 100 cm, an average of seven times a year; if the height is 110 cm, an average of three to five times a year), compressed air is introduced into the gates to expel the water. They thus rise, rotating around the axis of the hinges, to emerge and block the tidal flow. The inlets remain closed for the duration of the high water only and for the time required to manoeuvre the gates (on average, a total of four and a half hours). The feasibility and effectiveness of the proposed solution has been studied for years with tests using mathematical and physical models involving the collaboration of world leading hydraulic modelling laboratories. A life-sized prototype of a gate has been constructed (*Modulo Sperimentale Elettromeccanico* – MoSE) and subjected to four years of testing, enabling operation of the gates to be optimized and the individual components of the system, the constructional materials and their reaction to the chemically and biologically aggressive marine environment to be verified.

Time scale, costs and jobs

The mobile barriers could be completed within eight years from contracting the work (and activities could begin on the first sites two years from the start of the final design phase) with a gross cost of about 1,900 million Euro (unforeseen events and cap excluded). Construction of the barriers would create a total of about 1,000 direct jobs per year, plus a further 4,000 jobs in indirectly related or spin-off sectors, in addition to jobs associated with work to raise the level of paved areas. In the operational phase, maintenance and running of the mobile barriers, and regulation of the lagoon ecosystem in association with defence from flooding, would generate a further 150 jobs per year. This type of project requires a high percentage of unusual work and advanced technology. This suggests that it could lead to local development of specialized new activities and professions. In short, execution of the design, with massive funding concentrated in a short space of time and limited geographic area, would have major repercussions on the economy of Venice, representing an important objective for the targeted conversion of production in the city and industrial area.

31 • Venice mobile barriers project: Barrier caissons construction details

Y. EPRIM

INTRODUCTION

The Venice flood prevention system consists of four mobile barriers which will be erected in the three inlet channels that connect the lagoon to the Adriatic Sea, namely, Lido, Malamocco and Chioggia inlets. Because of its width, the Lido inlet channel will be defended by two barriers (S. Nicolò and Treporti), separated by an artificial island.

Table 31.1 *Mobile barrier characteristics.*

Mobile barrier	Length (m)	Water depth (m)
Lido - Treporti	420	6
Lido - S. Nicolò	400	12
Malamocco	380	14
Chioggia	360	11

The barriers consist of 18 to 21 buoyant steel gates supported on 6 to 7 reinforced concrete foundation caissons (Figs. 31.1 and 31.2).

The proper functioning of the mobile barrier system requires that the gates be prefabricated and their foundation caissons be cast and positioned with a greater accuracy than is normally required for conventional civil engineering works, such as submerged road tunnels.

The gates (Fig. 31.3) are 18 to 30 m long, 3.6 to 5 m thick and they all have a width of about 20 m. Typically, the concrete caissons are approximately 60 m long, each supporting three gates.

In order to achieve the designed performance of the flood defence system, the gaps between the gates must be kept to a minimum (average value of the gaps ≤ 100 mm). This ensures that, after closing the barriers, the rate of rise of the lagoon water level due to inflow through the gaps does not exceed 3 mm h^{-1}. This is one of the criteria of the barriers' operation system.

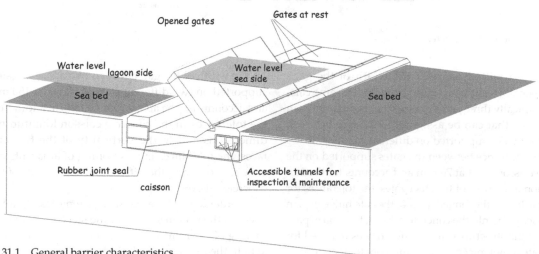

Fig. 31.1   General barrier characteristics.

Flooding and Environmental Challenges for Venice and its Lagoon: State of Knowledge, ed. C. A. Fletcher and T. Spencer.
Published by Cambridge University Press, © Cambridge University Press 2005.

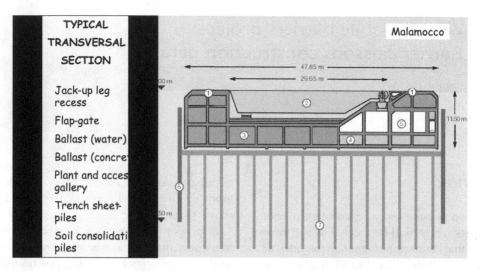

Fig. 31.2.   Typical section of the barrier structure.

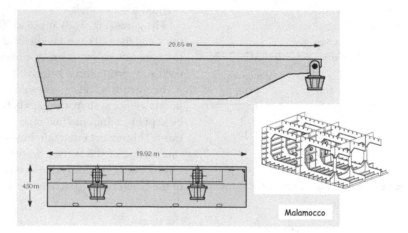

Fig. 31.3   Buoyant flap gate.

The permitted value of the width of the gaps between adjacent gates supported on the same caisson (typically three gates per caisson) is different than the value that can be assigned to the gap between adjacent gates supported on different caissons.

The clearance between the gates supported on the same caisson is set at 70 mm and accounts for the fabrication tolerances of the steel gates, the tolerances for embedding of the female units of the gate hinge in plan and elevation into the concrete caisson, thermal expansion of the structures, and the tolerances required for the gate removing/placing equipment to operate.

The width of the gap between two adjacent gates, supported on two different caissons is set at 150 mm and accounts for the above four factors, in addition to the tolerances in the concrete caisson longitudinal dimensions and in the verticality of the two end faces, the tolerances in positioning of adjacent concrete caissons, and the differential settlements of concrete caissons.

In order to meet the above requirements, the tolerances in the prefabrication of the steel gates and the casting of the reinforced concrete caissons must be kept to the minimum technically practical.

Also the placing of the caissons in their correct position, in plan and vertically, must be achieved with minimum tolerances.

This chapter outlines the construction methods and details that will be used in order to attain the required tolerances of the concrete caissons of the barriers and to achieve the necessary durability of the submerged structures.

MOBILE BARRIERS' METHOD OF CONSTRUCTION

The method of construction of the mobile barriers has been developed with the objective of minimizing interference with navigation, the *in-situ* work to minimize traffic due to movement of workers, the effects on the hydrodynamics of the inlet channels during construction and the dredging/excavation volumes.

To achieve this, the construction method is based on prefabrication of most of the barriers main elements in basins away from the inlet channels and the use of temporary or permanent sheet piling to reduce the dredging/excavation volumes. Thus, the main concrete structures are precast in appropriate dry basins, floated, transported to site, sunk into their final position and permanently ballasted.

After sinking to their final positions, some of these structures will need some limited finishing work; work that can be carried out under dry conditions.

Almost all structural and construction details are conventional and are in accordance with well consolidated standard engineering and construction practices.

However, special attention is required to aspects such as: the durability of the concrete structures (100 years useful life); the installation of the female connector units on the barriers caissons (precision necessary for the hinge anchoring of the flap gates); the installation of the rubber (Gina type) joints on one end of the barriers caissons and the contact plates on the other end (accuracy required to achieve full efficiency of the water tightness); and the positioning of the precast barriers caissons in their final locations (accuracy is required for the proper functioning of the flap gates and the water tightness elements of the service tunnels).

DURABILITY OF THE BARRIERS' CONCRETE CAISSONS

The fact that the structures will be precast will allow work under controlled conditions, hence accurate control of the results, i.e. minimum cover and concrete quality. The concrete grade will be at least 45 MPa. A high concrete grade and controlled curing will produce a very dense and durable concrete.

The high durability of the reinforced concrete structures will also be achieved by providing 50 mm minimum cover of the reinforcement and appropriate detailing. The structures will also be pre-stressed to prevent crack formation occurring in the concrete during the hardening and curing process.

INSTALLATION OF THE GATES' HINGE CONNECTORS – FEMALE PART

Each gate is connected to the foundation caisson by means of two hinge connector assemblies.

The connector assembly (Fig. 31.4) consists of a spigot (male part) and a socket (female part).

The spigot is bolted to the base plate of the hinge fork and by means of the hinge pin it is connected to the gate.

The socket (the female part of the connector assembly) is cone-shaped and is embedded in the reinforced concrete elements of the foundation caisson. The female connector elements will be installed at least three to four months after the caisson concreting has been completed, so that their correct position (horizontally and vertically) can be controlled. In this period, the movements of the caisson (due to hardening, temperature, creep) will be continuously monitored. The positions of the female connectors will then be accurately determined and the socket parts of the connectors embedded.

INSTALLATION OF THE GINA PROFILE AND THE CONTRAST PLATES

The installation of the Gina type rubber joints (Fig. 31.5) on one end face of the caisson and the contact

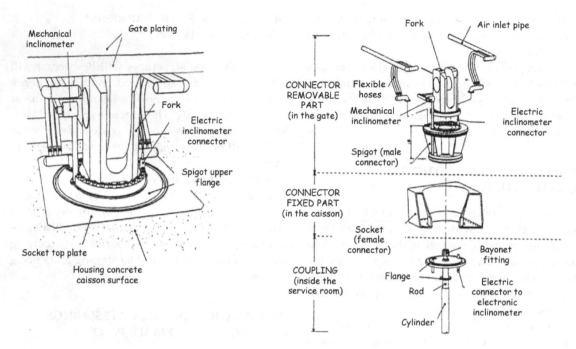

Fig. 31.4   Buoyant flap gate – caisson connector assembly.

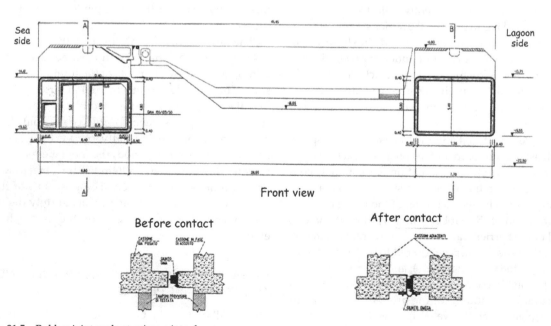

Fig. 31.5   Rubber joint seals at caisson interfaces.

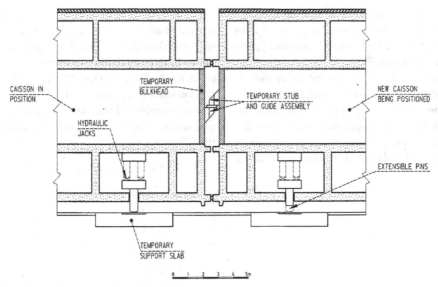

Fig. 31.6   Details of temporary supporting of caissons.

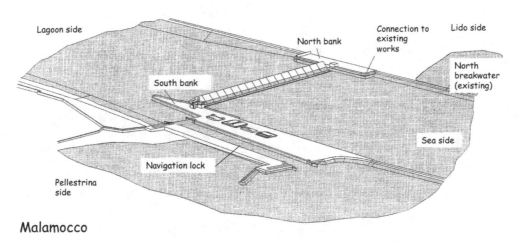

Malamocco

Fig. 31.7   Malamocco barrier overview.

plates on the other end of the caisson will be carried out three to four months after the caisson concreting has been completed. All the relevant dimensions and shapes will be accurately measured. Epoxy mortar will be used to achieve the correct vertical planes, on which will be installed a rubber joint base plate on one end and the contact plate on the other end of the caisson.

PLACING OF THE BARRIERS CONCRETE CAISSONS

Each caisson is fitted with a stub element on one end and a guide element on the other end (Fig. 31.6). These elements are of trapezoidal shape. Also each caisson is fitted with hydraulic jacks and pin assemblies at the four corners. All these assemblies are removed after assembly.

So that the correct position of the barriers caissons can be controlled, they will be sunk in position with the aid of removable wedge-shaped stub and guide assemblies, and temporarily supported on pins activated by hydraulic jacks. The stub and guide assemblies will be dismantled, partial ballasting of the caissons will be carried out and the elevation of the caissons will be adjusted as necessary by means of the hydraulic jacks and extensible pins, operated from within the compartments of the concrete caissons.

Only after two other caissons have been installed and partially ballasted and connected to the first one, can the latter then be completed by injecting cement mortar to fill the underside gap. Once the mortar is hardened, the hydraulic jacks will be de-activated and the caisson ballasting completed.

An overview of the Malamocco barrier (Fig. 31.7) and some of the details of the barriers are shown in the following figure.

32 • Venice port activities and the challenges of safeguarding Venice

A. RAZZINI

INTRODUCTION: THE PORT OF VENICE

The port of Venice is one of the major Italian ports and is included in the list of the leading European ports which are located on the strategic nodes of trans-European networks.

The port of Venice is located within the Venice Lagoon and consists of three different sections:

- *Marittima*, in the Venice town centre, essentially devoted to passenger traffic (cruise ships and ferries);
- *Marghera*, in the hinterland, close to Venice, devoted to commercial and industrial goods traffic; and
- *S. Leonardo*, in the central part of the lagoon, devoted to oil traffic.

From the territorial and infrastructural point of view, the port of Venice is characterized by:

- 2000 ha surface area;
- 30 km of wharves;
- 163 operational moorings;
- a 205-km internal railway network;
- a 70-km port road network; and,
- a 12-km optical fibres network.

The 'Venice port community' generates a turnover of more than €1 300 000.00 per year and consists of:

- more than 18 500 staff, supplemented by 270 000 seafarers in transit each year;
- 14 industrial terminals;
- 6 commercial terminals;
- 5 service companies;
- 37 port operators;
- 138 shipping and forwarding agencies,
- 45 logistics and transport companies, and
- 500 operators of other types.

In the year 2002 the port traffic:

- about 5000 ships in transit;
- the overall handling of about 30 000 000 tons of goods (mainly commercial traffic);
- the handling of 262 337 containers (largest Adriatic port);
- passenger traffic reaching *c.* 1 000 000 persons.

THE SPECIAL LEGISLATION FOR VENICE AND THE PORT

Because of its social and economic context, the law has seen the safeguarding of the port as an important objective and has set out specific indicators for social and economic revitalisation (Art. 1, Act of Parliament 171/73, the first Special Law for Venice), as well as infrastructural policies related to the removal of oil traffic and the strengthening of commercial traffic and its related activities (governmental guidelines of 27th March 1975), along with the rebalancing and re-launching of port activities (Art. 17, Act of Parliament 798/84, second Special Law for Venice). It is essential to know the relationship between laws, responsibilities and tasks of the Venice Port Authority (which is also called APV, the acronym for Autorità Portuale di Venezia), which manages the port of Venice, and, along with the Magistrato alle Acque, is in charge of dredging port canals (the Magistrato alle Acque is concerned with environmental issues in the lagoon).

As far as the physical safeguarding of the lagoon is concerned, the Venice Port Authority, in compliance with the previously mentioned Acts of Parliament, has no direct competence. The Government, through the Magistrato alle Acque, proposes projects that the Venice Port Authority assesses according to their

Flooding and Environmental Challenges for Venice and its Lagoon: State of Knowledge, ed. C. A. Fletcher and T. Spencer.
Published by Cambridge University Press, © Cambridge University Press 2005.

impact on port activities. As a matter of fact, the difficulties which port activities would undergo are clear, should the mobile barriers at the port mouths be built, with the consequent reduction in full accessibility for vessels (which is already limited by the insufficient dredging of canals, which are threatened by both polluted and non-polluted sediments). Such difficulties would damage not only the port, but also, and above all, the related economic activities, with serious consequences and repercussions for the whole of the Venetian economy.

As far as safeguarding is concerned, the Venice Port Authority has never proposed projects, because, among other things, it does not participate in, unless invited, the meetings of the organisms provided for by the Special Legislation (the Comitatone, the Commissione di Salvaguardia, etc.). Here follows a historical perspective on the standpoints of the Venice Port Authority as regards the planned mobile barriers at the lagoon inlets.

HISTORY OF THE VENICE PORT AUTHORITY'S STANDPOINT CONCERNING THE MOBILE BARRIERS PROJECT

REA project – Riequilibrio e ambiente (Rebalancing and the environment)

First, the Venice Port Authority did not exist at the time and the body in charge of the management of the sea-lagoon state property was the Provveditorato al Porto di Venezia, which was solely in charge of managing commercial traffic:

- from a quantitative point of view, commercial traffic was at a much lower level than industrial and oil traffic;
- container traffic was still at its beginning and 'hub ports', as well as 'feeder ships', did not exist;
- there were no regular ferry connections; and
- cruise traffic was just beginning.

The standpoint of the Provveditorato al Porto as regards the REA project was as follows:

- keeping the sea bed at the levels provided for in the port master-plan (including those leading to

S.Leonardo), which meant that the base of the mobile barriers should be at:
- 15 m below mean sea level at Malamocco inlet;
- 11 m below mean sea level at Lido inlet; and
- offsetting the delays caused by mobile barriers at the lagoon inlets, with a temporal extension of port operations (construction of the new illuminated route)

Building plans and the EIS

The observations of the Venice Port Authority on the building plans and the Environmental Impact Study (EIS) have focussed mainly on aspects related to the economic impact of the works on the port system and the management of the mobile barriers at the lagoon inlets.

The Venice Port Authority considered that the economic losses for the port, as indicated by the Environmental Impact Study, were largely underestimated. It highlighted that the Environmental Impact Study neglected the impact of the mobile barriers on the port economy as a whole ('halo effect') and on 'scheduled' traffic in particular, without taking into consideration that most of this traffic could be lost. The Venice Port Authority also highlighted the need to evaluate and minimize interferences with the port system during the long construction phase of the mobile barriers.

'Ufficio di piano' (Working group, letter of 22 March 1999 Italian Ministry of Public Works)

Important points regarding the level of the sea bed at the inlets indicated by the Ministry for Transport and Navigation were clarified during the meetings of the working group (draughts for ships):

- Lido mouth 9.5-10 m; and
- Malamocco mouth 12 m.

These figures (considering the 'under-keel clearance' indicated by the Venice Harbour Master's Office) mean a minimum depth at the inlets of:

- 12 m at Lido; and
- 14 m at Malamocco.

Deliberation of the Council of Ministers of 15 March 2001 – plan of supplementary works

The planned interventions in the framework of the so-called supplementary works are:

- outer breakwaters;
- navigation lock at Malamocco. The Venice Port Authority demanded that the construction of the navigation lock at Malamocco is built (already indicated as a fundamental work by the President of the Venice Port Authority with letter no. 266 of 15 January 2000) at the beginning of the works, so that it might be used during the construction phase; and
- raising the ground level in Venice town centre to +120 cm which will have consequences on the closure level of mobile barriers (+110 cm) and will mean a reduction in the annual number of closures.

Navigation lock at Malamocco

The Venice Port Authority has asked for the navigation lock with regard to the following planning data provided by the Environmental Impact Study:

- for the construction phase: 17 non-consecutive days of total interruption of navigation (executive summary, paragraph 4.5.3); and
- for the functioning of the barriers when the works have been completed, an average annual number of closures, in the presence of supplementary

works, of between 3 (current conditions) and 25 (sea level increase of 22 cm) times per year.

In the case of different indications as to the level to be protected (+100 cm) and of an increase in annual closures, utilisation of the navigation lock would change, with consequent problems in terms of delays and costs. The use of the lock would be also detrimental should it be used as a 'permanent access structure' for both incoming and outgoing ships whose draught is between 10 and 12 m.

Construction phase – repercussions on navigation

Repercussions on navigation during the construction phase have been assessed in the building plans as regards the number of days on which the complete stopping of navigation is planned. These situations are being closely examined in the definitive plans as well. The Venice Port Authority asks that:

- the reliability of preliminary forecasts on port incoming and outgoing routes is checked during the definitive planning stage, according to detailed construction-site layouts; and
- detailed simulations are done for any critical situation demanding the advice of the Harbour Master's Office, as well as of Pilots.

33 • Gates strategies and storm surge forecasting system developed for the Venice flood management

Y. EPRIM, M. DI DONATO AND G. CECCONI

SUMMARY

Storm surge prediction models for Venice have been continuously improving over the last 10 years, mainly motivated by the increasing occurrences of high water events and interest in their frequent application to the optimisation of the mobile gates *closure strategies*.

The first part of this chapter describes the adopted classification method of flood events, in terms of *return period* on the basis of maximum forecasted water level, including forecast rain and measures of forecast wind. Flood gate management is based on a combination of meteo and water level forecasts and observations and it is optimized to give warning of flood events. This chapter also describes the different *threshold levels* used for determining the reference time to alert the Port Authorities and to determine the subsequent decisions to be taken before the last closure input is undertaken.

The second part of the chapter describes the Venice water level forecast system, co-developed and operated by Danish Hydraulic Institute (DHI) and Venice Water Authority-Consorzio Venezia Nuova (CVN), that provides the necessary operational information for the proposed mobile gates located at the lagoon inlets. The system has been continuously improved, both in terms of system architecture and of automatic data acquisition, and it now provides the experimental basis for supporting the final design of the mobile barriers. The forecast system consists of two different sets of models (statistical and deterministic) operating on different theoretical bases, and using meteo and water level observations and meteo forecasts respectively. A general database for the acquisition and retrieval of the relevant parameters is also part of the forecast system. Long-term opera-

tional flood forecast systems, linked with databases of past model performance and historical environmental information, provide an invaluable tool for the evaluation of past performance, for simulating new assimilation techniques, and for the efficient re-analysis of the data. The database re-analysis can ultimately be used to provide updated information on short-term corrections and to provide statistics on uncertainty. One of the most interesting aspects of the analysis is the ability to achieve a good and well-defined data assimilation scheme. It is therefore important that the analysis system is able to support the experimentation and analysis of different assimilation schemes.

This chapter also describes a new procedure implemented in DVAAFS (Deterministic Venice Acqua Alta Forecast System) that allows hourly updates of the results, similar to the statistical model. With this procedure, it will be possible to test different correction algorithms. The main aim here is to improve the timing of the prediction of future water levels (say 4 to 6 hours) on the basis of the raw output from the model and with the assimilation of the most recently observed levels. Despite its complexity, the deterministic forecast is quite flexible when new scenarios need to be implemented in the model. The scenarios can be derived from new physical algorithms or simply from modified forcing functions. The improvement of a modified wind field for the meteo forecast of the event that occurred on the 6 November 2000 is shown.

MOBILE BARRIERS CLOSURE OPERATION

The Venice Mobile Barrier system is a unique flood defence system. It will normally be called upon several times a year to defend the inhabited areas of the city against a few decimetres of flooding and only exceptionally against water levels up to 220 cm above the datum of the Punta della Salute (PdS) tide gauge, which is the reference level for Venice. The safeguarding level is fixed at +110 cm with respect to this reference. The available freeboard of the city (pavement level – lagoon level) is only a few decimetres. After closing the barriers, the lagoon will become a closed basin and its level will increase from any water that may flow into, or fall onto, the lagoon. Hence the barriers' closure strategy and procedures must also account for all the factors that may contribute to the rise in the lagoon water level during and after the closing of the barriers. The factors that have been evaluated and accounted for are:

- direct rainfall on the lagoon;
- inflow from the surrounding watershed;
- wind set-up in the lagoon;
- flow through the three lagoon mouths during the closure operation; and
- flow through gaps in the barriers and from overtopping after closure.

Since all the factors above are functions of the storm surge and meteorological conditions, the forecasting uncertainties and errors must also be evaluated and accounted for defining the closure strategy and procedures.

Closure criteria

The closure strategy laid down in this phase of the project development is aimed at providing a relatively simple and straightforward method by which all the high water events can be grouped into a few classes of closure criteria. The procedures to be used for each class are then pre-set, covering all the probable combinations of the lagoon level influencing factors and their uncertainties. While the closure criteria for exceptional events (RP[1] > 10 years) can be laid down on the conservative side because the events are less frequent and therefore their effect on ship traffic and on the environment may be considered negligi-

---

[1] RP = Return Period in years

ble, the criteria and procedure for the more frequent events must consider the requirement that the number and duration of these closures be kept to the minimum but without jeopardising the effectiveness of flood prevention. The main classification of the closure events is based on the storm surge maximum height and duration. Since the level in the lagoon after closing can increase in time due to all the other factors, the duration parameter is considered more important. Analysis of the duration of the storm surge events that occurred during the period 1955 to 2002 gives the following results (duration is referred to as the time the level remains above +110 cm PdS):

mean value	RP 10 years	Extreme value
3 hours	11 hours	24 hours

The 10-year Return Period event duration has been selected to define the criteria separating the two main closure classes:

Class 1   for all the events with RP < 10 years, also called frequent events; and

Class 2   for all the events with RP > 10 years, also called extreme events.

These separating criteria are defined as 150 cm PdS for the storm surge maximum height and/or 11 hours for storm surge duration. Considering the probable error in water level forecasting, the separating criteria with a lead time of 24 and 3 hours become (Table 33.1):

A sub-classification of the events is required to cover the effects and uncertainties of all the other factors (rainfall, watershed inflow, wind set-up). However, this sub-classification of the closure events, which is done to reduce the number of false alarm closures and to keep the duration of closures to the minimum (with an adequate margin of safety), is relevant only to the frequent Class 1 events. In fact, extreme Class 2 events, having a minimum forecasted surge height of 130 cm (at a lead time of 24 hours) would not provoke any false alarm closures. Furthermore, such events are normally accompanied by intense persistent rainfall and strong winds before and during the closures. These conditions must be

Table 33.1   *Separating criteria for definition of Class events on the basis of forecast errors.*

Forecast storm surge	Lead time (hours)	
	24	3
Max. height (cm)	130[2]	140[3]
Duration (hours)	9	10

taken into account in the definition of the standard closure procedure of all the Class 2 events, regardless of intensity appearing in the meteorological forecasts. Therefore Class 1 events, which deal with frequent high water events which have a RP < 10 years are sub-classified as follows:

Class 1A   to cover all frequent events with forecasted rain 'weak' or nil and an expected wind velocity of less than 10 m s⁻¹; and

Class 1B   to cover all frequent events with forecasted rainfall greater than 'weak' (> 1 mm h⁻¹) and/or with the watershed having experienced greater than 'weak' rain on the preceding day but with a forecast wind velocity of less than 10 m s⁻¹.

A second sub-classification is required to cover the relatively rare events in which a sustained wind velocity may exceed 10 m s⁻¹ over the lagoon for at least 6 hours. Hence:

Class 1AV   to cover all 1A events with a forecasted wind of higher than 10 m s⁻¹ for more than 6 hours; and

Class 1BV   to cover all 1B events with a forecasted wind of higher than 10 m s⁻¹ for more than 6 hours.

The separating criteria for high winds are fixed at 10 m s⁻¹ and 6 hours duration because the local set-

up at this wind velocity and duration does not exceed 10 cm at Vigo Chioggia for *bora* winds and 1 cm at Punta della Salute (PdS) for *sirocco* winds. Since the safeguarding level for Chioggia is 10 cm higher than that for Venice (i.e.: +120 cm PdS), wind set-ups at Chioggia lower than 10 cm do not influence the closure procedures, which must protect Venice from flooding.

Method of defining the closure criteria

The method adopted to define the barriers' closure criteria consists of evaluating the probable rise in the lagoon level for each of the contributing factors listed previously above relative to the frequent events (return period less than 10 years) and those for having higher return periods to represent the exceptional events. The probable total rises in the lagoon so calculated are then used to establish the critical levels at which barrier closing operations must be initiated for each class of events, with sufficient margin to cover the uncertainties inherent in the forecasting instruments currently available. As a measure of safety, the maximum contributions of a 10-year return period event are used to define the Class 1 criteria for all events having a return period < 10 years. Similarly, for all the events having a return period > 10 years (Class 2), the contribution of the various factors have been entered with the values corresponding to the extreme 200 year return period event for the local meteorological conditions and the 1000 year high water event.

The factors contributing to the rise in the lagoon level have been analysed and their effects evaluated for the frequent and extreme events based on 70 years of data. Extreme value analysis was used to

---

[2] In the 95% of cases for 24 hours lead time the under prediction error is limited to –18.6 cm

[3] In the 95% of cases for 3 hours lead time the under prediction error is limited to –8.7 cm

Table 33.2   *Total rise of lagoon water level for storm surge events under sirocco and bora conditions.*

RP (years)	Sirocco (for Venice) (cm)	Bora (for Chioggia) (cm)
10	23	31
200	39	48

Table 33.3   *Total rise of lagoon level for 10 and 200 years return period storm surge events of sirocco (for Venice) and bora (for Chioggia).*[4]

Influencing factor	Sirocco			Bora		
	Return Period			Return Period		
	10 years	100 years	200 years	10 years	100 years	200 years
Duration of high water (hours)	11	19	24	6.5	10	11
Direct rainfall (cm)	7	12	13	6	10	12
Watershed inflow (cm)	4	11	13	2	5	6
Wind set-up In Venice (cm)	3 (4)	2 (7)	2 (8)	14	21	22
Flow during closing (cm)	6	5	4	7	5	5
Flow through barriers gaps (cm)	3	5	7	2	3	3
Total rise in Venice (cm)	**23**	**35**	**39**			
Total rise In Chioggia (cm)				**31**	**44**	**48**

define the characteristics of the 10, 100 and 200 years return period events. Summing up the contribution of all the factors influencing the lagoon water level which have the same return periods, the total rise becomes as shown in Table 33.2.

Table 33.3 illustrates the contribution of each of the five factors for *sirocco* and *bora* events.

Maximum level at which closing must be initiated

The fundamental element of the closing strategy is the level at which closure must be initiated for each class of event. In defining these levels, besides accommodating the freeboard necessary to cope with the lagoon level rises established above, appropriate consideration must also be given to the errors and uncertainties of the forecasting instruments. Two other important aspects must be clarified. The first concerns the fact that at the moment of closure, the water level in the lagoon is not horizontal (due to the entering tidal gradient) and therefore there will be changes in water level after barrier closure. For

example, if for the closing level reference is made to the tide gauge at Punta della Salute (PdS), the relative horizontal level of the lagoon after barriers closure will be 4 to 8 cm lower than the PdS reading. The second aspect is that it is possible at any moment to establish the horizontal level of the lagoon (that is the level to which the lagoon surface will settle after the closure of the mobile barriers) by applying a transfer function to the readings of two water level gauges located in appropriate locations in the lagoon, such as PdS Venice and Vigo Chioggia. These two aspects, which indicate that the true freeboard at any time can be known and that the freeboard after closing is greater than the difference between pavement levels and the PdS gauge reading at the time of closure, are important for the evaluation of the volume of incom-

Table 33.4   *Closure level criteria for each class event. Levels in cm referred to PdS datum.*

Class	Closure level	Class	Closure level
Class 1A	≈ +100	Class 1AV	≈ +80
Class 1B	≈ +90	Class 1BV	≈ +75
		Class 2	≈ +65

---

[4] The number in brackets for the wind set-up value refers to the maximum value not concurrent with the other maximum values.

ing water that can be contained in the lagoon without causing flooding. Table 33. 4 shows the criteria for closing, in terms of PdS water level readings.

To check the validity of the Class 1 closures in also covering the uncertainties of the forecasting instruments or technology, the events that occurred during the 1955 to 2002 period with a return period not higher than 10 years, along with their meteorological conditions, were first modified as follows:

The rainfall intensity of all the events with 1 to 1.5 mm h^{-1} rain was increased up to 5 mm h^{-1}, which represents the 10 year return period rainfall intensity;

The wind velocities were increased by 20% based on the results of the analyses of the variations of the wind growth gradients for 140 events; and

The storm surge curves and peak levels were increased by 20% and consequently the duration of the events by about 1.0 hour to account for the uncertainty in forecasting.

The simulation of the closure of Class 1 events (1955 to 2002) yielded the following results (Table 33.5):

The levels in the lagoon and the freeboard margins are referred to PdS because the resulting freeboards at Chioggia Vigo will be much higher (the safeguarding level at Chioggia is 10 cm higher than that for Venice). As can be observed, even after increasing the factors that contribute to the lagoon level rise to account for the uncertainties of forecasting, the freeboard margin for classes 1AV and 1BV seems to be high. Table 33.6 shows the total duration of the closure events that would have taken place in the period 1955 to 2002 had the barriers been operative. It can be observed, even having artificially increased the intensity of the meteorological conditions of the 144 events, the total closure as a percentage of time is less than 0.2%.

Various scientists (IPCC, 2001) have estimated that mean sea level may rise by 20 cm in the coming 30 to 50 years. Assuming that the mean sea level were higher by 20 cm, the total closure time would be in the order of 1%. To check the validity of Class 2 procedures, closure simulations were carried out for the 1966 event and the 100 and 200-year return period *sirocco* and *bora* events. The results are summarized in Table 33.7.

Table 33.5   *Results of closure simulation for Class 1 events in the period (1955 to 2002).*

Closure class	Frequency of occurrence (%)	Closure level (cm)	Levels in the lagoon mean/max (cm)	Freeboard margin mean/min (cm)
1A	50	100	101 - 102	9 / 8
1B	46	90	94 - 98	16 / 12
1AV	2	80	89 - 92	21 / 18
1BV	2	75	86 - 90	24 / 20

Table 33.6   *Duration of closure for frequent Class 1 events.*

Closure class	Number of events	Average closure duration (hours)	Total duration in 48 years (hours)	Closure as % of time
Class 1A	76	3.6	274	0.07
Class 1B	64	4.8	307	0.07
Class 1AV	2	6.2	12	< 0.01
Class 1BV	2	7.5	15	< 0.01
All Class 1 events	144	4.3	608	0.14

Table 33.7   *Closure simulation of the extreme events (Class 2).*

Storm surge event	1966 event	*Sirocco* RP 100 years	*Sirocco* RP 200 years	*Bora* RP 100 years	*Bora* RP 200 years
Surge peak level (cm)	194	175	210	160	185
Closure duration (hours)	24.0	20.0	24.3	10.1	11.0
Closure level (cm)	65	65	65	65	65
Max. level at Venice (cm)	102	97	103	70	75
Max. level at Chioggia (cm)	96	89	97	104	108
Freeboard (cm)	8	13	7	16	12

Operational procedures

The application of the closure strategy and procedures is based both on forecasting instruments and/or technology that concerns the assigning of an event to a particular closure class, taking into account a conservative measure of the forecasting uncertainties.

The decision to close is made 3 hours before the expected time of closure. The execution of the closing procedure itself is based on the real levels at PdS – Venice, Vigo – Chioggia, Lido and at the CNR platform[5] offshore. The proposed procedures set down a standard closing level for each class of events, namely +100, 90, 80, 75, 65 cm as referred to the datum of the PdS tide gauge (mean tide in 1897). These closing levels also account for the uncertainties in the meteorological/hydrodynamic forecasts and in the duration of the events. A pre-alarm will be given 24 hours in advance of a probable high water event. At this moment, if the forecast is for a high water peak level > 130 cm (= 150 – error) and/or for a duration > 9 hours (= 11 – error), the event will be classified Class 2, otherwise the event will be classified Class 1. Approximately 9 to 6 hours before the time it is foreseen that the level at PdS may reach +100 level, the storm surge forecast will be re-examined and if Class 2 conditions are still foreseen, an alarm will be given to stop navigation 3 hours in advance of the expected time of starting to close the

gates. When the decision is made to give the alarm to stop navigation, the procedure requires that the levels outside and inside the lagoon be continuously monitored to allow the operator to initiate the closing operations at the pre-set level.

If the event has been classified as Class 2 (extreme event), the operator will commence closing operations when the level at PdS Venice reaches +65 cm. If the event is classified Class 1, the monitoring of levels will commence 6 hours in advance, using the knowledge of the meteorological forecasts and the actual measurements of wind over the last few hours and of rainfall in the last 24 hours. The event will be sub-classified as 1A, 1B, 1AV or 1BV. Each of these sub-classes requires a standard closing level (+100, 90, 80, 75 cm), which will be measured at PdS Venice. When reached, the closing operations will commence. The decision to close will be taken on the basis of the actual levels required for each class. In terms of time, it means that the closure will commence some time in advance of reaching the +110 level at Venice. For Class 1A closures which account for almost 50% of events, the average time before reaching the +110 PdS level will be 45 min, for Class 1B which accounts for about 45% of events, it will be 1h 15 min and for classes 1AV and 1BV which account for the rest of the frequent events the average time will be 2h and 2h 15 min respectively. A very important fact is that between the tide levels at the CNR platform and at PdS in Venice there is a time lag of 45 to 60 minutes. This means that when the level at CNR platform is + 100 the level at PdS is still below +90. This difference allows for the avoidance of false

---

[5] CNR Platform is located outside Venice Lagoon in the Adriatic Sea

alarm closures for all those events which have been classified 1A (requiring closing to be commenced when the level at PdS reaches +100). In fact, by continuously observing the levels at the CNR platform and at PdS, in all those events classified 1A that do not exceed + 99 at CNR platform closure can be avoided. Hence, the eventual false alarm closures (events with a peak level between 100 and 109 cm) will be limited to those Class 1 events that have been classified as requiring closing levels lower than the + 90 level, i.e. less than 50% of the events. In terms of frequency or average number of closures per year this means that the true closures will be 2 to 3 and the false alarm closures about 1 to 2.

Considering the possibility that the mean sea level in future, i.e. over the next 30 to 50 years may rise by 20 cm due to climate change and global warming, the average number of closures per year at that time may become 25, of which 10 will be false alarms, assuming that the closure criteria remain unchanged during this period.

STORM SURGE SYSTEM IN VENICE

The operational flood forecast system was developed for Magistrato alle Acque and has been in operation for the last seven years. The main reason for developing and operating the system is to provide water levels predictions for optimising the gate closure procedures. The forecast modelling system comprises five main components: a data acquisition and archiving system; a statistical model component; a deterministic model component; an error analysis set of procedures; and a web server.

Data acquisition and archiving system

Numerous data sources are used in the production of the forecasts. They are all collected automatically using standard data communication methods. Data fall into two categories: measurements and forecasts. The measurements are mainly used to establish the reference or boundary conditions, to apply corrections to forecast results, and for analysis of the results in order to access forecast performance. The meteorological forecasts are used to provide forecasted predictors values for the statistical models and to force the hydrodynamic models

to simulate the storm surge forecast. The forecast data used for the modelling consists of pressure and wind fields covering the Mediterranean. The system operates with input from two different meteorological models, one from the European Centre for Medium-Range Weather Forecasting (ECMWF) and another from the Emilia-Romagna Regional Meteo Service (LAMI). All results of the forecast, as well as all measurement data, are entered into a common SQL Server database system to access the data in a very flexible way during the analysis phases of the project.

Statistical modelling system

The Statistical model complex has been operating automatically since late 1995. This system is based on a number of models that differ according to the number and the nature of the pressure predictors (Cecconi et al., 1997). The Extended Complete (EXCO2) model is based on a multiple correlation among the residual water levels (meteorological contribution) in Venice and the atmospheric pressure forecasted at the station of the SYNOP network located on the coast

Fig. 33.1 Location of predictors for the statistical modelling system.

of the Adriatic Sea and at two external stations (Genova and Alghero) shown in Fig. 33.1. When new data is available, the model is executed and a new forecast produced. For the EXCO2 model, the execution is repeated every hour to update the forecast with the latest measurement values. The EXCO2 model consists of one unique multiple regression algorithm with 78 constant coefficients (Kerper *et al.*, 2002). EXCO2 produces water level forecasts for 72 or 120 hours, depending on the meteo input time lead (72 for LAMI and 120 for ECMWF). The calibration of this model was based on measurements in the period 1981 through 1990 and was verified for the period 1996 through 1998. During the calibration, EXCO2 was optimized for strong storms detecting five different meteo conditions. A mean error correction proportional to the maximum water level has also been implemented. Results from the model are currently available at the address http://cvnws.port.venice.it.

Deterministic modelling system

The Deterministic Venice Acqua Alta Forecast System (DVAAFS) was first implemented in October 1996. The system is based around DHI's MIKE 21 HD numerical model including grids for the Mediterranean Sea, Adriatic Sea, and sub-grids of the northern Adriatic and the Venice Lagoon (Cecconi *et al.*, 1997). A total of four grids are used in the DVAAFS system and are shown in Fig. 33.2. The Mediterranean module operates de-coupled from the Adriatic Sea grids and the Venice grid, to provide external effect boundary conditions from systems arising in the Mediterranean that are not captured in an Adriatic Sea model. The model set-up covers the entire Mediterranean Sea because (Cecconi *et al.*, 1997) even small water level variations at the entrance to the Adriatic Sea, 800 kilometres south of Venice, are amplified due to the geometry of the basin. The Adriatic Sea and Venice Lagoon grids are dynamically coupled, or nested, so that boundary information is transferred between grids dynamically at each computational time step. One of the primary driving forces of the computational model is the applied time-varying water level boundary condition at the Otranto Strait, where the Adriatic Sea meets the Mediterranean Sea. Four main effects are combined into the single boundary condition: these are the astro-

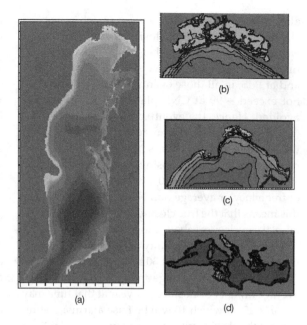

Fig. 33.2   Example gridded wind and pressure field from ECMWF: (a) nested Adriatic grid; (b) Venice Lagoon grid; (c) Northern Adriatic grid; and (d) Mediterranean grid

nomical tide generated from known harmonic constituents from near Otranto; external surge from the Mediterranean HD model; 'short-term' or seasonal mean sea level variations (i.e. global effects) from filtered measurements at Brindisi (near Otranto); and an inverted barometric condition by adjusting approximately 1 cm of water level per hPa of pressure. Wind and pressure fields produced by the ECMWF and the Emilia Romagna Meteorological Service are used to produce two independent forecasts.

The DVAAFS system is automated to run 120-hour forecasts every 24 hours using ECMWF forecast wind and pressure at the arrival of the new forecast data. At the completion of the forecast simulation, the predicted water levels in and around Venice are automatically corrected using the daily average at five water level recordings available in and around the lagoon. A new procedure has been implemented in DVAAFS which allows hourly updates of the results, similar to the statistical model (Kerper *et al.*, 2002). With this procedure, it is possible now to test different correction algorithms to better predict the next levels (4 to 6 hours) on the basis of the raw output from the model and on the assimilation of the most recently observed levels, providing water level forecasts suitable for gate management. The resulting time series forecast is plotted and sent to the web page, results are stored to the database and then finally 2D animations are created.

Performance, uncertainty and re-analysis

In order to assess the accuracy of the operational Forecast System in relation to the mobile barriers project, the general database has been used for both the statistical and deterministic models. The structure of the database considers all the parameters

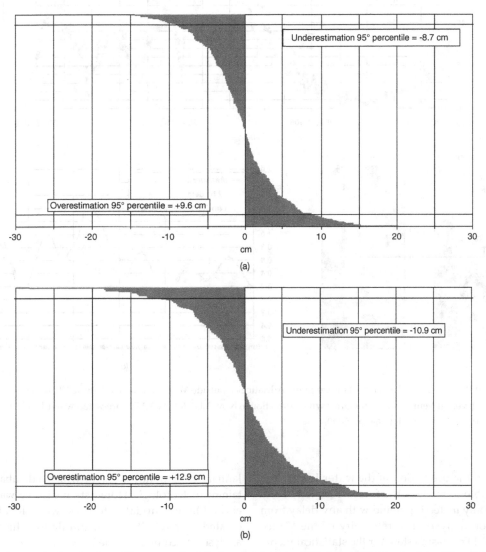

Fig. 33.3   Percentile distribution of the errors. Statistical model. January 1996–May 2003. Events above 80 cm. (a) forecast issued 3 hours before the maximum observed level; (b) forecast issued 3 hours before the time of upcrossing level 80 cm (3 to 9 hours in advance).

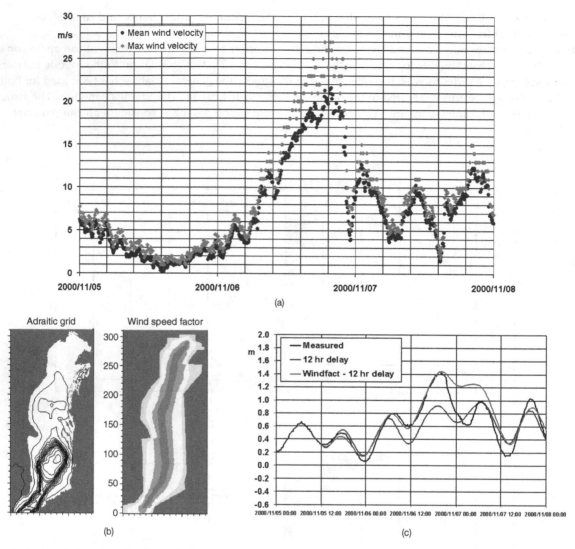

Fig. 33.4   (a) Wind measured at CNR platform in the Adriatic Sea outside Venice Lagoon. 6 Nov., 2000; (b) DVAAFS fore-casted water levels at Venice before and after wind correction. 6 Nov. 2000; (c) DVAAFS forecasted water levels at Venice before and after wind correction. 6 Nov. 2000.

involved in the prediction of the water levels. It is now possible to retrieve the forecast issued at any time with any meteo input and with any delay from the time of analysis. The reliability of the Venice Water Level Forecast System for the statistical model has already been accurately tested by comparing the forecast and observed values for different lead times. The analysis has been performed for the period

January 1996 to May 2003, for which a database containing meteorological forecasts, measured water levels and hourly updated forecast water levels were created. The results of the simulations have been analysed by comparing the Punta Salute forecast and observed values for different lead times considering all the simulations as in the real case (i.e. using the forecasts that would have been available in an operational context). All the runs issued with all the meteorological forecast inputs and with time leads of 72 hours have been considered.

The forecast model with the most extended forecast series is the hourly updated EXCO2 statistical model with the ECMWF meteorological forecast. The accuracy of EXCO2 has been extensively investigated for the events of high water in the period 1996 to 2003 (Consorzio Venezia Nuova, 2003). According to specific design needs, the analysis was performed considering the errors with forecasts issued with 24, 12, 9, 8, 6, 5, 4, 3, 2 and 1 hours and with two different start times. In the first case, the start time was computed as the number of hours before the maximum observed level, while in the second case as the number of hours before the time of upcrossing of a specified reference level. The first method is used for assessing model accuracy, while the second is used mainly for detecting the criteria for the management of the gates at lagoon entrances. Two examples illustrating this type of analysis are shown in Figs. 33.3a and 33.3b.

New scenarios

Recently the effect of a modified wind field was tested for the forecast of a critical event, which occurred on the 6 November 2000. The wind field forecast showed a strong *sirocco* wind starting at 3:00 pm. The wind direction was rather well reproduced, but the forecasted speed lower than that observed.

Figure 33.4a shows the wind speed time series recorded at the CNR platform. It should be noted that the maximum measured wind speed was close to 30 m s^{-1}, while the forecasted one was close to half that speed. According to the measurements at the SYNOP stations in the Adriatic Sea, it was possible to detect a shape factor for the wind speed distribu-

tion. This was then applied to DVAAFS to simulate this modified scenario. Different hypothesis were tested, with varying intensity and distribution. Fig. 33.4b shows the situation with the wind field intensity multiplied by a factor 1.5 (as the record often reports) and lasting until the end of the day. The example shows that after this modification the maximum water level was nearly perfectly reproduced. The overprediction shown in Figure 33.4c, occurring after the maximum level occurred, is due to the test scenario. The wind continued to blow nearly 2 to 3 hours longer than the observed one for matching the wind field 3-hourly forecast structure. These results are important because they offer new solutions for one of the more critical problems in water level forecasts and show that the deterministic forecast can be used in new ways to improve the forecast.

REFERENCES

Cecconi, G., Canestrelli, P., Di Donato, M., and M. Cecchinato. 1997. *I nuovi sistemi per la previsione dell'acqua alta a Venezia*. Venice: Consorzio Venezia Nuova – Quaderni trimestrali Anno V, nn. 3 and 4.

Consorzio Venezia Nuova. 2003. *Sistema di previsione delle acque alte 3° trimestre*. Venice: Magistrato alle Acque, Studio C.1.6 – Sviluppo di uno strumento operativo informatizzato (Modello Decisionale) per la gestione delle chiusure delle opere mobili alle Bocche di Porto, Parte B.

IPCC, 2001. *Climate Change 2001 – Synthesis Report of the IPCC Third Assessment Report*. Cambridge; Cambridge University Press, 397pp.

Kerper, D., Larsen, L.C., Di Donato, M., and G. Cecconi. 2002. Storm surge forecasting and flood management system in Venice, Italy. *Proceedings of Solutions in Coastal Disasters '02, San Diego, CA*. Reston, VA: American Society for Civil Engineers, pp. 238–52.

34 • Mobile barriers as a management tool for water quality and lagoon flushing

D. R. F. HARLEMAN

INTRODUCTION

There is general, although not universal, agreement that the only way to save Venice from the increasing frequency and height of flooding during meteorologically induced high tides is to construct four sets of movable gates – two at the Lido opening and one set each at Malamocco and Chioggia – to give a total of 78 gates that can be closed to temporarily isolate the Venice Lagoon from the Adriatic Sea. One of the recurrent themes in the discussions on how to save Venice from increasing flood damage is the claim that gate closure, even for a few hours, would interfere with tidal flushing and cause significant damage to the lagoon's ecosystem. Others point to the consequent build-up of sewage. There is rarely any mention that the solution to the latter problem is the collection and treatment of Venice's sewage or that the mitigation of the former problem is in the use of the gates as a management tool for improving lagoon water quality through enhanced circulation when the gates are not being used for flood control.

This chapter has four objectives:

- to summarize the temporal trend of nutrient concentrations and water quality in the lagoon over the past forty years (1962–98);
- to predict the average duration of flooding events and gate closures per year under various scenarios of sea-level rise during the next hundred years;
- to demonstrate the use of the movable gates as a management tool for enhancing lagoon circulation and improving water quality; and
- to summarize ecological changes from construction of flood control gates and dams in The Netherlands' Rhine River delta and discuss the relevance of these changes to the Venice Lagoon.

LAGOON NUTRIENT LEVELS AND WATER QUALITY HISTORY

The temporal history of concentration of dissolved nutrients that have contributed to the eutrophication of the lagoon over the past 40 years (Fig. 34.1) shows a downward trend. Ammonia (NH_4) nitrogen has decreased primarily because of wastewater treatment of industrial and municipal effluents in the vicinity of the Port, in Mestre and other mainland areas. However, further reduction is necessary and this probably will not happen until Venice collects and treats its sewage. It is noted that nitrate (NO_3) nitrogen has not decreased significantly. The major source of nitrate in the lagoon is agricultural run-off from the large drainage basin and this loading is the most difficult to control because of the diffuse sources.

One of the most important results of the decreased nutrient inflows to the lagoon has been the dramatic drop in the macroalgae *Ulva*. This organism, which had noxious summer blooms in the latter part of the 1980s, appears to be under control. However, given its preference for the ammonia nitrogen coming from Venice's sewage, it may reappear under the right climatic conditions. From about 1998 there has been a significant increase in water quality monitoring activity in the lagoon by the Venice Water Authority (MAV) – Consorzio Venezia Nuova (CVN), the Veneto Region and CORILA. Of particular interest is the tracking of transient nutrient loadings from the lagoon watershed during rainfall periods and measuring inflow/outflow rates and water quality exchanges between the three lagoon inlets and the Adriatic. The monitoring results are being used to calibrate and verify a new multi-dimensional ecological model for the lagoon.

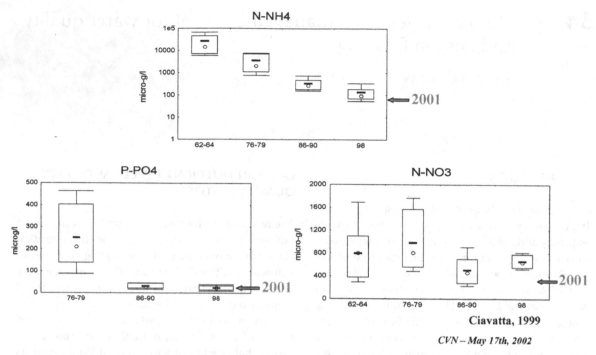

Fig. 34.1   Temporal trend of dissolved nutrients in Venice Lagoon (CVN/Thetis, 2002).

PREDICTION OF GATE CLOSURE FREQUENCY UNDER VARIOUS SEA-LEVEL RISE SCENARIOS

The prediction of future sea-level rise during a period of anthropogenic global warming is not an exact science. The most respected predictions come from the Intergovernmental Panel on Climate Change (IPCC) Their central value, within a wide range, is 50 cm of sea-level rise by 2100.

A three-dimensional, state of the art, hydrodynamic numerical model is being used to predict yearly average gate closure flooding events and durations under various sea-level rise scenarios ranging from present conditions to 50 cm in 10 cm increments (MAV-CVN unpublished report, 2002). The gate closure frequencies are based on historical water level and meteorological conditions during the past 50 years. Each sea-level rise scenario is generated by adding the incremental rise to the historical water level and climatic record. In addition, extensive research is underway to develop protocols for forecasting gate closure events. The number and duration of closures/year is greater than the number and duration of floodings/year because imperfect forecasts result in a number of false alarms. It is expected that by the time the gates are operational the accuracy of the closure forecasting will have been improved.

Figure 34.2 shows the average duration in hours/year of floodings and closures under future sea-level rise scenarios up to 50 cm. In addition, three threshold levels – 100, 110 and 120 cm – for gate closures are shown. Present thinking is that the gates would be operated so as to protect Venice from flood levels greater than 110 cm above the official city datum. This reflects benefits of the extensive work that has been, and continues to be done, to raise pavements over most of the city to prevent flooding at elevations less than 110 cm (MAV-CVN unpublished report, 2001). For example, given an expected rise of 50 cm in mean sea level in 100 years, the pre-

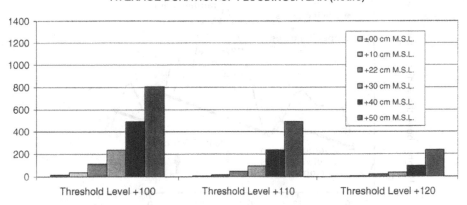

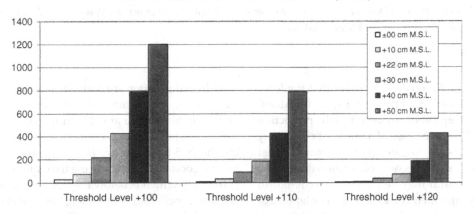

Fig. 34.2   Duration and gate closure frequency under various sea-level rise scenarios (MAV-CVN, 2002).

dicted average duration of closures/year at a threshold of 110 cm is about 800 hours or 10% of the total number of hours per year. Assuming that two-thirds of the *acqua alta* events commonly occur in the four months from October through January, the gates would be closed less than 20% of the time in that period.

Ammerman and McClennen (2000) have pointed out that 'such a high concentration of gate closures will limit the circulation of water that is essential to biological life in the lagoon, this could have negative impacts on levels of water pollution and the ecology of the lagoon.' They conclude, 'there will be a need for fresh thinking in the search for new, alternative

solutions.' However, they give no indication of an alternative solution. Pirazzoli (2002) expresses concern about the gates becoming rapidly obsolete because of future sea-level rise. The case study chosen to illustrate his point is the historical high water event of 29-30 October 1976 which reached a flood level of 124 cm above the city datum. Figure 34.3 shows the 1976 tide levels increased by 50 cm thus simulating the expected 100 year future event in which the city (without gates) would have been flooded with a peak tide of 174 cm – not far below the record flood of 194 cm in November 1966. The heavy line shows a water level rise of about 1 cm h^{-1} in the lagoon during his simulated time of gate closure of

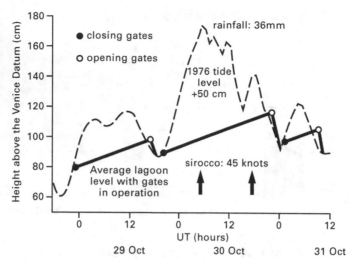

Fig. 34.3   Tide levels on 29–30 October 1976 in Venice increased by 50 cm. Average water levels in the lagoon have been obtained by simulating operation of MOSE flood gates.

about 26 hours. Pirazzoli concludes that the 'case study shows clearly that the very expensive MOSE gates would represent an incomplete protection against flooding in the case of a sea-level rise of only (!) 50 cm'. He goes on to state, 'there is a clear risk that the MOSE gates will become obsolete within a few decades and that it may be necessary to demolish them shortly after their construction, in order to separate the lagoon from the sea in a more effective way'. In a rebuttal, Bras et al. (2002) point out that during Pirazzoli's simulation period (with gates) water levels in the lagoon were below the flooding level of 110 cm except for about three hours when they reached a peak of 115 cm above the city datum. They also questioned the magnitude of Pirazzoli's lagoon water level rise during gate closure. He cites excessive gate leakage due to adjacent gate oscillation angles of 20° to 30° due to incident waves. However, the gates have been carefully designed to keep oscillations below 10° under foreseeable wave conditions. (MAV-CVN unpublished report, 2003). Recent studies (CVN, personal communication 2003) have analysed the lagoon water level rise for the 29-30 October 1976 flood event with a gate closure at a water level of 88 cm above datum and a duration of 26.5 hours. The water level rise is a result of: (1) direct

rainfall on the lagoon at 1.5 mm h^{-1} or 4 cm; (2) inflow from the lagoon watershed at 300 m^3 s^{-1} or 7.1 cm; and (3) gate leakage at 0.28 cm h^{-1} (based on 1:30 model tests) or 7.5 cm. The total water level rise during the 26.5 hour closure is 18.6 cm and the maximum lagoon water level during the entire simulation is 106.6 cm which is below the flooding level of 110 cm. The conclusion is that the gates will effectively protect Venice from extreme flood events within a time frame of 100 years.

MOVABLE GATES AS A MANAGEMENT TOOL FOR ENHANCING LAGOON CIRCULATION AND WATER QUALITY

In 1993, a team of experts were asked by CVN to consider the widely expressed opinion that gate closures to prevent flooding would diminish communication with the Adriatic and therefore aggravate pollution within the lagoon. The team, consisting of B.Battaglia, R. Frassetto, W. Munk, G. di Sciara and G. Puppi, prepared an interesting one page document entitled 'Let the moon sweep the lagoon'.

The essence of their response was that the water quality impact of gate closures can be mitigated by opening and closing the sets of gates at the three openings at appropriate tidal phases so as to enhance the internal circulation in the lagoon. For example, closure of the Malamocco gates during ebbing tide would induce a circulation and a net outflow towards both Lido to the north and Chioggia to the south. This leads to a convective ventilation of the lagoon waters which is much more effective than the diffuse ventilation associated with the present localized tidal in-and-out flow through all three openings (see Fig. 34.4).

The induced circulation effect is illustrated graphically by the following two examples from lagoon circulation model studies. (CVN/TCH unpublished reports, 1993, 1999) shown in Fig. 34.5. The upper pair of figures compare dispersion patterns for the polluting source from the city of Venice – the left with all the gates normally open and the right after four tidal cycles with Malamocco closed during inflow and Lido and Chioggia always open. The lower pair show polluting sources from the Dese River and the city of Venice – the left with all gates open and the

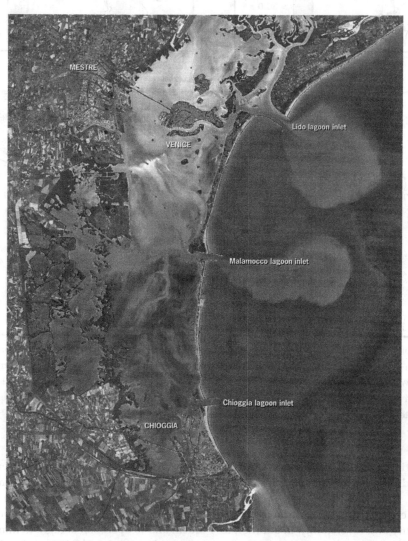

Fig. 34.4   Venice Lagoon and its three inlets.

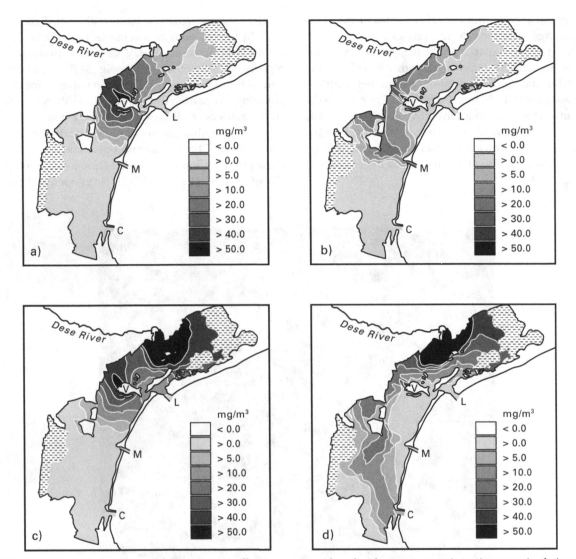

Fig. 34.5   Dispersion patterns under one or two polluting sources with and without gate operation to improve circulation:
(a) normal situation with one pollution source (Venice (V)); (b) as (a) but after four tidal cycles with Malamocco (M) gates
closed during inflow and Lido (L) and Chioggia (C) gates always open; (c) normal situation with two pollution sources
(Venice and Dese River); (d) as (c) but after seven tidal cycles with Malamocco gates closed, Chioggia gates closed during
inflow and Lido gates closed during outflow. Light areas on lagoon margins = fish farm areas.

right after seven tidal cycles with Malamocco closed,
Chioggia closed during inflow and Lido closed during outflow.

The induced circulation scenarios have a high
degree of flexibility and reversibility. In addition to
changing the order in which the sets of gates are

opened and closed, other possible scenarios relate to
operation of the gates during spring and neap tides,
increased operation during summer periods when
pollution impacts are high and floods are less frequent and to operation at night to minimize interference with shipping. The latter is less important now

that a large ship lock will be built at Malamocco to avoid shipping delays to the port of Venice when the gates are closed for any reason. Residual currents generated by differential closure of the gates are able to spread the tidal flushing over the lagoon more effectively, thus reducing the residence time of polluted water even in areas located far from the inlets. Additional studies of induced circulation should be made as soon as newly calibrated and verified ecological and morphological models become available.

ECOLOGICAL CHANGES RESULTING FROM CONSTRUCTION OF FLOOD CONTROL GATES AND DAMS IN THE RHINE RIVER DELTA

The great North Sea storm of 30 January 1953 flooded a third of The Netherlands and resulted in the death of more than 2000 people, mostly in the delta region. In addition to the estuaries of the Eastern and Western Scheldt, there are three main estuaries of the Rhine. A Delta Committee was formed immediately and in a report a year later advised that the mouths of the Dutch estuaries be closed by dams to prevent future flooding. The Western Scheldt, which serves the port of Antwerp, was to remain open with flood protection provided by raising the dikes around its perimeter.

Flooding in the most northerly Port of Rotterdam waterway is controlled by a massive horizontal sector gate that is normally open (see Bol, chapter 38). The original plan called for the remaining two Rhine River estuaries (Haringvliet and Grevelingen) and the Eastern Scheldt to be permanently separated from the North Sea by dams with small, normally closed, sluices used only to discharge river water to the North Sea. This would have created an enormous inland freshwater lake from the three salt-tidal estuaries. The plan was modified in the 1970s in response to environmental concerns by adding 62 normally open gates in the Eastern Scheldt dam so as to retain some aspects of a salt-tidal estuary (see Saeijs and Guerts van Kessel, chapter 39).

REFERENCES

Ammerman, A. J., and C. E. McClennen. 2000. Saving Venice. *Science* **289**, 1301–2.

Bras, R. L., Harleman, D. R. F., Rinaldo, A., and P. Rizzoli. 2002. Obsolete? No. Necessary? Yes. The Gates Will Save Venice. *EOS* (Transactions of the American Geophysical Union) **83**, 20, 218.

Pirazzoli, P. A. 2002. Did the Italian government approve an obsolete project to save Venice? *EOS* (Transactions of the American Geophysical Union) **83**, 20, 217–18.

35 • The Thames Barrier – now and in the future

D. WILKES AND S. LAVERY

BACKGROUND

Much of central London has developed on low-lying marshland alongside the tidal Thames. Archaeological evidence and written records are witness to man's struggle against rising sea levels and increasingly severe tidal flood conditions over the past 2000 years. Despite this, progressively sophisticated tidal defence arrangements have proved effective and the United Kingdom's capital city has continued to thrive. Historically, the Thames' tidal defences have been improved in response to tidal flooding disasters, the most recent of these being the devastating flood of 1953. This acted as the catalyst for construction of the current system of Thames tidal defences. These defences, which include the Thames Barrier, 32 additional moveable gates and 337 km of riparian works provide one of the best standards of flood defence in the United Kingdom. With over 20 years of reliable operations, they have provided confidence to investors and reassurance to the people of London. The defences provide a high standard of flood protection, with an allowance for sea-level rise being built into the design, giving at the year 2030 a standard of protection of 1:1000 years or 0.1 % risk of flooding. It took 30 years to plan and build the current system of defences following the 1953 flood. With less than 30 years remaining before the design standard is reached in 2030, the time is right now to commence planning for the next generation of defences for London and the Thames Estuary.

It could be argued that these seemingly impregnable defences have distorted public and institutional perceptions about flood risk from the tidal Thames. Development continues apace in the confident understanding that the likelihood of flooding is very low. But there is very little appreciation of the conse-quences of tidal flooding should the defences fail. Therefore, the next generation of tidal defences are being planned against a background of inadequate perception of flood risk and even complacency. Today, in the early part of the twenty first century, London continues to expand, and substantial growth is planned through the UK government's major regeneration initiative known as the 'Thames Gateway'. London is thus moving to the east, with extensive residential and commercial property being built in the natural Thames tidal floodplain. These new developments will fundamentally change the developed footprint in the floodplain, and will be in place for at least the next 50 to 100 years. This presents us with a dual challenge of planning the next generation of defences, and ensuring that the right decisions are made now concerning the nature and location of new building in the tidal floodplain. The Environment Agency is planning a long-term flood risk management strategy for London and the Thames estuary which will include the proposals for how the Thames Barrier is managed now and in the future. This chapter describes the Agency's approach to long-term management of the Thames Barrier as our single largest flood defence asset, although the context of the wider Thames estuary is also described here for completeness. In the 1970s and 1980s when the Thames Barrier was constructed, the vocabulary of sustainability was not then in use. It was recognized however, that the new works would provide a high standard of tidal defence to London for at least 60 years. Removal of the threat of imminent flooding has provided us with 'breathing space' to consider future options for longterm sustainable flood management. An understanding of sustainable development is therefore key to our current development of flood management strategy for the Thames Barrier and associated defences.

Flooding and Environmental Challenges for Venice and its Lagoon: State of Knowledge, ed. C. A. Fletcher and T. Spencer.
Published by Cambridge University Press, © Cambridge University Press 2005.

INTRODUCTION

Sustainable development

The management of estuaries requires complex and often conflicting issues to be balanced and this requires a real commitment to sustainable development. The Brundtland Report (World Commission on the Environment, 1987) defines this as: 'The development that meets the needs of the present without compromising the ability of future generations to meet their own needs.'

Sustainable development is not just concerned with balancing economic, social and environmental interests; it seeks to enhance all three components, or at least to ensure that there are no overall adverse effects. Compensatory activity may thus be needed to offset an unavoidable impact. For example, on the Thames estuary, new intertidal habitat could be created to replace that unavoidably lost by engineering works. Across-the-board raising of tidal defences on existing alignments to cope with sea-level rise is not a sustainable solution, although in some areas there will be no suitable alternative option. A mix of long term flood risk management solutions, together with appropriate compensatory measures will provide the best possible sustainable flood risk management. The opportunities for creating morphological change to accommodate anticipated rising sea levels, or at least to delay the need for otherwise needed improvements, are limited on the Thames. There are, however, some significant opportunities for controlled inundation of areas to contain an element of the North Sea surge, which merit further investigation. Where new defences are planned on the tidal Thames, a key element of the design brief will be to 'future proof' the design so that the defence can easily be raised to accommodate sea-level rise and is adaptable for future land use. The challenge is to build in this 'future proofing' over a long planning horizon and against a background of considerable uncertainty.

Climate change and uncertainty

Climate change is one of the key uncertainties against which major decisions must be made. Climate change threatens an increase in flood risk. The Thames Estuary is vulnerable to three types of flooding: the inundation of floodplains by river water; local flooding when the drainage network is overwhelmed by intense rainstorms; and by tidal surges from the North Sea. Sea level is rising relative to the land as a result of global warming and the downward land tilt associated with post-glacial recovery. The latest UK climate scenarios suggest extremes may change: in particular, the frequency and magnitude of rainfall. The amount of change will depend on the greenhouse gases emitted, but by the 2080s, seasons could have changed and become more intense and many coastal locations could well be experiencing extreme water levels more frequently. For some UK east coast locations such as the Thames, current extreme sea levels could occur between 10 and 20 times more frequently than they do now and new extremes will occur. It is anticipated that sea level will continue to rise. These changes will vary from a few centimetres in the north west to up to 90 cm around London under the worst-case scenario. Sea-level rise and tidal surge could give rise to new and greater flood risk problems although the extent is as yet unknown.

The Environment Agency sees climate change and related effects as a major driver in the need to develop long term flood risk management strategies for areas at risk. Through the Thames Estuary Flood Risk Management Plan, strategies are being developed to counter the heightened flood risk, the associated increase in erosion with loss of mudflats and intertidal habitat, and to foster the general well-being of the environmental and morphological regimes of the estuary.

Managing flood risk in a defended tidal flood plain

Climate change brings with it much greater uncertainty over the base line conditions that govern flood risk. It also means that in the much longer term, post-2100 sea level will almost certainly continue to rise. Therefore if development is placed in tidal flood risk areas, a high level of flood risk management is needed. Part of this will be a need to recognize that extreme events will always pose some risk to development no matter how good the defences. Developers and planners must be made aware of this from the outset and develop resilience of systems into their plans. Adherence to the recommendations of PPG25 (Government guidance on Development and Flood Risk) is essential to ensure that flood risk management is factored into developments at planning stage. There is, however, some need for further interpretation of the guidance when a floodplain is defended to a standard greater than 0.1% risk of flooding in any one year, as is the case on the Thames Estuary. There is also a need for more widespread understanding of the two elements of flood risk; namely *probability* and *consequence*, and the different ways in which they are managed. In parallel, there is a need to move away from a mind set that sees developments within defended tidal flood plains as being at negligible risk of flooding just because the probability of flooding is low. In the Thames tidal floodplain there are opportunities for regeneration, a key objective of the Environment Agency being to ensure that such development is undertaken in a manner and with sufficient education so that the communities and businesses that settle have sufficient awareness and appropriate infrastructure to manage flood risk in the most effective and cost efficient manner.

Other drivers

In addition to the factors described above, key drivers for the development of a long term flood risk management strategy for the Thames are: the ageing flood defence infrastructure; the need to plan for changing socio-economic needs on the estuary; and compliance with environmental legislation. In particular, future plans must incorporate the requirements of relevant European Union (EU) Directives including the Habitats and Birds Directive, the Water Framework Directive and the Strategic Environmental Assessment/Environmental Impact Assessment Directive.

THE THAMES ESTUARY

Estuary characteristics

The Thames Estuary has a significant defended floodplain of 35 000ha (Fig. 35.1). Large-scale reclamation of marshes and mudflats commenced in the seventeenth century and this has continued until the major tidal defence works of the 1970s and 1980s. There is a rich archaeological and historical heritage and once defended from tidal inundation, the floodplains were rapidly occupied by industrial and agricultural uses and by the expansion of London and Thames-side towns in the Thames Estuary. This continues today. The estuary is dynamic, with a tidal range of up to 7 m and, although consisting of a relatively stable single-channel system in the upper middle and lower reaches, the outer estuary is composed of continually shifting channels, sandbanks and mudflats. The mobile bed sediments along the estuary vary from coarse sands and gravel to very fine mud which, where suspended in the flow, belies the dramatically improving water quality. Although more than 60% of the Thames estuary flood plain is either heavily urbanized or has development planned, the lower estuary floodplain supports a considerable acreage of agricultural land and there are a number of important SSSI and sites designated under the EU Habitats and Birds Directive.

The Thames tidal defences

For the purposes of this chapter, the Thames Estuary is taken as extending from Teddington in west London to Sheerness/Shoeburyness in the east. Tidal defence is provided by the Thames Barrier at Woolwich, nine other major barriers owned and operated by the Environment Agency and 337 km of tidal walls and embankments which protect an estimated 1.25 million people living and working in the tidal floodplain (Fig. 35.1). Thames flood defences have developed since Roman times, being improved with successive and increasingly severe tidal flood events. The most recent major improvements of the 1970s and 1980s included the Thames Barrier and associated defences, constructed in response to the devastating tidal flood of 1953.

The present generation of flood defences were made possible when the Thames Barrier Act (1973) passed into law. Amongst other issues the act stipulated that the barrier should be reliable in operation, should offer minimal disruption to the passage of shipping and should be aesthetically pleasing. The Environment Agency continues to maintain and

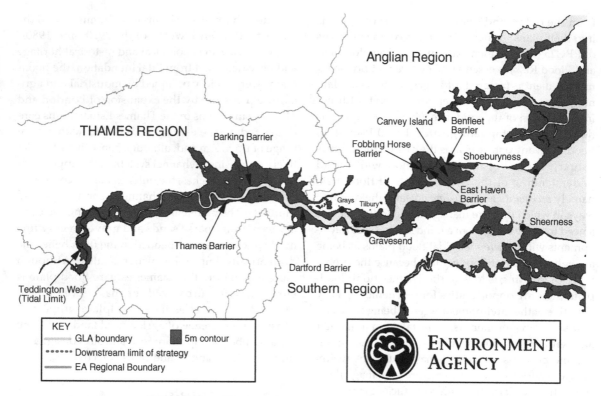

Fig. 35.1   The defended Thames tidal floodplain.

operate the barrier with these points in mind. During the first twenty years of barrier operation the Agency and its predecessor organizations commissioned two formal independent reviews of reliability. In summary these reviews concluded that the greatest risks of malfunction clustered around the electrical and electronic control systems and around people factors. With the passage of time and possible deterioration due to wear and tear and ageing, the risk of failure profile is expected to change.

These current defences provide London and the Thames estuary with a high standard of defence against tidal flooding with an allowance for sea-level rise of 8 mm per year built into the design standard. This design standard will be reached soon after the year 2030 so the time is right now, irrespective of climate change or other drivers, to start the planning for the next generation of tidal defences. With the future sustainability of London and the highly developed Thames corridor at stake, there is a compelling case

for production without delay of a long-term strategy. There are twelve major tidal tributaries discharging into the tidal Thames downstream of Teddington in addition to the many 'lost rivers' of London which have been culverted during the past 300 years and now discharge through tidal flaps. Low-lying land has pumped-drainage systems at Rainham and Canvey Island on the north bank and at Thamesmead on the south bank, much of this land being up to 4 metres below design high tide level. Elsewhere in London, surface water runoff is drained via a combined sewer outfall system to the Becton Sewage treatment works on the north of the river and Crossness STW on the south bank. London's Victorian sewage system has served the city well for over 120 years, but urgent improvements are now being planned through the 'Tideway Strategy' which is looking at the problem of raw sewage overflow into the tidal Thames during times of heavy rainfall. Declining industrial use combined with better control and regulation of discharges to the Thames has resulted in an impressive recovery of the water quality. From being biologically 'dead' in the 1950s, the tidal Thames now supports more than 120 species of fish, 350 species of invertebrates and hundreds of thousands of visiting wildfowl and waders each year. The Thames estuary also provides an important nursery ground for marine fish and supports fisheries throughout the North Sea.

A vision for the Thames

The Environment Agency's approach to its tasks is based on providing a better quality of life and is set out in 'The vision for our environment: making it happen'. This may be translated into the following vision:

- an unpolluted estuary
- a healthy fishery with a sustainable run of migratory fish to and from the inland catchments;
- thriving wildlife with an expanded inter-tidal habitat;
- flood defences that provide an assured level of protection while maintaining the estuary's natural processes;
- prosperous ports, industry and agriculture, which have minimized their environmental impacts and benefit from the diverse and healthy river, and the rivers and canals running to it;
- a landscape with outstanding archaeological and historic features that are enhanced by the Environment Agency's activities;
- a tourist economy that benefits from the river but is sensitive to its wildlife and historic riches;
- a vibrant community that understands, cares for and enjoys the river.

THE NEXT-GENERATION FLOOD MANAGEMENT PROJECT

Project objectives

The Thames Estuary Flood Risk Management Plan is being developed by the Environment Agency to develop a strategy for flood risk management for London and the Thames estuary for the next 100 years that addresses the management of the estuary from a risk perspective and with a whole estuary approach. The work is funded by the Thames, Essex and Kent Flood Defence Committees and application is being made for grant aid from Defra (Department of Environment, Food and Rural Affairs). Support has been obtained from the Interreg 3B European funding programme and from the UK's Office of the Deputy Prime Minister (ODPM) to fund a sub project which examines public and official attitudes to creative floodplain management jointly with partners from Germany, Belgium and the Netherlands.

Programme, progress and outputs

Having completed extensive scoping studies the project team is developing an approach to the management of 'priority' flood defence works, which cannot wait for the completion of the Thames flood risk management plan. Through historic legacy and implementation of major schemes, such as the construction of the Thames Barrier, the tidal Thames has a disparate collection of more than 33 different management plans and strategies covering tidal defence across Anglian, Southern and Thames regions (Fig. 35.2). There is a pressing need for an overarching management framework to reconcile these various plans and strategies and build them into a coherent whole. The advent of the Catchment Flood Management Plan programme in the Thames catchment further emphasizes the need for this whole estuary management framework. The Thames study is in its early stages, and it is anticipated that work on the strategy will be completed in five phases over a six-year period. Implementation of construction on the Thames is likely to commence around 2015 and be carried out over a 20-year period, although it is recommended that this is preceded by a period of collaboration with the Thames Gateway development, to ensure appropriate sustainable flood defences are built into the new riverside construction due for completion by 2016.

Stakeholders

In the Thames Estuary, establishment of partnerships and liaison with stakeholders is still in its early stages. The Agency is a partner in the Thames

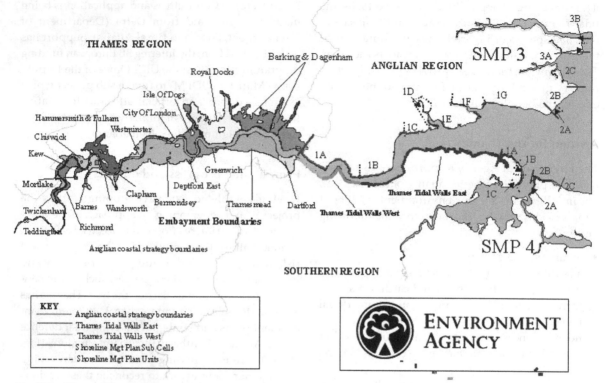

Fig. 35.2   Thames Flood Management framework.

Estuary Partnership – a neutral body which brings together over 100 different stakeholder groups to work towards common aims and sustainable development of the Thames Estuary. Principal Thames Estuary stakeholders are the Port of London Authority and the Thames Gateway partnerships and there is urgency for early engagement with these organizations, because of the interrelation of their plans with the Agency's flood defence interests. The Thames Gateway is one of the government's key areas for regeneration in the south-east of England, which includes proposals for construction of additional housing for a population equivalent to a new city the size of Leeds (approximately 700 000 people).

The future of the Thames Barrier

The Thames Barrier is just one part of the overall Thames tidal defences system – albeit the single largest structure. All major structures are being examined for robustness and sustainability under climate change and other changing environment scenarios. Early indications are that, subject to appropriate modifications, the useful life of the Thames Barrier could be extended by a further 100 years. Figure 35.3 shows how the gates of the Thames Barrier may be modified to provide protection against an additional 1.2 metres of flood level. This, in conjunction with structural and other improvements, may cost many millions of pounds but would be considerably cheaper than the construction of a new barrier and might 'buy' a further 60 to 100 years of useful life. An increasing number of closures because of higher tides and higher fluvial flows may have implications for reliability and maintenance and an impact on sediment morphology.

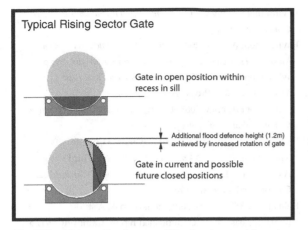

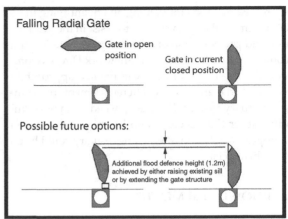

Fig. 35.3   Thames Barrier – modification to provide an increased level of protection.

MAKING THE RIGHT DECISIONS

Sharing experience

The Thames project team was established in January 2002. The team has gained valuable experience from the pioneering work of colleagues on managed setback on the Essex estuaries, where there is now a track record of over ten years of practical setback experience, and on the Humber initiative which is trailblazing techniques and policies, having started over seven years ago. The successful bid for European funding to examine elements of the work opens up opportunities for exchanging views and developing solutions to common problems with European partners.

Decision making

Climate change raises difficult questions for all organizations including the Environment Agency. Uncertainties are inherent, and thus it requires a capacity to manage risk. The crosscutting nature of the problem necessitates the ability to manage the bigger picture and work between different policy areas over long time-scales.

Interpretation and compliance with the Water Framework Directive provides an incentive for a fully integrated approach to valuing and managing the different elements of complex estuary systems. Early work on an appraisal methodology for Heavily Modified Water Bodies (HMWB) indicates that the Thames Estuary – which falls under the description of HMWB – could usefully adopt this approach in conjunction with other decision testing tools.

CONCLUSIONS

There may be a possibility that some of the effects of sea-level rise can be accommodated by managed morphological change although somewhat limited in the Thames Estuary. Similarly there are some significant opportunities for controlled inundation of areas to contain an element of the North Sea surge. Climate change adaptation measures are likely to include significant improvements to existing defences in conjunction with a new approach to sustainable flood management. Ideally, these design solutions must be produced in partnership with planning authorities and developers. Support of government is required to ensure that planning guidance and developer responsibility is understood and consistently applied. There is an urgency to establish these partnerships with strategic planners and developers to ensure that imminent widescale development within the Thames Gateway does not prejudice future flood risk management options, and that practical partnership arrangements are established to ensure that best practice flood risk management is factored in to the planning at an early stage. There is excellent potential for habitat protection, creation and improvement, but there is a need for robust methods for valuing and quantifying these environmental improvements to assist in the justification and prioritization of schemes. Interpretation and compliance with the Water Framework Directive provides an incentive for a fully integrated approach to valuing and managing the different elements of a complex estuary system. If this is grasped as an opportunity rather than seen as another piece of imposed European legislation, the Thames estuary could be an early beneficiary.

ACKNOWLEDGEMENTS

We are grateful to colleagues who have provided information. The views expressed are those of the authors and are not necessarily those of the Environment Agency.

BIBLIOGRAPHY

Ash, J., Fenn, T., Venables, J., and S. Lavery. 2003. *Planning for Flood Risk Management in the Thames Estuary*. CIWEM Rivers & Coastal Group meeting January 2003.

Ash, J., Risk and Policy Analysts. 2002. Humber Estuary decision framework – use of HMWB appraisal methodology (unpublished advisory note).

Department for Environment, Food and Rural Affairs. 2001. *Development and Flood Risk*. Planning Policy Guidance No. 25, The Stationery Office, London, 52pp.

Department of Environment, Transport and the Regions. 1999. *A Better Quality of Life: Strategy for Sustainable Development for the UK*. The Stationery Office, London, 96 pp.

Environment Agency. 1998. *Humber Estuary Action Plan: Our Commitment*. 58 pp.

Environment Agency. 2002. *Our Vision For The Environment: Making it Happen – Draft Corporate Strategy*. Bristol, 59pp.

Environment Agency. 2003. *Thames Estuary Flood Risk Management Plan*. Project Appraisal Report.

Environment Agency. 2003. *The Implications of Climate Change for the Environment Agency*.

IPCC (Intergovernmental Panel on Climate Change). 2001. *Summary for Policy Makers*, First, Second and Third Assessments, Third Assessment, March 2001. Cambridge: Cambridge University Press.

Kelman, I. 2000. Coastal settlements at risk: A study of England's east coast. Unpublished report, Cambridge : The Martin Centre, University of Cambridge.

Meyer, A., and T. Cooper. 2000. *Why Convergence and Contraction are Key*. Environmental Finance, May 2000. Environment Agency briefings from A. Meyer.

Reeder, T. 2003. Climate change and development in the Thames gateway. Unpublished advisory note.

UK Climate Impacts Programme. 2000. *Socio-economic Scenarios for Climate Change Assessment*. UK Climate Impacts Programme. Oxford: UK CIP, Oxford.

UK Climate Impacts Programme. 2002. Scientific Report, Tyndall Centre for Climate Change Research, School of Environmental Sciences, University of East Anglia, Norwich, UK.

UK Met Office.1999. *The Greenhouse Effect and Climate Change*. A briefing from the Hadley Centre and the Met Office.

UK Met Office. 2000. *Climate Change*. An update of recent research from the Hadley Centre.

36 • Cardiff Bay Barrage – lessons learnt 1990–2003

P. HUNTER

INTRODUCTION

This chapter briefly describes the main features of the Cardiff Bay Barrage and discusses some of the lessons learnt during its design, construction and operation for the period 1990–2003.

In 1986 the UK Government decided that the Cardiff Bay area was in urgent need of urban renewal. The keystone to this strategy was the construction of the Cardiff Bay Barrage, to impound the estuaries of the rivers Taff and Ely, create a 200 ha freshwater lake and about 12 km of attractive waterfront. To achieve this, solutions had to be found for five key objectives, which can be summarized as follows:

Objectives	Solutions
Create lagoon	Build barrage/embankment
Not increase flooding risk	Provide sluices/ exclude high tides
Allow navigation	Provide locks and harbour of refuge
Encourage fish	Provide fish pass
Satisfy the public	Maximise access, interest and delight

The 1,100-metre-long barrage includes sluices, locks, an outer harbour, fish pass, embankment and bridges.

INNOVATION

The Cardiff Bay Barrage (Fig. 36.1) has many innovative design features which combine to create a unique construction project built in a maritime environment with the second highest tidal range in the world. Among these are:

- economical optimized composite sand/rock embankment;
- largest European fish pass;
- new filling method for locks in hypertidal estuaries;
- float-in caisson breakwaters with double-wall 'Jarlan' wave absorbing cross-section; and
- unique integration of engineering, architecture and landscaping to give maximum public access throughout this engineering structure.

THE BARRAGE

The embankment is about 800 m long, and 100 m wide at its base. The design was dominated by:

- foundation conditions: 2 to 7 m of soft mud overlying gravels; and
- stability of the embankment during construction and final closure.

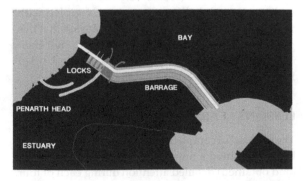

Fig. 36.1   Plan of the Cardiff Bay Barrage.

Flooding and Environmental Challenges for Venice and its Lagoon: State of Knowledge, ed. C. A. Fletcher and T. Spencer.
Published by Cambridge University Press, © Cambridge University Press 2005.

A large fish pass was required to allow fish to migrate between the sea and the bay and then on to the rivers. The fish pass was designed for a flow of 10 m³/s, plus an additional gravity flow of 5 m³/s water in a culvert beneath the fish pass to attract fish to the area of the fish pass.

The five sluice gates are each 9 m wide and 7.5 m high and have a total capacity of 2300 m³/s. The gates close to exclude the highest tides, allowing bay water levels to be controlled even in extreme conditions.

Three navigation locks allow safe passage between the inland bay and the sea. The locks have to withstand head differences in either direction.

The impounded Cardiff Bay has now become a freshwater bay and saline intrusion is highly undesirable for ecological reasons. Flows of salt water into the bay are therefore controlled and minimized

Two breakwaters enclose the Outer Harbour, each about 20 m high because of the high tidal range. The breakwaters are an innovative float-in 'Jarlan' screen design, consisting of a permeable seaward screen formed with concrete 'planks' with a solid screen on the near (harbour) side. Waves hitting the breakwater are partly dissipated when passing through the front screen, and then reflected from the solid near screen. The reflected wave is then out of phase with the incoming wave. The locally increased velocities and turbulence maximize energy dissipation and minimize the overall wave reflection from the structure.

The architectural and engineering design of the scheme ensures a unified design and character and allows maximum public access across the structure.

Details of the design of the barrage have been described elsewhere (Hunter and Gander, 2002). The wide variety of challenges to be met, coupled with the unique tidal conditions of Cardiff, ensure that this project ranks as one of the most interesting civil engineering projects recently completed in the United Kingdom.

LESSONS LEARNT

The successful completion of the Cardiff Bay Barrage has highlighted a number of topics that demanded (and obtained) detailed attention during each of four distinct stages of the project, i.e. Planning and Design, Procurement, Construction and Operation. These are discussed below:

Planning and Design

During this stage the following activities were fundamental to laying the foundations for the success of the project:

- Clarifying objectives;
- Identifying all benefits and drawbacks;
- Identifying all potential uses and users;
- Assessing and mitigating project risks of all types
- Developing a robust design concept to avoid later changes; and
- Consultation and communication with all project stakeholders.

Procurement

Key objectives during this stage were:

- Maximizing communication between designers and contractors before finalizing design;
- Freezing the design before starting construction; and
- Selecting experienced organizations to construct the barrage. (Selection of subcontractors can be as important as selection of main contractors.)

Construction

During the four-year construction period, it was important to:

- Plan construction methods to minimize exposure to risks;
- Minimize environmental and social impacts of the construction process; and
- Keep stakeholders informed of operations and issues.

Operation

Finally during the operation (and maintenance) of the barrage it is essential to have:

- Adequate staffing;
- Spare parts policy (consumables and provision for breakages);

- Rigorous adherence to planned testing and maintenance schedule;
- Maintaining planned environmental monitoring; and
- Maintaining effective two-way communication with the public and all other stakeholders.

The barrage operators have planned all these activities from the beginning and have successfully implemented their plans.

PERFORMANCE

Cardiff Bay was impounded in 1999, and has achieved its objectives. Sluices have been subject to high river flows, the maximum being 535 m^3/s when the bay level rose by 2.2 m as forecast. On 12 August 2001 when a flash shower occurred that resulted in 200 m^3/s flowing down the rivers in a very short time, the bay was maintained at the required level of 4.5 m OD.

The locks have accommodated up to 230 boats passing through in one day. Fish numbers are in line with expectations. The saline sump is designed to contain salt water that enters the bay via the locks, whilst the flushing pipe discharges sea water at low tide. At times the containment of seawater has proved problematic and alternative methods to avoid lock restrictions are being examined. More stringent requirements (relating to saline intrusion) since the barrage was designed have resulted in more monitoring and control being needed than was originally envisaged.

Other demands on fresh water flows into the bay include the fish pass, limited industrial extraction, extraction by the Port of Cardiff for maintaining water levels in the commercial docks, and also an allowance for seepage from the bay. In the periods of normal river flow, there will be ample fresh water for all purposes, but in periods of low flow then conservation measures are necessary, particularly to ensure adequate flows down the fish pass. Two of the locks are linked by cross-culverts to allow filling water to be re-used in adjacent locks at successively lower levels. This can reduce the outflow from the bay by about 35%.

A topic requiring detailed investigation during the design of the barrage was the effect of the barrage on siltation, which was initially considered as a potentially major environmental impact of the project. The main source of siltation in Cardiff Bay is the Severn Estuary, and river-borne sediments are insignificant. Therefore no significant siltation within the Bay is expected to occur. Substantial maintenance dredging had been required in the access channel for Cardiff Docks, and this requirement is expected to fall now that the barrage is complete. This is because the main sediment transport mechanism into the access channel occurred during the last phase of the ebb tide, when sediments deposited during high tide are washed back into the dredged channel. This does not occur now, when the barrage is present, although this benefit is partly offset by the reduced scouring currents in the access channel.

The impoundment causes a rise in groundwater levels in the area surrounding the bay. This was studied in great detail during the planning of the barrage. As well as pre-impoundment surveys, 16,000 surveys of properties have been completed as a precautionary monitoring measure since the bay was impounded. Only one has shown that the predicted rise in ground water has caused problems and this was where it was expected, at the pumping station in Penarth Road next to the River Ely.

The architectural and design attention to all engineering and landscape aspects of the scheme has ensured a unified design and character, with potential for further enhancement as the bay develops. The embankment design varies from the rugged rocky seaward edge to the landscaped lagoon side, with the stone crest of the barrage dividing the principal access paths. A water edge walk follows the scalloped edge of the barrage, providing a constantly changing view of the bay and the barrage.

CONCLUSIONS

The barrage has been maintaining an impounded bay for over three years, and during that period it has produced the expected flood control benefits as well as providing an attractive waterside environment that is encouraging substantial redevelopment of the extensive waterfront of the city of Cardiff.

The final construction cost was £120m, compared with the 1993 budget of £114m. Private sector

Fig. 36.2   The completed Cardiff Bay Barrage.

investment in the bay area of £1 billion has already been secured, and the Development Corporation calculated that at least £170 million a year was being returned to the public purse in 2000 in terms of business rates and other taxes directly attributable to the developments in Cardiff Bay. The barrage has therefore proved to be successful both as an engineering project and in its role of developing the waterfront economy of the city of Cardiff.

REFERENCE

Hunter, P. D. and H. C. W. Gander. 2002. Cardiff Bay Barrage: planning and design. *Proceedings of the Institution of Civil Engineers – Water and Maritime Engineering*, **154**, 117–128.

37 • The modelling of Cardiff Bay Barrage control system: the revised automatic control logic for the sluice gates

E. FAGANELLO AND S. DUNTHORNE

INTRODUCTION

Cardiff Bay Barrage is a tidal exclusion barrier forming a 200 ha freshwater lake by impounding the river flows from the Rivers Ely and Taff. The flood control structure built is designed to pass river floods and exclude high tides, allow passage for migrating fish and continued navigation between the bay and the Severn Estuary as well as to provide a public amenity with 12 km of attractive waterfront (Hunter and Gander, 2002).

The principal structures and ancillary works associated with the Barrage comprise five 9 m wide × 7.5 m high automatically controlled sluices with double-leaf vertical lifting gates, an 8 m wide fish pass and three navigation locks (8 m to 10.5 m wide) with hydraulically operated sector gates. The 700 m long embankment, formed from rock and marine dredged sand, not only impounds the freshwater lake but also prevents both the loss of freshwater and contamination by seawater. The outer harbour (formed by two breakwater arms comprising a series of pre-cast reinforced concrete caisson units) provides a tidal shelter and 'harbour of refuge' for small boats waiting to use the locks.

Prior to the construction of the Cardiff Bay Barrage the risk of flooding of the Cardiff Bay waterfront was mainly from inundation by surge tides. The 1 in 100 year tide is +7.87 m OD and the highest recorded sea level is +7.96 m OD. The Severn Estuary hosts the world's second highest tidal range of almost 14 m: from +7.17 m OD (Highest Astronomical Tide) to −6.30 m OD (Lowest Astronomical Tide). The Barrage has now eliminated the probability of tidal flooding and the residual flood risk depends upon the ability of the inland bay to provide sufficient storage for river floodwater during the tidelock period, as well as the capacity of the sluices to discharge all the stored volume when the sea level drops below the bay level and before the next tidelock occurs.

A fully automatic system controls the gate movements and maintains the normal bay level between +4.0 m OD and +4.5 m OD by responding to and/or predicting rises in bay and tidal sea water levels. Sensors within the bay and sea, downstream of the gates, continually record the water levels (with an accuracy of ±5 mm). The accurate measurements are stored in a database. The rate of change of the bay level and the sea level are then calculated and the Programmable Logic Controllers (PLCs) dictate, as necessary, the raising or lowering of a gate or series of gates.

Following comments received from the Environment Agency, Cardiff City Council appointed GIBB Ltd (now Jacobs Infrastructure) in February 2000 to construct a mathematical hydraulic model of Cardiff Bay. The model was used to examine the reasons for frequent gate movements experienced in the prototype during the sluice gate commissioning tests. It was also used to explore measures which could be taken to reduce the number of movements, whilst maintaining adequate discharge capacity during extreme flood events. Initial indications were that the programming of the original automated control system would have to be revised to accomplish these aims.

This chapter describes the construction, calibration and verification of the mathematical hydraulic model, and in particular the way in which the discharge coefficients of the sluice gates are dependent on the number of gates in operation and the reasons for frequent gate movements. In addition the revised sluice gate control logic is presented and the use of the hydraulic model to verify the logic is detailed.

Flooding and Environmental Challenges for Venice and its Lagoon: State of Knowledge, ed. C. A. Fletcher and T. Spencer.
Published by Cambridge University Press, © Cambridge University Press 2005.

MODEL CONCEPTUALIZATION AND CONSTRUCTION

In order to define a new set of operating rules for the sluice gates, a one-dimensional hydrodynamic model was constructed using ISIS Flow, a widely used software package developed by Halcrow in partnership with HR Wallingford. ISIS Flow solves the Saint-Venant equations (for unsteady flow in open channels) using an implicit finite difference scheme: the Preissmann four-point scheme (*ISIS Flow User Manual*, 1997). The programme computes flow depths and discharges for each node (cross section) of the channel network. A large variety of hydraulic structures (such as weirs and sluices) can also be modelled using a combination of empirical and theoretical equations. A detailed bathymetric survey, covering the bay and the tidal reaches of the River Ely and River Taff (carried out in 1998), was used to derive the model cross-sections and chainages.

When the barrage operates, water levels in the bay can fluctuate between approximately +4.0 m OD and +7.5 m OD. Thus, the storage volume between these two levels (approx 6 500 000 m^3) is of primary concern in the hydraulic model as it represents the active storage of the bay. The plan area of the bay does not vary significantly in this water level range and remains virtually constant at about 2.0 km^2.

A check was therefore made between the storage capacity represented by the one dimensional model and that calculated from the 1998 bathymetric survey to verify the consistency of the geometric boundary utilized in the hydraulic model. This check confirmed that the volumetric representation of the bay system was correctly reproduced in the model.

The schematic diagram (Fig. 37.1) shows the model representation of Cardiff Bay. The labels *LVSXU* and *LVSXD* (X = 1, … 5) represent the five operating vertical sluices. The fishpass and the navigation locks have not been included in the model as they are not affected by the operating rules governing the discharge capacity of the bay through the gates. Bay cross-sections are represented by labels *BayX* (X = 1, … 8). The River Ely and the River Taff hydraulic nodes are represented by labels *ElyX* (X = 0, 1, 2) and *taffX* (X = 0, 1, 2). The hydraulic roughness of all sections has been represented by Manning's *n* values varying from 0.02 to 0.03.

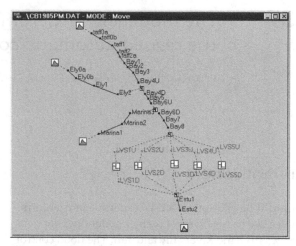

Fig. 37.1   ISIS model schematic. Extract from the ISIS Model – HR Wallingford Software Ltd.

The water level gauge, which is used to control the sluices, is located adjacent to the lock gates (Penarth Marina) and this location is shown in the model as node *Marina2*.

The tidal boundary conditions are attached to the label *Estu2*. Inflow hydrographs (from the Rivers Ely and Taff) are attached to the labels *taff0a* and *Ely0a* (at *Marina1* a constant inflow hydrograph has been introduced as a sweetening flow – 1.0 m^3/s – just to model the little Penarth Marina area and thus the water level gauge). Junction units have been used to link the bay to the tidal channels (Rivers Ely and Taff).

MODEL CALIBRATION AND VERIFICATION

Five drawdown tests, with several different gate opening combinations (2,3,4 and 5 gates), were carried out in April and May 2000 at the barrage by site staff. The test results have proved to be very useful for model calibration although the measured combined fluvial inflows were on the whole modest (10 to 46 m s^{-1}). River flows were recorded by the flow gauging stations situated upstream on the River Taff and River Ely and operated by the Environment Agency. Seawater levels were recorded by sensors located downstream of the barrier and bay water levels were recorded at several locations around the bay.

Description of the sluices and discharge equations utilized in the model

The sluices comprise five automatically operated sluice gates controlling discharges through the 9 m wide by 7.5 m high openings. The gates are a double leaf hook type. A key aspect considered for the design choice was the overshot facility. Although relatively expensive, there are several advantages to having an overshot arrangement (Pope, 2002):

- Debris can be passed over the gate without hindrance;
- The considerable turbulence that can be generated close to the fish pass entrance provides a powerful attraction factor and increases the effectiveness of the fish pass; and
- They have less visual intrusion when raised than a number of other designs.

In flood conditions the gates operate in undershot mode. The gates are housed within a reinforced concrete structure. The sill level is –2.50 m OD (the piers are integral with the base of each sluice) and the soffit level is +5.00 m OD. They discharge into a 60 m long stepped, reinforced concrete apron with concrete faced guide walls either side (see Fig. 37.2). Physical model tests were used to determine the discharge coefficients for the main sluice gates and these

Fig. 37.2   View of sluices and stilling basin under construction.

were used in the original design of the barrage (HR Wallingford Ltd, 1991).

The five identically sized vertical sluice gates have been modelled as separate and independent Vertical Sluice ISIS units in parallel. Dimensions and coefficients are the same for each gate. Gate openings have been imposed on the model in order to replicate exactly the movements that occurred during the tests.

During the simulation the operating modes change depending on upstream and downstream water levels as well as the gate settings.

The operating modes that occur in the modelled scenarios are listed below:

1) gate closed
2) free weir flow under gate
3) drowned weir flow under gate
4) free gate flow
5) drowned gate flow

Each operating mode is described by a specific discharge equation, which applies only if specific hydraulic conditions are satisfied simultaneously. The equations used in the ISIS model are a strict application of those found in Bos (1989).

Theoretical background

Flow beneath the sluice gates, where the underside of the gate is clear of the water surface flow (condition (2) above), is described by the round nosed horizontal broad crested weir equation (Equation 37.1). In the case where the weir flow is drowned under gate (condition (3) above), the discharge is described by the Bernoulli equation constrained to a smooth transition between free and drowned flow at the modular limit (Equation 37.2).

The modular limit m for weir flow is taken from Harrison (1967) and it is a function of $\log_{10}(h_1/p_2)$, where p_2 is the height of crest above bed of downstream channel. The modular limit is the value of the submergence ratio when the flow just begins to be affected by the downstream level, that is, when the weir begins to be drowned. For both these conditions two coefficients apply – the discharge coefficient C_d and the coefficient of approach velocity for weir flow C_{vw}:

$$Q = C_d C_{vw} \left(\frac{2}{3}\right)^{1.5} \sqrt{g}bh_1^{1.5} \quad (\text{m}^3/\text{s})$$

free weir flow under gate (Eq. 37.1)

Conditions:

$$h_0 \geq 0.001, \, h_2/h_1 \leq m, \, 0.005(L-r) < h_1 < 1.5h_0, \, h_2 < h_0$$

$$Q = C_d C_{vw} \left(\frac{2}{3}\right)^{1.5} \sqrt{g}bh_1 \left[\frac{(h_1 - h_2)}{(1-m)}\right]^{0.5} \quad (\text{m}^3/\text{s})$$

drowned weir flow under gate (Eq. 37.2)

Conditions:

$$h_0 \geq 0.001, \quad h_2/h_1 > m, \quad 0.005(L-r) < h_1 < h_0$$

Free or drowned orifice equations are used when the underside of the gate interferes with the flow (Equations 37.3 and 37.4, conditions (4) and (5) above). Here also two coefficients apply: the discharge coefficients C_e and the coefficient of approach velocity for under gate flow C_{vg}.

$$Q = 0.60 C_{vg} \sqrt{2g}bh_0^{1.5} \left[\left(\frac{h_1}{h_0}\right) - \alpha\right]^{0.5} \quad (\text{m}^3/\text{s})$$

free gate flow (Eq. 37.3)

Conditions:

$$h_0 \geq 0.001, \, h_1 \geq 1.5h_0, \, \frac{h_2}{h_0} < \frac{\alpha}{2}\left\{\sqrt{1+16\left[\left(\frac{h_1}{h_0\alpha}\right)-1\right]}-1\right\}$$

$$Q = C_e C_{vg} \sqrt{2g}bh_0[(h_1 - h_2)]^{0.5} \quad (\text{m}^3/\text{s})$$

drowned gate flow (Eq. 37.4)

Conditions:

$$h_0 \geq 0.001, \, h_1 \geq 1.5h_0, \, \frac{h_2}{h_0} \geq \frac{\alpha}{2}\left\{\sqrt{1+16\left[\left(\frac{h_1}{h_0\alpha}\right)-1\right]}-1\right\}$$

where

$$C_d = \left[1 - \delta\frac{(L-r)}{b}\right]\left[1 - \frac{\delta(L-r)}{2h_1}\right]^{1.5}$$

$$C_e = 0.61\left[1 + 0.15\frac{(b+2h_0)}{(2b+2h_0)}\right]$$

and

C_d discharge coefficient for weir flow ($C_d \approx 0.96$)
C_e discharge coefficient for under-gate flow ($C_e \approx 0.66$)
C_{vw} coefficient of approach velocity for weir flow
C_{vg} coefficient of approach velocity for under-gate flow
g gravitational acceleration (m s^{-2})
b breadth of each sluice gate at control section (m)
h_1 upstream water level over the crest (m)
h_2 downstream water level over the crest (m)
h_0 gate opening (m)
L length of weir crest in the direction of flow (m)
δ coefficient which accounts for boundary layer effects ($\delta = 0.01$)
r radius of curvature at the leading edge of the weir crest ($r = 0.1$)
α contraction coefficient ($\alpha \approx 0.63$)
m modular limit ($m = h_2/h_1$)

Reverse flow (from sea to bay) through the gates was allowed and assumed the same coefficients' values used for positive flow.

Calibration and verification of the drawdown tests

The model has been calibrated and verified using the data collected during the sluice gate commissioning

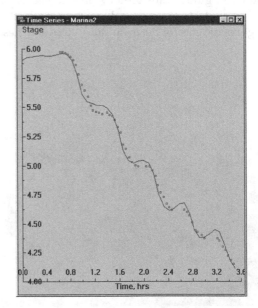

Fig. 37.3   11/04/00 calibration test (gates operating = 5, $C_{vw} = 1$, $C_{vg} = 1.1$). Extract from the ISIS Model – HR Wallingford Software Ltd.

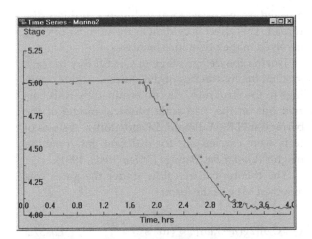

Fig. 37.4   19/05/00 AM calibration test (gates operating = 3, C_{vw} = 1, C_{vg} = 1.3). Extract from the ISIS Model – HR Wallingford Software Ltd.

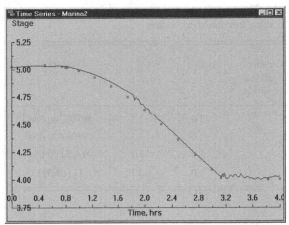

Fig. 37.5   19/05/00 PM calibration test (gates operating = 2, C_{vw} = 1, C_{vg} = 1.5). Extract from the ISIS Model – HR Wallingford Software Ltd.

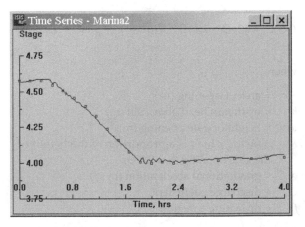

Fig. 37.6   09/05/00 AM calibration test (gates operating = 4, C_{vw} = 1, C_{vg} = 1.1). Extract from the ISIS Model – HR Wallingford Software Ltd.

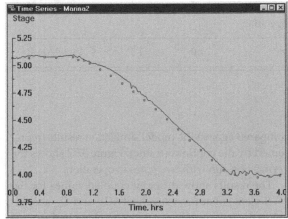

Fig. 37.7   18/05/00 verification test (gates operating = 2, C_{vw} = 1 and C_{vg} = 1.5). Extract from the ISIS Model – HR Wallingford Software Ltd.

tests. Four tests were used for model calibration and one for model verification. Calibration of the model initially focussed on the bay roughness coefficients (Manning's n) and a sensitivity analysis was carried out. However, the calculated bay levels appeared to be quite insensitive to the bay roughness being dominated by the storage in the system. The focus of the calibration then became that of reproducing the discharge capacity of the gates (by using different combinations of discharge and approach velocity coefficients). By

modifying these two principle parameters it was possible to achieve a very high degree of accuracy in both the bay water level and its timing. The model outputs clearly show that the bay levels are particularly sensitive to the approach velocity under-gate coefficient C_{vg} (but not to C_{vw} in weir flow mode). In fact the results suggest different values for C_{vg} depending on the number of the gates in operation. Figures 37.3, 37.4, 37.5 and 37.6 show the calibration runs: the recorded bay levels at Penarth Marina (indicated by the dotted line) are

Table 37.1 *Combination of discharge coefficients resulting from model calibration.*

Number of gates in operation	C_{vw}	C_{vg}	Test date
2	1.0	1.5	19/05/00 PM
3	1.0	1.3	19/05/00 AM
4	1.0	1.1	09/05/00 PM
5	1.0	1.1	11/04/00

Table 37.2 *Combination of discharge coefficients resulting from model verification.*

Number of gates in operation	C_{vw}	C_{vg}	Test date
2	1.0	1.5	18/05/00

compared against the model simulation results (represented by the continuous line). Figure 37.7 shows the verification run with two gates in operation.

The results show different coupled values for the coefficients of approach velocity C_{vw} and C_{vg} according to the total number of gates involved in the operations (and regardless the sluice gate combination considered):

- $C_{vw} = 1.0$ $C_{vg} = 1.1$ for 4 and 5 gates
- $C_{vw} = 1.0$ $C_{vg} = 1.3$ for 3 gates
- $C_{vw} = 1.0$ $C_{vg} = 1.5$ for 2 gates

The different combinations of discharge coefficient values are mainly caused by the effects of approach flow (particularly significant in the gate flow modes (3) and (4)).

Another important factor influencing the discharge capacity of the gates system is the presence of abutments and piers. They constrain the flow through the openings and cause curvature as well as the contraction of the streamlines. The flow is then

affected by increased local turbulence and it is disturbed. This effect is evident as the number of gates involved in the operations increases.

During the design stage of Cardiff Bay Barrage system, the hydraulics of the sluices i.e. gate openings in the structure, was determined by tests carried out on the 1:40 scale physical model of the barrage at H.R. Wallingford Laboratories. A series of tests were carried out for different gate opening heights (*Cardiff Bay Barrage Design Study*, 1991).

The theoretical free flow under the gates was assumed to be of the form:

$$Q = C_d A[2g(h - \lambda w)]^{0.5} \quad (m^3\ s^{-1}) \qquad Eq.\ 37.5$$

The drowned jet flow under the gates was assumed to be of a form similar to equation 37.5:

$$Q = C_d f A[2g(h - \lambda w)]^{0.5} \quad (m^3\ s^{-1}) \qquad Eq.\ 37.6$$

where

A	= area of opening (m²)
h	= upstream head above sill (m)
w	= height of gate opening (m)
λ	= factor which is a proportion of the height of the gate opening ($\lambda = 0.853$)
g	= gravitational acceleration (m s⁻²)
C_d	= discharge coefficient
f	= drowning function

Parameters λ, C_d and f were derived from the physical model tests. No consideration was however given to the use of multiple gates and therefore a fixed value for the discharge coefficient (C_d) was utilized, equal to 0.634.

REVISED GATE OPERATION DURING SPATE

Once the ISIS hydrodynamic model was calibrated and verified, the work focussed on the analysis of potential flood scenarios and the optimization/improvement of the gate rules. This was necessarily an iterative procedure where changes were proposed to the logic and tested using the hydraulic model.

The original sluice gate control logic allowed the gates to open or close sequentially dependent on the water level in the bay and its rate of fall. The application of this logic was shown to cause the sluice gate settings to oscillate or 'hunt'. This effect was reproduced in the model and was found to be as a result of seiching of bay water levels, being instigated by initial gate operation when the water levels in the bay were drawn down during flood conditions. The hydraulic model has also indicated that the present location of water level sensors accurately represents the water level in the bay and that there is no need or advantage to be gained by relocating them.

In order to reduce the number of gate movements and avoid any consequent peak flood water level increase experienced in the bay, the gate control logic had to be amended. After many trials using the hydraulic model and critical peer review the following sequential procedure was determined:

1) The water level rise in the bay during the tidelock period and the duration of the tidelock period is recorded.
2) The water level rise during the tidelock period is used to estimate the inflow to the bay via a look-up table. This table is an array and because the bay storage curve is known it produces the estimated total inflow as a simple extrapolation from the bay water level change and tide lock duration.
3) The inflow estimate is then doubled and the resultant discharge used to determine a gate pre-set position to achieve this discharge.
4) The sluice gates move in a single movement to this pre-set (P1) at the end of tidelock and hence control the fall in water level in the bay.
5) The logic checks the rate at which the bay water level drops. Once the water level is within 200 mm above the target bay level then the pre-set gate position switches (to P2) to slow the rate of bay drawdown.

The logic includes checks to allow additional gate stage openings if the bay water level does not fall within a predetermined time after the end of tidelock.

These criteria have been applied to determine the maximum water level experienced in the bay. The design flood conditions used to analyse this situation and the simulations undertaken using the hydraulic model of the bay are presented below.

Table 37.3 indicates the gate preset positions P1 and P2.

By applying the free gate flow mode formula it is estimated that the 5 gates can reach a combined discharge capacity of up to 2300 m^3/s with the bay level at +7.50 m OD.

DESIGN FLOOD CONDITIONS

Flood hydrographs for the River Taff and the River Ely discharging to the bay for different return periods have been derived by applying the Flood Studies Report (FSR) Unit Hydrograph Method to both rivers. The methodology has been calibrated using the gauged flow data available on both rivers. These are the hydrographs agreed for the project and represent those initially defined by the Institute of Hydrology (1991) (Table 37.4).

The main flood event considered was the 1 in 100 year river flood event, with a combined peak discharge of 930 m^3 s^{-1}, combined with the Highest Astronomical Tide at +7.17 m AOD. This event has been shown by H. R. Wallingford Limited to have a return period of approximately 1 in 70 700 years at the present time (Samuels and Burt, 2002). In addition, in order to make some allowance for the critical flood events occurring during maintenance periods, only four of the five sluice gates have been assumed to be available to pass discharges from the bay. During these conditions the gates would fully open to pre-set gate stage 16 (see Table 37.3). Additional fluvial design events have also been investigated in order to analyse a wider range of possible scenarios and better understand the results of the pre-set openings proposed.

BAY FLOOD WATER LEVEL SIMULATIONS

The assumptions made are that the four gates are initially fully open, the initial bay water level is +4.00 m OD and the initial sea level is +3.46 m OD. Above this level it is estimated, using the calibrated model, that the discharge capacity of the four operating gates becomes less than 930 m^3 s^{-1}, the peak inflow. As soon as the sea-level rises above this threshold, the bay

Table 37.3 *Description of the gate pre-set positions (P1 and P2) for 4 gates in operation.*

Condition (m³/s)	Stage	Sluice 1 opening (m)	Sluice 2 opening (m)	Sluice 3 opening (m)	Sluice 4 opening (m)	Sluice 5 opening (m)
Preset P1 (approximately 2.0*river inflows)						
Inflow > 823	16	0.00	7.50	7.50	7.50	7.50
705 < Inflow ≤ 823	15	0.00	7.50	7.50	7.50	5.50
646 < Inflow ≤ 705	14	0.00	5.50	7.50	5.50	5.50
529 < Inflow ≤ 646	13	0.00	5.50	5.50	5.50	5.50
412 < Inflow ≤ 529	12	0.00	3.50	5.50	3.50	5.50
375 < Inflow ≤ 412	11	0.00	3.50	3.50	3.50	3.50
301 < Inflow ≤ 375	10	0.00	3.50	3.50	3.50	2.25
264 < Inflow ≤ 301	9	0.00	2.25	3.50	2.25	2.25
220 < Inflow ≤ 264	8	0.00	2.25	2.25	2.25	2.25
176 < Inflow ≤ 220	7	0.00	1.50	2.25	1.50	1.50
154 < Inflow ≤ 176	6	0.00	1.50	1.50	1.50	1.50
110 < Inflow ≤ 154	5	0.00	1.50	1.50	1.50	0.75
88 < Inflow ≤ 110	4	0.00	0.75	1.50	0.75	0.75
44 < Inflow ≤ 88	3	0.00	0.75	0.75	0.75	0.75
Inflow ≤ 44	2	0.00	0.00	0.75	0.00	0.75
Preset P2 (approximately 1.2*river inflows)						
Inflow > 823	13	0.00	5.50	5.50	5.50	5.50
705 < Inflow ≤ 823	12	0.00	3.50	5.50	3.50	5.50
646 < Inflow ≤ 705	12	0.00	3.50	5.50	3.50	5.50
529 < Inflow ≤ 646	11	0.00	3.50	3.50	3.50	3.50
412 < Inflow ≤ 529	10	0.00	3.50	3.50	3.50	2.25
375 < Inflow ≤ 412	8	0.00	2.25	2.25	2.25	2.25
301 < Inflow ≤ 375	8	0.00	2.25	2.25	2.25	2.25
264 < Inflow ≤ 301	7	0.00	1.50	2.25	1.50	1.50
220 < Inflow ≤ 264	6	0.00	1.50	1.50	1.50	1.50
176 < Inflow ≤ 220	5	0.00	1.50	1.50	1.50	0.75
154 < Inflow ≤ 176	4	0.00	0.75	1.50	0.75	0.75
110 < Inflow ≤ 154	4	0.00	0.75	1.50	0.75	0.75
88 < Inflow ≤ 110	3	0.00	0.75	0.75	0.75	0.75
44 < Inflow ≤ 88	3	0.00	0.75	0.75	0.75	0.75
Inflow ≤ 44	2	0.00	0.00	0.75	0.00	0.75

Table 37.4   *Cardiff Bay Barrage: fluvial flood frequency.*

Return period (years)	Flood peak (m³/s)		
	Ely	Taff	Total
2.33	44	365	409
5	64	500	564
10	78	560	638
20	92	635	727
50	108	735	843
100	122	808	930

level starts to increase and when the difference between the two levels becomes less than 300 mm, the gates close. Tidelock ends when the sea level falls to the same level as the bay. At that time the gates open fully to 7.5 m with an opening velocity of 1m/min. Changing the timing of the peak flow to that of the peak in the tidal cycle reveals that the maximum water level in the bay occurs when the peak flow is simultaneous with the peak tide level and produces an estimated bay level of +7.42 m OD. The results for the critical case, tidal peak coincident with peak inflow, are presented graphically as Figure 37.8. The

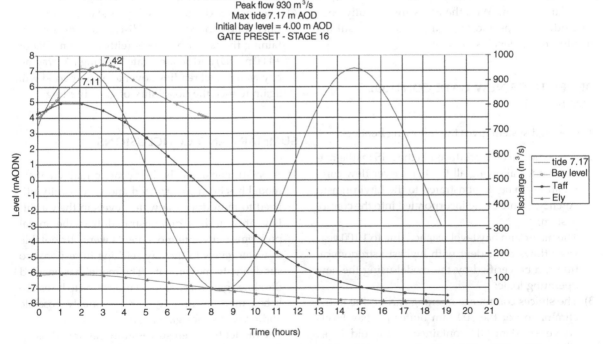

Fig. 37.8   Resulting bay water level from 1 in 100 year fluvial event (initial bay water level = +4.00 m OD) and four gates fully open.

Table 37.5 *Summary of maximum water levels in the bay.*

Peak inflow (m³/s)	Peak bay water level (m OD) with peak tide = +7.17 m OD (H.A.T.)					
	Three gates operation		Four gates operation		Five gates operation	
	Initial bay water level +4.0 m OD	Initial bay water level +4.5 m OD	Initial bay water level +4.0 m OD	Initial bay water level +4.5 m OD	Initial bay water level +4.0 m OD	Initial bay water level +4.5 m OD
930	+7.48	+7.52	+7.42	+7.45	+7.27	+7.32

initial water level condition of +4.0 m OD is that which would occur if the water level in the bay had been above a level of +4.9 m OD in the previous tidal cycle. This represents the condition expected to occur in the extreme tidal conditions expected in such an event. However, the sensitivity of the flood water level to the initial bay water level has been examined.

A similar analysis has also been carried out with three and five gates in operation. Table 37.5 contains the summary of all the model results for the 1 in 100 year fluvial event. With the gates always fully open (including the period with high tide) the resulting maximum bay level is +7.56 m OD.

BENEFITS OF NEW GATE CONTROL LOGIC

The revised system gives the following benefits:

1) The logic can be reproduced in the ISIS mathematical model, allowing all tidal and river flow combinations to be demonstrated to the Environment Agency, before being incorporated into the control system;

2) The maximum bay level reached due to 1:100 year river flows coincident with highest astronomical tide can be verified by the model using the new operating logic;

3) The sluices commence opening earlier due to the elimination of the 300 mm buffer for bay levels above +6.00 m OD contained in the old logic. (This will assist in reducing the maximum bay level reached during this period.);

4) The time necessary to open the gates after tide lock is reduced by:
   – opening the gates to the pre-set position in one movement, not through the stages;
   – moving all gates together instead of in pairs as in the old logic.
   This will assist in reducing the maximum bay level reached during this period;

5) The number of sluice gate movements is reduced in comparison with the previous logic; and

6) The increased number of gate stages allow for finer control of bay water discharge whilst maintaining the target bay level (either +4.5 m OD or +4.00 m OD) and so limit gate 'hunting'. Averaging bay water level readings over a longer period will again reduce the possibility of gate 'hunting'.

SUMMARY AND CONCLUSIONS

1) A one-dimensional hydrodynamic mathematical model has been constructed and successfully utilized to reproduce the water levels in the Cardiff Bay area. It has demonstrated that the initial problems with gate movements were attributable to water levels in the bay seiching in response to sudden gate movements. The model has enabled the sluice gate discharge relationships to be established more accurately than previously possible and revised sluice gate logic to be tested.

2) The model has been successfully calibrated using on site tests, which recorded water level variation in response to predetermined gate operations.

The results of the calibrations demonstrate that there is a tendency for discharge coefficients of multiple sluice-gated systems to vary dependent on the number of gates in operation. This effect can result in significant variation in the discharge capacity per gate. Accurate analysis is therefore required for any type of barrage outlet system that comprises multiple gates.

3) The use of one-dimensional mathematical models seems to offer an acceptable compromise in terms of costs and accuracy when compared to more expensive and difficult to calibrate two and three-dimensional models for such applications especially if used in conjunction with physical models.

ACKNOWLEDGEMENTS

The authors wish to thank the Cardiff Harbour Authority, particularly Roger Thorney (Barrage Manager) and Peter Hunter, Jacobs Project Director, for their knowledge and support.

REFERENCES

Bos, M. G. 1989. *Discharge Measurement Structures*, The Netherlands: ILRI.

H. R. Wallingford Ltd. 1991. *Cardiff Bay Barrage Design Study*. Report EX 2297.

Harrison, A. J. M. 1967. The streamlined broad-crested weir, *Proceedings of the Institution of Civil Engineers* **38**, 657–78.

Hunter, P.D., and H.C.W. Gander. 2002. Cardiff Bay Barrage: planning and design. *Proceedings of the Institution of Civil Engineers – Water and Maritime Engineering* **154**, 2, 117–28.

ISIS Flow User Manual. 1997. H. R. Wallingford & Halcrow.

Pope, N. J. 2002. Hydromechanical equipment for Cardiff Bay Barrage. *Proceedings of the Institution of Civil Engineers – Water and Maritime Engineering* **154**, 149–56.

Samuels and Burt. 2002. A new joint probability appraisal of flood risk. *Proceedings of the Institution of Civil Engineers – Water and Maritime Engineering* **154**, 2, 109–15.

38 • Operation of the 'Maeslant Barrier': (storm surge barrier in the Rotterdam New Waterway)

R. BOL

INTRODUCTION

Over the night of 1 February 1953 the south-western part of The Netherlands was threatened by a severe north-westerly storm. The seawater reached previously unknown high levels and the dikes broke; a disaster resulted in which a large area was flooded with salt water. 1835 people died. After this disaster a study was started on how to prevent such flooding in the future. This study resulted in the 'Delta Plan', a plan to reinforce the dikes and to close many of the existing estuaries with dams to shorten the length of coastal defence. Only two of the estuaries had to be left open: the Western Scheldt Estuary, which leads to the Port of Antwerp, and the New Waterway which is the entrance for the Port of Rotterdam. Along these estuaries the dikes had to be strengthened to reach the so-called 'Delta Level' i.e. the need to have a protection level against storm surges that statistically occur only once every 10 000 years.

The idea to build a storm surge barrier in the New Waterway to give the area behind it the demanded safety level had been suggested in the 1960s. However at that time the reinforcement of the dikes was a cheaper option. In 1985, based on advances in knowledge and improved calculations of flooding risk, it was concluded that an even further reinforcement of the dikes was needed. This led to problems, especially in the cities of Rotterdam and Dordrecht where the existing high water protection (the 'dike') is situated in the heart of the city. Reinforcement would have required an enormous construction project in each city, with many constraints for local businesses and transport. Therefore the possibility of a storm surge barrier was studied again.

THE MAESLANT BARRIER

This time the storm surge barrier proved to be an acceptable alternative. It would be cheaper and much quicker to be realized than previously. The only constraint was that the Port Authorities feared the negative effects of the barrier on the image of the Port of Rotterdam as an 'open port'. It was thus agreed that navigation must be maintained and that more specifically the closure of the port by the storm surge barrier should occur no more than once every 10 years.

In 1987 the design of the barrier started. Out of six designs, made by consortia of private building companies, a design consisting of two sector-gates was chosen. While at rest these gates are 'parked' in docks. During severe storms the gates, floating in the water, are transported to the centre of the waterway to close the navigation channel of 360 m width (Fig. 38.1). In 1991 the building started, and in the spring of 1997 the building of the Maeslant Barrier was completed. An enormous construction emerged; in size the two gates are comparable to the Eiffel Tower in Paris, but comprising four times more steel.

The Maeslant barrier is situated in the south-western part of the Netherlands near the Hook of Holland, 30 km west of the centre of Rotterdam (Fig. 38.2).

THE OPERATION OF THE BARRIER

The Maeslant Barrier has been operational since autumn 1987. A unique procedure exists to manage the closure of the barrier exists here as the decision is fully automated by a computer. For that reason the BOS has been built (BOS is the Dutch acronym for a DSS (Decision and Supporting System)). This BOS guides the closure of both the Maeslant Barrier in the

Flooding and Environmental Challenges for Venice and its Lagoon: State of Knowledge, ed. C. A. Fletcher and T. Spencer. Published by Cambridge University Press, © Cambridge University Press 2005.

Fig. 38.1   The Maeslant Barrier; closed (under normal conditions).

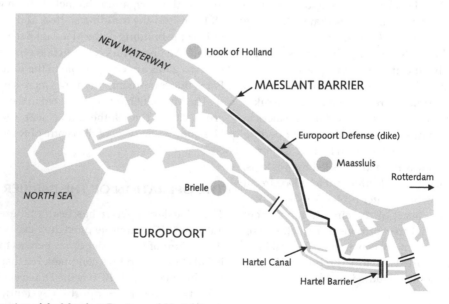

Fig. 38.2   Location of the Maeslant Barrier and Hartel Barrier.

New Waterway, as well as the Hartel Barrier in the Hartel canal. The two barriers together constitute the protection of the area behind it in case of extreme storm surges (Fig. 38.3).

Thus the 'BOS':

• decides if and when the barrier has to be closed; and
• how the closure, and the subsequent opening, is executed.

The only task 'man' has to do is to watch. To watch and see how the BOS undertakes its actions. Intervention by man is not possible as long as the BOS is in operation. The BOS is the Boss! Only when the BOS should fail, will it be down to human control to close and open the barrier 'by hand', and, to make the decisions. The chance of such a failure of the BOS is estimated at once in 100 000 times.

THE ACTIONS OF THE BOS

The BOS can be in 2 main states: 'at rest' or 'operational'. When the BOS is 'at rest' this doesn't mean that the BOS is not active. On the contrary, every 10 minutes the BOS checks if the pre-warning level or the closing level will be passed in the next 24 hours.

Therefore every 10 minutes a new prediction is made of the expected water levels at Rotterdam and Dordrecht for the next 24 hours. Such a prediction consists of:

• A prediction of the expected sea water level at Hook of Holland and at the Haringvliet sluices. This starts with the prediction of the weather for the next 24 hours by the Royal Dutch Meteorological Institute (KNMI). With this information calculations are made for the resulting hydraulics in the North Sea and resulting water levels at Hook of Holland and Haringvliet sluices. In these predictions two organizations are involved: the Hydro-Meteo Centre Rijnmond (HMR) and the Storm Surge Warning Authority (in Dutch SVSD). These two organizations estimate the expected development of the sea water level during storm surges.
• Subsequently the BOS makes a calculation with the 1-dimensional hydrodynamic computer model SOBEK for the Northern Delta Area. This is the estuary where the rivers Rhine and Meuse meet

the North Sea. For this calculation the following information is used:

– the predicted sea water levels at Hook of Holland and the Haringvliet sluices;
– the discharge of the rivers Lek and Waal (downstream parts of the Rhine) and the Meuse; and
– the openings of the Haringvliet sluices.

The result of this calculation is the predicted water levels at Rotterdam and Dordrecht.

THE BOS IS 'OPERATIONAL'

When the predicted water level at one of the stations Rotterdam or Dordrecht exceeds the following levels; Calling level (Dutch: O-level) / Pre-warning level (Dutch: V-level):

Rotterdam   2.60 m above sea level (Dutch: NAP)
Dordrecht   2.30 m above sea level (Dutch: NAP)

the BOS changes from the state 'at rest' to the state 'operational'.

This is a signal that a closure of the Maeslant Barrier can be expected. At that time the BOS initiates the following actions:

• the operational leaders, the people that observe the performance of the BOS and have to act if the BOS fails, are called to duty;
• the HMR/SVSD is requested to make a prediction of the sea water level at Hook of Holland and Haringvliet sluices every 3 hours instead of every 6 hours;
• the expected moment of the beginning of the closure is determined;
• a pre-warning is sent to the Harbour Coordination Centre (HCC) to inform it that a closure may take place (this pre-warning is sent 8 hours before the expected beginning of the closure); and
• 8 hours before the beginning of the closure the filling with water of the docks is started in order to make the gates float. This filling takes about 15 minutes.

The BOS continues predicting and checking the expected water level at Rotterdam and Dordrecht. The Maeslant Barrier is prepared for closure in the case when one of the following levels is exceeded:

Rotterdam   3.00 m above sea level (Dutch: NAP)
Dordrecht   2.90 m above sea level (Dutch: NAP)

The BOS then initiates the following actions:

- 4 hours before the beginning of the closure, the Harbour Coordination Centre receives the request to stop all shipping 2 hours before the start of the closure; and
- at the same time the gates of the docks are opened.

Two hours before the expected beginning of the closure the navigation signals are set to 'red'-alert, which means that all navigation is forbidden. It is then a matter of waiting for the closure.

Every ten minutes the BOS still undertakes the prediction of the water levels and the determination of the moment to start the closure. The predicted starting moment can therefore change during the preparation of the closure.

THE CLOSURE OF THE MAESLANT BARRIER

The determination of the starting moment for the closure of the barrier is different under varying hydrological circumstances. First of all the discharge of the rivers is important.

When the discharge on the Upper Rhine, measured at Lobith, is less than 6000 m^3/s (this is 98% of the time) a so-called 'Level-Closure' is executed; the barrier is closed at the moment that the water level at the barrier for the last time before the expected peak level exceeds the level of 2.00 m above sea level (Dutch: NAP).

In the case when the water level at the barrier is already higher than 2.00 m above sea level (NAP) and does not fall under this level, the barrier is closed at the moment that the water level is at its lowest level, just before the expected peak level. This is called a 'Lowest Level-Closure'.

When the discharge of the Upper Rhine, measured at Lobith, is more than 6000 m^3/s (2% of the time) then the closure is started at the moment that, at the barrier, for the last time before the expected top level the tide turns (changes from outflow to inflow). This is called a 'Turn of the Tide Closure'.

A closure of the Maeslant Barrier starts with the transport of the gates to the centre of the river. This takes about half an hour. After that, the valves in the gates are opened to let water flow into the hollow gates; this starts the sinking of the gates. This sinking is also guided by the BOS. The process takes about 2 hours. At about 1 m above the concrete sill the sinking of the gates is temporarily stopped to clean the sill. The strong current under the gates will remove the sediment from the sill. When the gates have 'landed' on the clean sill, extra water is let into the gates to bring the barrier 'under pressure'. In this way the barrier can resist water levels seaward of more than 5.00 m above sea level; a water level that statistically will occur once in 10 000 years.

When the sea water level drops again, after the storm surge, the water in the gates of the barrier is pumped out, so that the gates start floating again. This process starts at the moment a negative head difference is detected; that means the moment the water level at the river side of the barrier is higher than the water level at sea side. When the gates are fully lifted, they are moved back into the docks. These processes take 2.5 hours. All the actions are still controlled by the BOS.

When the gates are back in the docks, the dock gates are closed and the docks are pumped dry again. The shipping can start again and the operational people are sent back home. Then the BOS returns back into the state 'at rest', ready to detect another storm surge.

All together the closing procedure of the Maeslant Barrier depends on many information lines, moments of decisions and of control. These actions are all taken completely automatically by the computer system, in the presence of the BOS.

SOME REMARKS

The Maeslant Barrier has been in operation for 7 years now. In this period the Barrier has not needed to be closed because of a storm surge. However, it has been closed yearly, just before the beginning of the each storm season, to check it works correctly.

The closing procedure of the Maeslant Barrier is a complex matter. A lot of research has been executed to develop the procedure in the BOS. Now the BOS is operational, the people that executed the research have started working in other jobs, and the knowledge soon disappears. It is hard to maintain this knowledge and to keep it directly available, if needed, in the future. Therefore, a Knowledge Management System for the Maeslant Barrier was built. It remains hard, however, to keep the knowledge available for the future when there is no immediate need for its application.

The life cycle of a computer-system, on which the BOS is implemented, is about 10 years. Nowadays, both software and hardware systems change rapidly, which means that the computer system has to be overhauled every ten years.

The closing procedure has very high demands of reliability. The chance of failure needs to be less than once per thousand (storm) closures. Because of these high demands, an extensive testing procedure is applied to all software components, similar to those applied at nuclear plants. This testing procedure takes months of time. At each change of the software or procedure, this testing has to be executed again, to meet the required high standard of reliability.

During development of the closing procedure, while the barrier was already under construction,

new information upon the phenomena of seiches became known (seiches are long waves with a period of 15 to 90 minutes). It was recognized that strong seiches with amplitudes up to 2 m might occur. These seiches may result in a large negative head over the gate of about 1.5 m, while the construction was only designed to resist a small negative head. A solution was found by changing the sinking and lifting procedure, but this has made the operation much more complex.

The Maeslant barrier was built as a sophisticated, high-tech barrier. The people that now have to operate and maintain it are not used to such a high-tech construction. It takes a long time to get used to the demands and vulnerability of this high-tech construction compared to other barriers. An important lesson learnt, therefore, is to be aware of the needs and capabilities of the organization and people that have to execute the operation and its maintenance.

More information on the barrier is available at: www.keringhuis.nl

39 • The Oosterschelde, a changing ecosystem after completion of the delta works

H. L. F. SAEIJS AND A. J. M. GEURTS VAN KESSEL

SUMMARY

The Oosterschelde in southwest Netherlands, as part of the Delta Project, was to be 'dammed off' for safety reasons and transformed into a fresh water lake. Emotional protests from environmentalists and fishermen culminated in a reconsideration. After conscientiously weighing up interests, costs/benefits, and risks based on a policy analysis with alternatives, the Dutch Government decided to build one storm surge barrier (SSB) and two compartment dams and to reinforce the existing dikes, making the works six times more expensive. A delay of seven years was accepted. Safety or environment had become safety and environment. This chapter highlights experiences with the decision making process, the changes in the ecosystem of the Oosterschelde over 17 years and illustrates important lessons learned by trial and error, with examples.

INTRODUCTION

Large parts of the Netherlands actually consist of a delta plain formed by the rivers Rhine, Meuse and Scheldt. Today, the nation is being defended against floods and more than 60 per cent on average lies 4 m below mean sea level.

The delta works in the southwest of the Netherlands were completed in 1987. They brought about considerable changes in the tidal area. Today the Oosterschelde covers approximately 450 km². It was the last estuary in southwest Netherlands to be 'blocked off' with a dam. Blocked off however, is the wrong phrase. The construction of the SSB, the Oesterdam and the Philipsdam (Fig. 39.1) divided the estuary into three parts: the part on the seaward side with full tidal exchange, the middle part with reduced tidal exchange and the eastern part without tidal influence and where saltwater has been replaced by freshwater. These structures considerably increased human impact and drastically changed the life sustaining conditions of flora and fauna. It has completely changed human interaction with nature. In the past, forces of nature would dominate, man was not able to control them and therefore was not to blame when things went wrong. Nowadays man can control natural forces within certain bounds. As a result of the attempt to control natural forces, man is held responsible when things go wrong and has to act before it is too late. This requires an understanding of how the ecosystem works. However, we have to admit that our current knowledge of ecosystem structure and functioning is limited. This now raises some important questions. How do we guide changes into the right direction? How did the system adapt after completion of the delta works? What lessons can be learned?

HISTORICAL SUMMARY

A historic misunderstanding

Roughly two thirds of the Netherlands lies about 4 m below mean sea level. These lowlands are organized in polders. Polders are pieces of land, situated below sea level and surrounded by dikes. They are kept dry by pumps. The Dutch are very proud of their polders. Is that justifiable? Eventually polders turn out to be the demonstration of a thousand years of misunderstanding of the sneaky processes that lead to a sinking of the land. How could the Dutch have manoeuvred themselves into such a precarious position?

During the Holocene global temperatures have been rising. As a result, large parts of the world's ice-

Flooding and Environmental Challenges for Venice and its Lagoon: State of Knowledge, ed. C. A. Fletcher and T. Spencer.
Published by Cambridge University Press, © Cambridge University Press 2005.

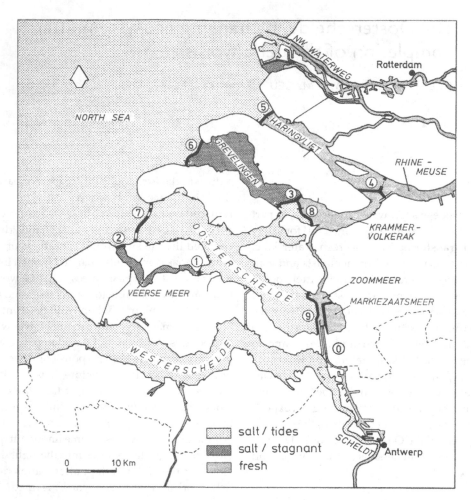

Fig. 39.1   Map of the Delta area of the rivers Rhine, Meuse and Scheldt in the SW Netherlands, with various waterbodies as resulting from the Delta Project engineering scheme. 0 = Kreekrakdam, 1867; 1 = Zandkreekdam, 1960; 2 = Veersegatdam, 1961; 3 = Grevelingendam, 1964; 4 = Volkerakdam, 1969; 5 = Haringvlietdam, 1970; 6 = Brouwersdam, 1971; 7 = Oosterschelde stormsurge barrier, 1986; 8 = Philipsdam, 1987; 9 = Oesterdam, 1986; Markietzaatsmeer was closed off from Zoommeer by Markiezaatsdam in 1983 (Nienhuijs and Smaal, 1994).

sheets and polar icecaps have melted. The impact has been an impressive sea-level rise of more than 35 m in the last 10 000 years (van de Ven, 1993; Mulder *et al.*, 2003). Mean sea-level rise during the Holocene has been about 45 cm per century. This has decreased to 5 cm in the last few centuries. The North Sea has filled up. About 6000 years BC the area where the Netherlands is situated today was threatened with flooding. However, by 1000 years

AD, when man had settled and began to have an impact on the area, there was land. The reason for this was the fact that the Netherlands has been rising along with the sea-level rise by the building up of peat layers (up to 60 metres of peat are found in this area) and through sedimentation.

When man entered the delta area he implicitly chose two strategies to make the land acceptable for settlement. The marshlands were drained, by digging small ditches and using pumps (the famous Dutch windmills), and dikes were constructed to prevent flooding. The impact of these measures turned out to be quite dramatic in the long term; drainage led to irreversible settling of the soil and thus a lowering of the land, whilst the construction of dikes prevented the natural compensating process of sedimentation (Pons, 1992). We cannot blame our ancestors because they did not understand what they were doing. But surprisingly, the present day government seems to persist in this behaviour, by installing additional pumps as a reaction to global climatic change and consequent sea-level rise, and continues to facilitate the urbanization of polders with peat soils (van de Ven, 1993). In the case of the Oosterschelde, the ultimate result of a thousand year struggle between man and nature is that the estuary gradually turned into an erosion basin (1950: net sediment export 5 million m^3 yr^{-1}), surrounded by neglected dikes which defended vulnerable low-lying polders against the sea (van Veen, 1953).

Then the 1953 flooding occurred. One month after the flood the Delta Commission was set up to advise the Government. In 1958, Parliament passed the Delta Law and, in doing so, accepted the recommendations of the Delta Commission (1962). The principal argument of the recommendations was to shorten the coastline (from approximately 700 to 100 km), provide protection against storm surges and prevent further salinization (a major concern for the agricultural sector). The minor aims of the recommendations were to open up the area for socio-economic development by creating facilities for horticulture, establishment of industries, increasing recreational facilities, gaining some land and creating freshwater reservoirs for drinking water supplies. The main disadvantage mentioned was the disappearance of the shellfish industry. At that time the loss of a unique saline tidal system was not mentioned at all.

LESSONS FROM THE LARGE SCALE PROJECTS IN THE TWENTIETH CENTURY

The Oosterschelde was the last and most difficult estuary to be closed off. For this closure, experience with former coastal engineering projects, the Zuiderzee project in the north (van de Ven, 1993) and the other estuaries of the Delta project in the south, was used. What lessons could be learned from the large scale transformations undertaken to this point? (Saeijs, 1982, 1994).

A step by step approach has its advantages

By dividing an area into separate compartments, a phased realization and differentiation per system was possible. This approach is called the 'compartmentalization strategy'. When the differences between the original environment (tidal estuaries, dunes, salt and freshwater marshes) are compared with the current situation with chosen and developed environments (polders and fresh-, brackish- and saltwater lakes), it becomes clear that the process of creating a compartmentalized region has the potential to set new ecological conditions, which in turn can provide new opportunities. The chain of results as a consequence of such major human adaptations of the environment is illustrated in Fig. 39.2.

An important lesson is that compartmentalization allows for a phased approach in space and time and thereby increases the flexibility in land use and decision making. However, a major drawback is the irreversible loss of the original environment, and the fact that it is hard to predict and control the development of these newly created water systems (also see conclusions below).

Water reclamation instead of land reclamation

The 'Markerwaard', would actually form the fifth and last land reclamation in the Zuiderzee Project. After reassessing whether to proceed with land reclamation, the decision was made to preserve the water system of the Markermeer, rather than gaining new land. This represented a significant change in traditional thinking. Today the creation of terrestrial systems is no longer considered to be the only important issue. Freshwater systems are also being seen as valuable alternatives. This trend is actively being pursued in the delta area, but including new elements such as saline and brackish waters and controlled tidal systems. Land reclamation developed into land and water reclamation. It has become evident that water reclamation is as important as land reclamation.

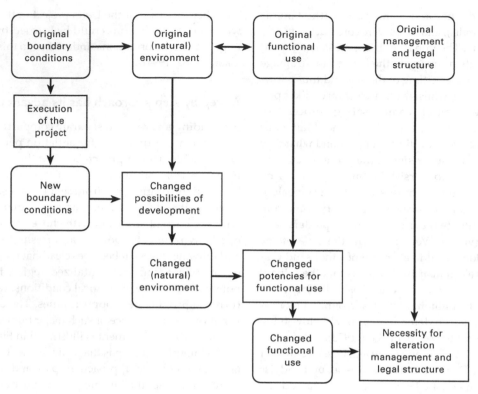

Fig. 39.2   Chain of results as a consequence of a major human action in the environment (Saeijs and Bannink, 1978).

Purifying effect of a succession

The opportunities offered by the aquatic infrastructure are allowing a new ecological perspective to be developed. The following example illustrates this point. The River IJssel, which mainly contains water from the River Rhine, flows into the freshwater Ketelmeer. The bulk of the sediment it carries, being highly contaminated with pollutants such as heavy metals and organic micro pollutants, settles in the Ketelmeer. The accumulation of toxic sludge in the IJsseldelta has highly polluted the floor of the Ketelmeer. However, as a result, the quality of the water that flows on into the much larger Ysselmeer has improved considerably. A substantial amount of organic micro pollutants and heavy metals are retained and settle on the floor of

the Ketelmeer.[1] Further improvements in quality are also occurring in the IJsselmeer due to physical, chemical and biological processes. Since the Marketmeer receives its water from the IJsselmeer, in addition to supplies of rainwater, this lake derives substantial benefits from its relative position within the succession of lakes. The Gouwzee is situated even more favourably in this respect. It is not surprising that this lake was the first in this series of freshwater lakes where abundant submerged vegetation, previously lost due to eutroph-

---

[1] The long-term effect on the polluted lakebed has been eliminated by the construction of a depot in which the contaminated sludge is stored temporarily until purification is possible.

ication, recovered. In the Netherlands the order in which relatively contaminated riverine water passes through this series of interconnected lakes is managed in such a way that the profit gained from this purifying effect of a succession is optimized. The long term effects and possible future implications are being studied. We have identified two interesting management strategies, the 'strategy of concentrating problems in specified areas' and the 'strategy of interconnected surface waters'.

A saltwater lake, a bold experiment

The Grevelingenmeer was formed when the Grevelingendam (1968) and the Brouwersdam (1971) had been completed (Fig. 39.1). The Grevelingen estuary has been transformed into a saltwater lake. In the first few years after transformation, the water in the lake was very clear (secchi classification 10 m),

even in the summer months. Nowadays, the visibility is still relatively acceptable (2m). In spite of high concentrations of nitrogen and phosphates, there are no eutrophication problems. In response to the environmental changes the bird population has undergone major changes (Fig. 39.3).

The properties and processes of a saltwater lake ecosystem were unknown but by using the approach of 'adaptive environmental management techniques' (Holling, 1980), the final result is quite acceptable. The main management actions concerned were control of the water level, residence time, salinity, stratification and added nutrients (Saeijs and Stortelder, 1982). In order to facilitate the exchange of living organisms, the link with the sea was restored by constructing and opening an inlet / outlet sluice in the Brouwersdam. The biological developments taking place were studied and monitored carefully, using techniques such as ecological modelling (RAND, 1977d, 1977e). In

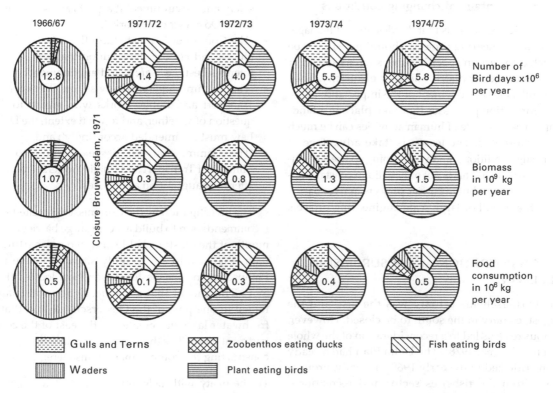

Fig. 39.3   The response of bird in Lake Grevelingen on the closure of the Brouwersdam in 1971. Distribution of presence (in bird-days and biomass) and food consumption (Saeijs and Baptist, 1976).

doing so, it was possible to have a kind of 'dialogue' with the system (a cycle of monitoring – modelling – evaluation). This water management strategy can be described as the strategy of allowing spontaneous development within certain environmental boundary conditions as mentioned above. Oysters, mussels and denitrifying bacteria play a vital part in this success. The lake is appreciated now as an important wetland from the point of view of nature conservation (EU Bird Directive, Ramsar Convention), recreation and fisheries. The main lessons are that saltwater lakes offer wide potential, in terms of both economic and ecological valuation. However, these water systems are vulnerable, mainly due to variations in salinity and the occurrence of stratification and oxygen depletion (Hoeksema, 2002). Therefore, it is essential that when changes occur, choices are made in time and priority is given to the management of these transformation processes.

Taking advantage of changing conditions

The conclusion so far is that the design and management of engineering projects should not simply be directed to certain specific aims, such as safety, or water storage, or distribution, or electricity. They need full acknowledgement of the importance of the transformation processes that take place in a landscape. Then the role of human activities can be much more profitable. The lesson is 'take advantage of changing conditions' and when doing so use a multi-sector approach. This is not only valid in coastal engineering projects, but in barrier projects in river basins and other small or large scale engineering projects as well.

RECONSIDERATION AND A SURPRISING DECISION

The Oosterschelde estuary was the last, and the largest, estuary in the south to be closed. However, various reasons led to a reconsideration of the whole scheme (Parma, 1978). The Delta Plan had already been criticized in the early 1960s. Initially, protests arose from the fisheries sector and recreational sailors but after the confrontation with the impact of the closure of the Hollands Diep, and later on the

Grevelingenmeer, people became increasingly aware of the enormous scale of the environmental effects of large-scale hydraulic engineering works. Fierce protests soon led to a social conflict. The original Delta Plan envisaged total closure of the Oosterschelde in 1978, followed by desalination and the creation of a fresh-water lake. Much of the water contained in this lake would be supplied by the rivers Rhine and Meuse. The deterioration of the water quality of the Rhine in the last few decades also made a significant contribution to the need for reappraisal. The discussion in the 1960s was concentrated on two alternatives: either total closure with desalination or leaving the Oosterschelde open with reinforcement of dikes (as in the Westerschelde). Many arguments were put pro and contra. In 1973, the `Klaasesz Commission' was set up and given the following brief (Commissie Oosterschelde, 1974):

- 'The Commission will inform the Minister of Transport and Public Works with regard to all safety and environmental aspects at issue related to the Oosterschelde works';
- 'The Commission must study the existing possibilities and choose the solution that serves both interests best, i.e., safety and environment';
- 'The Commission will advise the Minister of Transport and Public Works with regard to the question of whether, and to what extent, the Delta Law must be amended accordingly'; and
- 'The Commission will issue a report to the Minister of Transport and Public Works six months after its inauguration at the latest'.

The investigation by the Commission resulted in a recommendation to build a storm surge barrier at the mouth of the Oosterschelde and two compartment dams that would be, roughly situated where the Philipsdam and the Oesterdam are now located (Fig. 39.1). This meant that a (reduced) tide was maintained in one part of the Oosterschelde and that a freshwater lake was created to the east of the compartment dams. The three most important reasons for constructing the compartment dams were:

1) The treaty with Belgium, in which it was agreed that the shipping route from Rotterdam to Antwerp would be tide-free;

2) The effect of the salinity on the Hollands Diep (Fig. 39.1) must not increase, in the interests of drinking water supplies and agriculture. The Volkerak sluices were not constructed for an effective salt/fresh water separation; and

3) In order to maintain an adequate tidal range in the (remaining) Oosterschelde (minimum 2.70 m at Yerseke) and to remain within a certain budget, the tidal area must not be too large.

In 1974, the Government expressed its intention to close off the Oosterschelde with a storm surge barrier (SSB). The works were halted. After a brief study period, the following was put to Parliament:

1) the Oosterschelde would be closed off by 1985 at the latest by a sluice dam (storm surge barrier);
2) dikes at risk would be raised to an acceptable level (frequency exceeding 2.10^{-3} times yr^{-1}); and
3) the Ministry of Transport and Public Works was given 18 months in order to study the technical aspects of this solution.

Parliament accepted the proposals subject to three preconditions:

1) The technical feasibility of the scheme;
2) The solution must not cost more than DFl 3.8 billion (at 1974 prices); and
3) The barrier must be ready no later than 1985.

A policy analysis for the different Oosterschelde alternatives was prepared and reported in 1976 (RAND, 1977b, 1977c). This study also looked into the environmental aspects of the scheme. During the study, it was assumed that the preconditions made by Parliament could be met. In 1976, the recommended technical concept was accepted by Parliament. Based on an analysis entitled 'wet cross-section variants in the Oosterschelde SSB', it was decided to adopt a minimal wet cross-section in the SSB of 14 000 m^2 (IDWDO, 1977). However, in order to guarantee a tidal range of 2.70 m at Yerseke and to keep the tidal volume as large as possible, a wet cross-section in the storm surge barrier of 19 000 m^2 was finally realized.

The location of the Philipsdam, the Oesterdam, and the Markiezaatdam and the sluice to the Westerschelde were also chosen based on policy

analysis. The SSB is a flexible instrument to obtain safety now and in the future, a real breakthrough in flexible design. For example, the SSB can withstand sea-level rise, although this will result in an increased number of closures.

ADMINISTRATIVE APPROACH TO 'AFTERCARE'

Physical planning and management

Before we describe the adaptation process of the Oosterschelde, we briefly highlight the administrative approach (for more details see Saeijs and de Jong, 1981). Authorities like the Government, Provinces, Municipalities and Waterboards all have their own responsibilities with regard to the institution and management of the Oosterschelde. Moreover there are many other stakeholders, each with their own initiatives and interests. The challenge is how to attune the decisions of all these stakeholders with their respective responsibilities and interests. Particular attention was paid to the areas which were affected by the construction and to the functioning of the engineering structures. To get all the authorities involved, a consultation/working structure was founded for the Oosterschelde (Fig. 39.4). Within this consultation/working structure, decisions were based on consensus. The most important task of the Oosterschelde consultation structure can be regarded as the preparation of a policy for organizing and managing the Oosterschelde basin to the west of the Philipsdam and the Oesterdam. This policy consisted of the following components:

1) A description of the general policy to be followed, with regard to the relevant functions;
2) An organization plan, investigating the development opportunities of projects;
3) A management policy coordinated with the organization policy;
4) An investigation plan in order to be able to fill in gaps in knowledge relevant for policy (after-care or feed-back studies; effects on (parts of) the policy plan; and
5) A feed-back procedure (evaluation) and adjustment of the policy plan.

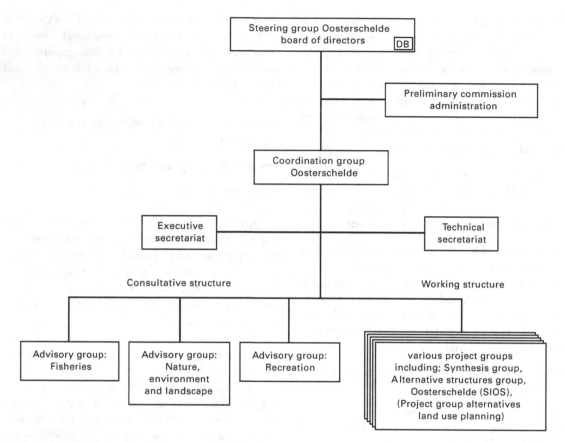

Fig. 39.4   Consultation and working structure for drawing up a plan of action for the Oosterschelde. Central, provincial and local government and water authorities are involved in this structure. Source: Rijkswaterstaat.

One essential difference in the administrative approach of the Oosterschelde compared to other areas such as the Lake Grevelingen (Fig. 39.1), is that in that case the government solely, without participation of other stakeholders, made plans for management and land use (Segeren *et al.*, 1975). Of course other authorities and stakeholders put forward a lot of objections and did not accept the governmental plans. Other differences in approach were:

1) The Oosterschelde policy plan was not just a national plan but was a combined plan involving local, provincial and central Government;
2) A clear administrative working relationship and structure made it possible to make decisions and

facilitated integration with regional and budget allocation plans;
3) The 'process approach' of the plan, rather than aiming for a final plan, made it possible to respond quickly to future changes;
4) The integral nature of the plan meant that the physical planning and managerial aspects were viewed in terms of their mutual relationship; and
5) The Policy plan produced was translated into a Plan of Action.

The choice of management of the SSB was prepared within the scope of another project called BARCON or Barrier Control (RAND, 1979a; Saeijs, 1982). However, it was clear that the decision with regard to the management of the Oosterschelde barrier should not conflict with the formulated policy in the 'Plan of Action' for the Oosterschelde. And, as we have learned, there has not been a conflict of interests.

Oosterschelde Plan of Action

Many interests are involved. The main interests are: fisheries, nature conservation, aquatic sports and inland navigation. Everyone has his own ideas about the use of the new facilities. Moreover, there are many uncertainties about future developments. In order to prevent a situation where everyone puts his own particular ideas into practice in an uncoordinated way, it is essential to determine a clearly defined plan of action and to make the necessary choices from the outset. A strategy must be agreed upon. In the case of the Oosterschelde, the strategy has been established in a plan of action in which the aspects of physical planning and management of the area have been co-ordinated. The method of approach which has been used to formulate the plan of action for the Oosterschelde has already been discussed at the beginning of this paper. The report entitled 'The Oosterschelde – a summary of possible courses of action' (Rijkswaterstaat, 1976; SGO, 1980a) forms an important step in the development towards the final plan of action. The principal aim is of great significance for the contents of the physical planning and management policy for the Oosterschelde: 'The conservation and wherever possible, the improvement of the prevailing values, taking into consideration the basic principle for a proper

social functioning of the area, in particular with regard to the fisheries'. (SGO, 1980c). This principal aim is based on the government decision on how the Oosterschelde works have been carried out and the considerations on which they were based. Implementation of the main aim must not divert from the primary aim of the Delta works which is the safety of the population (for more details see Saeijs, 1982: ch. 6).

ADAPTATIONS AND MODIFICATIONS AFTER 17 YEARS

In the case of the Oosterschelde it is possible to evaluate the effects that the construction of the Delta works has had on the environment over the 17 years after completion. Generally, the effects can be subdivided into primary, secondary and tertiary type effects, in which the primary effects consist of hydrodynamical and geomorphological changes, secondary effects embody the ecological responses to a changing environment and tertiary effects consist of the human reactions to a changing ecosystem.

Primary effects: changes in hydrodynamics and morphology

As a result of a decreased tidal volume, the tidal channels in the Oosterschelde are transporting less water and therefore tend to adapt themselves towards a smaller cross-section. In order to do so, the tidal channels require sediment and have a so called 'hunger for

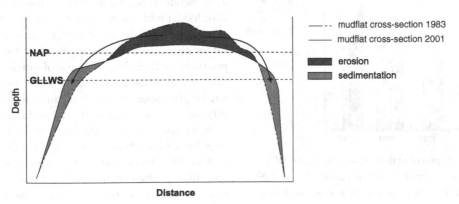

Fig. 39.5   Schematic drawing of the erosion process in the Oosterschelde after the completion of the Delta works. Note patterns of erosion and sedimentation (source: Geurts van Kessel *et al.*, 2003).

sand'. Since the SSB itself forms a barrier for the import of sand from the foredelta and sedimentary input from the riverine areas has been cut off by the compartmentalization dams, the only material available for this process is located in the Oosterschelde itself: the intertidal mudflats. Figure 39.5 shows a schematic drawing of the erosion process of intertidal mudflats in the Oosterschelde. Sediment is eroding from the higher areas of the mudflats and is being deposited near the Mean Low Water Level. As a result, the geomorphologic profile of the intertidal mudflats is gradually changing into what can be described as a flat, 'pancake-like' shape.

Before the construction of the works it was predicted that between 400 and 600 million m^3 of sediment would be required in order to reach a new morphological equilibrium in the Oosterschelde. This suggests that eventually nearly all intertidal areas will turn into shallow areas, over a period of several hundred years. Although it will take such a long period before the tidal flats will disappear entirely, the time of exposure (the period that the tidal flats are exposed to the air) decreases much faster, in the order of decades. In the period after completion the exposure time class of 40–60 per cent has decreased while the exposure time class of 0–20 per cent has increased (Fig. 39.6).

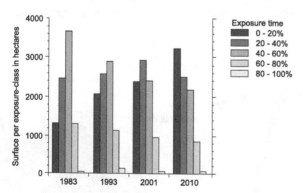

Fig. 39.6   Development of the exposure time of tidal mudflats in the Oosterschelde since the completion of the Delta works. The exposure time is decreasing much faster (decades) than the total loss of surface of tidal flats (source: Geurts van Kessel et al., 2003).

Secondary effects: ecological responses to a changing environment

A broad range of ecological adaptation processes were triggered by the construction of the Delta works. We can give an example of a well-predicted, and an unexpected, development.

Erosion of salt marshes

Some of these responses turn out to have been predicted in advance quite well, like the decline of salt marshes. The main reason for the decline of these areas which harbour specific salt resistant vegetation is the decreased inundation frequency, as a result of the smaller tidal amplitude. Drying out of these marshes makes them vulnerable to erosion. Furthermore there is no silt import from the rivers anymore to stimulate natural expansion by sedimentation. At the current rate the loss is c. 4 hectares of salt marsh per year due to erosion.

Less time for birds to forage

The decrease of exposure time in the Oosterschelde is a process that was not explicitly put forward before the construction of the SSB. It turns out, however, to be the determining factor for its carrying capacity of waders. A decreasing exposure time translates directly into less time for birds to forage on these mudflats. Especially in winter, when birds require more food daily in order to maintain their body temperature, and, therefore, need almost twice the time to gather their food, problems occur first (Figs. 39.7, 39.8).

This process does not affect a single bird, but rather all waders depending on the Oosterschelde mudflats for their food. The possibilities for adaptation by individual birds and the availability of alternative prey species will determine to what extent different bird species are affected and in what order.

In the whole of the Netherlands the surface of estuaries has steadily been decreasing from 866 100 ha in 1900 to 393 000 ha in 2000 by transformation into lakes and through the construction of dikes and polders (Saeijs, 1999). In the southwest, the Oosterschelde and Westerschelde are still the most important contributors to intertidal surface area. In both cases, however, these

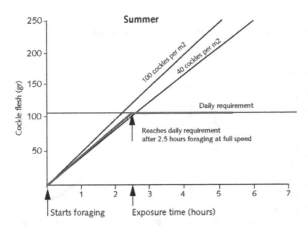

Fig. 39.7   Schematic presentation of Oystercatcher foraging in summer. Source: Alterra.

Tertiary effects: the reactions of man to a changing ecosystem

Human efforts to adapt to a changing ecosystem are vast as alternatives and opportunities for existing use are explored and, if possible, developed. The Oosterschelde has a history of providing fishing grounds as well as the cultivation of shellfish. Some examples of the response of this sector illustrates how tertiary effects can determine for the final outcome.

Adaptation of mechanical suction-dredge fishing for cockles to changes in cockle habitat induced by the Deltaworks

As a result of the Deltaworks the best locations for cockles (*Cerastoderma edule*) have shifted from the Northern and Eastern part to the Central and Western part of the Oosterschelde (Plates 13, 14). The changes in cockle habitat have been induced directly by the decrease of current velocities and reduction of tidal amplitude. Long term effects on cockle habitats are most likely induced by the decreasing exposure time of tidal flats (Kater *et al.*, 2003).

In the Oosterschelde mechanical suction dredge fishing of cockles developed in the 1970s. Since cockles are the main food source for oystercatchers (*Haematopus ostralegus*) competition between these birds and fishermen became apparent. The policy on mechanical cockle fishing was only formulated in 1993 in response to the declining oystercatcher numbers in the Oosterschelde. A reservation policy was implemented, which means that a certain minimal number of cockles must be present in the system before cockle fishing is allowed. In addition, certain areas were permanently closed for cockle fishing, but, as it turns out, these areas were less suitable for cockles than the areas that were left open to cockle fishing. An important question is: 'What is the impact of this kind of fishery, besides the removal of cockles from the system?' One of the effects is illustrated in Fig. 39.9.

The left hand curve (1) shows the distribution of cockle biomass before fishing. The central box (2) shows the distribution of the intensity of the fisheries. The right hand curve (3) shows the distribution

areas are declining: in the Oosterschelde intertidal area is being lost as a result of the on-going erosion process, whilst in the Westerschelde dredging activities may be having the same result, albeit to a lesser extent. In the long term, the consequences for the birds from three different major north-south fly-routes that depend on the Dutch Wadden Sea and Delta area for their stopovers, will be dramatic. Because of these developments, we doubt whether in future we will be able to maintain present-day bird numbers.

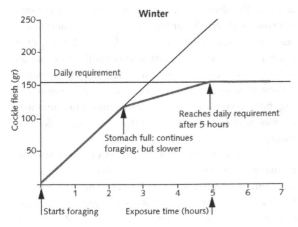

Fig. 39.8   Schematic presentation of Oystercatcher foraging in winter. Source: Alterra.

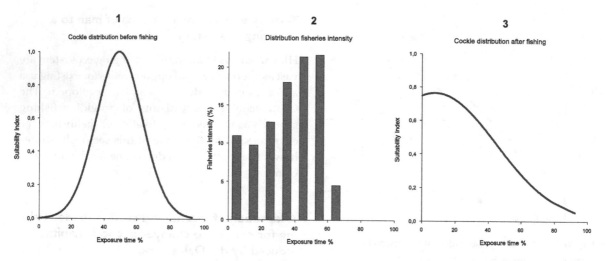

Fig. 39.9   Biomass distribution of cockles at different exposure time classes before and after cockle fishing (Geurts van Kessel *et al.*, 2003).

of cockles in the following year, after fishing. Of the remaining cockles, most are found low in the intertidal area, and are therefore submerged for most of the time (Geurts van Kessel *et al.*, 2003). As a result, birds are hardly able to reach the remaining cockles. The extent to which the oystercatchers can actually reach the cockles present in the mudflats (physiological versus ecological food requirement; Rappoldt *et al.*, 2003) is an extremely important factor, which will have to be taken into account during the next policy period by the Ministry of Agriculture, Nature Conservation and Fisheries.

UNCONTROLLED EXPANSION OF THE IMPORTED PACIFIC OYSTER (*CRASSOSTREA GIGAS*)

An example of an unforeseen phenomenon is the unlimited spreading of the Pacific oyster (*Crassostrea gigas*). This species was imported in 1964 as an alternative to the native flat oyster (*Ostrea edulis*). As at that time plans were focussed on turning the Oosterschelde into a fresh water lake by constructing a dam at the mouth, the introduction was considered a 'closed experiment' and risks of spreading were considered non-existent. As we have seen, the plans turned out differently: the Oosterschelde

remained saline. Possibly due to increased retention time of the water resulting in a quicker warming up of the water in spring, oyster spat fall was triggered. Today more than 640 ha of the intertidal area is covered with Pacific oysters (Plate 15, Fig. 39.10; Kater *et al.*, 2003).

In addition, an estimated 700 ha of the subtidal area is covered. All together this makes a covered area of approximately 15 km². Besides the threat these sharp oysters form for recreational activities (surfing, swimming), they act as a competitor for space with other macrobenthic species and can compete for food with filter feeders like cockles and mussels and predate on their larvae. Due to the fact that the explosive development of this exotic species has not appeared to slow down to date, and that no natural enemies seem to be able to control further spreading, both coastal managers and mussel farmers are increasingly concerned about the implications for the future.

With regard to the examples presented above, primary, secondary and tertiary effects can be summarized as follows (Fig. 39.11). The erosion of higher parts of the tidal flats is forcing the cockles downward, the oysters are expanding most in lower areas where they compete with cockles for space, and both fisheries and birds are becoming depend-

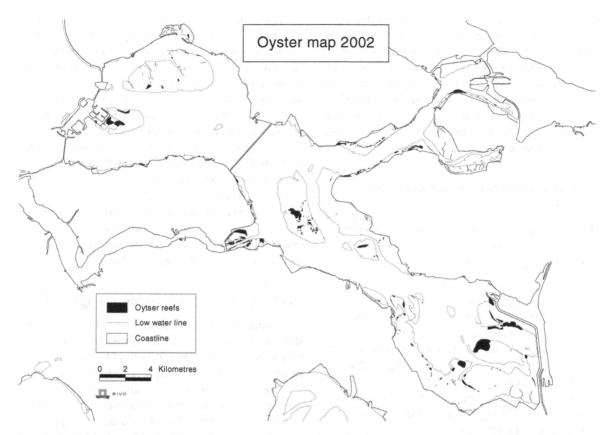

Fig. 39.10   Tidal area covered with Pacific oyster reefs in the Oosterschelde in 2002: more than 640 hectares (Kater *et al.*, 2003).

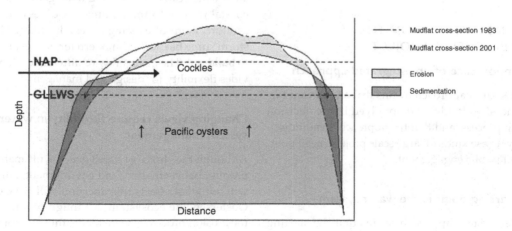

Fig. 39.11   Example of the combination of primary, secondary and tertiary effects in the Oosterschelde (Geurts van Kessel *et al.*, 2003).

ent on a decreasing amount of intertidal area. Clearly it takes a long period of time before all three types of effects following a large scale intervention become visible. Due to the fact that fisheries activities turn out to have a determining impact on the adapting Oosterschelde ecosystem as a whole, the need for the incorporation of fisheries management into the decision making processes following a large scale infrastructural project is underlined.

Conclusion: adaptations and modifications after 17 years

The net result of primary, secondary and tertiary responses to large construction works like the Delta works are very hard to predict:

- Because it takes a very long time before the final combined effects become visible, politicians should take the timescale of these adaptive processes into account;
- Hydro-morphological changes turn out to be an extremely important determining factor for future ecological potentials;
- Fisheries should be part of the decision making process as fishing pressure should be controlled in the period after construction; and
- The import of exotic species into a water system that is responding to a major intervention should be avoided as the ecosystems within the system tend to be more vulnerable in this situation.

LESSONS LEARNT FROM THE OOSTERSCHELDE PROJECT

The importance of an integrated approach

An integrated approach means that all relevant factors and stakeholders are involved in the decision making process, and that the approach is multidisciplinary. These kinds of large scale projects need political and public engagement.

The starting point is the water system

A water system approach means that the starting point is the water system (ecosystem), not the wishes of society. The wishes of society have to be attuned

to the possibilities and potentials provided by the water system that can be developed. When considering measures for a water system it is important to agree on the boundaries of the water system. The water system includes the seaward side of the dams, the water body, its floors and shores, its life communities and all the physical, chemical and biological processes in it. In the case of flood protection, special attention should be devoted to the seaward side (outside the planned barrier) and the water of the rivers that discharge into the system, as well as subsystems within the lagoonal area. The integrated water system approach has become the fundamental basis of the revolutionary developments in water management in the Netherlands, and in other, more distant locations, in the past few decades.

Flexibility of design is vital for management

Following the first lesson, an obstinate misunderstanding is that these kinds of projects are a matter for engineers alone. Engineers tend to focus on construction, budget and safety issues. From their point of view, a dam is the cheapest and simplest solution to cover these issues. But from a management point of view the construction should be an important management tool as well, necessary to manage modified and transformed coastal waters. Let us consider the Delta project at this point in time. There are, for example, many available solutions to this problem, all of which can be considered management tools for coastal waters. Examples are: new dikes and dams; reinforcement of existing dikes; discharge sluices; storm surge barrier and modern terps (elevated land to build on). The availability of different options provides flexibility for design and management.

Changing views require flexibility in water system management

Although the choice of a SSB was the ultimate compromise between safety and environment, only time will tell what effects this decision will have on the Oosterschelde ecosystem. Although predictions of the erosion processes were made, this did not play a significant part in the decision making process at the time. Today the view of society on this matter has changed as environmental awareness has increased considerably. Also, we have been able to evaluate the effects of the Delta works on the Oosterschelde some 17 years after completion. Because of the rigid solution that was chosen, modifications that have to be made in order to relieve some of the negative (side) effects of the Delta works have been accompanied by high costs.

Aftercare is an inevitable part of the project

The project is not finished with the completion of the civil technical works alone. This cannot be stressed enough, since there tends to be a mismatch between timescale of political decision-making (decades) and adaptive processes (centuries). Besides a proper determination of the starting point, on-going monitoring and evaluation is needed in order to adapt management to changing conditions within the water system in order to ensure that the areas continue to function as well as possible. Involvement of all relevant authorities (Ministries, Provinces, municipalities, etc.) in consultations, creation of a joint vision and the willingness to use their own authority for the project are all needed in the period after completion.

In the case of the Oosterschelde, a consultation and working group was specially devised for this task: the Steering Committee Oosterschelde. Unfortunately fisheries policy did not take part in the creation of a joint vision as this frequently led to conflicts. Eventually, aftercare should become embedded in water management.

Estuaries are irreplaceable

Although interesting new environments (like the Grevelingenmeer) can be created by compartmentalization, the net result is simply not comparable with the original estuarine environment. Estuaries are generally too valuable to be disrupted or destroyed as they fulfil functions of international importance. Therefore an international justification should be agreed upon when estuaries are to be modified. In the case of the Oosterschelde this never happened. Loss of estuarine environments cannot be compensated for by creating fresh or saltwater lakes.

However, by creating interesting environmental boundary conditions, new environments can be developed.

The collection of water bodies created in the Delta area is, without doubt, unique. But besides the Oosterschelde other water systems like the Veerse Meer (Holland, 2004), the Grevelingenmeer (Hoeksema, 2002) and the Volkerakmeer (Tosserams et al., 2000) are showing signs of instability (oxygen depletion, water quality, the presence of cyanobacteria). There is a growing feeling that something has to be done about these problems. Redesigns are necessary to keep the balance between various functions. Projects have started to inform the public about the necessity to take action in the coming years and experiments are being done to see if the restoration of estuarine gradients can provide the preferred solutions. Examples are the reconnection of the Veerse Meer to the Oosterschelde (established in June 2004), in order to improve water quality of the eutrophic Veerse Meer, and experiments to restore the estuarine gradient in the Haringvliet, by managing the Haringvliet Sluices as a Storm Surge Barrier. Bringing the erosion process of the tidal mudflats of the Oosterschelde to a halt however, would require much greater measures, such as an enlarged tidal prism and restoration of sediment import from the foredelta.

CONCLUSIONS

A translation into modern water management

The new knowledge was translated into the new water policy and water management of the Netherlands but how? Until the 1960s, water was only seen as a resource for human use and abuse.[2] In the 1970s, policy was concentrated on the protection of the water against pollution. In the 1980s, the general public became aware of the fact that surface waters are ecosystems. That is to say, where organ-

---

[2] This policy led to an enormous water pollution which forms the heritage of one generation's profit, to the next. Restoration will be accompanied by great costs and will take many years.

isms as well as physical, chemical and biological processes play an important part besides the development of infrastructure. Water is no longer considered as merely a resource for human use and abuse. The importance of a properly functioning aquatic ecosystem is acknowledged today. Developing a stronger ecological basis for water management, has been considered as being essential. This process started in the second Policy Document on Water Management (Ministerie van Verkeer en Waterstaat, 1984). In this document the key points are:

1) The water system as a whole is of primary importance. (A water system includes the water; the lake and river beds; the banks including the living community, the dams and dikes and sluices, the living organisms and all physical, chemical and biological processes);
2) All aspects are important for a balanced decision-making process. These include: safety; agriculture; residential property; industry; electricity supply; the services sector; shipping; fisheries; recreation; landscape and nature;
3) Cost/benefit considerations are necessary;
4) The wishes of society and the possibilities of the systems have to be brought into line;
5) Quantity and quality are considered as interrelated subjects; and
6) Groundwater and surface water are interrelated too, in lakes, estuaries and the sea.

The approach outlined above takes time and has had far-reaching consequences in terms of policy. It was implemented in the third national policy document on water management of 1990 (Ministerie van Verkeer en Waterstaat, 1989). To sum up the aspects of special attention of this document:

1) An integrated, national water policy, based on a watersystem approach is opted for;
2) Action-plans to clean up river basins are defined;
3) Groundwater receives special attention;
4) For each water system a harmonized policy is formulated;
5) Differentiation of emission for each water system is accepted. Special policies are formulated for polluted beds, non-point source pollution, banks, etc.;

6) Effective utilization of the infrastructure for planned ecosystem management is in full swing;
7) Innovation of administrative and legal machinery; and
8) A financing system is also included.

Currently the fourth national policy document on water management is being implemented in the Netherlands (Ministerie Verkeer en Waterstaat, 1998).

REFERENCES

Aalst, W., van Huis in 't Veld, J. C., Stuip, J., Walther, A. W., and J. M. van Westen . 1987. *The Closure of Tidal Basins*. Delft: Delft University Press, pp. 1–743.

Commissie Oosterschelde. 1974. *Rapport Commissie Oosterschelde*. The Hague: Staatsdrukkerij. (Dutch, English and Fremch.)

Delta Commission. 1962. *Final Report of the Delta Commission*. The Hague: Staatsdrukkerij. (Dutch, English and French.)

Geurts van Kessel, A. J. M., Kater, B. J., and T. C. Prins. 2003. *Veranderende draagkracht van de Oosterschelde voor kokkels. Rapportage van Thema's 2 en 3 uit het 'Lange Termijn Onderzoeksprogramma Voedselreservering Oosterschelde', in het kader van de Tweede Evaluatie van het Nederlands Schelpdiervisserijbeleid, EVA II*. Rapport RIKZ/2003.043, RIVO rapport C062/03. Rijksinstituut voor Kust en Zee, Middelburg.

Hoeksema, R. 2002. *Grevelingenmeer van Kwetsbaar naar Weerbaar? Een beschrijving van de ontwikkelingen van 1996 tot 2001 en een toetsing aan het beleid*. Rapport RIKZ/2002.033. Middelburg: Rijksinstituut voor Kust en Zee.

Holland, A. M. B. M. 2004. Veerse Meer aan de Oosterschelde. Toestand ecosysteem Veerse Meer vóór ingebruikname doorlaatmiddel, Rapport RIKZ/2004.007. Middelburg: Rijksinstituut voor Kust en Zee.

Holling, C. S. (ed.). 1980. *Adaptive Environmental Management*. Int Series on Applied System Analysis 3. New York: John Wiley & Sons Ltd.

IDWDO (Interdepartementale Werkgroep Doorstroom Opening Oosterschelde). 1977. *Varianten doorlaat opening stormvloedkering Oosterschelde*. The Hague, The Netherlands: Publ. Rijkswaterstaat, Koningskade 4.

Kater, B. J., and J. M. D. D. Baars. 2003. *Reconstructie van oppervlakten van litorale Japanse oesterbanken in de Oosterschelde in het verleden en een schatting van het huidig oppervlak*. Nederlands Instituut voor Visserij Onderzoek (RIVO) rapport C017/03. Yerseke: RIVO.

Kater, B. J., Brinkman, A. G., Baars, J. M. D. D., and G. Aarts. 2003. *Kokkelhabitatkaarten voor de Oosterschelde en de Waddenzee*. Nederlands Instituut voor Visserij Onderzoek (RIVO) rapport C060/03. Yerseke: RIVO.

Ministerie van Verkeer en Waterstaat. 1967. *1e Nota Waterhuishouding.*

Ministerie van Verkeer en Waterstaat. 1984. *2e Nota Waterhuishouding.*

Ministerie van Verkeer en Waterstaat. 1989. *3e Nota Waterhuishouding. Water voor nu en later.*

Ministerie van Verkeer en Waterstaat. 1998. *4e Nota Waterhuishouding. Water Kader.*

Morzer Bruyns, M. F., and R. J. Benthem. 1979. *Spectrum Atlas van Nederlandse landschappen*. Utrecht/Antwerp: Publ. Het Spectrum.

Mulder, F. J. and co-workers. 2003. *De ondergrond van Nederland. Geologie van Nederland*. Utrecht: Nederlands Instituut Voor Toegepaste Wetenschappen TNO.

Nienhuis, P. H. 1978. Lake Grevelingen. A case study of ecosystem changes in a closed estuary. *Hydrobiol. Bull.* **12**, 3/4, 246–59.

Nienhuis, P. H., and A. C. Smaal. (eds.). 1994. *The Easternscheldt Estuary. A Case Study of a Changing Ecosystem*. London, Boston and Dordrecht: Kluwer Academic Publisher, pp. 1–579.

NIG (Nederlands Instituut voor Toegepaste Wetenschappen). 1997. *De ontstaansgeschiedenis van het Zeeuwse kustlandschap.* (Audiovisual.) Delft, The Netherlands.

Parma, S.1978. Political aspects of the closure of the Easternscheldt. *Hydrobiol. Bull.* **12**, 3–4, 163–75.

Pons, L. J. 1992. Holocene peat formation in the lower parts of the Netherland. In J. T. A. Verhoeven (ed.), Fens and Bogs in the Netherlands, Vegetation, History, Nutrient, Dynamics and Conservation. *Geobotany* **18**, 7–79.

RAND. 1977a. *Protecting an Estuary from Floods. A Policy Analysis of the Oosterschelde*. Santa Monica, CA: RAND Corporation.

RAND. 1977b. *Summary*. Report R 2121/1. Netherlands I, 1–153. Santa Monica, CA: RAND Corporation.

RAND. 1977c. *Assessment of Security from Flooding*. Report R2121/2. Netherlands II 1–162. Santa Monica, CA: RAND Corporation.

RAND, 1977d. *Assessment of Long Run Ecological Balances*. R2121/3. Netherlands III, 1–215. Santa Monica, CA: RAND Corporation.

RAND, 1977e. *Assessment of Algae Blooms. A Potential Ecological Disturbance*. 2121/4. Netherlands IV, 1–81. Santa Monica, CA: RAND Corporation.

RAND, 1979a. *Controlling the Oosterschelde Storm Surge Barrier. A Policy Analysis of Alternative Strategies*. R2444/1. Santa Monica, CA: RAND Corporation.

RAND, 1979b. *Sensitivity Analysis*. R2444/2. Netherlands II, 1–54. Santa Monica, CA: RAND Corporation.

RAND, 1979c. *Basis Response to North Sea Waterlevels*. The BARCON Simplic model, R2444/4. Santa Monica, CA. RAND Corporation.

Rappoldt, K., Ens, B. J., Berrevoets, C. M., Geurts van Kessel, A. J. M., Bult, T. P., and E. Dijkman. 2003. *Scholeksters en hun voedsel in de Oosterschelde, Rapport voor deelproject D2 thema 1 van EVA II, de tweede fase van het evaluatieonderzoek naar de effecten van schelpdiervisserij op natuurwaarden in de Waddenzee en Oosterschelde 1999–2003*. Alterra rapport 883. Wageningen: Alterra.

Rijkswaterstaat. 1976. *Beleidsanalyse Oosterschelde Alternatieven (Policy analysis Easternscheldt Alternatives)*. The Hague, The Netherlands: Rijkswaterstaat, Koningskade 4.

Saeijs, H. L. F. 1982. Changing estuaries. A review and new strategy for management and design in coastal engineering projects. Thesis, University of Leiden, The Netherlands. *Rijkswaterstaat Communications* **32**, 1–413.

Saeijs, H. L. F. 1994. *Creative in a changing delta. Towards a controlled ecosystem management in the Netherlands.* Proceedings, 18th ICOLD-Congress, 6–11 November 1994, Durban, South-Africa (Question 69; R.250). Paris: International Commission of Large Dams, pp. 371–95.

Saeijs, H. L. F. 1999. *Levend water, goud waard*. Symposium Het verborgen vermogen. Rijkswaterstaat Directie Zeeland, p. 70

Saeijs, H. L. F., and B. A. Bannink. 1978. Environmental considerations in a coastal engineering project. The Delta Project in the south-western part of the Netherlands. *Hydrobiol. Bull.* **12**, 4–5, 180–202.

Saeijs, H. L. F., and H. J. M. Baptist. 1976. *Vogels Grevlingenmeer 1971–1975*. Nota 76–31 Rijkswaterstaat, Deltadienst. In Saeijs and Bannink (1978).

Saeijs, H. L. F., and A. De Jong. 1981. The Oosterschelde and the protection of the environment. A policy plan for a changing estuary. *Land and Water International* **46**, 15–34; *Rijkswaterstaat Communications* **32**, 21–278.

Saeijs, H. L. F., and P. B. M. Stortelder. 1982. Converting an estuary to Lake Grevelingen. Review of a coastal engineering project. *Environmental Management* (Springer Verlag, NY) **6**, 5, 377– 405.

Saeijs, H. L. F., Smits, A. J. M., Overmars, W., and D. Willems. (eds.). 2004. *Changing Estuaries, Changing Views*. Rotterdam: Erasmus Universiteit; Nijmegen: Radboud Universiteit.

Segeren, W. A., Brinkman, J. H., van Dordt, J. H., Feitsma, K. S., van der Hoeff, J. J., Reitsma, T., Saeijs, H. L. F., Smittenberg, J., Stiemer, D., Tiggelaar, L., and M. Zijlstra. 1975. *De Nieuwe Inrichtingsschets Grevelingenbekken. Rijksdienst voor de IJsselmeer polders.* The Hague: Cartoprint BV.

SGO (Stuurgroep Oosterschelde). 1980a. *De Oosterschelde. Een overzicht van de beleidsmogelijkheden.* The Hague, The Netherlands: Rijkswaterstaat, Koningskade 4.

SGO (Stuurgroep Oosterschelde). 1980b. *Studies over de Oosterschelde deel 1: Ontwikkelingsmogelijkheden van natuur-, visserij- en recreatie functies.* The Hague, The Netherlands: Rijkswaterstaat, Koningskade 4.

SGO (Stuurgroep Oosterschelde). 1980c. *Studies over de Oosterschelde. Deel 3: Beheersaspecten en instrumentarium Oosterschelde.* The Hague, The Netherlands: Rijkswaterstaat, Koningskade 4.

SGO (Stuurgroep Oosterschelde). 1980d. *Inrichting Oosterschelde: Deelrapport Inventarisatie.* The Hague, The Netherlands: Rijkswaterstaat, Koningskade 4.

Tosserams, M., Lammens, E. H. R. R., and M. Platteeuw. 2000. Het Volkerak-Zoommeer. *De ecologische ontwikkeling van een afgesloten zeearm.* RIZA rapport 2000.024. Lelystad: RIZA.

Veen, J. Van. 1953. *Land Below Sea-level.* The Hague: Publ. Trio.

Ven, G. P. van de (ed.). 1993. *Man Made Lowlands, Second Revised Edition.* Utrecht: Matrijs.

Plate 1   Satellite image of the Venice Lagoon.

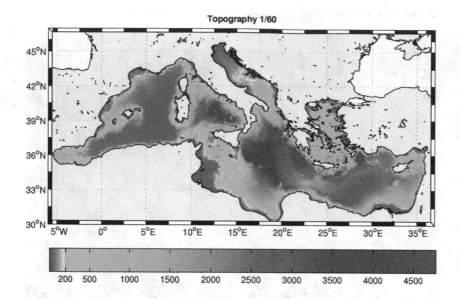

Plate 2 Bathymetry (water depths in metres) of the Mediterranean Basin (see Pinardi *et al.*, chapter 6).

Plate 3 Dynamic height anomaly at 5 m (with respect to 800 m reference level) in the eastern Mediterranean Sea for the month of August in selected years between 1981 (top left) and 2000 (bottom right). Units are in cm (see Pinardi *et al.*, chapter 6).

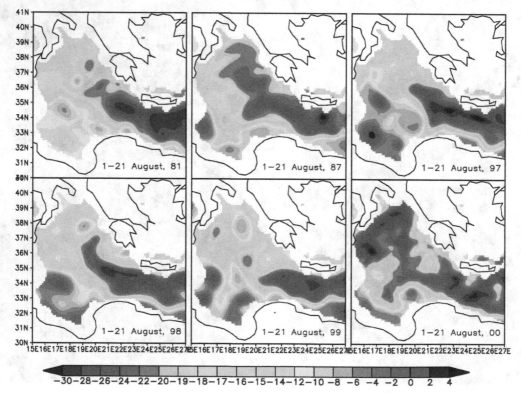

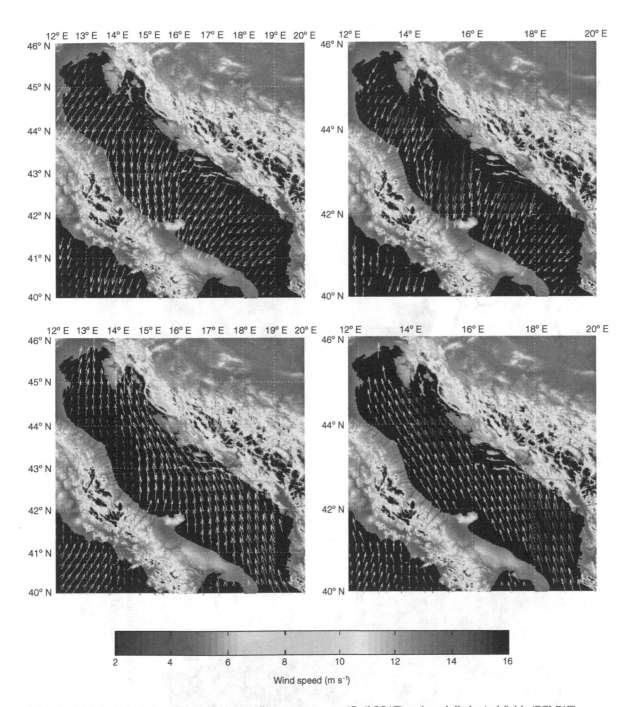

Plate 4   Adriatic Sea winds as described by satellite scatterometry (QuikSCAT) and modelled wind fields (ECMWF analysis). Top panels: *bora* wind (11 February 2001). Bottom panels: *sirocco* wind (25 January 2001). Left panels: wind field from the ECMWF analysis, interpolated over a 25 km by 25 km grid, respectively at 06:00 and 18:00 GMT. Right panels: QuikSCAT satellite scatterometer wind field at 04:08 and 16:56 GMT (see Zecchetto *et al.*, chapter 7).

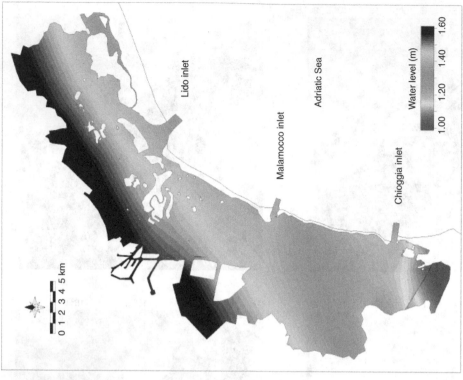

Plate 5   Spatial variability in maximum water level on 8 December 1992, showing the effect of a *bora* (NE) wind (see Ferla, chapter 12).

Plate 6.   Spatial variability in maximum water level on 6 November 2000, showing the effect of a *sirocco* (SE) wind (see Ferla, chapter 12).

Plate 7   An early photograph (1880–90) of the Dario Palace in Venice (left) and a present day view
of the same building (right). The observed algal displacement is 46 ± 10 cm (for the apparent
sea-level rise subtract 13 cm with reference to the 2003 algal level) (see Camuffo *et al.*, chapter 16).

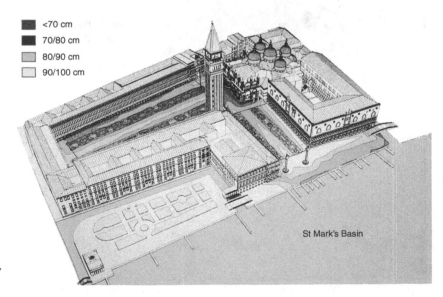

Plate 8   Flooding of St Mark's
Square. Water levels relate to
the 1897 reference level
(present mean sea level lies at
c. 25 cm above this level)
(see Vio, chapter 21, and Brotto,
chapter 23).

Plate 9 Fragment of a figurative fresco and detail in Venice (see Danzi *et al.*, chapter 24).

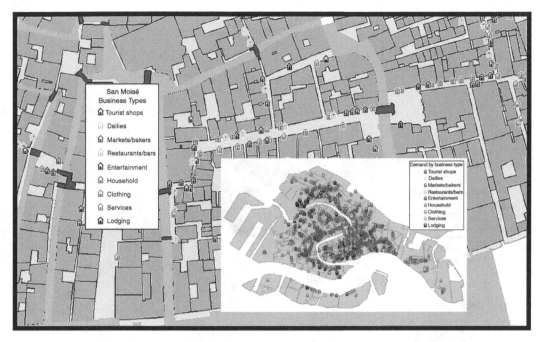

Plate 10   Inventory of all commercial establishments
in Venice by category (with detail for San Moisé)
(see Carrera, chapter 27).

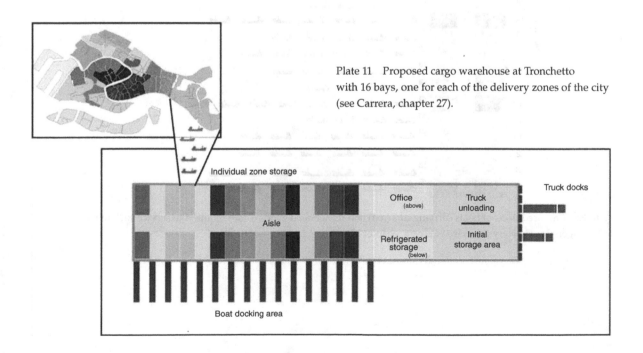

Plate 11   Proposed cargo warehouse at Tronchetto
with 16 bays, one for each of the delivery zones of the city
(see Carrera, chapter 27).

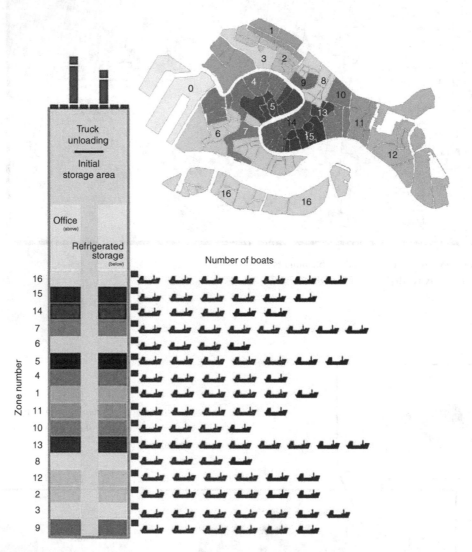

Plate 12 Proposed fleets of boats carrying dry (red) and refrigerated goods (blue) required to service each of the 16 homogeneous delivery zones (see Carrera, chapter 27).

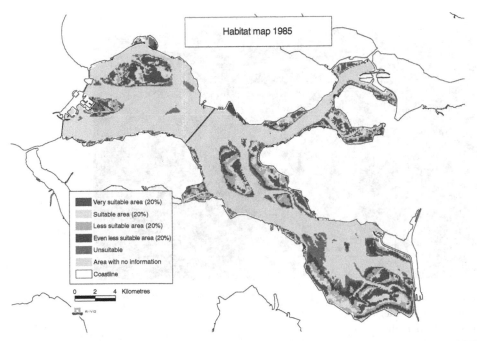

Plate 13   Cockle (*Cerastoderma edule*) habitat in the Oosterschelde, The Netherlands, in 1985 (source: RIVO-CSO) (see Saeijs and Guerts van Kessel, chapter 30).

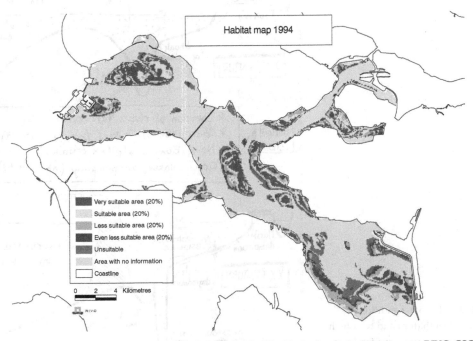

Plate 14   Cockle (*Cerastoderma edule*) habitat in the Oosterschelde, The Netherlands, in 1994 (source: RIVO-CSO) (see Saeijs and Guerts van Kessel, chapter 30).

Plate 15   Pacific oyster (*Crassostrea gigas*) reef in the Oosterschelde, The Netherlands (photograph: Dr N. Steins) (see Saeijs and Guerts van Kessel, chapter 30).

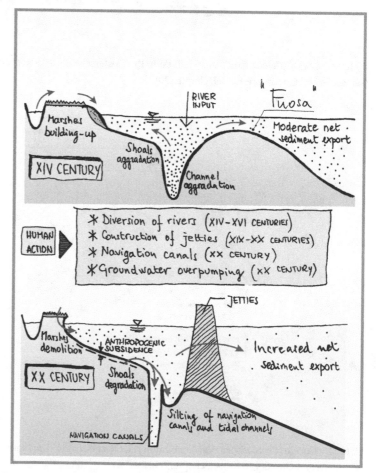

Plate 16   Lagoon conditions in the fourteenth (top) and twentieth (bottom) centuries (see Di Silvio, chapter 43).

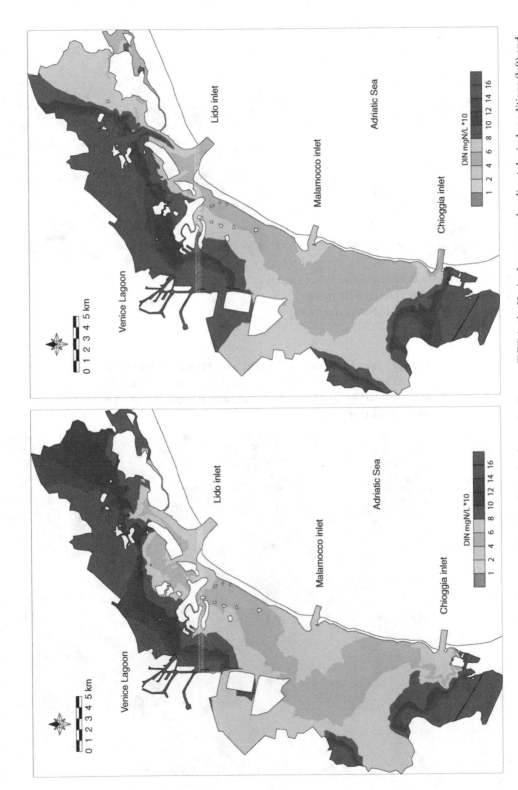

Plate 17 Spatial distribution of yearly averaged values of dissolved organic nitrogen (DIN) in the Venice Lagoon under climatological conditions (left) and realistic forcings (right) (see Umgiesser *et al.*, chapter 47).

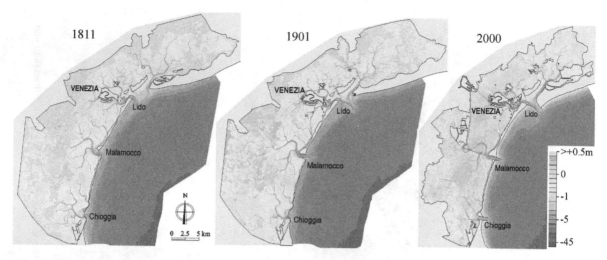

Plate 18   Bathymetric configurations of the Venice Lagoon surveyed in: (a) 1811; (b) 1901; and (c) 2000 (present configuration) (see D'Alpaos and Martini, chapter 48).

Plate 19   Examples of the velocity field at the Malamocco inlet to the Venice Lagoon. Flood phase (a, c) and ebb phase (b, d) in the present configuration (left) and for the 1811 configuration (right) (see D'Alpaos and Martini, chapter 48).

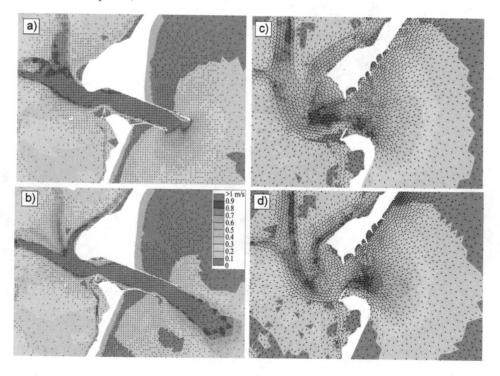

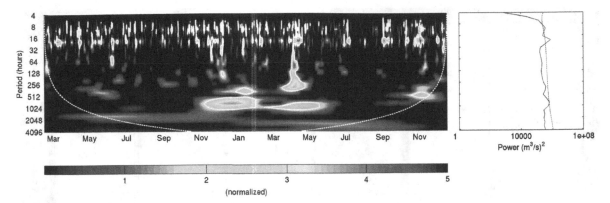

Plate 20 Wavelet spectrum of non-tidal water flux for Chioggia inlet (left panel). Wavelet spectrum values are normalized by the total variance, region under dashed line is affected by edge effects. On the right panel time-averaged wavelet spectrum is shown; dashed red line represents the background-noise spectrum with an autocorrelation of 0.14 for the lag 1 estimated from data (see Gačić, chapter 49).

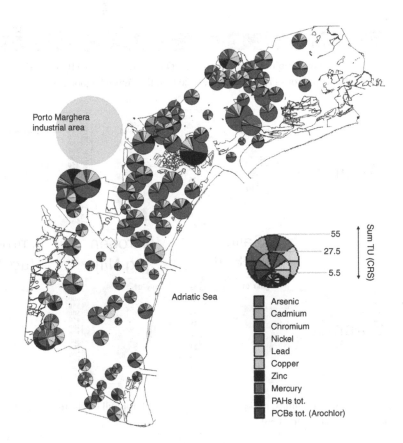

Plate 21 Pie chart visualization of the individual contaminant contribution to the cumulative risk (CRS) for the benthic community at sampling stations in the Venice Lagoon. The risk was estimated through the Toxic Units (TU) method and the ecotoxicological TEL benchmark (see Marcomini *et al.*, chapter 54).

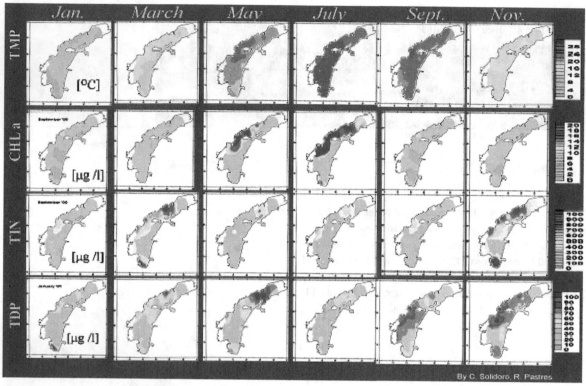

Plate 22 Temporal evolution of temperature (T), chlorophyll-a (CHL-a), total inorganic nitrogen (TIN) and total dissolved phosphorous (TDP) for the year 2001 in the Venice Lagoon (see Zirino, chapter 55).

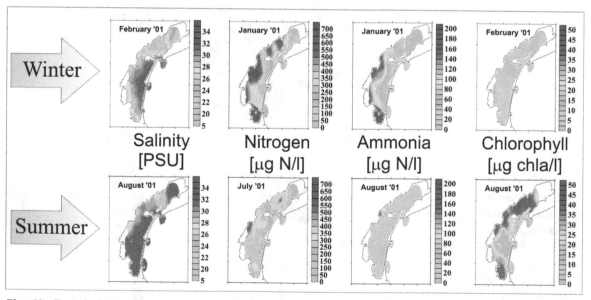

Plate 23 Typical salinity, nitrogen, ammonia and chlorophyll levels during winter and summer in the Venice Lagoon (see Penna *et al.*, chapter 56).

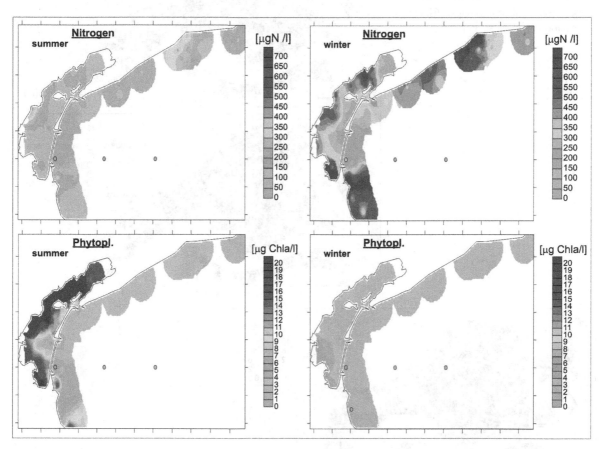

Plate 24 Examples of integrated data processing of
water quality of the lagoon (Venice Water Authority and
CVN) and the coastal area (ARPAV) in 2001 (see Penna
et al., chapter 56).

Plate 25   Distribution of (a) Zn and (b) Cu concentration in the upper sediment layer of the canal network. Colours indicate the relative magnitude of contaminant concentrations measured (mg/kg, dry weight) (see Zonta *et al.*, chapter 64, for explanation of white arrow).

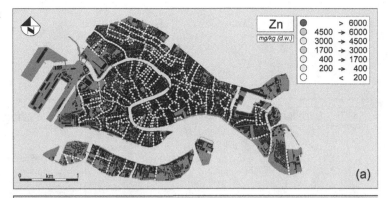

(a)

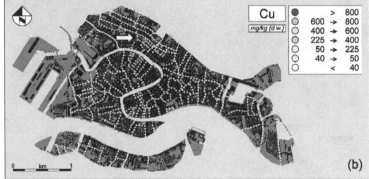

(b)

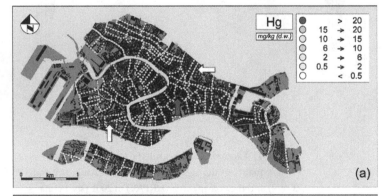

(a)

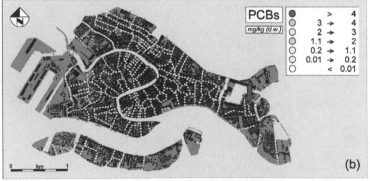

(b)

Plate 26   Distribution of (a) Hg and (b) PCBs concentration in the upper sediment layer of the canal network. Colours indicate the relative magnitude of contaminant concentrations measured (mg/kg, dry weight) (see Zonta *et al.*, chapter 64, for explanation of red and white arrows).

40 • Integrated water management for support of sustainable development of St Petersburg region

R. MIKHAILENKO, A. N. SAVIN, H. GERRITSEN AND H. VAN PAGEE

INTRODUCTION

St Petersburg is situated at the mouth of the River Neva where it flows into the Baltic Sea. Its location has strategic value for economic and social development. Presently, however, the various economic and social uses of the water resources have a negative influence on the water environment and are potential sources of ecological risk.

Significant population growth and the extension of the city territory since the 1960s, as well as an increase of the industrial sector in the city economy when water protection measures were not undertaken, have led to a deterioration of the ecological situation in the Neva river and Neva Bay. The water system Ladoga Lake – River Neva – Neva Bay – eastern Gulf of Finland is an important strategic region (Fig. 40.1). The complexity of ecological conditions in the ecosystems of these water bodies is determined by peculiarities in anthropogenic and natural factors. As a result of an analysis on the human impacts on Neva Bay, eleven types of effect have been identified: along-coast landfill plus construction of housing and infrastructure; recreation; agricultural activity; water-supply

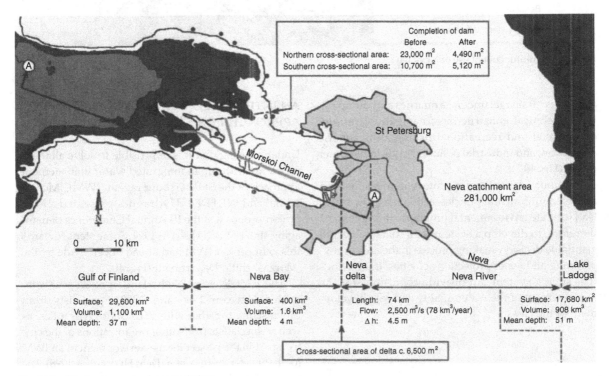

Fig. 40.1 Hydrodynamic characteristics.

Flooding and Environmental Challenges for Venice and its Lagoon: State of Knowledge, ed. C. A. Fletcher and T. Spencer.
Published by Cambridge University Press, © Cambridge University Press 2005.

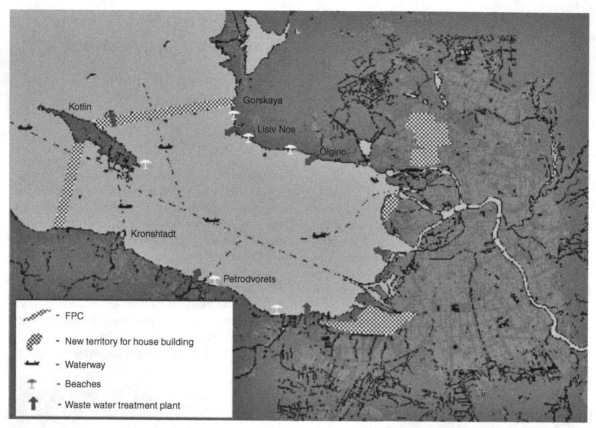

Fig. 40.2   Multifunctional use of Neva Bay.

and water-drainage; undersea quarries and dumping; hydrotechnical constructions; navigation; rafting; commercial and recreational fishing; atmospheric deposition; and industrial pollution loads in the water phase (Fig. 40.2).

Without relevant water protection measures and active environmental policies, economic growth will lead to acute environmental problems. It is necessary, therefore, to develop a system for a sustainable and justifiable decision-making process in the area of city-planning and water protection activities. It is obvious that such a system can provide conditions for the improvement of the water quality and safe development of the city.

AN INTEGRATED WATER MANAGEMENT APPROACH FOR ST PETERSBURG

Understanding that it is impossible to solve all these problems without an integrated water management approach of the St Petersburg region (IWM), 'Morza-schita' and WL | Delft Hydraulics presented a joint project proposal at the III Annual Conference of managing staff of EBRD held in 1994. Some steps towards this joint project IWM had already been made by the 'Morzashchita' Department (Fig. 40.3).

The joint Russian-Netherlands project was conducted between 1996 and 2000. Specialists from Morzaschita together with WL/Delft Hydraulics, as well as specialists from other organizations of the city, took part in the project for the development of an IWM for the St Petersburg region (Delft Hydraulics, 1998). The main technical components of this approach were: data-

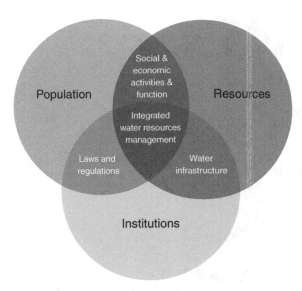

Fig. 40.3   Main principles of IWM.

- IWM-1 (1997):
  - focus on the definition of a Strategic Management Action Plan;
  - development of Decision Support Systems;
  - bringing together of monitoring data in databases; and
  - case studies.
- IWM-2 (1998):
  - publication of the Strategic Management Action Plan;
  - case study : cleaning up of water quality in the Okhta River;
  - start of development of model-based flood forecasting; and
  - installation of Coordinating Council for Water Management.
- IWM-3 (1999-2000):
  - development of a GIS for Flood Damage Assessment; and
  - further improvement and user training of flood forecasting model.

bases of system properties; parameters; pollution loads and other anthropogenic parameters; information systems about available properties of separate water environment constituents; of bottom sediments and biota; a load model of polluting substances for definition of a general sum of system pollution; and forecasting models regarding water quality and ecological parameters.

To study the influence of the manoeuvring of the water gates of the Flood Protection Barrier (FPB) on the ecological condition of Neva Bay, 'Morzaschita' had previously carried out a large-scale field experiment (Figs. 40.4, 40.5).

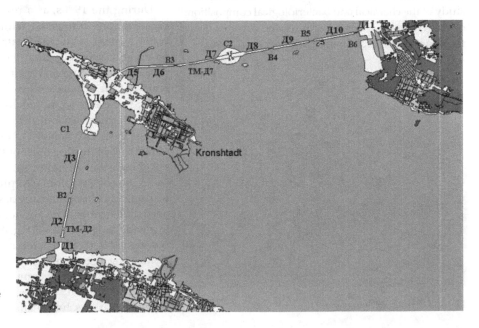

Fig. 40.4   Existing flood protection barrage (barrier).

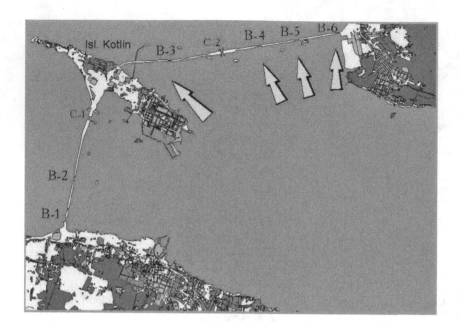

Fig. 40.5   Project flood
protection barrage (barrier).

The study programme included the following activities: hydrometeorological and hydrological observations; analysis of the water quality on many hydrochemical and bacteriological factors; hydrobiological and ichthyologic studies; remote sensing observations; tracer observations on the pollutant distribution from the city water purification plants; and study of the chemical and bacteriological composition of the sediments. Sixteen institutes and organizations took part in this experiment (Guralnik *et al.*, 1995).

Fig. 40.6   Water exchange gates.

The results of the experiment confirmed the possibility of purposefully influencing the hydrological regime and the ecological situation of the Neva Bay water area and the eastern part of the Gulf of Finland by means of closing specific section openings of the FPB and redistributing the water volumes passing the FPB (Fig. 40.6).

During the 1990s, as a result of a polling of 17 organizations connected with researches into the water system, a detailed catalogue of all natural observations (both regular monitoring and specific project observations) was made, and a detailed review of the databases documented (Mikhailenko, 1997). The tables in the catalogue contain observational information about all parts of the water system Ladoga Lake – River Neva – Neva Bay – eastern Gulf of Finland, including a list of the basic characteristics of the water environment.

As part of the joint IWM project, a GIS for flood warning and flood damage assessment was developed, that is now at an applied stage and of interest to a wide range of users (Fig. 40.7). This GIS model is designed both for macroeconomic damage assessment, using aggregate indices, and, if sufficient detail is available, for microeconomic assessment.

Results of the GIS are already being used by many city organizations. Information about flooded territo-

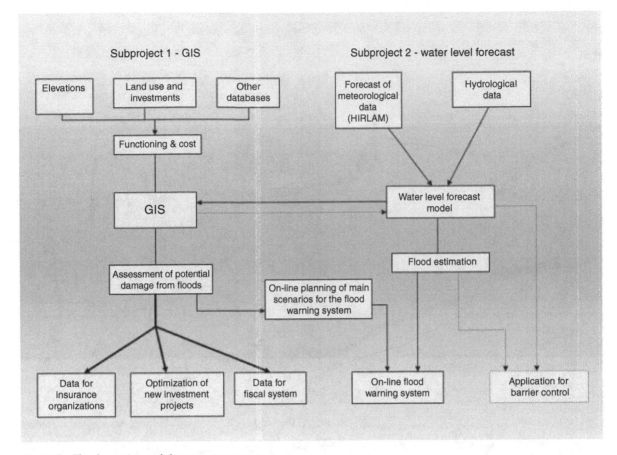

Fig. 40.7   Flood warning and damage assessment.

ries in the geographical information system can be useful for insurance, rent pricing and local taxation, development of plans of actions in emergency situations and for planning of investments and general urban planning (Fig. 40.8).

CONCLUSIONS

In summary, the implementation of the integrated water management programme will allow the creation of a regional system of water management. The managerial approach for development will be based on understanding the best balance between the various uses of water system, leading to a more stable development and rational use of the water system (Lake of Ladoga – Neva River – Neva Bay – eastern

part of the Gulf of Finland). It will also allow significant progress in the development of the flood early warning system to be achieved and a method to estimate the potential damage will be realized. Implementation of the programme will make the city more attractive for investments, as the threat of flooding of the historical centre will be eliminated.

REFERENCES

Delft Hydraulics. 1998. *Integrated Water Management of the Ladoga-Neva-Galf Water System*. Final Report, WL Delft Hydraulics, Delft.

Guralnik, D. L., Kassatsier, K. E., and R. R. Mikhailenko. 1995. Analysis of distribution of heavy metals pollution in the Neva Inlet. *Int. Quart. J. Ecol. Chem.* **4**.

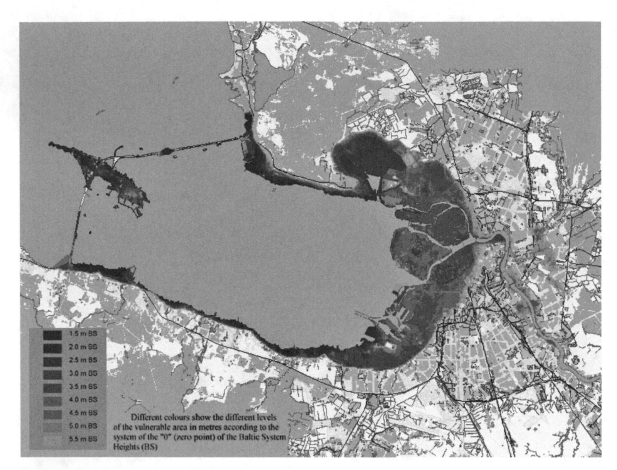

Fig. 40.8   Flooded areas of the city of St Petersburg.

Mikhailenko, R. R. 1997. *Integrated Water Management for St Petersburg Region*. Data Inventory, St Petersburg.

41 • Flood protection, environment and public participation – case study: St Petersburg Flood Protection Barrier

H. GERRITSEN, M. VIS, R. MIKHAILENKO AND M. HILTUNEN

SUMMARY

A public information and participation programme is a standard part of environmental impact assessment studies (EIAs) for large infrastructural works. This chapter illustrates its importance using the example of the Project of Completion of the Flood Protection Barrier for St Petersburg, Russia. During the initial construction stage of the Barrier in the 1980s, public concern on the environmental impacts effectively led to a halting of the construction works when the Barrier was two thirds completed. In 2002, a feasibility study for the completion of the Barrier was conducted. Public information and involvement was considered a key element in the 2002 study.

This chapter presents an overview of the relevant history of the Project. It discusses the requirements for EIA and public participation under the European Bank for Reconstruction and Development (EBRD) and the Russian Federation, and gives an overview of how the EIA and public participation were conducted. Attention is given to the information exchange during the First Public Consultation and the way in which the suggestions and concerns of the public were incorporated in the scope of the EIA. Based on the results of the Final Public Consultation for the environmental impact assessment, it is concluded that the public participation has resulted in an improved feasibility study, and also in increased local support for the Barrier. This is relevant for the sustainability of such a large infrastructure.

INTRODUCTION

The city of St Petersburg was founded in a low-lying area, where the delta of the river Neva meets the waters of the Gulf of Finland. These three elements (low-lying city, river delta and Gulf) define a key environmental problem for St Petersburg. High water in the Gulf of Finland causes frequent flooding of the low areas of the city, while the river outflow with its long-term average discharge of 2500 m^3/s carries the effluent of the population (approximately five million people) and industry of the city to the shallow Neva Bay and eastern part of the Gulf of Finland. The flooding of the city has been a concern since its foundation in 1703. A total of 296 floods have been recorded to date, which gives about one per year. When the water level reaches 160 cm above the reference datum BS (BS is approximately equal to local mean sea level), streets and basements in the low-lying parts of the city centre start to flood, and a flood is officially recorded. The most recent flood occurred on 30 December 2003 (199 cm). The five largest floods occurred in 1824 (421 cm), 1924 (380 cm), 1777 (321 cm), 1955 (293 cm) and 1975 (281 cm BS).

THE INITIAL CONSTRUCTION STAGE OF THE FLOOD PROTECTION BARRIER (1966-1987)

After the 1955 flood, options were investigated for providing the then called city of Leningrad with adequate structures for flood protection. From four original designs, a design with a flood protection barrier across the island of Kotlin was selected and taken further for study (see Mikhailenko, this volume). Sixteen major Russian institutes with relevant expertise participated in this large-scale feasibility study. The 1975 flood accelerated the studies and the decision making process. During the period 1973–7 some 100 reports were produced, analysing the present environmental and geological situation, design and construction and environmental protection issues. According to local

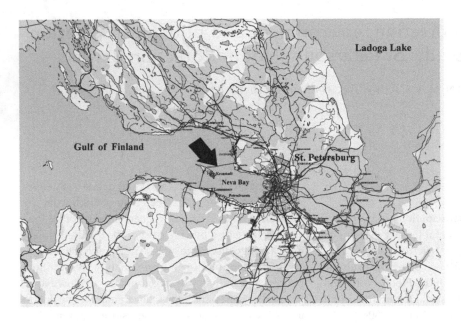

Fig. 41.1   Map of St Petersburg at the meeting point of Neva River and Gulf of Finland. Arrow shows the location of the Flood Protection Barrier in Neva Bay.

scientists, about 20% of the studies addressed the environmental situation and protection. In 1978, reviews by the USSR State Committee for Science and Technology and the Main State Expertising Commission concluded that preceding construction of the Barrier, treatment facilities to purify the waste water discharges to the Bay were required.

Based on the Project Report and supporting studies, the Government of the Soviet Union in 1979 approved the construction of the western design,

consisting of a 25.4km-long Flood Protection Barrier across the island of Kotlin, west of the city of Kronstadt (Fig. 41.1). The Barrier is composed of 11 dam sections D1–D11, 6 water exchange complexes B1–B6 distributed along the Barrier to allow free passage of water and 2 navigation passages C1 and C2. Figure 41.2 shows the concept of the Barrier.

Construction started in 1980 and by December 1984 Kotlin was connected to the northern coast via the Barrier. At the time, not all designed through flow openings in that section were immediately realized, though, to facilitate the construction activities and associated transport (Neva river water was still able to flow through the large openings in the southern section).

PUBLIC CONCERNS AND EXPERT ANALYSIS (1984–1995)

Around that time, local deterioration of water quality became apparent. Bathers experienced itching on their skin and recreational beaches along the northern coasts were more frequently closed for reasons of bacterial pollution. Recreational fish catch decreased, with more and more fish being observed to have sores. As the site was officially declared a closed military area, and concerns of the public and local com-

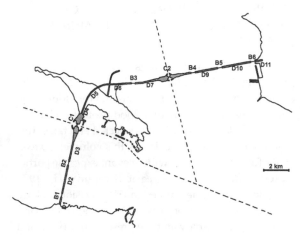

Fig. 41.2   Design state of the St Petersburg Flood Protection Barrier.

munities were not adequately addressed, public rumours and speculation resulted. In the autumn of 1987 concerned citizens and scientists appealed to Secretary General Gorbachev. They felt that the observed pollution of Neva Bay was caused and aggravated by the construction of the Barrier and they protested against its completion. As a result of these and similar concerns voiced by the Academy of Sciences of the USSR, the construction was temporarily halted.

In short succession, three commissions of Soviet experts were nominated to investigate the environmental state of Neva Bay and the possible role of the Barrier in its condition. The reports of the commissions all agreed on the deterioration in the environmental state of Neva Bay, but strongly disagreed as to the role of the Barrier in creating the water quality problem. To resolve the then highly politicized issue, in 1990 an International Commission of Experts was invited to study and evaluate the problems. A Technical Working Group of WL | Delft Hydraulics was asked to support the Commission. In its report, this Commission confirmed the serious environmental problem in Neva Bay, but also made it clear that the impact of the completed Barrier on the environmental state would be minimal (International Commission, 1990). It recommended that completion should proceed as planned, giving attention to minimizing impacts during the construction period. In parallel, measures to improve wastewater treatment should urgently be implemented as discharge of untreated or partially treated sewage was the main cause for the deteriorated environmental state of the Bay. Equally, a continuing active public information programme should be conducted by the authorities. The Commission actually started the latter by announcing and organizing its own public hearings and inviting people to comment on the Commission's assessments and recommendations. These meetings gave opportunities for a two-way exchange of information and proved to serve a strongly felt need by the public to have their concerns addressed and questions answered. Although the Commission's results and recommendations were generally accepted during the years that followed, political and financial changes allowed the work to continue at only a very minimal pace. A 1996-7 Pre-feasibility Report by Sir Alexander

Gibb and Partners again confirmed the earlier analysis and recommendations of the International Commission (Gibb, 1997). Due to the break-up of the Soviet Union and lack of finances, though, work between 1987 and 2002 remained largely limited to maintenance of works already completed, with the Barrier itself only about two thirds completed.

THE PROJECT FOR THE COMPLETION OF THE FLOOD PROTECTION BARRIER

By the turn of the millennium, political and financial stability in Russia had much improved. In 2002, a four-part Feasibility Study including an Environmental Impact Assessment was conducted within the framework of a possible international loan to the Russian Federation for the accelerated completion of the Barrier. This was commissioned by the State Construction Ministry (Gosstroy) and financed by the European Bank for Reconstruction and Development (EBRD). The study addressed three alternatives for completion (the Project):

- 'Do minimum'; reference situation; protection of the present infrastructure;
- Completion of the Barrier only; and
- Completion of the Barrier with provisions for a future highway (in the form of extension of ramps for a tunnel under the large navigation lock).

The Technical Study involved a design review, a construction review, an operational review, environmental modelling, cost assessment and implementation scheduling. Separate Economic and Cost Recovery studies were undertaken, as was the Environmental Impact Assessment Study. There was much interaction between all the studies. The main objectives of the EIA study was to assist Gosstroy and the EBRD to assess the environmental impacts of the Project, to provide a platform for a public discussion, and to ensure consistency with the Russian environmental and permitting requirements and EBRD's environmental policy and procedures.

While public discussion, information and participation is part of any EIA nowadays, the history of the present Project gave it special meaning. In the 1980s, the Project had been halted due to public concern, and it was expected that there was likely to be historic negative public feeling about the Project. The project had often been described as one of the last massive prestige projects associated with a political system and style which had since been replaced. As such the Barrier had become a highly politicized issue. While the existing materials from the original studies and the analysis of 1990 and 1997 facilitated the EIA, the credibility issue with the community was likely to be a complex one. Further complicating this were the different views and requirements on EIA procedures in the west and in Russia where they were still in the process of being adjusted from earlier Soviet plan economy requirements.

So, information, communication and participation aspects were seen as key issues by Gosstroy, EBRD and the independent Dutch consultant NEDECO/WL | Delft Hydraulics that had been commissioned to conduct the EIA. They would be vital in developing a proposal for the completion of the Barrier that would adequately balance the requirements of flood protection and an environmental healthy environment, and would moreover meet with sufficient public support to make it sustainable in the longer term.

EIA REQUIREMENTS OF EBRD AND OF RUSSIA

The EBRD considers an Environmental Impact Assessment as one of the main types of environmental appraisal work required in its lending process. Environmental investigations have to be carried out or commissioned by the initiator of the Project (Project Sponsor) in accordance with both the national, in this case Russian, and the Bank's EIA requirements (EBRD, 1992). Russian requirements are laid down in the Federal Law 'On Environmental Protection' (Russian Federation, 2002), the Law 'On the State Environmental Review' (Russian Federation, 1995) and in the Regulation 'On Impact Assessment of Proposed Operation on the Environment in the Russian Federation' (Russian Federation, 2000). According to these laws and regulations, any project in the Russian Federation, in order to obtain official authorization, has to pass environmental appraisal by State authorities. This process consists of 2 procedures: OVOS, the Russian form of EIA, the purpose of which is to evaluate the significance of the environ-

mental impacts of the project, consider alternative options, incorporate mitigation measures into the project design and address public concerns regarding the project impact; and State Environmental Review (SER), the purpose of which is to analyse OVOS materials together with other Project documentation and to take a decision on the environmental soundness of the proposed activity. The conclusion of the SER is legally binding for the Project.

PUBLIC CONSULTATION REQUIREMENTS

According to the EBRD's Environmental procedures, project sponsors must ensure that all national requirements for public consultation in the country where the Project will take place are met. In addition, they have to follow the Bank's guidelines for public consultation.

The Russian EIA procedure prescribes public participation in two phases of the EIA: in the initial project scoping phase when basic documentation and the scope or Terms of Reference (ToR) for the EIA study are prepared, and in a final phase, when the EIA report is available. A public participation plan has to be defined by the local authorities together with the Project Sponsor and interested public. The plan has to be laid down in an official document and time and location of the planned public consultations have to be announced by means of short notifications in official journals, by radio or TV broadcasts, or via the internet, at least 30 days before the meetings take place. The Project Sponsor has to make sure that those interested people who do not have access to the media are also informed.

In a final stage of the EIA the draft EIA report has to be made available to the public for at least 30 days before the final public consultation. The Project Sponsor has to react to any written comments on the draft EIA report within 30 days of the final public hearing.

All public hearings have to be documented in official protocols, in which the main issues discussed, and the main disagreements between the public and the Project Sponsor, are documented. These protocols have to be attached to the final EIA report. The public should have free access to the final EIA report from the moment of its approval until the time when the final decision on realisation of the Project is taken.

The Environmental Procedures of the EBRD require that the Project Sponsor meaningfully consults with the stakeholders on the preparation and results of the EIA. The results of the EIA process have to be disclosed to the public. Ongoing consultation is also required during the construction and operation phases of the Project. A Public Consultation and Disclosure Plan (PCDP) must be developed at the initial (scoping) stage of the EIA. It must define a technically sound and culturally appropriate approach to consultation and disclosure. The goal is to ensure that adequate and timely information is provided to people affected by the project and other stakeholders, and that these groups are given sufficient opportunity to voice their opinions and concerns.

The World Bank equally stresses the importance of public consultation. In a recent evaluation, the World Bank has assessed experience with public consultation on a number of occasions. Its overall conclusion has been that the quality of the resulting EIA has been improved significantly when information and views provided by the stakeholders in the EIA process are taken into account (Kjorven, 1997; World Bank, 2002).

OVERVIEW OF THE EIA

The EIA study was conducted from January to July 2002, with a formal completion in September that year. Its activities were grouped into six stages:

- conducting a Scoping Study;
- conducting a Scoping Meeting or First Public Consultation;
- definition of the environmental baseline;
- assessment of environmental impacts of the Project, both positive and negative;
- conducting a Final Public Consultation; and
- reporting.

The actual EIA analysis followed the general framework for EIA (Fig. 41.3) and started with the collection and analysis of the available basic data, both on the Project and possible Project alternatives, and on the environment as far as it was likely to be affected. The collection and analysis of environmental data served to provide a description of the so?called environmental baseline conditions.

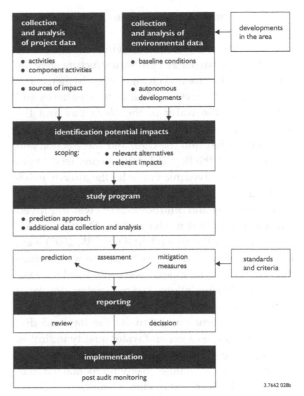

Fig. 41.3 General framework of Environmental Impact Assessment.

By combining the information on baseline conditions, sources of impacts, literature, checklists and cause-effect networks, a list of potential environmental impacts was made. This identification involved an estimate of the order of magnitude of the impacts.

In a Public Consultation Meeting with all involved and affected parties, key environmental issues and local concerns were further identified and discussed, prioritized and selected for further study, resulting in the final scope for the EIA. For the selected alternatives and impacts an assessment was executed which was largely based on available data and information. Two groups of Russian scientists, all having specific expertise of the Neva Bay/Gulf of Finland marine environment and the Flood Protection Barrier construction project, were involved along with the specialists of the EIA consultant. One group for the assessment of the impacts of the Barrier as a physical structure in its operational phase and one for the assessment of impacts related to the construction activities.

It was reasoned that the driving forces causing impacts of the Barrier as a physical structure were the occupation of space and changes in bathymetry. The changes in bathymetry influence flow patterns and velocities of the water in the Neva Bay and the nearby Gulf of Finland, the exchange of water between the Bay and the Gulf, and wave action and water levels in the Bay. To be able to predict changes in these hydro-dynamic variables the already existing local CARDINAL hydrodynamic model application (Klevanny and Smirnova, 2002) was improved and runs with that model and the Delft3D model application for the area (Delft3D-FLOW, 2003) were made for different river discharge and wind condition scenarios. The models were also used to make water quality simulations. The parameters simulated were total coliform concentration, as an indicator for sanitary/hygienic conditions in the Bay and dissolved oxygen, as a general water quality indicator. The outcome of the model simulations were interpreted in terms of changes in environmental parameters by the Russian and foreign scientists.

PREPARATION OF SCOPING MATERIAL AND PUBLIC CONSULTATION

In the Scoping Study, the Project was described, plus the characteristics of the environment in which the Project is located: Neva Bay and the Eastern Gulf of Finland. An overview was given of environmental norms and standards and Russian water quality standards. A comparison was made between EIA requirements according to Russian law, EBRD Environmental Procedures and European Union (EU) regulations. Based on existing literature, the key environmental issues and the potential environmental impacts to be studied in more detail in the actual impact assessment were identified. The Scoping Study (NEDECO, 2002a) was conducted during the first month of the study, January 2002.

Early in the project (January), the identification of stakeholders, the people who have a role in the Project, could be affected by the Project or who are interested in the Project, was undertaken in a series of meetings of representatives of Gosstroy, Morzashita

(the Flood Defence Agency of St Petersburg) and NEDECO. The majority of the people attending the meetings had a long history of engagement with the Barrier Project and knew the different parties in the field. Over 125 individuals and organizations from the following stakeholder groups were identified:

- local governmental authorities;
- national governmental ministries with authority in the project area;
- relevant environmental regulatory authorities;
- representatives of the potentially affected local population, including local community organizations and groups who might be affected by the project;
- representatives of non-governmental organizations (NGOs);
- representatives of relevant academic or scientific bodies;
- representatives of the Finnish Ministry of the Environment;
- representatives of the EBRD; and
- representatives of the press.

A brochure, in Russian and English, was prepared to provide the stakeholders with background information on the Project. The document contained a brief description of the proposed Project and possible Project alternatives, the existing environment and the objective and approach of the environmental impact study. It contained a list of potential environmental impacts, an annotated Table of Contents of the proposed environmental impact statement, and a timetable for the remaining part of the study and further public consultation and information disclosure activities (NEDECO, 2002a). The information brochure was distributed before the public consultation meeting, together with an inquiry form on which comments or suggestions could be presented. Announcements of the public consultation were made in the local papers and in a 15-minute radio interview on the local radio.

THE FIRST PUBLIC CONSULTATION MEETING

The first Public Consultation Meeting took place on 31 January 2002 in a central location in the city. All identified stakeholder groups were well represented. At the start, a number of short presentations were given in which the project was further introduced to the public. More detailed information on the project and its history was provided by means of a small exhibition, consisting of poster boards displaying information on the Project and past environmental studies.

After the introductory presentations, the floor was given to the public and eleven people made use of the opportunity to express their worries and concerns about the Project. In addition, thirty written comments were received. During a break a group of specialists organized the remarks made into a summary and made a list ranking the key environmental issues and concerns as brought forward by the public. This list was presented and the audience was asked to comment on the summary and conclusions. At the end of the meeting a press conference was held.

Most participants in the discussions were in favour of the completion of the Flood Protection Barrier and expressed the view 'the sooner the better'. Leaving the Barrier in its present state was considered to be an environmental risk. None of the comments given expressed the wish to stop construction or to remove the existing works. Many presentations and comments stressed the point that a very large number of environmental studies had already been carried out since the 1970s, culminating in the Report of the International Commission (International Commission, 1990) and the Pre-feasibility Report (Gibb, 1997). It was strongly argued that the available studies should be used as much as possible. It was advised that the available expertise and experience concerning the Barrier and the environmental state of Neva Bay be optimally used by involving local scientists in the impact assessment as much as possible.

Environmental impacts that were of most concern to the public and that were specifically mentioned during presentations and in written comments were, in order of importance: impacts related to sedimentation and displacement of polluted sediments, changes in water quality in the Bay and the adjacent Gulf of Finland and the possible resulting impacts on the recreational beaches and fish stock/fish migration routes. A further concern raised was the much delayed completion and improvement of the waste water treatment of the city per se, while its urgency

had already been identified in the original design studies and had been confirmed in later studies. One alternative to the proposed Project was proposed, consisting of leaving out the gates in the main navigation lock C1, with the arguments that the Barrier with a remaining permanent gap of, e.g., 5000 m^2 would already reduce flood levels to manageable heights, the flows would be affected less, and the cost of completion could be significantly reduced.

During the meeting specific attention was paid to the flow of information. It was considered very important by the public, but also by the press, to have timely and sufficient access to the relevant information.

IMPACT OF THE FIRST PUBLIC CONSULTATION MEETING

The discussion during the First Public Consultation showed a concern for both adequate flood protection and protection of the environment. For some, the image of the environmental problems of the 1980s and the associated mistrust and emotions regarding the Barrier were still very much alive, even though studies since then had shown that insufficient water treatment and not the Barrier as such was the main cause of the deteriorated environment of Neva Bay. However, the provision of easily accessible material to the audience, the open discussion and the clear intention to take the concerns and suggestions of the public serious in the definition of the actual EIA scope helped to further increase confidence in the proposed study.

Based on the outcome of the meeting, the scope of the EIA was finalized. Some 20 Russian scientists with proven expertise on specific subjects of study were involved in the EIA, with the explicit understanding that the overall responsibility for the analysis remained with the EIA consultant. The alternative was included in the study by simulating the damping characteristics of the barrier with openings of various size and different flood rise and fall intensities. Care was taken that the analysis addressed the specific potential problems mentioned and provided explanations, answers and possible mitigation, whenever applicable. As part of the continuing information process, the Scoping Report and the Meeting Report were made publicly available through the internet (NEDECO, 2002a). This was published in local papers and communicated to interested parties.

The extended team evaluated the environmental baseline and then worked in two groups, one assessing the environmental impact of the Barrier during operation (in completed state), the other the environmental impacts during construction. With the involvement of the local specialists and the regular meetings with the full team, the knowledge base clearly increased, which improved the overall analysis and increased its comprehensiveness. The tuning with the activities that were being carried out in the parallel Technical Study, in particular the environmental modelling, further enhanced the completeness and depth of the assessment. During the period of work, relevant information on the EIA and its progress was provided by several articles in the local and national press, while further information was provided through the internet. With the relevant expertise more readily available, the analysis and assessments were conducted in about four and a half months following the First Public Consultation.

SUMMARY OF THE EIA RESULTS

More extensive conclusions of the EIA are contained in (NEDECO, 2002c). For completeness, the main conclusions are listed below (NEDECO, 2002b).

- The completion and the operation of the Barrier are designed to meet relevant Russian and EU environmental and health and safety standards.
- Completion of the Barrier will have no significant negative impacts on the environment of the Neva Bay and adjacent Gulf of Finland, particularly relating to water quality.
- There are at present serious environmental problems in Neva Bay and the adjacent part of the Gulf of Finland, but these are related to other activities in and around the City of St Petersburg. To remedy these, the capacity and level of waste water treatment in the St Petersburg region should be increased, for both domestic and industrial effluents.
- There will be a partial return to Pre-Barrier conditions in the distribution of the water flows north and south of Kotlin and the general circulation.

- The safety of navigation to and from the ports will increase due to separation of short distance and long distance (large-vessel) navigation lanes via C2 and C1, respectively.
- Through appropriate mitigating measures, adverse impacts during construction can be adequately minimized.
- These conclusions agree with those of earlier studies, notably the Report of the International Commission (International Commission, 1990) and the Pre-feasibility Report (Gibb, 1997).

PUBLICATION AND INFORMATION PROCESS OF THE EIA RESULTS

On 19 June 2002 the Draft Environmental Impact Report was made available to the public, through publication on the web sites of Gosstroy and Morzashita. It included a separate Executive Summary (NEDECO, 2002b), specifically intended for non-specialist stakeholders. On the latter web site, materials on the project and the First Public Consultation had already been available since early March 2002. The Final EIA Report (NEDECO, 2002c) was accessible to the public for comments as of Friday 5 July 2002. For a period of several months hardcopies of the report could also be read freely and commented upon in the local offices of Gosstroy in St Petersburg and Moscow, at the Morzashita office in St Petersburg, in four key public libraries in St Petersburg and Kronstadt, at the Business Information Centre of the EBRD in London and at the local EBRD office in St Petersburg. Forms for comments were provided in parallel, with an invitation to submit comments.

On 17 and 18 July a well-announced exposition of the results of the EIA was held in the Hall of Investment Projects in St Petersburg. Besides posters, computer animation and brochures were used to illustrate the results of the study. Again written comments were solicited on special forms.

FINAL PUBLIC CONSULTATION

The Final Public Consultation on the project took place on Friday, 19 July 2002 in the House of Scientists in St Petersburg, 30 days after the draft EIA report had been made public. The meeting was announced in advance in a local newspaper and in the official *St Petersburg City Bulletin*, while invitations were sent directly to the identified stakeholders like municipalities in the project area, representatives of NGOs, etc. The meeting was attended by over 140 people and, like the Scoping Meeting, had considerable press coverage. Those attending the meeting were provided with the printed Executive Summary of the EIA report, available in both Russian and English. In a number of short presentations the main results of the EIA study were presented to the public by members of the EIA study team.

After the presentations the public was invited to comment on the results of the EIA. Twenty people participated by presenting oral comments and questions, while a similar number provided comments and questions in writing. These reactions came from all types of involved parties: local municipalities, city government, NGOs, the scientific community, interested individuals and the press. During a break the written comments were organized and summarized by an ad-hoc committee. Replies to the comments were given after the break and the main outcomes of the afternoon's discussions were summarized. A press conference was held, which was broadcast on local and national TV and radio stations (Fig. 41.4).

A few voices claimed that flooding is a 'natural flushing feature' of Neva Bay and should not be interfered with, stating that the Barrier should therefore be

Fig. 41.4   Press coverage of the Final Public Consultation.

removed. Apart from this, the emotions which had dominated the discussion in the 1980s, were largely absent. Various issues were again raised, such as reduction of faecal pollution along the beaches, eutrophication in areas with low velocities, impacts of construction activities on fish migration, and the impact of resuspension of polluted bottom sediments. All these had been addressed in the analysis, though, including mitigating measures when appropriate.

Most parties noted that the existing and potential project related problems in the area had indeed been recognized and addressed, and agreed with the general conclusions of the EIA. It was specifically mentioned and appreciated that the ideas put forward during the first Scoping Meeting such as building on existing relevant studies, involving local expertise, prioritizing the key issues, and analysing a further alternative, had indeed been taken into consideration in the EIA. Equally, the readiness to openly publish the extensive EIA materials via the internet and the openness of the consultation meetings was explicitly appreciated.

FOLLOW UP AND PRESENT SITUATION

While the EIA materials are still publicly accessible for comments through the internet, the hard copy materials in the public libraries remained publicly accessible for information and commenting until Autumn 2002. An unknown number of people have made use of the opportunity to inspect the materials and only a limited number of people provided a written response. In September, the response submitted after the July meeting was analysed and evaluated and added to the EIA report by EBRD and Gosstroy to form an integral part of the project documentation.

The full documentation of the four part Feasibility Study, of which the EIA and Public Participation was one, was evaluated and approved by EBRD as the leading bank for a consortium consisting of EBRD, the Nordic Investment Bank and the European Investment Bank. In December 2002, on behalf of the consortium of banks and the Government of the Russian Federation, respectively, EBRD and Gosstroy signed a loan agreement for the Project for the accelerated completion of the St Petersburg Flood Protection Barrier. The expected construction time from the start of actual construction is estimated to be about 5 to 7 years, following contractual and implementation preparations in 2003.

CONCLUSIONS REGARDING THE PUBLIC PARTICIPATION

As part of the present EIA study, an extensive and open interaction process was conducted throughout the study, involving a First Public Consultation to identify the concerns and finalize the scope at the start of the EIA, and a full accounting of the results in an Exposition and Final Public Consultation at the end.

The following conclusions can be drawn:

• The First Public Consultation provided much useful feedback from the stakeholders in terms of suggestions for extending and prioritizing the EIA topics, suggestion of a further alternative and the recommendation to involve local expertise. This has improved the quality of the EIA process, the quality of the EIA results and has led to a much improved Project;

• The active information flow in the form of topical EIA brochures, news articles, publication of Meeting reports and the publication of the full EIA report on the internet and via libraries has provided the public and stakeholders with access to the factual information. This has served as a basis to renounce possible historic views resulting from rumours or a simple lack of information;

• In the Final Public Participation, the Project Sponsor Gosstroy and the EIA consultant openly presented the previously published results and discussed these with stakeholders. In the process, the stakeholders expressed their appreciation for the openness and the fact that their earlier comments had been acted upon by the Project Sponsor and EIA consultant. This active approach has led to enhanced support for the project, and also to its sustainability;

Our overall conclusion is that the key factor in the success of the public information process has been the open character of the process, the systematic analysis and the free and full availability of the analysis, whereby its results and conclusions have been available to all.

The loan agreement between the consortium of banks and the Russian Federation can be seen as an implicit further confirmation of the importance and success of the public participation to enhance the quality and sustainability of this large-scale infrastructure project.

ACKNOWLEDGEMENTS

The present study was conducted by a NEDECO consortium with WL | Delft Hydraulics as leading partner (EIA consultant). The Finnish Environment Institute (FEI) was second partner. It was commissioned by the Russian State Construction Ministry (Gosstroy, Project Sponsor) and funded by the Japan-Europe Cooperation Fund (JECF) at the EBRD. It was conducted in the period January–July 2002. The authors furthermore wish to thank all involved Russian scientists (see preface to the Final Report (NEDECO, 2002c) for their contribution to the EIA, and the head and staff of Morzashita for their technical and logistic support during the study.

REFERENCES

Delft3D-FLOW. 2003. *Validation document 3D hydrodynamics modelling software Version 1.0*. Delft: WL | Delft Hydraulics www.wldelft.nl/soft/d3d/intro/validation/valdoc_flow.pdf.

EBRD (European Bank for Reconstruction and Development). 1992. *Environmental Procedures (Revised 1996)*. London: EBRD.

Gibb. 1997. *St Petersburg Flood Protection Barrier. Pre-feasibility Study*. Final report prepared for EBRD. Reading: GIBB Ltd.

Kjorven, O. 1997. *The Impact of Environmental Assessment: A Review of World Bank Experience*. World Bank Technical Paper No. 363.

Klevanny, K. A., and E. V. Smirnova. 2002. Simulation of current and water pollution changes in the Neva Bay after completion of St Petersburg Flood Protection Barrier. *Environmental and Chemical Physics* **24**, 3, 144–50.

International Commission of Experts. 1990. *Leningrad Flood Protection Barrier*. Final Report, prepared for the USSR State Committee on Science and Technology and the USSR State Committee on Nature Protection. Delft: WL | Delft Hydraulics.

NEDECO. 2002a. *Initial Public Consultation – Scoping Meeting. Information documents and Reports*. www.morzashita.spb.ru/nedeco_en.html.

NEDECO. 2002b. *St Petersburg Flood Protection Barrier. EIA Executive Summary*. Document for Final Public Consultation (Russian and English), 22 pages. www.ebrd.com/projects/index.htm or www.morzashita.spb.ru/nedeco_en.html.

NEDECO. 2002c. *St Petersburg Flood Protection Barrier*. EIA Final Report. Prepared for Gosstroy and EBRD, fully available through www.morzashita.spb.ru/nedeco_en.html.

Russian Federation. 1995. Federal Law 'On the State Environmental Review'. Law No. 174-FZ of 23.11.1995, Moscow.

Russian Federation. 2000. Regulation 'On Impact Assessment of Proposed Operation on the Environment in the Russian Federation' (annex to the Order of the Russian Federation State Committee for Environmental Protection No. 327 of 16.05.2000). Moscow.

Russian Federation. 2002 Federal Law 'On Environmental Protection'. Law No. 7-FZ of 10.01.2002, Moscow.

World Bank. 2002. *Public Consultation in Environmental Assessment 1997-2000, Findings from the Third Environmental Assessment Review*. Environmental Department Working Paper, No. 87, Environmental Assessment Series. Washington: World Bank.

Part V · The Venice Lagoon: physical processes, sediments and morphology

42 • Introduction: physical processes, sediments and morphology of the Venice Lagoon

T. SPENCER, C. A. FLETCHER AND J. DA MOSTO

Venice Lagoon is the largest lagoon on the margins of the Adriatic Sea, having an area of about 550 km^2 (with a drainage basin of 1877 km^2) and an average depth of c. 1 m. Most of the lagoon is occupied by a large central water body (c. 350 km^2) and extensive intertidal salt marshes (c. 33.5 km^2 in 1999) of varying typologies (Bonometto, chapter 52). The subtidal lagoon, partially vegetated by macroalgae and sea-grasses, gives way to tidal flats (*velme*) which lie within +/-10 cm of mean sea level (Day *et al.*, chapter 50). A series of saltmarsh (*barene*) associations then characterize elevations between +10 and +32 cm above mean sea level (Cecconi, chapter 51; Bonometto, chapter 52). These wetlands are of international importance, being included in the Ramsar Convention (1971), and recognized through Natura 2000 and the Habitats Directive within the European Community.

The lagoon is connected to the Adriatic Sea through three relatively deep inlets. The most northerly, Lido, is the largest (900 m wide with a mean depth of c. 12 m); the central inlet, Malamocco, is the deepest (18 m mean depth) with a width of c. 450 m; and the most southerly, Chioggia, is c. 400 m wide, 8 m deep and has a tidal flux of about half that of the other two inlets. Over 90 per cent of the total variance in average flux through the inlets is due to tidal forcing (typical exchanges over an individual tidal cycle are c. 500 x 10^6 m^3). Low-frequency flux variations, at timescales from 20 to 40 days and max-imized in winter, are generated by local wind effects. Thus, for example, surface wind stress from the northeasterly *bora* creates a north–south lagoon sur-face slope. Under these conditions, higher water lev-els in the southern lagoon, and lower levels in the northern lagoon, compared to the open sea, strengthen outflow at the Chioggia inlet and rein-force inflow at the Lido (Gacic, chapter 49). The aver-age residence time of the whole basin is about 30 days but these times are much reduced under the southeasterly *sirocco* and, particularly, the *bora* when the residence time near the inlets can be as low as three days (Umgiesser *et al.*, chapter 47). Freshwater discharge into the lagoon has been estimated at c. 100 m^3 s^{-1} (40 m^3 s^{-1} river runoff, 60 m^3 s^{-1} industrial water discharge), although with significant seasonal and inter-annual variability (Gačić, chapter 49).

Understanding such a topographically, physically and ecologically complex system is a considerable challenge, particularly where lagoon-wide monitor-ing of many physical parameters has only been estab-lished in recent years. One approach is via box models, as typified by Di Silvio (chapter 43), in which a set of morphological elements (sea, channels, shoals, marshes) are allowed to exchange water and sedi-ments according to simplified relationships, assumed to represent the more complex processes occurring in reality. Alternatively a more mechanistic approach attempts at developing a thorough understanding of the hydrodynamics and morphodynamics of each morphological unit. Here one of the biggest chal-lenges is the difference in timescales of hydrodynamic and morphologic responses (Seminara *et al.*, chapter 44; Tambroni *et al.*, chapter 45). Hydrodynamic mod-els have been progressively applied to the 'Venice problem' since the 1970s, progressing from simple reproduction of city water levels to 1-D and 2-D mod-els able to model tidal propagation into channels and across tidal flats. Both finite difference and finite ele-ment models have been developed. An example of the former is Gozzi and Menel Lemos' (chapter 46) three-layer grid application of the model 'Delft 3-D Flow' to the city and the lagoon. Sub-lagoon scale grid models are required because wide tidal flats lie next to quite large salt marshes which periodically wet and

Flooding and Environmental Challenges for Venice and its Lagoon: State of Knowledge, ed. C. A. Fletcher and T. Spencer.
Published by Cambridge University Press, © Cambridge University Press 2005.

dry during the tidal cycle and because the network of tidal channels also supports a widespread network of smaller creeks; high resolution grids are required to resolve these complexities under low flow depths but wide area application produces large computational problems (D'Alpaos and Martini, chapter 48). Another way forward is through finite element modelling; this allows for more flexibility, with its subdivision of the system into triangles varying in form and size. Here Umgiesser *et al.* (chapter 47) describe the development, calibration and validation of the SHYFEM model for the Venice Lagoon. This model also shows a characteristic of many models developed from the 1990s, the coupling of hydrodynamic models to other modules in order to perform additional sediment transport, water quality and ecological assessments of this long-modified ecosystem.

Coastal lagoons are transitional environments between fully terrestrial and fully marine conditions. In the absence of intervention, their long-term tendency is to infill with sediments. However, in the case of the Venice Lagoon, long-term human manipulation has been to maintain the marine element in this balance. The Brenta and Piave rivers formerly discharged into the lagoon and formed large areas of deltaic wetlands, but these rivers, and others, were diverted around the lagoon, principally in the fifteenth to seventeenth centuries. The last great work of the Venice Republic ended in 1791 with the embankment of the entire lagoon margin (*conterminazione lagunare*), including the building of the great seawalls (*murazzi*) along the Adriatic coast. In chapter 52, Bonometto argues that these manipulations led to the establishment of a new balance between 'the natural and manipulated protective dynamics' and that it has been the interventions from the mid nineteenth century onwards that have been beyond the 'system's supporting capacity'. These later, cumulative processes have been (i) the construction of the jetties at the three inlets (Malamocco 1850, Lido 1890, Chioggia 1914); (ii) the separation from the lagoon of several fish breeding areas (1900–30); (iii) the reclamation of large areas of shoals and marshes (1920–50); (iv) the dredging of two large navigation canals (1925 and 1963–5); and (v) increased land subsidence (4 mm yr^{-1}) due to excessive groundwater withdrawal (1950–70) (Di Silvio, chapter 43).

Prior to 1850, the inlets were much wider and shallower than at present and the external part of the inlet (the *fuosa*), an important sediment source area, was shallower than the tidal channel inside the lagoon (Di Silvio, chapter 43). It is probable that the tidal flow fields were largely symmetrical, allowing the finer sediments carried out to sea with the ebb tide to be reintroduced into the lagoon during the next flood tide (D'Alpaos and Martini, chapter 48). The inlet deepening that accompanied the restriction of inlet width both removed the shallow water sediment sources (Di Silvio, chapter 43) and effectively moved the inlet mouths beyond the surf zone, where sediments could be resuspended for transport. The jetties have also interfered with the longshore movement of sediments, trapping material and promoting shoreline advance updrift of the inlets and accelerating coastal retreat downdrift to the south (Seminara *et al.*, chapter 44). At the same time as diminishing sediment inputs into the lagoon, the jetties now force the ebb tide into a seaward jet which is maintained out into the open sea, preventing the local recycling of fine sediments (D'Alpaos and Martini, chapter 48). Whilst there is dispute as to the magnitude of sediment losses, and the apportionment of those losses between coarse and fine sediments (see, for example, chapters by Seminara *et al.* (44); Day *et al.* (50) and Cecconi (51)), it is clear that the lagoon sediment balance is severely in deficit.

This deficit has given rise to a situation in which the lagoon floor has deepened and flattened but where the tidal channels are filling up by siltation. The 1901 hydrographic chart suggests that the central part of the lagoon was 50–100 cm shallower than today, in accordance with Denaix's early nineteenth-century observations that, during low tide, significant portions of the lagoon dried, showing the convoluted structure of the channel network (D'Alpaos and Martini, chapter 48). In addition in this region, excavation of two large canals (Canale Vittorio Emanuele, *c.* 1925, and Canale Malamocco Marghera, *c.* 1963) produced strong transversal currents across the original tidal network, with consequent siltation of natural and artificial channels and erosion of adjacent shallows (Di Silvio, chapter 43).

Saltmarsh edge retreat rates in the Venice Lagoon have accelerated in recent years. This appears to be due to both increased wave attack and tidal channel lengthening and widening. Both these processes may be related to the increase in tidal prism and amplitude of the tidal wave consequent upon sea level rise and inlet deepening. Wetland loss is also taking place as a result of insufficient sediment supply to meet the vertical surface accretion needed to keep pace with sea level rise; this problem will be exacerbated by near-future accelerations in sea level rise (Day *et al.*, chapter 50). These quasi-natural processes behind loss of intertidal habitat have been augmented by extensive reclamations of wetland for both agriculture and industry: the industrial harbour at Marghera (1919), the second industrial area of Marghera (1925), the last wetland reclamation for agriculture (1930), the impoundment of large areas for aquaculture (1940 and thereafter), the international airport at Tessera (1957), and the third industrial area of Marghera (1963). Overall, therefore, during the twentieth century wetland area in Venice Lagoon decreased by two-thirds to 4000 ha (Day *et al.*, chapter 50.)

These losses can be set alongside the creation of 500 ha of artificial islands, mudflats and marshes since the end of the 1980s, using sediments (totalling 6×10^6 m^3) recovered from the maintenance dredging of channels (Cecconi, chapter 51). This work has not been without its difficulties – the elevation window for creating marshland is very small under the tidal regime of the lagoon – or its critics, specifically on sediment emplacement methods and the design and construction of artificial marsh margins. It has been argued that there is a need to avoid standard, 'rigid' techniques and materials throughout in the lagoon in favour of more 'plastic' reconstructions, sensitive to the diverse characteristics of particular parts of the lagoon (Bonometto, chapter 52). The dredging and reconstruction programme has been combined with other measures to protect and restore natural marshes and mudflats, alongside attempts at regulation of erosion due to mechanized clam fishing and boat traffic in the lagoon. These include transplantation of vegetation onto mudflats and saltmarshes, additions of mineral and/or organic materials to marsh surfaces, installation of permeable sediment fences to reduce wave fetch lengths/retain sediments. However, calls for larger scale ecological engineering, and particularly the re-introduction of riverine sediments into the lagoon to create new areas of wetland, have been proposed, but contested, and thus remain subject to further evaluation and review at the present time (Day *et al.*, chapter 50; Cecconi, chapter 51).

43 • Sediment balance, morphodynamics and landscape restoration

G. Di SILVIO

SUMMARY

In its present condition, the Lagoon of Venice is subject to conspicuous erosion of shoals, reduction of salt marsh surfaces and siltation of tidal channels. The main reason for this evolution is the substantial *sediment deficit* of the lagoonal basin compared to 150 years ago (i.e. before the construction of the jetties at the inlets), and even more with respect to the fourteenth century (i.e. before the diversion of the rivers). Recent evolution is also dependent on other forcing factors, such as the anthropogenic modification of *tidal currents* (due to the construction of the large navigation canal in the central lagoon) and, up to a point, the long-term variations of *relative mean sea-level*, both in terms of eustasy and soil subsidence. While eustasy and soil subsidence tend to increase the flooding frequency of artificial intertidal surfaces (e.g. the street levels of Venice), as far as vegetated salt marshes are concerned the increase of relative sea-level does not give rise to a corresponding decrease in their elevation (which is somehow compensated by a corresponding continuous trapping of sediments) but rather to a decrease of their surface area.

Present and past morphology of the lagoon can be interpreted by a simple conceptual model which simulates morphodynamic processes. Besides relative mean sea-level change, the long-term conceptual model incorporates the effects of tidal currents (dominant in the channels), local wind waves (dominant in the shoals), sea wave climate (dominant in the littoral zone near the inlets) as well as anthropogenic interventions (dredging of canals, sediment dumping, wave breakwaters and other constructions). The same model indicates the measures to be taken for slowing down or even halting degradation processes. In fact, landscape restoration projects being carried on in the Lagoon of Venice in recent years basically consist of the reconstruction of shoals and marshes by the mechanical relocation of sediments. For mitigating the present erosive tendencies, however, attention should also be given to possible modifications to the prevailing boundary conditions.

INTRODUCTION

The Lagoon of Venice has experienced over recent centuries a progressive evolution from a predominantly terrestrial towards a predominantly marine landscape, due to a remarkable shift from a positive to a negative balance of sediments entering and leaving the lagoon. This balance inversion has been determined by radical changes in boundary conditions that took place first in the fifteenth and sixteenth centuries and later in the nineteenth and twentieth centuries.

The first change consisted in the re-direction away from the city, and eventually in the complete diversion from the basin, of all the major rivers (Brenta and Piave) flowing into the lagoon. This objective was attained through successive interventions by the government of the Serenissima Repubblica. They were concerned about the siltation of the lagoon which put in jeopardy the accessibility to the port; the requirements of military defence; and even the healthiness of some peripheral areas of the basin.

The second fundamental change, also imposed by the needs of navigation, was caused by the construction of jetties in the three inlets of the lagoon (Malamocco, Lido and Chioggia), accomplished in the period 1850-1920. The strong reduction of the inlet width determined by the jetties inevitably caused a corresponding increase of inlet depth. This was exactly as expected, in order to improve the accessibility to the port. On the other hand, the bot-

Flooding and Environmental Challenges for Venice and its Lagoon: State of Knowledge, ed. C. A. Fletcher and T. Spencer. Published by Cambridge University Press, © Cambridge University Press 2005.

tom deepening on the seaward side of the inlet produced a substantial decrease in sediment mobilisation by sea waves. That jetties have produced a remarkable drop of the suspended sediments at the seaward ends of the inlets can be inferred by comparing bottom profiles along the inlets before and after construction. While in 1810 (the Denaix survey), the maximum depth of the tidal channels (c. 10 m) was much deeper than the adjoining sea (c. 6 m), after the construction of the jetties the channel depths progressively increased towards the open sea (>10 m). This led to a strong reduction in the sediments entering the lagoon during the flood tide, which were no longer sufficient to compensate the sediments leaving the lagoon during the ebb tide. The consequent deficit in the sediment balance has given rise to a general erosion of the lagoonal basin which is still in progress. Erosion concerns especially the shallow areas (shoals and marshes) rather than the tidal channels, which on the contrary are still affected by a silting phase. The erosion of the shallow areas shows itself in a remarkable deepening of the shoals and a dramatic reduction of the surface area covered by halophytic vegetation (salt marshes). This implies a strong decrease of the intertidal areas (over the entire range of submegence frequencies) as well as a deterioration of the natural network of tidal channels, with a general tendency towards a progressively flatter basin.

The most recent erosion of shallow areas and progressive flattening of lagoon morphology have also been facilitated by anthropogenic changes which took place in the last century. Among them, were the excavation of two large canals (Canale Vittorio Emanuele, c. 1925, and Canale Malamocco Marghera, c. 1963) which produced strong transversal currents across the original tidal network, with consequent siltation of natural and artificial channels and erosion of adjacent shallows. Also the accelerated soil subsidence in the 1950s and 1960s, associated with excessive withdrawal of the groundwater, has contributed to increasing the depth of the shoals and to reducing the surface area of salt marshes. The larger uniformity of the lagoon morphology, produced by the recent interventions, is reflected in a lower biodiversity and even in the removal of several species from certain areas. Moreover, the large expanse of water

that is the present lagoon is very different from the traditional landscape bequeathed to us by old paintings and literature.

For all these reasons, great public pressure has been put on the Magistrato alle Acque to undertake the morphological restoration of the lagoon. An important part of the Venice project is devoted to this objective, while a certain number of interventions have been already carried out in several parts of the lagoon (see Cecconi, chapter 51). The causes of the recent deterioration of the lagoon morphology are numerous and their effects combine in complicated patterns. Thus it is often difficult to recognize the mitigation measures that need to be taken to invert the degradation processes in a coordinated and sustainable way, taking into account and assessing all the possible positive and negative interactions. In this chapter an effort is made to set forth rational criteria for choosing the necessary interventions and for evaluating their effectiveness. This will be made through a relatively simple conceptual model which incorporates the most important erosion/transport/deposition mechanisms and includes all the relevant boundary conditions.

THE TRANSPORT CONCENTRATION MODEL

A very useful tool (Di Silvio, 1989) which has been applied to the morphological modelling of the Venice Lagoon (Di Silvio, 1991), and then extended to other semi-empirical models (de Vriend, 1996; Wang et al., 1996; Stive et al., 1996), is the concept of *transport concentration* which controls the evolution and equilibrium configuration of tidal lagoons. The morphodynamics of tidal lagoons are regulated by long-term (net) sediment fluxes between littoral and tidal channels as well as between channels and shallows (shoals and salt marshes). Long-term fluxes can be expressed in advective and/or dispersive form as a function of the local transport concentration, namely long-term averaged concentration of sediments.

Long-term sediment fluxes are obtained by averaging the short-term transport-diffusion equation over a relatively long period (say, one year). In a two-dimensional form, long-term sediment fluxes in the direction x and y are expressed as:

$$T_x = -Dh_e \frac{\partial C}{\partial x} + UC \qquad\qquad Eq.\ 43.1$$

$$T_y = -Dh_e \frac{\partial C}{\partial y} + VC \qquad\qquad Eq.\ 43.2$$

where C is the long-term averaged concentration (transport concentration) in a given location of the lagoon; U and V are the components of the residual (Eulerian) waterflow respectively in the x and y directions; h_e is the local water depth and D a local dispersion coefficient which incorporates the residual (Lagrangian) transport, corresponding to all the non-linear deviations from the averaged quantities in the original short-term advective-diffusion equation.

Dispersive transport at tidal scale is due to a variety of mechanisms, ranging from turbulence to the non-uniform velocity distribution over the estuary's depth and width (Fisher et al., 1979). At intertidal scale, however, the long-term dispersive transport is basically due to the alternate tidal flow between channels and adjacent shoals (*trapping and pumping*). The scalar quantity D, called the *intertidal dispersion coefficient*, has the order of magnitude of several hundreds of [m²/s]. This is found to be proportional to the local squared tidal velocity (Dronkers, 1978) and can be evaluated in each location of a lagoon by an advective-diffusion tidal model. Long-term sediment transport may also be determined by the asymmetry of the tidal velocity during the ebb and flood phases. In a branching channel network like the Lagoon of Venice, however, this component appears to be much smaller that the trapping and pumping effect.

If the sediments of the lagoon are composed of non-uniform particles, having a different behaviour depending on grainsize, Eqs. (43.1) and (43.2) should be written for the concentration C_i of each grainsize class (i = 1, 2, 3, ..., N). Moreover, if the material is very fine, a distinction has to be made between the actual time-averaged concentration C_i present in a location and the *equilibrium* time-averaged concentration C_{eqi} in the same location (Di Silvio and Padovan, 1998). For sake of simplicity, however, in the present chapter the hypothesis will be made of uniform material (silt), coarse enough to assume that the actual concentration coincides with the equilibrium

concentration, $C = C_{eq}$. This is true when the settling time of particles is much shorter than the tidal cycle, so that the sediment concentration practically depends solely on the local hydrodynamics (short adaptation distance). In this hypothesis, the long-term evolution of the local depth h_e is given by the Exner equation:

$$\frac{\partial h_e}{\partial t} = \frac{\partial T_x}{\partial x} + \frac{\partial T_y}{\partial y} + \alpha \qquad\qquad Eq.\ 43.3$$

where $\alpha = \alpha_e + \alpha_s$ is the combined rate of eustasy (sea level rise) α_e and soil subsidence α_s. For solving Eq. (43.3) an expression for the long-term averaged concentration (transport concentration) C in Eqs. (43.1) and (43.2) is needed.

TRANSPORT CONCENTRATION IN CHANNELS AND SHOALS

Transport concentration C in a given location is obtained by averaging the instantaneous concentration c(t) over a long period of time. The instantaneous concentration can be computed by any appropriate transport formula, as a function of the local currents (especially active in the channels) and of the local waves (especially active on the shoals). As far as currents are concerned, a number of formulae, more or less theoretically founded, have been proposed in the last hundred years to express the sediment transport as a function of the shear stress or flow velocity. In a first category of formulae, sediment transport appears to be proportional to the difference, raised to a certain power m, between the actual flow velocity and a critical flow velocity (binomial formulae; e.g. Ackers and White, 1973). In a second category, sediment transport results appear directly proportional to the *n*-th power of the flow velocity, but with an exponent n much larger than m (monomial formulae; e.g. Brown, 1950). Let us consider any monomial transport formula for the total sediment transport and let us call q_s the volumetric sediment discharge per unit width produced by the corresponding water flow q = v × h, where v is the water velocity averaged over the depth h. As monomial formulae provide:

$$q_s \propto v^n \qquad\qquad Eq.\ 43.4$$

where n = 4 – 6, the instantaneous sediment concentration (n = 5) can be written:

$$c = \frac{q_s}{q} \propto \frac{v^n}{q} \propto \frac{q^{n-1}}{h^n} \propto \frac{q^4}{h^5} \propto \frac{Q^4}{B^4 h^5} \qquad Eq.\ 43.5$$

where Q is the water discharge in a channel of width B. In a 'short' lagoon like Venice, the instantaneous discharge q (or Q) in a given location (both in flood and ebb conditions) is substantially proportional to the area of the corresponding drainage basin, while the statistical distribution of q (or Q) over the year is practically linear. By averaging Eq. (43.5) over the year one finds the long-term average concentration produced by tidal currents in any location:

$$C_c = \frac{f_c}{h_e^5} \qquad\qquad Eq.\ 43.6$$

where h_e is the local tide-averaged depth and the coefficient f_c is proportional to the 4-th power of the tidal prism.

An expression similar to Eq. (43.6) can also be obtained for the long-term average concentration produced by waves by assuming that the proportionality (4) still holds if v is the instantaneous orbital velocity produced by the waves near to the bottom (shallow waters) and that v is proportional to q/h. This assumption implies that the alternate sediment flux q_s during the passage of a wave is in phase with the corresponding depth-integrated water flow q. Under the hypothesis made, the instantaneous concentration during the passage of a wave is:

$$c = \frac{q_s}{q} \propto \frac{v^n}{v \cdot h} \propto \frac{v^{n-1}}{h} \qquad Eq.\ 43.7$$

By evaluating the value of the maximum orbital velocity v_m by:

$$v_m = \frac{1}{2}\sqrt{\frac{g}{h}} H_w \qquad\qquad Eq.\ 43.8$$

where H_w is the significant wave height, one finds the concentration on the shoals averaged over the wave period:

$$c = k_1 \frac{H_w^{n-1}}{h^{\left(\frac{n-1}{2}+1\right)}} \qquad\qquad Eq.\ 43.9$$

with k_1 including all the constants of proportionality. In order to obtain the long-term average concentration in a given location, Eq. (43.9) should be averaged over, say, a few years.

The height of significant wave H_w in shallow water and within a limited range of length of fetch, F_e, average depth along the fetch, h_s and wind velocity, u_A(F_e =1000-5000 m, h_s = 0.5–2.0 m, u_A = 15–25 m s^{-1}) , can be computed through the formula (Dal Monte and Di Silvio, 2004):

$$H_w = 3.2 \cdot 10^{-3} \cdot F_e^{0.32} \cdot h_s^{0.5} \cdot u_A^{1.45} \qquad Eq.\ 43.10$$

which approximates the formula of Bretschneider (1969).

By substituting in Eq. (43.9), with n = 5, one finds, approximately:

$$c = k_3 \frac{h^2 F_e^{4/3} u_A^6}{h_s^3} \qquad\qquad Eq.\ 43.11$$

By assuming that the average depth along the fetch h_s is proportional to the local depth, h, the wave-averaged concentration is:

$$c \propto \frac{F_e^{4/3} u_A^6}{h} \qquad\qquad Eq.\ 43.12$$

Equation (43.12) indicates that the concentration produced by waves depends on the local depth h (varying with the tidal level) and on a combination of fetch length and wind velocity. For the sake of simplicity we shall consider tidal level and wind velocity, both time-dependent quantities, as mutually independent. The duration of tidal elevation is almost linearly distributed over the lagoon, around its average value. Wind and fetch both depend on direction; however, their statistical distributions are again mutually independent (Dal Monte and Di Silvio, 2004). In conclusion, Eq. (43.12) can be integrated over the time and over directions to get the averaged value:

$$C_w = \frac{f_w}{h_e} \qquad\qquad Eq.\ 43.13$$

where h_e is the local tide-averaged depth and f_w a wave resuspension coefficient. Note that waves put sediments in suspension but their contribution to long-term transport is negligible: long-term transport of suspended sediments (on the shoals as well as in the channels) is produced, via the long-term averaged concentration, by intertidal dispersion and residual flow (Eqs. 43.1 and 43.2).

The value of the resuspension coefficients f_c in Eq. (43.6) and of f_w in Eq. (43.13) depends upon the statistics of water flow and, respectively, upon wind velocity and fetch length. Moreover, as q_s in Eq. (43.4) depends on sediment grain size, the values of f_c and f_w tend to decrease for coarser bottom composition. An appropriately smaller value of f_w can also account for the enhanced resistance of the bottom due to vegetation, either seagrasses (e.g. *Zostera marina*) on the shoals or terrestrial halophytic species on the salt marshes. Even more so, the vegetation canopy on the salt marshes practically prevents sediment resuspension, so that one can put here directly $f_w = 0$.

The total transport concentration C in any location is obtained in principle by summing up expressions (43.6) and (43.13):

$$C = C_c + C_w \qquad\qquad Eq.\ 43.14$$

It has been observed above, however, that C is dominated either by waves (on the shallows, where $f_c \cong 0$), or by currents (in the channels, where C_w rapidly decreases as the depth increases).

MORPHOLOGICAL EVOLUTION OF THE LAGOON OF VENICE

For predicting the lagoon evolution Eq. (43.14) should be introduced in the long-term balance (Exner) equation of sediments (Eq. 43.3), where the sediment transport components, T_x and T_y, are expressed as a function of C (Eqs. 43.1 and 43.2). By integrating Eq. (43.3) in space and time, one obtains the distribution of the transport concentration $C(x,y,t) = C_c + C_w$ and, from Eqs. (43.6) and (43.13), the corresponding distribution of the water depth $h_e = h_e(x,y,t)$.

The initial conditions $C(x,y,0)$ are provided by the distribution at $t = 0$ of the *water depth* h_e and of the *resuspension coefficients* f_c and f_w. As the hydrodynamics of waves and currents depend on the climatic and planimetric characteristics of the lagoon (tidal prism, fetch length, wind velocity) the values of f_c and f_w are not affected by the evolution of water depth and therefore remain reasonably constant during the computation. The resuspension coefficients however, may change if some modification is introduced by man, as for example, excavation of navigation canals, change of the tidal prism, change of vegetation cover or construction of wave breakwaters.

Boundary conditions are provided by the sediment *input* of the rivers (now diverted) flowing into the lagoon, by the sediment *output* due to dredging and by the transport *concentration at the seaward side* of the inlets. This last value basically depends on the wave climate in the sea and on the inlet water depth.

If no perturbation is introduced in the boundary conditions or in the resuspension coefficients, the morphological evolution described by Eq. (43.3) will continue until a stationary condition is reached ($\partial h_e/\partial t = 0$ all over the lagoon). Note that the model also accounts for the formation and destruction of salt marshes (Di Silvio and Padovan, 1998). The transformation of a shoal into a salt marsh takes place as soon as the bottom of the shoal reaches a certain elevation (around mean sea level) where halophytic aerial vegetation starts to thrive and f_w becomes zero. By contrast, the destruction of a salt marsh occurs when the depth of the bare shoals increases too much and the cliffed margin of the adjoining vegetated surface collapses under the action of waves. In this way salt marshes expand or lessen their vegetated surface area but change very little in their elevation. It should be noted, however, that the destructive processes of the marsh margin are produced by waves but only when the adjacent shoals continue to deepen.

The morphological model described above explains the evolution undergone by the lagoon of Venice after the diversion of the rivers and the consequent reduction of sediment input (fourteenth to sixteenth centuries). A simplified (zero-dimensional) version of the model has been also applied to the three main basins of the lagoon (Di Silvio, 1991) to simulate their most recent evolution (nineteenth and twentieth centuries). The model was able to reproduce the major effects of (a) the construction of the jetties in the three inlets (Malamocco 1850, Lido 1890, Chioggia 1914); (b) the temporary reintroduction in the southern lagoon of the river Brenta (1850-90); (c) the separation from the lagoon of several fish breeding areas (1900-30); (d) the reclamation of large areas of shoals and marshes (1920-50); (e) the dredging of two large navigation canals (1925 and 1963-5); and (f) the increased soil subsidence (4 mm yr^{-1}) due to excessive groundwater withdrawal (1950-70).

All these perturbations had consequences for the lagoon morphology and have been duly replicated by the model. As already mentioned, the construction of the jetties through the original outer delta (*fuosa*) gave rise to a strong reduction of the suspended sediment at the seaward end of the inlet. Consequently jetties produced a strong deepening of the shoals (water depth become almost double), an even stronger reduction of the surface area covered by salt marshes and a much less dramatic increase of the channel cross-sections. This general picture, however, appears more complicated because of the other perturbations. In the Chioggia basin, for example, the reintroduction of the river Brenta's sediment input for 40 years before the construction of the jetties, initially fostered sedimentation of shoals and marshes (see Day *et al.*, chapter 50). Also the increase of tidal channels cross-section produced by the jetties was initially more than compensated for by the (temporary) transport of sediments from the shoals to the channels. A strong transfer of sediments from the shoals to the channels ('*zand-honger*' or sand-hunger of the channels, as it is called by the Dutch) is also displayed by the model as a consequence of other perturbations that reduced the flow in the original tidal network, such as the construction of new navigation canals.

A schematic picture of the lagoon conditions in the fourteenth and the twentieth centuries is given in Plate 16. The long-term sediment fluxes between the various components of the lagoon morphology (marshes, shoals and channels) are controlled by the transport concentration prevailing at the seaward side of the inlet and in the inner zones of the lagoon. The transport concentration, in its turn, depends on the local depth (Eqs. 43.3 and 43.13). As water depth appears as a denominator, the lagoon invariably tends to reach an equilibrium condition (when erosion takes place, local concentration decreases and more sediments are attracted to that location).

EQUILIBRIUM AND QUASI-EQUILIBRIUM CONDITION

When the morphology of the lagoon eventually reaches a stationary configuration ($\partial h_e / \partial t = 0$ in Eq. 43.3), we may say that the lagoon is in *equilibrium* with its boundary conditions (sediment input, transport concentration at the seaward side of the inlets, resuspension coefficients). In principle, depending upon the prevailing boundary conditions, a tidal lagoon may eventually evolve into a uniform deep bay (extreme erosion) or into a uniform salt marsh (extreme deposition). Strict equilibrium conditions are rarely met in a natural lagoon (climate is changing) and even less in the presence on human interventions.

If evolution is relatively slow, however, the lagoon can be considered in *quasi-equilibrium* and the term ($\partial h_e / \partial t$) can be neglected in Eq. (43.3). In the Lagoon of Venice (where the residual (Eulerian) waterflow is negligible and the rate of eustatism + subsidence α is much smaller than $\partial h_e / \partial t$), Eq. (43.3) can be written:

$$-\frac{\partial h_e}{\partial t} = \frac{\partial}{\partial x}\left(Dh_e \frac{\partial C}{\partial x}\right) + \frac{\partial}{\partial y}\left(Dh_e \frac{\partial C}{\partial y}\right) \qquad Eq.\ 43.15$$

If the term $\partial h_e / \partial t$ is negligible and no external input (rivers) is present, as in the modern Lagoon of Venice, the solution of Eq. (43.15) is $T_x = T_y = 0$ and $\partial C/\partial x = \partial C/\partial y = 0$, namely the long-term sediment concentration C is uniform all over the lagoon. It is possible to approximately evaluate the circumstances under which the term $\partial h_e / \partial t$ can be neglected. For a rectangular lagoon of uniform depth h_e, having its axis along the axis x, Eq. (43.15) can be integrated between the seaward end (x = L, where C = C_M) and the landward end (x = 0, where $\partial C/\partial x = 0$). By assuming ($Dh_e$) and ($\partial h_e / \partial t$) constant, the integration provides:

$$C(x) = C_M + \frac{1}{2}\frac{(\partial h_e / t)}{(Dh_e)}(L^2 - x^2) \qquad Eq.\ 43.16$$

with the concentration increasing landward from C_M to $C(0)$.

The hypothesis of quasi-equilibrium conditions ($\partial h_e / \partial t \cong 0$) holds if:

$$\frac{C(0) - C_M}{C_M} = \frac{1}{2} \frac{(\partial h_e / t)}{(Dh_e) C_M} L^2 \ll 1 \qquad \text{Eq. 43.17}$$

namely if the spatial differential of the sediment concentration over the lagoon is relatively small. This means, in other words, that the dispersion mechanism should be strong enough to propagate perturbations all over the basin in a rather short time.

For the southern and central part of the Lagoon of Venice under present conditions, the orders of magnitude of the quantities in Eq. (43.17) are as follows: transport concentration at the seaward end, $C_M = 20 \times 10^{-6}$; length of the lagoon, $L = 0.5 \times 10^4$ m; average erosion rate $(\partial h_e / \partial t) = 3 \times 10^{-10}$ m s^{-1}; product of dispersion coefficient and waterdepth $Dh_e = 10^3$ m^3 s^{-1}. The hypothesis of quasi equilibrium conditions is therefore satisfied even with the actual, relatively strong, rate of erosion. The northern part of the lagoon is longer ($L \cong 2 \times 10^4$ m), but the erosion rate is much smaller. In any case this means that the long-term concentration $C(x)$ can approximately be assumed as uniformly distributed all over the lagoon and equal to the concentration at the seaward end. Namely (Eq. 43.14):

$$C_M \cong C = C_c + C_W \qquad \text{Eq. 43.18}$$

and in particular:

$$C_M \cong \frac{f_W}{h_{es}} \qquad \text{Eq. 43.19}$$

$$C_M \cong \frac{f_c}{h_{ec}^5} \qquad \text{Eq. 43.20}$$

respectively on the shoals (wave dominated) of waterdepth h_{es} and in the channels (current dominated) of waterdepth h_{ec}.

INTERVENTIONS FOR RESTORING THE LAGOON'S MORPHOLOGY

The quasi-equilibrium conditions (Eqs. 43.18, 43.19 and 43.20) provide a relation between the present bathymetric configuration of the lagoon, present sediment concentration at the seaward side of the inlets and the present resuspension coefficients. In particular, we may express the water depth of the shoals in any given location, h_{es}, in the following way:

$$h_{es} \cong \frac{f_W}{C_M} = \frac{f_W h_{ec}^5}{f_c} \qquad \text{Eq. 43.21}$$

where f_W is the wave resuspension coefficient in that location, C_M is the present concentration at the seaward side, h_{ec} is the water depth of any channel cross-section and f_c the resuspension coefficient in the same cross-section (proportional to the 4th power of the corresponding tidal prism).

Equation (43.21) clearly indicates the measures to be taken in order to bring the lagoon towards the pristine configuration or, at least, to reduce the present rate of evolution from a 'terrestrial' towards a 'marine' landscape. This objective can obviously be attained by creating shallower shoals and increasing the area of salt marsh surfaces by the mechanical deposition of sediments. It should be noted, however, that the artificial reconstruction of these morphological features, as in fact being carried out by the Magistrato alle Acque (Cecconi, chapter 51), presents a number of issues that need to be carefully considered. First of all, the quantities of sediments involved and the relative costs are huge (hundreds of millions of cubic metres of sediments have been lost from the shoals to the canals and eventually to the sea in the last century). Secondly, the material dredged from the navigation canals (often below the quality standards in force for the Lagoon of Venice) can rarely be employed for the reconstruction of shoals and marshes but must be disposed of in the open sea. Finally, under the present conditions prevalent in the lagoon, the sediments mechanically deposited in the artificial shoals and marshes tend to be rapidly removed and lost again to canals and the open sea, unless they are adequately protected with revetments and other structures.

In conclusion, the morphological restoration of the lagoon cannot simply consist of replacing in the same locations the material lost in the past. To reduce the quantity of deposited sediments, and to minimize the artificial arrangements needed to keep it in place, attention should be given to the possibility of modifying the *boundary conditions* that control the present morphology.

This can be done by looking at Eq. (43.21). If we want to move in a 'natural' way towards a smaller water depth h_{es} (and, consequently, also towards a larger surface of vegetated salt marshes), there are basically three possible types of action:

1) *Increasing the concentration C_M at the seaward end of the inlets.* This could in principle be done by removing the jetties and by letting the inlets assuming again the configuration prior to 1850. At that time the inlets were much wider and shallower than in the present situation. The external part of the inlet (called *fuosa* in Venetian, or *foce* in Italian, i.e. 'mouth' in English) was shallower than the tidal channel inside the lagoon. It was a proper 'outer delta' with a broad and rather unstable main channel cut in the shallows. During sea storms (especially these generated by *sirocco* winds blowing from the south-east) waves were able to re-suspend high sediment volumes on the shallows of the *fuosa*, which were then trapped in the tidal channel. Later on, following the storm, the sediments were gradually removed by the tidal currents and deposited again on the shallows. In this way the net sediment flux from the lagoon to the sea was almost zero or even reversed. The presence of the outer delta before 1850 is accounted for in the long-term model by prescribing an appropriately larger value of C_M with respect to the present conditions. The evolution of the outer delta, however, can explicitily be simulated in the model by introducing an additional external element (Stive *et al.*, 1996). The introduction of the outer delta as a morphological component of the lagoon, permits the consideration of the concentration C_M as a dependent variable of the model. On the other hand, a realistic modelling of this new component (physically controlled by both the tidal currents and by the external wave climate) would probably require further investigation.

The complete removal of the jetties and the consequent new formation of the original *fuosa* is probably not feasible today because of the presence of settlements, infrastructure and, above all, touristic beaches on both sides of the inlets. The removal of the jetties might result in the retreat of the present coastline, as well as to a continuous net flux of sediments from the sea into the lagoon, to attain again the equilibrium configuration prior to 1850. As the amount of material required to reach the equilibrium condition of 1850 would be several hundreds of millions of cubic metres of sediment and the only sediments available could only originate from the adjacent beaches, these would be severely damaged by the restoration of the original *fuosa*.

A more feasible intervention, instead, might be a *partial* and *controlled* widening of the present inlet width. As a consequence, the depth of the outer inlet would correspondingly be reduced, but not like in the original *fuosa*. In this way, the value of C_M in Eq. (43.21) would increase, but much less than in 1850. The widening of the inlet should be selected in such a way that the present net flux of sediments from the lagoon to the sea would be strongly decreased, but not reversed.

2) *Decreasing the resuspension coefficients f_w in the shoals.* This goal could be achieved by either decreasing the 'erodibility' of the bottom, or by decreasing the 'erosivity' of the waves. The first approach might consist in expanding the seagrass beds (*Zostera marina*) and increasing their stem density. The relative reduction of f_w in the presence of seagrasses can be evaluated by comparing the depth-integrated concentration measured in two locations, respectively protected or non protected by vegetation, having the same depth and subject to the same resuspension action (e.g. wind waves or waves produced by motorboats).

The second approach consists in decreasing the shear stress of the wave motion by reducing the fetch length (Eq. 43.12). This is obtained by building wave breaks, either rigid or vegetated (salt marshes), across the direction of the strongest and most frequent winds.

3) *Increase the resuspension coefficients f_c in the channels*. The resuspension coefficients f_c (Eqs. 43.5 and 43.6) are proportional to the 4-th power of the water discharge in the channel; i.e. (for a short lagoon) to the 4-th power of the tidal prism. As a consequence, even a slight reduction of the tidal flow would be reflected in a strong decrease of f_c and on a much larger water depth in the shoals (the so-called 'sand hunger' of the channels mentioned above). By contrast, an increase of the water flow in the channels would bring sediments to the shoals. Water discharge in the channels could be increased, not only by augmenting the tidal prism (hardly possible in the present lagoon) but also by concentrating the flow within the tidal channels ('channelling'). Channelling could be fostered by bordering the existing channels by means of artificial salt marshes. Those artificial salt marshes appear to be similar to the natural ones created by deposition along channel banks in proximity to sediment sources.

CONCLUSIONS

The Lagoon of Venice has experienced a progressive deepening of the shallow areas, reduction of the vegetated surfaces (salt marshes) and siltation of the natural and artificial channels over a long period of time. This trend, probably initiated by the diversion of rivers out of the lagoon, has been accelerated in the last 150 years by the construction of the jetties at the inlets and other anthropogenic perturbations.

The restoration project being carried out over recent years by the Magistrato alle Acque basically consists in the reconstruction of artificial shoals and marshes via mechanical relocation, in the eroding shallows, of material dredged from the channels and/or imported from outside the basin. However, due to the extremely large amount of material eventually required to reconstruct the pristine morphology, attention should also be paid to alternative measures, namely to the modification of the boundary conditions that prevail today.

A first change of boundary conditions might be the reintroduction to the lagoon of the rivers diverted in the fourteenth to sixteenth centuries. Implementing

this measure, on the other hand, would present insuperable difficulties, because the sediment transport of these rivers has progressively become very depleted, due to the effects of reservoirs, soil conservation measures and sand mining. In any case, this scarce sediment is indispensable for feeding the valuable (but sediment-starved) beaches adjacent to the lagoon.

Other changes of boundary conditions might also be considered, to reach the effects predicted from Eq. (43.21). These changes might involve:

1) a controlled widening of the present inlets (by careful adjustment of the jetties) and consequent increase of the seaward sediment concentration C_M;
2) a reduction of the resuspension coefficient f_w in the shoals; and
3) an increase of the resuspension coefficient f_c in the channels.

As it appears from Eq. (43.21), all these changes would produce a reduction of the shoals depth h_{es} or, at least, reduce the present erosion rate.

REFERENCES

Ackers, H. B., and W. R. White. 1973. Sediment transport: new approach and analysis. *Proceedings ASCE, J. Hydr. Div.* **99**, HY11, 2041–60.

Bretschneider, C. L. 1969. *Forecasting Handbook of Ocean and Underwater Engineering*. New York: McGraw-Hill, ch. 11.

Brown, C. B. 1950. Sediment transportation. In H. Rouse (ed.), *Engineering Hydraulics*. New York: John Wiley and Sons.

Dal Monte, L., and G. Di Silvio. 2004. Sediment concentration in tidal lagoons. A contribution to long term morphological modelling. *Journal of Marine Systems*, 234–55.

De Vriend, H. J. 1996. Mathematical modelling of meso-tidal barrier island coast. Part I: Empirical and semi-empirical models. In P. L. F. Liu (ed.), *Advances in Coastal and Ocean Engineering* 2. Singapore: World Scientific, pp. 115–49.

Di Silvio, G. 1989. Modelling of the morphologic evolution of tidal lagoons and their equilibrium configurations. In Proceedings of the IAHR, XXIII Congress, Ottawa, Canada, 21–25 August.

Di Silvio, G. 1991. Averaging operations in sediment transport modelling: short-step versus long-step morphological simulations. In Proceedings of the International Symposium on the Transport of Suspended Sediments and its Mathematical Modelling. Florence, Italy.

Di Silvio, G., and A. Padovan. 1998. Interaction between marshes, channels and shoals in a tidal lagoon investigated by a 2-D morphological model. In Proceedings of the 3rd Int. Conference on Hydroscience and Engineering, Cottbus/Berlin, Germany.

Dronkers, J. 1978. Longitudinal dispersion in shallow well-mixed estuaries. *Coastal Engineering Conference* 3, 169.

Fischer, H. B., Imberger, J., List, E., Koh, R. C. Y., and N. H. Brooks. 1979. *Mixing in Inland and Coastal Waters*. Academic Press, 151pp.

Stive, M. J. F., Wang, Z. B., Capobianco, M., Ruol, P., and M. C. Bujsman. 1996. Morphodynamics of a tidal lagoon and adjacent coast. In Proceedings of the 8th Int. Biennial Conference on Physics of Estuaries and Coastal Seas. The Hague, Netherlands.

Wang, Z. B., Karsen, B., Fokkink, R. J., and A. Langerak. 1996. A dynamic-empirical model for estuarine morphology. In Proceedings of the 8th Int. Biennial Conference on Physics of Estuaries and Coastal Seas. The Hague, Netherlands.

44 • Open problems in modelling the long-term morphodynamic evolution of Venice Lagoon

G. SEMINARA, M. BOLLA PITTALUGA, N. TAMBRONI AND V. GAROTTA

SUMMARY

Through the centuries Venice Lagoon has undergone significant morphological changes. Assuming that its present morphology is a state of sickness, the diagnosis of the causes of the disease is not too difficult. Various well-known factors have contributed to this state, e.g. the diversion of rivers discharging into the lagoon, the construction of long jetties bounding the three inlets, the processes of sea-level rise and land subsidence. As a result, the lagoon has progressively deepened, and is commonly stated to export to the sea a considerable volume of sediments each year (Consorzio Venezia Nuova, 1992).

Two major consequences of the above changes are the source of a widespread concern in the international community on the fate of Venice:

- an increased frequency of high waters; and
- a progressive loss of salt marshes which tends to transform the Venice Lagoon into a bay.

This chapter starts with a short discussion of an ongoing debate, summarized here by the provocative slogan: *saving Venice or saving the lagoon?* We argue that this debate is ill-formulated as the former goal is quite distinct from the latter, and requires different measures which must be pursued on a timescale much shorter than the timescale on which one might reasonably expect to affect the morphodynamic evolution of the lagoon.

We then concentrate on the second issue and briefly review some of the available knowledge on the hydrodynamics and morphodynamics of morphological units composing the tidal system. We discuss successes and difficulties encountered in some of the attempts to model these analytically, numerically or experimentally and possible implications for

the pathological state of the Venice Lagoon. A related discussion on the research developments still needed in order to provide a reliable *prognosis* and *therapy* for Venice concludes the chapter.

SAVING VENICE OR SAVING VENICE LAGOON?

Describing the morphological changes undergone by the Venice Lagoon during recent centuries is not a new exercise. We simply briefly summarize the main issues here. The diversion of rivers discharging into the lagoon was an important decision of the Venetian Republic taken during the Renaissance: it produced the beneficial effect of stopping an ongoing aggradation process in the central part of the lagoon and it improved the health of Venetians, threatened by malaria. However, it also removed a significant fluvial input of sediments into the lagoon, a fact which has contributed to the progressive deepening of shoals. The construction of long jetties at the three lagoon inlets, completed at the beginning of the twentieth century, aimed at deepening the inlets to allow for steam navigation, achieved this goal but at some price. In fact, the inlets were moved a few hundred metres seaward, i.e. beyond the surf zone where sediment is strongly resuspended during storm events. This reduced the concentration of suspended sediments available to tidal currents in the near inlet regions. Furthermore, the degree of ebb – flood asymmetry of the flow field near the inlet was enhanced by the presence of the jetties. As a result, the import of sediment from the sea into the lagoon was progressively reduced, leading to progressive deepening of the inlets. Moreover the jetties intercepted the littoral circulation, which is dominantly counter-clockwise in the upper Adriatic Sea, and this

Flooding and Environmental Challenges for Venice and its Lagoon: State of Knowledge, ed. C. A. Fletcher and T. Spencer.
Published by Cambridge University Press, © Cambridge University Press 2005.

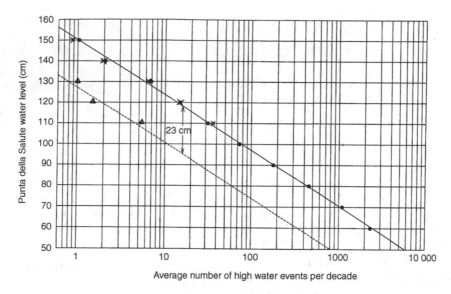

Fig. 44.1   The average number of high water events observed in a decade plotted as a function of the free surface elevation exceeded by those events (after Marchi, 1997). (x events occurring after 1997; • events occurring between 1966 and 1997; Δ events occurring between 1927 and 1946). The dotted line is obtained by simply shifting the solid line by 23 cm.

led to shore progradation updrift of the inlets and degradation downdrift. The industrial development of the Marghera area, as well as further actions related to the development of the city, have also been held responsible for the modification of the tidal regime of the lagoon. These included the construction of the Canale dei Petroli, which removed the oil ship traffic from the centre of the city, the construction of the international airport, the artificial filling of portions of the lagoon (*casse di colmata*) and the construction of fences to isolate large portions of the lagoon for fishing purposes (*valli da pesca*).

Along with the above man-induced processes, two natural phenomena, namely sea-level rise and subsidence, have produced significant relative lowering of the city. Note that subsidence was severely enhanced between 1950 and 1970 as a result of extensive groundwater pumping (stopped in the 1970s) for the industrial development of the Marghera area.

Research undertaken in the decades since 1970 (see Ghetti, 1990 for a systematic review of the most important contributions) has conclusively demonstrated that the sea-level rise and subsidence (amounting to an increased elevation of the free sur-

face relative to the city level of 23 cm from evidence at Punta della Salute) are mainly responsible for the increased frequency of high waters in the last century while other factors described above have had negligible influence. St Mark's square is as a result now flooded ten times more frequently than a century ago. Figure 44.1 (Marchi, 1997) is a clear demonstration of the above statement.

While the debate concerning the diagnosis of the disease is reasonably clear, the issues of an appropriate prognosis and an effective therapy have led to argument which still continues. It suffers from the mixture of political considerations with technical arguments and has generated two parties. The former party, which we will call for the sake of simplicity the party of Venice, supports the idea that ' …defending Venice from high waters can only be insured by regulation of the lagoon inlets by means of suitable mobile gates…'. (Vote of the General Assembly of The Council of Ministry of Public Works of Italy, 1999). The latter party, the party of the lagoon, is founded on the statement that '…defending Venice is a goal which cannot be pursued unless the whole lagoon system is simultaneously governed … In particular … the sediment deficit should be cut and it should be demonstrated that the system tends to a new state of dynamic equilibrium ...' (Commission for Environmental Impact of the Ministry for the Environment of Italy, 1998). In order to ascertain whether the above debate is well posed, let us first briefly review the state of knowledge on lagoon morphodynamics in the light of the need to achieve long term predictions of the evolution of the lagoon.

WHICH APPROACH TO TAKE TO LONG-TERM PREDICTIONS OF THE MORPHODYNAMIC EVOLUTION OF THE VENICE LAGOON?

Predicting the long term morphodynamic evolution of the lagoon is a major subject of research which can be approached from a variety of viewpoints of increasing complexity. The simplest, yet useful approach, employed by practical engineers, relies on a number of empirical relationships which grossly interpret the morphodynamic response of tidal basins to variations of external forcings: a well known example is that of O'Brien (1969) who established empirically a quasi-linear relationship between cross sectional area of tidal inlets or tidal channels at equilibrium and tidal prism. The rationale behind this approach is the idea that tidal inlets, as well as tidal channels and tidal basins, may reach an ideal state of statistical equilibrium. The above approach is usually employed to get some general guidance for the management of tidal basins.

A second, more advanced, level of understanding may be achieved by means of box models which describe the tidal system as a collection of morphological elements (sea, channels, shoals, marshes), the geometry of which is given in terms of empirical relationships: such morphological elements are allowed to exchange water and sediments according to simplified relationships which are assumed to interpret the complex processes occurring in reality (Di Silvio, 1989; Van Dongeren and de Vriend, 1994, Stive et al., 1998). These models, which are discussed elsewhere in this volume, are useful tools when guided by more detailed analyses of the type discussed below.

A third, mechanistic, approach attempts at developing a thorough understanding of the hydrodynamics and morphodynamics of each morphological unit,

namely inlets, channels, shoals and marshes. Firstly, one investigates the existence of a state of morphodynamic equilibrium of a straight channel closed at one end and connected at the other end with a tidal sea, progressively adding to the model further features of the real configuration, such as channel convergence, curvature, the presence of bedforms, sorting effects and cohesion. The second step is to investigate the morphodynamics of tidal inlets with the aim of ascertaining which mechanisms control the exchange of sediments between the sea and the lagoon. The last step is to investigate the morphodynamic interactions between channels, lateral shoals and marshes. The state of knowledge on the processes underlying each of the above problems is briefly reviewed below and can be respectively described as well developed, developing and poorly developed.

The fourth possible approach to lagoon morphodynamics is to pursue a fully numerical model of the whole system, attempting to describe all the above processes simultaneously. Such an approach has the obvious advantage of describing the geometry of the system with great accuracy and must be seen as the final step of the present research. However, the computational effort required by these numerical models is so large as to make reliable long term predictions presently not feasible.

THE MORPHODYNAMICS OF TIDAL CHANNELS

Strictly speaking, a channel is in a state of morphodynamic equilibrium provided its bottom profile does not undergo net variations within each tidal cycle. In practice, as the forcing tide is subject to variations on a variety of further time scales (that associated with storm events, seasonal cycles) the above notion must then be interpreted in an averaged sense. A number of factors listed below may affect this state:

1) The nature of the sediment present in the channel bed controls the sediment transport. Sediments are typically cohesionless in the larger tidal channels of Venice Lagoon and their size is small enough for transport to occur as suspended load as well as bed load. The progressively finer size of sediments encountered as one moves from the

inlet towards the inner lagoon reflects sorting effects, and leads to the possible role of cohesion.

2) The geometry of the channels (channel convergence, meandering and the presence of tidal flats) also affects significantly both the lateral and the longitudinal shape of the bed profile at equilibrium.

3) The third ingredient is the forcing effect of the tidal oscillation at the inlet as well as the ability of the sea to exchange sediments with the channel. The latter is essentially determined by the geometry of the inlet itself as well as by the sediment concentration at a distance far from the inlet section. This is in turn affected by the wave climate and by the presence of littoral currents.

4) Finally, non linear effects associated with the presence of both steady and migrating tidal bars may drive a net sediment flux able to affect the establishment of equilibrium. This is readily understood if one notes that, at a linear level, a harmonic oscillation of bed elevation drives a corresponding harmonic oscillation of sediment flux with the same frequency: at second order non linearity drives a second harmonic of sediment flux as well as a zero component which is essentially a contribution to the net sediment flux.

In the last decade progress in the understanding of some of the above processes has been made through the work of various groups. To our knowledge, the first fully mechanistic attempt to investigate the long-term 1-D equilibrium of straight, non-convergent tidal channels is due to the Utrecht group led by H. de Swart (Schuttelaars and de Swart, 1996, 2000). Though the importance of such a pioneer contribution should not be undermined, we recall that in these papers the modelled channel did not establish its own equilibrium depth at the inlet nor did it choose its own equilibrium length: these were fixed *a priori*. More recently, Lanzoni and Seminara (2002) considered the morphodynamic evolution of convergent channels showing that, starting from an initially horizontal bed profile, a sediment wave develops in the channel carrying sediments landward and leading to deposition at the inner boundary until a shore develops in the inner portion of the channel. A final equilibrium profile is established which is concave seaward and slightly convex landward; an equilibrium depth which depends only on channel geometry and tidal forcing is reached at the inlet section (Bolla Pittaluga and Seminara, 2002). This work was also based on simplifying assumptions. In particular, sediment transport was assumed to be in equilibrium with the local and instantaneous hydrodynamic conditions, an assumption which prevented a proper description of sediment exchange through the inlet. Bolla Pittaluga and Seminara (2002) have removed some of the latter restrictions. They have accounted for the effect of spatial and temporal variations of the hydrodynamics on the dynamics of suspended load with the help of a model of transport in suspension (Bolla Pittaluga and Seminara, 2003), which avoids the need to solve the classical advection – diffusion equation for concentration, an effort which would make computations of long term equilibrium prohibitively demanding. The ebb–flood asymmetry of sediment transport at the inlet was also accounted for by a simple model whereby the sediment flux during the ebb phase was controlled by the transport capacity of the flow in the channel, while, during the flood phase, the sediment flux was forced by an externally driven concentration. Finally, the effect of higher harmonics contributing to the forcing tidal oscillation was investigated. Each of the latter effects turned out to affect the long term morphodynamic evolution of tidal channels, though none of them prevented channels from reaching equilibrium: more precisely, the role of non-equilibrium sediment transport did not turn out to be crucially important, but higher harmonics with an appropriate phase affected the equilibrium profile significantly as did the presence of a non vanishing 'sea concentration of suspended sediments' which affected the exchange of sediments through the inlet. An important question may be answered based on the analysis of these model problems: *Does equilibrium imply that the instantaneous sediment transport must vanish at the inlet?* The answer is positive, provided the external forcing consists of a single harmonic and the concentration at infinity vanishes. Both the role of overtides and the presence of a sediment supply forced by a non vanishing sea concentration may lead to inlet equilibrium, i.e. to a net flux of sediment vanishing in a cycle, with non vanishing instantaneous values of the sediment flux. The latter result has interesting implications concerning the equilibrium of Venice inlets: they will be discussed in the next section.

Further effects which await to be explored, still in the context of 1-D formulations, are sediment sorting and the role of cohesion in the innermost channels. Other aspects cannot be explored in the context of 1-D formulations. The formation and non linear development of tidal bars, i.e. large scale bedforms (Dalrymple and Rhodes, 1995), has only recently been investigated with the help of 2-D or 3-D analyses, though in fairly idealized contexts (Seminara and Tubino, 2001; Schramkowski et al., 2003). The only attempt known to the authors to incorporate their role in the context of long term morphodynamic computations is that of Hibma et al. (2003). Finally, 3-D effects cannot be ignored if the role of channel curvature is investigated: the available knowledge on bars forced by tidal meandering is also fairly recent (Solari et al., 2002; Marani et al., 2002; Solari and Toffolon, 2001) and their possible effect on the longitudinal long-term equilibrium of the bottom profile is as yet unknown.

The morphodynamic coupling between inlet and channel as well as between channel and adjacent tidal flats also requires 2-D or 3-D approaches. Recent preliminary attempts in this direction are briefly discussed in sections 4 and 5 and more deeply in a companion chapter (Tambroni et al., chapter 45).

No attempts to investigate the long term equilibrium of tidal channels on the basis of controlled laboratory observations have been made until recently. This is not surprising as the time scale of morphodynamic evolution is typically much larger than the hydrodynamic time scale. Moreover, the evolution becomes slower and slower as equilibrium is approached: hence, the time required to approach equilibrium in the laboratory is of the order of days or weeks. In spite of its complexity, such an experimental exercise was recently undertaken by Tambroni et al. (2002) and has proved instructive. It allowed a check, under controlled conditions with simple channel geometries, of the main mechanisms emerging from the theoretical work mentioned above. Such a goal is more difficult to achieve on the basis of field observations whose interpretation is generally complicated by the simultaneous presence of a variety of natural features whose role cannot be readily isolated. The experimental observations confirm that tidal channels closed at one end and connected at the other end with a tidal sea tend to a morphodynamic equilibrium characterized by an upward concave bed profile seaward and by a convex profile landward, vanishing net sediment flux in a tidal cycle at each cross section and the formation of a 'shore' at the landward end of the channel. Also, the channel chooses its own equilibrium length and equilibrium depth at the inlet. Comparison between the experimental and theoretical results proves satisfactory except close to the inlet where the 1-D model of the flow field fails.

The morphodynamic evolution of the inlet region was also monitored, providing interesting observations on the formation of an outer delta (see below). Finally, the formation of large and small scale bedforms emerged, displaying features which confirmed the recent theoretical predictions of Seminara and Tubino (2001).

THE MORPHODYNAMICS OF TIDAL INLETS

The tidal inlets of Venice Lagoon have various major conflicting functions. They must be wide and deep as well as stable enough to allow for navigation; however, they control the exchange of water with the sea, and hence their width and depth also affect the amplitude of tidal oscillations in the enclosed basin, posing a constraint partly in conflict with the previous requirement. Control of the morphodynamic degradation of the lagoon would call for inlet shapes able to reduce the ebb-flood asymmetry of the flow field so as to encourage sediment carried by littoral currents to merge into the flood current at the inlet; however such a feature would also encourage siltation of the inlets! Finally, pollution control would call for large volumes of exchanged water and inlet shapes enhancing the ebb-flood asymmetry of the flow field, in contrast with the just mentioned requirements posed by the need to enhance sediment exchange.

A fundamental issue then arises: under what conditions are tidal inlets stable? This problem has been widely explored (see Bruun et al., 1978). In particular, empirical observations led O'Brien (1969) to propose a well-known relationship between the cross-sectional area of tidal inlets and the so-called tidal prism. Lanzoni and Seminara (2002) have provided some substantiation of O'Brien's (1969) relationship showing numerically that, indeed, as tidal channels (not just tidal inlets) tend to morphodynamic equilibrium an O'Brien type relationship is progressively satisfied at each cross section. Recently, Tambroni and Seminara (2004) have investigated the problem theoretically, revisiting the analysis of Marchi (1990). The latter approach leads to the following relationship for the maximum flow speed experienced in a tidal inlet consisting of a rectangular channel connecting a sea with prescribed tidal forcing with a tidal basin:

$$U_{max} = f(\delta_1, \delta_2) \frac{\omega Sa}{bY} \qquad Eq.\ 44.1$$

where ω is the angular frequency of the tide, S is the area of the surface of the basin, a is the amplitude of the fundamental component of the sea oscillation, b and Y are width and average flow depth of the tidal channel assumed to be of rectangular shape. Furthermore, f is a function of the harmonic content of the tidal forcing and of the following dimensionless parameters:

$$\delta_1 = \frac{L\omega^2 S}{gYb} \quad , \quad \delta_2 = \frac{La\omega^2 S}{gC^2 Y^2 b} \qquad Eq.\ 44.2a,\ b$$

with C flow conductance (i.e. the dimensionless Chezy coefficient) of the channel. Using the data employed by Marchi (1990), the latter relationships can be applied to the three inlets of Venice Lagoon and may be interpreted as a dimensional dependence of the maximum flow speed on the flow depth for given values of the remaining parameters. Such dependence is plotted in Figure 44.2 for a pure M_2 tidal oscillation. Note that the values of U_{max} fall in the correct range and agree fairly well with Marchi's predictions: they display a maximum for values of the flow depth smaller than those presently observed

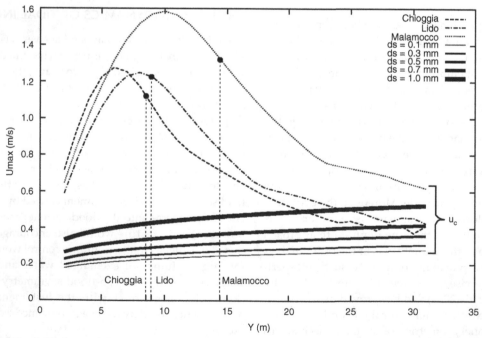

Fig. 44.2   The dependence of the maximum speed at each of the Venice Lagoon inlets is plotted versus the inlet depth for a pure M_2 forcing tidal oscillation with amplitude a = 0.5 m. Also plotted is the critical speed for sediment motion for sediment sizes in the range 0.1 – 1.0 mm.

at each inlet. The question of inlet stability can now be examined in the limit condition of no sediment supply from the sea, stipulating that the maximum speed must not exceed the critical speed U_c for sediment motion. Imposing Shields criterion one finds the threshold condition plotted in Figure 44.2 for a realistic range of sediment sizes. Figure 44.2 clearly shows that in the absence of a sediment supply from the sea exceeding the transport capacity of tidal currents, the present configuration of each of the Venice inlets is far from equilibrium and potentially prone to further erosion.

A qualitative justification of the observed erosive tendency of the Malamocco inlet as opposed to that of the Lido inlet is then obvious: the progradation of the coast on the north side of the Lido inlet has clearly enhanced the sediment supply to the inlet coming from the sea, a feature much weaker at Malamocco.

The analysis of Tambroni and Seminara (2004) also shows that the dominance of tide propagation in the inlet channel is strongly dependent on the harmonic content of the forcing sea oscillation. A pure M_2 forcing leads to a very slight flood dominance at each inlet. The role of a significant K_1 component typically observed in the Adriatic Sea, with a phase lag ranging about $-\pi/4$ relative to the M_2 component, leads to a weak ebb dominance in each inlet. An estimate of an upper bound for the loss of sediments experienced by each inlet through the years can then be simply obtained by calculating the sediment transported through the inlet during the sequence of tidal oscillations recorded for several years by the CNR gauge station located in the sea region adjacent to the Venice Lagoon. Assuming that the amount of sediment carried by both flood and ebb currents is determined by their transport capacity, the effect of a sediment overload forced by the sediment concentration in the far field is excluded: such a limit condition implies that the above estimate does indeed provide an upper bound for the sediment loss. Preliminary results of such an exercise (Tambroni and Seminara, 2004) suggest that:

1) the yearly loss of sand, ranging around 50 000 m³, is an order of magnitude smaller than it is usually claimed;
2) significant loss of sediments occurs only during high water events; and

3) a repetitive seasonal dependence of the net volume of the exchanged sediment is experienced at each inlet: more precisely, sediments invariably enter the lagoon during winter and summer while a net loss of sediment occurs during spring and autumn.

The result 1) suggests that the major contribution to sediment loss is not likely to be associated with an exchange of the sand available close to the inlets: attention should rather be focussed on the very fine material resuspended by wind in the shoals or produced by progressive bank collapse on salt marsh margins. These sediments may be fine enough to be transported by tidal currents as wash load: in other words they are not able to settle in the lagoon and are carried by tidal currents to the inlets where they are lost, merging into littoral currents. The result 2) has been obtained correlating the net exchange of sediments in a tidal cycle with the amplitude of the forcing sea oscillation in that tidal cycle. The above findings outline a picture which is somewhat more complex than usually described.

In order to investigate the effect of the sea climate and littoral currents on inlet morphodynamics, a more detailed analysis of the inlet hydrodynamics is required. The only early attempt known to us to tackle this complex problem is that of Blondeaux et al. (1982) who showed, both theoretically and experimentally, that, in the fixed bed case, the flow field induced in a neighbourhood of inlets bounded by jetties protruding into the sea exhibits a strong ebb–flood asymmetry, behaving as effectively irrotational during the flood phase and as a turbulent unsteady jet during the ebb phase. The latter gives rise to a pair of counter-rotating vortices leaving the inlet under the effect of their self induced velocities. The asymmetry of the flow field is one of the mechanisms which leads to a net flux of sediments through the inlet at each cycle, an effect which was much weaker before the construction of the jetties at the inlets of the Venice Lagoon. The recent experimental observations of Tambroni et al. (2002) have shown that, when the bed is cohesionless, the ebb jet gives rise to the excavation of a submerged channel and to the build up of an outer delta, which somewhat reduces the degree of ebb-flood asymmetry.

Preliminary attempts to model the latter process solving numerically the shallow water equations coupled with the Exner equation have been recently performed (see Tambroni *et al.*, chapter 45; D'Alpaos and Martini, chapter 48). However, modellers are confronted here with major problems: reproducing the large scale coherent structures shed from the ends of the jetties is a numerically heavy task which is hardly compatible with the aim to perform long term morphological predictions. Moreover, close to the ends of the jetties strong 3-D effects are present and lead to the development of deep scour holes, posing an even harder modelling problem. This notwithstanding, the preliminary results discussed in Tambroni *et al.* (chapter 45) suggest that coupling a 2-D model of the inlet region to a 1-D model for the channel allows a significant improvement of predictions concerning the flow depth at the inlet. This is encouraging and suggests the opportunity to extend the latter numerical simulations to account for the effects of a far-field concentration of suspended sediments forced by the sea climate and by littoral currents.

THE MORPHODYNAMICS OF SHOALS

The presence of shoals affects the hydrodynamics of tide propagation in tidal channels. A 1-D formulation to treat this effect was proposed by Speer and Aubrey (1985) who pointed out that the presence of flats may turn a flood dominant channel into an ebb dominant one, provided that the area of flats is sufficiently large. At closer scrutiny the picture appears to be more complicated. Bolla Pittaluga *et al.* (2003) have recently undertaken a systematic investigation which suggests that:

- in the absence of flats, and with a horizontal bottom, the tidal oscillation at the inlet typically changes from flood to ebb dominant when the wave amplitude increases beyond some threshold value;
- a sloping character to the bottom induces a prevailing ebb dominance; and
- the effect of flats is to accelerate the flow in the channel and further enhance ebb dominance, at least in the case of sloping channels.

However, note that a comparison between results obtained using 1-D and 2-D models suggests that the 1-D formulation is no longer adequate as the width of the flats becomes large. This finding calls for a systematic extension of the latter analysis to describe the case of wide flats using the 2-D formulation.

Enhancing the ebb dominance of tidal flow in the channels implies that channels tend to export sediments to the sea. However, what mechanisms control the sediment supply to the channels, i.e. the sediment exchanged between channels and flats? The analysis of these mechanisms still awaits sound investigations, but the ingredients which play a role are qualitatively fairly clear. Tidal currents in the shoals are typically too weak to resuspend sediments lying on the bottom, hence they act mainly to redistribute sediments exchanged with channels when shoals are flooded or sediments resuspended by the action of wind generated waves. The latter effect appears to be crucial and may have been strengthened by shoal deepening. However, modelling it is a major task. Firstly, one should model wind generated waves in small basins delimited by islands and marshes, where the action of wind is unable to fully determine its effects. (Observations performed in various estuaries suggest that for relatively shallow shoals a linear relationship can be established between the significant wave height and the shoal depth, Le Hir *et al.*, 2000). Secondly, a model is required for the distribution of sediment suspended by the action of wind generated waves. In this respect, it is well known that the wind generates a stress at the surface and a free surface set up: as a result of both these effects a wind generated current arises, with a velocity distribution which reverses its direction close to the bed where the wind current has a direction opposite to that of the wind stress. Sediment suspension is then driven by a complex interplay among the flow fields associated with the wind wave, the wind current and the tidal currents. The process is further complicated by the possible cohesive character of sediments lying in the shoals. Though various numerical models presented recently claim to be able to model all the above mechanisms (often using commercial codes), in our opinion a sound scientific exploration of the above process is as yet unavailable.

The processes which control the morphodynamic evolution of tidal marshes are even more complex, involving the crucial role of vegetation. This is a strongly interdisciplinary field of research which falls outside our present research activity: we can neither assess the available knowledge nor suggest research needs. Let us only point out that an important hydrodynamic process which appears to be insufficiently explored is the mechanics of marsh bank collapse, controlled by seepage through the wetting and drying banks, as well as by wave action and biological effects affecting marsh vegetation.

CONCLUDING REMARKS

The brief and admittedly superficial overview given above allows us to go back to the political debate described in our introductory section: saving Venice or saving the lagoon? Though a discussion of the hydrodynamics of high waters and of the choice of appropriate defence works for Venice was outside the scope of this paper, we can now clearly state that the above debate is ill formulated and misleading. In fact, as discussed above:

* cutting the sediment deficit can only be achieved by affecting the morphodynamic evolution of the lagoon, a process which has the time scale of decades, or even centuries; on the contrary, defending the city from the effect of high waters is a problem which can (and must) be confronted on a much shorter time scale;
* reversing the morphological degradation of the lagoon is definitely a major goal but its effect on the intensity of high waters is inevitably minor;
* the 'morphodynamical' therapy is far more complex and less understood than the therapy conceived for the high water problem: the latter is amenable to a surgical operation, the former relies on a variety of remedies far from being fully assessed, some of which of uncertain success.

In conclusion: 'saving Venice' and 'saving the lagoon' are aims to be both pursued, but they involve different time scales and require different tools, while their success is not equally certain.

Indeed, proposing technical solutions appropriate to reverse the morphodynamic degradation of Venice Lagoon besides the artificial reconstruction of *barene* already undertaken by Consorzio Venezia Nuova, can only be the outcome of further deep investigations able to clarify the as yet unexplored mechanisms discussed above. In particular, what would be the short term and long term outcome of modifying the shape of the inlets? Our calculations (Tambroni and Seminara, 2004) suggest that such an operation can only affect a very small fraction of the total yearly loss of sediments. Is the idea of reintroducing part of the discharge of the Brenta river into the lagoon feasible and effective? These are difficult questions which must be answered before any such suggestion may be offered for public scrutiny.

A relatively simple investigation could, however, help the understanding of the effect of the closure of the mobile gates on the loss of sediments experienced by the lagoon: one might reasonably expect that such an effect would not be negligible, as closures are associated with flood events which are responsible for most of the sediment exchange between the sea and the lagoon. In other words, can the mobile gates also be useful to reverse the morphodynamic degradation of the Venice Lagoon?

ACKNOWLEDGEMENTS

This work has been supported by CORILA and by the PRIN project 'Morfodinamica dei sistemi a marea' co-funded by the University of Genova and MIUR (Italian Ministry for Higher Education and Scientific Research).

REFERENCES

Blondeaux, P., de Bernardinis, B., and G. Seminara. 1982. Correnti di marea in prossimità di imboccature e loro influenza sul ricambio lagunare. In Proceedings of XVIII Conv. Idraulica e Costruzioni Idrauliche. Bologna: 21–23 September.

Bolla Pittaluga, M., and G. Seminara. 2002. Modelling suspended sediment transport in tidal flows: theory and application to long term equilibrium of tidal channels. In Scientific Research and Safeguarding of Venice: CORILA research program, 2001 results. Venice: CORILA, Instituto Veneto di Scienze, Lettere ed Arti, pp. 661–74.

Bolla Pittaluga, M., and G. Seminara. 2003. Depth integrated modelling of suspended sediment transport. *Water Resources Research* **39**, 5, 1137, 1/1–1/11.

Bolla Pittaluga, M., Lanzoni, S., and G. Seminara. 2003. The hydrodynamics of tidal channels: the role of tidal flats. In book of abstract of the Second Annual Workshop of CORILA, Venice: 31 March to 2 April.

Bruun, P., Metha, A. J., and I. G. Johnson. 1978. *Stability of Tidal Inlets: Theory and Engineering*. Amsterdam: Elsevier.

Consorzio Venezia Nuova. 1992. *Morphological Modelling of Venice Lagoon*. Final Report on Study A.2.14.

Dalrymple, R. W., and R. N. Rhodes. 1995. Estuarine dunes and bars. In G. M. E. Perillo (ed.), *Geomorphology and Sedimentology of Estuaries*. Amsterdam: Elsevier, Chapter 13.

Di Silvio, G., 1989. Modelling the morphological evolution of tidal lagoons and their equilibrium configurations. In Proceedings of the XXII Congress of IAHR, Ottawa, Canada, 21–25 August.

Dongeren, A. P. van, and H. J. de Vriend. 1994. A model of morphological behaviour of tidal basins. *Coastal Engineering* **22**, 3/4, 287–310.

Ghetti, A. 1990. Italian contributions to hydraulics of lagoons. *Excerpta* **5**, 7.

Hibma, A. H. J., de Vriend, H. J., and M. J. F. Stive. 2003. Numerical modelling of shoal pattern formation in well mixed elongated estuaries. *Estuarine and Coastal Shelf Science* **57**, S–b, 981–91.

Lanzoni, S., and G. Seminara. 2002. Long term evolution and morphodynamic equilibrium of tidal channels. *Journal of Geophysical Research* **107**, C1,1–13.

Le Hir, P., Roberts, W., Cazaillet, O., Christie, M., Bassoullet, P., and C. Bacher. 2000. Characterization of intertidal flat hydrodynamics. *Continental Shelf Research* **20**, 1433.

Marani, M., Lanzoni, S., Zandolin, D., Seminara, G., and A. Rinaldo. 2002. Tidal meanders. *Water Resources Research* **38**, 1225–39.

Marchi, E. 1990. Sulla stabilità delle bocche lagunari a marea. *Rend. Fis. Acc. Lincei* **s.9**, 1/2.

Marchi, E. 1997. Protection of Venice and of the other inhabited areas in the lagoon from high water. In Proceedings of the Symposium Venice and Florence: a Complex Dialogue with Water, 24 May, pp. 13–26.

O' Brien, M. P. 1969. Equilibrium flow areas of inlets on sandy coasts. *Journal of the Waterways and Harbour Div.*, *ASCE* **95**, WW1, 43–52.

Schramkowski, G. P., Schuttelaars, H. M., and H. E. de Swart. 2003. The effect of geometry and bottom friction on local bed forms in a tidal embayment. *Continental Shelf Research* **22**, 11–13, 1821–33.

Schuttelaars, H. M., and H. E. de Swart. 1996. An idealized long-term morphodynamic model of a tidal embayment. *Eur. J. Mech.*, *B/Fluids* **15**, 55–80.

Schuttelaars, H. M., and H. E. de Swart. 2000. Multiple morphodynamic equilibria in tidal embayments. *Journal of Geophysical Research* **105**, C10, 105–24.

Seminara, G., and M. Tubino. 2001. Sand bars in tidal channels. Part one: free bars. *Journal of Fluid Mechanics* **440**, 49–74.

Solari, L., and M. Toffolon. 2001. Equilibrium bottom topography in tidal meandering channel: preliminary results. Paper presented at IAHR-RCEM Symposium 2001, Japan, 1—14 September.

Solari, L., Seminara, G., Lanzoni, S., Marani, M., and A. Rinaldo. 2002. Sand bars in tidal channels. Part 2 Tidal meanders. *Journal of Fluid Mechanics* **451**, 203–38.

Speer, P. E., and D. G. Aubrey. 1985. A study of non-linear tidal propagation in shallow inlet/estuarine systems Part II: Theory. *Estuarine, Coastal and Shelf Science* **21**, 207–24.

Stive, M. J. F., Capobianco, M., Wang, Z.B., Ruol, P., and M.C. Buijsman. 1998. Morphodynamics of a tidal lagoon and the adjacent coast. In D. J. Dronkers and M. B. A. M. Scheffers (eds.), *Physics of Estuaries and Coastal Seas*. Rotterdam: Balkema, 397–407.

Tambroni, N., Bolla Pittaluga, M., and G. Seminara. 2002. Morphodynamics of tidal channels: experimental observations. In Scientific Research and Safeguarding of Venice: CORILA research program, 2001 results. Venice: CORILA, Instituto Veneto di Scienze, Lettere ed Arti, pp. 675–89.

Tambroni, N., and G. Seminara. 2004. What is the yearly loss of sediments through the inlets of Venice Lagoon? In Proceedings of XXIX Convegno di Idraulica e Costruzioni Idrauliche, Trento, 7–10 September.

Scientific paper

45 • Modelling the morphodynamics of tidal inlets

N. TAMBRONI, P. K. STANSBY AND G. SEMINARA

SUMMARY

The morphodynamic evolution of a model of tidal inlet, consisting of a straight channel closed at one end and connected at the other end with a 'sea' basin, has been investigated numerically. Results suggest that a 2-D depth-averaged model for the inlet region, coupled with a 1-D model for the channel, captures main features of the near inlet region. In particular, the flow field at the inlet turns out to be highly asymmetric throughout each tidal cycle, being characterised by a jet-like flow with the presence of large-scale coherent structures during the ebb phase and an accelerating, convergent and near irrotational flow into the channel during the flood phase. Coupling the hydrodynamics with the morphodynamics, the tendency to long term morphodynamics equilibrium has also been investigated. Preliminary results show qualitative agreement with the recent experimental measurements of Tambroni *et al.* (2002). Results also suggest that the 2-D nature of the flow field in the near inlet region significantly affects the equilibrium depth reached at the inlet cross section.

INTRODUCTION

The morphodynamic evolution of Venice Lagoon is ultimately due to the net exchange of sediments with the adjacent sea, occurring through tidal inlets. Such an exchange is partly associated with the transport of sand as bedload and suspended load with intensities which may be significantly asymmetric throughout the tidal cycle as a result of the harmonic content of the forcing oscillation, of the availability of sediments in the far field and of the asymmetry of the flow field in the flood and ebb phases driven by the inlet shape. Very fine material, resuspended by the action of wind in salt marshes and transported by tidal currents, is unable to settle in the channels and is directly lost into the sea. This was briefly discussed in a companion chapter (Seminara *et al.*, chapter 44) where the scene was set for this more focussed contribution. Making detailed short term predictions, given the complexity of the above problem, calls for the need to develop our understanding through a close interaction between theoretical and experimental investigations of the process reproduced under controlled conditions.

A set of experiments was then performed in the laboratory of the Department of Environmental Engineering of the University of Genova, where a model of a tidal channel was built. The channel was bounded by an inlet adjacent to a basin where a 'tidal' oscillation was generated. Both the channel and the basin had a mobile bottom consisting of light material (crushed hazelnuts) and the experiment was continued until the bottom of the channel and of the basin approached equilibrium. The major outcome of the experiment was indeed the observation that the bottom profile does approach equilibrium conditions characterised by vanishing net sediment flux in a tidal cycle: the bed profile at equilibrium developed a shore in the inner part of the channel and reached an equilibrium depth at the inlet, while an outer delta formed in the 'sea' region opposite the inlet. The experimental measurements (reported in detail elsewhere, Tambroni *et al.*, 2002) will be used as a test case for the validation of the numerical model presented herein.

The theoretical approach we employ to investigate the morphodynamics of the inlet region is based on the classical 2-D depth averaged shallow water equations for the fluid phase, coupled with the Exner equation and some semi-empirical closure relation-

Flooding and Environmental Challenges for Venice and its Lagoon: State of Knowledge, ed. C. A. Fletcher and T. Spencer.
Published by Cambridge University Press, © Cambridge University Press 2005.

ship for the solid phase. The problem is solved numerically by extending the recent approach of Stansby (2003) to the present mobile bed case. The 2-D model of the inlet region is then coupled to the 1-D model for the hydrodynamics and morphodynamics of the tidal channel developed by Bolla Pittaluga and Seminara (2002) and then applied to the test case.

The main limit of the present approach is connected to the inability of a 2-D analysis to reproduce 3-D effects associated with the occurrence of upwelling currents and the shear layers shed by the inlet corners. In spite of this, results described below are encouraging as the model appears to grasp the main features of the observed process and predict the equilibrium morphology fairly satisfactorily.

The chapter is organized as follows. The next section is devoted to the theoretical formulation of the model and is followed by a description of some hydrodynamic and preliminary morphodynamic results, as well as some comparisons with experiments. The chapter concludes with some discussion and final remarks.

MATHEMATICAL FORMULATION

The governing equations

The mathematical problem can be formulated taking advantage of the fact that phenomenon to be investigated is such that the horizontal scales are much larger than the vertical depth-limited scale, a feature which suggests that the shallow water equations can be safely employed to describe the hydrodynamics. The limits of the present approach are related to the local role of 3-D effects which 2-D models are obviously unable to reproduce. Moreover, the morphological time scale is typically much larger than the flow time scale, which implies that a quasi steady approach can be used for the flow field. Recalling that the evolution of the bed interface is governed by the 2-D form of Exner's (1925) equation, we end up with the following differential problem:

$$\frac{\partial h}{\partial t} + \frac{\partial (hu)}{\partial x} + \frac{\partial (hv)}{\partial y} = 0 \qquad Eq.\ 45.1$$

$$\frac{\partial (hu)}{\partial t} + \frac{\partial (hu^2)}{\partial t} + \frac{\partial (huv)}{\partial t} = -gh\frac{\partial h}{\partial x} - gh\frac{\partial z_b}{\partial x} - \frac{\tau_{bx}}{\rho}$$
$$+ \frac{\partial}{\partial x}\left(2v_e h\frac{\partial u}{\partial x}\right) + \frac{\partial}{\partial y}\left(v_e h\left(\frac{\partial u}{\partial y} + \frac{\partial v}{\partial x}\right)\right) \qquad Eq.\ 45.2$$

$$\frac{\partial (hv)}{\partial t} + \frac{\partial (huv)}{\partial x} + \frac{\partial (hv^2)}{\partial y} = -gh\frac{\partial h}{\partial y} - gh\frac{\partial z_b}{\partial y} - \frac{\tau_{by}}{\rho}$$
$$+ \frac{\partial}{\partial x}\left(v_e h\left(\frac{\partial v}{\partial x} + \frac{\partial u}{\partial y}\right)\right) + \frac{\partial}{\partial y}\left(2v_e h\frac{\partial v}{\partial y}\right) \qquad Eq.\ 45.3$$

$$\frac{\partial z_b}{\partial t} = -\frac{1}{1-p}\left(\frac{\partial q_{sx}}{\partial x} + \frac{\partial q_{sy}}{\partial y}\right) \qquad Eq.\ 45.4$$

Following the notation of Fig. 45.1, in Eqs. (45.1-45.4), t denotes time, h is the local water depth, z_b is bed elevation, u and v are the local depth-averaged components of velocity in the x and y directions respectively, q_{sx} and q_{sy} are the corresponding components of the depth averaged sediment flux which accounts for both bed load and suspended load, p is water density and v_e is the eddy viscosity obtained from a turbulence model given below.

The bottom stresses τ_{bx} and τ_{by} are evaluated based on the following classical formula:

$$(\tau_{bx}, \tau_{by}) = (u,v)\rho C_f \sqrt{u^2 + v^2} \qquad Eq.\ 45.5$$

where C_f is the friction coefficient which, for the case of plane undisturbed bed, may be given the classical logarithmic form:

$$C_f = \left[6 + 2.5\ln\left(\frac{h}{2.5d_s}\right)\right]^{-2} \qquad Eq.\ 45.6$$

Here d_s is the grain size assumed to be uniform and 2.5 d_s is used as roughness height after Engelund and Hansen (1967). Note that Eqs. (45.5) and (45.6) essentially extend the Chézy law (Chézy, 1776) to a slowly varying 2-D context. The above formulation requires that the governing equations be supplemented with closure relationships for the eddy viscosity and the sediment flux.

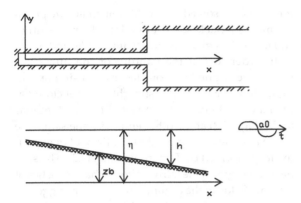

Fig. 45.1   Sketch of a channel connected to a rectangular basin and notation.

As concerns eddy viscosity, Stansby (2003) has recently proposed a general 3-D form for v_e with negligible vertical velocity, which has been reduced to a 2-D form for the present application in the following way:

$$v_e = \left(l_h{}^4 \left[2\left(\frac{\partial u}{\partial x}\right)^2 + 2\left(\frac{\partial v}{\partial y}\right)^2 + \left(\frac{\partial v}{\partial x} + \frac{\partial u}{\partial y}\right)^2 \right]^{\frac{1}{2}} + (\beta u_* h)^2 \right) + v$$

Eq. 45.7

where l_h is a horizontal length scale, u_* is the friction velocity, and β is a constant.

The term involving β accounts for vertical mixing and in parallel flow Stansby (2003) finds $\beta = 0.0067$. However, here β also accounts for dispersion and the effect of the horizontal length scale on vertical mixing which in turn affects the bed shear stress (and dispersion). There are thus three disposable parameters, l_h, β and C_f which describe complex coupled physical effects and may be tuned for a particular application. In this case we set l_h equal to a value typical of a two-dimensional jet (0.14 times the channel width) postulating that this is appropriate for eddy formation at the channel exit. C_f is defined by formula (45.6) and results are insensitive to β in the range 0.067 to 0.2. The sediment flux still remains to be defined. For the bedload contribution we assume:

$$(q^b_{sx}, q^b_{sy}) = (\cos\gamma, \sin\gamma)\Phi\sqrt{(s-1)gd_s^3}$$

Eq. 45.8

where s is the relative density of sediments, g is gravity, and γ is the deviation angle of the direction of the bed load flux relative to the longitudinal direction due to deviation of bottom stress and gravitational effects (Seminara, 1998).

Neglecting the effect of the longitudinal bed slope (a restriction which can be easily removed) the deviation angle may be given the following form:

$$(\cos\gamma, \sin\gamma) = \left(\frac{u}{\sqrt{u^2 + v^2}}, \frac{v}{\sqrt{u^2 + v^2}} - \frac{r}{\sqrt{\tau_*}} \frac{\partial z_b}{\partial y} \right)$$

Eq. 45.9

with r an empirical parameter ranging about 0.5-0.6 (0.56 according to Talmon et al.,1995).

Moreover, Φ is the intensity of bed load, a monotonically increasing function of the excess Shields stress which can be given the classical Meyer-Peter and Müller (1948) form:

$$\Phi = 8(\tau_* - \tau_{*c})^{\frac{3}{2}}$$

Eq. 45.10

where the Shields stress is defined as:

$$\tau_* = \frac{u_*^2}{(s-1)gd_s}$$

Eq. 45.11

The flux of suspended sediment is calculated assuming local equilibrium, i.e. employing the classical Rouse form of the vertical distribution of sediment concentration associated with the local and instantaneous characteristics of the flow field. The resulting expression for q^s_{sx} and q^s_{sy}, reads:

$$(q^s_{sx}, q^s_{sy}) = (u,v)h\psi$$

Eq. 45.12

with

$$\psi = \frac{\sqrt{C_f}}{k} C_e [I_2 + K_1 I_1]$$

Eq. 45.13

Here k is von Kàrmàn constant, C_e is the reference concentration for which various semi-empirical expressions have been proposed in the literature (e.g. van Rijn, 1984), while K_1, I_1 and I_2 are integral functions depending on two parameters, namely:

- ζ_R, the conventional dimensionless value of the reference elevation where the boundary condition is imposed under uniform conditions;
- the Rouse number Z defined in terms of the ratio between particle fall velocity W_s and friction velocity u_* as follows:

$$Z = \frac{W_s}{ku_*} \qquad\qquad Eq.\ 45.14$$

Note that the suitability of the assumption of local equilibrium for suspended sediment has been checked using the recent approach of Bolla Pittaluga and Seminara (2003) which provides a framework able to account for slow variations of the flow field both in space and in time. Preliminary tests suggest that non equilibrium effects are not crucial.

The numerical approach

The solution of the complete unsteady equations for the liquid (Eqs. 45.1–45.3) and solid phase (Eq. 45.4) must be obtained numerically.

The 2-D numerical approach will only be summarized here, being similar to that proposed for the 3-D shallow water flows by Stansby and Lloyd (2001). Quick upwind scheme is used to discretize the advection terms while all the other spatial discretizations are central difference, giving second order accuracy. Temporally the advection and diffusion terms are treated explicitly using second order Adams Bashforth. As regards the bed shear stress and the depth gradient, they are respectively discretized using the semi-implicit 1st order and the 2nd order Crank-Nicolson methods. The expressions for hu and hv, determined from the momentum equations (45.2, 45.3), are substituted into the mass conservation equation (2.1) forming a pentadiagonal equation set

for h which is solved by an efficient conjugate gradient method. Back substitution into the momentum equations then gives the new values of hu and hv.

In order to test the validity of the numerical model, we applied it to our test case: a simple configuration consisting of a straight channel closed at the landward end and connected to a rectangular basin (Fig. 45.2) at the other end, with the same size as the apparatus employed in the first set of experiments performed by Tambroni *et al.* (2002). The simple geometry and the fairly uniform characteristics of the flow fields in the channel cross section suggested the opportunity to couple the 2-D model used for the flow in the basin to a 1-D model sufficient to reproduce the morphodynamic evolution of the channel. The main advantage of the latter coupling is a substantial reduction of the computational effort when compared with a fully 2-D model. The one dimensional equations governing mass and momentum conservation in the channel, coupled with the one dimensional equation of mass conservation of the solid phase, were solved by means of a classical McCormack scheme; for details on the numerical model the reader is referred to Bolla Pittaluga and Seminara (2002). Clearly the equations must be supplemented by appropriate boundary conditions, through which the 1-D and 2-D models are actually coupled.

As regards the hydrodynamics, at the seaward boundary of the basin the free surface elevation is imposed, according to the following formulation:

$$h\big|_{x=Lt} = h_0 = a_0 \cos\left(\frac{2\pi t}{T}\right) \qquad Eq.\ 45.15$$

where h_0 is the initial mean flow depth, a_0 is the amplitude of the tidal wave and T is the tidal period.

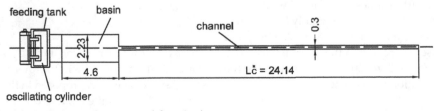

Fig. 45.2   Sketch of the experimental apparatus (plan view).

Note that, we have concentrated our attention on the hydrodynamical response of the tidal system to an M2 forcing oscillation, ignoring, for the moment, the effect of higher order harmonics. Since the basin has vertical sides, a 'log' law is assumed for cells adjacent to walls with an assumed roughness height of 1 mm. Furthermore, the normal component of velocity was set to vanish at the walls, hence:

$$v\big|_{y=0} = 0, \qquad v\big|_{y=Bb} = 0, \qquad u\big|_{x=0} = 0 \qquad Eq.\ 45.16$$

As regards the boundary conditions imposed at the inlet, the value of the longitudinal velocity obtained from the 1-D solution was forced to solve the 2-D flow field in the basin. Reciprocally the solution for the flow field in the basin provides the water elevation at the inlet forcing the tide propagation in the channel. The formulation of the morphodynamic problem is completed by imposing the condition that the instantaneous sediment flux at the closed ends of the system must vanish. The conditions on the sediment flux imposed at the inlet basin need further discussions. The exchange of sand through the channel inlet occurring in the field results from a complex phenomenon where various factors may play a significant role. In particular, two of them are of prevailing importance:

1) the hydrodynamics of the inlet region and in particular its ebb-flood asymmetry, which depends on the inlet shape; and
2) the availability of sediments in the far field, a feature depending on the wave climate and on the presence of littoral currents.

In the present work the latter effect has been ignored as it could not be reproduced in the experiments of Tambroni et al. (2002) which provide the test case for the numerical model. Extending the present approach in order to remove the above restriction will require some empirical correlation between wave climate and concentration of suspended sediments in the far field. On the contrary, the exchange of sediments between the channel and the basin has been taken in account through the following scheme: the sediment flux during the ebb phase has been assumed to be determined by the transport capacity

of the stream at the inlet section of the channel, while the sediment flux during the flood phase has been taken to be driven by the flow field evaluated in the inlet region of the basin. This approach will be shown to provide more accurate predictions, as compared with those of a simpler 1-D model for the morphodynamic evolution of a tidal channel. The computation was initiated with a flat bottom and the problem was solved by calculating at each time step, firstly the flow-field over the whole domain and then advancing the bed level. The numerical simulation is characterised by the same set of parameters used in the experiment: $h_0 = 0.082$ m, $a_0 = 0.024$ m and $T = 180$ s. The geometry of the channel and of the basin was chosen to be identical to that employed in the experiment. The flow field in the basin was evaluated by solving the shallow water equations on a staggered grid (159×63 nodes, mesh size 0.03×0.03 m^2) while the number of nodes along the channel was set equal to 50.

RESULTS

Hydrodynamics

As pointed out by Blondeaux et al. (1982) when the bed is fixed and plane, the flow field at tidal inlets is highly asymmetric throughout each tidal cycle. In fact, during the flood phase, the channel acts as an unsteady sink, which gives rise to a nearly irrotational 2-D flow pattern. During the ebb phase, vorticity is continuously shed from the separation points at the sharp edges of the inlet and gives rise to an unsteady turbulent jet, characterised by the formation of a pair of counter-rotating vortices. These leave the generation area under the effect of the induced velocity that each vortex determines on the other. An asymmetry of the flow field at the inlet of the tidal channel was displayed also in the present experiments. A comparison between the surface velocity measured in the basin by a particle tracking technique during the initial phase of the experiment (when the bottom was still flat) and the flow field obtained from the present computations is shown in Figure 45.3a, b.

Note that the numerical model is able to capture the qualitative behaviour of the flow field observed

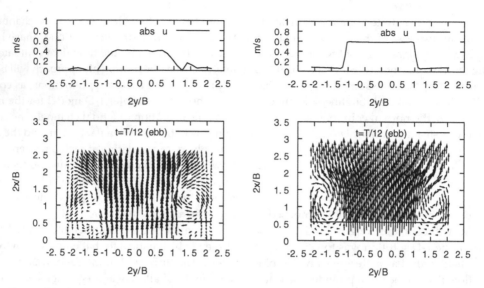

Fig. 45.3a   Comparison between experimental (left) and numerical (right) flow fields at the inlet during the ebb phase. The upper graphs show the velocity distribution at a cross section located at a given distance from the inlet ($x = 0.075$m). In the lower plots the vertical and horizontal axis are respectively the longitudinal and transversal axis of the basin scaled with half width of the channel.

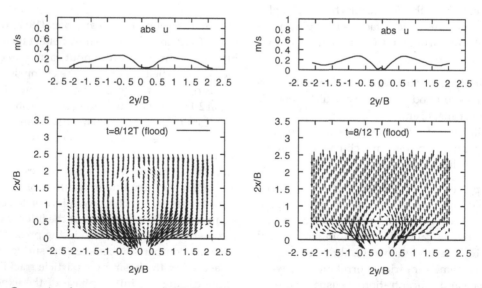

Fig. 45.3b   Comparison between experimental (left) and numerical (right) flow fields at the inlet during the flood phase. The upper graphs show velocity distribution at a cross section located at a given distance from the inlet ($x = 0.075$ m). In the lower plots the vertical and horizontal axis are respectively the longitudinal and transversal axis of the basin scaled with half width of the channel.

near the inlet during both the flood and the ebb phase, though the computations appear to somewhat overestimate the flow speed. This may be due to the inability of the present model to reproduce accurately the dissipative role of the shear layers.

Morphodynamics

A sequence of topography fields of the bottom of the basin region close to the inlet is shown at different times in Fig. 45.4a, b. It appears that, starting from an initial flat bottom, the strong ebb current coming out of the channel leads to an intense bed scour and to the excavation of a submerged channel which gradually deepens and progresses seaward.

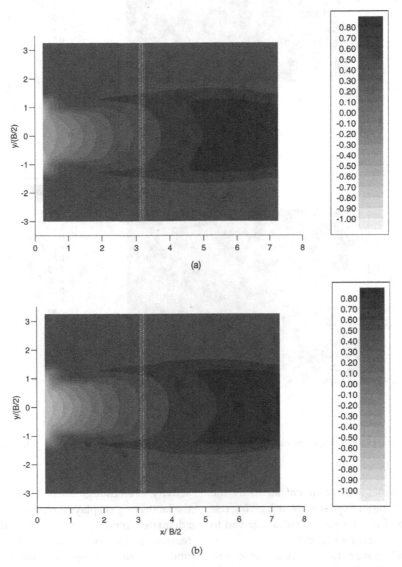

Fig. 45.4 Evolution of the topography field in the inlet region: (a) after 20; (b) after 80 tidal cycles. The vertical and horizontal axis are respectively the longitudinal and transversal axis of the basin scaled by half channel width (B/2), while the bottom elevation is scaled by the initial mean flow depth (h_0).

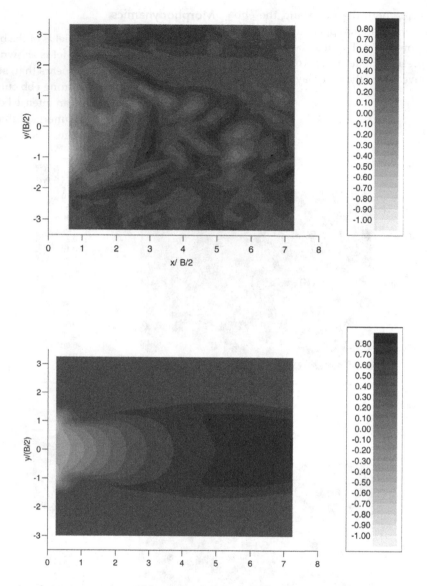

Fig. 45.5 Comparison between experimental (upper) and numerical (lower) topography field in the inlet region after 20 tidal cycles.

At the same time, sediments entrained by the scouring action of the jet deposit seaward, giving rise to the formation of an elongated central bar and to the generation of a sort of 'outer delta'.

A comparison between the numerical and experimental patterns of the bottom of the basin after 20 cycles (Fig. 45.5) shows a reasonable agreement as regards the magnitude and position of the scour activity, even though the bed evolution measured in the laboratory displays a patchy pattern which is not adequately reproduced by the calculations. The latter feature seemingly arises from the complex interaction between the large scale coherent structures generated by the shear layer issued from the sharp edges of the inlet and the mobile bed, an effect not easy to reproduce in a 2-D model. Obviously a shallow water

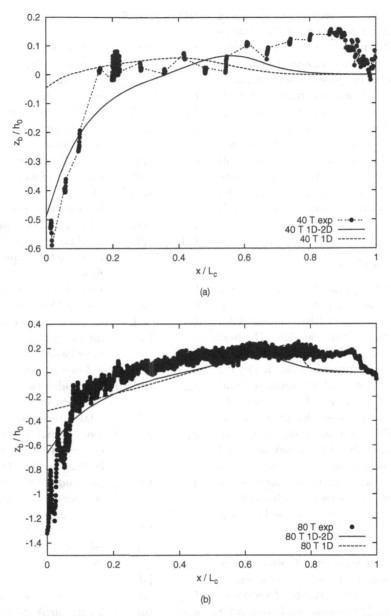

Fig. 45.6 Comparison between the longitudinal bed profiles after 40 (a) and 80 (b) tidal cycles. Observed in experiment calculated by the 1-D-2-D numerical model, which accounts for the exchange of sediments induced at the inlet by the 2-D character of the flow field in the basin; and obtained from a simple 1-D numerical model which ignores the latter effect.

model cannot fully capture the dispersive effects associated with the character of the vertical velocity distribution.

As regards the evolution of the bed profile in the channel, the resulting asymmetries in surface eleva-

tions and tidal currents lead to a net sediment flux directed landward at each tidal cycle. This mechanism is associated with the flood dominant character of the flow field in the initial stage of the morphodynamic evolution process. Figure 45.6 reports a comparison between the longitudinal bed profiles of the channel observed in the experiment, those calculated by the 1-D to 2-D numerical model described herein and those obtained from a simple 1-D numerical model which ignores the presence of the basin.

At the initial stage of the experiment sediments are scoured in the seaward portion of the channel. A fairly sharp front forms in the bed profile and migrates landward. Note that the present numerical approach, accounting for the exchange of sediments induced at the inlet by the 2-D character of the flow field in the basin, predicts the magnitude of the scour at the inlet section more accurately than the 1-D approach.

CONCLUDING REMARKS

The preliminary results reported here suggest that a shallow water model can be a useful tool to investigate the exchange of sediments driven through tidal inlets by the 2-D nature of the flow field. The next major step is to include the forcing effects associated with the concentration distribution generated by the action of breaking and non-breaking waves in the far field, as well as with the role of littoral currents. However, it must be emphasised that the computational effort required by the present approach, even under the somewhat idealized conditions analysed herein, is too large for long term predictions of the kind needed for applications to the long term evolution of Venice Lagoon. This notwithstanding, results obtained by the present model may be used for various purposes:

- firstly, the effect of possible modifications of the inlet setting can be at least qualitatively assessed;
- secondly, one may attempt to construct a relationship between the hydrodynamic conditions in the far field and the sediment exchange at the inlet, an ambitious goal which could be usefully incorporated in the context of simpler morphological models of the inlet (of the type discussed in the chapter by Seminara *et al.*, chapter 44);
- finally such a relationship would play the role of a simple boundary condition for morphological models of the whole lagoon, reducing the large computational effort they require.

NOTATIONS

a_0	[m]	amplitude of the tidal wave imposed at the seaward boundary of the basin.
B	[m]	width of the channel.
B_b	[m]	width of the basin.
C_e	[/]	equilibrium concentration at the reference elevation zR.
C_f	[/]	friction coefficient.
d_s	[m]	diameter of sediment particles.
g	[m s⁻²]	gravity.
h	[m]	local water depth.
h_0	[m]	initial mean flow depth.
I_1, I_2, K_1	[/]	coefficients appearing in the formula for the suspended sediment flux.
k	[/]	von Karman constant.
l_h	[m]	vorticity horizontal length scale.
L_c	[m]	length of the channel .
L_t	[m]	total length of the system (channel + basin).
p	[/]	sediment porosity.
q_{sx}, q_{sy}	[m² s⁻¹]	longitudinal and transversal components of depth averaged total sediment flux.
q^b_{sx}, q^b_{sy}	[m² s⁻¹]	longitudinal and transversal components of depth averaged bedload sediment flux.
q^s_{sx}, q^s_{sy}	[m² s⁻¹]	longitudinal and transversal components of depth averaged suspended sediment flux.
r	[/]	empirical factor appearing in (45.9)
s	[/]	relative density of sediment particles.
t	[s]	time.
T	[s]	period of the tidal wave.
$u \, v$	[m s⁻¹]	longitudinal and transversal local depth-averaged components of velocity.
u_*	[m s⁻¹]	friction velocity.
W_s	[/]	particle fall velocity.
$x \, y$	[m]	longitudinal and transversal coordinates.
z_b	[m]	local bed elevation.
Z	[/]	reference Rouse number.
β	[/]	vertical mixing constant.

Φ	[/]	dimensionless intensity of bedload.
γ	[/]	deviation angle of the direction of the bed load flux relative to longitudinal direction.
η	[m]	local free surface elevation.
ρ	[kg m^{-3}]	density of the fluid phase.
τ_*	[/]	Shields' stress.
τ_{*c}	[/]	critical Shields' stress value for sediment transport.
τ_{bx}, τ_{by}	[N m^{-2}]	longitudinal and transversal components of bottom stresses.
υ	[m^2 s^{-1}]	cinematic viscosity.
υ_e	[m^2 s^{-1}]	eddy viscosity.
Ψ	[/]	dimensionless suspended sediment flux.
ζ_R	[/]	conventional dimensionless reference elevation for concentration.

ACKNOWLEDGEMENTS

The work of the first and last authors has been supported by CORILA and by the PRIN project 'Morfodinamica dei sistemi a marea' co-funded by the University of Genova and MIUR (Italian Ministry for Higher Education and Scientific Research). This work is also part of the PhD dissertation of the first author, to be submitted to the University of Genoa in partial fulfilment of her degree. The authors wish to thank M. Bolla Pittaluga for making available his 1-D numerical code to model the morphodynamics of tidal channels.

REFERENCES

Blondeaux, P., De Bernardinis, B., and G. Seminara. 1982. Correnti a marea in prossimità di imboccature e loro influenza sul ricambio lagunare. In Proceedings of XVIII Conv. Idraulica e Costruzioni Idrauliche, Bologna: 21–23 September.

Bolla Pittaluga, M., and G. Seminara. 2002. Modelling suspended sediment transport in tidal flows: theory and application to long term equilibrium of tidal channels. In *Scientific Research And Safeguarding Of Venice: Corila research program, 2001 results*. Venice: CORILA, Istituto Veneto di Scienze, Lettere ed Arti, pp. 661–74.

Bolla Pittaluga, M., and G. Seminara. 2003. Depth integrated modelling of suspended sediment transport. *Water Resources Research* **39**, 5, 1137, 1/1–1/11.

Engelund, F., and E. Hansen. 1967. *A Monograph on Sediment Transport in Alluvial Streams*. Copenhagen: Danish Technical Press.

Exner, F. M. 1925. Uber die wechselwirkung zwisehen Wasser und geschiebe in Flussen. *Sitzer Akad. Wiss. Wien*, 165–80.

Lanzoni, S., and G. Seminara. 2002. Long term evolution and morphodynamic equilibrium of tidal channels. *Journal of Geophysical Research* **107**, C1, 1–13.

Meyer-Peter, E., and R. Müller. 1948. Formulas for bedload transport. *Conference of the International Association of Hydraulic Research*, Stockholm, Sweden, pp. 39–64.

Seminara, G. 1998. Stability and morphodynamics. *Meccanica* **33**, 1, 59–99.

Stansby, P. K. 2003. A mixing length model for shallow turbulent wakes. *Journal of Fluid Mechanics* **495**, 369–84.

Stansby, P. K., and P. M. Lloyd. 2001. Wakes formulation around islands in oscillatory laminar shallow-water flows: Part II Boundary layer modelling. *Journal of Fluid Mechanics* **429**, 239–54.

Talmon, A. M., Struiksma, N., and M. C. L. M. Van Mierlo. 1995. Laboratory measurements of the sediment transport on transverse alluvial-bed slopes. *Journal of Hydraulic Research* **33**, 519–34.

Tambroni, N., Bolla Pittaluga, M., and G. Seminara. 2002. Morphodynamics of tidal channels: experimental observations. In *Scientific Research and Safeguarding of Venice: CORILA research program, 2001 results*. Venice: CORILA, Istituto Veneto di Scienze, Lettere ed Arti, pp. 675–89.

Van Rijn, L. C. 1984. Sediment transport, Part II: suspended load transport. *Journal of Hydraulic Engineering ASCE* **110**, 11, 1613–41.

46 • Application of hydrodynamic and morphological models

A. GOZZI AND G. MENEL LEMOS

INTRODUCTION

Since first studies were developed for the safeguarding of the lagoon and the city of Venice from high tides (*acqua alta*), mathematical models have been used to foresee water levels in the lagoon induced by tidal events. In the past, *ad hoc* mathematical models were developed and applied but with the evolution of models and increased power of computers, more and more advanced mathematical models have been developed, beginning with 0-D and 1-D models and then most recently 2-D and 3-D models. Initially, these models were used only as an analysis tool in order to provide hydrodynamic parameters for engineering and design purposes. However, more recently, these models have been used as an integrated tool, to provide hydrodynamics in the study of more complex phenomena, such as water quality or the morphological evolution of estuarine systems.

Recently the Magistrato alle Acque di Venezia, the Water Authority for the Lagoon of Venice, decided to develop a hydrodynamic mathematical model of the Lagoon of Venice. This new model uses a curvilinear mesh with finite differences schematization that can be easily connected to water quality and/or morphological modules, thus providing a hydrodynamic base for environmental analysis. To achieve this result, a number of schematizations of the selected model had to be developed and the model had to be calibrated in the best possible way. Usually models are calibrated using water levels but, to evaluate environmental processes, it is important to know flow distribution. To achieve this aim, a number of intensive measurement campaigns were undertaken to collect flow data under controlled situations.

In this chapter we briefly described how the model was applied to the Lagoon of Venice and how measurement campaigns for calibration purposes were organized. The main calibration results are presented. One of the first applications of the model was to evaluate effects of some interventions at the inlets of the lagoon. Some examples of results achieved during this study are presented.

MODELS DEVELOPED FOR THE LAGOON OF VENICE

The Lagoon of Venice is one of the most studied estuarine environments in the world. Since 1970 many hydrodynamic models have been developed to analyse and to understand how tidal waves propagate inside the lagoon. The main purpose of these models has been to reproduce, in an efficient way, time series of water levels in the most important locations inside the lagoon itself.

The first model developed was a 0-D model that was able to reproduce in a very good way levels at the 'Punta della Salute' location, the most important tide gauge in front of St Mark's Square in the historical centre of Venice. After this first exercise, many other models were issued *ad hoc* for the Lagoon of Venice, with different schematizations and algorithms of solution. The most important ones to be noted here are the 1-D model (Fig. 46.1), the first model able to represent the network of channels and to evaluate the propagation of tides into it, and 2-D finite element (Fig. 46.2) and finite difference models, able to evaluate the propagation of tidal waves both into the channels and on tidal flats. These models were, and sometimes still are, used for many different purposes such as to foresee levels in the city of Venice and in other locations in the lagoon, to begin

Flooding and Environmental Challenges for Venice and its Lagoon: State of Knowledge, ed. C. A. Fletcher and T. Spencer. Published by Cambridge University Press, © Cambridge University Press 2005.

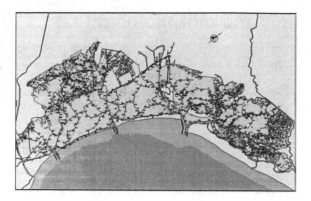

Fig. 46 1   1-D model.

to study the very complex hydrodynamics of the Lagoon of Venice and to evaluate effects of engineering interventions in Venice itself, in other historical centres of the lagoon (i.e. Murano, Burano, Chioggia, etc.) and all over the lagoon. The main type of analysis performed with these models has been related to water levels.

Since 1990 these models, mainly the 2-D finite elements model, have been coupled with specific

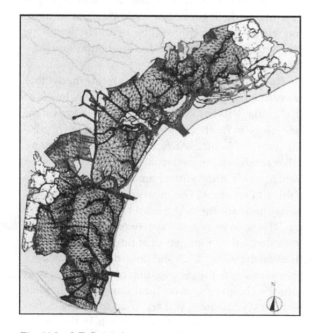

Fig. 46.2   2-D finite element model.

modules in order to perform morphological, water quality and ecological analyses, creating an integrated modelling approach with different models being constantly upgraded and calibrated.

Recently, morphodynamic and quality analysis has become more and more important and there has been a need to have available mathematical tools to evaluate local effects of restoration interventions. In this case the first model to be updated had to be the hydrodynamic model, the base for any morphological and ecological simulation.

This new model had to present the following requirements:

- to be a commercial model, internationally known, tested and certified;
- to have the possibility to perform simulations with schematizations both in 2-D and 3-D;
- to be usable with a user-friendly interface; and
- to have the possibility to be linked with morphological, water quality and ecological modules to perform complex simulations, trying to limit as much as possible the user's intervention during the coupling procedure.

The mathematical model, complying with all these requirements, chosen to develop the new hydrodynamic model of the Lagoon of Venice was Delft3D FLOW, the hydrodynamic module of a suite of models developed by the Delft Hydraulics laboratory (The Netherlands). This is a finite difference model working on a curvilinear grid. This suite also includes morphodynamic, water quality and ecological modules. The FLOW module of Delft3D is a multi-dimensional (2-D or 3-D) hydrodynamic (and transport) simulation programme which calculates non-steady flow and transport phenomena resulting from tidal and meteorological forcing on a curvilinear, boundary fitted grid. In 3-D simulations, the vertical grid is defined following the so-called sigma coordinate approach. This results in a high computing efficiency because of the constant number of vertical layers over the whole computational field.

SCOPE OF THE NEW MODEL

The objective of the hydrodynamic modelling of the Venice Lagoon and the Adriatic Sea is to obtain water levels and flow patterns in areas of interest inside the lagoon and near the inlets. Any environmental aspect of the lagoon is connected to tidal flows, to the amount of fresh water from the drainage basin and to the meteorological phenomena that directly act on the lagoon. To set-up a hydrodynamical model of the Venice Lagoon is a very complex matter, where the analysis of different physical systems, including the lagoon basin, the drainage basin and the Adriatic Sea, is required. To fulfil all the requirements needed by the Magistrato alle Acque in Venice, the water authority for which the model was developed, the lagoon had to be schematized in three different configurations.

Technical specifications

Technical specifications approved by the technical committee of the Magistrato alle Acque indicated the way that the morphological and emergent structures should be represented. Some requirements are indicated below:

- All the islands greater than 1–2 km² in area;
- All the main channels, for instance those which are connected to the inlets and all the secondary channels, meaning those channels more than 30 metres wide;
- The salt marshes and tidal flats (*barene* and *velme*) as they are represented in the last survey made by the Magistrato alle Acque;
- The main emergent structures, recognizable from the most recent surveys of the lagoon.

Schematization

The overall Venice Lagoon model consists of the following schematizations:

- *Basic Configuration* – the lagoon itself and the rivers discharging into it – the extent of the model is limited to the lagoon, with open boundaries at the entrances of the lagoon and the part of the rivers entering the lagoon which are influenced by the tides.
- *Complete System* – the lagoon itself, rivers discharging into it and part of the Adriatic Sea – the domain of the model is extended by including a

part of the Adriatic Sea covering a coastal stretch of approximately 40 km along shore and 15 km cross-shore.
- *City of Venice* – a detailed model of the channels of the Venice, where the domain is extended sufficiently far away from the city to include area of exchanges between the city itself and the lagoon. This model is connected to the general model of the lagoon using the domain decomposition techniques.

Numerical grid

The adopted mathematical model allows the schematization of the Lagoon of Venice with a curvilinear grid to reproduce the morphology and structures in the lagoon. The Venice Lagoon covers an area of about 400 km² with many irregular morphological elements (island, channels etc.) oriented in many different directions, distributed uniformly over the lagoon. To guarantee an accurate representation of the geometry of channels and shallow areas, the grid has been locally modified to follow channels and land boundaries as far as possible. The highest resolution is concentrated along the main channels, with dimensions of individual cells of about 25 to 50 m. In shallow water areas and where such high resolution was not necessary, the cell dimensions increase to about 150 m. To obtain the 'complete system', the model grid was extended including a part of the Adriatic Sea covering a coastal stretch of approximately 40 km along shore and 15 km cross-shore. For the 3 configurations of the computational grids, the number of cells is:

- Basic Configuration – ~75 000
- Complete System – ~85 000
- City of Venice – ~40 000

Bathymetry

The model bathymetry was generated using the data collected during the 2000 survey of the lagoon. For the sea, the bathymetry was generated using sample points derived from nautical maps (Istituto Idrografico della Marina – 1:50 000 scale) updated with the most recent alongshore survey (Study

Fig. 46.3   Grid and bathymetry of the complete system –Venice Lagoon model.

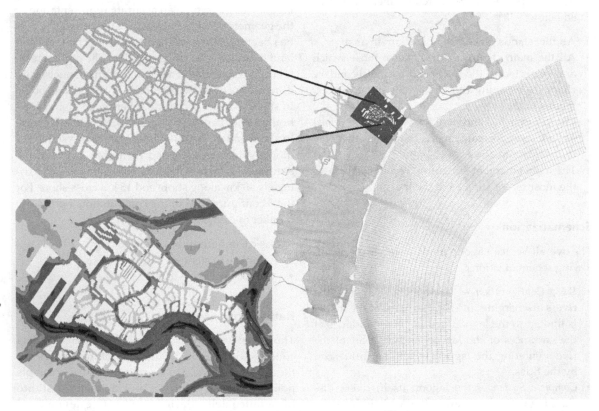

Fig. 46.4   Grid and bathymetry of the detailed model of the historical centre of Venice.

C.2.1/V – 'Monitoraggio dei litorali da Jesolo a Isola Verde'). The reference level for the bathymetry is Mean Sea Level (MSL).

INTENSIVE CAMPAIGNS

Beyond the geometrical requirements, technical specifications required that many hydrodynamic verifications had to be done. In fact, as stated above, the model must be able to reproduce not only water levels at various tide-gauges, but also to represent the flow distribution and velocity field in the main hydraulic nodes in the lagoon. To support the numerical modelling activities, a comprehensive field data collection programme was conducted. Some intensive campaigns were done at the eight main hydraulic junctions of the lagoon with the purpose of measuring, using a synoptic approach, water levels, temperature, salinity and turbidity in the channels and in the shallow waters that constitute the single hydraulic node. Location of the main hydraulic nodes and a scheme of measurement method, with parameters measured, are shown in Figs 46.5 and 46.6.

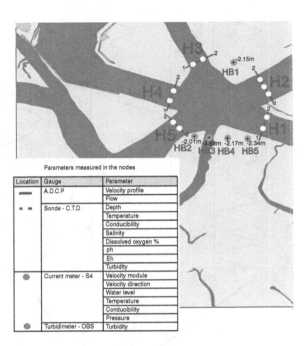

Location	Gauge	Parameter
—	A.D.C.P	Velocity profile
		Flow
– –	Sonde - C.T.D	Depth
		Temperature
		Conducibility
		Salinity
		Dissolved oxygen %
		ph
		Eh
		Turbidity
●	Current meter - S4	Velocity module
		Velocity direction
		Water level
		Temperature
		Conducibility
		Pressure
●	Turbidimeter - OBS	Turbidity

Fig. 46.6   Parameters measured in the nodes.

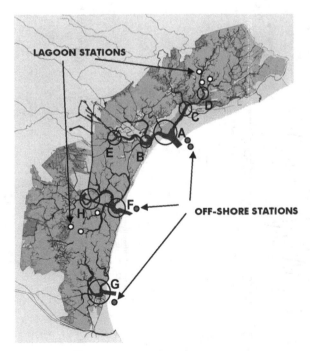

Fig. 46.5   Main hydraulic nodes and fixed stations.

The purpose of the field data collection was to characterize the physical processes in the Venice Lagoon, including the general circulation patterns and the specific circulation patterns in and near the inlets. In each node two different campaigns were carried out for the duration of 24 hours, for a total of 16 events. The campaigns took place under neap and spring tide conditions. The field data were collected as shipboard and *in situ* measurements, with real-time data acquisition.

In the channels, the measures were taken using three boats: two equipped with ADCP (Acoustic Doppler Current Profiler) for the measurement of the velocities and one equipped with a multiparametric CTD probe for the measurement of water quality parameters. In shallow water, the measurements were performed by the installation of S4 current meters and fixed OBS turbidimeters. To provide data to set-up the morphological model, 9 stations were installed (six in the lagoon and three along the littoral – see Fig. 46.5) equipped with automatic fixed instrumentation for the measurement of turbidity, temperature, salinity, current velocities and water levels.

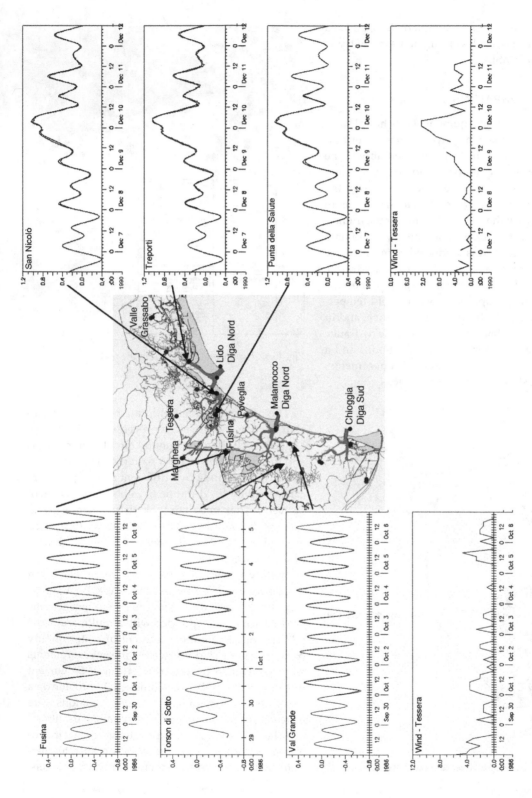

Fig. 46.7   Examples of simulated versus measured water levels (m) for calibration purposes.

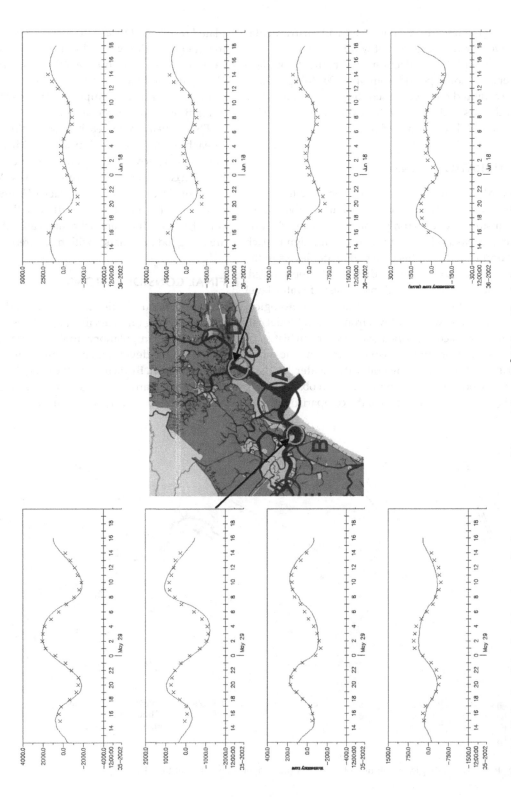

Fig. 46.8   Examples of simulated (continuous line) versus measured (crosses) fluxes (m^3/s) for calibration purposes.

Two of them were also equipped with instrumentation for the measurement of wind and waves.

The field data collection programme took place over a 12-month period from May 2002 to April 2003. The acquired data were used to define the boundary conditions and to calibrate/validate the hydrodynamic and dispersion models.

CALIBRATION RUNS

According to technical specifications, the hydrodynamic model had to be calibrated using five different historical events (only water levels) and the tide events measured during the field campaign in each hydraulic node. In this case both water levels in the lagoon and fluxes in the node had to be reproduced. The model forcing functions consist of water level elevation along the open boundary, meteorological data, and freshwater inflows from rivers. Model calibration was accomplished using data from historical events and from the intensive hydrodynamic monitoring programme. The computed water levels at various locations were compared with observations. In Fig. 46.7 some examples of the comparison of sim-

ulated results and measured water levels are shown, together with tide-gauge locations. The model was calibrated not only on water levels, but, as said above, also on fluxes. This was one of the main goals of the measurement campaigns at the main hydrodynamic junctions of the lagoon.

Fluxes measured in each cross-section belonging to each junction were simulated with the model. Figure 46.8 shows some examples of comparisons between computed and measured fluxes. As it can be seen, for water levels the accuracy of simulated values is very high and even for fluxes, that are more complex to evaluate and measure, calibration results are in a good accordance with measured values.

FINAL CONSIDERATIONS

The application of a finite difference hydrodynamic model working on a curvilinear grid has allowed the creation of a computational tool which can be used as a base for the development of future models of the Venice Lagoon. By increasing the spatial resolution of the grid schematization, it was possible to evaluate in a more detailed way internal water exchanges,

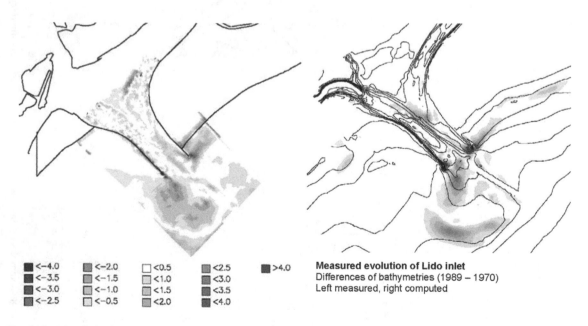

■ <−4.0	■ <−2.0	☐ <0.5	▦ <2.5	■ >4.0
■ <−3.5	▦ <−1.5	☐ <1.0	▦ <3.0	
■ <−3.0	▦ <−1.0	▦ <1.5	▦ <3.5	
■ <−2.5	☐ <−0.5	▦ <2.0	■ <4.0	

Measured evolution of Lido inlet
Differences of bathymetries (1989 – 1970)
Left measured, right computed

Fig. 46.9 Morphological evolution of Lido inlet (in co-operation with Delft Hydraulics).

mainly in the marginal zones of the lagoon. The evaluation of these processes of water exchange is very important, not least for the future models, both morphological and ecological, that will be prepared by the Magistrato alle Acque and the Consorzio Venezia Nuova. It must be underlined that the main goal of this model is the hydrodynamic description of Venice Lagoon as a whole. Detailed analysis in specific areas of the lagoon will be undertaken using domain decomposition techniques (2-way nesting); this is an existing capability of the current model.

The Venice Lagoon model has been verified against water level and flux data for several events. The overall reproduction of the tide is in good agreement with the observations, even in the most northern and most southern parts of the lagoon. The modelled total fluxes through the inlets have also been shown to be in good agreement with observations. The model, as it was defined during activities described above, is nearing completion. Further activities will be:

- Final calibration of the complete system and the city of Venice system; and
- Final calibration of the dispersive module.

At present the model is being used as the hydrodynamic base for the new morphodynamic model Consorzio Venezia Nuova-Magistrato alle Acque which is being developed.

At present a subset of the hydrodynamic model is being used as a base for studying the morphological evolution of the lagoon. Figure 46.9 shows calibration results for the Lido inlet.

47 • An open source model for the Venice Lagoon and other shallow water bodies

G. UMGIESSER, C. L. AMOS, E. CORACI, A. CUCCO, C. FERRARIN,
D. MELAKU CANU, I. SCROCCARO, C. SOLIDORO AND L. ZAMPATO

INTRODUCTION

In the past a consistent effort has been put into the modelling of the physical, chemical and biological aspects of the Venice Lagoon. Due to its complicated bathymetry, the Venice Lagoon is one of the most complex coastal systems in Italy, and due to this difficulty, along with extensive field campaigns, modelling techniques must be used if some understanding of this system is to be achieved.

Finite difference hydrodynamic models have been applied in the past to the Venice Lagoon by Chignoli and Rabagliati (1975), Volpi and Sguazzero (1977) and Sguazzero et al. (1978). In 1987 the Danish Hydraulic Institute set up an operational model for the Venice Lagoon that was used in planning the sluice gates that were to be built in the lagoon (Marchi et al., 1992). All these models used the finite difference method with a grid size of 300 m or more. Recently a higher resolution model with a grid size of 100 m was implemented by Casulli and Cheng (1992) and Casulli and Cattani (1993). This model can describe the channel network faithfully, but the computational demand of such a high resolution model is quite high. However, all finite difference models have suffered from the same problem: the need of a small grid size to resolve the narrow channels inside the lagoon that has to be imposed everywhere. A compromise between accuracy and computer time has to be made.

The finite element or finite volume method, on the other hand, allows for more flexibility with its subdivision of the system into triangles varying in form and size. One such approach has been used by Casulli and Zanolli (1998), where a model with an unstructured grid was presented. This finite volume approach was then extended to the resolution of the non-hydrostatic equations. It has also been applied by Casulli and Zanolli (2002) to the Venice Lagoon with promising results.

The Lagoon of Venice has been a case study for ecological models too. Contributions in the peer reviewed literature started to appear in the1980s, with a number of contributions on modelling macroalgae blooms, first by using a 0-D representation (Solidoro et al., 1997b; Bocci et al., 1997; Coffaro and Bocci, 1997) and then within a 3-D coupled model (Solidoro et al., 1997a). Subsequent contributions have dealt not only with the growth, fishing and rearing of clams (Solidoro et al., 2000; Pastres et al., 2001) but also with using a methodology for the identification of Maximum Permissible Loads compatible with a predefined Water Quality Target (Pastres et al., 2003). More recent contributions include the definition of VELFEEM, a finite element ecological model for the Lagoon of Venice (Umgiesser et al., 2003a) and the first applications of this coupled model (Melaku Canu et al., 2001; Melaku Canu et al., 2003).

At the Venice Institute of the CNR (National Research Council) a finite element model (SHYFEM) has been developed that was conceived as a model suitable for the application to basins with the characteristics of the Venice Lagoon. This model has been enhanced with modules that treat wave, sediment and ecological processes. This model has been applied not only to the Venice Lagoon but also to other Italian and Mediterranean lagoons, showing its general suitability for shallow water basins.

The model presented in this chapter is freely available to the scientific community. The source code has been released under the GPL (Gnu Public Licence) and is free to use for everybody who wants to use it. The model can be downloaded from the web page

Flooding and Environmental Challenges for Venice and its Lagoon: State of Knowledge, ed. C. A. Fletcher and T. Spencer.
Published by Cambridge University Press, © Cambridge University Press 2005.

http://www.ve.ismar.cnr.it/shyfem. Here not only the source code for the general model can be found but also the set-up for the Venice Lagoon (numerical grid, data, input files), in order to serve as a reference implementation of a numerical model of the Venice Lagoon for other scientists that work on these topics.

In the following sections the model framework is presented and its application to the Venice Lagoon and other shallow basins described. The main focus in this chapter is on hydrodynamic modelling; this is where most of the results are available.

THE MODELLING FRAMEWORK

In this section the general modelling framework of SHYFEM is presented. An overview is given that covers all the models that have been integrated into this modelling framework, without going into too much detail. References are given for the background of the equations, the resolution methods and all other questions concerning more specific topics.

The hydrodynamic model

The model presented here is a 2-D hydrodynamic finite element model that was developed initially (Umgiesser, 1986) as an application to the Venice Lagoon. References that deal with the resolution techniques and the details of the finite element method can be found elsewhere (Umgiesser and Bergamasco, 1993, 1995; Umgiesser *et al.*, 2003a; Umgiesser *et al.*, 2004a). Although a 3-D version of the model exists, only the 2-D version is described here, since this is the most suitable form for applications to very shallow basins like the Venice Lagoon.

The model uses staggered finite elements for spatial integration and a semi-implicit algorithm for integration in time. The terms treated implicitly are the water level gradient, the Coriolis and the bottom friction term in the momentum equation and the divergence term in the continuity equation; all other terms are treated explicitly. The model resolves the vertically integrated shallow water equations in their formulations with levels and transports:

$$\frac{\partial U}{\partial t} - fV + gH\frac{\partial \zeta}{\partial x} + RU + X = 0 \qquad Eq.\ 47.1a$$

$$\frac{\partial V}{\partial t} + fU + gH\frac{\partial \zeta}{\partial y} + RV + Y = 0 \qquad Eq.\ 47.1b$$

$$\frac{\partial \zeta}{\partial t} + \frac{\partial U}{\partial x} + \frac{\partial V}{\partial y} = 0 \qquad Eq.\ 47.1c$$

where ζ is the water level, U and V the vertically integrated velocities (total or barotropic transports) in x,y direction, f the Coriolis parameter, g the gravitational acceleration, $H = h+\zeta$ the total water depth, h the undisturbed water depth, t the time and R the frictional parameter.

The transports are related to their velocities u,v by vertical integration over the water column:

$$U = \int_{-h}^{\zeta} u\,dz \qquad Eq.\ 47.2a$$

and:

$$V = \int_{-h}^{\zeta} v\,dz \qquad Eq.\ 47.2b$$

The terms X and Y contain all other terms that need not to be treated implicitly in the time discretization. These are the wind stress, the non-linear advective terms and the horizontal viscosity terms. The advective terms are discretized with an upwind scheme; for the horizontal viscosity term either a constant viscosity or one computed by a Smagorinsky formula can be used. All details of the implementation are given in Umgiesser *et al.* (2004a).

The frictional parameter R can be expressed as:

$$R = c_B\sqrt{\frac{u^2 + v^2}{H}} \qquad Eq.\ 47.3$$

where c_B is a bottom drag coefficient (non dimensional) that may either be constant for deeper water bodies (typical value 2.5×10^{-3}) or may be computed through the empirical Chezy formula $c_B = g/C^2$, with C the Chezy coefficient. The Chezy term itself is not a constant but varies with the water depth as $C = k_s H^{1/6}$ where k_s is the Strickler coefficient.

At the open boundaries the water levels are prescribed in accordance with the Dirichlet condition, while at the closed boundaries the normal velocity is set to zero and the tangential velocity is a free parameter. This corresponds to a full slip condition.

The model deals also with intertidal flats that during a tidal cycle become both dry and wet. If the water level of the element is below a certain threshold, the element is taken out of the computation and is included again only if the surrounding elements reach a certain water level. These computations are done in a mass conserving way and no instability results from the switching off of the elements.

The transport and diffusion module

For the computation of the transport and diffusion of a passive tracer, as well as temperature and salinity, the model provides also a module that handles these processes. To simulate the behaviour of the tracer concentration the model solves the well known advection and diffusion equation, which, in the vertically integrated form, reads:

$$\frac{\partial S}{\partial t} + \frac{\partial uS}{\partial x} + \frac{\partial vS}{\partial y} = K_H \left(\frac{\partial^2 S}{\partial x^2} + \frac{\partial^2 S}{\partial y^2} \right) + Q \qquad Eq.\ 47.4$$

where $S = \int_{-h}^{\zeta} s\,dz$, u and v the barotropic velocities and K_H is the horizontal eddy diffusivity. The term Q represents a source term like a fresh water flux for the salinity or the heat flux through the water surface. Fluxes through the bottom have been neglected here.

The heat fluxes through the water surface can be computed with a radiative model that has been implemented in SHYFEM (Zampato *et al.*, 1998). This heat flux model is especially suited for applications to shallow lagoons.

The transport and diffusion equation is solved using an explicit time stepping scheme. The advective part uses an upwind discretization to ensure the exact conservation of mass. The mass consistency is checked for each time step and each finite volume centred on a node, both for the water mass and the mass of the dissolved substance. For small elements the time step could exceed the time step of the hydrodynamic model, and therefore fractional time steps may be used.

The wave module

Resuspension of sediments due to wind-generated waves is an important source of sediment for the water column. The wave sub-model utilizes empirical formulations which provide approximate estimates of significant wave height and wave period. This model does not take in account refraction, dispersion or wave breaking effects.

The formulation of the significant wave height (H_{m0}) and period (T_p) is based on the empirical prediction equations for shallow water (US Army Engineer Waterways Experiment Station, 1984) and depends on the wind speed (m s^{-1}), the water depth (m) and the wind fetch (m).

This empirical formulation assumes that the wind blows, with essentially constant direction, over a fetch for sufficient time to achieve steady-state fetch limited values. For shallow lagoons this is a reasonable approximation.

The ecological module

The ecological model is extracted from the EUTRO code of WASP, the Water Analysis Simulation Program released by the US EPA-Environmental Protection Agency (Ambrose *et al.*, 1993). It simulates the evolution of up to 9 state variables: ammonia NH3, nitrate NOX, phosphate OPO4, phytoplankton PHY, zooplankton ZOO, organic nitrogen ON, organic phosphorous OP, carbonaceous bio-geo-chemical oxygen demand CBOD, and dissolved oxygen DO. They are connected together in four interacting systems: the nitrogen cycle, the phosphorous cycle, the oxygen cycle and the carbon cycle (Fig. 47.1).

The evolution of phytoplankton, considered as a pool of primary producers, is driven by the nutrient concentration and by the dynamics of the grazers and therefore both top-down and bottom-up controlled dynamics can be reproduced. It is described by the growth term, the death term, and by grazing. The optimal growth is multiplied by the functional relations which simulate the growth limitation due to sub-optimal levels of radiation light intensity, temperature and nutrient concentration. The limiting factors are computed following the Michelis Menten -

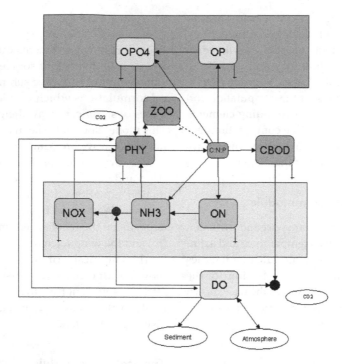

Fig. 47.1   Structure of the EUTRO eutrophication module. Shown are the Phosphorus cycle, the Nitrogen cycle and the Carbon cycle.

Monod equation for nutrient limitation, the Smith formulation for the limitation due to light intensity, and an exponential relation for temperature. The evolution of the zooplankton is described by the grazing term, upon which the growth depends, and by the mortality term. The grazing term is described by a Holling type II relationship between phytoplankton and zooplankton concentration and by the grazing parameter while the mortality term is described by a first order kinetics. Details on the implementation can be found in Umgiesser *et al.* (2003a).

The sediment transport model

Sedtrans96 (Li and Amos, 1995; Li and Amos, 2001) is a one-dimensional numerical computer model developed at the Geological Survey of Canada-Atlantic (GSCA) to deal with boundary layer dynamics and sediment transport on continental shelves and in coastal environments. Sedtrans 96 not only gives boundary layer parameters but also predicts bedform

development and bedload as well as suspended load transport rates for both sand and cohesive sediments. For given input data on wave, current and seabed conditions (i.e. grainsize, sediment density, bedform dimension), the model applies the Grant and Madsen (1986) continental shelf bottom boundary layer theory to derive near-bed velocity profile and bed shear stress. Then it calculates sediment transport for currents only or combined waves and currents over either cohesive and non-cohesive sediments. The calculations under pure wave, pure current and combined wave-current conditions are treated in separate algorithms. The model calculates the critical shear velocities for three distinctive modes of transport: bedload, suspension and sheet-flow transport. Also an explicit combined-flow ripple predictor is included in the model to provide time-dependent bed roughness.

Five sediment transport formulae are available in the model to integrate the instantaneous sediment transport through a wave cycle in order to obtain the time-averaged net sediment transport rate. Sedtrans also includes the predictions of the velocity and suspended concentration profiles: their product is integrated through depth to derive the suspended-load transport rate.

Integration into one model

The above mentioned models are integrated into one single modelling framework. The model first resolves the hydrodynamics, computing for every grid point the water level and the barotropic transports. After this step, the transport diffusion module is run and the tracers and/or heat and salt are advected and diffused.

The sediment transport and the ecological model are linked to the hydrodynamic model through generic interfaces and are kept in separate modules. They are run only as needed. The sediment transport module computes the erosion and deposition rates at every element and determines the sediment volume that is injected into the water column. After this step the sediments are advected with the transport and diffusion module described above. For the bedload component a direct advection scheme is used. The ecological module first computes all the reactions and transformations in every box. After this step the tracers are advected as with the sediments, taking into account a settling rate for the detritus.

All these computations run in parallel with the hydrodynamic module. This ensures an exact mass conservation of the variables and avoids problems of data exchange, where hydrodynamic variables must be saved and integrated over time, written to file and then read in by another model that runs the ecological processes.

Other utilities

The models that have been presented until now have the aim of resolving the differential equations that result from the various conservation formulas describing the processes that are resolved. However, in order to apply the model a numerical grid must be prepared on which the equations can be solved. Moreover, after the model run the results must be post-processed in order to visualize the important features. The model SHYFEM offers most of the tools that are needed to create the numerical mesh and bathymetry, manipulate it and visualize the results of the simulations. A short overview is given here.

The automatic mesh generator

The mesh generation for finite difference models with a Cartesian grid normally reduces to the choice of a suitable mesh size. After that, if no special features are needed like adaptive grids, the mesh generation is finished; only the bathymetry has to be interpolated onto the chosen numerical grid. On the other hand, finite elements need to be provided with a mesh that represents the geometry and bathymetry in an optimal way. Areas where higher resolution is needed must have a higher density mesh. The elements should also faithfully follow the coastline and any other important linear features inside the basin.

The SHYFEM model includes an automatic mesh generator that may be used to generate a numerical mesh, starting from the boundary line of a basin. The overall density of points may be given, as well as any internal lines (fault lines) that should not be crossed by the generated elements. Through the use of a background grid, different levels of resolution may be obtained, varying from area to area. The mesh generator uses a Delauny algorithm for the insertion of points and the re-gridding of the elements. The resulting grid, which will still show an irregular appearance, can then be treated with a post-processor that tries to smooth and regularize the grid, a feature important for the hydrodynamic model since a regular grid ensures good condition numbers for the inversion of the linear system matrix.

The graphical editor

The model also includes a graphical editor that allows an easy browsing of any grid that has been created with the mesh generator. It also allows for manual creation and manipulation of the grid in case of those areas that need special treatment. The programme can work on nodes and elements of the grid, changing the type and depth values of the elements, move and create new nodes and change the meshing of the triangles. With a colour table, it can also show the bathymetry of the grid and helps to find rough spots and errors in the interpolation.

Post-processing of the results

Equally important as the computation of the results through the model is the ability to render the results graphically. Only through the use of graphical tools is it possible to identify important patterns that are present in the simulations. The SHYFEM model includes a post-processor for the results that allows for an easy generation of maps of the most important variables, both vectorial (velocities, transports) and scalar (water level, temperature, salinity, biolog-

ical variables, suspended matter). The programme creates files in Postscript format that can then easily be transformed into PDF files. Black and white plots and colour plots with different colour scales are supported. For the plotting of time series, the programme relies on the tool gnuplot. This tool accepts a variable number of time series files and plots temporal evolution, also creating Postscript files.

APPLICATIONS

The model SHYFEM was originally developed as an application optimized for the Venice Lagoon. However, it turns out that the hydrodynamic code can also be successfully applied to other basins with

Fig. 47.2   Grid of the FEM model. The grid consists of 7842 triangular elements and 4359 nodes.

similar characteristics, like shallow lagoons and semi-enclosed bays. Here we give an overview of its application to the Venice Lagoon and then to some other lagoons, mainly in Italy.

Application to the Venice Lagoon

As its main application the model SHYFEM has been applied to the Venice Lagoon. An overview of applications of the hydrodynamic model to the Venice Lagoon is given in Umgiesser (1997).

The numerical computation has been carried out on a spatial domain that represents the entire basin through a finite element grid. The grid contains 4359 nodes and 7842 triangular elements (Fig. 47.2). The model considers as open boundaries the three lagoon inlets of Lido, Malamocco and Chioggia, elsewhere as closed boundary the whole perimeter of the basin. The bathymetry represented by the model can be seen in Fig. 47.3.

To reproduce the tidal wave propagation inside the lagoon, a calibration process had to be carried out. Measured data of phase shift and amplification rate with respect to Diga Sud Lido of semi-diurnal and diurnal principal tidal constituents were collected by 12 tide gauges situated inside the lagoon (Goldmann *et al.*, 1975). These data have been used for comparison with the calibration runs. By varying the bottom friction coefficient and comparing the numerical results with the observed data the model

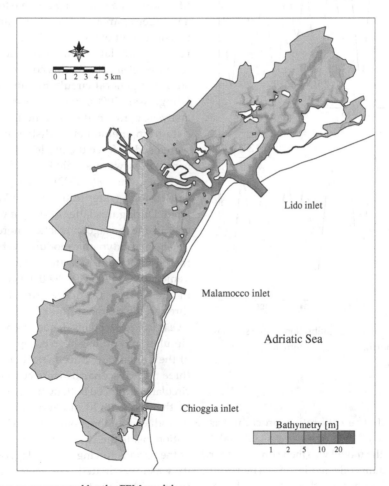

Fig. 47.3 Bathymetry as represented by the FEM model.

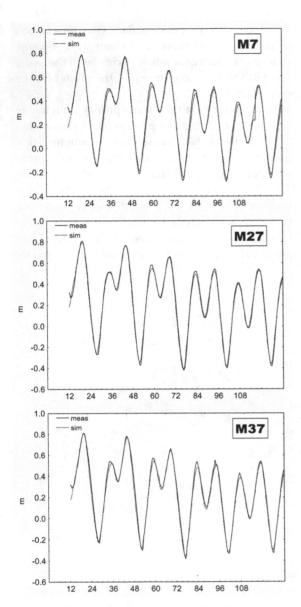

Fig. 47.4   Results of water level calibration. Three stations are shown from the north, centre and south. Measured and modelled results.

could be calibrated. The calibration procedure is described in more detail in Umgiesser *et al.* (2004a).

Once calibrated, the model was then compared for validation to tidal data collected in 1992. The result for three of the tide gauges (north, centre, south) are shown in Fig. 47.4. As can be seen, the model represents faithfully the water level variations in the lagoon.

The model has not only been validated with sea level data, but also with available temperature and salinity data during the year 2001. The agreement with temperature is good, and with salinity is acceptable, since a high uncertainty about the river discharge existed for this year. Details are given in Umgiesser *et al.* (2004a). Moreover, computed fluxes between the Venice Lagoon and the Adriatic Sea have been compared to available observed data, and the results were also satisfactory. This application is described later and can also be found in Cucco and Umgiesser (2002).

The calibrated model has been used for the identification of a partitioning scheme for the lagoon into a number of areas internally homogeneous from a physical point of view, and for a detailed analysis of residual circulation under different wind regimes (Solidoro *et al.*, 2004). This complements a previous study on general circulation and residual current (Umgiesser, 2000), and a storm surge modelling analysis, performed by Zecchetto *et al.* (1977) where data from anemometers distributed in the lagoon were used to force the model. In this paper mainly *bora* winds were studied. On the other hand, in Melaku Canu *et al.* (2002) a strong *sirocco* event that caused high water in the city of Venice was simulated. During this latter event the water level difference inside the lagoon reached more than 60 cm.

The hydrodynamic module has been used also for the computation of water residence times in the Venice Lagoon (Cucco, 2000; Cucco and Umgiesser, in press). The residence time has been defined through the remnant time (Takeoka, 1984) of a passive tracer released inside the basin during the model simulations and has been computed for each point of the domain. In Cucco and Umgiesser (in press) three typical scenarios were considered. The basin circulation induced both by the astronomical tide and by the two main local winds, *bora* and *sirocco*, was reproduced by the model, the residence time distribution inside the basin was computed and the effect of the various forcings on the lagoon renewal capacity was investigated. The effect of the return flow from the Adriatic Sea was also studied. When tide

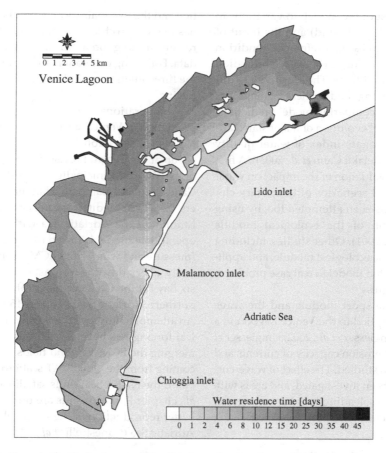

Fig. 47.5   Residence times in the Venice Lagoon. The situation refers to calm conditions.

only is forcing the basin, the residence time distribution is heterogeneous and mainly characterized by high values. The average residence time of the whole basin is about 30 days, certainly higher than the turnover time of the lagoon which is not more than 2 days (see Fig. 47.5). When the *sirocco* wind blows, the southern areas of the basin are subjected to a vigorous circulation that induces a high water renewal capacity. The average residence time is 13 days under these conditions. In the case of *bora* wind, the basin residence time is the lowest of all the cases, 3 days, and the distribution is highly homogeneous.

The ecological module, after several tests and refinements, has been coupled to the transport module, and the integrated model (VELFEEM) was then applied, as a first test, to the Venice Lagoon (Melaku Canu *et al.*, 2001; Umgiesser *et al.*, 2003a; Melaku Canu *et al.*, 2003). For this study idealized forcings (tide, meteorology) were used together with actual field data (river discharge, nutrient loadings) to run simulations over a one-year period. Analysis of the results shows that the model is numerically stable, variations in the state variables are consistent, and nutrients, plankton and oxygen evolve in space and time in an ecologically coherent way. The main features of the lagoon, namely DO seasonal evolution and dispersion of nutrients from the point sources are qualitatively reproduced (Umgiesser *et al.*, 2003a).

The coupled model was then used to derive some indications on the sensitivity of the Lagoon of Venice ecosystem (or at least of the model) to variations in physical forcing (tide and wind) and the input of macronutrients. To this aim, a reference condition was identified by running a one year simulation under climatologic conditions. The sensitivity to different scenarios of forcings and nutrient loadings was then investigated, by comparing model predictions of spatial and temporal evolution of major state variables and of an aggregate index of water quality (TRIX) (see Plate 17 (Melaku Canu *et al.*, 2003)). A first contribution to the evaluation of the impact on water quality under several scenarios of temporary closures of the inlets has been attempted too, by using a simplified version of the ecological module (Melaku Canu *et al.*, 2001). Other studies, including a full calibration of the ecological module, and applications of the calibrated model to real case problems, are currently in progress.

The sediment transport module and the wave module has been applied to the Venice Lagoon in a first application (Umgiesser *et al.*, 2002; Umgiesser *et al.*, 2004b) where the erosion capacity of currents and wind waves has been studied. The effect of wave-current interaction has been investigated, and areas with large bed-stress values, leading to enhanced erosion of the bottom sediments, have been identified.

Other applications to the Venice Lagoon

The Lagoon of Venice exchanges a huge amount of water with the Adriatic Sea (more than 500×10^6 m^3 in one tidal cycle). As a consequence, the definition of the boundary conditions at the inlets is critical in determining the behaviour of the model. For this reason, in a number of applications the model of the lagoon has been integrated with a larger model of the Adriatic Sea and, in one case, with a model (for surface elevation) of the whole Mediterranean Sea. In particular, in one of these applications the 2D model version of the Venice Lagoon and the Adriatic Sea was calibrated and validated by comparing the computed water levels with data measured by tide gauges along the Adriatic coast and discharge rates through the three lagoon inlets with ADCP measured fluxes (Cucco and Umgiesser, 2002). The instantaneous and residual sea-lagoon water exchange through the inlets induced both by the astronomical

tide and the two main local winds, bora and scirocco, has been reproduced by the model. The obtained results are in good accordance with observational data. Following calibration, the current system inside the three inlets was studied under varying external forcing.

An operational forecasting system for sea level prediction in Venice, based on the SHYFEM model, has also been established in the last two years, and set up at the Centro Previsioni e Segnalazioni Maree of the Venice Municipality (Canestrelli *et al.*, 2003b; Canestrelli and Zampato, chapter 11). The grid covers the whole Mediterranean Sea with a coarse resolution and the Adriatic Sea with a finer one. The operational model is forced by meteorological fields (pressure and wind) from ECMWF. It gives automatically, every day, the forecast sea level for the next six days at the CNR oceanographic platform, in the northern Adriatic Sea, situated 15 km off the coast. An attempt to improve the quality of the meteorological forcing has been carried out in a recent study, merging the ECMWF wind fields with satellite data, coming from the QuikSCAT scatterometer hosted on board the NASA SeaWinds satellite (see Zecchetto *et al.*, chapter 7). The results are promising, showing a better reproduction of the sea level in the Northern Adriatic Sea (Canestrelli *et al.*, 2003a).

Applications to other lagoons
The Sea of Taranto

The model SHYFEM has been used to study the hydrodynamics and the variability of temperature and salinity of the Taranto Sea (Scroccaro *et al.*, 2002a; Scroccaro *et al.*, 2002b). The Taranto Sea (situated in the Ionian Sea in the southern Italy) is composed of two parts: the Mar Grande and the Mar Piccolo. The Mar Grande (35 km^2) communicates with the Ionian Sea through two openings. The Mar Piccolo (20.7 km^2), connected to the Mar Grande by two narrow channels, is structured in two shelves, the 'First Seno' and the 'Second Seno'. In particular the Mar Piccolo is characterized by the presence of submarine freshwater springs, locally called '*citri*', the two most important being the citro Galeso and the citro Le Kopre.

A one-year simulation was set up taking into account different forcings such as tide, meteorological data and the freshwater discharges. The grid for the Taranto Sea was constructed with an automatic mesh generator and contains 4452 nodes and 8059 triangular elements. Temperature variability is well reproduced in the basin and the salinity is mainly determined by the varying contribution of freshwater inflow and evaporation (Fig. 47.6). Experimental data, collected by local research organizations during 2000–1 *in situ* campaigns, were used to calibrate and validate the model. Comparing the final results of the model with the corresponding data, a root mean square error equal to 1.37 °C for the temperature and a root mean square error of 0.85 PSU for the salinity were found (Scroccaro *et al.*, 2004).

The Cabras Lagoon

The SHYFEM model has been applied to the Cabras Lagoon to describe the water circulation and water residence time. Moreover, with the help of the radia-

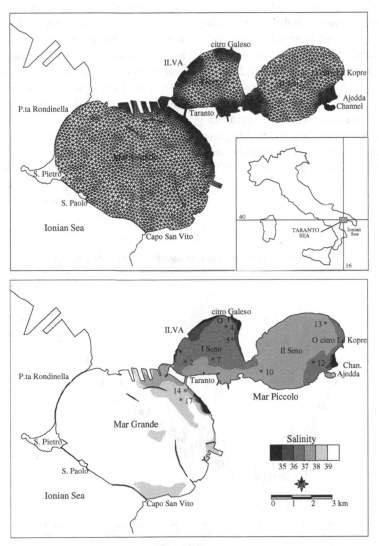

Fig. 47.6   The Sea of Taranto. Upper: grid of the model SHYFEM. Lower: instantaneous representation of the salinity field (30 July 2003).

tional transfer module of heat at the water surface and a transport diffusion model, the salinity and the water temperature have also been simulated. The hydrodynamic simulations indicated the wind as the main forcing for the water circulation of the Cabras Lagoon, while the tide determines the discharges through the inlets between the lagoon and the sea and modulates the circulation pattern set up by the wind (Ferrarin and Umgiesser, in press). The rivers do not influence significantly the hydrodynamic circulation but they determinate the salinity budget and the water residence times. The model reproduces well the temporal and spatial variability of the salinity data collected during a field campaign. The data range from close to 5 PSU during January, when river discharges were at a maximum, to nearly 25 PSU during July, when river discharge was close to zero and the lagoon slowly approached the salinity of the open sea (Fig. 47.7). The rivers have also great influence on the lagoon's renewal capacity, making the water residence times much lower than in the case without river inflow (Ferrarin and Umgiesser, 2003).

The Orbetello Lagoon

An idealized study of the circulation has been carried out with the SHYFEM model, which might be useful in the management and safeguarding of the Orbetello Lagoon (southern coast of Tuscany, Italy) which covers an area of 27 km^2 and is composed of an eastern lagoon (12 km^2) and a western lagoon (15 km^2). The two lagoons are connected by the Orbetello isthmus and separated by a dam. The lagoon exchanges water with the Thyrrenian Sea by the Nassa channel, Fibbia channel and Ansedonia channel. Each inlet has water pumps that may pump water from the sea into the lagoon. The lagoon is a semi-enclosed coastal basin characterized by shallow water (average depth 1 m), poor hydrodynamic activity and algal blooms. These characteristics may cause problems for the water quality and the conservation of fish population, the most important resource of this ecosystem.

Some typical situations have been simulated prescribing tidal level, water pumps and wind as forcing factors (Scroccaro *et al.*, 1999; Scroccaro *et al.*, 2001). Results show that the water movement is mainly

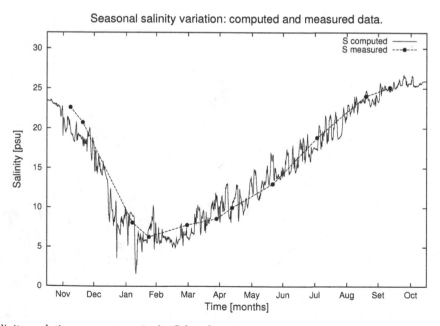

Fig. 47.7   Salinity evolution over one year in the Cabras lagoon.

induced by the wind action, whereas the tide has a quite negligible effect. Moreover the pumping system seems to be not efficient enough to improve the hydrodynamic activity.

Finally, different grids, corresponding to different openings of the dam, have been implemented for the simulations, showing that a partial opening of the dam might be a feasible compromise between a complete removal of the dam and the actual situation that blocks the water exchange (Scroccaro *et al.*, 2003).

The Nador Lagoon

The Nador Lagoon, situated on the Mediterranean coast of Morocco, is the largest lagoon in Morocco (115 km²). It has a maximum depth not exceeding 8 m and it communicates with the Mediterranean Sea through one channel. In the last fifty years this lagoon has been characterized by eutrophication and a dramatic increase of organic matter inputs and sedimentation, due to the increasing urbanization and the development of aquaculture activities.

The hydrodynamic circulation has been investigated with the finite element model SHYFEM. The grid of the lagoon is represented by 3464 nodes and 6486 triangular elements. The model was forced by prescribing an idealized tide at the open boundary and the prevailing regimes of wind (wind from ENE and wind from WSW). The study shows that the circulation is mainly induced by the wind and that the situation is also related to the unique communication channel with the Mediterranean Sea, which strongly affects the water exchange of the basin. The model needs still to be calibrated and validated with tidal and meteorological data. The first simulations with the finite element model are in good agreement with a circulation pattern found in the bibliography (Guelorget *et al.*, 1987), shown in Fig. 47.8. The lagoon seems to be divided into three parts, the western, the central and the southern part, with different hydrodynamic characteristics. The zones where the hydrodynamic activity is less prominent seem to be part of the central zone and the western area of the lagoon (Umgiesser *et al.*, 2003b).

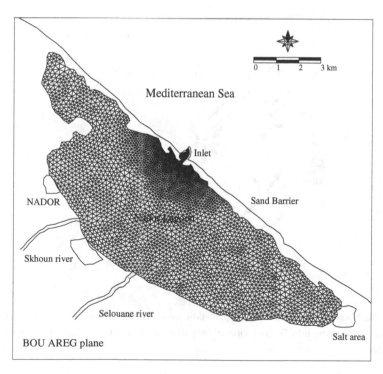

Fig. 47.8   The Nador Lagoon, Morocco: grid of the model (*continued overleaf*).

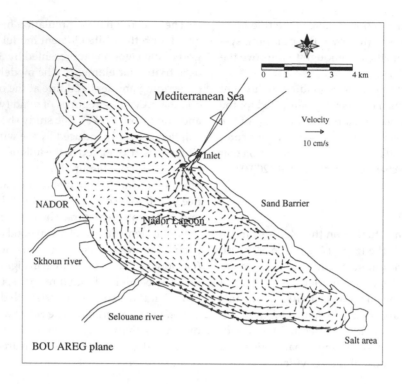

Fig. 47.8 (*continued*) The Nador Lagoon, Morocco. Top: instantaneous picture of the velocity field in the Nador Lagoon for wind from ENE and out-flowing tide. Bottom: circulation pattern proposed by Guelorget *et al.* (1987), based on hydrographic parameters.

CONCLUSIONS

A framework of numerical models (SHYFEM) has been presented that are well suited for application in very shallow lagoons and coastal seas. The model uses the finite element method for maximal adaptability to complicated geometries and bathymetries. The semi-implicit algorithm ensures unconditional stability for fast gravity waves.

The model has been extensively applied to the Venice Lagoon, for which all its modules have been developed and tested. The model has been calibrated with tide gauge data and has been verified with water level, temperature and salinity and discharge data through the inlets. It is capable of describing the processes occurring inside the lagoon to a high degree. Residual currents and residence times have also been computed. Other modules have been applied, including the wave and sediment transport module and the eutrophication module. Applications include the modelling of water fluxes through the inlets and the operational water level forecast for Venice.

The model has also been applied to other sites in Italy and in the Mediterranean, including the Orbetello Lagoon, the Sea of Taranto, the Cabras Lagoon and the Nador Lagoon. The model is available to the scientific community in the form of the source code and is licensed under the Gnu Public Licence (GPL). The model may be downloaded by following the link http://www.ve.ismar.cnr.it/shyfem. For the Venice Lagoon the actual set up of the model is also given.

ACKNOWLEDGEMENTS

This work has been partially funded by the CORILA project 3.2 'Hydrodynamics and morphology of the Venice lagoon' and project 3.5 'Quantity and quality of exchanges between lagoon and sea'.

REFERENCES

Ambrose, R. B., Wool, T. A., and J. L. Martin. 1993. *The Water Quality Analysis Simulation Program, Wasp5: Model documentation*. Technical report. Athens, Georgia: Environmental Research Laboratory.

Bocci, M., Coffaro, G., and G. Bendoricchio. 1997. Modeling biomass and nutrient dynamics in eelgrass (Zostera marina) applications to the lagoon of Venice (Italy) and Orensud (Denmark). *Ecological Modelling* 102, 67–81.

Canestrelli, P., Cucco, A., De Biasio, F., Umgiesser, G., Zampato, L., and S. Zecchetto. 2003a. The use of quikSCAT wind fields in water level modelling of the Adriatic Sea. In *Proceedings of the Sixth European Conference on Applications in Meteorology (ECAM)*. Rome, 15–19 Sept.

Canestrelli, P., Cucco, A., Umgiesser, G., and L. Zampato. 2003b. An operational forecasting system for the sea level in Venice based on a finite element hydrodynamic model. In *Proceedings of the Sixth European Conference on Applications in Meteorology (ECAM)*. Rome, 15–19 Sept.

Casulli, V., and E. Cattani. 1993. Stability, accuracy and efficiency of a semi-implicit method for three-dimensional shallow water flow. *Computers Math. Applic.* 27 (4), 99–112.

Casulli, V., and R.T. Cheng. 1992. Semi-implicit finite difference methods for three-dimensional shallow water flow. *International Journal for Numerical Methods in Fluids* 15, 629–48.

Casulli, V., and P. Zanolli. 1998. A three-dimensional semi-implicit algorithm for environmental flows on unstructured grids. In *Conference on Numerical Methods for Fluid Dynamics*. University of Oxford, pp. 57–70.

Casulli, V., and P. Zanolli. 2002. Semi-implicit numerical modeling of nonhydrostatic free-surface flows for environmental problems. *Mathematical and Computer Modelling* 36, 1131–49.

Chignoli, C., and R. Rabagliati. 1975. Un modello della idrodinamica lagunare. In *Venezia e i problemi dell'ambiente*. Bologna: Il Mulino, pp. 225–97.

Coffaro, G., and M. Bocci. 1997. Resource competition between Ulva rigida and Zostera marina: a quantitative approach applied to the lagoon of Venice. *Ecological Modelling* 102, 81–97.

Cucco, A. 2000. Modellizzazione degli scambi di materia tra la Laguna di Venezia e l'Alto Adriatico. Master's thesis, Università Ca' Foscari, Venice.

Cucco, A., and G. Umgiesser. 2002. Modeling the water exchanges between the Venice lagoon and the Adriatic sea. In P. Campostrini (ed.), Scientific Research and Safeguarding of Venice. *Proceedings of the annual meeting of the Corila Research program 2001 results*. Venice: Istituto Veneto SLA, pp. 499–514.

Cucco, A., and G. Umgiesser. In press. Modeling the Venice Lagoon residence time. *Ecological Modelling*.

Ferrarin, C., and G. Umgiesser. 2003b. Hydrodynamic modelling of the Cabras lagoon: water circulation and residence times. In Ozhan, E. (ed.), *MEDCOAST 03 – The Sixth International Conference on the Mediterranean Coastal Environment*, Vols. I–III. Ankara, Turkey: Middle East Technical University, pp. 2053–64.

Ferrarin, C., and G. Umgiesser. In press. Hydrodynamic modeling of a coastal lagoon: the Cabras lagoon in Sardinia, Italy. *Ecological Modelling.*

Goldmann, A., Rabagliati, R., and P. Sguazzero. 1975. *Characteristics of the Tidal Wave in the Lagoon of Venice.* Technical Report 47, Venice Scientific Center: IBM.

Grant, W. D., and O. S. Madsen. 1986. The continental shelf bottom boundary layer. *Annual Review of Fluid Mechanics* **18**, 265–305.

Guelorget, O., Perthuisot, J. P., Frisoni, G. F., and D. Monti. 1987. Le rôle du confinement dans l'organisation bio-geologique de la lagune de Nador (Maroc). *Oceanologica Acta* **10 (4)**, 435–44.

Li, M. Z., and C. L. Amos. 1995. Sedtrans92: a sediment transport model for continental shelves. *Computers & Geosciences* **21** (4), 533–54.

Li, M. Z., and C. L. Amos. 2001. Sedtrans96: the upgraded and better calibrated sediment-transport model for continental shelves. *Computers & Geosciences* **27**, 619–45.

Marchi, E., Adami, A., Caielli, A., and G. Cecconi. 1992. Water flow modelling of the Venice Lagoon. *Coastal Engineering* **20**, 1869–78.

Melaku Canu, D., Umgiesser, G., and C. Solidoro. 2001. Short term simulations under winter conditions in the lagoon of Venice: a contribution to the environmental impact assessment of a temporary closure of the inlets. *Ecological Modelling* **138 (1–3)**, 215–30.

Melaku Canu, D., Umgiesser, G., Bonato, N., and M. Ferla. 2002. Analysis of the circulation of the lagoon of Venice under scirocco wind conditions. In P. Campostrini (ed.), Scientific Research and Safeguarding of Venice. *Proceedings of the annual meeting of the Corila Research program 2001 results.* Venice: Istituto Veneto SLA, pp. 515–30.

Melaku Canu, D., Solidoro, C., and G. Umgiesser. 2003. Modelling the responses of the lagoon of Venice ecosystem to physical forcings variations. *Ecological Modelling* **170**, 265–89 (see also Erratum, Ecological Modelling (2004), 175, 197–216).

Pastres, R., Solidoro, C., Cossarini, G., Melaku Canu, D., and C. Dejak. 2001. Managing the rearing of *Tapes philippinarum* in the lagoon of Venice: a decision support system. *Ecological Modelling* **138**, 213–45.

Pastres, R., Ciavatta, S., Cossarini, G., and C. Solidoro. 2003. Sensitivity analysis as a tool for the implementation of a water quality regulation based on the maximum permissible load policy. Submitted to *Reliability Engineering and System Safety Journal.*

Scroccaro, I., Di Bitetto, M., and G. Umgiesser. 1999. *Sensibility Analysis of the Circulation in the Orbetello Lagoon.* Tech. Rep. 228. Venice: ISDGM-CNR.

Scroccaro, I., Di Bitetto, M., and G. Umgiesser. 2001. *Sensibility Analysis of the Circulation in the Orbetello Lagoon – Part 2.* Venice: Tech. Rep. 242, ISDGM-CNR.

Scroccaro, I., Matarrese, R., and G. Umgiesser. 2002a. *Sviluppo e applicazione di un modello agli elementi finiti al Mare di Taranto. Parte 1 – Idrodinamica.* Venice: Tech. Rep. 254, ISDGM-CNR.

Scroccaro, I., Matarrese, R., and G. Umgiesser. 2002b. *Sviluppo e applicazione di un modello agli elementi finiti al Mare di Taranto. Parte 2 – Temperatura e salinità.* Venice: Tech. Rep. 255, ISDGM-CNR.

Scroccaro, I., Cappelletti, A., and G. Umgiesser. 2003. An idealized circulation of the Orbetello lagoon. In E. Ozhan (ed.), *MEDCOAST 03 – Proceedings of the Sixth International Conference on the Mediterranean Coastal Environment*, Vols. I–III. Ankara, Turkey: Middle East Technical University, pp. 2087–98.

Scroccaro, I., Matarrese, R., and G. Umgiesser. 2004. Application of a finite element model to the Taranto Sea. *Chemistry & Ecology* **20 (Suppl. 1)**, S205–24.

Sguazzero, P., Chignoli, C., Rabagliati, R., and G. Volpi. 1978. Hydrodynamical modeling of the Lagoon of Venice. *IBM Journal of Research and Development* **22 (5)**, 472–80.

Solidoro, C., Brando, E. V., Dejak, C., Franco, D., Pastres, R., and G. Pecenik. 1997a. Simulation of the seasonal evolution of macroalgae in the lagoon of Venice. *Environmental Modelling and Assessment* **2**, 65–71.

Solidoro, C., Pecenik, G., Pastres, R., Franco, D., and C. Dejak. 1997b. Modelling Ulva rigida in the Venice lagoon: model structure identification and first parameters estimation. *Ecological Modelling* **94**, 191–206.

Solidoro, C., Melaku Canu, D., Cucco, A., and G. Umgiesser. 2004. A partition of the Venice Lagoon based on physical properties and analysis of general circulation. *Journal of Marine Systems* **51**, 147–60.

Solidoro, C., Pastres, R., Melaku Canu, D., Pellizzato, M., and R. Rossi. 2000. Modelling the growth of *Tapes philippinarum* in northern Adriatic lagoons. *Marine Ecology Progress Series* **199**, 137–48.

Takeoka, H. 1984. Fundamental concepts of exchange and transport time scales in a coastal sea. *Continental Shelf Research* **3 (3)**, 311–26.

Umgiesser, G. 1986. A Model for the Venice Lagoon. Master's thesis, University of Hamburg.

Umgiesser, G. 1997. Modelling the Venice lagoon. *International Journal of Salt Lake Research* **6**, 175–99.

Umgiesser, G. 2000. Modeling residual currents in the Venice lagoon. In T. Yanagi (ed.), *Interactions between Estuaries, Coastal Seas and Shelf Seas*. Tokyo: Terra Scientific Publication Company (TERRAPUB), pp. 107–24.

Umgiesser, G., and A. Bergamasco. 1993. A staggered grid finite element model of the Venice Lagoon. In K. Morgan, J. P. Ofiate and O. Zienkiewicz (eds.), *Finite Elements in Fluids*. Swansea: Pineridge Press, pp. 659–68.

Umgiesser, G., and A. Bergamasco. 1995. Outline of a primitive equation finite element model. In *Rapporto e Studi. Vol. XII. Istituto Veneto di Scienze, Lettere ed Arti*, Venice: Instituto Veneto di Scienze, Lettere ed Arti, pp. 291–320.

Umgiesser, G., Sclavo, M., and S. Carniel. 2002. Modeling the bottom stress distribution in the Venice lagoon. In P. Campostrini (ed.), *Scientific Research and Safeguarding of Venice. Proceedings of the annual meeting of the Corila Research program 2001 results*. Venice: Istituto Veneto SLA, pp. 287–99.

Umgiesser, G., Melaku Canu, D., Solidoro, C., and R. Ambrose. 2003a. A finite element ecological model: a first application to the Venice Lagoon. *Environmental Modelling and Software* **18 (2)**, 131–45.

Umgiesser, G., Scroccaro, I., and M. Snoussi. 2003b. Modeling the Nador lagoon, Morocco. In E. Ozhan (ed.), *MEDCOAST 03 – Proceedings of the Sixth International Conference on the Mediterranean Coastal Environment*, Vols. I–III. Ankara, Turkey: Middle East Technical University, pp. 2065–74.

Umgiesser, G., Melaku Canu, D., Cucco, A., and C. Solidoro. 2004a. A finite element model for the Venice Lagoon. Development, set up, calibration and validation. *Journal of Marine Systems* **51**, 123–45.

Umgiesser, G., Sclavo, M., Carniel, S., and A. Bergamasco. 2004b. Exploring the bottom stress variability in the Venice Lagoon. *Journal of Marine Systems* **51**, 161–78.

US Army Engineer Waterways Experiment Station. 1984. *Shore Protection Manual*. Washington DC: US Government Printing Office.

Volpi, G., and P. Sguazzero. 1977. La propagazione della marea nella Laguna di Venezia: un modello di simulazione e il suo impiego nella regolazione delle bocche di porto. *Rivista Italiana di Geofisica e Scienze Affini* **IV (1/2)**, 67–74.

Zampato, L., Umgiesser, G., and F. Peron. 1998. Simulazioni di scambio termico all'interfaccia acqua-aria nella Laguna di Venezia. In *53. Congresso Nazionale ATI*. Florence, pp. 205–16 (in Italian).

Zecchetto, S., Umgiesser, G., and M. Brocchini. 1977. Hindcast of a storm surge induced by local real wind fields in the Venice Lagoon. *Continental Shelf Research* **17 (12)**, 1513–38.

48 • The influence of the inlet configuration on sediment loss in the Venice Lagoon

L. D'ALPAOS AND P. MARTINI

INTRODUCTION

It is a widely accepted opinion (e.g. Day *et al.*, 1999) that for several years the Venice Lagoon has been subject to intense erosion processes which have significantly changed its morphology. Though the importance of this phenomenon is crucial to lagoon morphodynamics, it is not yet possible to realistically quantify sediment balance, owing to the numerous and complex physical mechanisms governing sediment dynamics and, moreover, the lack of reliable field data on sediment transport rates. In this chapter, we focus on the role that the geometry of the lagoon inlets plays in the propagation of tides within the lagoon and, consequently, in the sediment loss that characterizes the present exchanges with the sea. This analysis has been carried out using a mathematical model developed at the Department of Hydraulic, Maritime, Environmental and Geotechnical Engineering of the University of Padua. The model allows for the description of both the hydrodynamic behaviour of the entire lagoon and the interaction between tidal currents and bed sediments. In particular, the characteristics of the flow field in a neighbourhood of the inlets is investigated by considering: 1) the present (2000) configuration; 2) the configuration surveyed at the beginning of the nineteenth-century by the Napoleonic Captain A. Denaix (1811) and characterized by the absence of jetties; and 3) the configuration existing at the beginning of the twentieth century, when the jetties of the Malamocco inlet were already build up (precisely in 1865), while the Lido inlet had just been reconfigured (1901) by the installation of similar structures.

This investigation is unique in that for the first time the hydraulic characteristics of the inlets in various historical periods is studied, simulating the morphological behaviour of the entire lagoon rather than the local evolution of the inlet area. The results of the study show that the tidal current regime in the present lagoon is very different from the one that existed before the interventions carried out by man from the late nineteenth century. Compared to the Denaix's 1811 lagoon configuration, both the tidal prism (i.e., the flow exchange capacity between the sea and the lagoon) and the amplitude of the tidal wave propagating into the lagoon have significantly increased. Furthermore, the pattern of tidal currents in the sea just in front of the inlets has became strongly asymmetrical, thus leading to a systematic loss of suspended finer sediments. Indeed, only a minimal part of the finer material carried through the inlets during the ebb phase of the tide flows back to the lagoon during the next flood phase. A rough estimate of the huge volume of sediments that are lost as a consequence of this inlet asymmetric behaviour suggests the tendency towards the degradation of many of the morphological structures present in the lagoon (e.g., the salt marshes). Although the study of these problems has received less attention than those dealing with the defence of Venice from high tide floods, nevertheless the erosive processes described above are deemed to be quite relevant to the safeguarding of the entire lagoon ecosystem. Indeed, if the present significant loss of sediment is not counteracted, the lagoon will inevitably be transformed into a bay of the sea.

THE MATHEMATIC MODELLING OF THE BASIN OF THE LAGOON

Many mathematical models (e.g. D'Alpaos and Defina, 1993; Umgiesser and Bergamasco, 1993) have been proposed in the last three decades to examine both glob-

Flooding and Environmental Challenges for Venice and its Lagoon: State of Knowledge, ed. C. A. Fletcher and T. Spencer. Published by Cambridge University Press, © Cambridge University Press 2005.

ally and locally the characteristics of the hydrodynamic field within the Venice Lagoon. In this context, the contribution of researchers at the University of Padova has been mainly focussed on the attempt to include appropriate sub-grid models in the mathematical modelling of tide propagation. These sub-grid models are used to describe phenomena characterized by a length-scale which is smaller than the computational grid size used for the numerical solution of the governing equations. In particular, such a modelling procedure takes into account the rather complex morphology of the lagoon, where wide tidal flats lie next to quite large salt marshes which periodically wet and dry during the tidal cycle, and where a tight network of channels is intertwined with a widespread but not less important network of smaller creeks. Including such small elements in the computational grid would lead to a very huge computational problem, which can hardly be solved despite the powerful and fast computers available nowadays. The details of the solutions proposed to take into account the above problems are thoroughly discussed in D'Alpaos (1992), D'Alpaos and Defina (1993, 1995, 2004), Defina *et al.* (1994), Defina (2000) and Martini *et al.* (2004).

Here we report only a brief description of the fundamental aspects of the proposed mathematical model. The equations governing the two dimensional flow field are:

$$\rho \frac{d(\overline{U}_x)}{dt} - \frac{1}{Y}\left(\frac{\partial(YR_{xx})}{\partial x} + \frac{\partial(YR_{xy})}{\partial y}\right) + \frac{\tau_{bx}}{Y} - \frac{\tau_{sx}}{Y}$$

$$+ \rho g \frac{\partial h}{\partial x} = 0 \qquad \frac{\tau_{bx}}{\rho Y} = g \frac{|\vec{q}|q_x}{Ks^2 H^{10/3}}$$

$$\rho \frac{d(\overline{U}_y)}{dt} - \frac{1}{Y}\left(\frac{\partial(YR_{xy})}{\partial x} + \frac{\partial(YR_{yy})}{\partial y}\right) + \frac{\tau_{by}}{Y} - \frac{\tau_{sy}}{Y}$$

$$+ \rho g \frac{\partial h}{\partial y} = 0 \qquad \frac{\tau_{by}}{\rho Y} = g \frac{|\vec{q}|q_y}{Ks^2 H^{10/3}}$$

$$\eta \frac{\partial h}{\partial h} + \frac{\partial q_x}{\partial x} + \frac{\partial q_y}{\partial y} = 0 \qquad\qquad Eq.\ 48.1$$

where t is time, q_x and q_y are the discharges per unit of length in the two directions x and y, respectively,

R_{ij} are the Reynolds stresses (specified through the Smagorinsky turbulence closure model), ρ is the fluid density, τ_b and τ_s are the bottom and wind shear stresses, respectively, g is gravity, h is the free surface elevation, Y and H are two equivalent flow depths, modified when necessary to take into account wetting and drying processes, and η can be interpreted as a h-dependent storage coefficient, accounting for the actual area that can be wetted or dried during the tidal cycle. When dealing with quite small flow depths, in fact, the irregularities of the bottom surface of each grid element can significantly modify both the volume of water that accumulates on a grid element (accounted for through η and Y), as well as the flow resistances on the element itself (suitably modified through H) (D'Alpaos and Defina, 1993; 1995; Defina, 2000). In the numerical solution of Eq. (48.1), carried out using a finite element technique, the actual topography is taken into account through a suitable sub-grid model (Bates and Hervouet, 1999; Defina, 2000) to calculate the parameters η, Y and H. Note that when a given element of the computational mesh is completely wetted h tends to 1 while the equivalent flow depths H and Y tend to the geometric water depth. It is also worthwhile to point out that the introduction of such a sub-grid model not only allows for the use of a fixed computational grid, but also avoids numerical problems within computational domains where significant portions periodically wet/dry.

Another aspect of the proposed mathematical model is the possibility of coupling in a general way two- and one-dimensional elements of the computational grid (D'Alpaos and Defina, 1993), thus allowing a physically based description of smaller creeks, which can hardly be represented when using a purely two-dimensional grid. In fact, in this latter case it would be necessary to considerably reduce mesh size, with an unacceptable consequent increase in the computational effort. The proposed coupling is particularly efficient from an operational point of view: indeed, the local flow pattern can be reliably represented using a relatively coarse two-dimensional computational grid (for example modelling the entire Venice Lagoon currently requires about 80,000 two-dimensional elements and 1,500 one-dimensional ones).

Recently, the hydrodynamic model described above has been modified, by adding a new formulation of convective and turbulence terms and introducing a new sediment transport module (Martini *et al.*, 2004). In the latter model, the bed-load transport rate is estimated through the classical Meyer-Peter-Muller formula:

$$\frac{q_b}{\sqrt{g(s-1)d_{50}^3}} = 8 \cdot (\vartheta - k_{long} \cdot k_{trasv} \cdot \vartheta_{cro})^{3/2} \qquad Eq.\ 48.2$$

where q_b is the solid discharge, d_{50} is the average diameter of the cohesionless bed sediment, s is the relative sediment density, ϑ is the dimensionless shear stress, and ϑ_{cro} is the critical value of ϑ, calculated for a horizontal bottom according to the formula proposed by Van Rijn (1984). Moreover, the coefficients k_{long} and k_{trasv} are introduced to take into account the effects of longitudinal and transverse bottom slopes.

The suspended sediment transport rate is evaluated solving the concentration equation:

$$\frac{\partial(\overline{C}Y)}{\partial t} + \nabla \cdot (\overline{C}\mathbf{q}) - \nabla \cdot (Y\mathbf{D}\nabla\overline{C}) = W_s(C_{eq} - C_b)$$

$$C_b = \frac{\overline{C}}{k(z,\vartheta)} \qquad Eq.\ 48.3$$

where $\overline{C}$ is the depth averaged concentration (whose vertical distribution is assumed to follow the profile obtained by Rouse (1949)), and $\mathbf{D}$ is the dispersion-diffusion tensor, assumed to be isotropic. The right-hand side term of Eq. (48.3), expressing the exchange of sediment with the bottom, is assumed to be proportional, through the sediment particle settling velocity W_s, to the difference between the concentration of equilibrium C_{eq}, as expressed by Van Rijn (1984), and the actual concentration close to the bottom C_b.

Finally, the sediment balance equation governing the bottom evolution reads:

$$(1-n)\frac{\partial z_b}{\partial t} + \nabla \cdot \mathbf{q_b} = -W_s(C_{eq} - C_b) \qquad Eq.\ 48.4$$

where z_b denotes bottom elevation.

Equations (48.2), (48.3) and (48.4) are solved numerically using finite element volume techniques. In particular, the advective and diffusive-dispersive terms appearing in Eq. (48.3) are treated (Jasak, 1996) as the analogue terms that appear in the flow field equations (48.1). The validity of the above described mathematical model has been extensively tested in both experimental contexts, as well as considering case studies which are well-known in the literature (D'Alpaos and Defina, 1993; Martini *et al.*, 2004).

HISTORICAL ANALYSIS OF THE HYDRODYNAMIC BEHAVIOUR OF THE LAGOON

In this section we present the results of some studies whose aim was to analyse the similarities/differences between the 2000 tidal current regime and those characterising two different historical bathymetric lagoon configurations surveyed by Captain Denaix (1811), and by Genio Civile di Venezia (1901), when the jetties had already been built at the Malamocco and Lido inlets. The morphological features of the three configurations analysed shows significant differences (Plate 18).

According to Denaix's survey, the areas of the lagoon subject to tidal propagation were much wider (about 100 km² more) than today. The wide land surfaces that were created during the second half of the 1900s did not exist and only a few fish farms were dammed and completely excluded from tidal flow. Wide salt-marsh areas, whose extension is now greatly reduced, were interposed between the intertidal areas closer to the inlets (the so-called '*laguna viva*') and the shallow tidal flats located just before the fish farms. The channel network dissecting the tidal basin was much more developed than it is now, even though water depths of the large channels were quite similar to the present ones. Moreover, the 1901 hydrographic chart of the lagoon suggests that tidal flats were not as deep as they are today. The central part of the lagoon, located between the Lido and Malamocco inlets, was in fact 50-100 cm shallower than today, in accordance with Denaix's observations that, during low tide, significant portions of the lagoon dried, showing the structure of the channel network. All these observations highlight the signifi-

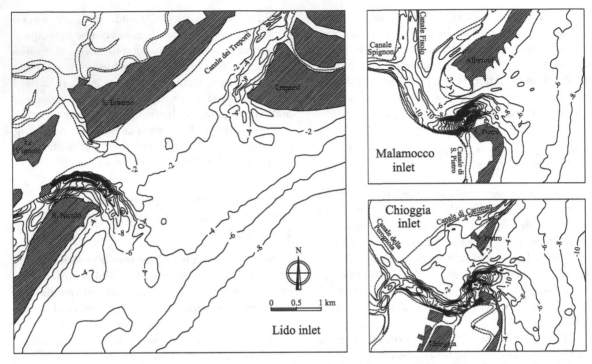

Fig. 48.1   Bathymetric configurations of the three lagoon inlets (Lido, Malamocco and Chioggia), according to Denaix's survey (1811).

cant erosion experienced by the lagoon itself in the twentieth century.

The profound change in inlet morphology that took place after the construction of the jetties had a significant effect on the propagation of the tidal wave. Compared to the present-day situation, at the beginning of 1800s the inlets were much wider and much shallower (Fig. 48.1), so that modern boats could not pass through the Lido inlet and also had difficulty entering the lagoon through the Malamocco inlet.

The calculations that can be made based on the 1811 and 1901 lagoon configurations described above are of particular interest for the current debate on the safeguarding of the lagoon. It is clear that at the beginning of the 1800s the tidal propagation inside the lagoon was mainly controlled by flow resistances, while the inertial phenomena were negligible. The tidal phase lag varied significantly in different areas of the lagoon and the tidal amplitude experienced a much stronger damping than today. For example, assuming a semidiurnal sinusoidal forcing tide with amplitude 1 m and oscillating around 0.30 m a.m.s.l.

(hereafter denoted as the reference tide), the maximum tide elevation calculated at Punta della Salute was about 10 cm lower than that of the open sea (Fig. 48.2), in comparison with the slight increase observed for the present lagoon configuration.

Moreover, even if the intertidal areas subject to tidal propagation were much greater than today, in 1811 the tidal prism was much smaller. In fact, the strong dissipation experienced by tidal wave as it propagated through the salt marshes located at the landward boundary of the so-called 'laguna viva' led to a remarkable damping of the water oscillations in the tidal flats just beyond these marsh areas. Note that these results are in agreement with both the analyses carried out some years ago with a different mathematical model using a coarser computational mesh (D'Alpaos, 1992), and the relevant field observations made by de Bernardi (1843), pointing out the difficulty of circulating waters in the channels crossing the fish farms.

The general distribution of maximum water levels (Fig. 48.3) in the 1901 lagoon configuration was not

A - Punta della Salute
B - Sacca Piccola
C - Porto Marghera
D - Valle Grassabò

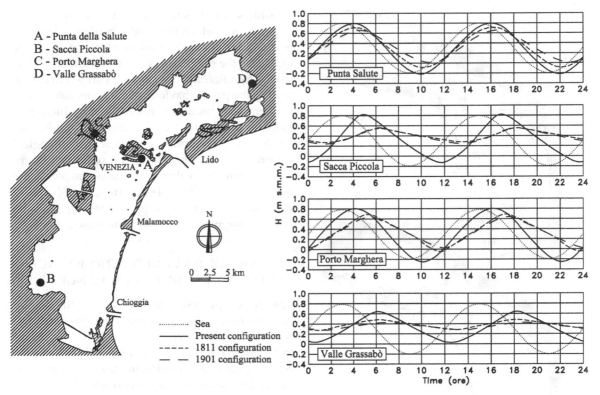

Fig. 48.2   Time evolution of the water surface elevation at four different locations within the lagoon (Punta della Salute, Sacca Piccola, Porto Marghera, Valle Grassabò) for the three lagoon configurations analysed. All the calculations have been carried out by assuming a reference semidiurnal, sinusoidal forcing tide, with amplitude 1 m, and oscillating around 0.30 m a.m.s.l.

very different from the case just considered of Denaix's lagoon. Nonetheless, one important difference was that by the early 1900s, jetties had already been built at the Lido and Malamocco inlets. Another difference worth pointing out is that maximum tidal elevation along the path of the historical centre was even smaller than it was in 1811: for the reference tide considered above, the maximum 1901 tide elevation at Punta della Salute was 15 cm lower than that of the open sea (Fig. 48.2). This was principally due to the increase in the flow resistances of the new Lido inlet, just a few decades after the S. Nicolò, S. Erasmo and Treporti passages had been unified into one (1892).

The construction of jetties also affected the exchange of flows with the sea. In fact, the maximum discharges in the past were much lower than they are today. For a typical spring tide, at the beginning of the

nineteenth century the maximum discharge through the Malamocco inlet was nearly half (~5000 m³ s⁻¹) of what it is today (~10 000 m³ s⁻¹), and even the maximum discharge through the Lido inlet was about 1,000 m³ s⁻¹ less than it is today (Fig. 48.4).

These results, and in particular the frictionally dominated character of the 1811 lagoon configuration, lead to some considerations which are extremely important and relevant to the issues regarding the project for building mobile gates at the inlets in order to prevent Venice from flooding. Firstly, restoring the hydrodynamic conditions characterizing the 1811 and 1901 morphological configurations, require a suitable increase of the flow resistance at the inlets and within the lagoon. However, since it is impracticable to restore the past morphological features within the lagoon, the only

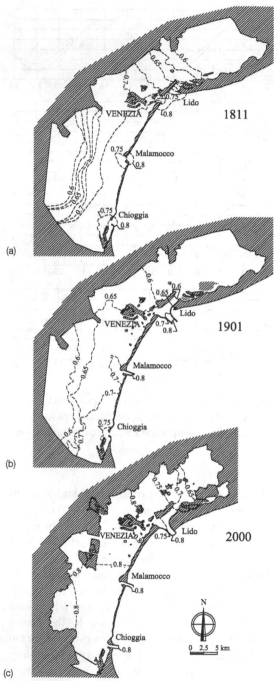

Fig. 48.3 Contours of maximum tide elevations for:
(a) 1811; (b) 1901; (c) 2000 lagoon configurations. The
calculations have been carried out by assuming a reference
semidiurnal, sinusoidal forcing tide, with amplitude 1 m,
and oscillating around 0.30 m a.m.s.l.

possible solution is to concentrate the flow resistances at the inlets. A non-negligible damping of the normal high tides can thus be obtained by either increasing the roughness of the inlet channels or narrowing significantly their sections. Furthermore, simulating the past behaviour of the inlets using local mathematical models requires the introduction of suitable boundary conditions which, obviously, cannot be deduced from the present morphological lagoon configuration. In fact, the results briefly summarized here, show that, for the same sea levels, the behaviour of the channels departing from the inlets was quite different in the past, regarding both the flow discharge and the water surface oscillations.

THE ROLE OF INLET GEOMETRY IN THE SEA-LAGOON SEDIMENT EXCHANGE

The erosion processes taking place in the lagoon basin are the result of many causes which often interact with each other. Nonetheless, there is no doubt that a decisive role is played by the wave action, generating shear stresses which are strong enough to re-suspend the sediments deposited on tidal flats and to induce the collapse of the banks bordering salt marshes. The sediment entrainment into suspension becomes much more intense when the waves generated by wind or boat wakes break and dissipate energy. Once the wave motion attenuates, the suspended sediments are then re-deposited on the bottom. However, during both the entrainment and settling phases, a significant amount of sediments flushes towards the channels; in particular, the sediments can be carried towards the inlets and then to the sea during the ebb tides. This leads to a continuous exchange of sediments between the lagoon and the sea, which is greatly influenced by the flow field at the inlets. It is therefore worthwhile discussing some aspects of this process.

Under present-day conditions, the velocity distribution at the inlets during the ebb and flood phases of the tide exhibits a strong asymmetry. In a situation with no coastal currents, the velocity field in the flood phase does not differ very much from that of a typical potential flow (Plate 19a): as sea water flows into the lagoon, the coastal areas near each inlet contribute to the total discharge. Completely different

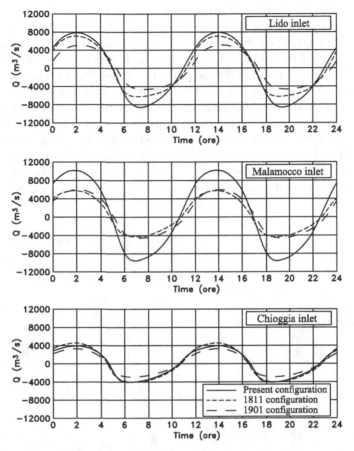

Fig. 48.4   Time evolution of flow discharge through the lagoon inlets for the three lagoon configurations analysed (i.e., 1811, 1901 and 2000). All the calculations have been carried out by assuming a reference semidiurnal, sinusoidal forcing tide, with amplitude 1 m, and oscillating around 0.30 m a.m.s.l.

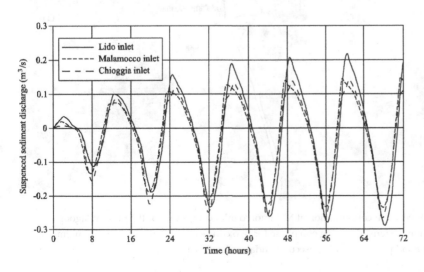

Fig. 48.5   Time evolution of the suspended sediment discharge through the three lagoon inlets for the present lagoon configuration. The calculations have been carried out by assuming a reference semidiurnal, sinusoidal forcing tide, with amplitude 1 m, and oscillating around 0.30 m a.m.s.l.

characteristics are exhibited by the ebb tide, when the current is forced in a given direction by the jetties which laterally delimit the inlets. Once the current has moved beyond the heads of the jetties, it maintains a jet structure that moves out towards the sea, preserving its own identity. This jet is characterized by a weak diffusion and penetrates for quite a long distance into the open sea. Wide vortices form on the sides of the jet and move away from the heads of the jetties towards the sea (Plate 19b).

These present-day conditions are completely different from the conditions that existed before the con-

struction of the jetties. In fact, if we consider the 1811 lagoon configuration, the temporal evolution of the flow field near the inlets appears to be essentially symmetrical, as shown in Plate 19 (c) and (d). During the flood tide (Plate 19c), all the coastal areas near each inlet contribute evenly to the total discharge, as they do today (Plate 19a). On the other hand, differently from today, during the ebb phase the current spreads out in the areas in front of the inlets and quickly mixes with the sea water (Plate 19d).

The effects of the different hydrodynamic behaviour of the inlets on the sediment balance are clearly

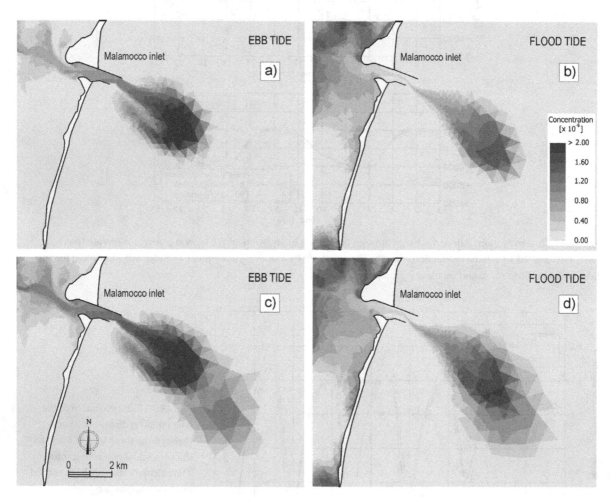

Fig. 48.6   Examples of the suspended sediment concentration at Malamocco inlet computed for the present lagoon configuration. Maximum sediment plume extension during ebb phase for two consecutive tidal cycles (a, c); minimum sediment plume extension during the flood phase for two consecutive tidal cycles (b, d).

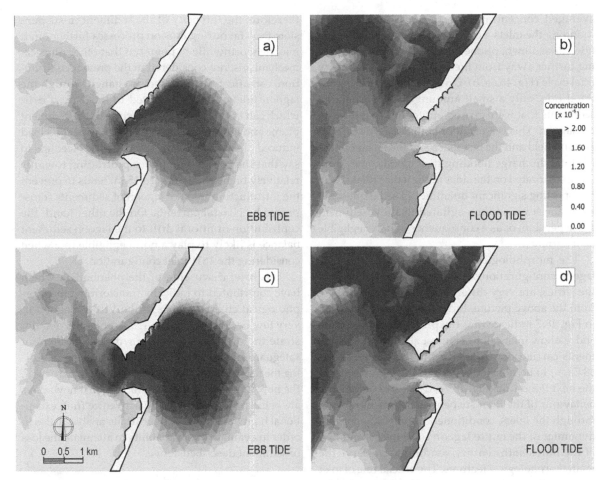

Fig. 48.7   Examples of the suspended sediment concentration at Malamocco inlet computed for the 1811 lagoon configuration. Maximum sediment plume extension during ebb phase for two consecutive tidal cycles (a, c); minimum sediment plume extension during the flood phase for two consecutive tidal cycles (b, d).

demonstrated by the results, shown in Figs. 48.6 and 48.7, obtained using the non-fixed bed model of the whole lagoon, considering, as before, a 1 m wide semidiurnal sinusoidal forcing tide oscillating at an average value of 0.30 m a.m.s.l. Moreover, as indicated by the results of a granulometric analysis carried out in some marshy areas of the lagoon, we assume a sediment grain size distribution characterized by a d_{50} of 0.050 mm and a d_{90} of 0.200 mm.

The resulting estimates of the solid discharge exchanged through the inlets in the present conditions suggest a significant loss of finer fractions. More specifically, as shown in Fig. 48.5, the net sediment vol-

ume delivered to the sea every tidal cycle turns out to be about 3000 m³. If we assume, for the sake of simplicity, that over one year the lagoon is subject to about one hundred tidal cycles with characteristics similar to the reference tide just considered, we obtain an overall loss of finer sediments of about 300 000 m³ yr⁻¹. This value is similar to the average estimated and obtained on the basis of the most recent lagoon bathymetries available for the lagoon.

This loss of sediments is favoured by the jetties at the inlets, as illustrated by the suspended sediment patterns of Fig. 48.6, where a grey colour scale depicts the instantaneous spatial contours of the averaged concentration in the sea areas just surrounding the inlets. For a given inlet, during the ebb tide, the sediment plume flows out into the sea, moving further away from the heads of the jetties every tidal cycle (Fig. 48.6a, b). On the contrary, during the flood tide, only a small amount of the suspended sediments that were carried out to the sea returns to the lagoon. This is a result of both the asymmetry of velocity field and the fact that a significant amount of the total discharge flooding the lagoon comes from the areas located on the sides of the jetties (Plate 19 a), transporting significant quantities of sediments only when the sea is enough agitated to allow sediment re-suspension, or as a consequence of non-negligible drift currents.

The morphological characteristics of the 1811 lagoon configuration, i.e. before the construction of the jetties, are very different from the ones arising from the above picture. Indeed, the results reported in Fig. 48.7 indicate that the presence of very wide and shallow inlets allows almost all of the finer sediments carried out to sea during the ebb tide (Fig. 48.7a, b) to be reintroduced into the lagoon during the next flood tide (Fig. 48.7c, d). Such a symmetric behaviour of the finer suspended sediment flowing through the inlets, conditioned heavily the morphodynamics of the entire lagoon until the second half of the nineteenth century, tending to preserve the main morphological features. The crucial role of inlet configuration is also substantiated by the fact that the lagoon bathymetry was scarcely influenced by net loss of incoming sediment consequent to the artificial changes carried out on the final reaches of the Brenta and Piave rivers, originally flowing into the lagoon, which were diverted directly to the sea during the sixteenth and seventeenth centuries, respectively.

After the construction of the jetties at each inlet, brought to an end in 1930, the systematic net loss of finer sediment fractions experienced through the inlets as a consequence of the above mentioned asymmetry, enforced an irreversible erosion of the lagoon. Obviously, other anthropic interventions contributed to modify the sediment budget, such as the dredging of industrial channels. Moreover, the flow depth increase experienced by the intertidal areas, resulting from erosive processes, subsidence and eustatism, enhances the wind wave amplitude and,

therefore, the intensity of the sediment re-suspension, thus favouring erosion processes further on. It may be worthwhile to point out that other physical mechanisms, not considered in the present contribution, can affect the sediment exchange through the lagoon inlets: they are related to the littoral drift which can both modify the structure of the sediment plume and feed inside the lagoon the suspended load induced along the coast by wind waves. Needless to say that the presence of jetties protruding towards relatively high sea depths (10-15 m) tends to prevent the introduction into the lagoon of sediments transported by coastal currents. On the other hand, the contribution of littoral drift to the lagoon sediment balance is likely to play a non-negligible role when considering the 1811 inlet configuration.

This brief discussion of the phenomena which have contributed to radically transforming the morphological characteristics of the Venice Lagoon in very few decades (i.e., from 1930 until now) demonstrate the urgency of dealing with the problem of safeguarding the lagoon and, where possible, restoring the morphology of the lagoon basin. In particular, the new interventions at the inlets which nowadays are being proposed to safeguard Venice from exceptional high tides, need to be critically analysed also in order to evaluate their capability to attenuate the loss of sediment described above.

CONCLUSIONS

The numerical analyses carried out to compare the present morphodynamic behaviour of the Venice Lagoon with those resulting from two historical configurations documented by the technical cartography of 1811 and 1901, have highlighted some important changes in the tidal current regime inside the lagoon and alterations that have taken place in the sea-lagoon exchange of sediment. Until 1865, tidal propagation was strongly influenced by dissipative phenomena, and both the maximum tidal elevation and the tidal prism were much smaller than they are today. At the same time, the particular morphology of the inlets, which were very wide and shallow, and the typical morphology of the inner portions of the lagoon implied that tidal propagation was frictionally dominated.

The numerical results presented above suggest that a significant reduction of both the amplitude and the peak of the tidal waves propagating inside the lagoon, similar to the one characterising the very recent past, can be achieved by narrowing the inlets and increasing their flow resistance. For example, the latter effect could be achieved by building fixed structures, which locally reduce the width of the inlet channels and increase considerably the local flow resistance. Actions aimed at suppressing the intense erosion processes currently taking place inside the lagoon are deemed to be as urgent as those proposed to defend Venice from flooding. One of the main causes of the erosion processes is the significant loss of finer sediments through the inlets during each tidal cycle, as a consequence of the construction, started in the second half of the 1800s, of jetties to enable modern large vessels to enter the lagoon. Putting off interventions to limit the erosive phenomena currently in progress in the entire lagoon should no longer be considered an option. Finally, the results obtained from investigating various inlet configurations suggest that the design of new structures within/outside the inlets requires a careful evaluation of the effects that such works will have on the sediment balance between the sea and the lagoon.

The complexity of the morphological problems pointed out by the numerical analyses described in previous Sections suggests the intrinsic weakness of those conceptual models which introduce a too simplified description of the physical processes shaping lagoon environments. Indeed, when the boundary conditions of the problem are likely to rapidly change, as is the case of Venice, these models lose their ability to make effective predictions since the empirical coefficients derived for a particular configuration are no longer valid.

REFERENCES

Bates, P. D., and J. M. Hervuoet. 1999. A new method for moving boundary hydrodynamic problems in shallow water. *Proceedings of the Royal Society of London*, Series A **455**, 3107–28.

D'Alpaos, L. 1992. Evoluzione morfologica della laguna di Venezia dal tempo del Denaix ad oggi e sue conseguenze sul regime idrodinamico. In *Conterminazione lagunare: storia,*

ingegneria, politica e diritto nella Laguna di Venezia. Venice: Istituto Veneto di Scienze, Lettere ed Arti, pp. 327–58.

D'Alpaos, L., and A. Defina. 1993. Venice Lagoon Hydrodynamic Simulation by Coupling 2D and 1D Finite Elements Models. In K. Morgan *et al.* (eds.), *Proceedings of the 8th Conference on Finite Elements in Fluids. New Trends and Applications*. Barcelona: Pineridge Press, pp. 917–26.

D'Alpaos, L., and A. Defina. 1995. Modellazione matematica del comportamento idrodinamico delle zone di barena solcate da una rete di canali minori. *Istituto Veneto di Scienze, Lettere ed Arti, Rapporti e studi* **12**, 353–72.

D'Alpaos, L., and A. Defina. 2004 (submitted). Contribution to mathematical modelling of Venice lagoon.

Day, J. W., Rybczyk, J., Scarton, F., Rismondo, A., Are, D., and G. Cecconi. 1999. Soil accretionary dynamics, sea-level rise and the survival of wetlands in Venice Lagoon: a field and modelling approach. *Estuarine, Coastal and Shelf Science* **49**, 607–28.

De Bernardi, A. 1843. *Inventario delle opere esistenti in Laguna in opposizione alle prescrizioni del Regolamento di Polizia Lagunare 20 dicembre 1841*. Venice.

Defina, A. 2000. Two dimensional shallow flow equations for partially dry areas. *Water Resources Research* **36**, 11, 3251–64.

Defina, A., D'Alpaos, L., and B. Matticchio. 1994. A new set of equations for very shallow water and partially dry areas suitable to 2D numerical models. In P. Molinaro and L. Natale (eds.), *Proceedings of the Special Conference on Modelling of flood propagation over initially dry areas*. Milan: ASCE, pp. 72–81.

Denaix, A. 1811. *Carta topografica idrografica militare della Laguna di Venezia*. Venice.

Genio Civile di Venezia, Ufficio del. 1901. *Carta Idrografica della Laguna di Venezia*. Venice: Ministero dei Laron Pubblicci.

Jasak, H. 1996. *Error analysis and estimation for the finite volume method with applications to fluid flow*. Unpublished Ph.D. thesis, London: Imperial College.

Martini, P., D'Alpaos, L., and L. Carniello. 2004 (in press). Un modello matematico bidimensionale per lo studio del trasporto di sedimenti nella laguna di Venezia. *Corila Research Program 2002–2003 results*. Istituto Veneto di Scienze, Lettere ed Arti. Venice.

Rouse, H. 1949. *Engineering Hydraulics*. New York, London and Sydney: John Wiley & Sons.

Umgiesser, G., and A. Bergamasco. 1993. A staggered grid finite element model of the Venice Lagoon. In K. Morgan *et al.* (eds.), *Proceedings of the Conference on Finite Elements in Fluids*.

New Trends and Applications. Barcelona: Pineridge Press, pp. 659–68.

Van Rijn, L. C. 1984a. Sediment transport, Part I: bed load transport. *Journal of Hydraulic Engineering ASCE* **110**, 10, 1431–56.

Van Rijn, L. C. 1984b. Sediment transport, Part II: suspend load transport. *Journal of Hydraulic Engineering ASCE* **110**, 11, 1613–41.

Van Rijn, L. C. 1984c. Sediment transport, Part III: bed forms and alluvial roughness. *Journal of Hydraulic Engineering ASCE* **110**, 12, 1733–51.

49 • Water fluxes between the Venice Lagoon and the Adriatic Sea

M. GACIC, V. KOVACEVIC, I. MANCERO MOSQUERA, A. MAZZOLDI AND S. COSOLI

SUMMARY

Water flux variations in the Venice Lagoon inlets have been analysed. Of the total variance of the flux, 92 per cent is due, on average, to tidal forcing. The remaining variance is associated with non-tidal local and remote forcing. Remote forcing acts through the lagoon inlets and is mainly related to the Adriatic seiches at the semidiurnal and diurnal timescales (11 and 22 hour periods respectively). Local wind generates an interior longitudinal sea-level slope that is responsible for the differential pressure gradient and the low-frequency flux variations. This then results in out-of-phase variations of fluxes in the Lido and Chioggia inlets. At longer time-scales (seasonal and year-to-year) the inlets' fluxes are due to the freshwater balance in the lagoon, with a net outflow of the order of 100 $m^3 s^{-1}$. On these long timescales, the Lido inlet shows an inflow while the Chioggia inlet shows an outflow of the same magnitude. Calculations of the flushing half-life using the tidal prism method demonstrate that the lagoon exchanges water with the open sea rather quickly. Flushing half-time ranges between 6 and 12 hours during spring and neap tides respectively.

INTRODUCTION

Venice Lagoon is the largest Adriatic lagoon, having an area of about 550 km^2 and an average depth of

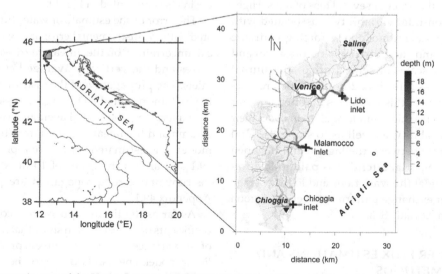

Fig. 49.1 Bathymetric map of the Venice Lagoon. ADCP mooring sites in inlets are denoted by crosses, triangles show tide gauge stations at Saline and Chioggia, the square denotes a rain gauge station in the city of Venice, while the Oceanographic Tower 'Acqua Alta' is denoted by an open circle.

Flooding and Environmental Challenges for Venice and its Lagoon: State of Knowledge, ed. C. A. Fletcher and T. Spencer.
Published by Cambridge University Press, © Cambridge University Press 2005.

about 0.5 m (Fig. 49.1). It communicates with the open Adriatic Sea through three relatively deep inlets: Lido, Malamocco and Chioggia. Lido is the largest (900 m wide with the average depth of about 12 m), Malamocco is the deepest (18 m average depth) with a width of about 450 m, while Chioggia is about 400 m wide and 8 m deep.

One of the important parameters in the management of semi-enclosed bays, lagoons and estuaries is the flushing or residence time (Zimmerman, 1981; Knoppers *et al.*, 1991). The flushing time concept in the Venice Lagoon becomes even more significant from the perspective of the construction of mobile gates at its inlets and temporal closures of the lagoon to mitigate flooding events (locally called *acqua alta*). The flushing time can be determined only after water fluxes and their variability have been determined. Detailed knowledge of water fluxes between the Venice Lagoon and the Adriatic Sea is also very important for understanding the generation of these flooding events and predicting them. Temporal variability of the inlets' fluxes in the Venice Lagoon is mainly associated with tidal forcing – as has already been shown for currents (Gačić *et al.*, 2004). Local forcing by the most frequent Adriatic winds, *bora* and *sirocco*, generates different response patterns in the inlets' low-frequency flows and fluxes at timescales on the order of several tens of days. High-frequency non-tidal variability is associated with remote large-scale atmospheric forcing (Adriatic basin-scale winds and atmospheric pressure) and occurs at timescales of semi-diurnal and diurnal Adriatic seiches (11- and 22-hour periods respectively). The aim of this work was to estimate water fluxes in the three inlets and to analyse in detail their temporal variability, as well as to study the total water flow rate between the lagoon and the open Adriatic Sea. Special attention is paid to low-frequency, non-tidal flux variations and to the average annual water exchange pattern between the lagoon and the open Adriatic Sea.

DATA, WATER FLUX ESTIMATIONS AND ANALYSIS METHODS

Water flow in lagoon inlets, in addition to prominent high-frequency variations, is subject to seasonal and year-to-year variability. Thus, current measurements in the inlets of the Venice Lagoon have been planned so as to be made over several years. Continuous current measurements have been carried out using bottom-mounted 600 kHz Acoustic Doppler Current Profilers (ADCP) with a recording interval of 10 minutes. One ADCP has been positioned per inlet; a detailed description of the experiment is given in Gačić *et al.* (2004). As mentioned in that paper, the position of the ADCP in each inlet was determined on the basis of a series of profilings undertaken at various phases of tides by a ship-borne ADCP. On average, around 100 profilings were carried out per inlet at different flood, ebb and slack tide conditions. For each profile the water flux was determined using WinRiver, the ADCP post-processing software provided by RD Instruments. Subsequently, linear regression was calculated between the vertically averaged currents at various points of the transect and the total water flux. The bottom-mounted ADCPs were then positioned at a point on the transect in each inlet where the linear correlation between the vertically averaged current and the water transport rate was maximized. In this way it was possible to obtain time-series not only of currents but also of water fluxes for all three inlets. The linear relationship between the water transport rates (Q) and vertically averaged current magnitude (V_m) in the inlets is presented in Fig. 49.2.

The error of the estimate of water fluxes is associated with the instrumental error and with the statistical uncertainty of the linear regression between fluxes and the vertically averaged currents at the inlets. The propagation of the instrumental error of the bottom-mounted ADCP was assessed across the time-averaging, the spatial averaging over the water column and the linear equation from the regression model. The estimated errors are reported in Table 49.1 and are on the order of 10 m^3 s^{-1}. Details on instrumental error propagation are presented in Appendix 49.1.

As far as the linear regression is concerned, the coefficients of determination show that more than 98% of the variance of measured flux is predicted via the linear model. The standard error of the estimation is:

$$S_e = \sqrt{\frac{1}{(n-2)}\sum_{k=1}^{n}(Q_k - \hat{Q}_k)^2}$$ *Eq. 49.1*

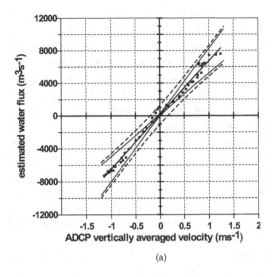

(a)

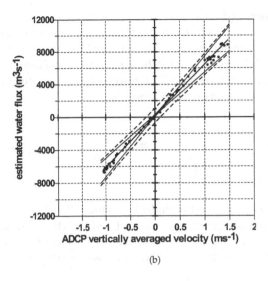

(b)

Fig. 49.2   Dispersion diagram showing linear relationship between the water flux and vertically averaged inlet currents for (a) Lido, (b) Malamocco and (c) Chioggia. Negative/positive signs denote the transport into the lagoon/out of the lagoon. Solid and dashed lines denote 99% confidence limits for individual and mean predicted values respectively.

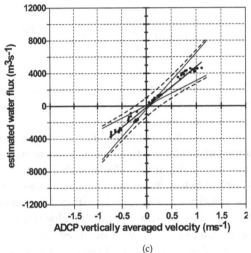

(c)

Table 49.1   *Regression coefficients and the correlation squared between the vertically averaged speed and flux for three inlets. Nb represents the number of vertical bins. The estimated error of water fluxes is due to instrumental errors.*

| | $Q = aV_m + b$ | | | # bins | Estimated error |
Location	a	b	R^2	Nb	(m^3/s)
Lido	6594.4	123.99	0.9931	9	13.24
Malamocco	6301.6	176.48	0.9962	16	9.49
Chioggia	4969.2	-159.61	0.9808	6	12.22

where Q_k is an observed value, while $\hat{Q}_k$ is its respective estimation from the linear equation; n is the number of observations. S_e is used to compute the confidence interval for the expected value of the estimator $\hat{Q}_i$ corresponding to some arbitrary value of V_{mi}, as given by the expression:

$$\hat{Q}_i \pm t_{(n-2,\alpha/2)} S_e \sqrt{\frac{1}{n} + \frac{(V_{mi} - \overline{V}_m)^2}{\sum\limits_{k=1}^{n}(V_{mk} - \overline{V}_m)^2}} \qquad Eq.\ 49.2$$

where $t_{(n-2,\alpha/2)}$ is the student's t value with n-2 degrees of freedom, α significance level, and V_m is the average value of V_{mk} , k = 1,..., n. Likewise the confidence interval for the individual estimated value $\hat{Q}_i$ is:

$$\hat{Q}_i \pm t_{(n-2,\alpha/2)} S_e \sqrt{1 + \frac{1}{n} + \frac{(V_{mi} - \overline{V}_m)^2}{\sum\limits_{k=1}^{n}(V_{mk} - \overline{V}_m)^2}} \qquad Eq.\ 49.3$$

These limits are not parallel to the regression line but hyperbolic, as presented in Fig. 49.2 for the 99 per cent confidence level.

Simpson and Bland (2000) had a slightly different approach; they took the index velocity measurements with an ultrasonic velocity meter positioned at about 2 m below mean sea level, thus measuring the line velocity at that depth. Then they determined the rating curve relating the index velocity to the average channel velocity as obtained from ship-borne ADCP measurements. Subsequently, instantaneous channel discharge was determined by multiplying the average velocity with the cross-sectional area of the channel.

In our case, we directly relate the index velocity taken with the bottom-mounted ADCP to the inlet flux obtained from the ship-borne ADCP surveys. The obtained relationship remains valid over much longer time-scales since the prevailing variability in the flow field is associated with the tidal forcing, while the meteorological forcing, on average, accounts only for less than 10 per cent of the total variance (as shown by Gačić *et al.* (2004)). The estimated flow rate reaches up to 10,000 m^3 s^{-1} in the Lido and Malamocco inlets and to about 5,000 m^3 s^{-1} in the Chioggia inlet.

The wind data were obtained from the oceanographic tower '*Acqua alta*' (Fig. 49.1). Hourly means of the wind vector were obtained from the 5-minute data record. The instrument height is 15 m above mean sea level. Subsequently from the wind velocity, the wind stress was calculated using a wind speed-dependent drag-coefficient (Large and Pond, 1981). The wind stress vectors were decomposed into components in the coordinate system with one axis in the direction of the maximum wind variance and the other one in the direction of the variance minimum. The major variance axis is parallel to the bora-wind direction (wind from NE), and the wind stress components are defined as positive when oriented in the direction of 54 deg. clockwise from north.

Sea level variations as a function of inlet fluxes and different wind patterns were analysed using tide-gauge data from stations positioned at two extremes inside the lagoon along its longitudinal axis (Fig. 49.1). This tide-gauge setting enabled the calculation of longitudinal (north–south) sea-level slope. All the analysed data were hourly averages obtained from recordings at a 5 min. interval.

The study period was from 12 February 2002 until 15 December 2003. Simultaneous ADCP measurements at all three inlets were carried out over this period. In the flux and sea-level time-series the tidal signal was determined by applying a harmonic analysis technique based on Foreman (1978). This fits the signal by the least squares method to the sum of sinusoidal terms of fixed frequencies. The non-tidal part of the flux time-series was then obtained by subtracting the tidal signal from the original time-series of hourly flux data. Only tidal components having a signal-to-noise ratio greater than one have been taken into consideration. The prevailing variability of the non-tidal flux was obtained from wavelet spectra (Torrence and Compo, 1998) of the detided time-series. Wavelet analysis provides information on how the localized and dominant frequency-dependent modes vary in time (Tokmakian and McClean, 2003). The method has been widely used to analyse variety of geophysical signals (Gu and Philander, 1995; Meyers *et al.*, 1993, Liu, 1994) since it enables to extract information on non-stationary features of time-series and to localize in time transient events, which are often present in geophysical processes.

RESULTS AND DISCUSSION

Tidal variations

Results of the harmonic analysis show that, as in the case of the vertically averaged currents (Gačić *et al.*, 2004), the major portion of the water flux variance in the inlets is associated with tidal forcing (Table 49.2). More than 80 per cent of the water flux variance is due to the four semi-diurnal constituents (M2, S2, K2 and N2), while the most energetic daily constituents (K1, P1 and O1) are responsible for about 10 per cent of the flux variance. The remaining variance is due mainly to a non-tidal forcing. The semi-diurnal constituent M2 shows the largest amplitude of more than 4000 $m^3 s^{-1}$ for the Lido and Malamocco inlets and more than 2000 $m^3 s^{-1}$ for the Chioggia inlet. Typically for all tidal constituents, the flux in the Chioggia inlet has an amplitude representing about 50% of those at Malamocco or Lido. As far as the phase-lag between different inlet fluxes is concerned, Chioggia leads both Malamocco and Lido for semi-diurnal constituents. Interesting,

although relatively weak, is a signal associated with a solar annual tidal constituent, SA. Despite its low amplitude of about 100 $m^3 s^{-1}$ in Lido and Chioggia, it is statistically significant, i.e. larger than the error. Only at Malamocco is the SA amplitude smaller than the error. Moreover, at that frequency the Lido and Chioggia fluxes are out of phase – so that an inflow at Lido is compensated by an outflow at Chioggia and vice versa. This exchange pattern is clearly evident from the monthly averages of tidal fluxes (Fig. 49.3). These were obtained by considering only those tidal constituents that have signal-to-noise ratio larger than one. By calculating the monthly means obviously all the constituents with a period smaller than a month were averaged out so we were left only with a signal associated with the SA tide.

Low-frequency non-tidal flux variations

Low-frequency, non-tidal flux variations were analysed over a time-scale range between 512 and

Table 49.2. *Tidal harmonic constants of the water flux for three inlets, including the solar annual SA in addition to the seven principal constituents. Analysed period: 12 February 2002 10h00 – 15 December 2003 11h00 (continued overleaf).*

EV	Explained variance (%) with respect to total variance (tidal+nontidal)
Amp	Amplitudes (m^3/s)
eAmp	error of Amplitudes (m^3/s)
Pha	Phases (°)
ePha	error of Phases (°)

Lido

constituent	EV	Amp	eAmp	Pha	ePha
O1	0.6	452.7	84.8	173.4	10.2
P1	1.1	588.0	87.2	179.2	8.0
K1	10.2	1827.8	85.9	179.9	2.8
N2	1.6	720.7	54.5	36.2	3.9
M2	58.7	4384.9	52.8	36.4	0.7
S2	19.3	2517.6	48.7	46.0	1.2
K2	2.0	805.5	43.3	40.7	3.2
SA	0.03	93.4	77.0	169.8	45.4
sum	93.5				

Table 49.2.   *continued*

Malamocco

constituent	EV	Amp	eAmp	Pha	ePha
O1	0.5	443.0	76.4	161.5	11.3
P1	0.9	574.0	81.4	170.3	8.9
K1	8.6	1794.9	77.1	170.2	2.7
N2	1.5	756.4	56.5	24.5	4.2
M2	60.0	4747.8	54.7	23.4	0.7
S2	21.1	2814.5	52.6	34.3	1.1
K2	2.2	905.0	52.9	27.9	3.3
SA	0.0	4.6	30.4	315.0	255.7
sum	94.7				

Chioggia

constituent	EV	Amp	eAmp	Pha	ePha
O1	0.6	240.8	41.3	162.7	8.4
P1	0.9	301.9	44.5	174.4	8.9
K1	8.3	901.3	44.0	166.5	2.6
N2	1.4	364.7	41.3	14.9	6.1
M2	58.1	2378.2	32.9	14.4	0.8
S2	20.6	1416.9	33.9	24.4	1.5
K2	1.9	426.4	31.9	20.6	3.9
SA	0.1	97.8	36.2	345.6	21.7
sum	92.0				

1024 hours (20 and 40 days respectively). The low-frequency non-tidal flux variations at this time-scale have a local variance maximum in winter months from November 2002 through January 2003, as determined from wavelet analysis and shown in Plate 20. The same feature was observed in the case of currents (Gačić *et al.*, 2004). The flux variations at the Lido inlet for this time-scale range are out-of-phase with respect to those in the Chioggia inlet (Fig. 49.4). Also, from the figure it is evident that the average flux in Lido is negative (inflow), while in Chioggia it is positive (outflow). The net water flux at this time scale for the entire study period (Fig. 49.5) shows an average outflow of about 88 m^3 s^{-1}around which oscillations with a maximum amplitude of about 500 m^3 s^{-1} occur. In addition, strong wind forcing events generate in-phase net flux variations; negative departures of the wind stress component along the major variance axis from the mean (i.e. strengthening of the *bora* wind) generate an increase in net inflow, mainly due

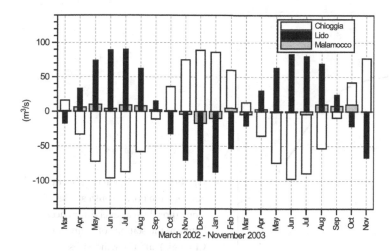

Fig. 49.3   Monthly average tidal fluxes at Lido, Malamocco and Chioggia.

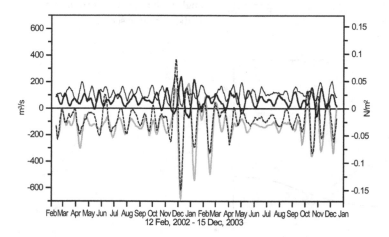

Fig. 49.4.   Time-series of non-tidal fluxes for the time-scale range between 512 and 1024 hours for Lido (dash), Chioggia (bold line), Malamocco (thin line) and wind stress component along the major variance axis (solid gray line).

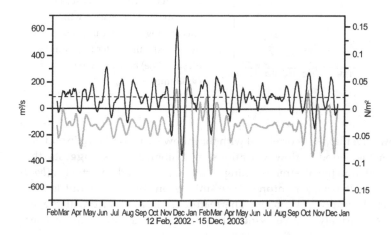

Fig. 49.5   Non-tidal net water flux (sum of corresponding three inlets' fluxes) at time-scale range 512–1024 hours as a function of time (thin line) and wind stress component along the major variance axis (solid gray line). Horizontal dashed line denotes the average net flux (88 m^3 s^{-1}) for the study period (12 February 2002 – 15 December 2003).

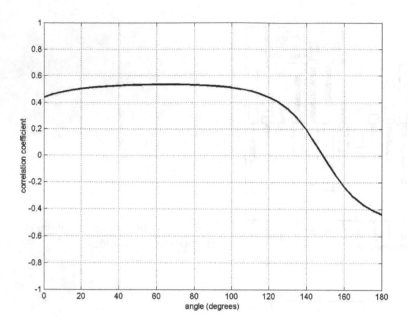

Fig. 49.6   Correlation coefficient between the north–south sea-level slope (hourly sea-level differences between the stations of Saline and Chioggia) and the wind stress component along axes with a varying angle with respect to north increasing in the clockwise sense.

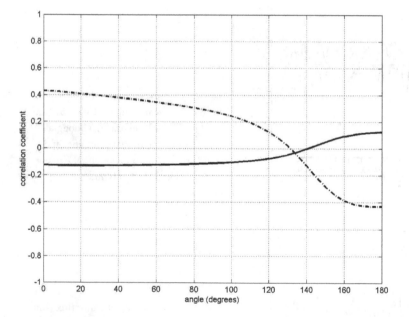

Fig. 49.7   Correlation coefficient between the wind stress projection (a component along axes with varying angle, increasing clockwise from north) and water fluxes: at Lido (dashed line) and Chioggia (solid line).

to the intensification of the Lido inflow. The maximum amplitude of the net flux appears in the period November/January, as already documented from the wavelet spectra of individual time-series. It was shown by Gačić *et al.* (2004), that the wind component along the axis of major variance is positively correlated with the inflow variations in Lido and outflow variations in Chioggia. This suggests that strengthening of the wind from the northeast (*bora*) reinforces the inflow in Lido and outflow in Chioggia. This feature can be associated with the sea-level wind set-up inside the lagoon creating a sea-level slope from north to south. This results in a higher sea level in the southern portion of the lagoon than in the open sea. On the other hand, in the northern portion of the lagoon the sea-level is lower inside the lagoon than outside it. These differences create a differential pressure gradient between the lagoon and the open sea that reinforces the inflow in Lido and the outflow in Chioggia. Longitudinal (north-south) sea-level slope was correlated with wind stress components in different directions taking into account only winds stronger than 5 m s^{-1}. The correlation coefficient as a function of the wind direction is displayed in Fig. 49.6. These calculations confirm that the wind blowing from the direction of about 60 deg. clockwise from north is the most efficient in generating the longitudinal sea-level slope in the lagoon. This coincides with the direction of the *bora* wind, and for the same direction a maximum positive correlation with the flux in Lido and negative one with

that in Chioggia was documented (Fig. 49.7). Also, this wind direction is quite close to the maximum wind variance axis.

Annual mean fluxes

From all available hourly flux data, moving average annual fluxes at each inlet, with a time step of one month, were calculated. They are presented as a function of time in Table 49.3, while their sum (net total flux) is plotted in Fig. 49.8. We calculated moving averages of the total water fluxes as well as of their tidal and non-tidal parts. The mean annual tidal fluxes are practically zero since most of the tidal constituents were averaged out. As far as the non-tidal annual flow is concerned, at Lido the pattern is one of inflow while at Chioggia outflow takes place. At Lido the flux varies between 60 and 120 m^3 s^{-1} and at Chioggia it changes to a lesser extent, varying from

Table 49.3 *Moving annual average of the water fluxes in m^3 s^{-1} (total, tidal, non-tidal) in each inlet and net total flux (sum of the total fluxes for three inlets).*

Annual averaging period	Lido			Malamocco			Chioggia			Net total
	total	tidal	nontidal	total	tidal	nontidal	total	tidal	nontidal	
1: Mar/2002 - Feb/2003	-59	1	-62	105	1	123	53	0	59	99
2: Apr/2002 - Mar/2003	-59	0	-63	110	1	123	49	0	55	100
3: May/2002 - Apr/2003	-60	0	-62	112	0	120	49	0	55	101
4: Jun/2002 - May/2003	-66	-1	-67	104	0	114	47	-1	53	85
5: Jul/2002 - Jun/2003	-73	-1	-72	95	-1	108	47	-1	53	69
6: Aug/2002 - Jul/2003	-69	-2	-68	109	-2	111	51	-1	54	91
7: Sep/2002 - Aug/2003	-78	-2	-78	101	-2	102	62	0	57	85
8: Oct/2002 - Sep/2003	-81	-1	-79	102	-1	105	62	0	57	83
9: Nov/2002 - Oct/2003	-99	0	-97	94	-1	101	61	0	56	56
10: Dec/2002 - Nov/2003	-115	0	-114	92	0	99	67	1	61	44

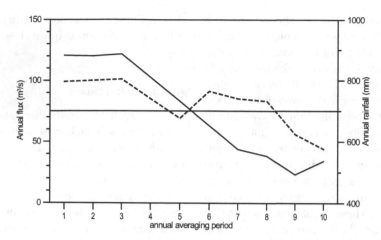

Fig. 49.8   Moving annual average net flux (dash) and the corresponding total rainfall (solid line) as a function of time according to Table 49.3.

about 50 to about 60 m³ s⁻¹. In both cases there is a continuous increase in the absolute value of the flux over the studied period. Malamocco, on the contrary, shows a decrease in outflow from about 120 to 100 m³ s⁻¹. The net water transport rate between the lagoon and the open sea (sum of mean annual total fluxes for all three inlets) presents an outflow occurring over the entire studied period with a decreasing trend from 100 m³ s⁻¹ to less than 50 m³ s⁻¹ that is dictated by the freshwater balance in the lagoon. The estimated freshwater discharge into the lagoon consists of about 40 m³ s⁻¹ of the river run-off (based on 1999 data; Zuliani *et al.*, 2001) and 60 m³ s⁻¹ of industrial water discharge (Magistrato alle Acque, 2002). The river run-off varies on an interannual time-scale and depends to a large extent on the rainfall. Our data show that minimum values of the average net water outflow from the lagoon occur when the averaging period contains to a major extent the year 2003, an extremely dry year in the area. In fact, we have calculated the moving annual rainfall with one-month time step for the entire current measurement period at a station in the city of Venice (see Fig. 49.1 for the location of the meteorological station). Data have been collected by Istituto di Scienze Marine/Biologia del Mare, Consiglio Nazionale delle Ricerche, Venice (URL http://www.ibm.ve.cnr.it). Annual rainfall data were then compared with the annual net fluxes (Fig. 49.8). It is clear that the documented decrease

in the net water outflow, as well as the Malamocco outflow, can be associated with a decrease in rainfall. It is, however, interesting that in parallel to a decrease in rainfall, an increase in the Lido inflow and, to a lesser extent, in the Chioggia outflow takes place. This may be explained in terms of the dependence of river run-off on rainfall. Thus, strengthening of the horizontal pressure gradient between the open sea and the lagoon interior can be related to reduced river run-off as a result of rainfall decrease. The stronger pressure gradient reinforces the inflow at Lido, while the net outflow decreases, presumably due to the lower run-off. Vice versa, an increase in the freshwater discharge reduces the pressure gradient between the lagoon interior and the open sea, reducing consequently the inflow to the lagoon that mainly takes place at Lido. On the other hand, in that situation the net water outflow increases due to the higher river run-off.

Flushing time

The steady state flushing time was calculated using the concept of flushing half-life or the time that it takes to renew half of the lagoon water volume. This definition is based on a rather restrictive assumption, i.e. it is assumed that complete mixing occurs rapidly compared to flushing half-time (Kjerfve *et al.*, 1996). Therefore, the obtained estimate of the flushing time represents the lower limit for water exchange. This estimate is also based on the assumption that the water that enters within one tidal cycle is not the same water that left during the previous tidal cycle. This is quite a reasonable assumption since it was shown that the quasi-steady coastal current outside the lagoon is rather strong, reaching in some months 20 cm s^{-1} (Kovačević *et al.*, 2004). Assuming first order kinetics (Pritchard, 1961), the time (τ) needed to renew half of the lagoon water is:

$$\tau = \frac{0.69}{\kappa} \qquad\qquad Eq.\ 49.4$$

where k is the ratio between the total water input rate and the volume of the lagoon. Water input rate consists of run-off, precipitation and tidal flushing rate. For the Venice Lagoon the precipitation of 800 mm/year is equal to a flux of 14 m^3 s^{-1}, while the river run-off is 40 m^3 s^{-1} and the industrial water discharge 60 m^3 s^{-1}. We were able to obtain quite reliable tidal prism estimates from our water flux time-series integrating the water volume entering during a flooding tide through all three inlets together. This term is the most important as it amounts to 11 900 m^3 s^{-1} during spring tide, to about 5500 m^3 s^{-1} during neap tide, or in average over one lunar month to about 8700 m^3 s^{-1} (i.e. one or even two orders of magnitude larger than the other water flux terms). Based on these estimates, the flushing half-time τ ranges between 6 hours during spring tides to about 12 hours during neap tides. The obtained values suggest that the Venice Lagoon renews water very quickly and it has quite a short flushing time as compared with estimates obtained for a number of other coastal lagoons (Kjerfve *et al.*, 1996; David and Kjerfve, 1998).

CONCLUSIONS

Long-term monitoring of the water flow in the Venice Lagoon inlets has enabled us to study the relative importance of tides and other forcings in driving the water exchange between the lagoon and the open sea. By setting up the linear regression between currents measured with bottom-mounted ADCP and fluxes, it was possible to estimate the total water flux as well as to study its temporal variation over variety of time-scales from tidal to sub-inertial and seasonal ones. The tidal flow is responsible for more than 90 per cent of the water flux through the inlets with the semidiurnal components being dominant. Phase-lag between different inlets shows that Chioggia tidal fluxes lead both those in Malamocco and Lido. Low-frequency flux variations at time-scales from 20 to 40 days, with a maximum variance in winter months, are generated by the local wind that creates an interior longitudinal (north-south) sea-level slope and subsequently a pressure gradient between the lagoon and the open sea of opposite sign in Chioggia and Lido. Thus, flux variations at these time-scales in the two inlets are out-of-phase, while Malamocco shows very weak correlation with the wind. Long-term average net flux (sum of all three inlets' fluxes) suggests an average annual outflow of the order of 100 m^3 s^{-1} that corresponds well with the total freshwater discharge into the lagoon (river run-off and industrial discharges). This may suggest that in the freshwater balance of the lagoon evaporation balances precipitation quite well. On a year-to-year basis, the net outflow changes appreciably. It was documented that it varies parallel to the precipitation rate, presumably due to the river run-off changes generated by changes in precipitation. Having detailed knowledge of the tidal prism, flushing half-life (the time needed to replace half of the lagoon water volume) was computed and the results show that it ranges between 6 and 12 hours for the spring and neap tides respectively. This suggests that the Venice Lagoon exchanges water faster than other similar water bodies.

ACKNOWLEDGEMENTS

This research was supported by CORILA, a consortium for promoting and coordinating the scientific research on Venice and its lagoon between the University of Venice, the University of Padua, the Architectural Institute of Venice, and the National Research Council of Italy. Financial support was also obtained from 'Centro Previsioni e Segnalazioni Maree' of the Municipality of Venice. I. Mancero Mosquera participated in the work with the support of the 'Programme for Training and Research in Italian Laboratories (TRIL)' of the Abdus Salam International Centre for Theoretical Physics, Trieste, Italy.

APPENDIX 49.1
ERROR PROPAGATION IN FLUX CALCULATIONS

Fundamentals

Let $x = (x_1, x_2, ..., x_N)$ be a random vector with mean vector $\mu_x = (\mu_1, \mu_2, ..., \mu_N)$ and standard deviation $\sigma_x = (\sigma_1, \sigma_2, ..., \sigma_N)$. Let $y = f(x_1, x_2, ..., x_N)$. Finding the distribution of y means to assess the propagation of statistical parameters of x_i across f. If f is a non-linear function, it becomes a very complex problem. However an approximation can be obtained with the propagation of the first two statistical moments, i.e. mean and variance; this is good enough under assumption of gaussianity. Also a linearization of f via first-order Taylor approximation can be done by expanding f about μ_x as shown:

$$y = f(x) \cong f(\mu_x) + \sum_{i=1}^{N}\left[\frac{\partial f}{\partial x_i}\bigg|_{x=\mu_x}\right](x_i - \mu_i)$$

Then, it can be shown (Arras, 1998) that:

$$\mu_y = E(y) = f(\mu_x) \tag{1}$$

$$\sigma_y^2 = E[(y-\mu_y)^2] = \sum_{i=1}^{N}\left[\frac{\partial f}{\partial x_i}\right]^2\sigma_i^2 + \sum_{i=1}^{N}\sum_{\substack{j=1 \\ j\neq i}}^{N}\left(\frac{\partial f}{\partial x_i}\right)\left(\frac{\partial f}{\partial x_j}\right)\sigma_{ij} \tag{2}$$

where the partial derivatives are evaluated in the mean vector. If x_i's are i.i.d. (independent, identically distributed) random variates, then the covariance $\sigma_{ij} = 0$ and the equation (2) becomes:

$$\sigma_y^2 = \sum_{i=1}^{N}\left[\frac{\partial f}{\partial x_i}\right]^2\sigma_i^2$$

On operation of the Acoustic Doppler Current Profiler (ADCP)

The Workhorse Sentinel ADCP's used in the inlets (Lido, Malamocco and Chioggia) were set up to work at 600Khz and vertical resolution of 1 m. The associated error to this configuration is $\sigma = 0.066$ m s^{-1}

(single-ping standard deviation), this error can be reduced by ensemble averaging. Since measurements are obtained every ten-minutes by averaging 60 pings, the associated error is reduced according to Central Limit Theorem:

$$\sigma_s = \frac{\sigma}{\sqrt{Np}}$$

Np is number of pings. The random error is uncorrelated from ping to ping, thus the ten-minutes data have an error of $\sigma = 0.0085$ m s^{-1}.

Error propagation on the computing of flux

A linear correlation was found between vertically-averaged hourly currents from the bottom-mounted ADCPs and hourly fluxes estimated on the basis of simultaneous measurements with a vessel-mounted ADCP. The propagation of instrumental error of bottom-mounted ADCP is first determined across the time averaging and the spatial averaging; then across the linear regression model.

Time averaging

Six ten-minutes data are averaged to yield one hourly value. This is expressed by the function:

$$V_h = \left(\frac{1}{Ne}\right)\sum_{i=1}^{Ne}x_i, \qquad Ne = 6$$

V_h is the hourly value, the x_i's are ten-minutes data. As a consequence of uncorrelated errors from ping to ping, the assumption of uncorrelated errors in ten-minutes data is justified, thus equations for the mean, μ_h, and the standard error of hourly values, σ_h, are derived by applying the formulae (1) and (2):

$$\mu_h = \left(\frac{1}{Ne}\right)\sum_{i=1}^{Ne}\mu_i \qquad \text{(the mean of means)}$$

$$\sigma_h = \left(\frac{1}{Ne}\right)\sqrt{\sum_{i=1}^{Ne}\sigma_i^2}, \qquad \text{since } \frac{\partial V_h}{\partial x_i} = \frac{1}{Ne}$$

Hence the standard error of hourly values is $\sigma_h = 0.0035$ m s^{-1}.

Spatial averaging

In this step a vertically-averaged velocity, V_m, is computed from hourly values:

$$V_m = \left(\frac{1}{Nb}\right)\sum_{i=1}^{Nb} V_{hi}$$

Nb represents the number of vertical bins (it depends on the depth of channel) and V_{hi} are the hourly values along the vertical profile. There are 9 bins monitored in the Lido inlet, 16 in Malamocco and 6 in Chioggia, so Nb takes values accordingly. With no knowledge about the spatial correlation of instrumental errors, the next approach is adopted:

a. Decompose the covariance from equation (2) as: $\sigma_{ij} = \rho_{ij}\sigma_i\sigma_j$, where ρ_{ij} is the correlation coefficient between the bins i and j.
b. Make some assumptions about the correlation coefficient.

Regarding step b, the best case would be no spatial correlation between bins so that $\rho_{ij} = 0$ and the Central Limit Theorem applies; the worst case would be a 100% bin-to-bin correlation ($\rho_{ij} = 1$), which implies that no decrease of error is reached by the vertical averaging of currents. Simpson and Bland (2000) use a 15% correlation between adjacent bins only. In general, for a constant correlation between adjacent bins, equation (2) becomes:

$$\sigma_m^2 = \frac{\sigma_h^2}{Nb} + \frac{2\,(Nb-1)}{Nb^2}\rho\sigma_h^2 \cong \frac{(1+2\rho)}{Nb}\sigma_h^2$$

where σ_h is the hourly standard error from the previous step ($\sigma_h = 0.0035$ m s^{-1}, constant). Hence, the vertical mean current has an associated standard error of σ_m.

The linear regression model

The instrumental error of bottom mounted ADCP is also propagated across linear regression model used to estimate fluxes, Q, from the vertically averaged velocity V_m. The linear relationship is modeled in the form of the equation Q= f(V_m) = aV_m + b. Applying equation (2) yields:

$$\sigma_Q^2 = a^2\sigma_m^2$$

where σ_m is the standard error in vertically averaged currents and σ_Q is the standard error in flux estimates.

REFERENCES

Arras, K. O. 1998. *An Introduction to Error Propagation: Derivation, Meaning and Examples of Equation $C_Y = F_X C_X F_X{}^T$.* Swiss Federal Institute of Technology Lausanne (EPFL), Technical Report N° EPFL-ASL-TR-98-01 R3.

David, L. T., and B. Kjerfve. 1998. Tides and currents in a two-inlet coastal lagoon: Laguna di Terminos, Mexico. *Continental Shelf Research* **18**, 1057–79.

Foreman, M. G. G. 1978. *Manual for Tidal Currents Analysis and Prediction. Pacific Marine Science Report 78–6*, Patricia Bay, Sidney, British Columbia: Institute of Ocean Sciences. Revised edn 1996.

Gačić, M., Mancero Mosquera, I., Kovačević, V., Mazzoldi, A., Cardin, V., Arena, F., and G. Gelsi. 2004 (in press). Temporal variations of water flow between the Venetian Lagoon and the open sea. *Journal of Marine Systems.*

Gu, D., and S. G. H. Philander. 1995. Secular changes of annual and interannual variability in the Tropics during the past century. *Journal of Climate* **8**, 864–76.

Kjerfve, B., Schettini, C. A. F., Knoppers, B., Lessa, G., and H. O. Ferriera. 1996. Hydrology and salt balance in a large, hypersaline coastal lagoon: Lagoa de Araruama, Brazil. *Estuarine, Coastal and Shelf Science* **42**, 701–25.

Knoppers, B., Kjerfve, B., and J. P. Carmouze. 1991. Trophic state and water turn-over time in six choked coastal lagoons of Brazil. *Biogeochemistry* **14**, 149–66.

Kovačević, V., Gačić, M., Mancero Mosquera, I., Mazzoldi, A., and S. Marinetti. 2004 (in press). HF radar observations in the Northern Adriatic: surface current field in front of the Venetian Lagoon. *Journal of Marine Systems.*

Large, W. G., and S. Pond. 1981. Open ocean momentum flux measurements in moderate to strong winds. *Journal of Physical Oceanography* **11**, 3, 324–36.

Liu, P. C. 1994. Wavelet spectrum analysis and ocean wind waves. In E. Foufoula-Georgiou, and P. Kumar (eds.), *Wavelets in Geophysics.* San Diego: Academic, pp. 151–66.

Magistrato alle Acque. 2002. *Relazione sulle caratteristiche degli scarichi idrici dell'area di Porto Marghera*, Dati relativi al 2000. Venice, 142 pp.

Meyers, S. D., Kelly, B. G., and J. J. O'Brien. 1993. An introduction to wavelet analysis in oceanography and meteorology: with application to the dispersion of Yanai waves. *Monthly Weather Review* 121, 2858–66.

Pritchard, D. W. 1961. Salt balance and exchange rate for Chincoteague Bay, Chesapeake. *Science* 1, 48–57.

Simpson, M. R., and R. Bland. 2000. Methods for accurate estimation of net discharge in a tidal channel. *IEEE Journal of Oceanic Engineering* 25, 4, 437–45.

Tokmakian, R., and J. L. McClean. 2003. How realistic is the high-frequency signal of a 0.1° resolution ocean model? *Journal of Geophysical Research* 108, C4, 3115, doi:10.1029/2002JC001446.

Torrence, R. T., and G.P. Compo. 1998. A practical guide to wavelet analysis. *Bulletin of American Meteorological Society* 79, 1, 61–78.

Zimmerman, J. T. F. 1981. The flushing of well-mixed tidal lagoons and its seasonal fluctuation. In *Coastal Lagoon Research, Present and Future*. Technical Papers in Marine Science, 33. Paris: UNESCO, pp. 15–26.

Zuliani, A., Fagarazzi, O. E., Zonta, R., and L. Zaggia. 2001. Fresh water transfer from the drainage basin to the Venice Lagoon. In *Preliminary Proceedings, DRAIN Project Workshop*. Venice: Ministero dei Lavori Pubblici, pp. 1–8.

50 • Venice Lagoon and the Po Delta: system functioning as a basis for sustainable management

J. W. DAY JR, G. ABRAMI, J. RYBCZYK AND W. MITSCH

SUMMARY

Over the past several thousand years, a number of important natural processes and anthropogenic impacts have influenced the formation, evolution and current status of the northwestern Adriatic shoreline, including the Po Delta and Venice Lagoon. The sandy coast, that was characterized by river distributary channels, deltaic marshes and estuaries and associated dunes, was relatively stable between 4000 and 2000 years ago. Beginning in Roman times, human interventions aimed at stabilizing the coast, inner delta, and Venice Lagoon gave rise to significant changes in the area. By the end of the twentieth century, the cumulative impact of these changes had resulted in profound changes in these coastal systems. In the Po Delta, river distributary channels are largely diked and much of the delta lies 1–3 m below sea level. The area of deltaic wetlands has been greatly reduced. During the twentieth century, wetland area in Venice Lagoon decreased from about 12 000 ha to 4000 ha. This loss was due to redirecting rivers around the lagoon to the sea, natural and human-induced subsidence, reclamation, and a strong net loss of sediments to the sea. There is a general crisis of instability and of unsustainable management of water and land resources. The unsustainability of present management is especially apparent considering predicted acceleration of sea-level rise, subsidence of the deltaic plain, and planned construction of gates in the lagoon inlets. Sustainable management will require that the rivers be re-incorporated into deltaic and lagoon functioning. This will require the utilization of pulsing events that maintained the delta and lagoon, especially the re-introduction of riverine inputs through river diversions.

INTRODUCTION

The coastline of the northwestern Adriatic Sea, which includes a broad plain of coastal lowlands, lagoons, coastal marshes, and river mouths that extend from lower Emilia-Romagna through the Veneto and Friuli-Venezia-Giulia regions has been greatly affected by human activity for over two thousand years. There was once a well-developed system of river mouths, beach–dune barriers, lagoons, salt marshes and freshwater wetlands. After the low stand of sea level about 17 000 years ago, water levels increased and stabilized between 4000 and 6000 years ago. At that time the coastline included small cuspate deltas and estuaries of the Rivers Po, Adige, Brenta and Piave. Since then, these coastal systems have been continuously modified by river flooding and marine processes and by human activities.

The present Po Delta has 5 principal distributaries and 14 secondary branches. Its area is 73 000 ha; about 60 000 ha are reclaimed land (*la bonifica*) and the rest is brackish lagoons and impoundments used for aquaculture (*valli*). There is a discontinuous sandy ridge of small dunes (between 2 and 4 m in elevation) parallel to the shoreline. The Adige and Brenta rivers discharge between the Po Delta and Venice Lagoon. The lagoon has an area of about 550 km² of brackish waters and wetlands and is connected to the sea by the Lido, Malamocco and Chioggia inlets. Venice Lagoon has been greatly altered by the development of urban areas (Venice being the most prominent) on the marshes and islands of Venice Lagoon and along the beach-dune system. The Brenta and Piave discharged into the lagoon and formed large areas of deltaic wetlands, but these rivers have been diverted around the lagoon and large areas of wetlands have been lost, both to reclamation and erosion, and to impoundment for fish ponds (*valli*). An embankment now completely surrounds Venice Lagoon and there is only local drainage into the lagoon. The interconnected series of lagoons in the northwestern Adriatic once formed an almost continuous network of deltaic systems and lagoons, which can be considered as deltaic lagoons (in the sense of Lankford, 1976).

In this chapter, we address impacts of human activities on the functioning of the Po Delta and Venice Lagoon. We specifically address the coastal management problems that have arisen due to these problems, including sustainable management in light of accelerated sea-level rise and the construction of the MOSE, the gates in the lagoon inlets. Our general hypothesis is that sustainable management of the delta and lagoon ecosystems must be based on system functioning.

THE DEVELOPMENT OF THE PO DELTA AND VENICE LAGOON AND THE IMPACTS OF HUMAN ACTIVITIES ON THE MORPHOLOGY OF THE AREA

The Po is one of the most important rivers discharging to the Mediterranean; it is 650 km long, has a mean discharge of about 1500 m^3 s^{-1} and the delta covers about 70 000 ha. The delta formed over the past several thousand years as the river successively occupied a number of different river channels (Sestini, 1996; Fig. 50.1). Former freshwater marshes that covered most of the deltaic plain have been reclaimed over the past century for agriculture. Most of the delta is now 1 – 4 metres below sea level due to subsidence caused primarily by extraction of shallow deposits of natural gas with a high water content (Sestini, 1996). Near the sea, there are beaches and dunes, shallow lagoons, salt marshes, and extensive reed swamps (approximately 2500 ha) dominated by *Phragmites australis* bordering the ends of the main river channels. Geological subsidence in the delta is 1-3 mm yr^{-1} (Sestini, 1992; Bondesan *et al.*, 1995) but human-induced subsidence was 5-20 mm yr^{-1} in the western Po Delta, and was 100 cm between 1958-1962 (Bondesan *et al.*, 1995). Marshes of the Po Delta and Venice Lagoon are highly productive (Scarton *et al.*, 1998, 2002) and serve as important habitat for many species of fish, mammals and birds.

Strong modifications in the morphology of land and rivers and of the natural processes in the area have been going on since Roman times 2000 years ago, including canal construction, agricultural development in a system called '*centuriatio*', and the construction of important military and commercial roads (via Popilia and via Annia), often located on the more dominant and stable sandy ridges (reaching up to 12 m in height). There are no data on land reclamation works in this period, but there was extensive cutting of forests in the lower plain of the Po river valley. This must have resulted in the first consistent increase in sediment transport by the rivers as consequence of soil erosion. There was a prominent delta with several cuspate features built by the southern branches of the river, among which the Po di Volano was one of the most active (Fig. 50.1a).

During late medieval times, there was a period of more extensive reclamation works of wetlands in the delta and in the drainage basin in the Po Plain and in the Apennines and the lower part of the Alps mountains. Partially as a result of these activities, there was a hydrological event called 'the breach of Ficarolo' (twelfth century) that changed the course of the river. New branches consolidated northward, such as the 'Po di tramontana' which was connected with the Adige River. This resulted in the transport of a large amount of sediment towards the Venice Lagoon. At the end of the sixteenth century, a large lobate delta formed in this area (Fig. 50.1d). In this period (1599-1604), the Republic of Venice carried out the largest diversion ever on a Po distributary. In the south-central part of the Po Delta, the old dune ridge was breached (Taglio di Porto Viro, see Fig. 50.1a., 50.1e) in order to force the water to flow into a new channel called the Po di Venezia. After this, the Po Delta began to take on its modern shape. The Venetians were also involved in other hydrological works in the Veneto region, such as the channelization of the Brenta and Piave rivers and their diversion around Venice Lagoon. The Brenta and Piave rivers now discharge to the south and north of the lagoon, respectively. At the same time they learned how to take advantage of the floods that, with growing frequency, were damaging the land. Some of the largest wetlands areas along the river courses were filled in with material transported by directing the

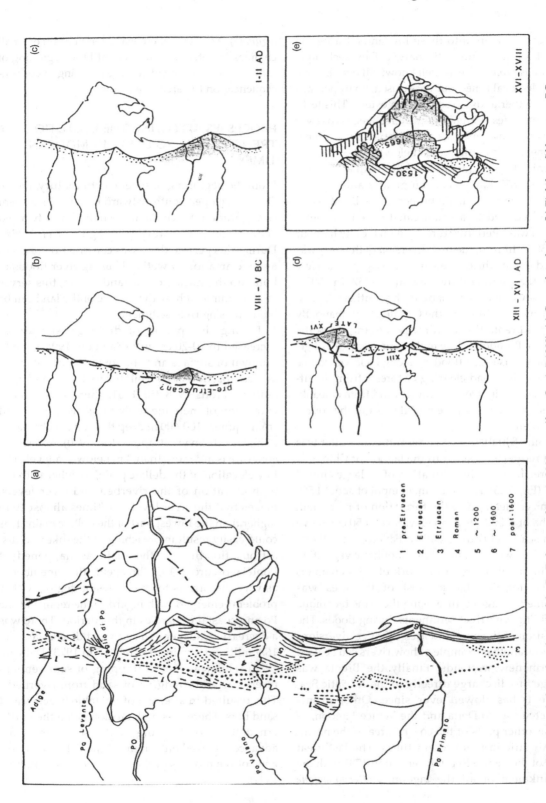

Fig. 50.1   Morphological evolution of the Po Delta over the past 3000 years (from Sestini, 1992). (a) General evolution. (b) Eighth to fifth century BC. (c) First to second century AD. (d) Eighth to the sixteenth century AD. (e) Sixteenth to eighteenth century AD.

flow of the rivers into these lowlands during the period of maximum discharge. This technique (*bonifica per colmata*) was well known all over Europe since Medieval times, but never used, apart perhaps in the Netherlands, to so large an extent. This technique provides evidence of the effectiveness of river diversions and serves as an example of what must be done now (as is discussed below).

Venice Lagoon (about 550 km^2, with a drainage basin of 1877 km^2) has been greatly altered. Discharge through the main branch of the Brenta river (today reduced to a small canal called 'Naviglio Brenta') was interfering directly with the city harbour (Fig. 50.2). In 1324, the first diversion of the river was carried out in which a new mouth was opened a few km to the south and then again in 1452. In 1507 a large, new outlet was made in the southern lagoon near Chioggia, through the Canale Montalbano. By 1540 the Brenta flowed directly into the sea at the mouth of Brondolo, close to the estuary of the Adige. But there was much flooding along its lower course due to widespread clearing of forests from the fifteenth to the eighteenth centuries and from embankments constructed in the middle part of the course of the Brenta.

From 1840 to 1896, the main Brenta outlet was again redirected back into the lagoon of Chioggia. This resulted in the formation of a 'large fluvial delta' (Fig. 50.3) due to sediment input of about 1 500 000 m^3 yr^{-1} leading to the formation of more than 2000 ha of tidal flats at an elevation 30-50 cm above mean sea level (Zunica, 1974; Favero *et al.*, 1988). These flats were colonized by halophytic vegetation (Beguinot, 1913) and a network of tidal channels (*ghebbi*) formed. Later, most of the area was reclaimed for agriculture using the same technique of infilling with river sediments during floods. The formation of this wetland area by river diversions serves as another example of how rivers can be used to form new wetlands. Finally the Brenta was changed to discharge directly to the Adriatic Sea, where it has flowed ever since. Other events impacted the Po Delta and the Venice Lagoon, as well as other parts of the coastal area of the northern Adriatic in more recent times. The last great work of the Venice Republic ended in 1791 with the embankment of all the lagoon (*conterminazione lagunare*). After this, there was a long period of small changes (nineteenth century) until the beginning of industrial development and the resulting major consequences on the area (see below).

PROCESSES AFFECTING THE EVOLUTIONARY TREND OF THE PO DELTA IN MODERN TIMES

From the beginning of seventeenth century, the Po Delta developed southeastward towards its current configuration. Since then, the delta has advanced about 7 km per century (Carbognin *et al.*, 1984). During this period, there was construction of dikes and reclamation of wetlands using river floods to build up the surface of the land. Again, this serves as an example of how the surface of the land can be built up using river sediments.

During this period of drainage, subsidence increased to 10-20 cm yr^{-1} (Zambon, 1994), due to removal of surface and shallow subsurface water and compaction of the soil due to drying and agriculture. From 1938 till 1961, there was intense extraction of methane with a high water content from aquifers 100-600 m deep that led to subsidence as high as 30 cm yr^{-1} (Caputo *et al.*, 1970). As a result, some areas are more than 3 m below sea level. The low elevation of the deltaic plain, together with the high elevation of the riverbed and river levels, means that the river level is at times almost 10 m higher than the lowest part of the deltaic plain. This could easily result in breaching of the dike and disastrous flooding of the plain as happened at Pontelagoscuro in 1951, an event still remembered as very disastrous for the Polesine region. Other problems emerged with regard to the regime of the Po and coastal dynamics in this period. There was a decrease in suspended sediment discharge from 16.9 million t yr^{-1} in 1965 to 10.5 in 1973 (Dal Cin, 1983). This was due to trapping of sediments in reservoirs and dredging of sand from riverbeds. This resulted in a retreat of unstable beaches and sand bars. There was high erosion along the northern deltaic coast, while in the south shoreline advance occurred only in the immediate down drift zone of the mouths of the distributaries (Zambon, 1994).

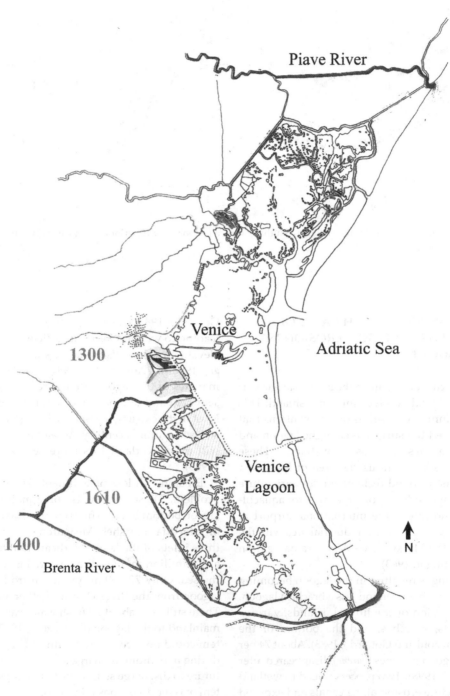

Fig. 50.2   The current status of the Lagoon of Venice, showing historical diversions of the Brenta River. The three different routes of the river are the historical course in AD 1300, the Brenta Nova which discharged into the lagoon and was excavated at the end of AD 1400, and the Taglio Nuovissimo, excavated in AD 1610.

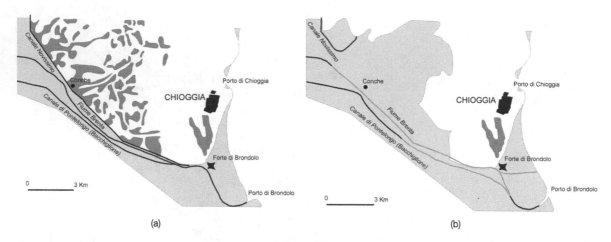

Fig. 50.3   Development of the Brenta Delta. (a) Southern Venice Lagoon in 1810, prior to river diversion. (b) Delta development in 1874.

THE EMERGENCE OF HIGH RATES OF WETLAND LOSS AND SOIL EROSION IN VENICE LAGOON

In 1919, construction began on the industrial harbour at Marghera, mostly on reclaimed marshland. This was the beginning of a series of interventions that greatly reduced the surface area of the lagoon and resulted in a massive loss of wetlands. These included the second industrial area of Marghera (1925), the last wetland reclamation for agriculture (1930), the impoundment of large areas for aquaculture (1940 and after), the international airport at Tessera (1957), and the third industrial area with the dredging of the Malamocco–Marghera navigation channel (starting in 1963).

These changes resulted in the lagoon becoming more marine in character and less a brackish lagoon. There was erosion of the littoral, islands, various embankments, marshes, and the bottom of the lagoon (Cavazzoni and Gottardo, 1983). About 74 per cent of the lagoon marshes (*barene*) disappeared after 1800 (Abrami, 1998). Two processes lead to wetland loss. One is edge erosion along canals and exposed shores due to wind generated waves and boat wakes. Exposed wetlands in the southern lagoon have shoreline retreat at rates of 1-2 m yr^{-1} (Cavazzoni and

Gottardo, 1983; Day *et al.*, 1998a). A second process leading to wetland loss is insufficient accretion and elevation gain to offset relative sea-level rise (Day *et al.*, 1998b). Day *et al.* (1999) showed that most marshes of the lagoon are not receiving sufficient sediment and will not survive future sea level rise. Most of the existing marshes will be gone or severely stressed within a century. Accelerated sea level rise will make wetland loss more severe (Day *et al.*, 1998b).

Erosion and loss of sediments is a very serious problem. These sediments come mainly from the lagoon bottom but also from edge erosion and retreat of intertidal marshes. Measurements made in the three inlets of the lagoon indicate that between 1.5 and 2 million m^3 of sediments are lost each year to the sea. Only 70 000 m^3 yr^{-1} are introduced to the lagoon from the River Dese and other small inputs. There still exist about 20 fresh water canals from the mainland to the lagoon (Cavazzoni, 1973). This system could be reconnected with old river branches during maximum discharge and could be used in the future to discharge sediments to the lagoon in a system of river diversions (Fig. 50.4).

The acceleration of sea-level rise will make the problems of sediment loss and wetland disappearance worse. Acceleration of eustatic sea-level rise

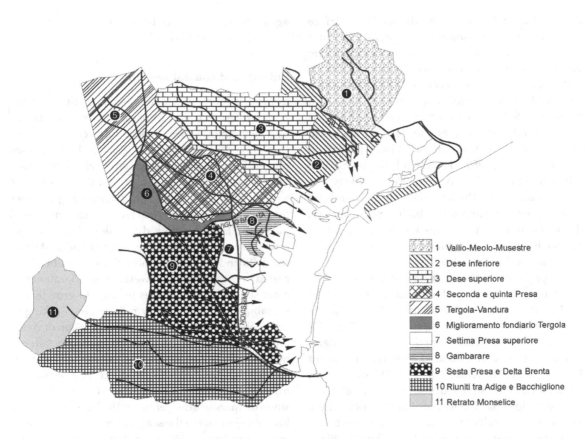

1 Vallio-Meolo-Musestre
2 Dese inferiore
3 Dese superiore
4 Seconda e quinta Presa
5 Tergola-Vandura
6 Miglioramento fondiario Tergola
7 Settima Presa superiore
8 Gambarare
9 Sesta Presa e Delta Brenta
10 Riuniti tra Adige e Bacchiglione
11 Retrato Monselice

Fig. 50.4   The Venice Lagoon and its drainage basin, with the subdivision into 11 water management areas. The arrows indicate the existing 20 outlets of freshwater inflow to the lagoon (from Cavazzoni, 1973).

combined with subsidence in increasing the rate of relative sea-level rise (RSLR). The background rate of geologic subsidence in Venice Lagoon during the twentieth century was 1.2-1.5 mm yr^{-1} (Pirazzoli, 1987; Carbognin et al., 1996), resulting in a RSLR between 2.4 and 3.0 mm yr^{-1} (Albani et al., 1983; Rusconi et al. 1993). In the Lagoon of Venice, the loss in relative elevation of the city was 23.5 cm during the last century, mostly due to ground water withdrawal. This increased water depths in the lagoon to an average of minus 35 cm for the same period. It means that a great quantity of sediment has already been lost to the sea and this is going to continue, resulting in a greater and greater sediment deficit. The increasing depth of the lagoon has resulted in an increase in tidal prism that is leading to an expan-

sion of the tidal channel network and further erosion of wetlands. The high rates of subsidence due to ground water withdrawal have ceased because pumping of ground water is now prohibited. But the rate of RSLR continues and will increase due the acceleration of eustatic sea-level rise (IPCC, 2001).

Another factor that has the potential to impact lagoon wetlands is the construction of gates, called MOSE, in the three lagoon inlets. These gates will be closed when high water levels threaten flooding in the city of Venice. With increasing sea level, the gates will be closed more and more often. Gate closure may cause an increase in wetland loss in the lagoon. The survival of wetlands in a rising water level scenario is strongly dependent on episodic inputs of sediments (e.g., Reed, 1989; Day et al., 1995). In Venice Lagoon, most sediment deposition on the wetland surface occurs during strong southerly storms when water levels are high and large amounts of sediments are resuspended from the bottom of the lagoon (Day *et al.* 1998a, 1999). It is precisely this kind of storm event that the MOSE will significantly reduce. Deposition of resuspended sediments on the marsh surface has been shown to be a major factor leading to coastal wetland survival in the Mississippi Delta (Baumann *et al.*, 1984; Reed, 1989). Thus, while the MOSE will reduce flooding in the city of Venice, it is likely that they will accelerate the loss of lagoon wetlands by reducing sediment deposition on the marsh surface. This will reduce accretion rates and thus the ability of the marshes to survive sea-level rise. Scenarios to offset this are addressed below.

The problems of the Po Delta and Venice Lagoon are not unique but are similar to other deltaic areas in the Mediterranean and elsewhere. In the Rhone Delta, isolation of the Rhone River from the delta plain has resulted in low accretion and low wetland productivity in areas isolated from riverine input (Ibañez *et al.*, 1999; Hensel *et al.*, 1998, 1999; Pont *et al.*, 2002). In the Ebro Delta, reduction of fresh water and sediment input has led to a sediment deficit. There are problems of saltwater intrusion, low vegetation productivity, and coastal retreat (Sanchez-Arcilla *et al.*, 1996; Ibañez *et al.*, 1997; Curco *et al.*, 2002). In both the Rhone and Ebro Deltas, the fringes of the delta are sinking below sea level. In the Mississippi Delta, isolation of the river by levees is primarily responsible for high rates of coastal wetland loss, which at times have been as high as 100 km^2 yr^{-1} (Day *et al.*, 2000a).

SYSTEM FUNCTIONING AS A BASIS FOR SUSTAINABLE MANAGEMENT OF THE PO DELTA AND VENICE LAGOON

If the trends of wetland loss in the Po Delta and Venice Lagoon are to be reversed, new management approaches must be adapted. What is needed to achieve sustainable management for Venice Lagoon and the Po Delta is a comprehensive, holistic, integrated approach that is based on the natural functioning of these coastal ecosystems. This section presents the conceptual basis for sustainable man-

agement and suggests how it can be applied to Venice Lagoon and Po Delta.

Pulsing and coastal systems

Coastal lagoons and deltas, such as those of the northwestern Adriatic, were formed and maintained by a hierarchical series of energetic forcings or pulses (see Day *et al.*, 1995, 1997, 2000b). These pulses ranged from changes of river channels on the order of every several hundred to over a thousand years to daily tides and include such forcings as great river floods, strong storms, annual river floods, and frontal passages. The foregoing discussion clearly shows how these energetic forcings formed and sustained the delta and lagoon, and how the systematic elimination of the pulses led to wetland deterioration and a net sediment loss. Forces leading to growth and sustainability were greatly reduced (e.g. diking of rivers eliminating riverine input, diversion of rivers around the lagoon, wetland reclamation, altered hydrology) while forces leading to deterioration were increased (e.g. enhanced subsidence, edge erosion of marshes due to boat traffic, reduced sediment deposition on marshes, deepening of the lagoon, and a net loss of sediments to the sea). Central to the problems of wetland loss in this region has been the elimination or reduction of riverine input and the re-integration of the rivers is an important key to sustainable management.

Basin-wide approach

Ideally, management of coastal systems should also include consideration of the entire drainage basin. Changes in the drainage basin such as reduced freshwater discharge, lower suspended sediment concentrations, and increased nutrient levels affect management in the receiving coastal system. For the Po River, there has been a reduction of sediment discharge due both to trapping of sediments in reservoirs and to illegal mining of sand from the river channel. There are also increased nutrient levels due to agricultural and urban inputs. In the Mississippi River basin, the creation and restoration of millions of hectares of wetlands has been proposed as a way to reduce nutrient levels in the river and to restore river and wetland habitats (Mitsch *et al.*, 2001; Day *et al.*, 2003). Such a basin-wide focus should be considered for the basins of the Po Delta and especially Venice Lagoon.

Ecotechnology, restoration ecology, and ecological engineering

The management approach proposed for Venice Lagoon and the Po Delta is ecotechnology, or more specifically ecological engineering. Ecological engineering is defined as 'the design of sustainable ecosystems that integrate human society with its natural environment for the benefit of both' (Mitsch, 1996, 1998; Mitsch and Jørgensen, 2003). Ecological engineering involves creating and restoring sustainable ecosystems that have value to both humans and nature. Ecological engineering combines basic and applied science for the restoration, design, and construction of aquatic and terrestrial ecosystems. Ecological engineering uses mainly the energies of nature with human energy used in design and control of key processes. The goals of ecological engineering are:

1) the restoration of ecosystems that have been substantially disturbed by human activities such as hydrological alteration or land disturbance; and
2) the development of new sustainable ecosystems that have both human and ecological value.

This is engineering in the sense that it involves the design of the natural environment using quantitative approaches grounded in the basic science of ecology. It is a technology where the primary tools are self-designing ecosystems. It is biology and ecology in the sense that the components are the biological species in these ecosystems. And, as the goals suggest, it is often ecological restoration. The concepts of ecological engineering and ecosystem restoration are intertwined. Ecological engineering is not a new field but an amalgam of several fields that deal with restoration and creation of ecosystems. Restoration ecology and the restoration fields (terrestrial, aquatic, wetland, etc.) have many features in common with ecological engineering. In fact, Bradshaw (1997) called ecosystem restoration 'ecological engineering of the best kind' because we are putting back ecosystems that used to exist, not creating new combinations of populations or systems.

Self-design and the related concept of self-organization must be understood as important properties of ecosystems in the context of their creation and restoration (Mitsch and Jørgensen, 2003). In fact, their application may be the most fundamental concept of ecological engineering. Self-organization is the property of systems in general to reorganize themselves, given an environment that is inherently unstable and non-homogeneous. Self organization is a systems property that applies very well to ecosystems where species are continually introduced and deleted, species interactions – e.g. predation, mutualism – change in dominance, and the environment itself changes. All of these activities go on at one degree or another all the time. In a sense, the organization is not derived from some outside force but from within the system itself. Self-organization manifests itself in microcosms and newly created ecosystems 'showing that after the first period of competitive colonization, the species prevailing are those that reinforce other species through nutrient cycles, aids to reproduction, control of spatial diversity, population regulation, and other means' (Odum, 1989).

All systems have some level of organization, but Pahl-Wostl (1995) argues that there are two general ways systems can be organized – by rigid top-down control or external influence (imposed organization) or by self-organization. These two types of organization are contrasted in Table 50.1. Imposed organization, such as implemented in many conventional engineering approaches, results in rigid structures and little potential for adapting to change. This of course is desirable for engineering design where predictability of safe and reliable structures are necessary, such as for bridges, furnaces, power plants, and sulphur scrubbers. Self-organization, on the other hand, develops flexible networks with a much higher potential for adaptation to new situations. It is thus the latter property that is desirable for solving many ecological problems, such as wetland loss and water quality deterioration. When biological systems are involved, the ability for the ecosystems to change, adapt, and grow according to their forcing functions and internal feedbacks is most important.

Mitsch and Jørgensen (2003) define self-design as 'the application of self-organization in the design of

Table 50.1 *Systems categorized by types of organization (modified from Pahl-Wostl, 1995).*

Characteristic	Imposed organization	Self-organization
Control	externally imposed; centralized control	endogenously imposed; distributed control
Rigidity	rigid networks	flexible networks
Potential for adaptation	little potential	high potential
Application	conventional engineering	ecological engineering
Examples	machine	organism
	fascist society	democratic society
	agriculture	natural ecosystem (coastal lagoons, river basin, and deltas)

ecosystems'. The presence and survival of species in ecosystems after their introduction by nature or humans is more up to nature than to humans. Self-design is an ecosystem function in which the chance introduction of species is analogous to the chance development of mutations necessary for evolution to proceed (Mitsch, 1998). In the context of ecosystem development, self-design means that if an ecosystem is open to allow 'seeding' (by river diversions for example) of enough species and their propagules, as well as fresh water, sediments and nutrients, through human or natural means, the system itself will optimize its design by selecting for the assemblage of plants, microbes, animals, chemical species, sediments, etc., that is best adapted for existing conditions. The ecosystem then 'designs a mix of man-made and ecological components in a pattern that maximizes performance, because it reinforces the strongest of alternative pathways that are provided by the variety of species and human initiatives' (Odum, 1989).

SUSTAINABLE MANAGEMENT OF VENICE LAGOON AND THE PO DELTA

The Po Delta and Venice Lagoon were formed and sustained by natural physical and ecological processes. These processes were generally not constant, but varied in both time and space and occurred episodically as a series of energetic forcings or pulses which included changes in river courses, river floods, and storms. The systematic elimination of these forcings is a major cause of the environmental problems of the lagoon and delta. In order to manage the lagoon and delta in a sustainable manner, management must be based on the natural functioning of the system, including the incorporation of pulsing events.

In the past, attention was given mostly to the effect of water pollution and eutrophication, as well as to the building of embankments for flood control and hard defense structures along the Adriatic littoral. This was necessary, especially after the disastrous sea surge of 1966 and as an answer to the growing tourist industry. These activities are concentrated north and south of the Lagoon of Venice and in the city itself, while the Po Delta is still not a major tourist destination. In the Po Delta, fishing remains the most important economic activity in the coastal fringe, the lagoon and in existing aquaculture ponds. Agriculture is the main activity in most of the Po deltaic plain. The recent formation of the Natural Park of the Po Delta is not yet having a real impact on the management of the area. But there is now a consensus among many in the region that one of the major problems in both the delta and lagoon, is the high rate of wetland loss and the net loss of sediments to the sea. Both of these problems are related to the pervasive hydrological changes in the region. Thus sustainable management of the northwest

Adriatic coastline will involve restoration, at least partially, of the natural hydrology of the region and retention and augmentation of sediments, especially within the coastal lagoons.

The problem of wetland deterioration and loss of sediments to the sea can be solved both by introduction of new sediments into the lagoon and delta system and the retention of sediments. The introduction of new sediments can be accomplished by river diversions. The reintroduction of the river back into the Po Delta and Venice Lagoon is not a new idea, but has been practiced for centuries in the region in the old, but very efficient, *bonifica di colmata*. As discussed earlier, this means the reconnection of the lagoons to the existing network of canals during periods of high river discharge. The canals of the *bonifica* could in turn to be easily reconnected with the old river distributaries as well as some of the large new canals such as the Idrovia Padova–Venezia Canal, which will be completed in the near future. Past actions in the Po Delta and Venice Lagoon have clearly shown that this approach is a very efficient method of building new wetlands as in the formation of wetlands by the Brenta River in the second half of the nineteenth century (Favero *et al.*, 1988; Fig. 50.3). This area of wetlands formed is equivalent to about one third of the total wetland loss during the twentieth century. Within the Po Delta, the river could be directly connected to the delta plain via diversions to build up the surface of the land, which could then be used for wetland habitat or agriculture. Internal dikes could be constructed so that the land surface could built up in very specific areas. River diversions are being used in the Mississippi Delta where it has been shown that they take up nutrients and reduce or reverse wetland loss (Lane *et al.* 1999, 2002, 2004; DeLaune and Pezeshki, 2003; DeLaune *et al.*, 2003).

Complementary to the introduction of new sediments from the river, measures should be undertaken to retain sediments in the lagoons. This involves managing wave energy that leads to erosion of bottom sediments by resuspension and to erosion of exposed marsh shorelines. In the open lagoon in areas with long fetches, wind generated waves are responsible for marsh erosion. In canals, wakes from boat traffic are the major factor leading to marsh edge erosion. A practicable way to reduce wave energy is to use permeable breakwaters or wave-stilling devices. These are also sometimes called sediment fences. Because they are permeable, they do not reduce sediment delivery to the marsh surface and would probably enhance accretion and revegetation in front of the marsh. Such breakwaters have been used successfully to establish marshes in The Netherlands and the Mississippi Delta (Schoot and de Jong, 1982; Boumans *et al.*, 1997). Sediment fences are currently in use in the Po Delta and Venice Lagoon where it has been shown that they reduce wave energy, protect exposed marsh shorelines and dramatically increase accretion in shallow subtidal and intertidal areas (Scarton *et al.*, 2000). It is possible that placement of such structures could be done in a way that waves are reduced and accretion is enhanced in both the immediate vicinity of the structure but also at some distance away in adjacent areas. Sediment fences could be used in conjunction with river diversions to enhance trapping of sediments. They could also be placed stratigically in the landscape to direct the development of marshes into desired areas.

The introduction and retention of sediments in the lagoons is important because it is critical to maintain important lagoon habitats. The ecosystem of this brackish water ecosystem is characterized by shallow subtidal areas and marshes, the *barene* (with an elevation from 10 cm to 40 cm above average sea level). The *barene* are closely interrelated with the tidal flats, the *velme* (with an elevation from plus 10 cm to minus 10 cm). The marshes, tidal flats, and shallow subtidal areas are in a balance with each other and are covered with different types of vegetation. In addition to its ecological role, the *barene* have many important effects on the human activities and the conservation of the historical islands and the city of Venice itself. It seems clear that in order to save the city of Venice, the natural habitats of the lagoon must also be saved.

The introduction and retention of sediments in the delta and lagoon will be critical to offset the predicted acceleration of sea-level rise. If marshes are to survive rising water levels, they must be able to accrete at a rate such that surface elevation gain is sufficient to offset the rate of water level rise (Cahoon *et al.*, 1995). Introduction of mineral sediments in river water increases accretion in two ways. First, it adds directly to vertical accretion. Second, nutrients associated with mineral sediments stimulate root growth and contribute to organic soil formation. The stimulation of accretion by river diversions will become all the more important because it is likely that the construction of the MOSE will reduce the input of sediments during storms. As mentioned earlier, most sediment deposition on the surface of marshes in Venice Lagoon occurs during strong southerly storms when water levels are high and large amounts of sediments are resuspended from the bottom lagoon (Day *et al.*, 1998a, 1999). It is this kind of storm event that the MOSE will significantly reduce. Thus, introduced riverine sediments will be very important in offsetting the reduction of resuspended sediment input due to the MOSE.

Water quality in many of the lagoons is poor due to nutrient enrichment leading to eutrophication. The introduction of river water has the potential to make these problems worse. The use of wetlands for water quality improvement is a widespread practice (Kadlec and Knight, 1996; Mitsch *et al.*, 2001; Day *et al.*, 2004) and diversions could be designed so that water quality improvement is optimized. In diversions in the Mississippi Delta, there was a strong reduction of nutrients, especially nitrate, as water flowed over wetlands and shallow water bodies (Lane *et al.*, 1999, 2002). Denitrification is an important permanent sink for nitrogen (Mitsch *et al.*, 2001). The use of wetland treatment can also be used for agricultural and municipal wastewater treatment in the watershed of Venice Lagoon as suggested by Mitsch *et al.* (2001) and Day *et al.* (2003, 2004).The capture of fish is an important economic activity in both Venice Lagoon and the Po Delta. There are both natural fisheries as well as the production of fish in aquaculture fish ponds. The fishing activities should be carefully integrated into comprehensive management plans for the area. New wetlands formed as a result of river diversions will become important fish nursery habitat.

Conservation of the visual landscape of the area is an important element of the restoration and management of the lagoon. Humans have been an important part of these ecosystems for thousands of years. Important cultural resources, such as Venice but many other areas also, are an integral part of the area.

The conservation of these cultural and visual resources is an important part of the overall management of the area. Such management actions should be taken as part of a holistic management strategy for the entire lagoon and delta ecosystem. A number of additional elements of such an approach have been proposed including construction of salt marshes and tidal flats with dredged material and vegetative plantings. Holistic management approaches have been proposed for the southeastern coast of the UK (Pethick, 1993), the Mississippi delta (Day and Templet, 1989; Day *et al.*, 1997), and for the entire Mississippi basin (Mitsch *et al.*, 2001; Day *et al.*, 2003). Similar management approaches have been suggested for other deltas and lagoons in the Mediterranean and elsewhere. Pont *et al.* (2002) analysed the environmental situation in the Rhone delta, France. They documented lowered wetland productivity and salt-water intrusion and reported that the fringes of the delta plain were slowing subsiding below sea level. These problems stem mainly from the isolation of most of the delta plain by levees from riverine input. They suggested that diversion of Rhone river water back into the delta was necessary for sustainable management. Ibañez *et al.* (1997) carried out a similar analysis for the Ebro Delta, Spain. Freshwater and sediment input to the delta have been reduced by greater than 90 per cent and there is a sediment deficit and loss of elevation of the deltaic plain with the lowest parts of the delta now below sea level. Similar to the Rhone, Ibañez *et al.* suggested that river diversions were necessary for sustainable management of the Ebro delta. In all of these deltas (Mississippi, Rhone, Po, and Ebro) river water is still flowing naturally into small areas the deltaic plain. In these areas there is rapid accretion, high wetland productivity, and little wetland loss.

ACKNOWLEDGEMENTS

Much of the research on which this paper is based was supported by the European Community through two projects: (1) the Environment Programme (Directorate General for Science, Research and Development, MedDelt Project. Impact of the Climate Change on North-western Mediterranean Deltas), contract no. EV5V-CT94-0465, Climatology and Natural Hazards; and (2) Gen. Dir. XI, Project LIFE-Barene 1999, LIFE Nat/IT/006246; and by the Consorzio Venezia Nuova, Venice, Italy. This paper was adapted from an earlier work (Abrami and Day, 2003) and is used by permission.

REFERENCES

Abrami, G. 1997. Coastal management and tourism development within the Po Delta and Venice Lagoon. In *Insular Coastal Area Risk Management in the Mediterranean, Workshop Proceedings*. Malta, 12pp.

Abrami, G. 1998. *Carta delle trasformazioni della Laguna Veneta: 1810–1992*. Venice: Magistrato alle Acque di Venezia.

Abrami, G., and J. Day. 2003. Historical and morphological aspects of the evolution of Venice Lagoon and of the Po Delta: new criteria for management of the area. In A. Sánchez-Arcilla and A. Bateman (eds), *Proceedings of the 3rd IAHR Symposium on River, Coastal and Estuarine Morphodynamics, Barcelona, 1–5 September 2003*. Madrid: IAHR, pp. 481–93

Albani, A., Favero, V., and R. Serandrei Barbero. 1983. Apparati intertidali della Laguna di Venezia. In *Laguna, fiumi, lidi: cinque secoli di gestione delle acque nelle Venezie. Atti del Convegno*. Venice: Magistrato alle Acque, pp. 221–8.

Baumann, R., Day, J., and C. Miller. 1984. Mississippi deltaic wetland survival: sedimentation versus coastal submergence. *Science* **224**, 1093–5.

Beguinot, A. 1913. *La vita delle piante superiori nella laguna di Venezia*. Publication 54 Ufficio Idrografico R. Venice: Magistrato alle Acque, 348pp.

Bondesan, M., and U. Simeoni. 1983. Dinamica e analisi morfologica statistica dei litorali del delta del Po e alle foci dell'Adige e del Brenta. *Mem. di Sc. Geo., Ist. Di Geo. E Min.* (Università di Padova), **36**, 1–46.

Bondesan, M., Castiglioni, G., Elmi, C., Gabbianelli, G., Marocco, R., Pirazzoli, P., and A. Tomasin. 1995. Coastal areas at risk from storm surges and sea-level rise in Northeastern Italy. *Journal of Coastal Research* **11**, 1354–79.

Bradshaw, A. 1997. Restoration of mined lands – using natural processes. *Ecological Engineering* **8**, 255–69.

Cahoon, D., Reed, D., and J. Day. 1995. Estimating shallow subsidence in microtidal salt marshes of the southeastern United States: Kaye and Barghoorn revisited. *Marine Geology* **128**, 1–9.

Caputo M., Pieri L., and M. Ungendoli. 1970. *Geometric investigations of the Subsidence in the Po Delta*. Venice:, CNR, Ist. Studio Dinamica Grandi Masse, TR 2, 37 pp.

Carbognin L., Gatto P., and F. Marabini. 1984. The city and the lagoon of Venice. A guidebook on the environment and land subsidence. In *Proceedings 3rd International Symposium on Land Subsidence, Venice, March 12–25, 1984*, pp. 1–81.

Carbognin, L., Gambolati, G., Marabini, F., Taroni, G., Teatini, P., and L. Tosi. 1996. *Analisi del processo di subsidenza nell'area Veneziana e sua simulazione con un modello tridimensionale non linear*. Ministero dell'Università e della Ricerca Scientifica e Tecnologica, Progetto Sistema Lagunare Veneziano.

Cavazzoni, S. 1973. *Acque dolci nella Laguna di Venezia*. Venice: CNR, Lab. Studio Dinamica Grandi Masse, 1–40pp.

Cavazzoni, S., and D. Gottardo. 1983. Processi evolutivi e morfologici nella laguna di Venezia. In Ministero dei Lavori Pubblici, Atti del Convegno, *Laguna, fiumi, lidi: cinque secoli di gestione delle acque nelle Venezie*. Venice: Magistrato alle Acque, pp. 2–18.

Ciabatti, M. 1979. Ricerche sull'evoluzione del Delta Padovano. *Giornale di Geologia*, Grandi Pr 2, 34, 26.

Curco, A., Ibañez, C., Day, J., and N. Prat. 2002. Net primary production and decomposition of salt marshes of the Ebre delta, Catalonia, Spain. *Estuaries* **25**, 309–24.

Dal Cin, R. 1983. I litorali del delta del Po alle foci dell'Adige e del Brenta, caratteri tessiturali e dispersione dei sedimenti, cause dell'arretramento e previsioni sull'evoluzione futura. *Boll. Soc. Geol. Ital.* **102**, 56.

Day, J., and P. Templet. 1989. Consequences of sea level rise: implications from the Mississippi Delta. *Coastal Management* **17**, 241–57.

Day, J., Pont, D., Hensel, P., and C. Ibañez. 1995. Impacts of sea-level rise on deltas in the Gulf of Mexico and the Mediterranean: the importance of pulsing events to sustainability. *Estuaries* **18**, 4, 636–47.

Day, J., Martin, J., Cardoch, L., and P. Templet. 1997. System functioning as a basis for sustainable management of deltaic ecosystems. *Coastal Management* **25**, 115–54.

Day, J., Scarton, F., Rismondo, A., and D. Are. 1998a. Rapid deterioration of a salt marsh in Venice Lagoon, Italy. *Journal of Coastal Research* **14**, 583–90.

Day, J., Rismondo, A., Scarton, F., Are, D., and G. Cecconi. 1998b. Relative sea level rise and Venice Lagoon wetlands. *Journal of Coastal Conservation* **4**, 27–34.

Day, J., Rybczyk, J., Scarton, F., Rismondo, A., Are, D., and G. Cecconi. 1999. Soil accretionary dynamics, sea level rise and the survival of wetlands in Venice Lagoon: a field and modeling approach. *Estuarine, Coastal and Shelf Science* **49**, 607–28.

458   J. W. Day Jr *et al.*

Day, J., Shaffer, G., Britsch, L., Reed, D., Hawes, S., and D. Cahoon. 2000a. Pattern and process of land loss in the Mississippi delta: a spatial and temporal analysis of wetland habitat change. *Estuaries* **23**, 425–38.

Day, J., Psuty, N., and B. Perez. 2000b. The role of pulsing events in the functioning of coastal barriers and wetlands: Implications for human impact, management and the response to sea level rise. In M. Weinstein and D. Dreeger (eds.), *Concepts and Controversies in Salt Marsh Ecology*. Dordrecht, Netherlands: Kluwer Academic Publishers, pp. 633–60.

Day, J. W., Ibañéz, A., Mitsch, W., Lara, A., Day, J. N., Ko, J., Lane, R., Lindsey, R., and D. Zarate. 2003. Using ecotechnology to address water quality and wetland habitat loss problems in the Mississippi Basin: a hierarchical approach. *Biotechnology Advances* **22**, 135–59.

Day, J. W. Ko, J.-Y, Rybczyk, J., Sabins, D., Bean, R., Berthelot, G., Brantley, C., Cardoch, L., Conner, W., Day, J. N., Englande, A J., Feagley, S., Hyfield, E., Lane, R., Lindsey, R., Mistich, L., Reyes E., and R. Twilley. 2004. In press. The use of wetlands in the Mississippi Delta for wastewater assimilation: a review. *Journal of Ocean and Coastal Management*.

DeLaune, R., and S. Pezeshki. 2003. The role of organic carbon in maintaining surface elevation in rapidly subsiding US Gulf of Mexico coastal marshes. *Water, Air and Soil Pollution* **3**, 167–79.

DeLaune, R., Jugsujinda, A., Peterson, G., and W. Patrick. 2003. Impact of Mississippi River freshwater reintroduction on enhancing marsh accretionary processes in a Louisiana estuary. *Estuarine, Coastal, and Shelf Science* **58**, 653–62.

Favero, V., Parolini, R., and M. Scattolin (eds.). 1988. *Morfologia storica della laguna di Venezia*. Venice: Arsenale Ed.

Gambolati, G., and Gonella, M. 1997. Analisi della vulnerabilità e del rischio di ingressione marina lungo la costa dell'Alto Adriatico sviluppata nell'ambito del progetto europeo CENAS. In *Cambiamenti climatici, energia e trasporti, Atti della Conf. Naz. strategie sviluppo sostenibile, Roma*.

Hensel, P., Day Jr, J. W., Pont, D., and J. N. Day. 1998. Short term sedimentation dynamics in the Rhone River delta, France: the importance of riverine pulsing. *Estuaries* **21**, 52–65.

Hensel, P., Day, J., and D. Pont. 1999. Wetland vertical accretion and soil elevation change in the Rhone delta, France: the importance of riverine flooding. *Journal of Coastal Research* **15**, 668–81.

Ibañéz, C., Canicio, A., Day, J. W., and A. Curco. 1997. Morphologic evolution, relative sea level rise and sustainable management of water and sediment in the Ebre Delta. *Journal of Coastal Conservation* **3**, 191–202.

Ibañez, C, Day, J., and D. Pont. 1999. Primary production and decomposition of wetlands of the Rhone Delta, France: interactive impacts of human modifications and relative sea level rise. *Journal of Coastal Research* **15**, 717–31.

Kadlec, R. H., and R. L.Knight. 1996. *Treatment Wetlands*. New York: Lewis Publishers.

Lane, R., Day, J., and B. Thibodeaux. 1999. Water quality analysis of a freshwater diversion at Caernarvon, Louisiana. *Estuaries* **2A**, 327–36.

Lane, R., Day, J., Kemp, G., and B. Marx. 2002. Seasonal and spatial water quality changes in the outflow plume of the Atchafalaya River, Louisiana, USA. 2001. *Estuaries* **25**, 1, 30–42.

Lane, R., Day, J. W., Justic, D., Reyes, E., Marx, B., Day, J. N., and E. Hyfield. 2004. Changes in stoichiometric Si, N and P ratios of Mississippi River water diverted through coastal wetlands to the Gulf of Mexico. *Estuarine, Coastal and Shelf Science* **60**, 1–10.

Lankford, R. 1976. Coastal lagoons of Mexico: their origin and classification. In M. L. Wiley (ed.), *Estuarine Processes*, Vol. II. New York: Academic Press, pp. 182–215.

Mitsch, W. J. 1996. Ecological engineering: a new paradigm for engineers and ecologists. In P. C. Schulze (ed.), *Engineering within Ecological Constraints*. Washington, DC: National Academy Press, pp. 111–28.

Mitsch, W. J. 1998. Ecological engineering – the seven-year itch. *Ecological Engineering* **10**, 119–38.

Mitsch, W. J., and S. E. Jørgensen. 2003. *Ecological Engineering and Ecosystem Restoration*. New York: John Wiley.

Mitsch, W., Day, J., Gilliam, J., Groffman, P., Hey, D., Randall, G., and N. Wang. 2001. Reducing nitrogen loading to the Gulf of Mexico from the Mississippi River basin: strategies to counter a persistent problem. *BioScience* **51**, 5, 373–88.

Odum, H. T. 1989. Ecological engineering and self-organization. In W. J. Mitsch and S. E. Jorgensen (eds.), *Ecological Engineering: An Introduction to Ecotechnology*. New York: John Wiley and Sons, pp. 79–101.

Pahl-Wostl, C. 1995. *The Dynamic Nature of Ecosystems: Chaos and Order Entwined*. New York: John Wiley and Sons, 267 pp.

Pethick, J. S. 1993. Shoreline adjustments and coastal management: physical and biological processes under accelerated sea-level rise. *Geographical Journal* **159**, 2, 162–8.

Pirazzoli, P. 1987. Recent sea-level changes and related engineering problems in the Lagoon of Venice, Italy. *Progress in Oceanography* **18**, 323–46.

Pont, D., Day, J., Hensel, P., Franquet, E., Torre, F., Rioual, P., Ibañez, C., and E. Coulet. 2002. Response scenarios for the deltaic plain of the Rhône in the face of an acceleration in the rate of sea level rise, with a special attention for Salicornia-type environments. *Estuaries* **25**, 337–58.

Reed, D. 1989. Patterns of sediment deposition in subsiding coastal marshes, Terrebonne Bay, Louisiana: The role of winter storms. *Estuaries* **12**, 222–7.

Rossetti, G., and M. Rossetti. 1977. Idrografia e idrologia della regione del Delta del Po in relazione con la degradazione delle aree tributarie. *Ist. Geologia* (University of Palma), 53.

Rusconi, A., Ferla, M., and M. Filippi. 1993. Tidal observations in the Venice lagoon: the variations in sea level observed in the last 120 years. *Proceedings of Sea Change 93*, The Hague, Netherlands: 19–23 April, 1993.

Sanchez-Arcilla, J., Stive, M., Ibañez, C., Pratt, N., Day, J., and M. Capobianco. 1996. Impacts of sea level rise on the Ebro Delta: a first approach. *Ocean and Coastal Management* **30**, 197–216.

Scarton, F., Rismondo, A., and J. Day. 1998. Above and below-ground production of Arthrocnemum fruticosum on a Venice Lagoon saltmarsh. *Bollettino del Museo Civico di Storia Naturale di Venezia* **48**, 237–45.

Scarton, F., Day, J., Rismondo, A., Cecconi, G., and D. Are. 2000. Effects of an intertidal sediment fence on sediment elevation and vegetation distribution in a Venice (Italy) Lagoon salt marsh. *Ecological Engineering* **16**, 223–33.

Scarton, F., Day, J., and A. Rismondo. 2002. Primary production and decomposition of Sarcocornia fruticosium (L.) and Phragmites australis Trin. Ex Steudel in the Po delta, Italy. *Estuaries* **25**, 325–36.

Schoot, P. M., and J. de Jong. 1982. *Sedimentatie en erosie metingen met behulp van de Sedi-Eros-tafel (Set)*. Rijkswaterstaat. The Hague: Notitie. DDMI-82, 401.

Sestini, G. 1992. Implication of climatic changes for the Po Delta and Venice Lagoon. In L. Jeftic, J. D. Milliman and G. Sestini (eds.), *Climatic Change and the Mediterranean*. UNEP, 428–94.

Sestini, G. 1996. Land subsidence and sea-level rise: The case of the Po delta region, Italy. In J. Milliman and B. Haq (eds.), *Sea-Level Rise and Coastal Subsidence*. Dordrecht, Netherlands: Kluwer Academic Publishers, pp. 235–48.

Zambon, M. 1994. Prevenzione e controllo della subsidenza. In *Il fiume e la sua terra*. Venice: Ist. Ven. Sc. Let. Arti e Acc. Concordi Rovigo, 77–104.

Zunica, M. 1974. La bonifica Delta Brenta. Un esempio di trasformazione del paesaggio nella laguna di Venezia. *Riv. Geo. It.* **81**, 3, 346–400.

Zunica, M. 1987. *Lo spazio costiero italiano*. Rome: V. Levi Ed., pp. 212.

51 • Morphological restoration techniques

G. CECCONI

SUMMARY

The safeguarding of the city of Venice from flooding is closely related to the protection of the lagoon morphology from erosion. Valuable components of the lagoon, the salt marshes and the mud flats, are disappearing at an alarming rate due to lack of sediments, wave attack and relative sea-level rise. Sediments from marshes and mud flats are eroded leading to channel siltation and marine dispersal and the lagoon is becoming deeper and flatter (i.e. losing its dentritic creek system). Since 1986 the Magistrato alle Acque – Consorzio Venezia Nuova have undertaken many morphological restoration works to maintain channel depths for navigation and to increase tidal flushing in the inner lagoon. Dredged material has been used to build artificial mud flats and marshes, re-creating 5 km² of fragile lagoon environment. The dredging and reconstruction programme has been combined with other measures to protect and restore natural marshes and mud flats.

This paper presents the problems affecting the morphological evolution of the lagoon and the techniques developed and implemented for the restoration of salt marshes and mud flats. The impacts of the works have been monitored in order to select the best practise for re-introducing self-preserving stabilising processes for each specific environment. In 2001 new lines of action have been defined as an upgrade of the 1992 initial plan. These new lines of action are aimed at reducing the re-suspension of sediments from the shallows, increasing the settling of sediments inside the lagoon, reducing the cross-flow of the sediments at the side of the channels and losses to the sea. The structures to reintroduce have to mimic the hydro-morphological functions of the shoals at the edges of the channels and the wind wave protection exerted by marshes and tidal flats. Other proposals such as the redirection of rivers into the lagoon or the permanent reduction of the tidal flow are shown to be not viable or counterproductive.

THE VENICE LAGOON

The Lagoon of Venice, the largest in Italy and one of the largest in the Mediterranean, is located in north-eastern Italy and has an area of approximately 550 km². It exchanges water with the sea through three large inlets. The sediment dynamics of the lagoon have been altered greatly over the past five centuries. Three rivers, the Brenta, Sile and Piave, which originally discharged into the lagoon, have been diverted from the lagoon to the sea since the sixteenth century. Presently, only a few small rivers (total discharge about 30 m³ s⁻¹) flow into the lagoon. Thus, riverine sediment input to the lagoon has been almost completely eliminated (from 700 000 m³ yr⁻¹ to 30 000 m³ yr⁻¹; Cecconi and Ardone, 1999). The import of coarse marine sediments into the lagoon has been greatly reduced (from 300 000 m³ yr⁻¹ to 30 000 m³ yr⁻¹) because of the construction of long breakwaters at the inlets at the end of the nineteenth century. Nowadays there is a net export of about 1.1 million m³ yr⁻¹ of sediments from the lagoon system into the sea. Most of the lagoon is occupied by a large central water body (about 350 km²) and extensive intertidal salt marshes (about 33.5 km² in 1999). The mean depth of the lagoon is 1.1 m and the tide range is 0.6-1.1 m. The subtidal areas are partially vegetated by macro-algae and sea grasses (such as *Zostera marina*, *Z. noltii* and *Cymodocea nodosa*) and the salt marsh species composition varies with elevation.[1] The marshes in the Lagoon of Venice are of European importance due to their areal extent, high productivity, and habitat value. The lagoon is among the sites included in the Ramsar Convention of 1971 and is of primary importance within the European Community for the conservation of species and habitats at a regional, national and European level under the provisions of Natura 2000 and the Habitat Directives.

THE MORPHOLOGICAL EVOLUTION

The elements which distinguish the Lagoon of Venice from other lagoons mainly relate to its conspicuous human presence. The Lagoon of Venice has always had a natural tendency to gradually silt up. This is supported by the fact that several neighbouring lagoons have disappeared, completely transformed into dry land and by the fact that for centuries the people of Venice have had to intervene to prevent the lagoon from silting up. Defence of the lagoon was therefore synonymous with the removal of sediments either transported by the rivers flowing directly into the lagoon or brought in by the sea through the inlets. Protection of the lagoon from river-borne sediments has been highly successful. This is clearly illustrated, for example, by the major works to divert the rivers Brenta, Piave, Po and Sile. Without doubt the most ambitious of these projects was the diversion of the River Po concluded in 1604, diverting it definitively to the south and creating the current delta. If the solid material which has since formed the Po Delta had continued to flow into the sea, the southern inlet would certainly have silted up. By diverting the rivers and relocating the mouths some distance away, the people of Venice countered the natural tendency of the lagoon to silt up. However, about a century after the diversion of the River Po, they were forced

to build embankments to defend the southern section of coastline which had started to suffer major erosion due to the lack of sediments carried by the river.

During the nineteenth century, the lagoon inlets had to be deepened to allow access to ships with ever greater draft with the construction of breakwaters, but this resulted in the obstruction of the flow of sand from the sea into the lagoon. Other works carried out during the nineteenth and twentieth centuries have had highly negative effects on the equilibrium of the lagoon. Excavation of navigable canals to allow ships access to the Marghera industrial zone and the formation of reclaimed areas have all had negative effects. The creation of the industrial zone has also had a major, albeit indirect, influence on the morphology of the lagoon. It has now been ascertained that soil subsidence in Venice was accelerated by the practice of draining water from the aquifers, leading to a general deepening of the water in the lagoon (Tosi and Carbognin, 2002).

Finally, the progressive deterioration of water quality must also be considered. Here it should be noted that morphology and the deterioration of water quality, far from being independent of one another, are in some ways linked in a self-perpetuating vicious circle. Indeed, on one hand, flattening of the lagoon bed increases water stagnation and consequent eutrophication. On the other hand, eutrophication of the water penalizes the growth of aquatic plants (eelgrass) on the bottom of the lagoon. All this makes the lagoon bed more vulnerable to erosion.

The area of salt marshes in the lagoon has fallen from about 115 km^2 in 1810 to about 33.5 km^2 at present, as a result of reclamation, erosion, pollution and natural and human-induced subsidence (Favero, 1992; MAV-CVN, 2000). These processes have not been counteracted by accretion due to the lack of sediment input from the basin and the sea. The overall effect of the current process of transformation is thus a marked evolutionary tendency towards erosion. The balance between sediment brought into the lagoon and that lost to the sea is highly negative: on average over one million cubic metres per annum in recent years (Fig. 51.1).

In turn, sediment dynamics within the lagoon are leading to the gradual disappearance of the morpho-

[1] A recent study has been carried out by Consorzio Venezia Nuova (MAV–CVN, 1999) to measure the elevations at which vegetation associations develop: *Puccinellio festuciformis-Sarcocornietum fruticosae* from 0.24 to 0.32m above mean sea-level (m a.s.l., frequency of inundation 16%); *Limonium narbonensis-Puccinellietum festuciformis* from 0.22 to 0.25m (20%); *Limonium narbonensis-Spartinetum maritimae* from 0.17 to 0.25m (25%); *Saliconornietum venetae* 0.10m (36%).

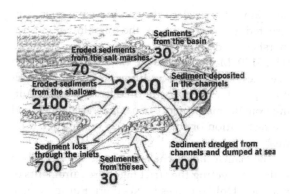

Fig. 51.1   The sediment balance of the Lagoon of Venice in thousands of cubic metres derived from bathymetric surveys of Magistrato alle Acque in 1970 and 1990 (MAV-CVN, 1992c).

logical elements peculiar to a lagoon environment. To summarize, the lagoon is becoming deeper (the combined result of sea-level rise and subsidence) and flatter (because of the way in which sediment is transported within the lagoon). The loss of features peculiar to a lagoon has extremely serious consequences both on the lagoon ecosystem, with the disappearance of plant and animal species, and on the systems for protecting the lagoon bulkheads, ever more exposed to wave motion.

THE RESTORATION WORKS

After the disastrous storm surge event of 4 November 1966, which brought home the precariousness of the entire lagoon basin, the need became all the more urgent for safeguarding interventions in the lagoon and in the unique and precious city of Venice (Cecconi and Ardone, 1999). These interventions, regulated by four Special Laws (1973, 1984, 1991, 1992), were shown to be indispensable. The State, through the Venice Water Authority and Consorzio Venezia Nuova, has elaborated a unitary plan of intervention, and numerous projects have already been completed. The safeguarding measures already implemented, or underway, include the nourishment of beaches and dunes (MAV-CVN, 1992a), the reconstruction and protection of wetlands, the local defence works in the historic city and other lagoon centres, as well as the

temporary closure of the three lagoon inlets with tidal barriers to stop storm surge flooding (MAV-CVN 1992b, 1997, 2002). The action plan to restore the lagoon environment aims to combat erosion and pollution. The working procedure has been broken down into two separate but integrated projects: the Plan for the morphological restoration (MAV-CVN, 1992c) and the Plan to improve the quality of the sediment and water (MAV-CVN, 1993). Their goal is to retain within the lagoon the sediments which up till now have been either naturally or artificially dispersed at sea, to restore the physical structures peculiar to the lagoon (channels, mud flats and salt marshes), reshaped so as to establish a mutual balance between the various elements, to block the dispersion into the lagoon of pollutants from dumps and to restore the lagoon bed, thereby reducing the quantity of industrial waste residues in the sediment. The anti-pollution measures include the clearing of abandoned dumps and the systematic transplantation of aquatic plants, such as eelgrass, to consolidate the lagoon bed and encourage biological diversity. The following sections describe the activities for the reconstruction and restoration of salt marshes and mud flats according to the updated Plan of 2001 (MAV-CVN, 2001).

The extension of salt marshes was drastically reduced during the last century. In the absence of interventions, they are likely to disappear completely in the near future. The causes of this reduction can be found in the reclamation which occurred up until the 1970s, in sea-level rise and the compacting process of clay and peaty soils, and in the erosion caused by wind generated waves and by ever-increasing motorboat traffic. However, the main historical cause of salt marsh disappearance is the lack of sediment input into the lagoon. In future only the salt marshes capable of keeping pace with the rising water levels by backfilling, that is, by trapping sediments and organic detritus, will be able to grow and thus survive.

A salt marsh monitoring programme was started in 1993 (Cecconi, 1997) in order to:

- measure accretion and soil elevation;
- document the processes that determine marsh accretion, bank erosion and 'ponding' (expansion of internal lakes); and

- monitor marsh management techniques, such as sediment fences and artificial nourishment with thin layers of sediments.

The Magistrato alle Acque and Consorzio Venezia Nuova have carried out several activities for the protection and the restoration of the lagoon morphology. These activities can be divided into three groups:

a Control and mitigation of adverse anthropic effects:
   - regulation of fishing with mechanical tools (manila clam), regulation of hunting;
   - regulation of navigation in the lagoon (harbour traffic, commercial traffic, pleasure boats);
   - works for dumping wave energy along the banks; and
   - works for confining and protecting the shallows, where manila clam is harvested.
b Reconstruction works:
   - reuse of dredged material from lagoon maintenance works; and
   - use of sandy sediments dredged at sea and near tidal inlets.
c Protection and restoration works:
   Protection of salt marsh edges and tidal flat from wave action
   - wooden piling and gabion; and
   - sand-works (construction of salt marshes, beaches, shoals and sandbars).
d System for promoting sedimentation and improving water quality
   - sediment fences ;
   - thin layer marsh nourishment;
   - deposition of organic material;
   - transplantation of vegetation on mud flats and on salt marshes;
   - dredging of tidal creeks and marsh ponds in tidal flats and constructed wetlands; and
   - capping of tidal flats.

SALT MARSH CONSTRUCTION WITH DREDGED SEDIMENTS

These works are aimed at the restoration of the lagoon salt marshes through the beneficial use of sediments (which meet local quality standards) recovered from maintenance dredging of the channels. So

far 5 km² of artificial salt marshes have been constructed (15 per cent of the remaining natural marshes) and 75 km of canals have been dredged (Fig. 51.2). More marshes are expected to be created over the course of the next 10 years as part of an optimum plan for the re-use of dredged material resulting from the maintenance of navigation channels and the restoration of small channels to improve tidal flushing.

The main technical challenge in the Venice Lagoon is that of meeting the strict height requirements of marshes. Unlike places in Northern Europe which experience a tidal excursion of more than 2 m, that of the Lagoon of Venice is limited to 0.6 to 1.1 m. Thus the elevation window for creating marshland is very small. Determining how much dredged material to stock in a marsh cell, so that when it decants it will still be affected by tides, has been a major undertaking. It has been observed that the elevation tends to stabilize at around 0.3 m above mean sea level after 1-2 years. Containment cells are constructed by using wooden piles and a hydraulic net to contain the mud and reduce turbidity. Sediments are pumped into the cells up to the 1.0 m elevation of the wooden piles and allowed to settle and drain to the optimal elevation range of 0.3 m.

Primary attention has been devoted to the start-up of the colonization process and other important factors, such as soil internal structure, avifauna, moisture content and plant root biomass, have been monitored for assessing naturalization processes. Restoration and consolidation work has been carried out on the edges to stabilize and maintain these artificial structures, as have naturalization works, including laying sand-based material upon the surface to encourage the growth of vegetation and creating openings along the edges for the formation of tidal creeks (Fig. 51.3).

RECONSTRUCTION OF SALT MARSHES ALONG CHANNELS WITH SANDY SEDIMENTS

The use of marine sand, from maintenance dredging of river outlets and borrow areas, is a very effective measure for stabilizing the tidal flats with shoal formations at the side of the channels. These projects are

Figs. 51.2 and 51.3   Constructed salt marshes and an example of colonization by *Salicornia spp*. and formation of a tidal creek in *Tezze Fonde* one year after filling.

under way, as required by the Inter-ministerial Committee, for mitigating the impact of artificial channels. This intervention consists of using a more stable mixture of pelitic and sandy sediments for the reconstruction of salt marsh structures along the edges of navigable channels in order to prevent silting up as a result of the cross-flow transport of sediment eroded in the shallows by wind waves and motorboat traffic. The edges of the constructed salt marshes bordering on the channel are protected against the waves caused by motorboats by small breakwaters or gabions, whereas the other edges not impacted by the boat traffic are not armoured and the filling will develop into a strip of tidal flats and beaches. Figure 51.4 illustrates

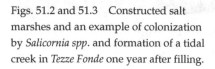

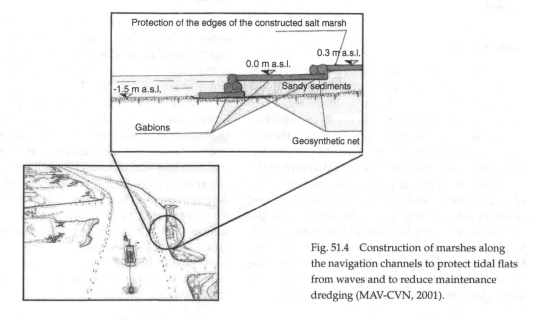

Fig. 51.4   Construction of marshes along the navigation channels to protect tidal flats from waves and to reduce maintenance dredging (MAV-CVN, 2001).

this kind of salt marsh reconstruction works along the Malamocco-Marghera harbour channel.

SALT MARSH RESTORATION AND PROTECTION

These activities include both the structures for local protection of the edges of the salt marshes and innovative techniques for the local management of sedimentation, erosion and water quality. The first group consists of different ways of armouring the edge of the salt marshes to avoid erosion, using wooden piles, gabions, artificial sandbars and beaches. The second group consists of different measures to guide and to aid sedimentation (sediment fences, artificial nourishment of marsh surfaces, depositing of organic materials, transplanting of vegetation) and to improve water quality (reopening of tidal creeks, capping polluted mud flats using clean sand). These interventions have been located in more than 20 different areas of the lagoon. The armouring of the border with wood and stone works has been designed for places in which currents and waves cannot be reduced by other measures, such as the regulation of motor boat traffic or the insertion of stable sandbars to absorb waves. This intervention is particularly necessary where the marsh risks being breached through the direct contact of ponds with the main channel.

Wooden piles

Many salt marsh edges are subject to intense erosion by tidal currents and waves generated by wind and motorboats. Erosion develops through the removal of the lower part of the bank and the consequent collapse of sods of vegetated soil. Subsequently the sediments are transported by the currents. The wooden pile screen along the border is erected to protect the marshes from wave action and to consolidate the edge, preventing its collapse. The piling is placed along the edge of, and at the same elevation as, the salt marsh, following the main indentations and correcting the contour where the damage is greater (Fig. 51.5). The gaps between the piling and the edge of the salt marsh are filled using low water content sediments dredged from the nearby channel. Piling has

Fig. 51.5   Wooden pile protection of salt marsh.

proved to be an effective protection against wave action generated by motorboats, halting erosion of the edges of the salt marshes. As early as 12 months after completion of the works, the material used to fill the gaps behind the piling can be seen to have been colonized by halophytic vegetation. The main problem with the use of wooden piles is the short duration of the portion above mean sea level because of the deterioration due to the colonization by *teredo navalis*.

Gabions

Armouring with small stones contained inside protected steel or geo-synthetic nets has been adopted as an alternative to wooden piling (Fig. 51.6). This alternative provides a better dissipation of wave energy and is of longer duration. The landscape impact of these works is still under evaluation by the Soprintendenza ai Beni Ambientali e Paesaggistici, and the Commissione di Salvaguardia, a State and a Regional institution charged with its approval. These works consist of geo-synthetic net gabions filled with stones of various sizes and placed on the mud flats at a depth of no more than 1 m, reaching an elevation of 0.30-0.5 m. Gabions have proven to be an excellent solution both for the improved dissipation of wave action, particularly in channels with a high level of motorboat traffic, and for their virtually unlimited duration. Costs are lower, and they can be removed should they no longer be necessary, due to reduced

Fig. 51.6   Gabion protection of salt marsh.

wave action or to natural compacting of the edge. Over the long term, gabions also become well integrated within the lagoon landscape as colonization of the gaps by encrustating organisms confers a greenish colour to the stones (MAV-CVN, 2001).

Shoals and sandbars

Some salt marshes in the lagoon present edges which have receded considerably due to waves and currents, with maximum values reaching 10 m yr^{-1} (Cavazzoni and Gottardo, 1983). These works consist of constructing a sand strip 10-50 m wide for the dissipation of wave energy, and of creating sandbars and shoals by raising the bed of the lagoon from the present depth of more than 1.5 metres to a depth corresponding to the low tide level (-0.5 m). Depending on how far the lagoon bed is raised with respect to sea level, the energy from wave action absorbed by this raising (and thus not transmitted to the edge of the salt marsh to be protected) varies between 80% and 40% of incident energy, according to experiments carried out on a physical model (D'Agremond et al., 1996). Figure 51.7 shows the sandbars created to date using a total of 300 000 m^3 of sand, as well as further planned interventions in the central area of the lagoon for the protection of a salt marsh front of around 10 km, employing a total of approximately 1.8 million m^3 of sand. The works carried out to date have allowed the technique to be perfected, testing the jet-spraying technique, which was seen to cause

neither the erosion of the lagoon bed nor an excessive dispersion of sediment. The wave reworking of the sand deployed near the marshes has induced a limited long-shore transport without the 'sand inundation' of the marsh. The surface of the detached sand bars instead has been eroded from a level of –0.2 m a.s.l. to –0.5 m a.s.l. by breaking waves (MAV-CVN, 2001).

WORKS FOR THE MANAGEMENT OF SEDIMENTATION, EROSION AND WATER QUALITY

Processes of sedimentation, accretion and repopulating of the shallows, mud flats and salt marshes can be triggered by a series of integrated and complementary interventions, including sediment fences, nourishment, depositing of organic material, transplanting of vegetation, the dredging of tidal creeks and capping. Since they result in the extension of salt marsh surfaces or the improvement of vegetation cover, these works create more areas in the lagoon which present a natural capacity for accretion.

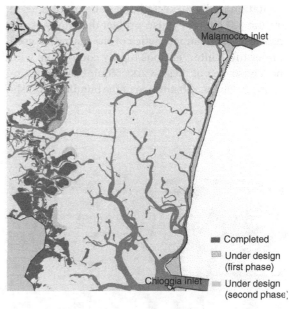

Fig. 51.7   Wave protection of salt marshes by sandbars and beaches.

Sediment fences

The lagoon shallows least affected by erosion are those located in sheltered areas where the wind-generated waves are very low. As a matter of fact, in these areas the sediments can be deposited after their re-suspension and transport by currents. The sediment fence is formed by large bundles of brushwood held in place by a double row of piles, and it is an artificial way of creating sheltered areas where the sediments can be deposited before they reach the channels. This technique was originally applied in the Netherlands and in Germany to reclaim land from the sea. A few years ago it was widely adopted in the coastal areas of the United States as a means of managing river deltas (Day *et al.*, 1999 and Scarton *et al.*, 2000). In the Lagoon of Venice, the first sediment fence was tested in 1993 in front of the Tessera marshlands. During the following three year monitoring period, the elevation of the lagoon bottom rose by about 15 cm. The new elevation of the bottom allowed the formation of an intertidal strip of land which has been colonized by vegetation (Fig. 51.8). A further 16 sediment fences have been built since then. The sediment fences have proved to be efficient in reducing water turbulence and in favouring sedimentation in the protected areas at annual rates of around 1-3 cm. They also protect the edges of salt marshes. The characteristics of the materials used render these interventions highly compatible with the landscape but these works require annual maintenance with the substitution of the bundles of brush-

wood. Even though the benefit to the sediment budget at the lagoon scale is limited (deposition of about 10^4 m^3 yr^{-1}), at the local level the sediment fences contribute to the expansion of the intertidal areas. In the future this benefit will increase with the extension of the areas that will be protected.

Nourishment of the marshes

Due to the reduced sediment loads of the rivers and sea, the marshes are not able to keep pace with the loss of elevation due to soil compaction and to the combined action of subsidence and eustasy. Vast salt marsh areas become permanently submerged, leading to the development of internal lakes (marsh ponds) which ultimately determine the complete disappearance of the marsh due to fragmentation ('ponding'). In fact, the loss of elevation causes the degradation of the halophytic vegetation, with drastic regression and a drop in the number of species. This eventually leads to the substitution of the original associations by more simple and pioneering compositions and, finally, to the formation of a mud flat if relative sea-level rise continues. Nourishment interventions aim to slow down surface erosion with a thin sediment layer to encourage accretion and to reduce stress on vegetation, taking care not to cause permanent damage to existing plants. A limited quantity of sediment is taken from neighbouring shallows and channels. A mixture of sediment and water is then sprayed mechanically over the surface of the salt marsh using an air jet. The artificial placing

Fig. 51.8   *Laghi* sediment fence after one year.

Fig. 51.9   Jet-spraying of a salt marsh.

of thin sediment layers was developed and tested in the USA as a technique of minimal impact in river delta channel dredging (Cahoon and Cowan, 1988). The resulting layer of soil, only a few centimetres thick, encourages the growth of vegetation. This activity has been seen to be effective at many test sites in the Venice Lagoon (Fig. 51.9). The positive effects include an increase in elevation in the order of 3-8 cm, with rapid compacting of the added material, lack of damage to existing vegetation and development of perennial vegetation 24 months after the completion of works with a 50% increase in plant root biomass.

Deposition of organic material

Waves and currents often cause erosion along the edges of the marshes, leading to the loss of portions of emerged land and the retreat of vegetation. Some marshes are seen to be less prone to erosion because of the formation of layers of organic material in the exposed zone (fragments of bivalves, eelgrass leaves and roots). These materials form a strip which, on one hand, protects the edge of the marshes and, on the other, forms an ideal microhabitat for the development of animal species (edge-dwelling benthos) and a nesting site for interesting bird species. This intervention, which aims to extend the zones naturally affected by the depositing of bivalves and eelgrass, was tested at six sites in the southern lagoon where it proved successful in attracting birds.

Transplanting of vegetation

The presence of meadows of eelgrass represents a good indicator of the environmental quality of the lagoon. Intertidal zones (shallows and marshes), characterized by loamy-clayey sediments, can be colonized by populations of *Zostera noltii* and *Spartina maritima*. These plants help stabilize the lagoon bottom, with the strengthening action of their rootstocks and to the ability of their long leaves to capture suspended sediments. The transplanting of *Zostera noltii* turfs in degraded areas which do not present natural colonization has favoured the development of broad expanses of eelgrass. Turf transplants have proved

to be an effective intervention, with an extension of the turf noted near the transplant site after only 12 months and coalescence after 24 months. Turf triggers the development of vegetation both in the shallows (*Zostera noltii*) and the mud flats (*Spartina maritima*).

Dredging of tidal creeks

Large zones of the lagoon mud flat have been eroded and have lost their network of tidal creeks, which guaranteed a higher rate of tidal flushing. The shallows most resistant to erosion are in fact those which have retained their meandering network of small channels. Using the *scomenzera*[2] principle, this type of intervention aims to trigger the natural development of tidal creeks linked to the main channels serving the shallows. As an example, the intervention in Palude Maggiore, a wide area of shallows in the northern lagoon, has led to the creation of creeks which connected the major channel with the shallows. The increased tidal flushing has proved to be beneficial for water quality. A noticeable reduction in the proliferation of macro-algae has been noted in the areas of intervention, together with a rise in the number of fish species present.

Capping of tidal flats

Many tidal flats in the Venice Lagoon have been affected by conditions of environmental decay related to the input of pollution loads. In areas with low tidal flushing, the inflow of fresh water from the mainland and the effluent from the industrial port and the city (containing a high load of organic substance and nutrients) has caused the accumulation of organic substances and nutrients in the sediments, risking anoxic crises and a superabundance of macro-algae. Three such areas in the Lagoon of Venice have been restored by covering the lagoon bottom with a layer of clean sand. For instance, the lagoon bottom of the Lago dei Teneri tidal flat was

---

[2] *Scomenzera* is a man-made inducement for currents to determine the path of a new channel through natural erosion.

raised by 0.5 m (up to a level of 0.3-0.4 m below sea level) by placing a layer of sandy sediments. A monitoring programme was conducted for two years to evaluate the effects of the intervention. This programme consisted of the collection of samples for toxicity tests, chemical analyses as well as analyses performed on the benthic communities in order to detect variations in different components of the ecosystem (Bona *et al.*, 1999). Further useful information was also obtained by means of Sediment Profile Imaging. This method uses an underwater camera, which acquires images of the vertical section of the first 0.2 m of the sediment. Data can be extracted from these images, concerning grain size, depth of the apparent discontinuity of the redox potential, succession stages of the infauna and the presence of reduced gases. All these parameters have been considered for the evaluation of an index of environmental quality of the benthic habitat (Organism Sediment Index). Capping has improved the benthic habitat and allowed the recolonization of the benthic community. This result has been partly confirmed by the toxicity tests, which confirmed the capability of 'Microtox' to discriminate between samples, while the Amphipods mortality test showed that the chemical concentrations in the sediment were always below the effect threshold. Nitrogen content and *Ulva* presence fell in the capped inner area. Capping is a good solution in those areas where the hydrodynamics is poor and where the contaminant level is not too high. Under these conditions the cap is stable and represents an efficient permanent barrier between the contaminated sediment and the environment where the level of oxygenation is sufficiently consistent to sustain benthic colonization.

ALTERNATIVE MEASURES

To restore the lagoon morphology other alternatives have been considered, but not implemented for different reasons. The redirection of the river Piave, Brenta, Adige, for instance, has been considered not feasible because of the pollution of water and pelitic sediments, the interruption of sand transport to the eroding littoral, the required landscape transformations and impacts of new convey-

ing channels. According to the decision of the Inter-ministerial Committee on 4 April 2003, this measure will be re-evaluated starting from an accurate evaluation of the amount and the quality of the sediment transport and the quality of the sediments intercepted by the hydro reservoir of Santa Croce Lake (Belluno). The permanent reduction of flow at the lagoon inlets can reduce the loss of sediments into the sea but it is counterproductive because the erosion of the tidal flats and the channel sedimentation will increase. Because of the recent increased depth of the tidal flat (due to erosion and relative sea-level rise) there is still a need to impede the cross-flow transfer of sediments from the tidal flats to the channels. This alternative can reduce the number of barrier closures but it will also produce a permanent reduction in tidal flushing. For these reasons it will be re-considered in association with a series of mitigation works inside the lagoon, as required by the Inter-ministerial Committee. The removal of the inlet breakwaters in general at the lagoon entrances is not feasible because it will produce the erosion of nearby littorals and sedimentation in the navigation channels. Instead the sand by-pass from Cavallino littoral into the Northern lagoon is effective and has been included into the restoration plan of 1992. It will be re-evaluated, according to the request of the Inter-ministerial Committee, when considering the experience gained from the on-going morphological restoration works of dredging and barge transport of sand from the Lido entrance channel.

CONCLUSIONS

Over the last 15 years, 6 million cubic metres of dredged sediments have been re-utilized within the Lagoon of Venice for the reconstruction of salt marshes and tidal flats, thereby avoiding the greater cost and the negative environmental impact of dumping the sediment at sea. The extension of the reconstructed marshes, which now account for 15% of the total marsh area, is comparable to the area of eroded natural marshes over the same time period. In the next 10 years, the same amount of sediments from navigation channel maintenance and the dredging of minor lagoon channels will be available to continue the re-construction and restoration of salt marshes and tidal flats.

From the above experience it is clear that the restoration of the lagoon morphology is an ongoing continuous effort of works of increasing complexity, aimed at reintroducing self-preserving hydro-morphological and biological processes. At the same time the activities for regulating and controlling fishing activities, motorboat speed and for mitigating their unavoidable impacts with protective structures must be implemented. Special attention has to be devoted to the choice of protection structures with limited landscape impact, extending the use of biodegradable materials (wood piles, bundles of brushwood, shells, natural geo-textiles) and of resistant structures that can be removed in future when their function may not be needed.

Other measures, such as diverting the Adige, Brenta and Piave rivers back into the lagoon to restore the huge sediment deficit of the lagoon, although very attractive in theory, are not feasible in the near future because the water and the pelitic sediments are polluted and the sandy sediments are needed for the natural nourishment of the shorelines. Also the narrowing the tidal inlets for reducing the ebb flow, is ineffective because it increases the tidal flat erosion and channel siltation, and produces a permanent reduction of the tidal flushing.

REFERENCES

Bona, F., Cecconi, G., and A. Maffiotti. 1999. An integrated approach to assess the benthic quality after sediment capping in Venice Lagoon. *Journal of Aquatic Ecosystems Health and Managment Society* (Canada), December, 1999.

Cahoon, D. R., and J. H. Cowan Jr. 1988. Environmental impacts and regulatory policy implications of spray disposal of dredged material in Louisiana wetlands. *Coastal Management* 16, 341–62.

Cavazzoni, S., and D. Gottardo. 1983. Processi evolutivi e morfologici nella Laguna di Venezia. *Laguna, fiumi, lidi: cinque secoli di gestione delle acqua a Venezia*, Proceedings, Venice: 10–12 June 1983.

Cecconi, G. 1997. Beneficial use of dredged material for recreating marshes in the Venice Lagoon. *In Proceedings of International Conference on Contamined Sediments*, Rotterdam: Sept.1997, pp. 150–7.

Cecconi, G., and V. Ardone. 1999. La protezione dell'ambiente lagunare e costiero veneziano. In *Atti della XVII Giornata sull'Ambiente 'Venezia: città a rischio'* (Accademia Nazionale dei Lincei, Rome, 4 June 1999), pp. 157–73.

D'Agremond, K., Van Der Meer, J. W., and R. J. De Jong. 1996. Wave transmission at low-crested structures. Paper presented at 25th Int. Conf. on Coastal Eng., Orlando, Florida.

Day, J. W., Rybczyk, J., Scarton, F., Rismondo, A., Are, D., and G. Cecconi. 1999. Soil accretionary dynamics, sea-level rise and the survival of wetlands in the Venice Lagoon: a field and modeling approach. *Estuarine, Coastal and Shelf Science* **49**, 607–28.

Favero, V. 1992. Evoluzione morfologica e trasformazioni ambientali dalla conterminazione lagunare al nostro secolo. In *Conterminazione lagunare: storia ingegneria, politica e diritto nella Laguna di Venezia*. Istituto Veneto di Scienze, Lettere ed Arti, Venice: pp. 165–84.

Magistrato alle Acque di Venezia, Consorzio Venezia Nuova (MAV-CVN). 1992a. *Progetto di Massima dei litorali veneti*. Venice.

MAV-CVN. 1992b. *Progetto di Massima degli interventi alle bocche lagunari per la regolazione dei flussi di marea*. Venice.

MAV-CVN. 1992c. *Interventi per il recupero morfologico della laguna. Progetto di Massima. Relazione finale*. Venice.

MAV-CVN. 1993. *Progetto generale interventi arresto e inversione del degrado lagunare. Rapporto finale*. Venice.

MAV-CVN. 1997. *Studio di Impatto Ambientale del progetto di massima delle opere mobili alle bocche di porto*. Venice.

MAV-CVN. 1999. Studio C.8.2: *Monitoraggio dell'erosione delle barene e dei bassifondi. Rapporto intermedio*. Venice.

MAV-CVN. 2000. *Valutazione dell'evoluzione morfologica della Laguna Veneta sulla base dei più recenti rilievi batimetrici*. Progetto delle opere mobili alle bocche di porto della laguna di Venezia per la regolazione dei flussi di marea, Attività svolte a seguito della delibera del 8/3/1999 del Comitato ex art. 4 L.798/84, Parte 2, Attività 5.

MAV-CVN. 2001. *Aggiornamento del piano degli interventi morfologici in laguna*. Venice: Ministero delle Infrastructure e dei Transporti–Magistrato alle Acque di Venezia.

MAV-CVN. 2002. *Progetto preliminare delle opere mobili alle bocche di porto*. Ministero delle Infrastructure e dei Transporti–Magistrato alle Acque di Venezia.

Scarton, F., Day, J., Rismondo, A., Cecconi, G., and D. Are. 2000. Effects of an intertidal sediment fence on sediment elevation and vegetation distribution in a Venice (Italy) lagoon salt marsh. *Ecological Engineering* **16**, 223–33.

472   G. Cecconi

Tosi, L., and L. Carbognin. 2002. Evidence of the present relative land stability of Venice, Italy, from land, sea and space observations. *Geophysical Research Letters* **29**, 12.

52 • Functional characteristics of salt marshes (*barene*) in the Venice Lagoon and environmental restoration scenarios

L. BONOMETTO

BACKGROUND

In the debate on the future of the Venice Lagoon there seems to be a common aim: guaranteeing the conservation of the quality, identity and functionality of this unique system. As a consequence of the floods in 1966 and the civil and scientific debate that ensued, this aim was also laid down in specific Italian laws such as the imperative '*the protection of the natural environment, the preservation of the ecological and physical unit, the preservation of the salt marshes and the exclusion of further acts of reclamation*'. This imperative was repeated by further special laws (1984 and 1992) requesting 'the hydro-geological equilibration of the lagoon, the halt and reversal of the degradation process', ... 'the elimination of the causes', ... 'the restoration of the lagoon morphology'.

The application of these laws requires the clearest possible definition of the meanings of the words and phrases and a verification of if, and to what extent, this decree has been respected. This then involves not only the 'state of knowledge' but also the 'state of application' of the aforementioned. It is therefore necessary to understand how this knowledge is actually applied in environmental management, in this case, the lagoon functional morphology.

These considerations pay particular attention to intertidal morphology (the salt marsh systems, locally known as *barene*), with just a brief mention of the submerged morphology to which the intertidal morphology is linked by permanent and feedback relationships. This chapter provides a summary of the extensive research carried out for the Ministry of Environment and ICRAM (Central Institute for marine scientific research and applied technology) and distributed by the Comune di Venezia.[1]

The overall situation that emerges can be summarized as follows:

- The particular environments of the Venice Lagoon are the result of the marine and fluvial inputs in a context defined by historic human activities as a fundamental component in directing the processes and equilibria. Actions and uses of the past century, and current ones, are responsible for the dynamics which represent a grave threat to the future of the lagoon.
- The artificialization and degradation of the lagoon are very different in different areas of the lagoon, and therefore require management, protection and restoration criteria which are differentiated likewise.
- The *barene* (salt marshes) in the Venice Lagoon are of different origins; their characteristics derive from the nature of their tidal flats, with typological and functional levels and margins, and this should represent the primary reference for protection and reconstruction works (which hasn't always happened in the interventions implemented to date).
- The forms and localization of pre-existing *barene*, and especially their geographical and functional significance, should represent the reference models in restoration works. These must be adapted to current situations, especially where the hydraulic

---

[1] Bonometto, L . 2003. *Analisi e classificazione funzionale delle barene e delle tipologie di intervento sulle barene. Analysis and functional classification of saltmarshes (barene) and types of intervention on saltmarshes.* Study produced under the agreement between the Environment Ministry and ICRAM (Institute for marine scientific research and applied technologies), distributed by Comune di Venezia.

Flooding and Environmental Challenges for Venice and its Lagoon: State of Knowledge, ed. C. A. Fletcher and T. Spencer.
Published by Cambridge University Press, © Cambridge University Press 2005.

and sedimentological dynamics have changed, and they must be implemented using techniques and instruments which relate to their inter-tidal characteristics.

- In the protection of the *barene* margins the use of rigid elements and materials must be limited as far as possible (albeit that these are currently the norm), since these go against the 'plastic' nature, permeability and ecotonal functionality of lagoon intertidal systems.

CHARACTERISTICS AND DIVERSITY OF THE LAGOON MORPHOLOGY

As for all lagoons, that of Venice is essentially a vast environment with 'ecotone' characteristics – an area of contact and transition between different environments – those of the terrestrial world above the water and the marine world below. All ecotones are characterised by intense dynamics, rapid evolutional transformations and considerable diversity over short space time-scales. This is because ecotones receive elements and energy from both the contexts that they are in contact with, thus placing themselves in a zone of interference and exchange between one another. Furthermore, ecotones are the site of specific and unique events. In the typical surroundings of the Venice Lagoon, both emerged and submerged, diversity and functionality are the result of marine and fluvial contributions with the addition of the interface dynamics due to, or induced by, human activity in both the present and the past.

Despite such a dynamic context, in nature salt marsh functionality is guaranteed over time by a surprising stability, linked to the roles of the few plant species present which, in a system with reduced energies, ensure a highly efficient ability of protection, construction and reconstruction. These abilities are largely linked to the refined differentiation present in natural salt marshes, the knowledge of which is a primary requisite for the planning and development of techniques for correct reconstruction or restoration interventions.

During the thousand-year history of Venice, the lagoon has been subject to stabilising activities that mainly concerned the highly evolutionary surroundings that were subjected to the unbalanced impacts of

continental and marine contributions. Man's actions thwarted the processes of silting of the lagoon areas by rediverting rivers and the erosive processes of the *lidos* and this led to a new condition due to the integration between the natural and manipulated protective dynamics. However, during the last century aggressive and degrading interventions and uses prevailed in the lagoon (drainage of areas for land reclamation, digging of canals incompatible with the characteristics of the lagoon, heavy industrial and agricultural pollutant loads, fishing destructive of the lagoon bed, erosion by waves from boat traffic) and were favoured by the investment of enormous new energies that favoured methods alien to the system's supporting capacity.

The impact of past human actions and recent aggressions varies greatly in the different areas of the lagoon. The lagoon is approximately differentiated into the following three areas (Fig. 52.1), and it is these distinctions that should guide protective intervention and management.

- **North lagoon** (A), which still has an extremely high level of original features, conserving both morphological elements and dynamics due to the relationship between marine and fluvial waters. The management principles here must be inspired by faithful conservation of identity, and any morphological restoration interventions, as defined by special laws, must, above all, be directed at environmental restoration and on the removal of anomalous factors, without introducing artificial elements unless in extremely localized areas where this is inevitable.
- **Central lagoon** (B), between Venice's Malamocco-Marghera Canal (known as 'Canale dei Petroli') and the Casse di Colmata, clearly dominated by twentieth-century interventions. First and foremost the flattening and deepening of the shallows resulting from the excavation of Canale dei Petroli and consequent erosion. The area requires a revised canal structure that will at least restore lagoon functionality, without the formal restoration of past features as the main objective.
- **South lagoon** (C) south of Canale dei Petroli, also including the areas between Casse di Colmata and the Giare peninsula, presenting an extremely

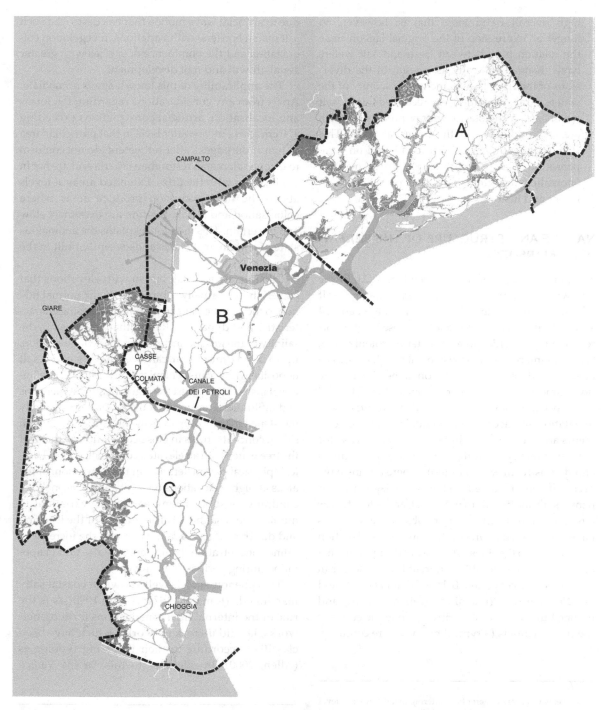

Fig. 52.1   Lagoon areas characterized in terms of human impacts: A. North Lagoon – highly conserved original characteristics and dynamics; B. Central Lagoon – areas dominated by the urban centre and twentieth-century interventions; C. South Lagoon – significantly natural, albeit that the freshwater/saltwater relationships have changed radically following the diversion of the rivers.

high overall naturalness that is, however, no longer an expression of the original lagoon since the relationships between fresh and salt waters have changed radically as a result of the diversions of the river Brenta, increasing salinity of the waters, transformation of reed beds into salt marshes and the start of decay processes which are still underway in the intertidal areas. This state requires managerial criteria based on compensation measures, with respect to degenerative anomalies resulting from human action, both recent and historic.

NATURE AND STRUCTURE OF THE *BARENE* (SALT MARSHES)

No matter what their origin (see Fig. 52.2), in the overwhelming majority of the Venice Lagoon salt marshes, ground elevation is mainly between +24 and +30 cm above mean sea level. These heights correspond to the widespread vegetation colonization, typical of internal and extensive salt marsh areas. The factor that determines elevation is the effect of the tide. Almost all the salt marshes are of the 'tidal flat' type since they have all originated as a result of sedimentation and accretion from tidally supplied sediments and fluvial contributions (this is the case for the salt marsh systems in the north lagoon upstream and downstream w.r.t. Burano). Where origins have been different (reed beds in the central lagoon, continental soils in the salt marshes of Campalto, etc.), as a result of the compaction or subsidence, marshes have also arrived at an equilibrium, stable elevation with respect to the tides.[2] It is only at the margins that one can see considerably different levels – lower in relation to the vegetative belt of *Salicornia veneta* and *Spartina maritima* (reaching levels near zero) and higher, but still less than +50 cm, along the edges of the tidal channels (a typical but now rare condition

due to artificial wave motion that has destroyed such salt marsh borders) with variations in vegetation colonization and the dominance of species with greater aerial growth and root development.

The applicability of this knowledge is immediate. Apart from any considerations regarding the forms and localisation, artificial constructions exceeding +30 cm above mean water level in that particular area (50 cm at the edges), after settlement, do not conform to lagoon salt marsh restoration criteria and are not in equilibrium with the tides. Extended areas at levels above the tidal excursion produce areas where colonisation and soil maturation are extremely slow and uncertain. This therefore explains the anomalous state of certain artificial islands recently built in the lagoon.

The rare areas of salt marsh with elevations that exceed those normally submerged by the normal tide due to their particular origins have surprising characteristics and dynamics. Rather than decreasing, the salinity of the ground increases in line with elevation, up to a maximum that can even have surface salt deposits, which cause chemical–physical and vegetational instability.[3] Whilst the regularity of submersions and infiltrations ensures that the ground is constantly moistened in the lower areas by interstitial water with saline concentrations in close correlation to those of the free water with an elevated constancy in the chemical/physical parameters and in the vegetation, in the areas at higher elevations the rarity and seasonal discontinuity of submersions leads to big variations in humidity and salinity. This is a result of the leaching and diluting effects of rain water and the increase in saline concentration due to capillary rise and evaporation during periods of aridity.

This phenomenon is well known in coastal salty marshlands (Ranwell, 1972; Adam, 1990) as is the rule of the internal tidal canalizations forming networks, i.e. 'tidal-creek networks' which have been classified according to their recurring typologies (Allen, 2000). In reference to this, in the Venice

---

[2] These elevations are coherent considering that the average level of spring tides is around +60–65 cm and also given that the accretion role of the tide weakens until it stops at the elevation where significant submersions no longer take place. As is known in literature, this stabilizes tidal flats at levels that indicate they have reached maturity (Pethick, 1984).

---

[3] This has already been observed by Marcello and Pignatti in their distinction between 'wet' salt marshes (low levels, in equilibrium with the tides) and 'dry' salt marshes, that are higher (Marcello and Pignatti, 1963; Pignatti, 1966).

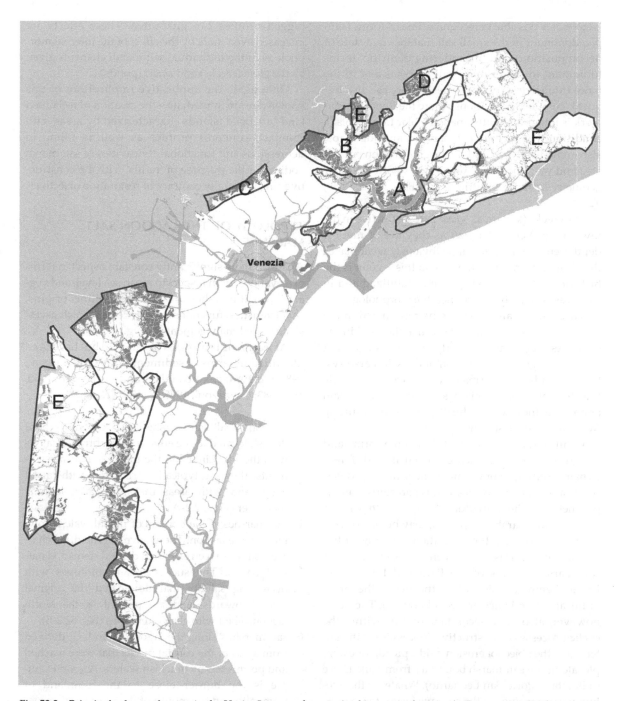

Fig. 52.2   Principal salt marsh areas in the Venice Lagoon, characterized in terms of origin: A. 'Lagoon canal' salt marshes, derived from marine sediment imports during incoming tides; B. Salt marshes at the borders of fluvial waters still entering the lagoon or at least partially active; C. Salt marshes originating on continental ground; D. Salt marshes originating from pre-existing brackish environments or salinised freshwaters arising from river diversions; E. Halophyte environments in the fish farms (*valli da pesca*), with characteristics directly dependent on human management.

Lagoon it is possible to recognize 'linear' forms (simple, dominant in the small salt marshes in a state of decomposition) and 'meandering dentritic' forms (dominant in the primary salt marshes and in the more extensive secondary ones). There are also frequent 'complex' formations where the dendritic network is integrated with small alveolar tidal pools (a condition that seemed extensive in antique maps, also immediately beyond the original lagoon boundary) and with local 'reticulate' types with a regular geometry of anthropic origins (see Figs. 52.3a, 52.3b, 52.3c).

The creeks (known as '*ghebi*') are distinctive in the low-energy Venice Lagoon as they are of modest depth (tens of cm up to 1 m). Obviously what makes the clear and unique difference in this lagoon is the historic context that has guided, stabilized and in some cases actively determined the morphology.

Three main categories can be recognized in the forms of the diverse natural salt marshes and in the elements they contain. Firstly, the forms resulting from the original construction of the salt marsh system; secondly, the forms resulting from the cyclic transformation which allow such systems to sustain and renew themselves; thirdly, the forms resulting from erosive or destructive processes.

Of importance is the fact that some forms and structures tend to coincide – even if in different dynamic settings. For example, the concave profiles, which are just as widespread and characteristic in the perimeters of the individual elements as they are in those of salt marsh complexes, may be of primary origin, derived from the deposition of marine or fluvial sediments (this can be seen, for example, in the salt marsh systems of the Palude del Tralo and Palude Centrega which follow the lines of the canals of Burano, San Felice and San Lorenzo). They can, however, also have secondary origins within the cyclic processes or destructive processes (in the latter being the object of erosion, widespread, for example, along the salt marsh boundary from Valle Dolce di Brenta to Porto San Leonardo). Whatever the evolutionary meanings of the concave forms, the expression of both conservative and destructive processes, these increase the perimeters of contact and exchange between the salt marshes and the water with all the biological, depurative and dissipative effects the

ragged profiles can guarantee. These effects are increased even more by the effect of the internal morphology of the individual salt marsh elements given by the *ghebi* (creeks) and *chiari* (ponds).

Once again the applicative implications of this knowledge are immediate – the creation of reclaimed land / artificial islands characterized by areas with compact, converse profiles as well as filling in depressions and functional concavities of salt marsh bodies, for the purpose of 'refills' with a reconstructive aim, is actually contrary to restoration objectives.

TYPOLOGY OF THE LAGOON SALT MARSHES

Apart from the significantly constant aspect in terms of elevations, with respect to mean sea level, and vegetation, salt marshes present very distinct origins, characteristics, functions and processes which necessitate varied management strategies.

At least four typologies of salt marshes have been identified in terms of sedimentology (Albani *et al.*, 1983 and 1984; Favero and Serandrei-Barbero, 1980, and 1983; Cavazzoni, 1995; Caniglia *et al.*, 1997):

A. 'lagoon canal' salt marshes, also known as 'tidal deltas', located at the edges of the entrance canals with the sea. Linked to the influx of marine sediments, they are typically structured with raised edges and with areas that gradually degrade thereafter (Fig. 52.2-A).

B. salt marshes at the borders of fluvial waters flowing into the lagoon. These have intermediary characteristics between extended delta systems and natural banks. The successions of biotopes with decreasing salinity that dominated the original lagoon towards land are present in the North lagoon, albeit with clear alterations (Fig. 52.2-B).

C. salt marshes along the lagoon boundary, derived from areas of the coastal plains that were reached and permeated by brackish waters. A good example is the border close to the mainland of Campalto and the mouth of the river Dese, recognisable for the superficial, or almost superficial, presence of compact sediments, a sign of continental origin, guaranteeing these salt marshes greater stability (Fig. 52.2-C).

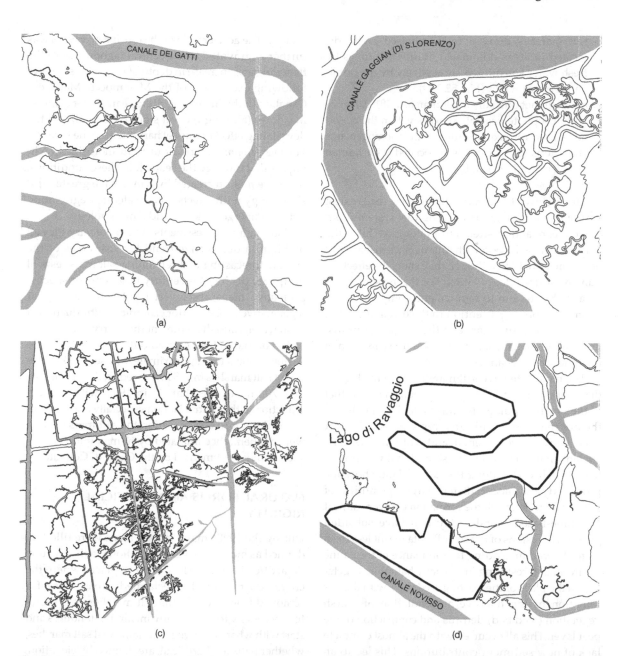

Fig. 52.3   Salt marshes types according to form and internal canalization (derived from Regional Technical Cartography): (a) 'linear' form, most frequent in the salt marshes which are breaking down due to increasing salinization of previously low salt environments (the salt marshes between Punta Cane and Cason delle Pescine, in the central-south lagoon are shown here); (b) 'meandering dentritic' and 'complex' form, most common in salt marshes which have evolved according to original characteristics (the salt marsh leading to Palude del Tralo, in the North Lagoon in shown here); (c) 'reticulate' forms with geometric man-made pools (the central lagoon example is shown, in the area beyond the reclaimed islands or Casse di Colmata, dominated by historic canalizations); (d) undifferentiated forms without internal canalizations, nebulous forms, represented by the recent artificially reclaimed areas ('artificial salt marshes' in the area of Lake Ravaggio are shown, in the central lagoon: thin lines delimit existing marshes).

D. salt marshes derived from salinised wetlands, located in areas dominated at different times by fresh water substituted successively by the dominance of sea water (Fig. 52.2-D).

E. although not officially a salt marsh, a fifth category with halophyte vegetation exists within the closed areas used for acquaculture, with inhomogeneous characteristics specifically associated with human intervention (Fig. 52.2-E).

The first two typologies characterize the greatest part of the north lagoon, forming vast systems both downstream and upstream from Burano. This is the only area in the lagoon with salt marshes that are still of a primary nature, although they show the effects of human-induced degradation. For this reason, maintenance has to aim to remove and compensate the anomalies, paying attention to the original values, aimed at the maintenance of the unique naturalistic/environmental characteristics and reactivation of the functional dynamics.

The most extensive salt marsh area in the lagoon that characterizes the central and south lagoon is that of the fourth typology (salinised environments). In the original lagoon this area represented a border zone between continental and salty waters. The ancient character of the marshland that was prevalently fresh water along the lagoon-land shoreline produced and accumulated large quantities of organic material, with the formation of layers of peat and lime-based marshy clay that have subsided greatly. Diversions of the river Brenta meant less fresh water flowed in, leading to dominance of the saline environment, resulting in a strictly halophile vegetation in the place of the existing reed beds, with the termination of the massive accumulation of marsh vegetation (reed bed) detritus and compaction of the peat layer. This all occurred with the almost complete lack of new sediment contributions. This led to an extremely rapid destructive dynamic such that the area, which had originally been populated by extensive reed beds with a submerged environment made of confined and detached valleys and 'lakes', was turned into a stretch of salt marshes where the submerged environment soon dominated and confined the intertidal and emerged environment to isolated patches – a process that greatly accelerated in the last century. The acceleration of the erosive processes was further aggravated in the central lagoon due to the flattening and increase in depth of extensive shallows following excavation of the Malamocco–Marghera Canal, with the increasing disappearance of eelgrass meadows (although recent signs of revival) and the devastating effect of mussel harvesting of the invasive *Vongole filippine*, carried out with motorized techniques that damage the seabed. In these conditions, the wave motion created by winds have greater erosive energy with effects that accelerate degradation of the submerged and intertidal morphology.

Although the assessments carried out vary, they all highlight a destructive trend for vast areas of the lagoon, forecasting that remaining salt marshes will disappear during this century unless major specific protective measures invert the trend and trigger regenerative processes (for example, reintroduction of fresh waters, and the sediment they carry). A quantification of marsh loss, carried out by Magistrato alle Acque,[4] shows that approximately 72 square kilometres of salt marshes in 1930 were reduced to around 47 in 1970 and to less than 45 in 1992, with marked reduction in the central lagoon especially and without even taking into consideration the salt marshes that were deliberately eliminated by human intervention, such as those submerged by the Casse di Colmata.

NATURAL FORMS AND ARTIFICIAL RIGIDITY

During the 1990s many lagoon areas were filled in, defined as morphological restoration via the creation of 'artificial salt marshes' even in places where the characteristics of the reclaimed areas were far removed from those of the salt marshes. It is therefore necessary to remove any misunderstandings and start with what is known for certain. Real salt marshes, whether natural or artificial, are defined by elevation, geomorphological and sedimentological structure and their ecosystem roles. When the functional mor-

---

[4] Magistrato alle Acque di Venezia. 1992. Nuovi interventi per la salvaguardia di Venezia (New interventions for Safeguarding Venice). Morphological restoration in the lagoon. Project coordinated by Consorzio Venezia Nuova. Venezia: Outline project.

phology is ignored, and when the elevations are such as to let the chemical-physical variables of the emerged environment prevail, the infills have the characteristics of an 'artificial island' and not of salt-marsh creation/'repristination'.[5] At best, one can refer to land reclamation via infilling using sediment laden waters, while bearing in mind that differences between a 'salt marsh' and 'land reclamation' are fundamental and concern the geomorphological processes and the ground characteristics, unstructured in the case of land reclamation but clearly structured, layered and developed for salt marshes. Vegetation dynamics consist of defined and lasting associations in salt marshes, whereas these dynamics are relatively undefined and variable in the reclaimed areas (Pignatti 1966; Visintini *et al.,* 2000).

The natural forms of mudflats, salt marshes and canals are the result of evolving conditions that, in non-rigid systems, are derived from the relationship between the forcing energies, the dissipative processes and sediment dynamics. Furthermore, due to their very origin, they are the forms that have the greatest capacity to coexist with the variables at work and to rapidly absorb or compensate the usual factors of natural disturbance, thus guaranteeing their self-preservation. They are 'plastic' structures and it is this very 'plasticity' that is linked to the ability to attain and conserve arrangements of optimal balance with respect to energy, salinity and sediment levels. The forms, elevations and the structural arrangements of salt marshes and mudflats are derived from gradual morphogenetic processes, that can be activated, accelerated and supported. These processes pass through evolutionary phases which have spe-

cific timings. This means that attention must be given to the temporal articulation of any reconstruction interventions. Works that do not reflect this knowledge and do not respect these natural principles do not create true, even if artificial, salt marshes and mudflats. This is shown by interventions that have filled in areas of the lagoon, selected according to criteria which favour infill capacity in relation to the expansion of the perimeter; in interventions in which new/reclaimed areas are created with undifferentiated techniques or too quickly; and in interventions in which the margins are reinforced using extraneous materials. Anyone who is acquainted with the lagoon morphology interventions in the 1990s knows that artificial reinforcement, nebulous forms and the undifferentiated implementation of measures were all the norm.

Further interventions using forms and positioning based on a logic that is extraneous to those of morphological and functional restoration must be avoided. However, it is also important not to go to the other extreme in the belief that a replication of the past is synonymous with design quality. 'Copying' may be easier than 'understanding and reinterpreting' but the restoration of an equilibrium first and foremost requires solutions that address functionality, particularly when working in contexts in which environmental factors (currents, fetch, depth, chemical-physical characteristics of the waters and sediments, etc.) have undergone profound changes, as is the case in vast areas of the central lagoon. Although reproduction of the forms[6] should be pursued where possible, it should not always be considered necessarily correct. It is the meanings of the forms that should serve as a ref-

---

[5] Despite this evidence, acts of 'repristination' or 'reconstruction' of salt marsh elements and forms have so far been described quantitatively by area only (hectares) rather than by functions/quality. The same comparison carried out by the Magistrato alle Acque between the forecasts of the 1992 general plan and the results of the 1999 assessment (activities requested by the Interministerial Committee established by art. 4 of the Law 24 Nov. 1984 n. 789 and deliberated on 8 April 1999. First report. Coordinated by Consorzio Venezia Nuova) confirm this simplification and make no reference to the quality of the infills carried out. Consequently, only in very few cases do the reconstructed elements conform to the original morphology of the areas in question, while in the majority of areas further environmental anomalies have been introduced.

[6] The lagoon has changed over the centuries as well as over the past decades. Thus 'repristination' interventions need to explicitly identify temporal reference models. It is a common opinion that the beginning of the 1800s is a prime reference point. It represents the situation of the lagoon immediately after the fall of the Republic and is therefore a model produced by the culture and technical and managerial expertise which were the very expression of Venice and its lagoon. It is the oldest period for which there is a modern cartographic document (the Denaix chart), thus allowing methods of study and comparison with successive phases based on scientifically credible elements (D'Alpaos, 1992; D'Alpaos and Martini, chapter 48). From a more practical point of view however, the most

(continued overleaf)

erence in a design scheme that is measured against actual dynamics. This scheme must also pay attention to structural characteristics, elevation, internal differentiations and the evolving and self-sustaining dynamics. For example, the creation of historic forms, even if reproduced correctly in their profiles, but demarcated with heaps of stones or using non-degradable materials means negating the restoration of the structural and dynamic characteristics of the salt marsh border, and understanding the reference forms as models that are rigid over time, whereas natural models are notoriously changeable over time.

Precisely because it is 'plastic', how the lagoon environment reacts to the application of rigid and unnatural elements is well-known. The erosion that occurs at the base of the margin due to the reflected waves or the enlarging of holes in the perimeter that often occurs to the rear of protective piles, is due to the imbalance in equilibria between resuspension and sedimentation. These reactions were well-known in the past – the famous saying 'palo fa palude' (a pile creates a marshland) expresses the knowledge that even the most minimally rigid materials can disturb hydraulic balances and sedimentological reactions which are capable of leading to changes even at a system level. The principle of favouring the plasticity of the salt marsh forms means accepting possible unforeseen spontaneous re-modelling. There are always too many variables involved to allow forecasts that are certain down to the finest details. Furthermore, the variability of settling rates – also in nature – are an essential part of a geomorphological functionality that tends towards optimal configurations in relation to the energies involved.

---

[6] (continued) important temporal reference terms are shown from 1931 until 1955, the years just before the great changes after the Second World War supported by cartographic surveys and aerial photographs. The lagoon reference that is closest to today's is therefore that of the mid twentieth century and this is why it is a more realistic model for repristination. But it must be clear that precisely because it preceded the excavation of the Canale dei Petroli and the creation of the Casse di Colmata that such a reference term must be understood flexibly in the areas involved in projects that created separations between lagoon bodies and important changes in the hydraulic regime. The re-proposal of the original forms has to be the principal objective, but that is not always possible and would not always lead to functional arrangements in transformed contexts.

INTERVENTIONS OF THE 1990S

Since the end of the 1980s, through their concessionary (Consorzio Venezia Nuova) and therefore through the companies and consultancies connected to it, the Magistrato alle Acque has been dredging the lagoon canals and placing the sediment thus obtained in designated areas. In these areas the waters mixed with sediments are distributed via jet-flow in such a way that the sediments accumulate and the waters drain away. The elevation reached once the sediments have settled is one of the decisive factors to see if the interventions actually do meet the objectives of restoring salt marsh conditions in terms of equilibrium and morphological features – and therefore complying with the special laws.

Interventions have been carried out throughout the lagoon but with a greater concentration near Chioggia and in the central lagoon, to the south of the Malamocco–Marghera Canal. These infill projects, technically 'reclaimed land/artificial islands', were given the name 'artificial salt marshes' to underline the constructive aspects of the interventions as opposed to loss of lagoon salt marshes due to erosion. From a naturalistic, geographical and geo-morphological point of view this represents a misunderstanding. The turbulent and almost undifferentiated manner in which these areas were created are in clear contrast to the aims of recreating environments derived in nature from gradual, slow and differentiated processes, thanks to which the relationships between energies and re-deposited materials create structured and stabilised ground and bottoms. The basic creations, which were carried out with simplified and standardized techniques, instruments and economised timing, as well as frequently incorrect geographical localization,[7] resulted in nearly all infill-interventions being

---

[7] Several projects detailed the criteria used to locate the reclaimed areas (Magistrato alle Acque di Venezia, Consorzio Venezia Nuova 2000. *Realizzazione di barene per il recapito di materiali di classe 'A' e 'B' provenienti dal dragaggio dei canali lagunari. Construction of saltmarshes with class 'A' and 'B' dredged materials originating from the lagoon canals. Barena Otregan and Barena Cornio. Venice definitive plans*). These include: 'the possibility of access by boats at a distance that is compatible with a direct filling in of the new morphological structures, *(continued overleaf)*

out of keeping with the morphological and functional characteristics of the real lagoon. Instead, the levels of information and differentiation that characterize the natural salt marshes in the sites where work is being carried out must correspond to the interventions of filling in (via flows of muddy-sandy water), paying particular attention to the morphological, structural and dimensional aspects as well as the temporal sequences needed in nature to carry out the constructive and evolutionary processes.

A great legacy of indications, useful comparisons (Cecconi, 1995; Cecconi *et al.*, 1998) and the verification of errors can be derived from more than a decade of interventions and these can serve as a reference base for future plans and the development of criteria and techniques that are genuinely aimed at morphological and functional restoration.

One characteristic that immediately distinguishes the 'artificial salt marshes' created so far is the anomaly of the forms compared to typical lagoon morphology (see Fig. 52.3d). They were created as compact bodies, with containment perimeters that reduce the length and therefore the costs. The resulting ratio area/perimeters is therefore greatly different from their natural counterparts which, as a rule, have an ample presence of concave, non-linear and irregular forms (analogous to intestinal villi). It is important to underline how these differences, reducing the ecotonal functions of the salt marsh environments, go way beyond the morphological aspects. Other characteristics that differentiate the in-fills in question from

the natural salt marshes are the internal morphology and the micro-altimetric structuring. In nature these are the fruit of the primary formation processes and successive evolutionary dynamics while the 'artificial' salt marshes have not yet been able to differentiate themselves and will still evolve into forms that differ from the spontaneous forms typical of this location. This is especially clear for the reclaimed areas at a higher elevation that maintain their amorphous aspect over time, without any lagoon functionality.

One area that was particularly affected by such work is the '*Laghi*' (lakes) near the Casse di Colmata (Lago dei Teneri and Lago di Ravaggio), where the positioning of the interventions led to further upheaval in a geography that had already been disrupted by work carried out in the 1960s. The sites filled with dredged mud were placed mainly in the historic 'lakes' that technically were amenable to being filled in since they were naturally relatively deep and capacious. With just one recent exception, reclaimed lands were created which are completely extraneous to their context in terms of form, location, dimensional relationships and sedimentological structuring. This resulted in the loss of geographical and morphological characteristics that gave the area its identity and functionality.

ARTIFICIAL INFILLING AND NATURAL PROCESSES

The halophile infills (low elevation, in constant rapport with tidal waters) and the 'artificial salt marshes'; as well as the restored margins of the degraded salt marshes would have had a much more stable structure, as well as a closer correspondence to the original characteristics, if they had been designed and carried out in synchrony with the constructive processes found in nature. This means creating artificial salt marshes by rapidly building as far as the elevation of the lowest salt marshes and then gradually proceeding with further additions of fine-grained material[8] where possible with sediments that are rich

---

[7] (*continued*) therefore foreseeing its location in the proximity of canals with sea bottoms also suitable for vessels with a draught up to 3.00 –3.50 m'; 'compatibility with other restrictions such as areas destined for fishing and hunting grounds …' This leads to the reflection that the location of sites which was of primary importance to achieve the aims of repristination and restoring the equilibrium in the lagoon were explicitly subordinate to other interests, some of which were clearly laughable in comparison, for example the position of a hunting ground. Above all, it should not be forgotten that if the aims of a correct morphological management is to be achieved, certain interventions require the capacity to proceed on the shallows and mudflats as well and to operate in both a refined and controlled fashion. However, here there has been a restriction that led to localisation and dimensioning of the projects in reference to the machines and vessels in use, with the consequence that it always tended towards simplifications, thus excluding the sites that would have required an updating of the instrumentation.

---

[8] In operations of filling in, the orientation and intensity of the flows are of great importance for the repristination of structures close to natural ones. At the input points of the fine grained material, there

(continued overleaf)

in organic particulate matter following colonization by the typical reinforcing pioneer vegetation. In this manner, it is possible to mimic the essence of the salt marshes, i.e. that of structured 'tidal flats' made tabular by the tide and redrawn in their internal forms by its ebb and flow.

Compared to other conditions, the slower, more gradual and articulated the creation of the artificial salt marshes, the more similar they will be to true salt marshes, with relatively low additional cost. Artificial lands created directly at their final height do not even partially reproduce the gradual pedogenetic processes that lead to a balanced structure and relative stability in nature. This is why they result in anomalies due to the lack of sedimentological stratification that are favoured at lower levels by regular submersions (Visintini et al., 2000) and benefit at a much slower rate from consolidating and protective effects (maintaining moisture and therefore cohesion and plasticity, and colonization by vegetation). The higher areas are crusty and powdery due to insolation without the presence of organic substances[9] that appear in nature (dead parts of the covering vegetation and the deposition of particulate matter); they also lack porosity due to the modest rooting by colonizing plants.

A further point that requires attention is the sedimentological characteristics (granulometry, texture, and organic content) of the sediments used in the infills. For example, the introduction of high quantities of sandy sediments from canals near the inlets in areas with a high silt-clay or peat bottom substrate –

which is the case for the more internal areas – means changing their identity considerably. Today, there is still no plan that allows the identification of the destinations most suited to the aims and characteristics of the sites for the various types of sediment.

These approaches can be seen in the way the work is usually organized with the patchy methodologies that are still dominant and in the decision to accelerate the disposal times of the mud; initially this can also lead to a higher and more detailed financial commitment. However, from a technical-naturalistic point of view, these are the terms of the matter – the need to apply this knowledge (at least in the most precious areas) and to move away from the creation of mechanical artificial islands that are undifferentiated and towards techniques that reintroduce natural processes of infilling by means of peat contributions as much as possible; this means exploiting the levelling and stratifying effects of the tide and using the jets in such a way that they come closer to imitating the effect of overtopping in nature that creates the characteristic raised margins of the salt marshes. From a financial point of view, the initially higher costs are not only justified by the quality of the results and their adherence to the aims of restoration, but can also be suitably compensated by the costs necessary for successive corrective interventions. These will be much lower, if necessary at all.

Until now, the reclaimed areas created have been stabilised with rigid protection at their margins. In many cases, however, the planning and creation of reclaimed areas destined to disappear rapidly may be more useful – creating them as relatively consolidated sediment deposits that will gradually be put into circulation and redistributed in the lagoon (the sediments may also simply be accumulated in such areas to intercept the currents or wave motion in demarcated areas with light protection to contain the turbidity produced by the filling in. In this manner, the artificial islands become 'disposable artificial islands', containers of sediments destined to successive redistribution events and therefore to remodellings shaped by the waves and currents.[10] This will

---

[8] (continued) is an accumulation of coarser particulate matter that results in a higher level. By simply paying attention to the positioning, orientation and consistency of the jets, it is possible to reproduce the constructive dynamics very effectively, thus coming closer in terms of margin sedimentological structures to their natural structure and their protective role.

[9] Some recent projects foresee creation of artificial areas via a final phase consisting of sediments with a higher organic content to allow a more rapid covering of vegetation, thus conforming more to natural structures as well as paying attention to the orientation of the jets (Magistrato alle Acque di Venezia, Consorzio Venezia Nuova 2002). *Recupero ambientale e morfologico. Environmental and morphological reatoration*. Ripristino morfologico dell'area Lago dei Teneri. Venice: general report and survey). These are signs of a transformation that should undoubtedly be welcomed, in the hope that the changes are actually put into practice.

---

[10] An analogous but simplified operation could consist of the simple dumping of suitable sediments for the creation of turbid waters carried with the currents, where hydraulic conditions and energies allow.

achieve the reformation of the existing salt marshes by means of constructive processes similar to natural ones with greater sedimentological and morphological stability. A further advantage of these techniques is that sediments with anomalous granulometry or sediments with origins outside the lagoon (from quarries etc.), can rapidly be taken up by the hydrodynamics, reworked and integrated in the organic particulate matter and returned to the bottoms in line with the true processes of the lagoon.

PROTECTION OF *BARENE* MARGINS

In morphology management, one fundamental aspect concerns the natural and artificial protection of the salt marsh margins that are subject to destructive forces via the actions of man, in both the past and the present. The solutions regarding consolidation vary according to a multitude of factors and variables: the integrity and functionality of the margins; the destructive processes that they are subjected to; the vulnerability they show; the location and type of salt marsh; and the potential to reactivate natural protective dynamics. Initial considerations are based on several certainties: the margins of the natural salt marshes are plastic borders that are subject to strengthening, destructive and reconstructive dynamics; they are permeable borders, functional ecotones responsible for exchanges with the aqueous environment; together with the submerged bodies they are fundamental factors of containment, orientation or deceleration of the aqueous flows. They are of primary importance in capturing and releasing sediments, in fixing floating detritus, in the dissipation of energy, and in their filtering and autopurification capacity of the lagoon system. The management and protection of the salt marsh margins must guarantee these dynamic processes as much as possible. It is these very processes that are interrupted or reduced by the rigidity of interventions. It is for these reasons that where possible, rigid protections should not be placed directly on the margins, but at a considerable distance, thus delimiting areas of 'calm waters' where natural processes can be reactivated with their consolidating and conservative effects. It would be even better if they consisted of temporary protections based on elements that can be easily installed and removed (e.g. resting diaphragms or floating barriers).

For more than a decade both the protection of salt marsh margins and the perimeters of the reclaimed areas were created by the planting of grouped piles, with the addition of geo-textile fencing, hydraulic nets or plastic film on the inside. This technique is generally extremely expensive, both financially and environmentally. The process of pile fixation requires cumbersome equipment and the associated vessels draught thus becomes a planning restraint or requires the opening of specific work site canals. The traditional pile fence creates reflected waves that cause depressions and start destructive processes in the perimeter mudflats. In addition, the piles and the associated materials hinder the salt marsh margin regenerative processes and the infiltration of water, thus opposing the normal dynamics and therefore also the restoration of protective and self-stabilising capabilities.

Today there is a clear tendency to follow different paths, favouring solutions suggested by environmental engineering that are closer to the natural structure (which still, however, require prudence and close attention since the best techniques can also lead to mistakes if they are not guided by an adequate knowledge of the values and functionality of the systems in question). These solutions consist in the use of various types of modular structure – *materasses*, *burghe* and *buzzoni*, the former having a flat rectangular profile while the other two are cylindrical, with the possibility of assuming lenticular profiles that can be modulated by filling with different materials and placement on the lagoon bottom or along the margins of the salt marsh.

These techniques offer precise applicative and environmental advantages. The forms of the elements used, whether individually or together, can recreate profiles that are more similar to those of natural salt marsh margins, remodelling the eroded borders and absorbing at least a part of the energy of the breaking waves. By using materials that are homogeneous to those of the intervention area, it is possible to replace and reactivate the spontaneous reformation processes of the marginal structures; during realization, further sediment can spontaneously accumulate in these elements, with a rapid development of a halophytic vegetation cover; the positioning and successive interventions of rearrangement do not require bulky barges and invasive worksite procedures; they are possible on any type of bed with extremely flexible modes of intervention, and facilitate both temporary and removable interventions. Finally, they present no serious problems of interference with the archaeological sites that are so widespread in the lagoon.

The structural resistance and naturalness of such elements depends on the types of materials used for the container and its filling – for example, the containers of the *burghe* can be made of either organic fibre (e.g. coconut) or plastic material (e.g. polyester); the filler can be earth or sediments that are analogous to those of the site in question and possibly integrated with other naturally degradable filling materials, whether sand, gravel or rubble. It is clear how the *burghe* in degradable containers and filling materials can be integrated with the evolutionary processes of the flexible systems while those in rubble and synthetics oppose the dynamic characteristics and are therefore only admissible in very limited situations that do not allow true restoration of the natural functions of the margins, e.g. where boat traffic is intense.

REFERENCES

Adam, P. 1990. *Saltmarsh Ecology*. Cambridge : Cambridge University Press.

Albani, A., Favero, V., and R. Serandrei Barbero. 1983. Apparati intertidali della laguna di venezia. Convegno: *Laguna, fiumi, lidi; cinque secoli di gestione delle acque nelle Venezia*. Venice: Magistrato alle Acque di Venezia.

Albani, A., Favero, V. and R. Serandrei Barbero. 1984. Appaati intertidali ai margini dei canali lagunari, studio morfologico, micropaleontologico e sedimentologico. *Ist. Ven. SS.LL., Rapporti e studi* 9, 137–62, Venezia.

Allen, J. R. L. 2000. Morphodynamics of Holocene salt marshes. a review sketch from the Atlantic and Southern North Sea coasts of Europe. *Quaternary Reviews* 19, 1155–231.

Caniglia, G., Contin, G., Fusco, M., Anoè, N., and A. Zanaboni. 1997. Confronto su base vegetazionale tra due *barene* della Laguna di Venezia. *Fitosociologia* 34, 118–19.

Cavazzoni, S. 1995. La Laguna: origine ed evoluzione. In G. Caniato, E. Turri, and M. Zanetti (eds.), *La laguna di Venezia*, Verona: UNESCO, Cierre Ed., pp. 41–67.

Cecconi, G. 1995. Un programma integrato per proteggere le barene. *Quaderni trimestrali del Consorzio Venezia Nuova*, (Venice) 5, 95.

Cecconi, G., Codato, F., Nascimbeni, P., and F. Mattarolo. 1998. Valore ambientale delle barene artificiali. *Quaderni trimestrali del Consorzio Venezia Nuova* (Venice) 1, 98.

D'Alpaos, L. 1992. Evoluzione morfologica della laguna di Venezia dal tempo del Denaix ad oggi, e sue conseguenze sul regime idrodinamico. In *Conterminazione Lagunare*, I.V.S.L.A., Atti conv. bicent. corterminaz., 1991, Padua: La Garangola, pp. 327–58.

Favero, V., and R. Serandrei Barbero. 1980. Origine ed evoluzione della laguna di Venezia – Bacino meridionale. *Lavori della Società Veneziana di Scienze Naturali* 5, 49–71.

Favero, V., and R. Serandrei Barbero. 1983. Oscillazioni del livello del mare ed evoluzione paleoambientale della Laguna di Venezia tra Torcello e il margine lagunare. *Lavori della Società Veneziana di Scienze Naturali* 8, 83–102.

Marcello, A., and S. Pignatti. 1963. Fenoantesi caratteristica sulle barene nella Laguna di Venezia. *Memorie di Biogeografia Adriatica* 5, Fantoni, Venezia.

Pethick, J. 1984. *An Introduction to Coastal Geomorphology*. London: Edward Arnold.

Pignatti, S. 1966. La vegetazione alofila della Laguna Veneta. *Istituto Veneto di Scienze, Lettere ed Arti* (Venice), Memorie 33, 1.

Ranwell, D. S. 1972. *Ecology of Salt Marshes and Sand Dunes*. London: Chapman and Hall.

Visintini Romanin, M., Rismondo, A., Scarton, F., and L. Leita. 2000. Interventi per il recupero morfologico della laguna di Venezia. La barena Fossei Est in Laguna Sud. *Quaderni trimestrali del Consorzio Venezia Nuova* (Venice) 3/4, 00, 1–33.

Part VI · The Venice Lagoon: ecological processes and environmental quality

53 • Introduction: ecological processes and environmental quality

C. A. FLETCHER, J. DA MOSTO AND T. SPENCER

The Lagoon of Venice is a complex ecosystem where natural and anthropogenic factors have interacted for many centuries. The shallow system experiences a diversity of physical, chemical and biological interactions across a range of space and time scales. For example, salinity varies across the lagoon, and over time: values recorded range from greater than 30 PSU at the inlets to a few PSU nearer the mainland at the freshwater inputs. The system currently shows signs of severe environmental degradation. In particular, agricultural, industrial and urban wastes have caused major water quality problems in the Venice Lagoon. Remedial action requires that the sources of the problem are identified clearly, but identification and quantification are difficult. The scale of the problem is substantial. The drainage basin measures approximately 1877 km^2 and is home to some 1 million inhabitants, to the industrial district of Porto Marghera – one of the largest industrial districts in the Mediterranean – and to two large wastewater treatment plants that operate for 350 000 inhabitants. Venice itself (with its 66 000 inhabitants, 10 000 overnight stay tourists and over 20 000 daily tourists and a net daily flux of 46 000 commuters) adds to the problem, particularly given that it has never provided for a sewage collection and treatment system. Untreated effluents from a large number of inputs are discharged directly into the canals and, ultimately, into the lagoon. Contaminants that were historically stored in sediments within the canals and the lagoon may now be released, adding further to the water quality problem. Finally, a number of toxic waste disposal sites were created in the 1970s within the lagoon to store contaminated wastes and these continue to need to be kept isolated from the system.

The ensuing environmental problems for the lagoon are multi-faceted. Nevertheless, important individual examples over the last 30 years point to the following three key issues. The first problem concerns high levels of nutrients in the Venice Lagoon. In the 1970s and 1980s eutrophication was the main water quality concern and inputs of nutrients caused algal blooms (most notably of the macroalgae *Ulva rigida*) and anoxic conditions in parts of the lagoon. The eutrophication problem started to decline by the 1990s but the exact reasons are still not clearly understood, due to a number of concurrent processes. Some researchers propose that the reduction in phosphate levels, due to a detergent ban in the late 1980s, and lower ammonium levels in the sediment have played a role (Marcomini *et al.*, chapter 54). Indeed, a comparison of dissolved nutrient and plankton distributions in the lagoon over time by Acri *et al.* (chapter 59) indicates that the ecosystem is changing in response to changes in the water quality caused by human activity. Others argue instead that the increase in dissolved inorganic nitrogen levels in the lagoon waters, especially in spring, may be due to considerable resuspension, greatly enhanced at the present time because of increasing erosion and more intensive fishing activities (i.e. mechanized dredging for clams (*Tapes philipinarum*)) which cause nutrient remobilization from sediments (Acri *et al.*, chapter 59). At present, high nitrate levels persist and Zirino (chapter 55) warns that eutrophic conditions could be seen again. Some scientists report that the seasonal fluctuation in nutrients means that in summer the lagoon waters are 'cleaner' than the sea and rivers. Also different parts of the lagoon are 'healthier' than others and display different patterns of species distribution over time. Thus, for example, areas of seagrass beds are increasing in the central lagoon (Rismondo *et al.*, chapter 60).

Flooding and Environmental Challenges for Venice and its Lagoon: State of Knowledge, ed. C. A. Fletcher and T. Spencer.
Published by Cambridge University Press, © Cambridge University Press 2005.

A second key problem is chemical pollution by persistent toxic substances (e.g. chlorinated organics such as dioxins, polychlorinated biphenols and the metals arsenic, cadmium and mercury). It is well known that chlorinated organics, micropollutants and metals from historic discharges may be stored in sediments as a 'toxic time bomb' with the potential to bioaccumulate and/or biomagnify in the food chain. These micropollutants can then cause adverse effects in benthos filter feeders and in the food web. Diffuse pollutants such as herbicides have also received increasing attention, in particular those herbicides which have replaced banned compounds (e.g. atrazine has been banned and replaced by ter-butylazine (Zonta *et al.*, chapter 62)).

The third, and emerging, problem is that of pollution by 'endocrine disruptor' substances which interfere with an organism's hormonal system even at extremely low concentrations (ppb and even ppt levels). Dabala *et al.* (chapter 65) present studies measuring selected endocrine disruptor substances and genetic damage and compare their findings with known values in the literature. Their conclusion is that Venice Lagoon waters show a significant capability to affect the endocrine system of aquatic organisms.

Developing a detailed understanding of the key issues is a pre-condition to proposing solutions. The complexity of the task should not be under-estimated; relating observations to natural and anthropic perturbations of such a system is fraught, as highlighted by Pérez Ruzafa and Marcos (chapter 58) in the context of the Mar Menor, a hypersaline coastal lagoon located on the southwestern Mediterranean coastline. Efforts in this respect are nevertheless underway for the Venice Lagoon. Monitoring programmes have been set up involving a number of institutions and a large amount of data has been acquired, including the UNESCO Venice Lagoon Ecosystem Project (Laserre and Marzollo, 2000), MAP, DRAIN, Project 2023 and MELa 1 and 2. DRAIN was set up to determine the freshwater and pollution discharges to the lagoon from the drainage basin (1998 to 2000). MELa, started in 2000, is the first major lagoon-wide monitoring programme, and demonstrates how institutions in Venice have worked together in a coordinated monitoring pro-

gramme (Penna *et al.*, chapter 56). The MELa programmes were also the first to approach systematically the topic of persistent toxic substances in the lagoon.

The monitoring programmes mentioned above, as well as other research projects, have yielded important information. They give new results, including the quantification of loadings from the rivers within the drainage basin (Zonta *et al.*, chapter 62), the lagoon sediments and the atmosphere (Capadaglio *et al.*, chapter 63). Here we draw attention to five areas of interest. First, the importance of sediments within the shallow lagoon to the quality of lagoon water has been highlighted by Zirino (chapter 55), who shows that bottom sediments dominate the recycling processes within the shallow lagoon; by Capadaglio *et al.* (chapter 63), who discuss the complex processes governing the transport and release of metals from sediments; and by Dabala *et al.*, (chapter 65), who investigate the scavenging role of humic substances in natural recovery processes.

Secondly, it can be demonstrated that tidal gradients are to be found in the 40 km long canal network of the city of Venice where large amounts of organic and inorganic matter accumulate from untreated waste waters and urban run off. However, these gradients can also be small, resulting in low velocities, and insufficient scour, in some of the canals. This has been shown to cause an accumulation of sediments that are strongly reduced and which store contaminants. Historically the canals of Venice were dredged for sanitary purposes but this was halted from the 1960s to 1990s, giving a 30-year sediment record and associated pollution profile. It is encouraging that a decrease in certain contaminants has been observed, with upper sediment layers being less contaminated than lower ones (Zonta *et al.*, chapter 64). Resuspension of contaminants from these sediments to the water column is also being researched. Preliminary results are discussed by Dabala *et al.* (chapter 65).

Thirdly, the distribution of the seagrasses *Zostera marina*, *Zostera noltii* and *Cymodocea nodosa* (which cover *c.* 4500 ha of the lagoon) was mapped in detail for the first time in 1990 and results have been compared to a more recent 2002 survey to determine changes in distribution over a decade (Rismondo *et al.*, chapter 60). The comparison shows an overall retreat in *Z. noltii* most strongly emphasized in the southern basin. However the central basin shows increases, mostly of *Z. marina* but also for *Z. noltii*. The different trends in coverage of these species highlight the importance of understanding the distinct pressures and processes occurring in the northern, central and southern parts of the lagoon (and see Bonometto, chapter 52 for saltmarshes). Zirino (chapter 55) discusses a greater flushing and better water quality for the central lagoon as a possible reason for the observed increase in seagrass population in that area.

Fourthly, the Venice Lagoon is of particular importance for bird species. It hosts the largest concentration of waterfowl in Italy and 80–90 per cent of the Italian populations of Redshank (*Tringa tetanus*) and Sandwich Tern (*Sterna sandvicensis*) are found in the lagoon area. Vegetation and wetland bird populations at recreated marsh sites (composed of dredged materials) in the Venice Lagoon have been made by Scarton (chapter 61). Dredge islands can play an important role for the feeding of wetland birds, although physical diversity, in the form of a network of tidal ponds and creeks, is needed at each site if a higher number of breeding species are to be encouraged. Older sites display lower diversity in terms of number of breeding species, demonstrating that further management such as vegetation cutting may be needed. This is one example of a more general conclusion, namely that remedies and conservation activities need to be monitored and managed over an extended period to achieve their goals.

Finally, controlling point sources is required because sewage is still discharged directly into the lagoon from some parts of the city of Venice. Tackling diffuse sources is also a challenge. Controlling the inputs from the most densely populated part of the lagoon basin is discussed by Casarin *et al.* (chapter 66). The Fusina Integrated Project includes new plans for waste water treatment within a strategic approach to water supply and strict final discharge limits.

The information obtained by the various monitoring programmes has greatly improved present understanding of the environmental issues facing Venice and forms the basis for further work, ultimately to assist decision making. Modelling is a pow-

erful tool in this context. Solidoro *et al.* (chapter 57) review the water quality and ecological modelling for Venice Lagoon to date. A screening of these papers indicates that most of them deal with nutrients and eutrophication but importantly the review examines the modelling of the mobile barrier closures. Opponents to the mobile barriers have raised concerns that, if closed too frequently, the barrier could have a severe impact on the water quality of the lagoon. However, the preliminary results of at least one modelling effort suggest that changes in the load of inputs to the lagoon have a far greater influence on adverse water quality conditions than short term hydrodynamic flow reductions. Reliable water quality models will be needed to evaluate the impact of the mobile barriers and the feasibility of the proposed use of the gates for water quality management purposes (Harleman, chapter 34). Thus water quality models with sediment interaction processes included need to be developed.

One major challenge for the future is how to manage the large amount of information already collected, and to be collected, by the monitoring programmes, and how to use it effectively in order to make it usable to stakeholders and decision makers. This is a problem, by no means unique to the issues facing Venice and the lagoon, of general application: how does one move from science, data collection and analysis to workable, manageable solutions for good ecological quality on a sustainable basis? One answer, within the framework of the new EU Water Framework Directive structure, is the application of the European Environment Agency's 'Driving forces, Pressures, States, Impacts, Responses' (DPSIR) model, coupled with risk assessment and environmental indicators, as discussed by Marcomini *et al.* (chapter 54).

REFERENCE

Lasserre, P., and Marzollo, A. (eds). 2000. *The Venice Lagoon Ecosystem: Inputs and Interactions between Land and Sea.* Man and the Biosphere Series 25. Paris/New York: UNESCO/Parthenon Publishing Group.

54 • Environmental quality issues in the perspective of risk assessment and management in the Venice Lagoon

A. MARCOMINI, A. CRITTO, C. MICHELETTI AND A. SFRISO

ENVIRONMENTAL QUALITY AND REGULATORY FRAMEWORKS

In the framework of the European Water Framework Directive (European Union, 2000), the definition of environmental quality of waters (i.e. surface water, groundwater) refers to the concept of ecological, physical, chemical and biological integrity of environmental resources. The measurement of ecological quality requires a reference area, which is a non-altered habitat of high ecological status. Accordingly, the environmental quality should be assessed by hydro-geo-morphological (tides, currents, sediment fluxes, etc.), physico-chemical (e.g. oxygenation, light, salinity, temperature, nutrient concentrations), and biological integrity (e.g. community health, species richness, individual abundance and dominance) parameters, as well as by the concentration of priority pollutants in biotic and abiotic environmental media (i.e. water, sediment, particulate matter and organism tissues).

Until recently, the estimation of water quality focussed mainly on nutrients and chemicals, assessing the environmental quality by using water quality criteria, and water quality standards. According to the US Clean Water Act (Public Law 92-500), a quality criterion is defined as a 'numeric or narrative expression that specifies concentrations of water (or sediment, [particulate matter, organic tissues]) constituents (such as toxic chemicals or heavy metals) which, if not exceed, are expected to support an ecosystem suitable for protecting life'. Environmental quality criteria are generally derived by applying different approaches: (1) Effects or weight-of-evidence database from laboratory or field exposures to contaminated media, for example, screening level concentrations for sediment (effect range approach,

Long and Morgan, 1990); threshold effect level (MacDonald *et al.*, 1996); (2) Equilibrium partitioning approach (e.g. derivation of sediment quality criteria from water quality criteria, Jones *et al.*, 1996); (3) Background levels, or some multiple of background levels, in the affected region (Oslo and Paris Commissions, 1993); (4) Bioaccumulation and biomagnification-based guidelines, based on bioconcentration factors (Van der Kooij *et al.*, 1991); (5) Biotic indices based on the presence-absence of target taxa (Occhipinti and Ferni, 2003).

A more suitable and integrated approach to evaluate environmental quality and to obtain environmental quality criteria is the risk-based approach, which takes into account the site-specificity of ecosystem, different exposure and effect measurements, and provides monitoring and management decision options by the involvement of stakeholders (ANZECC and ARMCANZ, 2000). The risk-based approach, using the weight of evidence method, is consistent with the most recent concept of environmental quality, such as adopted by the European Water Framework Directive (EU, 2000). In this chapter a summary of issues concerning the environmental quality of the Lagoon of Venice is presented.

HISTORICAL DEVELOPMENT

The Lagoon of Venice is a complex ecosystem where natural and anthropogenic factors have interacted for many centuries. Over the last 30 years, several environmental issues have emerged as a result of the increasing pressure of human activities (e.g. industry, urbanization, agriculture, fishing activities).

During the 1970s and 1980s, major attention was drawn by the evident eutrophication of the lagoon causing spectacular macroalgae (*Ulva rigida* C. Ag.)

Flooding and Environmental Challenges for Venice and its Lagoon: State of Knowledge, ed. C. A. Fletcher and T. Spencer. Published by Cambridge University Press, © Cambridge University Press 2005.

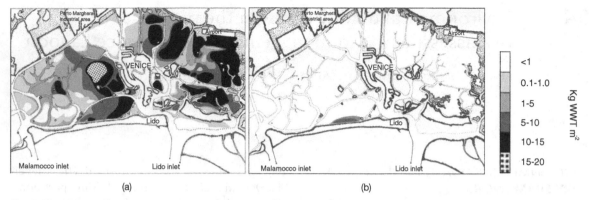

Fig. 54.1   Temporal trend of macroalgal biomass (kg m⁻², wet weight) in the central lagoon ((a) June 1987; and (b) June 1998, respectively).

blooms (Fig. 54.1) followed by frequent anoxic crises. Changes of climatic conditions, increased turbidity caused by fine sediment resuspension, increased grazing pressure, seagrass harvesting and the reduction of nutrient loads, especially phosphorus, led to the almost complete disappearance of macroalgal biomass (Sfriso and Marcomini, 1996; Sfriso *et al.*, 2003a).

In the 1990s, after the reduction of macroalgal biomass, attention was paid especially to nutrient availability, to the contamination by persistent and toxic pollutants (e.g. chlorinated compounds and pesticides, metals and metalloids) associated with sediment and water redistribution (e.g. erosion and sedimentation processes, changes in sediment and grain-size composition) possibly affecting economic

(e.g. fishing) and recreational (e.g. bathing) activities, as well as the biodiversity of the lagoon.

Figure 54.2 shows the spatial distribution of organic phosphorus (OP) in 1987 during the period of macroalgal dominance and in 1998 during intense clam fishing activities. Organic phosphorus decreased significantly, and was found to be homogeneously distributed over the whole central lagoon in 1998 (Sfriso *et al.*, 2003b).

Similar results were obtained for total nitrogen (Sfriso *et al.*, in press a) and many toxic and persistent pollutants such as DDTs and PCBs (Secco *et al.*, 2005). Over the last decade, concentrations in the sediment have been seen to reduce for all these chemicals and presently they appear to be more homogeneously spread in the central lagoon than in the past.

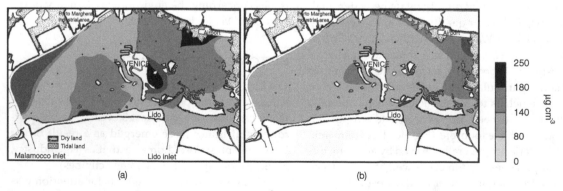

Fig. 54.2   Temporal trend of organic phosphorus (g cm⁻³, dry weight) in the 5 cm sediment top layer in the central lagoon ((a) June 1987; and (b) June 1998, respectively).

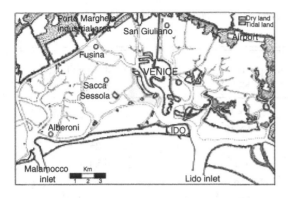
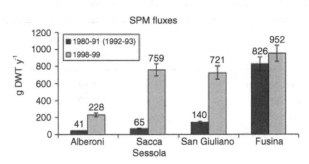

Fig. 54.3   Settled Particulate Matter (SPM) fluxes in four sampling stations in the central lagoon (from outermost, Alberoni, to innermost, S. Giuliano and Fusina, stations). Bars show the sampling errors.

These marked changes were related mainly to the reduction of macroalgae biomass in the lagoon and the bottom colonization by the clam *Tapes philipinarum*, a bivalve introduced in the lagoon for aquaculture purposes (Orel *et al.*, 2000). The catching of this species by hydraulic and mechanical dredges caused the re-suspension of high amounts of fine sediments. Figure 54.3 highlights the temporal trend of the settled particulate matter (SPM) fluxes in four sampling stations in the central lagoon during the period of macroalgae bloom and in the season of clam fishing.

SPM fluxes changed significantly especially in areas such as Sacca Sessola and Alberoni, which are more affected by clam-fishing. Clam harvesting affected especially the stations of Alberoni, Sacca Sessola and San Giuliano, whereas Fusina showed high SPM fluxes also in 1990-93 because of the transit of cargo boats in the Malamocco–Marghera canal which is very close to that area (Sfriso *et al.*, in press a; Sfriso *et al.*, in press b).

Large projects (MAP, DRAIN and 2023) have been undertaken in the Lagoon of Venice over the last five years with the goal of linking hydrodynamics with chemo-dynamics, comparing non-point (agricultural runoff) pollution versus point (discharges of municipal and industrial effluents) pollution sources, identifying pollution time trends and supporting the design of monitoring programmes. The MAP project provided the spatial distribution of organic and inorganic pollutant concentrations in sediment and in

aquatic organisms (MAV-CVN, 1999), while the DRAIN project concerned the transport and loads of nutrients and organic and inorganic chemicals from the catchment area (i.e. riverine transport) to the lagoon (MAV-CVN, 2001). Finally, the 2023 project estimated the atmospheric deposition of organic and inorganic pollutants, as well as the dynamics and contamination of lagoon sediments, and the temporal trend of nutrients in the central lagoon (MAV-CVN, 2000). Table 54.1 presents some results from the 2023 and DRAIN projects, i.e. the annual riverine loads, the Porto Marghera loads and atmospheric depositions of polychlorodibenzodioxins/furans (PCDD/Fs), polichlorobyphenils (PCBs), policyclic aromatic hydrocarbons (PAHs) and hexachlorobenzene (HCB).

These projects were conceived to support the regulatory implementation of water quality objectives (i.e. imperative and guide values for both river basin and lagoon water, Ronchi-Costa decree 23/04/1998), some of which are presented in Table 54.2. The quality objectives were defined in terms of chemical concentration of nutrients and pollutants dissolved in water. An additional decree has linked the individual concentrations to the flow of rivers and discharges into the lagoon by fixing maximum allowable annual loads (30/07/99 decree).

Based on the monitoring work carried out so far by a number of different institutions (e.g. University Ca' Foscari of Venice, CORILA, MAV-CVN) over the last five years, regulatory concentration limits were

Table 54.1   *Overview of major point and non-point (riverine, Porto Marghera and atmospheric loads) annual loads of nutrients and Persistent Organic Pollutants (POPs) in the Lagoon of Venice.*

	Riverine loads (kg y^{-1})	Atmospheric deposition (kg y^{-1})	Porto Maghera loads (kg y^{-1})
Total Nitrogen	3.3×10^{6} [a]	1.1×10^{6} [a]	
Total Phosphorus	2.0×10^{5} [a]	4.4×10^{4} [a]	
PAHs	< 100 [a]	100 [a]	25 [c]
Coplanar PCBs	0.4 [b]	0.4 [c]	$0.5\text{-}1.0$ [c]
HCB	0.9 [b]	0.8 [c]	0.3 [c]
PCDD/Fs (TEQ)	5.8×10^{-5} [b]	2.7×10^{-4} [c]	$0.5\text{-}1.0 \times 10^{-4}$ [c]

[a] MAV-CVN, 2000
[b] Collavini *et al*., in press
[c] Guerzoni and Raccanelli, 2003

Table 54.2   *Regulatory water quality criteria (guide values and imperative values) of selected pollutants in river basin and lagoon water (Ronchi-Costa decree, 23/04/1998).*

CHEMICALS (dissolved concentration)	River basin Guide value	Lagoon Imperative value µg/l	Guide value
NUTRIENTS			
Total nitrogen	400	350	200
Total phosphorus	30	25	10
METALS			
Arsenic	0.9	1.6	1.2
Cadmium	0.01	0.03	0.01
Chromium	0.2	0.7	0.2
Lead	0.03	0.15	0.03
Mercury	0.005	0.003	0.001
ORGANIC			
PCBs	0	0.00004	0
HCB	0	0.0008	0
Dioxins (TEQ)	0	1.30E-08	0
Tributilin	0.01	0.1	0.01

systematically found to exceed parameters, such as total Nitrogen and total Phosphorus, which showed concentrations two to three times higher than the imperative values reported in Table 54.2. Among inorganic chemicals, cadmium and lead have exceeded the imperative values by two to three orders of magnitude, while among organic pollutants, PCDD/Fs and PCBs exceeded the imperative values by about three orders of magnitude (MAV-CVN, 2002).

INTEGRATION OF ECOLOGICAL RISK ASSESSMENT AND MANAGEMENT

On the basis of the most recently acquired monitoring results (i.e. those on chemical contamination, primary and secondary producer distribution, species abundance etc.), an integrated analysis of environmental quality is envisaged in order to develop tools and approaches for the sustainable management of the Venice Lagoon. A sustainable environmental management process is expected to rely on an effective environmental conceptual model integrating several stress/impact factors, allowing a variety of information and data (e.g. physical, chemical, biological, ecological data) to be assembled together in a suitable database. This process requires also the interaction of different stakeholders and interested partners, in order to include all management interests. Several authors (Menzie, 2002; Tarazona, 2002) pointed out that a methodological approach capable of responding to those requirements is the environmental risk assessment coupled with environmental risk management. Risk assessment and management have been extensively applied for designing regulations (e.g. the EU regulation concerning the introduction of new hazardous substances), for providing the basis for site-specific decisions (e.g. brownfields remediation), and for ranking and comparing risks.

According to Menzie (2002), the development of a risk-based sustainable management support system requires dealing with five main challenges:

1) The integration of environmental risk assessment and environmental risk management, aiming to inform and to solve problems, dealing with stakeholder's interests and taking into account a provi-

sional diagnosis of the decision scenario. An efficient integration of environmental risk assessment and management will be stimulated and supported by the development of risk-based decision support systems (DSSs);

2) A better definition of the problem formulation phase, which is the preliminary and fundamental step of the risk assessment (US-EPA, 1998), to allow the clear identification of assessment endpoints (e.g. the ecological resources to be protected) and the conceptual model, to obtain a decision-driven procedure;

3) To include spatial, temporal and effects scales in the risk assessment framework, incorporating GIS tools, dedicated statistics procedures (i.e. spatial statistics), and environmental process modelling (e.g., pollutants transport and transformation, populations recovery);

4) The need of managing uncertainty, in a tiered risk assessment framework, from a screening-level assessment to an in-depth analysis, including probabilistic risk assessment; and

5) To improve the communication of results and decisions, especially for stakeholders, which should be included in the risk assessment procedure according to their potential to affect the decision and to contain important information for assessments and decisions, as well as their level of interest.

ENVIRONMENTAL RISK ASSESSMENT APPLIED TO THE VENICE LAGOON

According to the development of a sustainable environmental quality management and taking into account the conceptual basis of the Water Framework Directive, several research activities have been undertaken for the Lagoon of Venice. In particular, the ecological risk assessment (ERA) procedure was applied to the Lagoon of Venice to identify the stressors of concern, the lagoon areas and ecological resources affected by human activities and at high risk, with the aim of addressing the monitoring activities and remediation interventions required.

According to the EPA (1998), the ERA includes three phases: (1) problem formulation phase, where

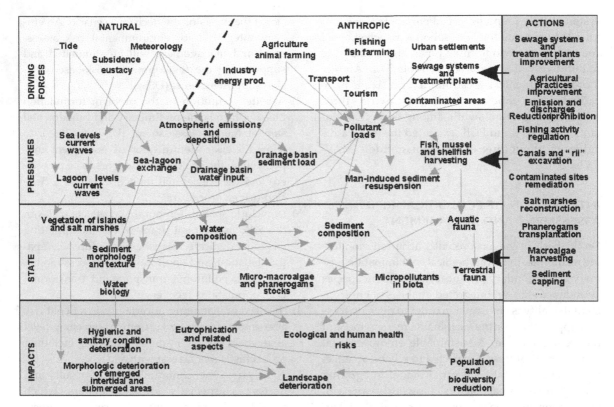

Fig. 54.4   DPSIR scheme for the Lagoon of Venice (Thetis, 2000).

the available information is integrated to develop a conceptual model, summarizing the risk hypotheses as well as the assessment endpoints (ecological resources to protect); (2) analysis phase, where exposure and effects for the assessment endpoints are characterized; and (3) risk characterization, where exposure and effects profiles are compared to estimate the ecological risk. In the ERA problem formulation phase, a conceptual model of the Lagoon of Venice was developed according to the Driving Forces-Pressures-State-Impacts-Response (DPSIR) scheme (Project 2023; Thetis, 2000) (Fig. 54.4), and according to US-EPA ecological risk assessment guidelines (Critto and Marcomini, 2001) (Fig. 54.5), respectively.

Figures 54.4 and 54.5 show the main sources (anthropogenic activities, i.e. driving forces) of stressors (i.e. pressures) affecting the lagoon environment, the stressor pathways, as well as the probable

adverse effects (i.e. impacts) on the ecological resources (targets). The conceptual model allowed the selection of three risk hypotheses associated with the accumulation of persistent and toxic pollutants in the lagoon sediments, concerning ecological entities such as the benthic community, the edible clam *Tapes philipinarum*, and the lagoon's food web. The benthic community may be affected by contaminants stored in sediments because of direct contact, while bioaccumulation and biomagnification can cause adverse effects in benthos filter feeders (i.e. clam *Tapes philipinarum*) and in the aquatic food web, resulting in ecological impacts and possible commercial limitations.

In order to characterize the potential risk to the benthic community a screening ERA, based on the quotient method (Jones *et al.*, 1999), was developed. The exposure to different classes of pollutants (metals, chlorinated organic compounds and polynuclear aromatic hydrocarbons) was investigated in more

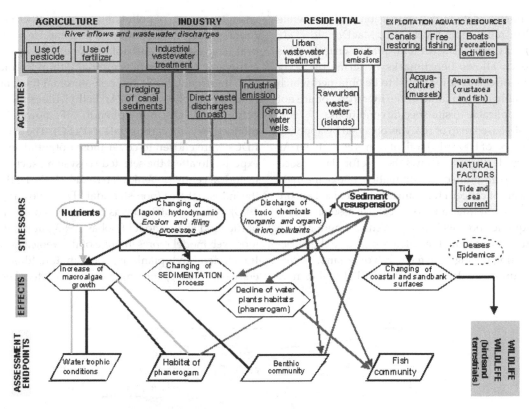

Figure 54.5   Conceptual model according to US-EPA ERA guidelines (Critto and Marcomini, 2001).

than 100 sediment stations, distributed across the whole lagoon. Geostatistical techniques and GIS (Geographic Information System) tools were used to draw the exposure maps and to estimate the pollutant loading in the top 15 cm of the lagoon sediment. Finally, a map of the spatial distribution of the estimated risk for the benthic community was drawn. The relative contribution of each pollutant to the cumulative risk at several sites was represented by pie charts in a GIS framework (Plate 21) (Critto and Marcomini, 2001; Critto *et al.*, in press).

Subsequently, the ecological risk associated with bioaccumulation of both total and dioxin-like polychlorobiphenyls (PCBs) and several inorganic pollutants (As, Cd, Cr, Cu, Hg, Ni, Pb, Zn) in the clam *Tapes philipinarum* was estimated (Micheletti *et al.*, 2004). Exposure was characterized by means of bioaccumulation regression models, applying spatial regression analysis (Cressie, 1993), a statistical

technique addressing the spatial issues associated with georeferenced data. Linear regression models were defined using explanatory variables concerning both sediment (e.g. pollutant concentration in sediment, total organic carbon) and organism (e.g. lipid fraction). Spatial regression and related statistical routines were calculated using Splus© spatial statistics software (MathSoft Inc., 1996). Bioaccumulation estimated by regression models was compared with bioaccumulation estimated by a partitioning-based model, the food chain (FC) model (Gobas, 1993). FC estimates the bioaccumulation of hydrophobic pollutants in benthic organisms by the pollutant partitioning between organisms tissues lipid fraction and sediment organic carbon fraction. Effects concentrations were estimated calculating the tissue screening concentrations, TSCs (Shephard, 1998), multiplying the bioaccumulation factors (BAFs) by the sediment quality criteria (SQC). The BAFs were estimated from experimental data with a regression analysis, while the threshold effect level (TEL, MacDonald *et al.*, 1996) was used as sediment quality criteria. This approach allows benchmarks in terms of 'body residue-effects relationships' to be obtained which are usually not available in the literature, or are not easily applicable to site-specific conditions.

A GIS-based map of risk was obtained by applying the method of Hazard Quotient (Jones *et al.*, 1999). As an example, Fig. 54.6 shows the risk for clam associated with cadmium bioaccumulation, one of the main chemicals of concern for clams in the Venice Lagoon. At the present time, an ecological risk assessment for the aquatic food web of the Venice Lagoon is in progress, to estimate the risk associated to the bioaccumulation through the food web of organochlorine compounds (i.e. PCBs and PCDD/Fs) and of new

emerging classes of pollutants, i.e. the endocrine disrupting compounds (EDCs). EDCs comprise both natural (e.g. estradiol, estriol, estrone) and synthetic estrogenic compounds, steroidal (ethinylestradiol, EE2; mestranol, MES) and non-steroidal (benzophenone, BP; bisphenol-A, BPA; diethylstilbestrol, DES; octylphenol, OP; nonylphenol, NP; nonylphenol monoethoxylate carboxylate, NP1EC). The selected EDCs were chosen to cover most potential sources expected to affect the selected ecosystem, such as the historical centre of Venice (steroidal EDCs, DES) and the industrial area (non-steroidal EDCs) (Pojana *et al.*, 2003). Bioaccumulation was characterized by means of the food chain model (Gobas, 1993), applied to a food web model representative of the Venice Lagoon. It included phytoplankton, zooplankton, filter feeders, detritivorous, and omnivorous-predators benthic

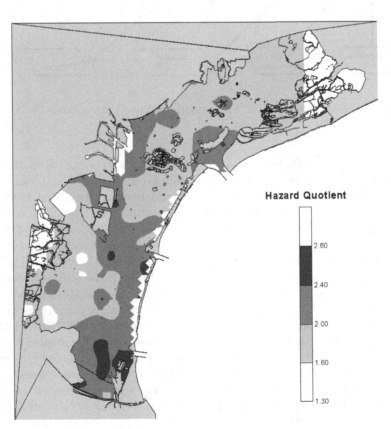

Fig. 54.6   Spatial distribution map of risk (Hazard Quotient) for the clam *Tapes philipinarum*, associated with cadmium bioaccumulation.

organisms, and various fish species (Libralato et al., 2002).

According to the critical body burden approach (McArty and Mackay, 1993), ecological effects will be characterized by tissue screening concentrations (TSCs, Shephard, 1998), and by the application of uncertainty factors (Calabrese and Baldwin, 1993) to the literature effects data.

Other activities being undertaken are aimed at standardising and validating procedures concerned with bio-assessment (Burbor et al., 1999), such as bio-markers and bioassays, for use in exposure and effects characterization. Site-specific and detailed toxicity data for the Lagoon of Venice have been obtained, especially those concerning spermyotoxicity and embryotoxicity bioassay on sea urchin (Volpi et al., 1999), and selected biomarkers (e.g. neutral red test) for Mytilus sp. (stress indices) (Nasci et al., 1998).

Such activities are useful in order to set up a site-specific risk assessment process based on the weight-of-evidence approach (Massachusetts weight-of-evidence workgroup, 1995). Finally, spatial analysis tools (e.g. GIS, geostatistics) will allow all these activities to be integrated and will provide an integrated multi-dimensional database and a spatially explicit exposure characterization.

DEFINITION OF A SUSTAINABLE MANAGEMENT SUPPORT SYSTEM FOR THE LAGOON OF VENICE

In the near future, efforts will be oriented to obtain a sustainable management support system (SMSS) for the Lagoon of Venice able to: (1) integrate environmental risk assessment and management; (2) improve stakeholders participation in the risk assessment process; (3) reduce the uncertainty associated with management decisions; and (4) increase the communication between the stakeholders concerned with lagoon management and the risk assessors.

Risk management involves decision makers and technological, economical, social, and political issues. A major scientific and technical task is to provide all relevant information in a manageable form for decision-making. A SMSS – sustainable management support system – is not designed to provide solutions, i.e. it does not replace decision makers, but it

is aimed at identifying realistic choices and integrating information into a coherent framework suitable for analysis and selection of alternatives. This effort requires close collaboration between different experts, such as environmental chemists and biologists, GIS specialists, risk assessors, multi-criteria analysts and economists (Beinat and Nijkamp, 1998).

The development of a SMSS can be supported by the integration of TRIAD-based site-specific risk assessment with weight-of-evidence approaches. TRIAD is a procedure that integrates exposure and effects data (i.e. lines of evidence), grouped into three disciplines: chemistry (e.g. bioavailability), toxicology (e.g. biomarker, bioassay) and ecology (e.g. trophic index) (Vollenweider et al., 1998), integrity biotic index (Karr and Chu, 1997)). The integration of results from the TRIAD activities, selected according to appropriate criteria (e.g. sensitivities, economic cost, etc.), in a weight of evidence framework, permits the user to evaluate, integrate and weigh exposure and effects lines of evidence, thus providing risk estimation. The main components of the weight-of-evidence approach are: (1) weight assigned to each measurement endpoint; (2) magnitude of response in the measurement endpoint; (3) concurrence among measurement endpoints. The weighting scores, evidence of harm, and magnitudes of response are integrated for each measurement endpoint in the 'weight of evidence matrix'. It is possible to integrate multi-criteria analysis tools with this information in specific risk indices (Massachusetts weight-of-evidence workgroup, 1995). The sustainable management support system (SMSS) requires the definition of trophic state and ecological integrity indicators/indices. This current activity in Venice aims to develop the ecological quality indices from biological (e.g. structure of the biotic communities), ecological (e.g. relations between biological and environmental variables), ecotoxicological (e.g. toxicity bioassays, biomarkers) and chemical (e.g. indicators of exposure and toxic potentiality) indicators. After a habitat classification, based on morphological, hydrodynamic, biological and chemical–physical characteristics of lagoon areas, a 'reference area' displaying high ecological conditions will be identified for each habitat class. Subsequently, the application of multi criteria analysis (Beinat and Nijkamp, 1998), and the weight-of-evidence approach will allow the integration of the indices into an environmental quality multimetric index (US-EPA, 2000), suitable for defining the environmental quality of selected lagoon habitats.

ACKNOWLEDGEMENTS

We are grateful to the Health Authority of the Veneto Region, the Venice Water Authority (MAV) and CORILA, for the sharing of data and partial financial support. Finally, we acknowledge the Venice in Peril Fund and the Cambridge Coastal Research Unit, University of Cambridge, for their remarkable efforts in the organization of the conference (Churchill College, Cambridge, September 2003), and publication of the conference proceedings.

REFERENCES

ANZECC and ARMCANZ. 2000. *Australian and New Zealand Guidelines for Fresh and Marine Water Quality*. National Water Quality Management Strategy Paper No. 4, Australian and New Zealand Environment and Conservation Council & Agriculture and Resource Management Council of Australia and New Zealand, Canberra, Australia.

Beinat, E., and P. Nijkamp (eds.) 1998. *Multi-Criteria Evaluation in Land-Use Management*. Dordrecht: Kluwer Academic Publishers.

Burbor, M. T., Gerritsen, J., Snyder, B. D., and J.B. Stribling. 1999. *Rapid Bioassessment Protocols for Use in Streams and Wadeable Rivers: Periphyton, Benthic Macroinvertebrates and Fish*, 2nd edn. EPA 841-B-99-002. Washington, DC: US Environmental Protection Agency, Office of Water.

Calabrese, J. E., and A. L. Baldwin. 1993. *Performing Ecological Risk Assessment*. Chelsea, MI, USA: Lewis Publisher.

Collavini, F., Bettiol, C., Zaggia, L., and R. Zonta. In press. Pollutant loads from the drainage basin to the Venice lagoon (Italy). *Environmental International*.

Cressie, A. C. N. 1993. *Statistics for Spatial Data*. New York: John Wiley & Sons.

Critto, A., and A. Marcomini. 2001. *Ecological risk and lagoon contamination*. (In Italian.) Venice: Cà Foscarina Ed.

Critto, A., Carlon, C., and A. Marcomini. In press. Screening Ecological Risk Assessment for the Benthic Community in the Venice Lagoon (Italy). *Environment International*.

European Union. 2000. *Directive 2000/60/EC of the European Parliament and of the Council Establishing a Framework for the Community Action in the Field of Water Policy*.

Gobas, A. P. C. F. 1993. A model for predicting the bioaccumulation of hydrophobic organic chemicals in aquatic food-webs: application to Lake Ontario. *Ecological Modelling* **69**, 1–17.

Guerzoni, S., and S. Racanelli. 2003. *The Wounded Lagoon*. (In Italian.) Venice: Ed. Cafoscarina Italy.

Jones, D. S., Hull, R. N., and G. W. Suter. 1996. *Toxicological Benchmarks for Screening Contaminants of Potential Concern for Effects on Sediment-associated Biota*. Lockheed Martin Energy Research Corporation Report ES/ER/TM-95/R2.

Jones, D.S., Barnthouse, L.W., Suter II, G.W., Efroymson, R.A., Field, J.M., and J.J. Beauchamp. 1999. Ecological risk assessment in a large reservoir: 3.Benthic invertebrates. *Environmental Toxicology and Chemistry* **18**, 599–609.

Karr, J. R., and E. W. Chu. 1997. *Biological Monitoring and Assessment: Using Multimetric Indexes Effectively*. EPA 235-R97-001. University of Washington, Seattle.

Libralato, S., Pastres, R., Pranovi, F., Raicevich, S., Granzotto, A., Giovanardi, O., and P. Torricelli. 2002. Comparison between the energy flow networks of two habitats in the Venice Lagoon. *Marine Ecology, PSZN* **23**, Suppl. 1, 228–36.

Long, E. R., and L. G. Morgan. 1990. *The Potential for Biological Effects of Sediment-sorbed Contaminants Tested in the National Status and Trends Program*. In NOAA Technical Memorandum NOSOMA 1990, 52, 8.

MacDonald, D. D., Carr, R. S., Calder, F. D., Long, E. R., and C. G. Ingersoll. 1996. Development and evaluation of sediment quality guidelines for Florida coastal waters. *Ecotoxicology* **5**, 253–78.

MAV-CVN. 1999. *Mapping the Lagoon Bottom Contamination*. (In Italian.) Final Report. Venice: Consorzio Venezia Nuova.

MAV-CVN. 2000. *Project 2023 Results*. (In Italian.) Final report. Venice: Consorzio Venezia Nuova.

MAV-CVN. 2001. *Determination of the pollutant load discharged into the Venice Lagoon by the drainage basin*. DRAIN project workshop, 14–15 June 2001, Venice, Italy.

MAV-CVN. 2002. *Monitoring the Venice Lagoon Waters. 2001–2002 data*. Venice: Consorzio Venezia Nuova, December 2002.

Massachusetts Weight-of-Evidence Workgroup. 1995 *A Weight-of-Evidence Approach for Evaluating Ecological Risks*. Draft Report. 2 November, 1995.

McArty, L. S., and D. Mackay. 1993. Enhancing ecotoxicological modelling and assessment: body residue and modes of toxic action. *Environ. Sci. Technol.* **27**, 1719–28.

MathSoft Inc. 1996. *S+SpatialStats. User's Manual for Windows and Unix*. Seattle, Washington: Mathsoft Inc.

Menzie, C. A. 2002. The evolution of ecological risk assessment during the 1990s: challenges and opportunities. *Environmental Analysis of Contaminated Sites*. Chichester: John Wiley & Sons.

Micheletti, C., Critto, A., Carlon, C., and A. Marcomini. 2004. Ecological risk assessment of persistent toxic substances for the clam *Tapes philipinarum* in the lagoon of Venice. *Environmental Toxicology and Chemistry* **23**, 6, 1575–82.

Nasci, C., Da Ros, L., Campesan, G., and V. U. Fossato. 1998. Assessment of the impact of chemical pollutants on mussel, *Mytilus galloprovincialis*, from the Venice Lagoon, Italy. *Marine Environmental Research* **46**, 279–82.

Occhipinti Ambrogi, A., and G. Forni. 2003. The biotic indices. *Biol. Mar. Med.* **10** (Suppl.), 577–604. (In Italian.)

Orel, G., Boatto, V., Sfriso, A. and M. Pellizzato. 2000. *Fishing Sources Management Plan for the Lagoons of the Venice Province*. Venice: Province of Venice. (In Italian.)

Oslo and Paris Commissions. 1993. *Guidelines for the management of dredged material, Annex 2*. Report to LC Scientific Group 17th Meeting, 18–22 July.

Pojana, G., Critto, A., Micheletti, C., Carlon, C., Busetti, F., and A. Marcomini. 2003. Analytical and environmental chemistry in the framework of risk assessment and management: the Lagoon of Venice as a case study. *Chimia* **57**, 542–9.

Secco, T., Pelizzato, F., Sfriso, A., and B. Pavoni. 2005. The changing state of contamination in the Lagoon of Venice. Part 1: organic pollutants. *Chemosphere* **58**, 3, 279–290.

Sfriso, A., and A. Marcomini. 1996. Decline of *Ulva* growth in the lagoon of Venice. *Bioresource Technology* **58**, 299–307.

Sfriso, A., Facca, A., and P. F. Ghetti. 2003. Temporal and spatial changes of macroalgae and phytoplankton in a Mediterranean coastal area: the Venice lagoon as a case study. *Marine Environmental Research* **56**, 617–36.

Sfriso, A., Facca, C., Ceoldo, S., Silvestri, S., and P. F. Ghetti. 2003b. Role of macroalgal biomass and clam fishing on spatial and temporal changes in N and P sedimentary pools in the central part of the Venice lagoon. *Oceanologica Acta* **26**, 3–13.

Sfriso, A., Facca, C., and A. Marcomini. In press a. Recording the occurrence of trophic level changes in the lagoon of Venice over the last 15 years. *Environment International*.

Sfriso A, Facca, C., and A. Marcomini. In press b. Sedimentation fluxes and erosion processes in the lagoon of Venice. *Environment International*.

Shephard, B. K. 1998. Quantification of Ecological Risks to Aquatic Biota from Bioaccumulated Chemicals. In *Proceedings, National Sediment Bioaccumulation Conference*, Bethesda, Maryland, USA, 11–16 September 1996, 2.31–2.52.

Tarazona, J. V. 2002. Trends in the risk assessment strategy of persistent organic pollutants. The European perspective. *Proceedings of the 22nd International Symposium on Halogenated Environmental Organic Pollutants and POPs*. Barcelona: CSIC, 59, 107–110.

Thetis. 2000. Project 2023. *Activity F. Preliminary assessment of state of Venice lagoon ecosystem*. (In Italian.)

US-EPA. 1998. *Risk Assessment Forum. Guidelines for Ecological Risk Assessment*. EPA/630/R-95/002F. Final Report. Washington, DC.: US Environmental Protection Agency.

US-EPA. 2000. *Estuarine and Coastal Marine Waters: Bioassessment and Biocriteria Technical Guidance*. December 2000. EPA 822-B-00-024. Washington, DC.: US Environmental Protection Agency, Office of Water.

Van Der Kooij, L. A., Van De Meent, D., Van Leeuwen, C. J., and W. A. Bruggeman. 1991. Deriving quality criteria for water and sediment from the results of aquatic toxicity tests and product standards: application of the equilibrium partitioning method. *Water Research* **25**, 697–705.

Vollenweider, R. A., Giovanardi, F., Montanari, G., and A. Rinaldi. 1998. Characterization of the trophic conditions of marine coastal waters, with special reference to the nNW Adriatic Sea: proposal for a trophic scale, turbidity and generalized Water Quality Index. *Environmetrics-Chichester* **9**, 329–57.

Volpi Ghirardini, A., Arizzi Novelli, A., Likar, B., Pojana, G., Ghetti, P. F., and A. Marcomini. 1999. Sperm cell toxicity test using sea urchin *Paracentrotus lividus* Lamarck [Echinodermata: Echinodea]: sensitivity and discriminatory ability towards anionic and nonionic surfactants. *Environmental and Toxicological Chemistry* **32**, 711–18.

55 • The monitoring programme in the Venice Lagoon: striving towards a comprehensive knowledge of the lagoon ecosystem

A. ZIRINO

SUMMARY

Although many, many studies have been conducted in the Venice Lagoon, in general, past efforts have focussed on a few or individual descriptors of its ecosystem, limited in both temporal and spatial coverage. However, under the sponsorship of the Magistrato alle Acque and its concessionary, the Consorzio Venezia Nuova, the last three years have witnessed the initiation and activities of MELa – (Monitoraggio Ecosistema Lagunare), a major programme designed to provide a comprehensive, synoptic, and lagoon-wide coverage of important water quality parameters, and provide insights into the major processes that affect the concentrations and distributions of ecological variables, e.g. primary production, growth and distribution of macroalgae and sea grasses, release rates of heavy metals from sediments, transfer and accumulation of toxic substances in the food web. Additionally, MELa provides water quality data in compliance with statutory requisites, baseline data against which the effects of new construction may be assessed, and also, data for mathematical modelling of the lagoon ecosystem.

Among many results, the MELa monitoring effort has provided a holistic vision of the processes occurring in the lagoon ecosystem. It has yielded an understanding of the spatial and temporal variability of primary production (phytoplankton, macroalgae, and seagrasses) and of the factors that 'drive' the concomitant nutrient field and physico-chemical parameters. The present state of the lagoon appears to be in a stable, mesotrophic condition, typical of similar environments. In sediments, several trace substances (Pb, Hg, As) are high near the industrial areas. However, a MELa sub-study, ARTISTA, which deals with the mobility and availability of contaminants from sediments (as well as with their distribution and concentration throughout the food web), shows little release from sediments, a condition largely attributable to the presence of excess acid volatile sulfides (AVSs) and binding by refractory organic matter. A study of trophic relationships at five stations dispersed throughout the lagoon, shows that they are influenced largely by physical parameters that determine the environment and not by the concentration of micropollutants in the sediments.

These observations strongly support the view that the ecosystem of the Venice Lagoon must be considered in its entirety and that its ecological 'state' must be assessed from a comprehensive knowledge of the interactions among multiple variables, rather than from the measurements of a single parameter, as is often done.

INTRODUCTION

The unusually high degree of international, national, and regional interest in the historical city of Venice has caused the Venice Lagoon to be one of the most studied ecosystems in the world.[1] However, most of the studies conducted prior to the last decade have tended to be highly individualized in that they reflected the particular scientific interests of individual investigators or organizations. Given the large area of the lagoon, high degree of ecological variability, and limited sampling, these studies could be made to reach virtually any conclusion about its ecological condition. There is, in fact, a popular tendency to equate the ecological state of the lagoon to that of

---

[1] See Lasserre and Marzollo (2000), Campostrini (2002).

the frequently observed and highly polluted canals in the historical centre.

The storm surge and subsequent flooding of Venice on 4 November 1966, forced a holistic perspective of the lagoon, in that the measures proposed to save Venice from future batterings, namely the closing of the entire lagoon for short periods of time with seawalls and mobile dikes (*dighe mobili*), would also affect the lagoon as a whole. It is for this reason that the Italian government, via the Ministry of the Interior and the local Venice Water Authority (Magistrato alle Acque Venezia, or MAV), via its concessionary, the Consorzio Venezia Nuova (CVN), decided to sponsor a series of studies designed to improve the knowledge of the state of the lagoon in its entirety, in so far as possible. These studies were designed to augment the already existing monitoring efforts carried out by MAV near the historical centre and by the laboratories of the Regione Veneto (ARPAV) which concentrated principally on the inputs to the lagoon from its drainage basin and the Adriatic coastal area adjoining it (see Penna *et al.*, chapter 56). These studies were also motivated by the passage of the so-called 'Ronchi-Costa' decree by the Italian government in 1998. Ronchi-Costa established water quality guidelines for Italian embayments.

The first study, termed Orizzonte 2023, ended in 1998 (MAV-CVN, 2000a). Its purpose was to attempt to estimate the water quality and ecological health of the lagoon from all of the available data, as well as additional measurements, to aid the Ministry of the Environment and the Ministry of Labour in the establishment of water quality criteria, and to define gaps in the knowledge of the lagoon ecosystem. The second study of the series, termed DRAIN, an Italian acronym for determination of drainage basin influx (Zonta *et al.*, 2001), attempted to establish reasonably accurate estimates of the fluxes of the major and minor contaminants (nitrogen, phosphorous, trace metals and organic pollutants, respectively) into the lagoon from the land, including inputs from rivers, ground waters and the air.

The third study was to be a comprehensive monitoring of the lagoon. It was called MELa, for Monitoraggio dell'Ecosistema Lagunare and its initial phase, MELa1, was launched in September 2000. It was followed by MELa2 and is continuing to 2006 as MELa3. The purpose of this chapter is to present the most pertinent aspects and results of MELa1 obtained during the last three years, as described in the internal reports. It is also an attempt to give an overview of the ecological condition of the lagoon, based on these initial data.

Physical setting

The lagoon is extremely shallow for its area. If, temporarily, we assume a square lagoon, its area to depth ratio would be 23 000 to 1, much thinner than an ordinary sheet of paper. This alone suggests that exchanges occurring at the lagoon floor as well as airborne inputs will be important in any budgetary calculations of its chemical constituents. Because of its shallowness, the lagoon is well flushed and well mixed vertically, by both bottom friction-induced turbulence and wind action. Hydrodynamic residence times vary from 1-3 days near the entrances to 20-30 days in the least flushed sectors adjoining the mainland (MAV-CVN, 1998). The lagoon floor is etched with many canals; the main ones, leading from each of the entrances to the mainland, are quite deep (*c.* 20 m) and wide enough to accommodate supertankers and the largest ocean liners. It is estimated that approximately 50 per cent of the water flow in and out of the lagoon goes via the canals. Throughout the lagoon, there are extensive subtidal flats (*velme*) and salt marshes (*barene*) that become exposed at low tide. Average tidal range is approximately +/- 35 cm, with large excursions occurring under seasonal *bora* (NE) or *sirocco* (S-SW) wind conditions. Freshwater input into the lagoon occurs year round and is greatest during winter and spring. Due to the historical diversion of much of the river inflow around the lagoon, the average freshwater input rate is only about 2 per cent of the input rate of Gulf of Venice water (200 $m^3 s^{-1}$ v. 10 000 $m^3 s^{-1}$).

Recent history

In the recent past, and even at present, the lagoon has been impacted by a very high input of nitrogen and phosphorous, coming predominantly from the agricultural drainage basin (*c.* 85 per cent) and the historical centre (*c.* 15 per cent). Up until *c.* 14 years ago,

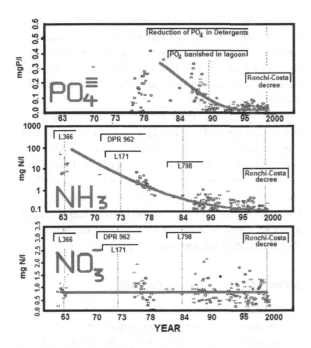

Fig. 55.1   Nitrate (NO_3^-), ammonium (NH_4^+), and phosphate (PO_4^{3-}) concentrations in the lagoon from the mid-sixties to the year 2000. Note: NH_4^+ concentration is on a log scale. Annotations in the figure refer to various laws and decrees issued to control pollution.

the banning of phosphorous from detergents and from direct discharges into the lagoon resulted in the present near disappearance of *Ulva* and of the return of large beds of *Zostera*. Figure 55.1 shows the time trend of nitrate (NO_3^-), ammonium (NH_4^+), and phosphate (PO_4^{3-}) concentrations in the lagoon from the mid-1960s to the year 2000.

Several things may be noted from Fig. 55.1. The large decrease in ammonium is linked to PO_4^{3-} and not to NO_3^-. NO_3^- concentration was very high and continues to be high. On a mol basis, and considering the Redfield-Richards ratio (Leibes, 1992), prior to 1990 there was a four fold excess of PO_4^{3-} over NO_3^-. Today the ratio has nearly reversed. The high values of ammonium appear linked to the very high primary productivity present during the period of phosphorous abundance, namely to the heterotrophic activity associated with that productivity. It is likely that much of the ammonium originates from respiration at the lagoon floor, pointing out the importance of that component in the over-all nutrient budget of the lagoon. Table 55.1 represents the present estimated annual total nitrogen and phosphorous load into the lagoon (MAV-CVN, 2003a).

The sediments

The granulometry of the lagoon sediments mirrors the residence time pattern fairly closely, in that fine, organic-rich, sediments are found in the back areas adjoined to land, while coarse, organic poor, sandy sediments are found near the lagoon entrances. At least two lagoon-wide studies of trace element distributions in the lagoon have been made: one by Albani *et al.* (1995) and one by Degetto *et al.* (see MAV-CVN 2000b). Both show high concentrations

these inputs lead to massive macroalgal blooms (*Ulva* sp.) during the summer months, displacement and reduction of the seagrasses (*Zostera* sp.), and to occasional localized anoxia near the historical centre, principally in August. These events, caused by the rapid microbial degradation of *Ulva*, occurred practically overnight, despite the shallowness of the floor and the good vertical mixing characteristics. In 1990,

Table 55.1   *Present total nitrogen and phosphorous loading in the Venice Lagoon (tons yr^{-1}).*

Source	Rivers	Direct discharge		Atmosphere	TOTAL
		Porto Marghera	Venice Centre		
Nitrogen$_{Tot.}$	4.0×10^3	1.3×10^3	5×10^2	1.1×10^3	6.4×10^3
Phosphorous$_{Tot.}$	2.3×10^2	1×10^2	6×10^1	4.4×10^1	4.5×10^2

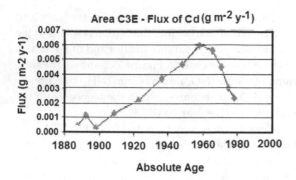

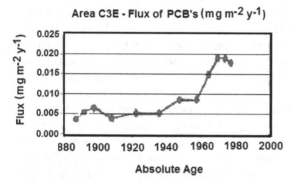

Fig. 55.2   Time series of Cd and PCB fluxes in a core taken from the industrial area in the central lagoon (MAV-CVN, 2000b).

and clustering of certain elements deemed to be of anthropogenic origin, namely Cr, Cu, Zn, Hg, As, and Pb, around the industrial zone of Porto Marghera and in some cases, Chioggia. According to Albani *et al.* (1995), 'The restricted occurrences of the anthropogenic sources suggest a very limited mobility of the bottom sediment within the lagoon' a conclusion reached independently by other workers (MAV-CVN, 1998). Be that as it may, the area of high concentration lies between Porto Marghera and the historical city in a zone where the hydrodynamic residence time is about 10 – 15 days (CVN, 1998; Zirino *et al.*, 2002). This suggests that this period is sufficiently long for anthropogenic pollutants to bind to particulates and settle close to the source.

Recent analyses of cores by Degetto *et al.* (MAV-CVN, 2000b) shows that chlorinated organic pollutants such as PCBs and dioxins follow the same

distribution pattern as the heavy metals, e.g., are concentrated and clustered around the industrial centres. Furthermore, the dating of the cores from the industrial area shows that the rates of deposition of many pollutants peaked between about 1960 and 1970 (Fig. 55.2).

This suggests that either (1) pollution has decreased since 1960 or (2) erosion in the central lagoon has increased, or (3) most likely, both have occurred. Also, the recent scouring of the floor of the central lagoon by clam fishermen tends to mobilize the sediment and causes pollutants to disperse throughout the body of water. Indeed, the recent work by Degetto *et al.* (MAV-CVN 2000b) suggests that the central lagoon is eroding.

MONITORING THE LAGOON

Penna *et al.* (chapter 56) present a thorough description of the several monitoring programmes that are taking place in the lagoon and of the efforts that are being made to integrate them. Here, only the first cycle (2000-03) of MELa, (MELa1) is discussed, a continuing, comprehensive, monitoring programme, launched by the Magistrato alle Acque via its concessionary, the Consorzio Venezia Nuova.

MELa1

MELa, essentially a chemical/biological monitoring programme established to monitor the lagoon environments in accordance with statutory requirements (Penna *et al.*, chapter 56), attempts to answer the following questions:

1) What are the long-term trends of the lagoon ecosystem?
2) Which physical, chemical, biological, parameters best determine the trends?
3) Which baseline values should be used to determine trends?
4) What are the expected ecological results of these trends?
5) What are the ecological effects of new construction?

MELa endeavours to answer these queries by monitoring water quality and benthic conditions throughout the lagoon, via measurements of water

Table 55.2   *Water quality parameters measured on MELa1.*

Parameter	CTD	Discrete	Parameter	Discrete
Temperature	*		PO_4^{3-}	*
Cond./Salinity	*		Total P	*
Diss. O_2	*		Total Org. C	*
PH	*		Part. Org. C	*
Eh	*		Diss. Org. C	*
Turbidity	*		Chlorophyll-a	*
Chlorphyll-a* (Fluor.)	*		Feopigments	*
Tit. Alkalinity		*		
Tot. Susp. Solids		*	Arsenic (As)	* #
NH_3		*	Copper (Cu)	* #
NO_3^-		*	Mercury (Hg)	* #
NO_2^-		*	Lead (pb)	* #
Tot. In. Nitrogen		*	Zinc (Zn)	*#
Tot. Diss. Nitrog.		* #	Cadmium (Cd)	*#
Total Nitrogen		*	Chromium (Cr)	* #
Total Diss. P		* #	Nickel (Ni)	* #

Consistently over the WQ guidelines set by 'Ronchi-Costa' decree of 4/23/1998.

and sediment chemical and biological properties, and attempts to determine how sediment concentrations impact on water quality and associated biota. It also archives and makes available time series of lagoon-wide, water quality data that provides (and will continue to provide) input and verification material for numerical water quality models.

Several sub-programmes comprise MELa1. The first of these, the hydrographic component, is based on a series of surveys occurring approximately every 29 days (at neap tide). Sampling is at 28 stations relatively uniformly distributed throughout the lagoon and at two stations just outside the Lido and Malamocco entrances. A relatively complete set of 'standard' water quality parameters is measured via discrete sampling at one or two depths, (depending on whether or not the station is located in a canal) and 'continuous' vertical profiles are also made with a CTD probe (Idronaut, Milano). 15 stations are also measured for seven heavy metals and As. All of the measurements are made using 'standard' oceano-

graphic techniques (Grasshoff *et al.*, 1999); metals are measured using inductively coupled plasma, mass spectrometry-based methods. The measured parameters are listed in Table 55.2.

Of the 28 stations sampled for water quality, ten also participated in the subprogramme ARTISTA (MAV-CVN, 2003b). The aim of ARTISTA was to study the rate, amount, and manner of transference of sedimentary pollutants into the biota. For ARTISTA, the designated stations were sampled during June and November 2001 for the seven metals and As listed in Table 55.2. Particulate fluxes were determined with sediment traps. Additionally, the sediments at these stations were measured for organic content, colloids, acid volatile sulfides (AVSs), and persistent organic pollutants (POPs) such as pesticides, polychlorinated-biphenyls (PCBs) and dioxins (PCDDs, TCDDs). Partitioning of metals and pollutants among the various physico-chemical species was also determined.

Of the ten ARTISTA stations, five were chosen for the sub-subprogramme ECOTOX, a study of the

510   A. Zirino

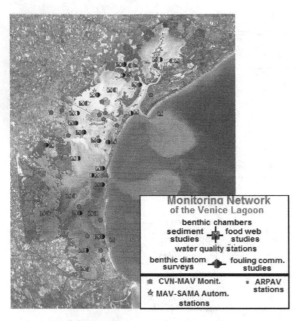

Fig. 55.3   Layout of MELa1 and SAMA stations superimposed on a Landsat image of the lagoon. (Plumes show the ejection of particulate-rich waters from the lagoon into the Adriatic Sea at ebb tide.)

transference and accumulation of these pollutants into the food web for the ultimate purpose of developing a working ecotoxicological model. Two of these 5 stations were equipped with benthic chambers in order to assess the rate of release of materials out of the sediments. Figure 55.3 show the location of the 30 MELa1 stations, along with the stations designated by the Venice Water Authority's SAMA programme and the stations sampled by the Regione Veneto (ARPAV).

ECOTOX involved taking a seasonal census of the biota at the five stations located to give a semi-representative coverage of the entire lagoon: two in the lower basin, two in the central basin, and one in the northern basin, in a region impacted by fresh water flows. Table 55.3 shows the organisms used in the trophic model of energy and material transfer that was set up with the programme ECOPATH (US EPA) into which field data from ARTISTA was introduced.

Once the efficiency of energy and material transference in the trophic web was established, an ecotoxicological model for the lagoon was constructed by coupling it to 'CRUP', a 516 cell, hydrodynamic-dispersive model of the lagoon (MAV-CVN-SI, 1995). CRUP calculates the steady-state distributions of 1 to 516 substances originating from any or all of the 516 cells, assuming conservative behaviour. From

Table 55.3   Ecological compartments considered in the trophic web model.

Compartment	Trophic level	Compartment	Trophic level
1. Carnivorous fishes	4	11. Micro+Mesobenthos	2
2. Gobius ophiocephalus	4	12. Meso+Macrozooplankton	2
3. Omnivorous fishes	4	13. Microzooplankton	2
4. Liza sp.	2	14. Phytoplankton	1
5. Carcinus mediterraneus	3	15. Epiphytes	1
6. Macrobenthos (predatory)	3	16. Sea Grasses	1
7. Tapes philippinarum	2	17. Benthic Microalgae	1
8. Macrobenthos (suspension feeders)	2	18. Macroalgae	1
9. Macrobenthos (detritus/herbivorous feeders)	2	19. Planktonic detritus	1
10. Polychaetes	2.5	20. Benthic detritus	1

there, by knowing the amount and location of the input, the potential concentration of a pollutant in a particular organism can be calculated and the risk to that organism estimated from toxicity studies available in the literature.

MELa2

A second monitoring programme, designated MELa2, was established to monitor benthic communities. This includes taking a census of soft-bottom benthic communities and, to survey sea grass coverage in the lagoon (Penna *et al.*, chapter 56; Rismondo *et al.*, chapter 60). The latter is a critical indicator of the ecological well being of the lagoon because seagrasses serve to anchor otherwise moving sea beds and provide a habitat for communities of desirable organisms. This programme consists aerial surveys supported by field sampling. A terrestrial component of MELa2 also surveys mudflat vegetation and terrestrial microvertebrates.

RESULTS AND DISCUSSION

Correlations with salinity

As the field components MELa1 and MELa2 have just terminated, only two years of data have been analysed (September 2000 to September 2002). However, a cursory survey of the third year's data does not appear to alter present understanding of lagoon processes. For the hydrographic parameters in Table 55.2, and discounting the heavy metals, the annual concentration means show a strong correlation to salinity, suggesting that their concentrations may be explained by the mixing of waters from two distinct sources: river waters and seawater from the Gulf of Venice (Fig. 55.4). This also suggests that physical mixing is occurring on time scales that are shorter than those of *in situ* biogeochemical processes. As expected, the degree of correlation is strongest in the winter, when primary production is at a minimum, and weaker in the summer, when primary production is much greater, yet the overall correlation is still present. (Further evidence to the

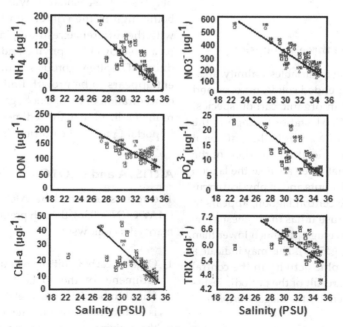

Fig. 55.4 Annual means of some macrovariables at each of the 30 stations, plotted as a function of annual mean salinity at the same stations (MAV-CVN, 2002a). Lines indicate trends, not statistical regressions.

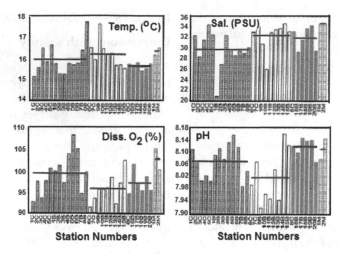

Fig. 55.5 Yearly averages of temperature, salinity, dissolved oxygen and pH in the three areas of the Venice Lagoon. Station numbers are on the X-axis. From left: southern, central and northern. The two stations on the far right of the graphs are from the Gulf of Venice (MAV-CVA, 2002a).

strength of mixing (vertical, in this case) is given by the fact that there is no statistical difference between values obtained by sampling at shallow and deep levels).

Differences among the three sub-basins

Statistical analysis of macrovariables (salinity, alkalinity, pH, turbidity, suspended solids, nitrogen and phosphate species and in organic carbon species) measured at the 28 MELa1 stations supports the observation of general differences in the seafloor of the southern, central and northern portions of the lagoon. This suggests that three areas of the lagoon have distinct patterns of inputs and of physical circulation. For example, the central lagoon, appears to be the best flushed because it has the highest average salinity, yet has an average pH much lower than the entering seawater (Fig. 55.5). This may indicate a much greater heterotrophic activity in the central lagoon, perhaps a direct result of the periodic resuspension of its the sediment by the illegal *vongolari*.

Seasonal cycles

As expected, the macrovariables in the lagoon display a strong seasonality, with the major nutrients being high in the winter and low in the summer, with the difference resulting in phytoplankton, as measured by chlorophyll-a and other primary producers. Despite vigorous summer growth, no nutrient appears to be growth limiting, e.g., reaches a concentration of zero. Once again, this suggests that re-mineralization in the sediments is rapid, and is an important part of the nutrient budget of the lagoon.

ARTISTA and ECOTOX

The results obtained by ARTISTA and ECOTOX (MAV-CVN 2003b), though preliminary, may be summarized as follows:

1) Organic carbon, nitrogen and phosphorous in the sediments of the 10 lagoon-wide stations was found to be distributed relatively uniformly. This is in disagreement with the water data that shows that mixing processes, and therefore, residence times, predominate and would control sediment

concentration. It may very well be that a gradient does exist between the inner and outer portions of the lagoon but has been masked by natural and experimental variability.

2) Sediments just below the surface are reducing (low REDOX values) with sulfides present and with some seasonal variability (more reducing in summer). This is as expected for an organic-rich environment.

3) Fluxes of materials from the bottom show a large degree of temporal variability but the values are comparable to those found in the literature. Fluxes are higher in the central lagoon and are wind-influenced.

4) Total metal concentrations and sediment simultaneously extractable metals (SEMs – DiToro et al., 1991; DiToro et al., 1992) are higher in the central lagoon and do not show seasonal variations. This is in agreement with the observation made earlier that heavy metals have little mobility and are found close to their sources. Throughout the lagoon the ratio SEM/AVS is always less than one. Thereby it may be inferred that no metal toxicity exists (McGrath et al., 2002).

5) There is no clear evidence of a relationship between metal concentrations in the sediments and the metal concentrations in the benthos.

6) At all five ECOTOX stations, carbon flow takes place predominantly between the first two trophic levels, e.g. polychaetes, nematodes, ciliates (and clams) are feeding off the detritus resulting from all primary producers. This is a condition typical of an organic-rich environment.

CONCLUSIONS

The limited survey of the MELa results conducted here indicates that, at present, and as a whole, the Venice Lagoon is not much different from other shallow, mesotrophic environments. The present lagoon is very different from the one that existed before 1990, that was choked with *Ulva* and risked anoxia in the summer. The present balance between nutrient input, solar irradiation, tidal flushing and wind-mixing, appears to be sufficient to also main-

tain the lagoon in its present, well oxygenated condition in the future. However, it must be remembered that the lagoon still contains a large excess of nitrogen (in various forms) over phosphorous, and that the balance with dissolved oxygen is controlled by the availability of phosphorous. Should it increase, the lagoon could well revert to its previous state.

It is estimated from hydrodynamic models that placement of the mobile barriers at the three entrances would reduce tidal inflow by about 5 per cent, even when the barriers are completely submerged. This does not appear to be a reduction that would have significant ecological consequences, particularly in view of the excellent vertical mixing that occurs in the lagoon. However, this conclusion is tenuous because sufficiently accurate water quality models of the lagoon have yet to be made. The data collected in MELa is a significant first step towards the development of these models. Nevertheless, more data are needed, particularly of the rates of primary production, oxygen consumption (biological oxygen demand) and of fluxes in and out of the sediments. Indeed, the important role of the lagoon floor in determining the lagoon ecosystem must still be evaluated. The central lagoon appears to be in a state of erosion, a phenomenon caused or aggravated by the illegal dredging of clams. More than pollution, this particular activity probably has had the most significant ecological impact on the central lagoon, especially on its infauna.

The limited results of ARTISTA indicate that, lagoon-wide, there is no clear evidence of toxicity caused by micropollutants (including metals and POPs). This might well be because of the presence of an excess of AVSs in the sediments and the complexing agents in the water column, as well as an abundance of suspended particles and dissolved and particulate organic carbon, all of which serve to limit the mobility of pollutants. In this context, the principal problem of the lagoon, e.g. the estimated high nutrient input and resulting high summer productivity is also its salvation. A decrease in nutrient input without a parallel decrease in heavy metal or organic pollutant input might result in possible toxicity to organisms. However, the present and continuing efforts by the citizens and government of Venice and the Veneto region, the Italian government, and the international community, to protect the lagoon, to understand its ecological processes, and to reduce all pollution into it, augur well for its future.

ACKNOWLEDGEMENTS

This brief ecological perspective of the Lagoon of Venice represents the work of the MELa Programme work group: A. Barbanti, A. Berton, M. Bocci, S. Carrer, E. Andreoli, A. Carlin, C. Castellani, M. Ciri, A. Regazzi, E. Delaney and E. Ramieri (Thetis), G. Bendoricchio, G. Caniglia and P. Venier (University of Padua) A. Bernstein, L. Montobbio and S. Bardino (CVN environmental engineering department), R. Rosselli and S. Silvestri (Venice Water Authority information service), G. Ferrari (Venice Water Authority pollution control section), R. Pastres, S. Ciavatta, A. Petrizzo and S. Libralato (INCA-Venice), A. Volpi, A. Arizzi-Novelli, P. Ghetti and G. Sburlino (University of Venice) A. Rismondo, D. Curiel, F. Scarton, D. Smania, G. Pessa, D. Mion, M. Cazzin and R. Zaja (SELC), C. Solidoro and G. Cossarini (OGS-Trieste), L. Albertotanza, A. Rabitti, G. Socal, L. Da Ros, C. Nasci, V. Moschino, S. Salviato, F. Braga and D. Tagliapietra (ISMAR-CNR), M. Bon (Venice Natural Sciences Museum), B. Chadwick (Zirino & Associates, San Diego USA), A. Viarengo (Eastern Piedmont University) and S. Focardi (Siena University). The writer gratefully acknowledges their work and the efforts of many of them to educate him in the mysteries of the lagoon. Many thanks also to O. Holm-Hansen, F. Delgadillo-Hinojosa and F. Azam, all of the Scripps Institution, for many fruitful discussions on this topic.

BIBLIOGRAPHY

Albani, A. D., Rickwood, P.C., Favero, V. M., and R. Serandrei-Barbero. 1995. The geochemistry of recent sediments in the lagoon of Venice: environmental implications. *Atti del Istituto Veneto di Science, Lettere, ed Arti* **153**, 2–3, 235–320.

Campostrini, P. (ed.). 2002. *Scientific Research and Safeguarding of Venice: CORILA research program, 2001 results*. Venice: CORILA, Instituto Veneto di Scienze, Lettere ed Arti.

DiToro, D. M., Zarba, C. S., Hansen, D. J., Berry, W. J., Swartz, R. C., Cowan, C. E., Pavlou, S. P., Allen, H. E., Thomas, N. A., and P. R. Paquin. 1991. Technical basis for establishing sediment quality criteria for nonionic organic chemicals using equilibrium partitioning. *Environmental Toxicology and Chemistry* **10**, 1541–83.

DiToro, D. M., Mahony, J. D., Hansen, D. J., Scott, K. J., Carlson, A. R., and G. T. Ankley. 1992. Acid volatile sulfide predicts the acute toxicity of Cd and Ni in sediments. *Environmental Science and Technology* **26**, 96–101.

Grasshof, K., Kremling, K.. and M. Erhardt. 1999. *Methods of Seawater Analysis*. Chichester: John Wiley and Sons Ltd.

Lassere, P., and A. Marzollo (eds.). 2000. The Venice Lagoon Ecosystem, Inputs and Interactions Between Land and Sea. Paris and New York: UNESCO/Partenon Publishers.

Leibes, S. M. 1992. *An Itroduction to Marine Biogeochemistry*. Chichester: John Wiley and Sons Ltd, 734pp.

MAV-CVN. 1998. *Interventi per l'arresto del degrado connesso alla proliferatione delle macroalgae in Laguna di Venezia – Studio delle cinetiche idrauliche connesse ai processi di immissione e trasporto degli inquinanti nella Laguna di Venezia*. Relazione finale.

MAV-CVN. 2000a. *Progetto 'Orizzonte 2023'. Rapporto sullo stato attuale dell' ecosistema lagunare veneziano – Attivita' A–F*. Rapporto finale.

MAV-CVN. 2000b. *Progetto 'Orizzonte 2023', Attivita' B3. Metodologia radiochimica per la determinazione dello sbilancio dei materiali particolati sospesi scambiati alle bocche di porto della Laguna di Venezia*.

MAV-CVN. 2002a. *Attività di Monitoraggio Ambientale Della Laguna di Venezia, Attività A.2: Trattamento ed analisi dei dati, elaborazione in linea, ed interpretazioni spazio temporali*.

MAV-CVN. 2002b. *Progetto 'Orizzonte 2023', Attività B (B1 e B2). Indagine radiochimica susedimentirecenti della Laguna di Venezia*.

MAV-CVN. 2003a. *Attività di Monitoraggio Ambientale Della Laguna di Venezia, Attività 3F.2.1: Aggiornamento al 2001 delle stime dei carichi inquinanti esterni ed'interni*.

MAV-CVN. 2003b. *Attività di Monitoraggio Ambientale Della Laguna di Venezia, Attività C.2.1. Indagini sperimental sui rilasci dei sedimenti e biodisponibilita' dei microinquinanti*.

MAV-CVN-SI. 1995. *Nuovi interventi per la salvaguardia di Venezia – Sezione 14, 4o Stralcio: Perfezionamento del sistema CRUP*. Relazione finale.

McGrath, J. A., Paquin, P. R., and D. M. DiToro. 2002. Use of the SEM and AVS approach in predicting metal toxicity in sediments, *ICMM Fact Sheet on Environmental Risk Assessment* **10**, February, 2002.

Zirino, A. *et al.* 2002. *Development of Procedures to Improve the CRUP System*. Final report.

Zonta, R. *et al.* 2001. *Progetto DRAIN, fase 4*. Venice: CNR-ISDGM.

56 • Institutional monitoring of the Venice Lagoon, its watershed and coastal waters: a continuous updating of their ecological quality

G. PENNA, A. BARBANTI, A. G. BERNSTEIN, S. BOATO, R. CASARIN, G. FERRARI, L. MONTOBBIO, P. PARATI, M. G. PIVA AND M. VAZZOLER

INTRODUCTION

Every visit to Venice refreshes the humanistic ideas we receive from literature, art, and other media. Lord Byron once wrote (Gordon, 1817) *'that nature, with its beauty, would survive in a long time the wonderful cultural and artistic heritage of Venice'*. While the shape of man's effect on the environment may be quite obvious to almost every wanderer into the Venetian urban maze, only careful and regular monitoring may reveal the state and dynamics of the lagoon environment. Huge resources have been spent to safeguard the lagoon and to control, restore, and preserve Venice's monuments, paintings and historical heritage so that future generations will still benefit from them. But, how does one control and keep in a good state the lagoon itself, its environment, its precious and different ecosystems? Understanding and perceiving the ecological quality of the lagoon is a complex and demanding process. A huge effort is required to carry out regular and detailed measurements, to acquire elements of judgement, and to build a shared knowledge base and perspective for design and planning. In particular, huge amounts of data, purposively collected and analysed, are required to support the design, construction, and implementation of the planned engineering and environmental protection and restoration work. The public authorities are implementing these monitoring efforts. However, local universities and research institutions, coordinated through CORILA (a consortium of local university and research institutes) may find a specific role.

THE PRESENT FRAMEWORK OF STUDIES, INVESTIGATIONS AND MONITORING ACTIVITIES

Several administrations, agencies and research institutions have been studying the hydrodynamics, morphology, chemistry and biology of the Lagoon of Venice for a long time, often with different aims and purposes. It is a general belief that the Lagoon of Venice is one of the most studied environments in the world; however, the very complex relations that exist between the physical/biological environment and the socio-economic and cultural issues that affect it limit an understanding of its dynamics.

Management of the lagoon has a long tradition. The present Venice Water Authority (VWA) (Magistrato alle Acque), was founded in 1501 (as the Council of Water Sages) to address and manage the lagoon environment, and its precursor, the Magistrato ai Lidi was founded even earlier, in the thirteenth century. Nowadays, the Special Law for Venice provides management tools for the purpose of safeguarding and restoring Venice and its lagoon, and reducing pollution from its drainage basin. Laws and regulations are continuously evolving; often influenced by politics, they can change in a short time. Nevertheless, safeguarding actions must to be planned for a long, useful, lifetime. The work being carried out today was approved yesterday and designed the day before, or earlier. Its effects will start tomorrow and last for decades. Administrations planning and carrying out environmental restoration programmes must therefore rely on a reliable base of information, suitable to support the various tasks to be performed. The information should, therefore, be based as much as possible on sound science rather than on regulatory compliance. The way under

Flooding and Environmental Challenges for Venice and its Lagoon: State of Knowledge, ed. C. A. Fletcher and T. Spencer.
Published by Cambridge University Press, © Cambridge University Press 2005.

which studies, investigations and monitoring activities for the Venice Lagoon correlate is currently under assessment and revision. Laws and official regulations set up objectives for environmental protection and restoration action and name the responsible authority. The authority then generates plans and projects to be developed on the basis of a sound understanding of the state of the lagoon. Therefore, it must have a clear understanding of the problems to be solved and/or of the impacts to be corrected. This knowledge is obtained from monitoring activities, specific investigations and more general studies.

Monitoring activities, carried out routinely on a basin-wide scale, for long periods of time, provide basic information and understanding of the lagoon environment and its dynamics. This information is useful for general planning, specific studies and investigations. On the other hand, specific investigations provide the database for specific works and projects. The relationship between the regulatory authorities and the research community is important and noticeable; the regulatory activities supply the broad-based knowledge of the lagoon and alert the research community to its needs. The latter, in turn, supplies a deeper and more refined understanding of lagoon processes and problems to the agencies, as well as new technologies for measurement and new assessment criteria.

KEY SUBJECTS AND REFERENCE LAWS

The protection and the restoration of the Lagoon of Venice are the shared responsibility of the Veneto Region and of the national government through the Venice Water Authority. As the Lagoon of Venice is a 'special' environment, the distribution of administrative tasks and duties is particularly complex and intertwined due to a double system of laws: there are 'special' laws and ordinary laws. Also, the political and ecological complexities of Venice result in major responsibilities and tasks being assigned to the national and local authorities. However, because of the worldwide attention to Venice, there are more financial resources available here than in other sites in Italy. This opportunity has been seized and Venice can be considered an advanced laboratory for environmental monitoring, centred on, but not limited to, the water system, both for the technical aspects and for the excellent collaboration between two different administrations: the Veneto Region and Venice Water Authority.

The main task of the Veneto Region, which carries out its environmental functions through its technical agency, ARPAV, is to develop plans to reduce and/or prevent pollution of the lagoon, to improve the quality of inland waters, the drainage basin, the coastal waters, and to manage human health aspects. The Venice Water Authority (VWA), with support from its concessionary, the Consorzio Venezia Nuova (CVN), is charged by the Italian Ministry for Infrastructure and Transportation to restore the hydrogeological equilibrium of the lagoon and its habitats, to improve its environmental quality, to preserve its ecosystems, and to protect Venice from high tides (Law No. 171, 1973), (Law. No. 798, 1984).

THE INTEGRATED SYSTEM OF CONTROL AND MONITORING OF THE LAGOON, ITS DRAINAGE BASIN AND THE COASTAL WATER BODY

The Veneto Region and Venice Water Authority, have set up together an integrated system to control and monitor the Lagoon of Venice, its watershed and the coastal water body. It is a melding of many different programmes, both new and existing, and of specialized activities. There are several individual monitoring programmes because national laws and regulations tend to provide funds for specific activities, not for a Lagoon-wide monitoring plan. The integrated monitoring system of the lagoon is thus the result of the convergence of different projects born with different aims and with different timings. The present integrated water quality monitoring network is shown in Fig. 56.1.

The Veneto Region, through its Agency ARPAV, is carrying out the following programmes:

- Monitoring quantity and quality of inland waters (rivers and ground waters) of the lagoon drainage basin;
- Health targeted controls for mussel farming inside the lagoon and in coastal waters, including the monitoring of organotin compounds;

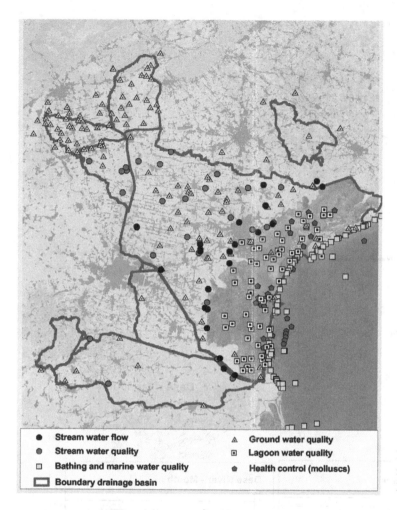

●	Stream water flow
●	Stream water quality
□	Bathing and marine water quality
▭	Boundary drainage basin
▵	Ground water quality
▣	Lagoon water quality
⬠	Health control (molluscs)

Fig. 56.1   Integrated monitoring network of water quality of Veneto Region/ARPAV and MAV-CVN (2001).

- Monitoring the quality of the coastal water; and
- Quality assessments and environmental studies.

The VWA through its pollution control service (SAMA), and supported by its concessionary (CVN), is carrying out the monitoring of the lagoon for sediments, waters, biota, and specialized and interdisciplinary studies.

The current level of integration of the monitoring activities of the two different authorities is to (1) coordinate a network of water monitoring stations, (2) coordinate sampling times and frequencies, (3) exchange data and information, and jointly process and analyse data. The next step will be the sharing and intercalibration of measurement techniques, the sharing of the list of parameters measured, and closer collaboration in the evaluation of monitoring results.

MONITORING THE DRAINAGE BASIN

Monitoring of the drainage basin includes the determination of river and stream flow rates and an assessment of the environmental water quality of surface and ground waters.

The Veneto Region has monitored surface fresh water quality since 1986. The availability of time series of water quality parameters is fundamental for evaluating the trend of pollutant loads coming into the

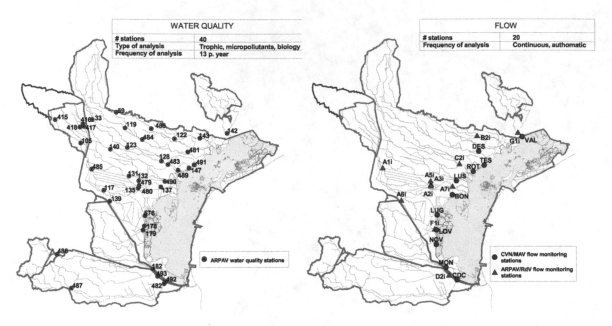

Fig. 56.2   The monitoring network of drainage basin (2002) of ARPAV.

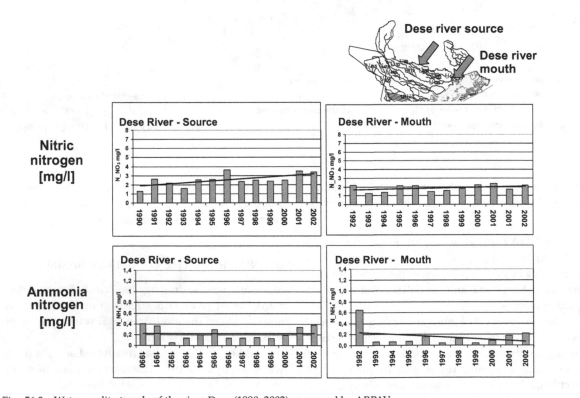

Fig. 56.3   Water quality trends of the river Dese (1990–2002) measured by ARPAV.

lagoon from the watershed and for checking the effectiveness of improvements to sewage treatments, as a result of reduction of the sources of pollution accomplished by the Veneto Region. Flows are currently measured by means of 11 automatic monitoring stations operated by VWA/CVN and of 9 stations directly operated by ARPAV (Fig. 56.2). Surface fresh water quality is measured at 40 stations distributed over the entire drainage basin. Monthly sampling and analyses are performed to determine the parameters required by the laws (Legislative Decree No.152, 1999), and the so called 'Ronchi-Costa' Decrees (Decree Ministry of the Environment, 1998). Extra measurements are performed during flood events. Currently, nearly 27 000 individual measurements are performed in the course of a year. This monitoring is supervised by the Veneto Region and coordinated by the ARPAV – Reference Centre for the Venice Lagoon Drainage Basin (Centro di Riferimento per il Bacino Scolante nella Laguna di Venezia). For example, Fig. 56.3 shows the yearly mean values of ammonia and nitrate concentrations of Dese river water, both at the source and at the mouth, since 1990 (ARPAV et al., 2002a). It is evident that the increase of nitrate in the upstream station

(source) during this period is (probably) due to the increased input of nitrogen by agriculture. On the other hand, the corresponding concentrations at the river mouth are lower, due to the *in situ* purification and dilution processes that occur while flowing downstream. Since 1990, ammonia has shown a significant decrease at the downstream station (mouth) due to the positive effects of waste water treatment plants, carried out under Veneto Region initiatives.

Water quality data are processed on the watershed basin scale to classify the environmental state of the rivers. Figure 56.4 (left) shows the distribution of the 75th percentile of nitrate in rivers in the year 2001. Higher values are measured upstream in the northern rivers where there is a high impact from cattle breeding and agriculture in the water table recharge area. The environmental state of river waters is shown in Fig. 56.4 (right) according to the classification required by Decree 152/1999, which combines the concentration of chemicals in the water with an extended biological index (Ghetti, 1995).

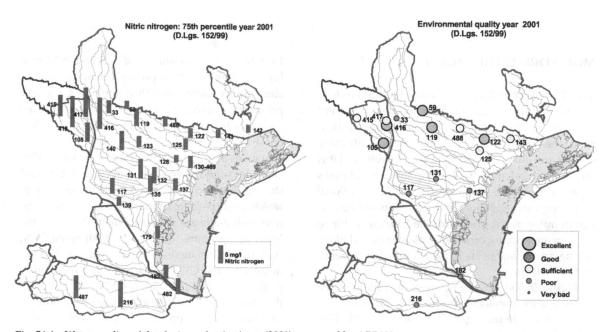

Fig. 56.4   Water quality of the drainage basin rivers (2001) measured by ARPAV.

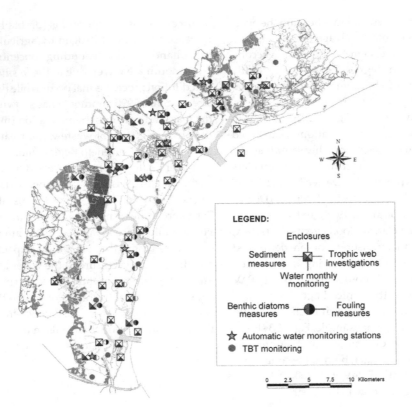

Fig. 56.5   The monitoring network of the lagoon water quality of Venice Water Authority and CVN (2001).

MONITORING THE LAGOON

The Venice Waters Authority (VWA) has been collecting data on the quality of the lagoon environment since the middle 1980s. A large number of investigations, measuring campaigns, problem identification studies and integrated evaluation of ecosystem status have been carried out to develop the general plans of restoration of the lagoon. As recent, local, national and European regulations started to demand ecological monitoring programmes, the VWA was able to plan the monitoring of waters, sediments, and biota, taking advantage of its background of experience and knowledge.

Figure 56.5 shows the network of stations of the water quality monitoring programme carried out by the VWA inside the lagoon for the year 2001. This network is currently composed of:

- 16 stations, close to sources of civil and industrial liquid wastes, in which physico–chemical parameters are measured monthly, and inorganic and persistent organic pollutants (POPs) are measured less frequently; these measures are carried out directly by SAMA (VWA); and
- 28 stations distributed all over the lagoon, not influenced by liquid wastes discharges, and 2 control stations (in the coastal area outside the lagoon), in which physico–chemical parameters and inorganic micropollutants are measured monthly within the MELa1 programme (MAV-CVN, 2000) by the CVN. At the same 28 stations, POPs are measured less frequently by the CVN as part of other monitoring programmes (MELa2) (MAV-CVN, 2001). Within the context of MELa1, benthic diatoms and fouling communities, both potential indicators of sediment and water quality, are also monitored.

The network composed of the 30 MELa1 stations serves as a reference, supplying base-line information to other specific studies and monitoring programmes designed with different objectives. Significant efforts are being made by CVN to guarantee operational coordination between different activities and programmes with the key participation of one of its specialized member, Thetis S.p.A.

A new network of monitoring stations is being implemented by the Veneto Region and ARPAV to carry out regulatory controls inside the lagoon. This new effort will involve 11 stations monitored for water quality, sediment and biota as requested by the 152/1999 Decree, and 24 additional stations monitored for health control of mussel farming.

After three years of monitoring by VWA (MELa1), the knowledge of water quality in the lagoon and its seasonal variability, in so far as physico-chemical and trophic parameters are concerned, has considerably improved (see Zirino, chapter 55). For example, maps of typical winter and summer salinity, nitrate, ammonia and chlorophyll distributions are shown in Plate 23. As shown therein, during winter, the nutrient distribution appears to be primarily controlled by river and industrial wastewater outflows, and by physical mixing. This is visible in the spatial gradient of nutrients, which decreases seaward. During summer, the nutrient concentrations in the water seems to be controlled by the primary production processes as well as by physical mixing.

Since the year 2000, the pollution control service of the VWA (SAMA) has been measuring micropollutants in water at 16 stations within the lagoon. Zn, Cd, Pb and Cu in water are measured directly in the field and in the laboratory using voltammetric techniques, whereas persistent organic micropollutants are measured by using preconcentrations techniques. As an example, typical distributions of PCDD/Fs in water are shown (Fig. 56.6). The higher concentrations in the waters close to the industrial area of Porto Marghera are evident and previously reported (MAV, 2002a).

As the Water Framework Directive (WFD) EU 2000/60 states, assessing the quality of a water body implies the monitoring of all the matrices of the water body, including the biota, which is highly sensitive to the quality of water and sediment. The VWA has also been monitoring aquatic vegetation in the lagoon since the 1990s, when dystrophic effects, caused by the overproduction of green macroalgae, had severe impacts on the ecosystem. Submerged aquatic vegetation is a key indicator of the ecological condition of the lagoon. The new recent mapping by VWA (MELa 2 programme) can be compared to

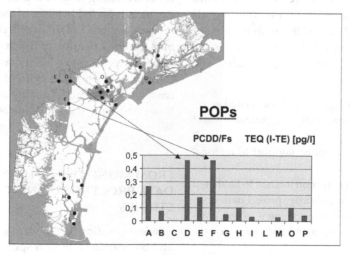

Fig. 56.6   Water quality of the lagoon close to the waste water outflows - typical PCDD/PCDF distributions in the 16 stations monitored by SAMA (2000).

the 1990 census, thereby allowing a cross comparison of the changes that have occurred in the past 12 years. The area colonized by seagrasses (*Cymodocea nodosa*, *Zostera marina* and *Zostera noltii*) appears to be stable at the whole lagoon scale; however, significant modifications of the distribution of specific populations have also occurred since the 1990s at sub-basin scale (see Rismondo, chapter 60). Specifically:

- large areas of the central basin are currently colonized by *Zostera marina*, unlike in the 1990s;
- substantial modifications have occurred in the southern basin: *Cymodocea nodosa* and *Zostera marina* pure meadows registered a strong increase, whereas *Zostera noltii* showed a large decrease;
- between 1990 and 2002 the northern basin has suffered a large decrease of *Zostera noltii*.

A more detailed evaluation of the VWA monitoring results is currently carried on in the MELa 2 programme (see Zirino, chapter 55).

MONITORING THE NORTH ADRIATIC SEA

Several countries face the Adriatic Sea and the activities of their settlements impact on the environmental quality of this semi-enclosed and very valuable resource. For this reason, the Veneto Region, through ARPAV, has recently set up the Northern Adriatic Observation Centre (Osservatorio Alto Adriatico or OAA) as part of the Interreg III Programme (ARPAV *et al.*, 2002b), with the goal of harmonizing different regional, national and transnational planning and monitoring activities. The main aims of the OAA are:

- Co-ordination and integration of sea-related activities within institutional bodies, organizations and research institutes;
- Providing technical-scientific support to regional policy makers;
- Establishing topics of common interest to be shared and to integrate the cross-border project partners;
- Common and efficient management of the environment by establishing a sea data bank for the Veneto Region;
- Decisional support aimed at restoration and protection of coastal water quality;

- Control and monitoring aimed at protection from pollution;
- Supporting tourism and improving value;
- Protection and improving the value of areas of naturalistic interest; and
- Disseminating information and increasing environmental knowledge.

A forthcoming action is to 'network' the different monitoring activities, including the offshore monitoring managed by CNR-ISMAR (Consiglio Nazionale delle Ricerche – Istituto di Scienze MARine) of Venice, and to implement a system of meteorological marine buoys to provide a well-organized database, accessible in real time to the stakeholders and operators.

The current (2003) Adriatic coastal waters monitoring programmes carried out by ARPAV for different regulatory purpose are:

- monitoring water quality at 90 littoral stations to check the quality of bathing waters as requested by law; controls are carried out every two weeks and more often during summer;
- monitoring the coastal water body (waters, sediments and biota) at 40 stations along 12 transects; and
- monitoring water quality for mussel farming at 33 stations.

For example, in Fig. 56.7, the quality of coastal waters resulting from year 2000, 2001 and 2002 monitoring is reported in terms of the trophic index TRIX for each station; TRIX (Vollenweider *et al.*, 1998) permits the classification of the trophic state of coastal areas, as prescribed by Italian laws (Legislative Decree 152, 1999). It is evident in Fig. 56.7 that the coastal waters of better quality are far from river outflows and near the shoreline along the Lagoon of Venice.

FROM MONITORING TO INTEGRATED DATA PROCESSING AND TARGETED STUDIES

Besides operating their own monitoring programmes and coordinating their respective data acquisition systems, the Veneto Region and the VWA collaborate to integrate data processing in order to have a shared

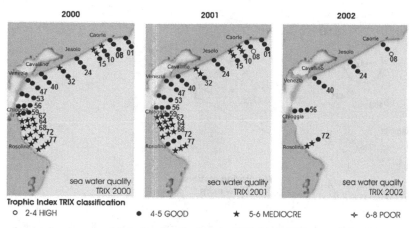

Fig. 56.7   Quality of coastal waters: trophic index distributions in 2000, 2001, 2002 years (ARPAV).

conceptual model of the functioning of the lagoon ecosystem(s).

INTEGRATED DATA PROCESSING

A preliminary integrated analysis of water quality data from the lagoon and of the coastal area has been recently carried out. Plate 24 shows the combined results of the seasonal averages (January–March = winter; July–September = summer) of trophic parameters in coastal waters measured by the combined ARPAV and CVN programme in 2001 (MAV-CVN, 2002).

Nitrate levels in the waters are similar inside and outside the lagoon, with higher concentrations close to the river outlets, especially in winter when there is greater rainfall and fresh water outflow. The map of chlorophyll 'a' in summer shows, as expected, the overall higher productivity of lagoon waters with respect to coastal waters, due to higher nutrient availability, higher temperatures and longer residence times (see Zirino, chapter 55).

POLLUTANT LOADS

One of the most significant cooperative activities is the periodic updating of pollutant loads into the lagoon within the programme MELa 2 (MAV-CVN, 2001). This activity includes the collection of data and

evaluations performed by different administrations, agencies and research institutes to produce an updated and shared reference framework of the lagoon's external and internal pollutant loads (nutrients, organic and inorganic micropollutants). It is a continuously evolving effort to make available a complete base of information needed for planning and designing depuration and restoration works. The estimate of annual loads into the lagoon for the year 2001 is available for a large number of parameters, but at different levels of confidence.

THEMATIC AND INTERDISCIPLINARY STUDIES

Background knowledge, and the large and shared database resulting from monitoring, allows for the design of specific and interdisciplinary studies to investigate particular processes or phenomena or to extend the results of local investigations to larger areas.

The ARTISTA study, carried out by the VWA through CVN within the MELa1 project, is an example. The main objectives of this study are to overcome the lack of information on (1) the mobility of micro pollutants (heavy metals and organic compounds) in the sediments of the lagoon, (2) the role of the sediments as sources of pollution, and (3) the amount of pollutants entering the trophic web. The

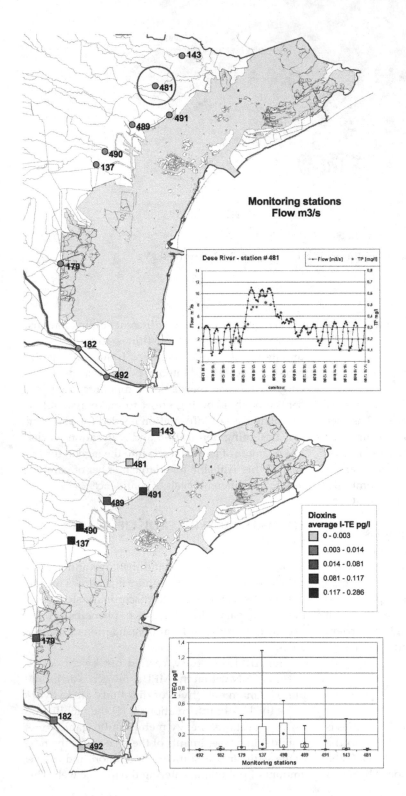

Fig. 56.8   ARPAV specific studies: investigation during flood events of the rivers and study on dioxin (PCDD & PCDF) in rivers conducted in 2002.

study integrates experimental investigations with the use of modelling, supported by a large database of existing data on contamination. The sites investigated are ten shallow areas with different environmental characteristics (Fig. 56.5).

A further example of specific studies and investigations carried out by Veneto Region through ARPAV is shown in Fig. 56.8, which shows some results from the research on flood events during severe rainfalls, carried out in 2002. This kind of study allows a more detailed description of the flows, and of the pollutant loads, into the lagoon. Specific measurement campaigns of POPs (PCDD/Fs, PCB and HCB) in the rivers have also been conducted (see Fig. 56.8) (ARPAV et al., 2002c).

OUTPUT AS DECISION TOOLS

Data, information and analyses deriving from monitoring studies and investigations carried out by the Veneto Region and VWA are available to the public at large. These are listed below:

- Data banks, airborne images, thematic mapping and GIS: available from the Information Service and CED (Data Elaboration Centre) of VWA, SIRAV Centre of Veneto Region/ARPAV for the coastal waters and the Venice Lagoon watershed;
- Publications and Technical Reports: Master Plan for pollution input reduction, prepared by Veneto Region (Regione del Veneto, 2000); *Quaderni trimestrali* edited by CVN/VWA; project Technical Reports (not published, but available on request), edited by CVN/VWA; publications edited by SAMA/VWA (Reports on the monitoring results of Industrial waste waters at Porto Marghera (MAV, 2002a), Reports on the monitoring results); publications of the monitoring results of rivers and watershed (ARPAV et al., 2002a) edited by ARPAV, North Adriatic Monitoring Centre bulletins (ARPAV et al., 2002d) by ARPAV; and
- Information to the public: *Puntolaguna* of the Venice Water Authority/CVN, accessible to anyone who asks, and the internet sites: www.regione.veneto.it, www.arpa.veneto.it, www.salve.it , www.magisacque.it

CONCLUSIONS AND PERSPECTIVES

A more complete and increasingly accurate knowledge base is required by the plans and projects aimed at protecting Venice, its lagoon and its valuable ecosystems. The Veneto Region, with its Regional Environmental Protection and Prevention Agency, and the State, through the Venice Water Authority, directly or with the support of its concessionary, the Consorzio Venezia Nuova, are carrying out monitoring activities in accordance with their respective institutional competencies. These agencies are striving to develop a more general, unified and integrated monitoring system, with the overall aim of setting up a common conceptual model and of developing a shared perception of the lagoon and its evolution.

The perspective, contained in the Water Framework Directive EU 2000/60, can only strengthen the cooperation between different administrations, authorities and research bodies operating on the lagoon. Its successful application requires that:

- the continuity of measurements and observations of long time series must be maintained in order to assess the evolution of the lagoon system and to evaluate the effectiveness of the restoring actions; and
- the integration of the monitoring activities, and in particular the data processing and the development of conceptual models, must be continued and improved to gain a better overall ecological perspective.

BIBLIOGRAPHY

ARPAV and Regione del Veneto. 2002a. *Bacino scolante nella laguna di Venezia. Rapporto sullo stato ambientale dei corpi idrici.* Anno 2001.

ARPAV and Regione del Veneto. 2002b. *Programma di iniziativa comunitaria Interreg III A/Phare CBC Italia Slovenia 2000–2006.*

ARPAV and Regione del Veneto. 2002c. *Caratterizzazione delle piogge intense sul bacino scolante nella laguna di Venezia.*

ARPAV and Regione del Veneto. 2002d. *Bollettini della costa veneta e Rapporti di balneazione*, website: www.arpa.veneto.it.

Decree of Ministry of the Environment, 23 April 1998 and others. *Requisiti di qualità delle acque e caratteristiche degli impianti di depurazione per la tutela della laguna di Venezia*, Official Gazette 18 June 1998, No. 140.

Ghetti, P. F. 1995. Indice Biotico Esteso (I.B.E.). *Metodi di analisi per ambienti di acque correnti, Istituto di Ricerca sulle Acque – CNR Notiziario dei metodi analitici*, Supplemento al Quaderno No. 10. Supplemento al No. 6/95 di Biol. Amb.

Gordon, G., Lord Byron. 1817. *Childe Harold's Pilgrimage.* London: Murray, canto 4.

Law No. 171. 1973. *Interventi per la Salvaguardia di Venezia.*

Law No. 798. 1984. *Nuovi interventi per la Salvaguardia di Venezia.*

Legislative Decree 152. 1999. *Disposizioni sulla tutela delle acque dall'inquinamento e recepimento della direttiva 91/271/CEE concernente il trattamento delle acque reflue urbane e della direttiva 91/676/CEE relativa alla protezione delle acque dall'inquinamento provocato dai nitrati provenienti da fonti agricole*, Official Gazette 29 May 1999, Ordinary Supplement, No. 124.

MAV. 2002a. *Sezione Antinquinamento del Magistrato alle Acque di Venezia. Relazione sulle caratteristiche degli scarichi idrici dell'area di Porto Marghera.* Dati relativi al 2000.

MAV. 2002b. *Sezione Antinquinamento del Magistrato alle Acque di Venezia. Monitoraggio delle acque della laguna di Venezia.* Dati relativi al 2000–2001.

MAV-CVN. 2000. *Attività di monitoraggio ambientale della laguna di Venezia.* Progetto esecutivo della 1^ fase triennale, MELa 1.

MAV-CVN. 2001. *Attività di monitoraggio ambientale della laguna di Venezia.* Progetto esecutivo della 2^ fase triennale, MELa 2.

MAV-CVN. 2002. *Attività di monitoraggio ambientale della laguna di Venezia.* Progetto esecutivo del 1° stralcio triennale, MELa 1. *Trattamento ed analisi dei dati, elaborazioni in linea ed interpretazioni spazio-temporalli, Pastres & Solidoro.*

Regione del Veneto. 2000. *Piano Direttore 2000 – Piano per la prevenzione dell'inquinamento e il risanamento delle acque del bacino idrografico immediatamente sversante nella laguna di Venezia.*

Vollenweider, R. A., Giovanardi, F., Montanari, F., and A. Rinaldi. 1998. Characterization of the trophic conditions of marine coastal waters, with special reference to the NW Adriatic Sea: proposal for a trophic scale, turbidity and generalized water quality index, *Environmetrics* 8, 329–57.

57 · Modelling water quality and ecological processes in the Venice Lagoon: a review

C. SOLIDORO, R. PASTRES AND D. MELAKU CANU

INTRODUCTION

In recent decades, the reputation of mathematical models has been steadily increasing in fields such as ecology and the earth and environmental sciences in which their introduction was welcomed with a certain dose of scepticism in the 1970s. With regard to the modelling of ecological and ecosystem processes, such an interest in the abstract realm of mathematics is justified for a number of very good reasons (Begon *et al.*, 1996). In our opinion, the two fundamental ones are: (1) models allow one to test the hypothesis concerning the main causal relationships about the dynamic of an ecosystem in a quantitative way and (2) models can be used to forecast the evolution of a given ecosystem, in relation to changes in the forcing functions.

In this chapter, we review the literature concerning the water quality, ecological and biogeochemical models which have been developed to investigate the Venice Lagoon ecosystem. Because of its peculiarity, this site has attracted the interest of both local and international scientific communities. As a result, over the last fifteen years, several specific modelling studies have been published in peer reviewed international scientific journals and presented in technical reports and conference proceedings. In spite of the fact that good scientific work may be found also in these latter forms, we only take into consideration here the peer reviewed papers as this choice gives one the possibility of consulting all the references reported. A screening of these papers indicates that most of them deal with the modelling of nutrient enrichment and eutrophication, whose consequences for the quality of the water and the sediment were acutely felt in the Venice Lagoon in the 1980s and early 1990s (Sfriso *et al.*, 1989, 1992). However, in the last five years the interest of the scientific community has shifted towards a more general understanding of biogeochemical cycles and ecosystem functioning. These topics have been investigated in a number of recent papers. As regards the model typology, an analysis of the peer reviewed literature shows that biogeochemical cycles and the dynamics of water quality and biological populations have been investigated by using deterministic models based, in general, on mass-balance equations.

This chapter is divided into two sections, in accordance with the two main goals of the modelling activity outlined above. The first section is dedicated to a description of the models proposed as research tools, while the second reviews the papers in which the models have been used for analysing policy alternatives, connected with species harvesting and ecosystem management issues.

MODELS AS RESEARCH TOOLS

The papers mentioned here present mathematical models which have been developed with the main aim of gaining insights into given biogeochemical and ecological processes, regardless of the relevance of the models for management purposes. In most cases, spatial variability has not been explicitly taken into consideration and, as a result, the investigations have been carried out using ordinary differential equation (ODE) models. However, attempts at coupling the above processes with the transport one, using partial differential equation (PDE) models can also be found.

In the 1980s, rising concern about the consequences of acute eutrophication prompted a number of studies on nutrient enrichment and primary production. The earliest attempt at modelling these processes in the Venice Lagoon dates back to the 1980s. The main findings were presented in a special issue (Dejak *et al.*, 1987), in which the authors presented a 3D reaction-diffusion model. The set of state variables included dissolved nitrogen and phosphorus and the densities of the phytoplankton and zooplankton. The structure of the reaction term was quite similar to those used in the classic water quality models of that time (Di Toro *et al.*, 1971; Kremer and Nixon, 1978). The growth of the phytoplankton pool depended on the water temperature, the incident light intensity and the concentrations of DIN, dissolved inorganic nitrogen and RP, reactive phosphorus, while the decay term explicitly included the grazing pressure exerted by the zooplankton pool. The kinetic parameters, such as the maximum phytoplankton growth rate and the nutrient half saturation constant, were estimated on the basis of site-specific data (Bertonati *et al.*, 1987). However, both the computational demands of this 3D coupled transport-reaction model and the lack of comprehensive calibration and validation data sets, severely limited its application to a description of the seasonal evolution of the ecosystem. In fact, the model was efficiently run on supercomputer in the early 1990s (Pastres *et al.*, 1995), even though the computational domain was limited to the central part of the lagoon.

Several field studies, performed in the 1990s (Sfriso *et al.*, 1992) indicated that the occurrence of repeated and extended anoxic crises that severely affected the sediment and water quality in the Venice Lagoon were related to the massive proliferation of the benthic macroalgae *Ulva rigida*, by far the dominant species in the 1980s. These findings prompted the initiation of modelling research activities, which led to the identification of a set of ODE models (Solidoro, 1994; Coffaro and Sfriso, 1997; Solidoro *et al.*, 1995; Solidoro *et al.*, 1997c). These were later applied to other coastal environments (Hull *et al.*, 2000; Martins and Marques, 2002; Zaldivar *et al.*, 2003; Biber *et al.*, 2004). The structures of the above models share some relevant common features, as is shown in Fig. 57.1. In the first place, the forcing functions are: the water temperature, the intensity of the solar radiation, and the concentrations of DIN and RP in the water column. Furthermore, these models describe the growth of *U. rigida* as a two-step process: in the first, the macronutrients in the water column

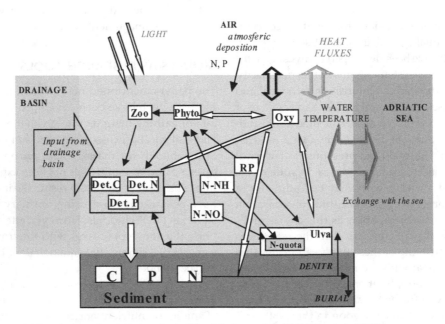

Fig. 57.1   Schematic description of nutrient cycles in *Ulva*/phyto models (Solidoro *et al.*, 1997).

are assimilated and, in the second, their intra-tissue storage is used for the synthesis of the new biomass. This two-step scheme was fully applied in the case of nitrogen, which was believed to be the most limiting macronutrient. As a consequence, intra-tissue concentration of nitrogen was explicitly considered as a state variable in these models. Conversely, based on the analysis of the specific field data, the influence of the dissolved inorganic phosphorus on the growth rate was simulated by using a one-step Monod kinetic (Solidoro, 1994; Coffaro and Sfriso, 1997; Solidoro et al., 1997c). The dependence of the growth rate on the set of potentially limiting factors, among which is the nitrogen intra-tissue concentration, is described by using a multiplicative model. However, the above models also presented some differences concerning the way in which the effects of light availability and water temperature on growth rate were simulated (Bendoricchio et al., 1994; Solidoro et al., 1996; Coffaro and Sfriso 1997) and regarding the parameterization of the mortality term. In fact, the first-order mortality term proposed in Coffaro et al. (1997) is characterized by a step-wise dependence on water temperature. Therefore, the mortality rate steeply increases if the water temperature exceeds a given threshold value, while, in accordance with the formulation proposed in Solidoro et al. (1996), the mortality rate increases below at low DO concentrations, regardless of the water temperature. The results presented in the above quoted papers demonstrate the fact that the introduction of the nitrogen intra-tissue concentration as a state variable is crucial for an accurate description of the seasonal evolution of U. rigida. In fact, the two-step kinetic allows one to simulate the variability in the apparent nitrogen half-saturation constants which are reported in Fujita et al. (1988). This mechanism could also explain the phase-lag between minimum concentration of dissolved nitrogen and maximum standing crop which characterized the time series of field data collected in the Venice Lagoon in the 1980s.

The competition between U. rigida and the other primary producers in the Venice Lagoon was studied in a number of papers published in the mid-1990s. In particular, the competition with seagrasses was investigated using an ODE model in which the equations proposed by Coffaro and Sfriso (1997)

were coupled with those used in Bocci et al. (1997) to simulate the dynamic of Zostera marina, the most abundant seagrass species. The presence/absence of the two species in different areas of the lagoon was explained in terms of the main forcings, i.e. the dissolved nutrient concentrations and the average kinetic energy of the water masses. The results, presented in Coffaro and Bocci (1997), indicate that Z. marina and U. rigida can co-exist only in the presence of low concentrations of dissolved inorganic nutrients and with medium to high bottom disturbances. Under these conditions, the ability of seagrasses to assimilate nitrogen and phosphorus through their root systems, and their resistance to bottom disturbances, counterbalance the intrinsically higher growth rate of the seaweed. Valuable insights into the competition mechanisms were also provided by the results presented in Coffaro et al. (1997). In this paper, the authors presented a simplified version of the coupled Z. marina – U. rigida model, in which some terms of the mass balance equations were parameterized as functions of two key parameters, the U. rigida surface to volume ratio and the density of below-ground biomass of Z. marina. These quantities were not treated as constant parameters but, rather, were dynamically adjusted, in order to maximize the system exergy in the course of the simulation (Bendoricchio and Jorgensen, 1997). The results were quite consistent with those obtained using the complete mass-balance model described in Coffaro and Bocci (1997) and pointed out the importance of those factors in determining the outcome of the competition. Therefore, this application of the maximum exergy principle provided a simple rationale for explaining the spatial distribution of these species in the lagoon. A later attempt at modelling the population dynamic of Z. marina can be found in Zharova et al. (2001). This model is characterized by a more accurate description of the relationships between biomass growth and water temperature, but the long term evolution of both the above and below ground biomasses, predicted by this model, is affected by instabilities, as shown by Pastres et al. (2004). The competition between U. rigida and phytoplankton was explored in two papers, Solidoro et al. (1994) and Solidoro et al. (1997c), using an ODE model, based on the formulations proposed by Bertonati et al.

(1987) and Solidoro et al. (1994, 1997c) (Fig. 57.1). These papers suggest that the while U. rigida is more efficient than phytoplankton in taking up nutrients, high densities of phytoplankton can significantly shade U. rigida mats and reduce the levels of light at the bottom to values too low for the efficient sustainability of seaweed growth. In addition, simulations indicate that if U. rigida biomass exceeds a critical threshold, the concentration of dissolved oxygen during the night drops to very low values, and, eventually, anoxic conditions occur. Such a condition causes an increase in the mortality of U. rigida and, in turn, in the concentration of easily degradable organic detritus and, ultimately, the DO demand. As a result, a collapse of U. rigida mats occurs, followed by the bloom of fast-growing phytoplankton species which take advantage of the sudden release of nitrogen and phosphorus from the remineralization of U. rigida tissues under reducing conditions. These predictions are qualitatively in agreement with the observations reported in Sfriso et al. (1989, 1992).

The reaction network proposed in Solidoro et al. (1997c) was then integrated into a 3D reaction-diffusion model. This model was used to analyse the seasonal evolution of the spatial distribution of U. rigida (Solidoro et al., 1997a) in the central part of the lagoon, and for the investigation of the inter-annual trend under different scenarios of meteorological conditions and nitrogen and phosphorus loads (Solidoro et al., 1997b). The explicit introduction of transport processes and bathymetric constraint gave new insights into the dynamic of an anoxic crisis. The analysis of the spatial distribution of dissolved nutrients and dissolved oxygen shows that the edge of a seaweed mat acts as a barrier for the diffusion of nutrients within the mat. Therefore, the concentrations of dissolved nutrients within the mat become lower and lower. This fact, together with the low level of light intensity near the seabed, slows down photosynthetic activity and triggers the onset of the crisis. In fact, in these sub-optimal conditions, the mechanisms of DO renewal can not compensate the DO demand exerted by seaweed respiration and an anoxic crisis can take place.

The relative importance of the initial conditions and parameters in determining the maximum yearly standing crop was studied from a more quantitative point of view by Pastres et al. (1999). In this paper, the authors present the results of a formal global sensitivity analysis which was performed using the model presented in Solidoro et al. (1997b). However, due to the computational burden of the global sensitivity analysis, the model was applied to a smaller computational grid, considered as representative of a shallow water basin crossed by deeper canals, such as the Venice Lagoon. The application of the Sobol variance-based method of sensitivity analysis (Saltelli et al., 2000) to this idealized system showed that the density of U. rigida in early spring was one of the key factors in regulating the maximum standing crop and, therefore, in the appearance of anoxic crises in late spring. These findings point to the role of the meteorological conditions in determining the maximum standing crop, since the biomass density in the early spring strongly depends on the levels of bottom disturbance and water temperature during the preceeding winter. In fact, a sharp decrease in the biomass of U. rigida occurred in the early 1990s. The decline in this population was probably due to the concurrence of three main factors: the enforcement of a more severe environmental legislation, which caused a marked reduction in the concentrations of ammonia and RP in the late 1980s (Pastres et al., 2004); the persistence of adverse meteorological conditions in the early 1990s (Sfriso and Marcomini, 1996); and the mechanical disturbance of the seabed. The latter was related to an increase in the use of shell-fishing devices. As a consequence of the collapse of the U. rigida population, the efforts of the local community of environmental modellers were directed, on the one hand, towards the analysis of the dynamics of certain species and, on the other, towards more general attempts at understanding and modelling both the nitrogen and phosphorus biogeochemical cycles and the structure of the whole trophic network.

With regard to the modelling of the dynamic of the ecosystem compartments or species, attention was focussed on the Tapes philippinarum bivalve, which was introduced into the lagoon in the mid-1980s and has become a remarkably profitable fishery target since the early 1990s. However, such intensive shell-fishing activity is a cause of concern from both ecological and morphological standpoints, because its impact on the benthic community and on the composition of the superficial sediment is very high. The first modelling study, which was aimed at understanding the reasons behind the spreading of this bivalve in the lagoon, led to the definition of an ODE bio-energetic model which simulates the growth of this filter feeder, in relation to water temperature, seston concentration and energetic content (Solidoro et al., 2000). This model was then used as a basis for developing a population dynamic model for the simulation of wild clam recruitment and survival (Solidoro et al., 2003).

The investigation of the nitrogen and phosphorus biogeochemical cycles was recently addressed in two papers (Cossarini et al., 2001; Solidoro et al., 2002) which present and discuss the results obtained using TDM (*Trophic Diffusive Model*), an improved version of the 3D reaction-diffusion model put forward in Dejak et al. (1987) and Pastres et al. (1995). The present version of the model includes the whole lagoon and can be run either on a high resolution (100 m × 100 m × 1 m), or medium resolution grid (300 m × 300 m × 1 m). In both cases, the integration time step is one hour. Advective terms are not considered, since the residual currents –averaged over the tidal cycle- are taken to be negligible (Dejak et al., 1998). Tidal mixing is implicitly taken into account by a space varying eddy-diffusion transport mechanism: the estimation of the anisotropic diffusivity tensor is described in Pastres et al. (2001). Therefore, this model can neither resolve time scales shorter than the tidal one nor provide suggestions on ecosystem response to variations in wind regimes or in other forcings capable of modifying the hydrodynamic fields. However, it can simulate, in a satisfactory way, climatological aspects and reproduce the main features of the observed seasonal evolution of DIN, RP and phytoplankton density (Solidoro et al., 2002). This model has been used for carrying out a detailed analysis of the nitrogen budgets of the lagoon and its sub-basins, and for estimating the annual fluxes among different biotic and abiotic compartments. The results, presented in the above-mentioned papers and in Solidoro et al. (2005), indicates that the lagoon, as a whole, exports nitrogen towards the sea and that the exchanges through the inlets play an important role in keeping nitrogen concentrations at an acceptable level. In fact, around one third of the dissolved inorganic nitrogen which

enters the water compartment from recycling within the system, net input from the drainage basin and from the other sources of nutrient, is used by primary producers, one third is transferred into the sediment compartment, and one third is exported towards the sea. The analysis of the model results also shows that the non-homogeneous spatial distribution of the tributary discharges and the nitrogen and phosphorus point-sources are the main cause of the differences in ecosystem response and water quality in the three sub-basins. Nutrient-poorer sub-basins fix a ration of available inorganic nutrient higher than nutrient rich ones. However, these sub-basins are more efficient in transferring the biomass to the highest trophic levels.

Another, more recent, attempt at analysing water quality and biogeochemical cycles in a spatially varying context led to the development of VELFEEM (VEnice Lagoon Finite Element Ecological Model) (Melaku Canu et al., 2001). This is a comprehensive transport-reaction model made by integrating a transport and a water quality module. The transport module is based on a 2D primitive equation finite element hydrodynamic model (Umgiesser et al., 1987), which was calibrated against a set of tidal wave propagation data and then validated by comparing its output with independent sets of water elevation, temperature and salinity data (Umgiesser et al., 2004). The water quality module is a purposely-modified version of the US EPA WASP water quality module (Umgiesser et al., 2003). Even though the coupled water quality-transport model has not yet been formally calibrated against water quality data, the effort is certainly interesting, since this model allows one to fully take into account the influence of the physical forcings on the dispersion processes and therefore on water quality. In fact, even if wind and tide are the major forcings driving the instantaneous general circulation, the actual transport is the result of the non-linear interactions between the actual wind field, tide, topographic constraint, inertia and the pre-existing set-up inside the lagoon (Solidoro et al., 2004). In Melaku Canu et al. (2003), the integrated model was used for analysing the relative importance of the wind intensity and directions; water elevation at the inlet; and nutrient loads from the drainage basin, in determining the spatial distribution of concentrations of nutrients and dissolved oxygen, as well as of the trophic indicator TRIX (Vollenweider *et al.*, 1998). Results indicate that the largest modifications are induced by variations in the nutrient loads, but that differences in wind regimes can cause significant differences in spatial distributions of water quality parameters too. Analysis of model results also points to the existence of high diel variabilities, superimposed on spatial and seasonal variabilities. This suggests caution in the use of averaged indexes for quantifying water quality status.

The modelling of the general structure of an ecosystem is becoming increasingly popular, since it provides a framework for integrating the results of specific population dynamic models. Such an approach has been applied to the Venice Lagoon in an attempt at estimating the energy fluxes through the ecosystem (Carrer and Optiz, 1999) and, more recently, for analyzing the changes in the ecosystem functioning which were brought about by the spreading of *Tapes philippinarum* (Libralato *et al.*, 2002; Pranovi *et al.*, 2003). These papers present the results of a steady-state analysis of the trophic network, simulated using the ECOPATH software. The results support the hypothesis that, in the year 1998, dissolved and particulate organic matter detritus accounted for a significant fraction of the energetic demand exerted by the *T. philippinarum* population. Since sediment resuspension represents, in general, an important source for organic matter, one can say that the high *T. philippinarum* standing stock was supported by the detritus which had been accumulating in the sediment during the acute eutrophication period. Trophic network models also provide a basis for the development of bioaccumulation and eco-toxicological models. A first example of the application of such models to the Venice Lagoon can be found in Carrer *et al.* (2000). This paper presents the estimation of the steady-state concentration of dioxin congeners in the different compartments of the trophic web.

MODELS AND ECOSYSTEM MANAGEMENT

In this section, we present three case-studies which illustrate how models have been used to address some of the complex management problems which

are still a source of concern regarding the Venice Lagoon. These problems were dealt with by using coupled transport-reaction models. In this regard, it must be stressed that the formal calibration/validation of such complex models is still an ongoing problem for two main reasons. On the one hand, the presence of many unknown input factors, i.e. the forcing functions, the initial and boundary conditions and the system parameters, makes it very difficult to choose the sub-set of factors to be tuned. On the other hand, the spatial and temporal resolution of these models is usually much higher than those of the time series of field data: as a result, there is no general agreement concerning the measurement of the misfit between the data and the model output (Evans, 2003). The models used in the selected examples given below were not formally calibrated, but their output had previously been compared with a multivariate space-time series of field data. These comparisons, which are reported in the papers quoted, showed that these models were able to capture the main features of the temporal evolution of the spatial distributions of the data. The first two case-studies represent specific applications concerning two general problems, i.e. species harvesting and the management of Nitrogen and Phosphorus loads. The third case-study concerns a very site-specific, but extremely important problem, i.e. the assessment of the environmental impact of the closures of the lagoon inlets.

Managing the fishing and rearing of manila clams

Manila clam (*Tapes philippinarum*) was introduced into the Venice Lagoon in the first half of the 1980s because the worsening of environmental conditions during the 1970s and 1980s had led to a drastic reduction in the population of the autochthonous *Tapes decussates*, one of the targets of the local artisanal fishery. The new species adapted extremely well to the local environment, quickly spread outside the experimental lots and colonized large areas of the lagoon. After a few years, the fishing community realized the potential economic value of the wild stocks, which from 1985 to 1990 had been growing at an exponential rate. As a consequence, the fishing pressure on

Tapes philippinarum suddenly increased at the beginning of the 1990s. However, this fishing effort was not subject to any regulatory policy. This was in spite of the fact that the fishing devices used disrupt the seabed and resuspend the sediment, thus causing a severe impact on both the benthic community and the lagoon morphology. Furthermore, cases were found in which the shellfish were released to the local market without having undergone any type of health checks. Nonetheless, the prospect of good revenues led to an increase in the number of fishermen who targeted *T. philippinarum* and, as a result, the Venice Lagoon has accounted for a large share of the Italian production of this species (around 80 per cent) over the last decade. However, as is well known in applied ecology, open access exploitation strategies are neither ecologically nor economically sustainable in the long run (Clark, 1976) and, therefore, appropriate regulatory policies needed to be introduced. This awareness prompted the initiation of a number of studies. The first step was the definition of individual growth and population dynamic models concerning *T. philippinarum* (Solidoro *et al.*, 2000; Solidoro *et al.*, 2003). These models were used in order to compare the ecological consequences and the revenues from controlled fishing activities (fixed quotas and fixed harvesting effort policies) versus the culture-based fishery regimes (Solidoro *et al.*, 2003). The results show that a culture-based fishery is to be preferred to a regulated type of fishing activity, and that ecologically conservative strategies (in which only large individuals are caught and always by using low efficiency fishing devices) are also more remunerative in the long run. The integration of the results of these studies in the 3D coupled water-quality transport model, TDM, required the characterization of the socio-economic context, in order to assess: (1) the economic income the local community received from this activity; (2) how many people were involved and where they lived; and (3) a first estimate of the total rearing area (3×10^7 m^2) which could support a production comparable to the one obtained in the open access fishing regime if a culture based fishery regime were implemented (Melaku Canu *et al.*, 2002). Finally, the potentialities of the MTD integrated model as a decision support system for the management of the rearing and/or culture-based fishery of *T.*

philippinarum in the Venice Lagoon have been illustrated by Pastres *et al.* (2001). This paper, quoted in the Integrated Strategic Design Plan for the Coastal Ocean Observations Module of the Global Ocean Observing System (UNESCO, 2003), represents a complete example of the application of an integrated modelling approach to the exploitation of a living marine resource within the framework of a classical Integrated Coastal Zone Management problem. In fact, the authors used the model for evaluating phytoplankton production in the different areas of the water basin, in relation to the average energy requirements of *T. philippinarum*. This information was then combined with a set of external constraints: the unhealthy areas were excluded first, then the bathymetry was taken into account. The locations of the lots to be devoted to clam rearing were then identified, by partitioning the total required rearing surface (3×10^7 m^2) in keeping with the current spatial distribution of the fishing effort and in accordance with the criteria outlined above. The results of the selection process compare quite well with the spatial distribution of the areas which the Local Authorities assigned to the *T. philippinarum* fisheries in 2001. The economic yields from different seeding density strategies in the selected rearing areas were computed and compared with the estimated yield for the year 1998. The model results indicated that the revenues from the rearing/culture-based fishery regimes in the selected areas would be similar to the present ones, but that the environmental impact would be much lower.

The estimation of Maximum Permissible Loads

In 1999 the Italian government issued an innovative law for the regulation of pollutant loads in waterbodies, which was based on the concept, in US-EPA terminology, of the Maximum Permissible Loads (MPL), or, the TDML (Total Daily Maximum Load) principle. Accordingly, the Local Authorities were entrusted with the drawing up of an inventory of the sources of pollution, with the subsequent settings of the required level of emission regarding each activity, in order to maintain the concentrations of potentially dangerous substances below the prescribed thresholds (known as the 'Quality Targets', 'QT', or 'Quality Standards', 'QS', in US EPA terminology). The implementation of this policy will clearly benefit from the use of mathematical models which can be used as tools both for (1) estimating the Maximum-Permissible-Loads (MPLs) which are compatible with the QTs; and (2) exploring the consequences of different input scenarios on ecosystem functioning. In fact, in mathematical terms, the loads are specified by a set of boundary conditions. Therefore, numerical models can be used to determine a functional relationship between the set of parameters which specify the boundary conditions and the output variables which are compared to the Quality Targets. Once this task has been accomplished, one can use the inverse of the above functional relationship, in order to estimate the MPLs which are compatible with the targets.

In Pastres *et al.* (2003), these applications were illustrated using the Nitrogen MPL in the Venice Lagoon as a case study, and the local sensitivity analysis with respect to each source of pollution as a tool. In fact, the alternative modelling approach consists of running the model repeatedly, in order to explore the response of the system to different load scenarios and then attempting to model the input-output relationship using statistical techniques. Such an approach is computationally expensive when dealing with large 3D models but rather interesting results can be obtained from an analysis of the effects of relatively small perturbations in the loads. These effects were computed using Local Sensitivity Analysis (LSA), which was performed in a computationally efficient way by means of an *ad hoc* implementation of the 'direct method' (Pastres *et al.*, 1997; Solidoro *et al.*, 2003). The MPL was related to the model output, i.e. the dissolved inorganic nitrogen concentration. The QT concerning the DIN was compared with the yearly averaged DIN concentration computed using the model, which was found to be linearly related to the nitrogen load. The results of the LSA were used to determine the percentage reduction in the nitrogen sources which would lead to the QT being respected. They were also used to explore the relationship between the DIN load and the percentage of the water basin in which the DIN yearly average is below the QT. Such a relationship was found to be highly non-linear regarding percentages of load reduction higher than 70 per cent. This means that, in this range, small reductions in the nutrient loads should bring about great benefits in terms of compliance with current legislation. In this regard, the paper also showed that a management policy based on the targeted reduction in the nutrient loads concerning specific sources would be more efficient in lowering such a percentage, rather than a general reduction of all of the sources. These findings point to the importance of a clear, and careful, definition of the QTs, since the identification of the optimal policy options greatly depend upon this definition.

Effects of the temporary closure of the inlets on water quality

Venice is famous all over the world because of its beauty, because it is a city built 'on the water', and because of the tidal flooding episodes, whose frequency has been increasing over the last decades. In order to prevent the flooding of the city, the Ministry of Public Works, almost 40 years after the worst flooding episode in 1966, has recently started a series of preliminary interventions, aimed at installing a set of mobile pneumatic gates, named MOSE, at the three inlets. Once in operation, the gates could be inflated, in concomitance with exceptionally high tide events, which usually occur in autumn and winter time. Such a solution to the flooding problem has been the subject of a heated debate, which is still far from over. One of the main reasons for concern is the possible impact of frequent and prolonged temporary closures on water quality, a concern which was also expressed within the framework of a formal Environmental Impact Assessment, commissioned by the Ministry of the Environment, in 1998. In particular, one of the main concerns was that the lagoon might no longer be able to self-purify its water at a rate high enough to cope with the input of pollutants from the historical city and the drainage basin, as a consequence of the reduction in the amount of water exchanged with the Adriatic Sea resulting from the closure.

VELFEEM, the model already quoted in the previous section (Umgiesser *et al.*, 2003), was used to test this particular hypothesis (Melaku Canu *et al.*, 2001) under several simplifying assumptions. In this paper, which represents one of the very few peer reviewed articles dealing with aspects relating to the MOSE project, the authors compared the time evolution of the spatial distributions of the BOD-DO concentration in the lagoon. The latter could be predicted on a no closure basis versus those resulting from several scenarios regarding temporary closures. The comparison was made firstly by assuming – regarding both the 'closure' and 'non-closure' runs – that the nutrients and BOD loads from industrial, agricultural, and domestic sources were those estimated in winter-time in the years 1995-2000, and then by running a second set of simulations, in which all these loads were doubled. The results of the simulations suggested that the closures do not significantly affect the water quality of the lagoon. In fact, even when the more critical situations, represented by the 'high-load' scenario, was considered, the simulated concentrations of BOD both during and after several repeated closures were only slightly higher than the values obtained without any temporary closures, and the DO concentrations never reached significantly low levels. Furthermore, the model showed that, in all cases, BOD and DO concentration patterns rapidly relaxed toward the unperturbed ones, once the inlets were opened again. Therefore, these findings suggested that repeated temporary closures in wintertime should not be a source of concern, as far as the DO-BOD dynamics are concerned. Moreover, the simulations showed that the water quality was more sensitive to the BOD loadings than to the short term perturbations in the hydrodynamical regime caused by the temporary closures.

Of course, these analyses were not meant to substitute an environmental impact assessment study, but merely to make a contribution to it. In fact, the construction of the MOSE is a very important intervention, with multifaceted aspects, which has to be considered from a much larger and more multidisciplinary perspective. Nevertheless, the paper shows that numerical models might be extremely useful as regards the environmental impact studies.

CONCLUSIONS

This paper presents a review of the modelling exercises which were applied to an investigation of the Venice Lagoon ecosystem (Table 57.1). The review shows that most of the modelling efforts were focussed on the analysis of the eutrophication problems and the macroalgae proliferation which characterized the ecosystem in the 1980s and early 1990s. As regards the above issues, the ODE models proved to be useful tools regarding the investigation of specific processes, such as the kinetics of nitrogen assimilation, the modality of intraspecific competition, the interaction among Dissolved Oxygen and the cycle of Nitrogen and Phosphorus, the importance of remineralization process. PDE models, instead, where particularly useful in the analysis of the influence of the physical forcing functions and the transport process on the ecosystem dynamics, and for assessing the environmental impact and effectiveness of alternative management strategies. These researches produced some original results, which were also applied to other eutrophic water basins. For example, U. rigida population dynamic models were applied to other coastal areas, both in Italy (Hull et al., 2000; Zaldivar et al., 2003) and abroad (Biber et al., 2004; Martins and Marques, 2002). Similarly, models on growth and population dynamic of T. philippinarum were incorporated in studies performed in other ecosystems (Bacher et al., 2002; Dowd, 2003; Gagnegy et al., 2004; Nakamura et al., 2001). Furthermore, it should be noted that these investigations represented a sources of ideas for the whole scientific community, and, in turn, stimulated field and laboratory researches, such as those reported in Sfriso and Marcomini (1999), in which experimental measurements of U. rigida growth rate as a function of intra-tissue concentration of N and P were performed, or Riccardi and Solidoro (1996) in which the functional response of Ulva growth to light intensity and water temperature was investigated. The definition of a methodology for the identification of MPL (TDML), which can be implemented in any coastal area, and of a modelling-based approach to the rational management of renewable resource, namely clam harvesting in a culture-based fishery context, also represents valuable contributions to the international scientific community (UNESCO, 2003).

More recently, the attention of the local scientific community of environmental modeller turned, on one side, toward a more comprehensive understanding of

Table 57.1   *Ecological, water quality and biogeochemical models applied to the Lagoon of Venice.*

authors	Dim	Temporal Scale	Spatial Scale	forcing	Physical submodel	Biological submodel								Tox	Objective
						phyto	ulva	zostera	Nutr	Detr	DO	zoo	tapes		
Trophic Network															
Carrer and Optiz, 1999	0D	steady state	—	—	—	Ecopath trophic network								—	trophic network
Carrer et al., 2000	0D	steady state	—	—	—	Ecopath trophic network + dioxyn ripartition								Dio	dioxyn fate
Libralato et al., 2002	0D	steady state	—	—	—	Ecopath trophic network								—	trophic network
Pranovi et al., 2003	0D	steady state	—	—	—	Ecopath trophic network								—	tapes diet
Single Species															
Solidoro et al., 2000	0D	year	—	T, phyto	—	Bioenergetic model-								—	tapes
Macroalgae															
Zharova et al., 2001	0D	year	—	N,P	—	—	—	zostera	—	—	—	—	—	—	zostera dynamic
Bocci et al., 1997	0D	year	—	I,T,N	—	—	—	zostera	—	—	—	—	—	—	zostera dynamic
Bendoricchio et al., 1994	0D	year	—	I and T	—	—	Ulva	—	—	—	—	—	—	—	ulva response T,I
Coffaro and Sfriso, 1997	0D	year	—	I,T,N,P, hydrodyn	—	—	Ulva	—	QN/QP	—	—	—	—	—	ulva dynamic
Solidoro et al., 1997c	0D	year	—	I,T	—	Phy	Ulva	-	NH/NO/QN/P	DN/DP	DO	Zoo	—	—	ulva dynamics
Competition															
Coffaro and Bocci, 1997	0D	year	—	I, T, N, P, hydrodyn	—	—	Ulva	zostera	—	—	—	—	—	—	ulva/zostera
Coffaro et al., 1997	0D	year	—	I, T, N, P, hydrodyn	—	—	Ulva	—	QN/QP	—	—	—	—	—	ulva/zostera
Solidoro et al., 1997c	0D	year	—	I,T	—	Phy	Ulva	—	NH/NO/QN/P	DN/DP	DO	Zoo	—	—	ulva/phyto
Ulva Analysis															
Solidoro et al., 1997a	3D	months	Central part	I, T	diffusive	Phy	Ulva	—	NH/NO/QN/P	DN/DP	DO	Zoo	—	—	ulva spreading
Solidoro et al., 1994	3D	1 year	Central part	I, T	diffusive	Phy	Ulva	—	NH/NO/QN/P	DN/DP	DO	Zoo	—	—	ulva/phyto dynamic
Solidoro et al., 1997b	3D	4 years	Central part	I, T	diffusive	Phy	Ulva	—	NH/NO/QN/P	DN/DP	DO	Zoo	—	—	ulva dynamic
Pastres et al., 1999	3D	year	Central part	I, T	diffusive	Phy	Ulva	—	NH/NO/QN/P	DN/DP	DO	Zoo	—	—	sens. analysis (global)

authors	Dim	Temporal Scale	Spatial Scale	forcing	Physical submodel	Biological submodel									Objective
						phyto	ulva	zostera	Nutr	Detr	DO	zoo	tapes	Tox	
Water Quality															
Pastres *et al.*, 1995	3D	1 year	central part	L, T	diffusive	phy	–	–	NH/NO/P	–	DO	zoo	–	–	phyto code parallelization
Pastres *et al.*, 1997	1D	year	channel	L, T	diffusive	phy	–	–	NH/NOP	–	DO	zoo	–	–	sens. analysis (local)
Cossarini *et al.*, 2001	3D	year	lagoon	L, T	diffusive	phy	–	–	NH/NO/QN/P	DN/DP	DO	zoo	–	–	N cycles
Solidoro *et al.*, 2002	3D	year	lagoon	L, T	diffusive	phy	–	–	NH/NO/QN/P	DN/DP	DO	zoo	–	–	N cycles
Melaku Canu *et al.*, 2003	2D	1 year	lagoon	L,T,tide	Prim. Eq. FEM	phy	–	–	NH/NO/P	DN/DP/DC	DO	zoo	–	–	Response to forcings
Umgiesser *et al.*, 2003	2D	winter	lagoon	L,T,tide	Prim. Eq. FEM	phy	–	–	NH/NO/P	DN/DP/DC	DO	zoo	–	–	Setup integrated model
Applied Studies															
Solidoro *et al.* 2003	0D	10 years	–	T, phyto	–		–	–			DO	zoo	tapes	–	Tapes fishing/ rearing
Melaku Canu *et al.*, 2001	2D	winter	lagoon	L,T,tide	Prim. Eq. FEM	phy	–	–	NH/NO/P	DN/DP/DC	DO	zoo	–	–	MOSE
Pastres *et al.*, 2001	3D	year	lagoon	L, T	diffusive	phy	Ulva	–	NH/NO/QN/P	DN/DP	DO	zoo	tapes	–	Tapes DSS
Pastres *et al.*, 2003	3D	year	lagoon	L, T	diffusive	phy	–	–	NH/NO/QN/P	DN/DP	DO	zoo	–	–	Max Permissible Loads

the dynamic of primary production and of the biogeochemical cycles and, on the other, toward the modelling of the structure of the trophic network. These two, complementary, research fields provided some basic results, which are currently being used in preliminary attempts at developing bioaccumulation and ecological risk analysis models. These model are usually based on the coupling of a food web, a physico-chemical partition model and a bioaccumulation sub-models. Once the pollutant inputs are known, the partition model allows one to compute its distribution among the different abiotic phases, such as air, water and surficial sediment, while the other two sub-models permit one to simulate the biomagnification of its concentration through the trophic web. The first applications of this modelling approach to the Venice Lagoon (Carrer *et al.*, 2000; Della Valle *et al.*, 2004; Micheletti *et al.*, 2004) were successful in describing the general trend of bioaccumulation and partitioning among abiotic phases. However, the results which have been obtained thus far can not be considered as entirely satisfactory from a quantitative point of view, in particular for heavy metals and some organics. This is certainly due also to the inherent complexity of these processes, and to the scarcity of field data. Further investigations are, however, still needed, in order to better understand the key processes and quantify the key factors that govern the interactions between water column and surficial sediment.

The analysis of Table 57.1 shows that the majority of the studies which were carried out using 3D models were performed by using the TDM model, in which advective terms are averaged out over the tidal cycle and – since residual currents are negligible – only turbulent diffusive processes concur to transport phenomena. For this reason, the results presented in this set of papers refer to processes which occur on the seasonal and inter-annual time scales. However, some researches addressed also processes which are characterized by shorter time scales, as in the study for the evaluation of the impact on water quality of different scenarios of temporary closures of the inlets in concomitance with exceptionally high tide events, or in analysing the influence of different wind regime and/or different tidal agitation on spatial distribution of nutrients, plankton and DO. In those cases, the simulations were performed by using

the model VELFEEM, in which the transport module is based on a primitive equation hydrodynamic model. As already remarked, both of these models have been corroborated by comparison against different sets of experimental findings, even if a full calibration/validation has not been accomplished as yet, because of the problems intrinsic in such a procedure and of the lack of experimental information. To this regard it must be stressed that even if the Venice Lagoon has been deeply investigated, until recently the large majority of the field investigations were carried out by small groups in a un-coordinated manner. In addition, an integrated monitoring program of the lagoon and its boundaries, which include loads from the drainage basins, exchanges with the sea, air, and sediment, is not operative as yet. As a consequence, until a few years ago there was a lack of basic knowledge both about space- time variabilities of water quality parameters and about fluxes of matter through different ecosystem components and among different areas. However, the implementation of the new monitoring programs, promoted by the Venice Water Autorithy (Magistrato alle Acque, MAV) and by Regional Authority (Regione Veneto), and the launch of several new research programmes, mostly managed by MAV and by the Consortium for Coordination of Research Activities concerning the Venice Lagoon System (CORILA), should improve this situation, and facilitate the development of an open source community model, a tool that has proved to be extremely valuable in most successful estuarine management programmes around the world. An attempt in this direction has been made by developers of VEELFEM model, who recently made available to the scientific community their model code together with pieces of documentation (Umgiesser *et al.*, chapter 47).

In closing this short review, we would like to express our belief that the ensemble of different modelling activities briefly summarized in this paper represents a rather complete example of modelling studies in a particularly complex transition ecosystem, and that the results which have been achieved thus far are important, since they contributed to enhance general understanding of the ecology of transition systems and not only that of the Venice Lagoon.

ACKNOWLEDGEMENTS

We would like to express our gratitude to Dr G. Umgiesser, who stressed that a review of ecological modelling in the Venice Lagoon would be of interest to the scientific community, and who encouraged the publication of this paper. We would like to thank Ms H. Maguire for the English revision of this article.

REFERENCES

Bacher, C., Grant, J., Hawkins, A. J. S., Fang, J., Zhu, M., and M. Besnard. 2002. Modelling the effect of food depletion on scallop growth in Sungo Bay (China). *Aquatic Living Resources* **16**, 1, 10–24.

Begon, M., Harper, J. L., and C. R. Towensend. 1996. *Ecology: Individuals, Populations and Communities*. Oxford: Blackwell.

Bendoricchio, G., and S. E. Jorgensen. 1997. Exergy as goal function of ecosystems dynamic. *Ecological Modelling* **102**, 5–15.

Bendoricchio, G., Coffaro, G., and C. De Marchi. 1994. A trophic model for Ulva rigida in the Lagoon of Venice. *Ecological Modelling* **75/76**, 485–96.

Bertonati, M., Dejak, C., Mazzei Lalatta, I., and G. Pecenik. 1987. Eutrophication model of the Venice Lagoon: statistical treatment of 'in situ' measurements of phytoplankton growth parameters. *Ecological Modelling* **37**, 1–2, 103–30.

Biber, P. D., Harwell, M. A., and W. P. Cropper Jnr. 2004. Modeling the dynamic of three functional groups of macroalgae in tropical seagrass habitats. *Ecological Modelling* **175**, 25–54.

Bocci, M., Coffaro, G., and G. Bendoricchio. 1997. Modelling biomass and nutrient dynamics in eelgrass (Zostera marina L.): applications to the Lagoon of Venice (Italy) and Øresund (Denmark). *Ecological Modelling* **102**, 67–80.

Carrer, S., and S. Opitz. 1999. Trophic network model of a shallow water area in the northern part of the Lagoon of Venice. *Ecological Modelling* **124**, 193–219.

Carrer, S., Halling-Sorensen, B., and G. Bendoricchio. 2000. Modelling the fate of dioxins in a trophic network by coupling an ecotoxicological and an Ecopath model. *Ecological Modelling* **126**, 201–23.

Clark, C.W. 1976. Mathematical Bioeconomics. *The Optimal Management of Renewable Resources*. New York: John Wiley & Sons.

Coffaro, G., and M. Bocci. 1997. Resources competition between Ulva rigida and Zostera marina: a quantitative approach applied to the Lagoon of Venice. *Ecological Modelling* **102**, 81–95.

Coffaro, G., and A. Sfriso. 1997. Simulation model of Ulva rigida growth in shallow water of the Lagoon of Venice. Ecological Modelling 102, 55–66.

Coffaro, G., Bocci, M., and G. Bendoricchio. 1997. Application of structural dynamic approach to estimate space variability of primary producers in shallow marine water. *Ecological Modelling* **102**, 97–114.

Cossarini, G., Solidoro, C., and R. Pastres. 2001. Modello tridimensionale di qualità dell'acqua della laguna di Venezia: analisi numerica del funzionamento dell'ecosistema e del ciclo dell'azoto. In *Atti 25. XI congresso nazionale della Società Italiana di Ecologia*.

Dejak, C., *et al.* 1987. Special issue on the Lagoon of Venice. *Ecological Modelling* **37**, 1–157.

Dejak, C., Pastres, R., Polenghi, I., Solidoro, C., and G. Pecenik. 1998. 3D modelling of water quality transport processes with time and space varying diffusivity. *Coastal Estuarine Studies* **54**, 645–62.

Della Valle, M., Marcomini, A., Sfriso, A., Sweetman, A. J., and K. C. Jones. 2004. Estimation of PCDD/F distribution and fluxes in the Venice Lagoon, Italy: combining measurement and modelling approaches. *Chemosphere* **52**, 603–16.

Di Toro, D. M., O'Connor, D. J., and R. V. Thomann. 1971. A dynamic model of the phytoplankton population in the Sacramento San Joaquin Delta. *Adv. Chem. Ser.* **106**, 131–80.

Dowd, M. 2003. Seston dynamics in a tidal inlet with shellfish aquaculture: a model study using tracer equations. *Estuarine Coastal and Shelf Science* **57**, 3, 523–37.

Evans, G.T. 2003. Defining misfit between biogeochemical models and data sets. *Journal of Marine Systems* **40–41**, 49–54.

Fujita, R. M., Wheeler, P. A., and R. L. Edwards. 1988. Metabolic regulation of ammonium uptake by Ulva rigida: a compartmental analysis of the rate-limiting steps for uptake. *Journal of Phycology* **24**, 560.

Gangnery, A., Bacher, C., and D. Buestel. 2004. Application of a population dynamic model to the Mediterranean mussel *Mytilus galloprovincialis*, reared in Thau lagoon (France). *Aquaculture* **229**, 1–4, 289–313.

Hull, V., Mocenni, C., Falcucci, M., and N. Marchettini. 2000. A trophodynamic model for the lagoon of Fogliano (Italy) with ecological dependent modifying parameters. *Ecological Modelling* **134**,. 2–3, 153–67.

Kremer, J. N., and S. W. Nixon. 1978. *A Coastal Marine Ecosystem: Simulation and Analysis*. New York: Springer-Verlag.

Libralato, S., Pastres, R., Pranovi, F., Raicevich, S., Granzotto, A., Giovanardi, O., and P. Torricelli. 2002. Comparison between the energy flow networks of two habitats in the Venice Lagoon. *Marine Ecology PSZN* **23**, 228–36.

Martins, I., and J. Marques. 2002. A model for the growth of opportunistic macroalgae (Enteromorpha sp.) in tidal estuaries. *Estuarine Coastal and Shelf Science* **55**, 2, 247–52.

Melaku Canu, D., Umgiesser, G., and C. Solidoro. 2001. Short-term simulations under winter conditions in the lagoon of Venice: a contribution to the environmental impact assessment of temporary closure of the inlets. *Ecological Modelling* **138**, 1–3, 215–30.

Melaku Canu, D., Solidoro, C., Pastres, R., and G. Umgiesser. 2002. *Tapes philippinarum* in the lagoon of Venice: a socio-economical-environmental analysis and suggestions for a more efficient harvesting strategy. In P. Duarte (ed.), *International Conference on Sustainable Management of Coastal Ecosystem* (Proceedings). Edicoes Universidade Fernando Pessoa, 247–255

Melaku Canu, D., Solidoro, C., and G. Umgiesser. 2003. Modelling the responses of the Lagoon of Venice ecosystem to variations in physical forcings. *Ecological Modelling* **170**, 2–3, 265–90.

Micheletti, C., Critto, A., Carlon, C., and A. Marcomini. 2004. Ecological Risk Assessment or persistent toxic substances for the clam *Tapes philippinarum* in the lagoon of Venice. *Environmental Toxicology and Chemistry*, in press.

Nakamura, Y. 2001. Filtration rates of the Manila clam, Ruditapes philippinarum: dependence on prey items including bacteria and picocyanobacteria. *Jnl Exp Mar Biol and Ecol* **266**, 2, 181–92.

Pastres, R., Franco, Davide, Pecenik, G., Solidoro, C., and C. Dejak. 1995. Using parallel computers in environmental modelling: a working example. *Ecological Modelling* **80**, 1, 69–86.

Pastres, R., Franco, D., Pecenik, G., Solidoro, C., and C. Dejak. 1997. Local sensitivity analysis of a distributed parameters water quality model. *Reliability Engineering & System Safety* **57**, 1, 21–30.

Pastres, R., Chan, K., Solidoro, C., and C. Dejak. 1999. Global sensitivity analysis of a shallow-water 3D eutrophication model. *Computational Physics Communication* **117**, 62–74.

Pastres, R., Solidoro, C., Cossarini, G., Melaku Canu, D., and C. Dejak. 2001. Managing the rearing of *Tapes philippinarum* in the lagoon of Venice: a decision support system. *Ecological Modelling* **138**, 1–3, 213–45.

Pastres, R., Ciavatta, S., Cossarini, G., and C. Solidoro. 2003. Sensitivity analysis as a tool for the implementation of a water quality regulation based on the Maximum Permissible Loads policy. *Reliability Engineering & System Safety* **79**, 2, 239–44.

Pastres, R., Brigolin, D., Petrizzo, A., and M. Zucchetta. 2004. Testing the robustness of primary production models in shallow coastal areas: a case study. *Ecological Modelling* **2**, 221–33

Pranovi, F., Libralato, S., Raicevich, S., Granzotto, A., Pastres, R., and O. Giovanardi. 2003. Mechanical clam dredging in Venice lagoon: ecosystem effects evaluated with a trophic mass-balance model. *Marine Biology* **1143**, 393–403.

Riccardi, N., and C. Solidoro. 1996. The influence of environmental variables on *Ulva rigida* C. Ag. growth and production. *Botanica Marina* **39**, 27–36.

Saltelli, A., Chan, K., and E. M. Scott. 2000. *Sensitivity Analysis*. Chichester: John Wiley & Son.

Sfriso, A., and A. Marcomini. 1996. Decline of *Ulva* growth in the Lagoon of Venice. *Bioresource Technology* **58**, 299–307.

Sfriso A., and Marcomini A. 1999.. Macrophyte production in a shallow coastal lagoon. Part II: Coupling with sediment, SPM and tissue carbon, nitrogen and phosphorus concentrations. *Marine Environmental Research* **47**, 285–309

Sfriso, A., Pavoni, B., and A. Marcomini. 1989. Macroalgae and phytoplankton standing crops in the central Venice lagoon: primary production and nutrient balance. *The Science of the Total Environment*, 139–59.

Sfriso, A., Pavoni, B., Marcomini, A., and A. A. Orio. 1992. Macroalgae, nutrient cycles and pollutants in the lagoon of Venice. *Estuaries*, 517–28.

Solidoro, C. 1994. Ampliamento di un modello 3D trofico diffusivo della Laguna di Venezia: simulazione della dinamica di popolazion macroalgale. Doctorate thesis, University of Ferrara.

Solidoro, C., Dejak, C., Franco, D., Pastres, R., and G. Pecenik. 1995. A model for macroalgae and phytoplankton growth in the Venice Lagoon environment. *International Enviroment* **21**, 5, 619–26.

Solidoro, C., Brando, E. V., Dejak, C., Franco, D., Pastres, R., and G. Pecenik. 1997a. Simulation of the seasonal evolution of macroalgae in the lagoon of Venice. *Environmental Modelling and Assessment* **2**, 65–71.

Solidoro, C., Brando, V. E., Dejak, C., Franco, D., Pastres, R., and G. Pecenik. 1997b. Long term simulations of population dynamics of *U. rigida* in the lagoon of Venice. *Ecological Modelling* **102**, 2–3, 259–72.

Solidoro, C., Pecenik, G., Pastres, R., Davide, Franco, and C. Dejak. 1997c. Modelling macroalgae (*Ulva rigida*) in the Venice lagoon: model structure identification and first parameters estimation. *Ecological Modelling* **94**, 2–3, 191–206.

Solidoro, C., Pastres, R., Melaku Canu, D., Pellizzato M., and R. Rossi. 2000. Modelling the growth of *Tapes philippinarum* in northern adriatic lagoons. *Marine Ecology Progress Series* **199**, 137–48.

Solidoro, C., Cossarini, G., and R. Pastres. 2002. Numerical analysis of the nutrient fluxes through the Venice lagoon inlets. In *Scientific Research and Safeguarding of Venice: CORILA research, program, 2001 results*. Venice: Istituto Veneto di Scienze, Lettere ed Arti, pp. 545–55.

Solidoro, C., Melaku Canu, D., and R. Rossi. 2003. Ecological and economic considerations on fishing and rearing of *Tapes phillipinarum* in the lagoon of Venice. *Ecological Modelling* **170**, 2–3, 303–19.

Solidoro, C., Melaku Canu, D., Cucco, A., and G. Umgiesser. 2004. A partition of the Venice Lagoon based on physical properties and analysis of general circulation. *Journal of Marine Systems* **51**, 147–60.

Solidoro, C., Pastres, R., and G. Gossanini. 2005. Nitrogen and plankton dynamics in the lagoon of Venice. *Ecological Modelling* **184**, 103–124.

Umgiesser, G., Melaku Canu, D., Solidoro, C., and R. Ambrose. 2003. A finite element ecological model: a first application to the Venice Lagoon. *Environmental Modelling & Software* **18**, 2, 131–45.

Umgiesser, G., Melaku Canu, D., Cucco, A., and C. Solidoro. 2004. A finite element model for the Venice Lagoon. Model set up, calibration and validation. *Journal of Marine Systems* **51**, 123–45.

UNESCO. 2003. *The Integrated, Strategic Design Plan for the Coastal Ocean Observations Module of the Global Ocean Observing System GOOS*, Report No. 125, IOC Information Documents Series No. 1183.

Vollenweider R. A., Giovanardi, F., Montanari, G., and A. Rinaldi. 1998. Characterization of the trophic conditions of marine coastal waters, with special reference to the NW Adriatic sea: proposal for a trophic scale. *Environmetrics* **9**, 329–57

Zaldivar, J. M., Cattaneo, E., Plus, M., Murray, C. N., Giordani, G., and P. Viaroli. 2003. Long-term simulation of main biogeochemical events in a coastal lagoon: Sacca Di Goro (Northern Adriatic Coast, Italy). *Continental Shelf Research* **23**, 1847–75.

Zharova, N., Sfriso, A., Voinov, A., and B. Pavoni. 2001. A simulation model for the annual fluctuation of *Zostera marina* biomass in the Venice lagoon. *Aquatic Botany* **70**, 135–50.

58 • Pressures on Mediterranean coastal lagoons as a consequence of human activities

A. PÉREZ-RUZAFA AND C. MARCOS

SUMMARY

The management of coastal zones involves developing suitable tools that allow human impact to be evaluated and, where possible, the establishment of relationships between the activity and the main ecological processes affected. Most Mediterranean coastal lagoons share both similar ecological processes, related to geomorphologic features and hydrographical conditions, and similar human impact and pressures. The large amount of information compiled over the last twenty years on the Mar Menor, a hypersaline coastal lagoon located on the south-western Mediterranean coastline of Spain, makes it a good reference point for analyzing the biological patterns and processes affected by changes in hydrographical conditions, nutrient inputs, biological assemblages and lagoon characteristics and allows for some general guidelines on cause-effect relationships between human activities and environmental indicators. Such knowledge can be utilized to formulate conceptual models for prioritising management activities and lines of research.

INTRODUCTION

In recent decades there has been growing interest in the integrated management of coastal zones (ICZM). Coastal lagoons and enclosed seas constitute one of the priorities because of their vulnerability to human impact and the concentration of competing uses (Sorensen, 1993). The management of coastal zones involves the development of tools which allow the consequences of human activities to be foreseen which in turn prompts ways of controlling those processes that lead to the degradation of coastal lagoons and other natural environments. An in-depth knowledge

of the nature of such processes is required, as is the knowledge of the causal relationships between them and the human activities that originate them.

However, such cause-effect relationships are, in general, unknown and establishing them requires detailed monitoring programmes. Sometimes they are incompatible with the accelerated evolution of the lagoon as forced by man-driven activities. In most cases, difficulties in establishing such cause-effect relationships have been worsened by the lack of detailed studies and data on the hydrographic conditions and biological assemblages before human intervention occurred. Baseline information on a system often does not exist, for example, before engineering work occurs. As a result, there is a need to share information and to know and understand the lessons learned from similar lagoons around the world (Sorensen, 1993). In this sense, the large amount of information compiled over recent years on Mediterranean and other European coastal lagoons should allow some general guidelines for cause-effect relationships to be established, in order to draw up conceptual models. These in turn should make the assignment of priorities regarding managerial actions and research lines possible.

THE MAIN FEATURES OF COASTAL LAGOONS

Coastal lagoons have historically been of great interest to humans because they offer high biological productivity and provide harbour and navigation facilities. At present, these ecosystems maintain important fisheries, aquaculture, urban development and tourist and recreational facilities.

Most lagoon properties come from their geomorphology and configuration. They are characterized

Flooding and Environmental Challenges for Venice and its Lagoon: State of Knowledge, ed. C. A. Fletcher and T. Spencer.
Published by Cambridge University Press, © Cambridge University Press 2005.

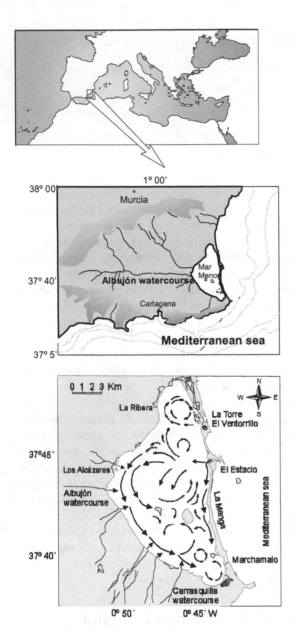

Fig. 58.1   Location of the Mar Menor lagoon.

by bottom topography and wind affects the entire water column, promoting the resuspension of materials, nutrients and small organisms from the sediment surface layer. Lagoon systems contain a high number of physical and ecological boundaries and gradients (between water and sediment, pelagic and benthic assemblages, lagoon-marine-fresh water and terrestrial systems, and with the atmosphere). So, most coastal lagoons correspond to the type of coastal ecosystems characterized by frequent environmental disturbances and environmental fluctuations (Barnes, 1980; UNESCO, 1981).

Thus, in spite of the high variability in size, form, salinity range, fresh versus marine water influence or trophic *status*, most of the variability between lagoons, including the number of species inhabiting them and

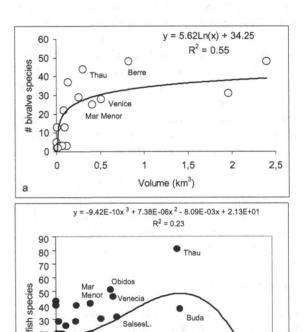

Fig. 58.2   Relation between some morphometric characteristics of Mediterranean lagoons and the number of species of bivalve molluscs (a) and fishes (b) that they shelter.

by being shallow and isolated from the open sea by coastal barriers that maintain some connecting channels or inlets (Fig. 58.1 shows the location and morphology of the Mar Menor lagoon). Due to their shallowness, lagoon beds are usually well irradiated, currents and hydrodynamics are closely conditioned ecological processes, can be explained by a set of parameters which describe the relationships through the aforementioned boundaries and their relative importance (Figs. 58.2, 58.3). The spatial organization of communities within each lagoon, species richness, phytoplanktonic versus benthic, algal or seaweed productivity and biomass have also been related to a confinement gradient according to the distance to the channels connecting with the open sea and the water renewal rates or colonization rates at each site (Guelorget and Perthuisot, 1983; Guelorget et al., 1983; Mariani, 2001; Pérez-Ruzafa, 1989; Pérez-Ruzafa and Marcos, 1992, 1993). The Venice Lagoon assemblages (Mainardi et al., 2002; Sconfietti, 1989; Sconfietti et al., 2003) are consistent with this model (Fig. 58.4). In this way, any activity which alters lagoon features such as their size, perimeter or coastal development, interchange of waters via inlets, renovation rates, bottom and sediment characteristics, will have important consequences on the biota and ecological processes.

THE MAR MENOR EXAMPLE

The Mar Menor is a hypersaline coastal lagoon, located on the south-western Mediterranean coastline. It has a surface area of 135 km^2, a mean depth of 3.6 m and a maximum depth of over 6 m. Currents are mainly wind driven (Fig. 58.1). There is no main river flowing into the lagoon, but the Mar Menor basin collects surface waters from more than 20 gullies or *ramblas* that contribute water and materials from agricultural runoff and from mountains, where mining activity took place over the centuries, during rainfall events. Amongst these wadis, the Albuján watercourse is the main collector in the drainage basin, and at present it maintains a regular flux of water due to changes in agricultural practices.

As an emblematic landscape of the region of Murcia, the Mar Menor has great socio-economic importance and supports a wide range of uses. It is considered a key factor in regional development plans providing key tourist and recreational services and maintaining important fisheries based on the quality and price of its natural products. The lagoon is, however, a subject of concern due to its high rate of change over recent decades, with detrimental impacts on the structure of its biological assemblages

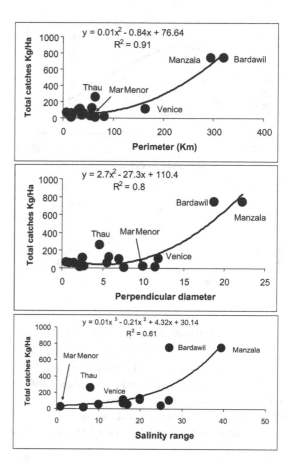

Fig. 58.3   Relation between fisheries production in Mediterranean coastal lagoons and morphometric and hydrologic characteristics. The fishing yield tends to increase with an increase in the total perimeter of the lagoon (a), the lagoon perpendicular diameter (with respect to the open sea coastline) (b), and the range of variation in salinity (c).

and dynamics. Some of the changes have been induced by coastal work on tourist facilities (land reclamation; opening, deepening and extending channels, urban development and associated wastes; the building of sports harbours; and artificial beach creation), whilst others relate to changes in agricultural practices in the watershed, with a change from extensive dry crop farming to intensively irrigated crops, increasing agricultural wastes and nutrient input into the lagoon. Some activities, such as mining, have existed since Roman times, whilst others,

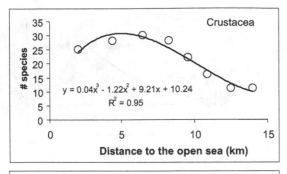

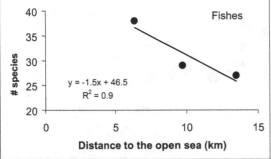

Fig. 58. 4   Gradients in species richness of Peracarid crustacea (upper; according to data from Sconfietti, 1989) and fishes (lower; according to data from Mainardi *et al.*, 2002) with distance to the mouth of the inlets that communicate the Venice Lagoon with the Adriatic.

such as salt mining or the opening of artificial channels for fisheries, were developed during the eighteenth and nineteenth centuries. Major urban development, land reclamations, widening and opening of channels and construction of sports harbours are the consequence of tourist growth since the early 1970s. This situation makes the Mar Menor a good reference case against which one can analyse the biological patterns and processes resulting from changes in hydrographical conditions, nutrient input, biological assemblages and lagoon characteristics forced by human activities.

Among historical activities, mining and the first agricultural developments produced the greatest impact, increasing sedimentation rates and introducing high amounts of heavy metals into the lagoon sediments. Some of them, such as lead, reach concentrations higher than 4000 ppm in the southern

basin (Simonneau, 1973). However, since mining stopped in the twentieth century, concentrations have remained constant (Pérez-Ruzafa *et al.*, 1987). At present, work related to the development of tourist facilities and changes in agricultural practices represent the main environmental problems for the lagoon.

The effect of changes in water renewal rates and the opening of inlets

One act, which probably represents the main impact on lagoon characteristics, has been the opening of artificial inlets. In the 1870s, in the southern basin, the Marchamalo channel was opened for the fishing of mugilids (grey mullet). It is now very shallow and is nearly closed. In 1973 El Estacio channel was deepened and enlarged to make it navigable and to build a sports harbour. The first consequence of this action was an increase in the interchange rates of water with the Mediterranean, and a reduction in the residence time of water. This led to a reduction of salinity and the easing of lower temperatures (Table 58.1) (Pérez-Ruzafa *et al.*, 1991). The main biological consequence of these changes was the colonization of the lagoon by new species, such as the algae *Caulerpa prolifera*. The muddy and somewhat sandy bed of the lagoon, with a low organic matter content and covered by scarce patches of the phanerogam *Cymodocea nodosa*, was replaced by anoxic bottoms with a high organic matter content (up to 8 per cent organic carbon) covered by a dense meadow of the algae which also colonized the rocky areas and displaced other photophilic algae (Pérez-Ruzafa and Marcos, 1987; Pérez-Ruzafa *et al.*, 1989). Other species, such as the jellyfish *Cotylorhiza tuberculata* and *Rhyzostoma pulmo*, are now new inhabitants of the lagoon, as part of a continuous process of colonization (Pérez-Ruzafa *et al.*, 2002).

In general, there has been an increase in biological diversity, both by raising the probability of colonization by Mediterranean species due to the higher renewal rates of waters and by increasing the possibility of establishing a stable population due to less extreme environmental conditions (Pérez-Ruzafa and Marcos, 1992, 1993). Figure 58.5 shows the number of species of fish and molluscs reported in the Mar

Table 58.1   *Hydrographic conditions at the Mar Menor before (1970) and after (1980, 1988) the enlargement of the El Estacio inlet.*

	1970	1980	1988
Outflow of water to the Mediterranean (m^3)	3.6×10^8	6.1×10^8	6.4×10^8
Inflow of water from the Mediterranean (m^3)	4.5×10^8	7.2×10^8	7.3×10^8
Residence time (years)	1.28	0.81	0.79
Temperature range (°C)	7.5–29	12–27.5	12–30.5
Salinity range (PSU)	48.5–53.4	43–46	42–45

Pérez-Ruzafa *et al.* (1991)

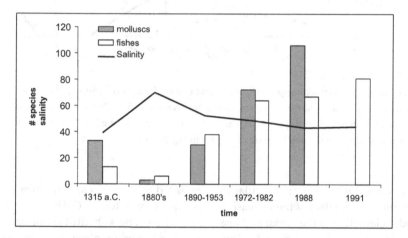

Fig. 58.5   Evolution of the number of species of fishes and molluscs in the Mar Menor lagoon through its recent history, related to changes in salinity as an indicator of the degree of communication with the Mediterranean Sea. High salinity in the late 1880s corresponds to an isolation period and the minimum in species richness. Through the twentieth century several works increased the number and size of inlets with a corresponding decrease in salinity and colonization rate of new species from the Mediterranean.

Menor at different stages in its recent evolution with regard to changes in salinity (an indicator of the degree of communication between the lagoon and the open sea). Such increases in biological diversity involve competition between alloctonous and lagoon species and a narrower partitioning of resources. This has had consequences for catches of commercial species. Figure 58.6 shows the falling catches of mugilids at the end of the nineteenth century, coinciding with the colonization of the lagoon by *Sparus aurata* and other fish species after the construction of the Marchamalo channel, and the decrease in catches

of both species after the enlargement of the El Estacio channel. The small increase observed over the last decade is related to changes in the trophic status of the lagoon over that time.

The impact of coastal works

Another factor having a greater impact on the Mar Menor lagoon has been the enlarging of beaches. This involves both the dredging of shallow sandy areas and the pumping of sediments along the coast. The stability of new beaches has been reinforced by the

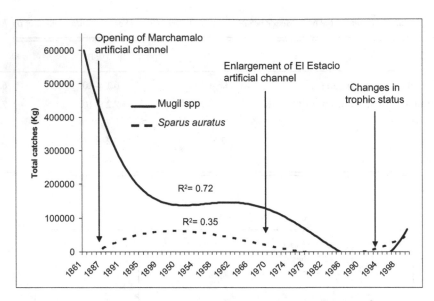

Fig. 58.6   Trends in fisheries of Mugilidae spp and *Sparus aurata* in the Mar Menor lagoon (R^2 values indicate the fit of the curves to real data). Increase in *Sparus aurata* catches relates to the decrease in Mugilidae fisheries after the opening of the Marchamalo artificial inlet. Catches of both species decreased after the enlargement of the El Estacio inlet in the early 1970s. (Compare these curves with data on the colonization of species in Fig. 58. 5.)

installation of rocky breakwaters perpendicular to the coastline. The immediate effect of the increasing turbidity and land reclamation is the burial and loss of traditional species and the loss of filter feeding assemblages and sciaphilic communities settled on the pylons under them (consisting mainly of sponges, cnidarians, bryozoans and ascidians) with losses in lagoon biodiversity (Pérez-Ruzafa, 1989). System instability produced by turbidity, nutrient reloading and increasing sedimentation has lead to the spread of the opportunistic algae *Caulerpa prolifera* and *Chaetomorpha linum* (Pérez-Ruzafa et al., 1991). This has produced an increase in organic matter, total nitrogen, lime and clay content in sediments. Figure 58.7 shows the significant differences (p< 0.05) found in these variables in dredged (D) and pumping (P) areas compared to unaffected *Caulerpa prolifera* (CM) and sandy bottoms (CS) before and after human action. At the same time as these physical-chemical changes in sediment properties occurred, fish assemblages changed in terms of species composition, richness and abundance. Sandy bottom assemblages, dominated by *Pomatoschistus marmoratus*, were

replaced by *Caulerpa prolifera* meadows on fine sediments, dominated by *Gobius niger* and *Hippocampus ramulosus*. The substitution of one assemblage by another has taken place according to ecological succession models after perturbations: increasing abundance during the initial months following interference and then decreasing again rapidly. Species richness and diversity decreased in the first instance. Species richness took at least one year to start to recover and diversity shows a pick up just some months after the end of the activity, to later decrease again and remain low according to the new established assemblage (Pérez-Ruzafa et al., in press).

On the other hand, the artificial rocky substrata which constitute breakwaters, experienced a rapid colonization process by acting as artificial reefs. Abundance has been higher than for natural bottoms, due to greater habitat complexity, and has remained constant with time. Meanwhile diversity and species richness have not shown significant differences to natural rocky bottoms and show a slower and progressive colonization process (Pérez-Ruzafa et al., in press).

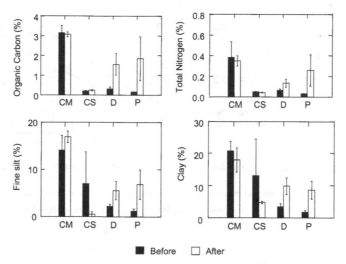

Fig. 58.7   Response of organic carbon, total nitrogen and fine silt and clay contents in sediments as a consequence of coastal works (before and after pumping or dredging actions to create new beaches). CM: *Caulerpa prolifera* meadows at control areas; CS: sandy bottoms at control areas; D: dredged areas; P: pumping areas.

The effects of increasing eutrophication

Another of the processes undergoing change in the Mar Menor over the last decade has been a generalized eutrophication process. During the 1990s, agriculture started to change from dry farming crops with low amounts of nitrogen fertilizers, to intensive irrigation crops with excessive nitrogen fertilization using waters diverted from the Tajo to the Segura river. Changes in the trophic status of lagoon waters was studied by Pérez-Ruzafa *et al.* (2002) through the comparison of two extensive time series (sampled weekly), one for 1988 and the other for 1997. The 1988 time series showed that nitrate concentrations were low throughout the year, contrasting with higher phosphate values. While nitrate concentration during 1988 was always under 1 µmol NO_3^- /l, much higher concentrations occurred in 1997, reaching concentrations of up to 8 µmol NO_3^- /l, particularly during spring and summer (Fig. 58.8), entering mainly via the main watercourse due to the rise of phreatic levels which have been increasing at a rate of 1 m per year (Pérez-Ruzafa and Aragón, 2002). As a consequence of changes in the nutrient input regime, the water column in the lagoon has changed from being moderately oligotrophic to being relatively eutrophic. During 1997, higher nitrate concentrations were usually found on the western coast of the lagoon, close to the mouths of the main watercourses, suggesting that nitrate input is related to agricultural activity.

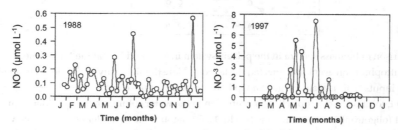

Fig. 58.8   Changes in nitrate concentrations in the water column at the Mar Menor lagoon, in 1988 and 1997.

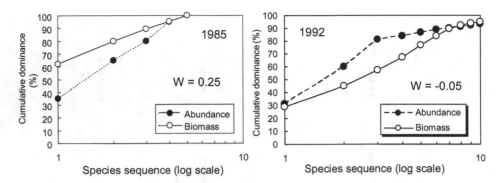

Fig. 58.9   Cumulative curves of abundance and biomass (ABC curves) for a coastal sandy bottom polychaete assemblage in the Mar Menor before (1985) and after (1992) the main eutrophication process that affected the lagoon.

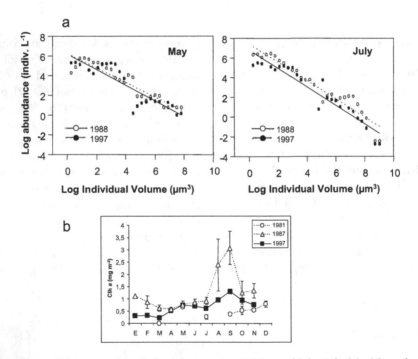

Fig. 58.10   (a) Comparison of biomass spectra in the pelagic system in May and July at the Mar Menor lagoon in 1988 and 1997 (after the main eutrophication event that affected the lagoon). Note the similarity in the slopes in spite of the differences in nutrient inputs. This is mainly due to a top-bottom control of the trophic web exerted by jellyfish (see Pérez-Ruzafa et al., 2002). (b) Evolution of chlorophyll a concentration in the water column in the Mar Menor in a sampling site in the north basin in 1981 (oligotrophic state without jellyfish), 1987 (oligotrophic to moderate tropic state without jellyfish), 1997 (eutrophic state with jellyfish).

The benthic communities, biological index as ABC curves (Warwick, 1986), or the related W parameter (Clarke, 1990) have shown changes in species composition structures (Fig. 58.9). Although an increase in the number of species has taken place in some cases (due to the combined effect of the increase in food resources and the continuing colonization process in the lagoon as mentioned above) only a few species have come to dominate the community which is constituted by a high number of individuals of smaller sizes and lower biomass.

In the pelagic assemblages, nutrient increases have provided conditions for the growth of larger phytoplankton cells and subsequent changes in trophic structure. Abundance of jellyfish has increased, reaching mean densities of 12 individuals per 100 m³. The total jellyfish population estimated in the lagoon by mid-summer 1997 was of the order of 40 million individuals. This population exerts a strong top-down control on the food web by selective grazing on large diatoms, ciliates, veliger larvae and copepods. The trade-off between competition for available resources (bottom-up) and predation (top-down) control mechanisms results in a planktonic size structure that is different to that which would be expected under eutrophic conditions. The interannual comparison of biomass spectra in the Mar Menor lagoon for 1988 and 1997 does not show significant differences. Results indicate that the seasonal trend was maintained over both years, independently from the nutrient load (Fig. 58.10a) (Pérez-Ruzafa et al., 2002). Chlorophyll has also been maintained at levels which existed prior to the eutrophication event (Fig. 58.10b).

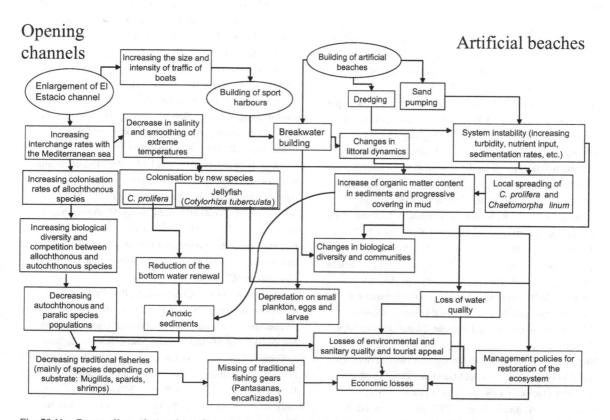

Fig. 58.11 Cause-effect relationships diagram suggested by the information mentioned in this chapter. Two main human activities are included: the enlargement of the El Estacio channel and pumping and dredging of sediments to build new beaches. Arrows connect each action or process involved to its immediate consequence.

BUILDING A NETWORK OF CAUSE-EFFECT RELATIONSHIPS

The combination of partial cause-effect relationships can give us a complex framework that conforms to a conceptual model, useful for the management of human activities in coastal lagoons (Fig. 58.11). As the figure shows, many of the consequences of these actions are losses in good water quality, invasion by allocthonous species, a decrease in fish catches and loss of artisanal fisheries, increasing the eutrophication or proliferation of jellyfish. These changes involve potential risks to human uses themselves, loss of tourist appeal and costs to the local economy. These factors must be taken into account as not only environmental and biological characteristics are at risk.

REFERENCES

Barnes, R. S. K. 1980. *Coastal Lagoons*. Cambridge: Cambridge University Press.

Clarke, K. R. 1990. Comparison of dominance curves. *Jnl.Exp.Mar.Biol.Ecol.* **138**, 1–2, 130–48.

Guelorget, O., and J. P. Perthuisot. 1983. Le domaine paralique. Expressions géologiques, biologiques et économiques du confinement. *Travaux du laboratoire de géologie* **16**, 1–136.

Guelorget, O., Frisoni, G. F., and J. P. Perthuisot. 1983. Zonation biologique des milieux lagunaires: definition d'une échelle de confinement dans le domaine paralique méditerranéen. *Journal de recherche océanographique* **8**, 1, 15–35.

Mainardi, D., Fiorin, R., Franco, A., Franzoi, P., Giovanardi, O., Granzotto, A., Libertini, A., Malavasi, S., Pranovi, F., Riccato, F., and P. Torricelli. 2002. Fish diversity in the Venice Lagoon: preliminary report. In P. Campostrini (ed.), *Scientific Research and Safeguarding of Venice*. Venice: Corila, pp. 583–94.

Mariani, S. 2001. Can spatial distribution of ichthyofauna describe marine influence on coastal lagoons? A central Mediterranean case study. *Estuarine, Coastal and Shelf Science* **52**, 261–7.

Pérez-Ruzafa, A. 1989. Estudio ecológico y bionómico de los poblamientos bentónicos del Mar Menor (Murcia, SE de España). Doctoral thesis, Universidad de Murcia.

Pérez-Ruzafa, A., and R. Aragón. 2002. Implicaciones de la gestión y el uso de las aguas subterráneas en el funcionamiento de la red trófica de una laguna costera. In J. M. Fornés and M. R. Llamas (eds.), *Conflictos entre el desarrollo de*

las aguas subterráneas y la conservación de los humedales: litoral mediterráneo. Madrid: Fundación Marcelino Botín – Ediciones Mundi-Prensa, pp. 215–45.

Pérez-Ruzafa, A., and C. Marcos. 1987. Los sustratos arenosos y fangosos del Mar Menor (Murcia), su cubierta vegetal y su posible relación con la disminución del mújol en la laguna. *Cuadernos Marisqueros Publicaciones Técnicas* **11**, 111–23.

Pérez-Ruzafa, A., and C. Marcos. 1992. Colonization rates and dispersal as essential parameters in the confinement theory to explain the structure and horizontal zonation of lagoon benthic assemblages. *Rapp. Comm. Int. Mer Méditerranée* **33**, 100.

Pérez-Ruzafa, A., and C. Marcos. 1993. La teoría del confinamiento como modelo para explicar la estructura y zonación horizontal de las comunidades bentónicas en las lagunas costeras. *Publicaciones Especiales del Instituto Español de Oceanografía* **11**, 347–58.

Pérez-Ruzafa, A., Marcos, C., Pérez-Ruzafa, I. M., and J. D. Ros. 1987. Evolución de las características ambientales y de los poblamientos del Mar Menor (Murcia, SE de España). *Anales de Biología* **12** (Biología Ambiental, 3), 53–65.

Pérez-Ruzafa, A., Ros, J. D., Marcos, C., Ballester, R., and I. Pérez-Ruzafa. 1989. Distribution and biomass of the macrophyte beds in a hypersaline coastal lagoon (the Mar Menor, SE Spain), and its recent evolution following major environmental changes. In C. F. Bouderesque, A. Meinesz, E, Fresi, and V. Gravez (eds.), *International Workshop on Posidonia Beds*, II, Marseille: GIS Posidonie, pp. 49–62.

Pérez-Ruzafa, A., Marcos, C., and J. Ros. 1991. Environmental and biological changes related to recent human activities in the Mar Menor. *Marine Pollution Bulletin* **23**, 747–51.

Pérez-Ruzafa, A., Gilabert, J., Gutiérrez, J. M., Fernández, A. I., Marcos, C., and S. Sabah-Mazzetta. 2002. Evidence of a planktonic food web response to changes in nutrient input dynamics in the Mar Menor coastal lagoon, Spain. *Hydrobiologia* **475**, 1, 359–69.

Pérez-Ruzafa, A., García-Charton, J. A., Barcala, E., and C. Marcos. In press. Changes in benthic fish assemblages as a consequence of coastal works in a coastal lagoon: the Mar Menor (Western Mediterranean). *Marine Pollution Bulletin*.

Sconfietti, R. 1989. Ecological zonation and dynamics of hard-bottom Peracarid communities along the Lagoon Estuary of the River Dese. *Riv. Idrobiol.* **28**, 1–2, 3–31.

Sconfietti, R., Marchini, A., Occhipinti, A., and C. F. Sacchi. 2003. The sessile benthic community patterns on hard bottoms in response to continental vs. Marine influence in northern Adriatic lagoons. *Oceanologica Acta* **26**, 47–56.

Simonneau, J. 1973. Mar Menor: évolution sédimentologique et géochimique récente en remplissage. Ph.D. thesis.

Sorensen, J. 1993. The management of enclosed coastal water bodies: the need for a framework for international information exchange. In J. Sorensen, F. Gable, and F. Bandarin (eds.), *The Management of Coastal Lagoons and Enclosed Bays.* New York: American Society of Civil Engineers, pp. 1–17.

UNESCO. 1981. *Coastal Lagoons Research, Present and Future.* Technical papers in Marine Science 33. Paris: UNESCO.

Warwick, R. M. 1986. A new method for detecting pollution effects on marine macrobenthic communities. *Mar.Biol.* **92**, 557–62.

Scientific paper

59 • Changes in nutrients and plankton communities in the Venice Lagoon

F. ACRI, F. BERNARDI AUBRY, A. BERTON, F. BIANCHI, E. CAMATTI, A. COMASCHI, S. RABITTI AND G. SOCAL

SUMMARY

A comparison between recent (1997–2002) and older data (1975–80) concerning the Venice Lagoon is presented. The investigations were carried out in the central and northern basins of the Lagoon of Venice, and the data, which refer to dissolved nutrient and plankton distributions, have been obtained within a framework of a number of different research projects. On comparing the two datasets, a general increase in orthosilicates was observed; a clear increase was also registered, especially in springtime, in dissolved inorganic nitrogen (DIN). These trends may be due to resuspension, which of late has been greatly enhanced because of the increased erosion induced by: (a) the tidal action or waves, and (b) the intensive dredging for molluscs, which causes nutrient mobilization from the sediments. Conversely, phosphates have shown a general decrease, due to the reduced amounts of phosphorus compounds in detergents, following Italian laws enforced in the 1980s. In the more recent dataset (1997-2002), the phytoplankton abundances have been higher than in the years 1975-1980 because of the more frequent diatom blooms; the increased resuspension phenomena probably favour a mobilization of benthic epipsamic diatoms from the sediments, resulting in a more pelagic life style. Therefore, the role of phytoplankton as primary producer has now been strengthened by the reduced competition between macroalgae and phytoplankton, with a reduction, in recent years, of the biomass of *Ulva*. Regarding zooplankton, a decrease in the standing stock, more marked in winter, has occurred. These results indicate that the Lagoon of Venice is subjected to anthropogenic pressures which continuously modify the ecosystem. For this reason, the lagoon is in need of continuing policies aimed at reducing the sources of pollution, as well as at controlling indiscriminate fishing.

INTRODUCTION

Brackish ecosystems, such as the Venice Lagoon, are subjected to great modifications in time and space because they are closely associated with drainage basin conditions, and with the action of seawater driven by tidal dynamics (Lara-Lara et al., 1980; Lapointe and Clark, 1992). In addition, anthropogenic influences may contribute to enhance natural trends, both directly and indirectly (Pavoni et al., 1992; Ravera, 2000). Domestic and industrial discharges from Venice, Marghera, Mestre and Chioggia, as well as run-off from rivers which contain many dissolved and solid materials (Dese, Naviglio del Brenta and others, see Zuliani et al., 2001), continually modify the status of the lagoon ecosystem (Perin, 1975; Cironi et al., 1993).

For nutrient budget considerations, some calculations of the nitrogen and phosphorus loads in the watershed have been made. The first estimates (Bendoricchio, 1996; Bendoricchio et al., 1999) reported values of 9000 t yr^{-1} of nitrogen and 1500 t yr^{-1} of phosphorus in the drainage basin, with c. 70 per cent reaching the lagoon. More recent estimates (Collavini et al., 2001) have reported lower discharges, of c. 4000 t yr^{-1} of nitrogen and 230 t yr^{-1} of phosphorus. The latest estimates (2001) have quoted a load of about 200 t yr^{-1} of phosphorus (ARPAV, 2002; Collavini et al., 2001). Recent studies on atmospheric deposition report values of about 1250 t yr^{-1} of DIN and 34 t yr^{-1} of phosphorus, showing the importance of the atmospheric input which has often been neglected (Guerzoni et al., in press).

Flooding and Environmental Challenges for Venice and its Lagoon: State of Knowledge, ed. C. A. Fletcher and T. Spencer.
Published by Cambridge University Press, © Cambridge University Press 2005.

In the last forty years, several changes have been observed in the composition of the primary producers of the Lagoon of Venice (Sfriso *et al.*, 1987; Sfriso *et al.*, 1992; Facca *et al.*, 2002). Since 1960, the seagrass meadows (*Zostera*, *Cymodocea*) which originally covered the whole lagoon have been disappearing (Sfriso *et al.*,

1987), while in the years up to 1990 the macroalga *Ulva rigida* C. Ag. has increased its coverage (Sfriso and Marcomini, 1996). During the entire period, several phytoplankton blooms have been observed, mainly near the industrial area of Marghera, in the deepest channels around Venice and in the marshes of the

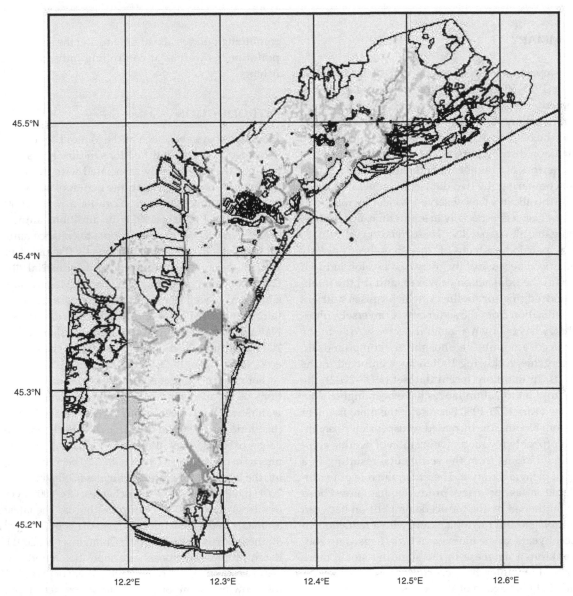

Fig. 59.1   Sampling sites during the two periods of investigation. Circles indicate the stations sampled 1975-80; triangles indicate the stations sampled 1997-2002.

northern basin (Voltolina, 1973; Alberighi et al., 1992; Sfriso and Pavoni, 1994; Socal et al., 1999; Facca et al., 2002). However, since 1990, the distribution and production of *Ulva rigida* has drastically decreased, mainly due to climatic changes (Sfriso and Marcomini, 1996). Other studies have demonstrated that the seagrass has increased (Rismondo et al., 1995, and chapter 60), covering several areas of the lagoon near the three inlets (Curiel et al. 1997). Regarding phytoplankton, comparing the data of 1998–9 with those of 1991–2 some authors have remarked on a general reduction in biomass, especially in the central basin (Facca et al., 2002), caused by increased turbidity. This change seems to have been generated by a remarkable resuspension of sediments, increased because of the intensive dredging for molluscs. As for the zooplankton community, the literature does not contain much information about the Lagoon of Venice other than our own dataset (see, for example, Brunetti et al., 1990). Most observations concern the fraction > 200 µm (mesozooplankton) while no data have yet been published about the smaller fractions (microzooplankton).

For three decades, the Istituto di Biologia del Mare has been carrying out research projects on hydrochemical and phytoplankton features of the Lagoon of Venice with the aim of demonstrating the changes in nutrients and plankton that have occurred throughout the years. A comparison of two periods of five years (1975-80 and 1997-2002) is presented and discussed here.

MATERIAL AND METHODS

Area and sampling periods

Two clusters of data deriving from different research projects on hydrochemical and plankton distributions in the Lagoon of Venice are considered here:

- from 1975 to 1980, fortnightly sampling was undertaken at six stations, located in the northern and central part of the lagoon (Fig. 59.1). Three sampling campaigns were carried out: from May 1975 until July 1976; from June 1977 until June 1978; and from August 1979 until September 1980 (Comaschi and Lombardo, 1977; Lombardo and Comaschi, 1977; Comaschi and Martino, 1978; Cioce et al., 1979; Socal, 1981; Comaschi and

Martino, 1981; Comaschi, 1987; Bianchi et al., 1999; Socal et al., 1985; Comaschi and Dalla Palma, 1988; Bianchi et al., 1990);
- from May 1997 to April 2002, monthly samples were collected at five stations in the northern and central part of the lagoon (Fig. 59.1; Comaschi et al., 2000; Bianchi et al. 2002; Bianchi et al., 2003).

Parameters and analytical methods

The following parameters are considered here: salinity by multiparametric probe (manual probe MC5, CTD Idronaut 801 in the years 1975–80 and 1997– 2002); dissolved nutrients (ammonia, nitrite, nitrate, orthosilicate, orthophosphate) according to the methods generally indicated by Strickland and Parsons (1972), Grasshoff and Ehrhardt (1976), Hansen and Koroleff (1999); phytoplankton abundance and species composition on samples fixed in hexametylentetramine-buffered formaldehyde and counted by means of an inverted microscope, according to Utermöl (1958); zooplankton on samples collected by a Clarke-Bumpus horizontal sampler equipped with a 200 µm net, fixed in neutralized formaline and counted both quantitatively and qualitatively under a stereomicroscope. All samples were collected at the surface (0.5 m depth).

Sampling procedures and analyses were constant during the research; thus data from different years and projects are homogeneous and comparable with each other.

Characterization of the datasets and statistical analysis

Because of the wide salinity data range, due to the presence of fresh as well as sea water in the lagoon, the whole dataset was divided into three subsets, according to the Venice system (1958), this being a well-accepted method of characterizing salinity zones, recently used to classify brackish waters (Healy, 1997). Three main classes were identified:

- mesohaline waters (5-18 PSU), that are not considered here because of the small amount of data;
- polyhaline waters (18-30 PSU); and
- euhaline waters (30-37 PSU).

Table 59.1   *Mean values of all parameters and results of t test (continued opposite).*

POLYHALINE WATERS	WINTER					SPRING					SUMMER					AUTUMN				
	1975-80		1997-2002			1975-80		1997-2002			1975-80		1997-2002			1975-80		1997-2002		
	n	mean	n	mean	p	n	mean	n	mean	p	n	mean	n	mean	p	n	mean	n	mean	p
DIN (μM)	27	80.56	32	96.52	-	34	23.57	52	59.86	**	47	32.41	66	29.53	-	47	61.73	38	63.56	-
Si-SiO$_4$ (μM)	30	25.31	32	54.00	**	40	27.37	52	48.69	**	51	26.54	66	48.94	**	47	32.02	38	65.50	**
P-PO$_4$ (μM)	30	1.02	32	1.10	-	39	0.68	52	0.97	-	51	1.06	66	0.79	-	47	2.34	38	1.15	**
Phytoplankton (cell/1 10^3)	37	518	32	1185	**	60	1314	51	7013	**	60	1429	60	12860	**	36	488	42	1467	**
Diatoms (cell/1 10^3)	37	247	32	660	**	60	466	51	3057	**	60	805	60	9330	**	36	227	42	743	-
Dinoflagellates (cell/1 10^3)	37	1	32	25	**	60	5	51	114	**	60	8	60	128	**	36	3	42	10	*
Nanoflagellates (cell/1 10^3)	37	238	32	485	**	60	806	51	3737	**	60	562	60	3100	**	36	221	42	608	**
Zooplankton (ind./m^3)	31	1270	32	114	**	53	4608	53	1369	**	58	9044	66	5288	**	46	797	41	480	**
Copepods (ind./m^3)	31	1255	32	105	**	53	4095	53	1198	**	58	5617	66	5000	*	46	713	41	450	**
Cladocerans (ind./m^3)	31	0	32	0	-	53	171	53	14	**	58	2670	66	18	**	46	36	41	5	**

n = number of observations; * = $p < 0.05$; ** = $p < 0.01$; - = not significant

Table 59.1 *continued.*

EUHALINE WATERS	WINTER					SPRING					SUMMER					AUTUMN				
	1975-80		1997-2002		p	1975-80		1997-2002		p	1975-80		1997-2002		p	1975-80		1997-2002		p
	n	mean	n	mean		n	mean	n	mean		n	mean	n	mean		n	mean	n	mean	
DIN (µM)	33	21.05	30	41.48	**	39	9.77	19	23.42	**	52	25.24	38	22.76	-	54	22.52	30	36.51	**
Si-SiO$_4$ (µM)	37	9.73	30	36.41	**	46	8.94	19	21.39	*	52	10.55	38	35.75	**	60	10.36	30	42.11	**
P-PO$_4$ (µM)	37	0.42	30	0.52	-	46	0.37	19	0.19	*	52	1.02	38	0.46	*	60	0.71	30	0.54	-
Phytoplankton (cell/1 10^3)	47	561	30	2148	**	48	727	18	2001	*	49	1927	37	4705	*	57	1011	31	903	-
Diatoms (cell/1 10^3)	47	384	30	1672	**	48	174	18	662	*	49	1388	37	2909	-	57	561	31	441	*
Dinoflagellates (cell/1 10^3)	47	4	30	27	**	48	13	18	49	-	49	38	37	61	*	57	13	31	10	-
Nanoflagellates (cell/1 10^3)	47	134	30	409	**	48	504	18	1274	-	49	381	37	1696	**	57	375	31	431	*
Zooplankton (ind/m^3)	64	2079	30	251	**	71	10700	20	2251	**	80	6884	38	2272	**	64	5187	31	522	**
Copepods (ind/m^3)	64	1982	30	230	**	71	8097	20	1915	**	80	4255	38	1577	**	64	3106	31	456	**
Cladocerans (ind/m^3)	64	5	30	1	*	71	692	20	61	**	80	1907	38	266	**	64	1708	31	17	**

n = number of observations; * = $p < 0.05$; ** = $p < 0.01$; - = not significant

Subsequently, the data obtained from polyhaline and euhaline waters were subdivided into four seasonal subsets. To statistically check the differences in averages between years, the t test was adopted. When the distribution of data was not normal, a logarithmic transformation was applied. Furthermore a modification of the t test (as developed by Welch, 1938) was applied when the Bartlett's test showed that the two compared populations of data did not have homoscedastic variance.

RESULTS AND DISCUSSION

Temporal trends in the polyhaline waters

The averages of polyhaline waters are reported in Table 59.1 and the following observations can be made:

- regarding nutrients: (i) the spring DIN concentrations were significantly lower in 1975–80 than in 1997–2002; (ii) the autumn orthophosphate levels were about two times greater in 1975–80 than in 1997–2002; (iii) silicate concentrations in all four seasons were lower in 1975–80 than in 1997–2002;

- phytoplankton abundances were higher in 1997–2002 than in 1975–80 (Fig. 59.2) in every season and the communities were mainly composed of diatoms and nanoflagellates; and
- in contrast, a clear decreasing trend was observed for zooplankton (Fig. 59.2): the abundances were distinctly lower in 1997–2002 and the decrease concerned both copepods and cladocerans.

Temporal trends in the euhaline waters

Means related to the euhaline waters are reported in Table 59.1 and the following observations can be made:

- regarding nutrients: (i) with the exclusion of summer, the mean DIN concentrations in 1997–2002 were significantly greater than in 1975–80; (ii) the orthophosphate levels were significantly lower in spring and summer 1997–2002 than in the corresponding seasons 1975–80; as in polyhaline waters, the silicate concentrations were always higher in 1997–2002 than in 1975–80;

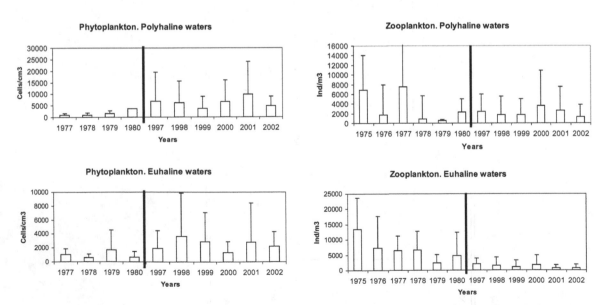

Fig. 59.2   Time trends of phytoplankton and zooplankton abundance in polyhaline and euhaline waters in the two sampling periods.

- except in autumn, phytoplankton abundances were significantly higher in 1997–2002 than in 1975–80 (Fig. 59.2), and the communities were dominated by diatoms and nanoflagellates; and
- zooplankton were distinctly more abundant in 1975–80 than in 1997–2002 (Fig. 59.2).

The natural and anthropogenic modifications which have taken place in the Lagoon of Venice have caused significant alterations in its hydrological and chemical states, and have probably influenced the biological communities in several ways. For this reason, describing and quantifying these hydrochemical and biological changes is a difficult task. Moreover, (i) although there has been much progress in data acquisition in recent years, the anthropogenic pressure on the whole lagoon (i.e. the qualities and quantities of wastes, sources of pollution, etc.) is not sufficiently well known; (ii) uninterrupted time series of samplings over large lagoonal areas are lacking; and (iii) when a few of these large-scale samplings have been carried out, this has been with differing sampling strategies, according to the various targets of each different project.

Bearing in mind these difficulties, we have tried nevertheless to make preliminary considerations about some general tendencies and/or changes which are occurring in the Lagoon of Venice. We have attempted this task throughout the comparison of data collected during two sampling periods. Two important events must be stressed: (i) the drastic decrease in distribution and production of *Ulva rigida* since 1990 (Sfriso and Marcomini, 1996); and (ii) the role of resuspension phenomena, recently highlighted (Facca *et al.*, 2002), and caused by increasing erosion, induced by tides and wind (Ravera, 1990), and by a more intensive dredging for molluscs (Orel *et al.*, 2000). Both these factors have produced several impacts on the distribution of nutrients and phytoplankton. In the latest sampling period (1997–2002), the increase of orthosilicate levels appears more evident in the whole lagoon, as a result of a major mobilization, through resuspension, of this nutrient from the interstitial waters to the water column. The concentrations of dissolved inorganic nitrogen (DIN) in the recent period have been higher than in the older period, particularly in spring (Table

59.1). In this season, the presently low biomass of macroalgae has a lesser nitrogen demand, with higher DIN quantities remaining in the waters; while its sequestration by *Ulva* was highly pronounced in the 1975–80 period, inducing much lower spring DIN concentrations in the water. The increase of phytoplankton abundance (especially of diatoms), observed between the two periods could be explained as follows: resuspension phenomena favour the mobilization of benthic epipsamic diatoms from sediments, returning them to a pelagic life style (see Bianchi *et al.*, 2000) in waters containing large concentrations of silica. Therefore the role of phytoplankton organisms, as primary producers, could now be strengthened by the reduced competition for nutrients and light between them and macroalgae. The increase of phytoplankton observed here is in apparent contrast with other data by some authors (Facca *et al.*, 2002), who describe a decrease of phytoplankton biomass from the early to the late 1990s. In fact, comparing these data (late 1990s), with our more recent dataset (1997–2002), shows that the phytoplankton standing stock appears to be very similar.

Conversely, orthophosphates have shown a general decrease over the years, especially in euhaline waters. This behaviour is related both to the reduced amounts of phosphorus compounds in detergents, following Italian laws enforced in the 1980s, and to the reduction of anoxic episodes in recent years, compared with 1970–80 (Sfriso *et al.*, 2003). In fact, it is documented that the reductive status of the sediments has favoured the release of phosphorus in the waters (Krom and Berner, 1980), while this nutrient has now been sequestered as an insoluble precipitate in the oxygenated sediments. Therefore, considering our data, it is suggested that this decreasing trend has not been sufficient to limit the growth of phytoplankton, over which the grazing pressure has been diminished, as shown by the negative trend demonstrated by the copepods and the cladocerans. However, the reasons for this marked decrease in zooplankton abundance are still not clear.

The analysis of both of these datasets confirms the high complexity of the Venice Lagoon from a hydrochemical and biological point of view. This environment is subjected to continued and relatively rapid anthropogenic and natural pressures, and for this reason it still needs continuing policies aimed at reducing the sources of pollution, as well as to a containment of indiscriminate fishing, both important instruments for its restoration and conservation.

ACKNOWLEDGEMENTS

The authors thank Dr T. Van Vleet and Dr Gloria Vallesi for assistance with the revision of the English version of this text and Dr Joan Coppola for help with diagrams.

REFERENCES

Alberighi, L., Bianchi, F., Cioce, F., and G. Socal. 1992. Osservazioni durante un bloom di *Skeletonema costatum* in prossimità della centrale termoelettrica E.N.E.L. di Fusina Porto-Marghera (Venezia). *Oebalia* **17** (suppl.), 321–2.

ARPAV (Agenzia Regionale per la Prevenzione e Protezione Ambientale del Veneto). 2002. *Bacino scolante nella laguna di Venezia*. Rapporto sullo stato ambientale dei corpi idrici, Anno 2001, Sintesi ottobre 2001, 15 pp.

Bendoricchio, G. 1996. Carichi di nutrienti nella laguna di Venezia e scenari di disinquinamento. In *Forum per la laguna ed.: La laguna di Venezia un patrimonio da riscoprire*. Venice: Filippi, pp. 75–9.

Bendoricchio, G., Calligaro, L., and G. M. Carrer. 1999. Consequences of diffuse pollution on the water quality of rivers in the watershed of the lagoon of Venice (Italy). *Water Science Technology* **39**, 113–20.

Bianchi, F., Cioce, F., Comaschi Scaramuzza, A., and G. Socal. 1990. Dissolved nutrients distribution in the central basin of the Venice lagoon. *Boll. Mus. Civ. Stor. Nat.* (Venice) **39**, 7–19.

Bianchi, F., Comaschi, A., and G. Socal. 1999. Ciclo annuale dei nutrienti, del materiale sospeso e del plancton nella laguna di Venezia. In M. Bon, G. Sburlino, and V. Zuccarello (eds.), *Aspetti ecologici e naturalistici dei sistemi lagunari e costieri*. Arsenale editrice, pp. 231–40.

Bianchi, F., Acri, F., Alberighi, L., Bastianini, M., Boldrin, A., Cavalloni, B., Cioce, F., Comaschi, A., Rabitti, S., Socal, G. and M. Turchetto. 2000. Biological variability in the Venice lagoon. In P. Lasserre, and A. Marzollo (eds.), *The Venice Lagoon Ecosystem. Inputs and Interactions between Land and Sea*. Man and the Biosphere Series, 25, Paris: UNESCO; Cornforth: Parthenon Publishing Group, pp. 97–125.

Bianchi, F., Acri, F., and A. Comaschi. 2002. Il trofismo degli ambienti lagunari: applicazione dell'indice TRIX nella laguna di Venezia (maggio 1997–aprile 1999). Studi trentini Sci. nat., *Acta biologica* **78**, 1, 195–200.

Bianchi, F., Acri, F., Bernardi Aubry, F., Berton, A., Boldrin, A., Camatti, E., Cassin, D., and A. Comaschi. 2003. Can plankton communities be considered as bio-indicators of water quality in the lagoon of Venice? *Marine Pollution* **46**, 8, 964–71.

Brunetti, R., Baiocchi, L., and M. Bressan. 1990. Seasonal distribution of Oikopleura (Larvacea) in the lagoon of Venice. *B. Zool.* **57**, 89–94.

Cioce, F., Comaschi Scaramuzza, A., Lombardo, A., and G. Socal. 1979. Hydrological and biological data from the northern basin of the Venice lagoon. June 1977–June 1978. *Atti Ist. Veneto Sci.* **137**, 309–42.

Cironi, R., Ioannilli, E., and R. Vitali. 1993. Assessment of effects of coastal power plants on marine biological resources in Italy. In *Coastal Ocean Space Utilization (COSU III)*, Santa Margherita Ligure, Portofino, Italy: 30 March – 2 April, 1993. London: Chapman and Hall, 17 pp.

Collavini, F., Zonta, R., Bettiol, C., Fagarazzi, O. E., and L. Zaggia. 2001. Metal and nutrient loads from the drainage basin to the Venice lagoon. In *Determination of the pollutant load discharged into the Venice Lagoon by Drainage Basin*. Venice: Ministero Lavori Pubblici, Consorzio Venezia Nuova, CNR Istituto per lo studio della dinamica delle Grandi masse, pp. 48–55.

Comaschi, A. 1987. Studio di popolazioni di copepodi planctonici nel bacino settentrionale della laguna di Venezia. Giugno 1977–giugno 1978. *Archo Oceanogr. Limnol.* **21**, 1–17.

Comaschi, A., and M. Dalla Palma. 1988. Studio di popolazioni di Copepodi planctonici nel bacino centrale della laguna veneta. Agosto 1979 – Settembre 1980. *Archo Oceanogr. Limnol.* **21**, 73–93.

Comaschi, A., and A. Lombardo. 1977. Hydrological data from the northern part of the Venice lagoon. May 1975 – July 1976. *Atti Ist. Veneto Sci.* **135**, 1–14.

Comaschi, A., and E. Martino. 1978. Studio del ciclo annuale dei Cladoceri nella laguna di Venezia mediante l'uso in componenti principali. *Archo Oceanogr. Limnol.* **19**, 99–120.

Comaschi, A., and E. Martino. 1981. Ciclo annuale dei Copepodi planctonici e loro distribuzione nel bacino settentrionale della laguna di Venezia. *Archo Oceanogr. Limnol.* **20**, 91–111.

Comaschi, A., Acri, F., Bianchi, F., Bressan, M., and E. Camatti. 2000. Temporal changes of species belonging to genus

Acartia (Copepoda, Calanoida) in the northern basin of the Venice lagoon. *Boll. Mus. St. Nat. Venezia* **50**, 189–93.

Curiel, D., Rismondo, F., Scarton, F., and M. Marzocchi. 1997. Flowering of Zostera marina in the lagoon of Venice (North Adriatic Italy). *Botanica Marina* **40**, 101–5.

Facca, C., Sfriso, A., and G. Socal. 2002. Phytoplankton changes and relationships with microphytobenthos and physicochemical variables in the central part of the Venice lagoon. *Estuarine Coastal Shelf Science* **54**, 5, 773–92.

Grasshoff, K., and M. Ehrhardt. 1976. Automated chemical analysis. In K. Grasshoff (ed.), *Methods of Seawater Analysis*. Weinheim: Verlag Chemie, pp. 263–97.

Guerzoni, S., Rossini, P., Molinaroli, E., Rampazzo, G., De Lazzari, A., and A. Zancanaro. In press. Atmospheric bulk deposition to the Venice lagoon: part I. Fluxes of metals, nutrients and organic contaminants. *Environmental International*.

Hansen, H. P., and F. Koroleff. 1999. Determination of nutrients. In K. Grasshoff, K. Cremling, and M. Erhardt (eds.), *Methods of Seawater Analysis*, Weinheim: Wiley-VCH Verlag, pp. 159–228.

Healy, B. 1997. Long term changes in a brackish lagoon, lady's island lake, south east Ireland. *Biology and Environment: Proceedings of the Royal Irish Academy* **97**, B 1, 33–51.

Krom, M. D., and R. A. Berner. 1980. Adsorption of phosphorus in anoxic marine sediments. *Limnol. Oceanogr.* **25**, 797–806.

Lapointe, B. E., and M. W. Clark. 1992. Nutrient input from the watershed and coastal eutrophication in the Florida Keys. *Estuaries* **15**, 465–76.

Lara-Lara, J. R., Alvarez Borrego, S., and L. F. Small. 1980. Variability and tidal exchange of ecological properties in a coastal lagoon. *Estuarine Coastal Shelf Science* **11**, 613–37.

Lombardo, A., and A. Comaschi. 1977. Biological data from the northern part of the Venice lagoon. *Atti Ist. Veneto Sci.* **135**, 133–48.

Orel, G., Boatto, V., Sfriso, A., and M. Pellizzato. 2000. *Fishing Sources Management Plant for the Lagoon of Venice Province*. Venice: Province of Venice, 188 pp.

Pavoni, B., Marcomini, A., Sfriso, A., Donazzolo, R., and A. Orio. 1992. Changes in an estuary ecosystem. The Lagoon of Venice as a case study. In D. A. Dunnette and R.J. O'Brien (eds.), *The Science of Global Change*. Washington, DC: American Chemical Society, pp. 285–305.

Perin, G. 1975. L'inquinamento chimico della Laguna di Venezia. In *Problemi dell'inquinamento lagunare*. Venice: Consorzio per la depurazione delle Acque della Zona Industriale di Porto Marghera, pp. 47–89.

Ravera, O. 2000. The lagoon of Venice: the result of both natural factors and human influence. *Journal of Limnology* **59**, 19–30.

Rismondo, A., Curiel, D., Solazzi, A., Marzocchi, M., Chiozzotto, E., and M. Scattolin. 1995. Sperimentazione di trapianto a fanerogame marine in laguna di Venezia: 1992–1994. *S.IT.E., Atti* **16**, 699–701.

Sfriso, A., and A. Marcomini. 1996. Decline of Ulva growth in the lagoon of Venice. *Bioresource Technology* **58**, 299–307.

Sfriso, A., and B. Pavoni. 1994. Macroalgae and phytoplankton competition in the central Venice lagoon. *Environmental Technology* **15**, 1–14.

Sfriso, A., Marcomini, A., and B. Pavoni. 1987. Relationship between macroalgal biomass and nutrient concentrations in a hypertrophic area of the Venice lagoon. *Marine Environmental Research* **22**, 297–312.

Sfriso, A., Pavoni, B., Marcomini, A., and A. Orio. 1992. Macroalgae nutrient cycles and pollutants in the lagoon of Venice. *Estuaries* **15**, 517–28.

Sfriso, A., Facca, C., and P. F. Ghetti. 2003. Temporal and spatial changes of macroalgae and phytoplankton in a Mediterranean coastal area: the Venice lagoon. *Marine Environmental Research* **56**, 617–36.

Socal, G. 1981. Nota sulla distribuzione quantitativa del fitoplancton nel bacino settentrionale della laguna di Venezia, giugno 1977 – giugno 1978. *Ist. Veneto Sci. Rappti Studi* **8**, 105–19.

Socal, G., Bianchi, F., and L. Alberighi. 1999. Effects of thermal pollution and nutrient discharges on a spring phytoplankton bloom in an industrial area of the lagoon of Venice. *Vie et Milieu* **49**, 1, 19–31.

Socal, G., Ghetti, L., Boldrin, A., and F. Bianchi. 1985. Ciclo annuale e diversità del fitoplancton nel porto-canale di Malamocco. Laguna di Venezia. *Atti Ist. Veneto Sci.* **143**, 15–30.

Strickland, J. D., and T. T. Parsons. 1972. A practical handbook of seawater analysis. *Journal of Fisheries Research Board of Canada* **167**, 2nd edn, pp. 310.

Utermöhl, H. 1958. Zur Vervollkommnung der quantitativen Phytoplankton-Methodik. *Mitteilungen der Internationalen Vereinigung für theoretische und angewandte Limnologie* **9**, 1–38.

Voltolina, D. 1973. A phytoplankton bloom in the lagoon of Venice. *Archo Oceanogr. Limnol.* **18**, 19–37.

Welch, A. 1938. The significance of the differences between two means when the population variances are unequal. *Biometrika* **29**, 350–62.

Zuliani, A., Zonta, R., Zaggia, L., and O. E. Fagarazzi. 2001. Fresh water transfer from drainage basin to the Venice lagoon. In *Determination of the Pollutant Load Discharged into the Venice Lagoon by Drainage Basin*. Venice: Ministero Lavori Pubblici, Consorzio Venezia Nuova, CNR Istituto per lo studio della dinamica delle Grandi masse, pp. 26–31.

60 • Distribution of *Zostera noltii*, *Zostera marina* and *Cymodocea nodosa* in Venice Lagoon

A. RISMONDO, D. CURIEL, F. SCARTON, D. MION, A. PIERINI AND G. CANIGLIA

INTRODUCTION

Coverage of marine seagrasses *Zostera marina* Linnaeus, *Zostera noltii* Hornemann and *Cymodocea nodosa* (Ucria) Ascherson occurring in the Lagoon of Venice was mapped for the first time in 1990 (Caniglia *et al.*, 1990, 1991; Magistrato alle Acque, 1991). In 2002 a new programme of monitoring was started in the framework of a project (MELa2) carried out by Consorzio Venezia Nuova, on behalf of Magistrato alle Acque (2003), the local State Water Authority in Venice. Updating of the seagrass map of Venice Lagoon was part of this project.

Field activities were performed during the summer of 2002, followed by data analysis and production of maps. All shallows, tidal flats and channels (not deeper than 2.5 m) of Venice Lagoon have been investigated, resulting in a highly detailed seagrass map. The principal objective consisted of the production of the seagrass map, in order to make a comparison with previous data possible. This activity permitted analysis of colonizing and regressing trends, relating them to other ecosystem components and functions, including their geographical location. ESRI ArcGis 8.2 software was used for the analysis.

METHODS

Field activities, performed from May to September 2002, and computer desk work were undertaken to obtain a complete digital map of Venice Lagoon seagrasses on GIS data files.

The DGPS global positioning system calibration, integrated with GIS software, was carried out in order to evaluate positional accuracy and to find the best way for data saving and transfer to GIS software. Calibration produced a GPS system with 1 m preci-

sion. Based on the results of this phase, a field data card, used by operators to collect coverage and other information, was set up. The aim of this phase was also to set and obtain a reference coverage scale (Fig. 60.1). Every field team, equipped with a GIS system installed on a small boat, was provided with photographs derived from IKONOS panchromatic images.

Surveys were carried out according to vegetation characteristics:

• along seagrass patch borders; and
• along transects to explore beds in the case of irregular coverage.

For each mapped seagrass patch, geographic and topologic information were collected, together with vegetation data (species and degree of coverage), using the estimation scale presented here. The data gathered in the field were than used to produce preliminary maps. These were later checked and processed with aerial photographs to produce the definitive cartographic data.

Based on these preliminary maps, an error evaluation was carried out. During surveying and map-

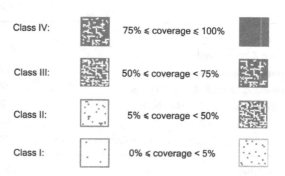

Fig. 60.1 Coverage scale used during field mapping.

ping, vector data are subject to gross, systematic and random measurement errors. Gross errors were identified and corrected using results obtained by comparisons between preliminary maps and aerial photographs, and later checked by field survey, as described above. Assuming there was no error in the identification and measurement of ecological attributes (species identification and coverage degree), and using aerial photographs with 1 m resolution in both axes, errors on positional data were first fixed in DGPS instrumental error (1 m on both latitude and longitude values). A refined evaluation of systematic and random errors will be carried out in the future, using data from the second and the third year of monitoring activities.

RESULTS AND DISCUSSION

Extension of all vegetation patches was calculated. For purposes of comparison (between 1990 and 2002), all the 1990 cartographic information (CAD Microstation software) was transferred to the same ArcGis 8.2 suite (Figs. 60.2, 60.3, 60.4).

Cymodocea nodosa now extends in the Lagoon of Venice over 2945 ha (1634 ha in 1990), *Zostera marina* over 3442 ha (3642 in 1990) and *Zostera noltii* over 633 ha (4141 in 1990). The comparison with 1990 seagrass distribution shows a very clear regression of *Z. noltii* in the inner areas, while *Z. marina* has been increasing its coverage in the central basin of the lagoon, where strong dystrophic events, caused by macro algae blooms, were frequently recorded until a few years ago. *C. nodosa* does not show a clear trend, apart from a strong loss close to Chioggia (southern basin), the result of artificial removal by clam collectors.

For the whole lagoon, we calculated a retreat of seagrass bed coverage of only 62 ha (-1.1%). Nevertheless, we observed high local variability, with a number of strong variations in extension, especially for mixed meadows, which indicates the strong dynamic nature of the system (Table 60.1).

The northern lagoon showed larger retreats in comparison with the 1990 data, with a loss of 584 ha of meadows. Higher losses are related to the disappearance of *Z. noltii* population (Fig. 60.5). This accounted for 88% of the total loss. This evidence indicates a most likely decrease in environmental quality in comparison with 1990.

In the central lagoon we found the most evident increase, leading to the hypothesis of improving water quality. The increase in *Z. marina* distribution amounts to 747 ha, with a high percentage coverage. This species is responsible for 97% of the whole increase in this basin. A similar trend has been observed for *Z. noltii* (Fig. 60.4) and *C. nodosa* (Fig. 60.2) with an increase of 88 and 60 ha respectively.

In the southern lagoon all three species showed distribution changes in comparison to 1990, although

Table 60.1   *Comparison between 1990 and 2002 species coverage.*

	1990 (ha)	2002 (ha)	Variations (ha)
pure *C. nodosa* beds	391	1777	+1386
pure *Z. marina* beds	266	2195	+1930
pure *Z. noltii* beds	1436	70	-1366
mixed *Z. marina* – *C. nodosa* beds	692	825	+133
mixed *Z. noltii* – *C. nodosa* beds	23	141	+118
mixed *Z. noltii* – *Z. marina* beds	2157	220	-1937
mixed *Z. noltii* – *Z. marina* – *C. nodosa* beds	528	203	-325
Total (ha)	5493	5430	-62

Fig. 60.2   Comparison of 1990 and 2002 *Cymodocea nodosa* distribution.

Fig. 60.3   Comparison of 1990 and 2002 *Zostera marina* distribution.

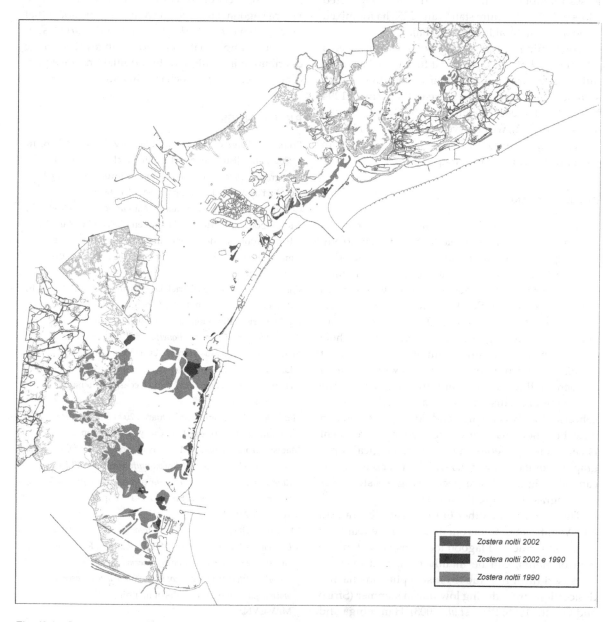

Fig. 60.4   Comparison of 1990 and 2002 *Zostera noltii* distribution.

less marked in comparison to the central lagoon. Net losses amount to 237 ha. *Z. noltii* (Fig. 60.4) registered a loss of 727 ha for pure stands and 2253 ha for mixed populations (a total loss of 2980 ha).

Concerning *Z. marina* (Fig. 60.3), we observed a retreat of 924 ha. It is important to note that a strong difference between pure and mixed meadows extension occurred, with an increase in the former (1294 ha) and a decrease in the latter (-2218 ha). For *C. nodosa* (Fig. 60.2), we observed an increase of 1202 ha. Only a few, limited meadows showed retreats in comparison with the 1990 map.

CONCLUSIONS

There are different trends in Venice Lagoon seagrass dynamics. Since 1990 *Z. noltii* has shown strong retreat, particularly in the southern basin. The other two species, at local scale, show different dynamics and high variability. A general overview of species dynamics points out that the differences in meadow extension between 1990 and 2002, about 62 ha, can not be considered significant at the scale of the whole lagoon. The most significant finding is the retreat of *Z. noltii*, both from mixed meadows with *Z. marina* in open shallow bottoms and from tidal flats in the whole lagoon. This represents a very critical issue, concerned with salt marsh and tidal flat erosion and retreat phenomena. The Venice Lagoon environment is very heterogeneous from a morphological-topographic point of view; thus each basin and each area can vary highly with respect to seagrass strategies and seagrass response to pressures.

The literature on other European Mediterranean and Atlantic sites points out that *Z. noltii* retreat from other estuarine and lagoon locations, in salt marsh environments, can be mainly explained by both increased turbidity due to erosion phenomena and desiccation stress during low tide in summer (Sfriso and Ghetti, 1998; Brun *et al.*, 2003). Hamminga and Duarte (2000) record that seagrasses may counteract turbidity to a high extent, by optimizing light absorption capacity. For this reason we believe that physical phenomena (erosion and consequent lost of shoots) are responsible for *Z.noltii* losses in Venice Lagoon. Other North-Adriatic lagoons we are monitoring have not experienced such high retreat phe-nomena; these locations have a more sheltered and protected intertidal environment by comparison to Venice Lagoon. However, it should also be noted that Frederiksen *et al.* (2004), referring to *Z. marina*, suggest more simply that even wide changes of coverage distribution in shallow water populations are equivocal and seem more stochastic in nature.

REFERENCES

Brun, F. G., Perez-Llorens, J. L., Hernandez, I., and J. J. Vergara. 2003. Patch distribution and within-patch dynamics of the seagrass *Zostera noltii* Hornem. In Los Toruños salt marsh, Cadiz Bay Natural Park, Spain. *Bot.Mar.* **46**, 513–24.

Caniglia, G., Borella, S., Curiel, D., Nascimbeni, P., Paloschi, F., Rismondo, A., Scarton, F., Tagliapietra, D., and L. Zanella. 1990. Cartografia della distribuzione delle fanerogame marine nella Laguna di Venezia. *Giornale Botanico Italiano* **124**, 1, 212.

Caniglia, G., Borella, S., Curiel, D., Nascimbeni, P., Paloschi, F., Rismondo, A., Scarton, F., Tagliapietra, D., and L. Zanella. 1991. Cartografia computerizzata delle fanerogame marine nella Laguna di Venezia. *Bollettino dell'A.I.C.* **81**, 1–4.

Frederiksen, M., Krause-Jansen, D., Holmer, M., and J. Sund Laursen. 2004. Long-term changes in area distribution of eelgrass (*Zostera marina*) in Danish coastal waters. *Aquatic Bot.* **78**, 167–81.

Hamminga, M. A., and C. M. Duarte. 2000. *Seagrass Ecology.* Cambridge: Cambridge University Press. 298pp.

Magistrato alle Acque. 1991. *Nuovi interventi per la salvaguardia di Venezia. Composizione delle comunità biologiche. 1ª Fase. Rilievi sui popolamenti delle barene ed aree circostanti e sulla vegetazione dei bassifondi*, Studio A.3.16. Rapporto finale. Venice: MAV-CVN.

Magistrato alle Acque. 2003. *Monitoraggio dell'Ecosistema Lagunare – 2° ciclo triennale (MELa2). Linea A: Rilievo delle fanerogame marine in Laguna di Venezia con taratura di un sistema di telerilevamento e completamento delle conoscenze sulle macroalghe.* Report on results of mapping – first year. Venice: MAV-CVN.

Sfriso, A., and P. F. Ghetti. 1998. Seasonal variation in biomass, morphometric parameters and production of seagrasses in the Lagoon of Venice. *Aquatic.Bot.* **61**, 207–23.

61 • Breeding birds and vegetation monitoring in recreated salt marshes of the Venice Lagoon

F. SCARTON

INTRODUCTION

In the Lagoon of Venice the area of salt marsh has fallen from about 12 000 ha to less than 4000 ha between 1900 and the present, due to reclamation, erosion and natural and man-induced subsidence. A dredging programme, undertaken for the Ministry of Public Works, has been under way since 1984 to maintain channel depths for the purposes of navigation and to increase tidal flushing in the inner lagoon. The resulting dredged material has been used to build artificial salt marshes (hereafter called dredge islands) and tidal flats (Cecconi, 2001). It was soon observed that after the end of the works, and following different patterns, dredge islands become covered with vegetation and become a suitable habitat for many species of wetland birds. The Lagoon of Venice is of particular importance for waterfowl, both wintering (it hosts the largest concentration in Italy, with about 130 000 birds counted in January; Zenatello *et al.*, 2002) and nesting. Among the nesting species, 80-90% of the Italian populations of Redshank and Sandwich Tern are found here, whereas for other species, i.e. Common Tern, Little Tern, Avocet, Black-winged Stilt, it hosts between 5% and 30% of the national totals (Valle and Scarton 1996, 1999).

Several studies have been published about birds or salt-marsh vegetation occurring at dredge islands, most of them dealing with sites in the USA (Delaney *et al.*, 2000; Erwin *et al.*, 2001, 2003; Neckles *et al.*, 2002; Shafer and Streever, 2000; Zedler, 2000) or the UK (ABP Southampton, 1998; Atkinson *et al.*, 2001) sites. Concerning the Lagoon of Venice, so far only a short review of birds has been published (Scarton and Valle, 1999) whereas information on vegetation has been presented only in unpublished reports. Here

data gathered during the years 1993-2002 is presented, with particular attention to the breeding birds and the vegetation succession observed at dredge islands.

STUDY AREA AND METHODS

The Venice Lagoon is a large (550 km^2) shallow coastal lagoon located on the north-eastern coast of the Adriatic Sea (with its centre at about 45° N 12° E). There are two barrier islands which separate the lagoon from the sea and water is exchanged through three large inlets. Most of the lagoon area is occupied by an open waterbody (about 400 km^2) which is partially vegetated by macroalgae (*Ulva* sp., *Chaetomorpha* sp., *Vaucheria* sp.) and seagrasses (*Zostera marina*, *Zostera noltii* and *Cymodocea nodosa*), these latter communities covering a surface of approximately 5430 ha. The mean depth of the lagoon is 1.1 m and the tidal range during spring tides is about one metre, 0.6 m being the mean range. There are extensive intertidal salt marshes, especially in the south-western and northern portions of the lagoon. Dominant marsh species include *Limonium narbonense*, *Salicornia veneta*, *Sarcocornia fruticosa*, *Atriplex portulacoides*, *Puccinellia palustris*, *Spartina maritima* and *Juncus maritimus*.

Up until 2003, 57 dredge islands had been created, ranging in size from 0.4 to 57.4 ha, with a mean of 10.5 ha and a total area of 600 ha. They are built by making a containment cell, using wooden piles and a hydraulic net to contain the mud and to reduce turbidity. The substrate of the islands is usually composed of clay-silty sediments (indicated here as 'silty islands'), but some islands have a prevalence of silty-sand or pure sand and are referred to here as 'sandy islands'. Elevation of the dredge islands is most of the

Flooding and Environmental Challenges for Venice and its Lagoon: State of Knowledge, ed. C. A. Fletcher and T. Spencer.
Published by Cambridge University Press, © Cambridge University Press 2005.

time around 0.5 m above sea level, which means they are flooded during high tides; nevertheless, some dredge islands are, entirely or for a large part of their area, completely above that level, which means they are rarely or never inundated.

Data on the use by breeding birds come from opportunistic observations from 1988-1992 and from regular visits, at least two to each site per season, from 1992 onwards. Some of these sites were studied in more detail, with bimonthly surveys. Data about nesting populations reported here refer to the 2000 period, unless specified otherwise. Vegetation surveys started in 1997, with a detailed mapping of each site being made in 2002.

BREEDING BIRDS

Overall, about 60 species (not considering passerines) were observed at least once at dredge islands, and gulls, waders and herons were the most abundant groups (Table 61.1). Most of the species used the islands as feeding sites (especially wintering waders such as Dunlins *Calidris alpina* and Curlews *Numenius arquata*, but also Little Egrets *Egretta garzetta* and Grey Herons *Ardea cinerea*) or resting sites (mostly gulls, i.e. Yellow-legged Gull *Larus michahellis* or Black-headed Gull *Larus ridibundus*, and herons), whereas nine species nested. The breeding species include rare or localised species throughout Italy, such as Shelduck *Tadorna tadorna*, Avocet *Recurvirostra avosetta*, Oystercatcher *Haematopus ostralegus*, Black-winged Stilt *Himantopus himantopus*, Redshank *Tringa totanus*, Kentish Plover *Charadrius alexandrinus* and Little Tern *Sterna albifrons*. Table 61.2 presents the number of pairs nesting at dredge islands, compared with that of the whole lagoon of Venice and with the most recent estimates available for Italy.

It can be seen that, for some species such as Oystercatcher or Little Tern, most of the pairs breeding in the lagoon are found at dredge islands. Moreover, for Oystercatcher, Shelduck and Redshank the number of pairs breeding at those sites make a significant percentage, more than 5%, of the total estimated Italian population. Colonies of Yellow-legged Gull occur each year at 10-15 dredge islands. The nesting of this species is of particular concern

as predation on chicks and eggs of other species, more important from a conservation point of view, has been reported for other Mediterranean sites (Vidal *et al.*, 1998; Hernandez-Matias and Ruiz, 2003). Despite the lack of a comprehensive census, the Yellow-legged Gull may be considered as increasing in the whole lagoon since in the last few years, it has also begun to nest on the roofs of historical buildings in Venice. Quite surprisingly, colonies of neither Common Tern *Sterna hirundo* nor Sandwich Tern *Sterna sandvicensis* have been observed so far at dredge islands; the two terns have comparable breeding populations in the Lagoon of Venice (800-1000 pairs each) and they are known to take advantage of dredge islands at other coastal sites, as happens along the USA Atlantic coast (Erwin *et al.*, 2003). Over the years 1996–2002, the breeding species exhibited different trends: Yellow-legged Gull and Oystercatcher increased markedly, Little Tern and Avocet showed strong fluctuations between years, whereas Black-winged Stilt, Kentish Plover and Redshank were more stable.

For nine sites for which more detailed data are available, it has been observed that ageing of sites reduces diversity, in terms of number of breeding species; maximum numbers are attained after five to six years, then the numbers start to decrease (Fig. 61.1). This is certainly due to vegetation encroachment, which has made most of the sites unsuitable for several species (such as Kentish Plover, Oystercatcher, Little Tern), which prefer bare substrates. Conversely, Redshank has become more abundant at the same sites. Quite different trends have been observed between sandy or silty islands (Fig. 61.2); the latter, which are also higher above sea level, become more rapidly covered with non halophytes (see also below) and even after a few years host very few species, sometimes only Yellow-legged Gull. Since other sandy sites, with an elevation below the high tide level, showed the occurrence of typical halophytes, it seems quite clear that elevation, not just the kind of substrate, is the driving force, or at least one of the most important factors, regulating vegetation growth. Within the nine dredge islands studied in more detail, the only morphological characteristic (among size, perimeter, elevation, distance from other salt-marshes, area of open water) which

Table 61.1   *Species, excluding passerines, recorded at dredge islands in the 1992–2002 period, and use of the sites (continued overleaf).*

	Nesting	Feeding	Resting
Podiceps cristatus		x	
Podiceps griseigena		x	
Podiceps nigricollis		x	
Phalacrocorax carbo			x
Egretta garzetta		x	x
Casmerodius albus		x	x
Ardea cinerea			x
Ardea purpurea		x	
Phenicopterus ruber		x	x
Anser fabalis			x
Tadorna tadorna	x	x	x
Anas crecca		x	
Anas platyrhynchos	x	x	x
Anas querquedula		x	
Somateria mollissima			x
Mergus merganser		x	
Circus aeruginosus		x	
Circus cyaneus		x	
Circus pygargus		x	
Buteo buteo			x
Falco columbarius		x	x
Falco peregrinus		x	x
Haematopus ostralegus	x	x	x
Himantopus himantopus	x	x	x
Glareola pratincola			x
Recurvirostra avosetta	x	x	x
Charadrius dubius		x	
Charadrius hiaticula		x	
Charadrius alexandrinus	x	x	
Pluvialis apricaria		x	
Pluvialis squatarola		x	x
Vanellus vanellus		x	
Calidris minuta		x	
Calidris ferruginea		x	x
Calidris alpina		x	x
Philomacus pugnax			x
Gallinago gallinago		x	

Table 61.1   *continued.*

	Nesting	Feeding	Resting
Numenius phaeopus		x	
Numenius arquata			x
Tringa erythropus		x	
Tringa totanus	x	x	x
Tringa glareola			
Tringa nebularia		x	
Actitis hypoleucos		x	
Arenaria interpres			x
Catharacta skua			x
Larus melanocephalus			x
Larus minutus			x
Larus ridibundus			x
Larus canus		x	x
Larus fuscus			x
Larus michahellis	x		x
Sterna sandvicensis			x
Sterna hirundo			x
Sterna albifrons	x		x
Chlidonias niger		x	
Cuculus canorus			x
Apus apus		x	
Alcedo atthis			x
Total: 58	9	38	35

correlated positively and significantly with the number of species was the area of tidal ponds and creeks excavated by the tides (Fig. 61.3). This suggests that tidal creeks and ponds, with the associated occurrence of invertebrates and juvenile forms of fish, are important feeding sites for birds breeding at those islands.

Human use of dredge islands is due to (1) fishermen, who lay nets over the surface to have them dried; (2) hunters, who build hides to shoot wintering birds and sometimes create ponds to attract more birds; and (3) sunbathers, especially during week-ends. Despite being limited in numbers, sunbathers posed a real threat to breeding birds, by causing desertion of nests or chicks at some sites.

The major findings can be summarized as follows:

- dredge islands are a new, important habitat feature of the Venice Lagoon; they host several species of particular concern, sometimes with populations of national importance (more than 1%);
- creation of a network of tidal ponds and creeks is needed at each site, if a higher number of breeding species is desired;

Table 61.2 *Number of pairs nesting at dredge islands compared to that for the whole Venice Lagoon and that for Italian populations, along with the status of each species in Italy.*

	Dredge islands	Lagoon of Venice[1]	Italy[2]	Italian Red List[3]
Shelduck	5–8	40–60	150–180	Endangered
Mallard	4–7	500–1000	10,000–20,000	Not evaluated
Oystercatcher	12–16	14–18	120–140[1]	Endangered
Black-winged Stilt	50–180	350–400	900–1,700	Lower risk
Avocet	30–140	100–200	1,200–1,800	Lower risk
Kentish Plover	20–30	70–90	1,500–2,000	Lower risk
Redshank	50–70	1,500–1,600	1,600–1,700[1]	Endangered
Yellow-legged Gull	200–400	5,000–6,000	24,000–27,000	Not evaluated
Little Tern	50–300	100–400	5,000–6,000	Lower risk

[1] Scarton and Valle, personal estimates, years 2000–03.

[2] Brichetti and Gariboldi, 1997.

[3] Bulgarini *et al.*, 1998.

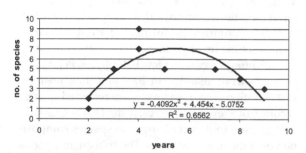

Fig. 61.1 Correlation between number of breeding species and years since the construction at nine dredge islands.

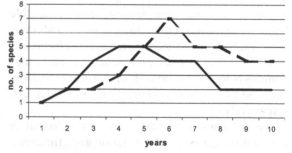

Fig. 61.2 Number of breeding species at sandy (solid line) and silty (broken line) sites.

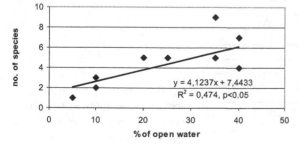

Fig. 61.3 Correlation between number of breeding species and percentage of tidal pond and creek areas inside dredge islands.

- ageing of sites and the concurrent vegetation growth reduces breeding species, so some management activities (such as vegetation cutting, or spraying of new sediments over the surface) are needed;
- sandy sites, if higher than high tide levels, become quickly less useful for breeding birds; and
- human use of the sites (especially by sunbathers) should be controlled and Yellow-legged Gull reproduction should be discouraged.

VEGETATION

Vegetation growth and species occurrence at dredge islands are clearly linked to (1) soil characteristics; (2) elevation above sea level; and (3) age of the sites. Silty-clay islands, regularly flooded, become covered with halophytes (mostly *Salicornia veneta, Puccinellia palustris, Spartina maritima*) after at least three years following the end of the works. Almost always the first two species to colonize the new sites are *Salicornia veneta* and *P. palustris*; the first species reaching in the colonized sites a coverage of some hectares at the end of the summer. After five to six years, about 90% of the islands are covered with vegetation, and all the sites develop a vegetation coverage after ten years or more. Bare areas do persist at several sites. Tidal ponds and creeks show only occasionally the occurrence of *Zostera noltii* and, even less frequently, that of *Ruppia maritima*. It is worth noting that at several sites vegetation cover appears well structured and different habitats (according to the 'Habitat' European Directive 92/43) were recorded during a recent mapping. Among the most valuable habitats, i.e. of community importance, 'Mediterranean and halophytic scrubs (Sarcocornetea fruticosi)', Natura 2000 code 1420, and 'Salicornia and other annuals colonising mud and sand', code 1310, were the most widespread, covering 72 ha and 82 ha each respectively.

Sandy islands, irregularly or never flooded, show a very different trend of vegetation growth; after a few years, halophytes (salt-tolerant) or truly psam-mophilous (sand-living) species become restricted to the less elevated areas, whereas most of the sites become covered with dense stands of nitrophilous or ruderal species (i.e. *Calamagrostis epigejos, Agropyron repens, Oenothera biennis, Conyza canadensis*). Small bushes like *Tamarix gallica* are also sometimes found. Since 2002 at a few dredge islands, and at natural marshes as well, we have observed the occurrence of *Spartina x townsendii*, which is the first record for Italy and most likely for the whole Mediterranean (Scarton *et al.*, 2003). We deem this hybrid has been introduced deliberately, maybe for erosion control, but no confirmed data is available. Almost circular clumps of this hybrid were found scattered on the surface of the islands, interspersed with *P. palustris* and *S. veneta*. Nevertheless, we have also observed this plant at lower elevation, on the fringe of the sites, on bare silty areas where no other terrestrial species occurs, the only other vascular plant being *Zostera noltii*. Preliminary data show that, as is known for other coastal areas (e.g. Denmark: Vinther *et al.*, 2001), *S. x townsendii* can grow at much lower elevations than any other halophyte (Fig. 61.4).

Other activities performed at dredge islands include setting up of permanent quadrats and testing transplanting success of a few halophytes. Permanent quadrats were installed in 2002 and their position and elevation above sea level were recorded with an accuracy of a few centimetres, using DGPS. Quadrats will be used to detect changes in species composition over the next few years. The transplanting tests were done at three sites, using different species at

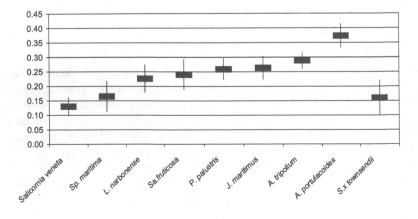

Fig 61.4   Vertical zonation (m above sea level; mean ±1 s.d.) of the most common native salt marsh species and of *S. x townsendii*, measured at Venice Lagoon salt-marshes (after Scarton *et al.*, 2003).

different elevation. *P. palustris* showed a good success at low (here defined as the 0.10–0.20 m range) and medium (0.20–0.30 m) elevation above sea level, whereas *Spartina maritima* grew vigorously only at low elevations, along tidal creeks and ponds. The results with *Sarcocornia fruticosa* (at medium or high, i.e. more than 0.30 m) and *Atriplex portulacoides* (only at high elevation) were on the contrary very poor.

ACKNOWLEDGEMENTS

Most of the data presented here were collected in the framework of surveys made on behalf of Magistrato alle Acque di Venezia – Consorzio Venezia Nuova (CVN). A. Bernstein, G. Cecconi and L. Montobbio (all from CVN) are especially thanked for encouraging biological monitoring at dredge islands.

BIBLIOGRAPHY

ABP Southampton. 1998. *Review of Coastal Habitat Creation, Restoration and Recharge Schemes*. Report no. R.909. 189 pp.

Atkinson, P. W., Crooks S., Grant A., and M. Rehlfish. 2001. *The Success of Creation and Restoration Schemes in Producing Intertidal Habitat Suitable for Waterbirds*. English Nature Research Reports, n. 425. 166 pp.

Brichetti, P., and Gariboldi A. 1997. *Manuale pratico di ornitologia*. Bologna: Edagricole.

Bulgarini F., Calvario, E., Fraticelli, F., Petretti, F., and S. Sarrocco. (eds). 1998. *Libro Rosso degli Animali d'Italia – Vertebrati*. Rome: WWF Italia.

Cecconi, G. 2001. Restoring saltmarshes in the Venice Lagoon. In L. Drévès and M. Chaussepied (eds.). *Restauration des écosystèmes cotiers*. Plauzané Editions Ifremer, Plouzané, vol. XXIX, 38–53.

Delaney, J., Webb, W., and T. Minello. 2000. Comparison of physical characteristics between created and natural estuarine marshes in Galveston Bay, Texas. *Wetlands Ecology and Management* **8**, 343–352, 2000.

Erwin, R. M., Truitt, B. R., and J. E. Jimenez. 2001. Ground nesting waterbirds and mammalian carnivores in the Virginia Barrier island region: running out of options. *Journal of Coastal Research* **17** (2), 292–296.

Erwin, R. M., Allen, D. H., and D. Jenkins. 2003. Created versus natural coastal islands: Atlantic waterbird populations, habitat choices, and management implications. *Estuaries* **26** (4), 949–955.

Hernandez-Matias, A., and X. Ruiz. 2003. Predation on common tern eggs by the yellow-legged gull at the Ebro Delta. *Scientia Marina* **67**, 95–101.

Neckles H. A., Dionne, M., Burdick, D. M., Roman, C. T., Bucksbaum, R., and E. Hutchins. 2002. A monitoring protocol to assess tidal restoration of salt marshes on local and regional scales. *Restoration Ecology* **10**, 556–563.

Scarton, F., and R. Valle. 1999. The use of dredge islands by birds in northern Adriatic lagoons. *Avocetta* **23**, 1, 75.

Scarton, F., Ghirelli L., Curiel D., and A. Rismondo. 2003. First data on *Spartina x townsendii* in the Lagoon of Venice (Italy). In: E. Özhan (ed.), *Proceedings of the Sixth International Conference on the Mediterranean Coastal Environment, MED-COAST 03* (7–11 October 2003, Ravenna, Italy). Ankara, Turkey: MEDCOAST, Middle East Technical University, vols. I–III. pp. 787–792.

Shafer, D. J., and W. J Streever. 2000. A comparison of 28 natural and dredged material salt marshes in Texas with an emphasis on geomorphological variables. *Wetlands Ecology and Management* **8**, 353–366.

Valle, R., and F. Scarton. 1996. Status and distribution of Redshanks breeding along Mediterranean coasts. *Wader Study Group Bulletin* **81**, 66–70.

Valle, R., and F. Scarton. 1999. The presence of conspicuous associates protects nesting Redshank *Tringa totanus* from aerial predators. *Ornis Fennica* **76**, 145–148.

Vidal, E., Medail, F., and T. Tatoni. 1998. Is the yellow-legged gull a superabundant bird species in the Mediterranean? Impact on fauna and flora, conservation measures and research priorities. *Biodiversity and Conservation* **7**, 1013–1026.

Vinther, N., Christiansen, C., and J. Bartholdy. 2001. Colonisation of Spartina on a tidal water divide, Danish Wadden Sea. *Danish Journal of Geography* **101**, 11–20.

Zedler, J. B. (ed.) 2000. *Handbook for Restoring Tidal Wetlands*. Boca Raton: CRC Press, 439 pp.

Zenatello, M., Baccetti, N., and L. Serra. 2002. Wintering waders in Italy: distribution, numbers and trends in 1991–2000. *Wader Study Group Bulletin* **99**, 18.

62 • Pollution in the Venice Lagoon (Italy): loads from the drainage basin

R. ZONTA, F. COLLAVINI, L. ZAGGIA AND A. ZULIANI

VENICE LAGOON AND DRAINAGE BASIN

The Lagoon of Venice and its drainage basin (Fig. 62.1) forms one of the most important lagoon systems in the world, where historical and recent cities, industrial districts and intensive agricultural activities coexist within a peculiar natural setting. The history of Venice and its lagoon is characterized by the difficult relationship between a particular and delicate environment and the constant effort of man to adapt it to their changing needs (Francia and Juhasz, 1993).

From the very first settlements and throughout the centuries the size, shape and structure of the lagoon have evolved significantly in consequence of natural and human actions. The main anthropic interventions (Avanzi *et al.*, 1984) have been concerned with the diversion of the rivers, modification and stabilisation of the sea inlets, dredging for the transportation infrastructure and fish farm development. In particular, a large channel was created in recent decades to carry shipments from the Adriatic Sea through the Malamocco inlet to the industrial area of Porto Marghera. The size, morphology and hydraulics of the drainage basin have been intensely modified as well (Cucchini, 1928), by the diversion to the sea of some large rivers (Piave (not shown in Fig. 62.1), Sile and Brenta), reclamation of a large belt of wetlands for agricultural and industrial purposes, and realisation of a network of artificial channels and a number of hydraulic devices (such as pumping plants, sluices and flood gates).

The present lagoon is a flat shallow water body (average depth is less than 1 m) accounting for a total surface area of about 550 km². It is *c.* 50 km long and 8–14 km wide, and is crossed by a network of tidal channels. Three inlets (Lido, Malamocco and Chioggia), 10–15 m deep and 450–900 m wide, connect the

lagoon to the northern Adriatic Sea. The average tidal excursion is 80 and 30 cm under spring and neap tide conditions respectively. As much as 110×10^9 m³ yr⁻¹ of water is exchanged through the inlet by the tide, whose large-scale circulation patterns identify three sectors: northern, central and southern lagoon.

The drainage basin consists of a low-gradient floodplain with a surface of about 1850 km². It encloses a population of more than a million inhabitants and includes some important urban centres and industrial areas. Porto Marghera, in particular, is one of the largest chemical industrial districts in the Mediterranean Sea. About 70% of the drainage basin surface is cultivated and relevant livestock activities are present. The catchment area encompasses several sub-basins characterized by different morphology and dimension, which are drained by streams with different hydraulic regimes, from natural to completely regulated. A belt of spring sources in the northern part of the drainage basin provide a permanent freshwater supply, which sustains the baseflow of the network in the driest periods. A considerable fraction of the basin surface (about 40%), particularly in the southern sectors and along the lagoon border, is below mean sea level. These areas are presently reclaimed lands, where the excess inflow is artificially drained by pumping plants, sluices and other hydraulic infrastructure. Finally, the lagoon receives additional freshwater inputs from two rivers (Sile and Brenta) that are external to the drainage basin.

The morphological features and the characteristics of the drainage, combined with the intense regulation of flow and the possibility of diversions, determine a different magnitude and time course of runoff in the various tributaries. Even where the natural slope governs the flow, a large number of connec-

Flooding and Environmental Challenges for Venice and its Lagoon: State of Knowledge, ed. C. A. Fletcher and T. Spencer.
Published by Cambridge University Press, © Cambridge University Press 2005.

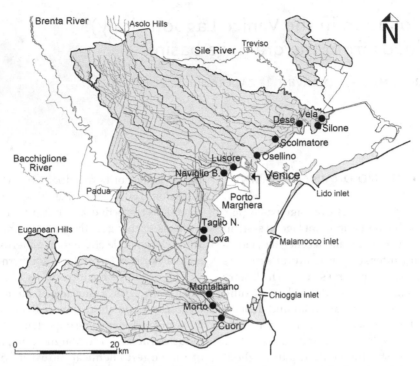

Fig. 62.1   The Venice Lagoon and its drainage basin. The twelve main tributaries and the location of the measurement sections are indicated.

tions permit water exchanges among the sub-basins, particularly during flood events. The hydraulic pathways in the drainage basin are therefore complex and not identifiable with any certainty. More detailed information on the characteristics of the lagoon and the drainage basin is available elsewhere (e.g., Avanzi *et al.*, 1984; Cossu and de Fraja Frangipane, 1985; Regione del Veneto, 2000; Magistrato alle Acque, 2000a; Zonta *et al.*, 2001).

THE DRAIN PROJECT

The load transferred by the drainage basin collectors represents a fundamental contribution from amongst the different sources of pollutants in the Venice Lagoon. The quantification and characterisation of this load, as well as the study of its time-variability and the definition of the relative importance of the different inputs, are essential in the planning of interventions for both the physical protection of the lagoon ecosystem and the improvement of water quality.

Regardless of their importance, in the past very few investigations have addressed the determination of the freshwater discharge and pollutant loads delivered to the lagoon. This issue was the objective of the DRAIN project,[1] which was undertaken to provide an estimate of the annual transfer of freshwater, heavy metals and nutrients from the entire drainage basin, as well as a preliminary assessment of the load of organic micropollutants.

The structure of the DRAIN project has been fundamental in dealing successfully with the complexity of the various aspects to be investigated, and to satisfy some important requirements (Zonta *et al.*, in press a). The expected accuracy and precision in the

---

[1] The DRAIN project (DeteRmination of pollutAnt INputs from the drainage basin into the Venice Lagoon) was carried out on behalf of Consorzio Venezia Nuova, concessionaire of the Italian Ministry of Public Works – Venice Water Authority, as part of the operations designed to safeguard the Venice Lagoon (Law 798/84).

evaluation of freshwater discharge and pollutant loads strictly depend on the percentage of the drainage basin surface intercepted by the monitoring. About 30 freshwater inputs are distributed on the lagoon perimeter. Among these, the DRAIN project considered the twelve main tributaries, accounting for about 90% of the whole drainage basin surface. The omitted surfaces (180 km^2) are mainly located along the lagoon border and mainly correspond to reclaimed lands, with a low specific discharge. As a result, the DRAIN monitoring accounted for about 97% of the total freshwater discharge from the drainage basin. To intercept the load from the territories close to the lagoon, the monitoring sections were located in the lower reaches of the tributaries. Most of them were therefore affected by the tide which complicated the evaluation of the freshwater outflow.

A wide set of chemicals were considered: total and dissolved metals (Fe, Mn, As, Cd, Cr, Cu, Hg, Ni, Pb, Zn) and nutrients (nitrogen and phosphorus species), organic carbon (POC and DOC) and organic micropollutants, namely polycyclic aromatic hydrocarbons (PAHs), dioxin-like polychlorinated biphenyls (PCBs), hexachlorobenzene (HCB), organochlorine pesticides, triazine herbicides, polychlorinated dibenzo-p-dioxins and dibenzofurans (PCDD/Fs). Furthermore, to investigate the flux of sediments in different flow conditions and to obtain a reference for the assessment of pollutant loads, an accurate investigation of suspended particle matter (SPM) was performed.

The project was characterized by intense field activity from May 1998 to August 2000. Discharge was continuously measured by self-recording current metres across known sections. Time series of the current speed recorded by the instruments were related to the velocity distribution in the entire section by means of a number of calibration measurements, under different flow rates and tidal conditions (Zuliani et al., in press). This allowed the hydrology of the sub-basins to be investigated and for an account of the effects of the different management practices on the stream runoff to be evaluated. To monitor pollutant concentrations, samples were collected on a biweekly basis. Numerous samplings were also performed in order to evaluate the short-term variability of concentrations and their distribution over the vertical profile of the water column. Furthermore, as floods greatly affect the overall budget of SPM and pollutants (Littlewood, 1992; Webb et al., 1997), every single event characterized by a noticeable increase in the flow rate was studied, with additional measurements and sampling. The hydrological response to flood events at the scale of the single tributary and the entire drainage basin was investigated, by comparing the average inflow of each morphological unit with the trend of runoff in the monitored sections. The pollutant loads were estimated for the year 1999 (Collavini et al., in press). In fact, this period experienced average characteristics with respect to both the total rainfall on the drainage basin and the temporal distribution of events.

RESULTS

Freshwater discharge

The mean annual discharge of the investigated tributaries referred to 1999 is reported in Table 62.1. Also shown in the table are the characteristics of the flow and the extension of the relative sub-basin surfaces, under both normal and flood conditions. The sum of individual discharges amounts to 34.5 m^3 s^{-1}. This value must be increased by 1.0 m^3 s^{-1} to account for the freshwater yield of the 180 km^2 unmonitored surfaces (Zuliani et al., in press). As a consequence, the value of 35.5 m^3 s^{-1} is assumed as the actual mean discharge from the entire drainage basin to the lagoon. The northern tributaries have a relatively high discharge, due to both the input from the spring sources and the external freshwater contribution from the Sile River. A relevant contribution (7.5 m^3 s^{-1}) comes from the Dese River, which may be considered the more 'natural' tributary of the drainage basin. It is in fact subjected to only limited flow regulation. The Vela and Silone, the latter completely deriving from the Sile, join upstream of the lagoon mouth, forming the largest freshwater input of the lagoon (8.2 m^3 s^{-1}). In the central part of the drainage basin, the permanently regulated flow of the system Naviglio Brenta – Taglio Nuovissimo accounts for a discharge of 9.8 m^3 s^{-1}. Naviglio Brenta originates from a lateral derivation of the Brenta River. It in turn originates the Taglio

Table 62.1   *Mean annual discharge (Q) and main characteristics of the DRAIN tributaries and sub-basins (from north to south), under both normal and flood conditions. The contributions from Sile (a) and Brenta (b) rivers do not correspond to specific sub-basin surfaces; (c): input from the 100 km² surface of the Asolo Hills territory; (d): diverted to the Brenta River; (e): diversion of the flow from the Euganean Hills to the Bacchiglione River.*

Tributary	Q (m³/s)	Flow type	Tide influence	Sub-basin surface (km²)	
				Normal flow	Flood
Silone Canal	4.7	intake	yes	(a)	
Vela Canal	3.5	natural	yes	106	106
Dese River	7.5	natural	yes	240	290[c]
Scolmatore Canal	0.7	regulated	prevalent	71	71
Osellino Canal	1.5	natural	yes	50	100[c]
Lusore Canal	2.4	natural	yes	155	155
Naviglio Brenta	5.1	regulated	no	251[b]	251
Taglio Nuovissimo	4.7	regulated	no		
Lova Canal	1.2	natural	yes	147	128[d]
Montalbano Canal	0.7	natural	yes	85	61[d]
Morto Canal	1.2	regulated	occasional	219	160[e]
Cuori Canal	1.3	mechanical	no	251	251

Nuovissimo by a diversion. Downstream of the diversion, the Naviglio Brenta also receives the discharge from two canals (Serraglio and Pionca), that drain an agricultural territory. It is important to note that the Silone, Dese and the system Naviglio – Taglio account for more than 70% of the whole drainage basin discharge (25.5 m³ s⁻¹). Due to the absence of spring inputs and the effective storage of water for irrigation purposes, the tributaries of the southern drainage basin are characterized by a comparatively low discharge.

Load relating to the year 1999

The estimated annual loads of the different contaminant species are reported in Table 62.2. The overall load, identified as 'Drain', is the sum of contributions from the individual tributaries. By referring to the value of 35.5 m³ s⁻¹, the loads from the entire drainage basin ('BASIN') are calculated from the DRAIN estimates by simply multiplying by the discharge ratio ('QR').

The estimates for organic micropollutants are less accurate, mainly because the role of floods in their transport was not studied with the same detail of the other contaminants. PCDD/Fs were monitored in only four representative tributaries. The load of the latter compounds is also expressed in terms of the Toxicity Equivalent (TEQ). It must be noted that the load of terbutrin, propazine, PAHs and organochlorine pesticides was not calculated

Table 62.2 (opposite)   *Annual loads (year 1999) of SPM, total and dissolved metals, nitrogen and phosphorus species, and organic micropollutants from the drainage basin. Mean annual discharges are also shown. The Scolmatore Canal is not included because data were insufficient for the load estimate. (a) Silone values also include the contribution of the Vela Canal; (b) ratio between the estimated discharge for the entire drainage basin (35.5 m³ s⁻¹) and the monitored discharge; (c) data were insufficient for the load estimate; (d) estimate based on data from four representative tributaries, accounting for 15 m³ s⁻¹; (e) data calculated on the basis of the ratio QR.*

Table 62.2

		Silone[a]	Dese	Osellino	Lusore	Naviglio	Taglio	Lova	Monta-lbano	Morto	Cuori	Drain	QR[b]	Basin
Discharge	m^3/s	8.2	7.5	1.5	2.4	5.1	4.7	1.2	0.7	1.2	1.3	33.8	1.05	35.5
SPM	t	4020	9428	1965	3229	6124	4461	1193	564	1409	1010	33403		35083
Fe-tot	t	120	256	58	109	175	119	45	17	34	26	958		1006
As-tot	kg	318	1032	246	819	868	556	232	129	175	205	4580		4811
Cd-tot	kg	11.6	16.1	4.2	4.9	11.8	6.8	2.4	1.0	1.5	2.9	63		66
Cr-tot	kg	582	615	134	250	917	600	104	60	95	130	3487		3662
Mn-tot	kg	5372	14017	2590	7945	7955	4997	3445	2393	4391	8328	61433		64523
Hg-tot	kg	6.9	6.6	1.0	2.2	3.3	1.6	0.8	0.7	0.6	0.7	24		26
Ni-tot	kg	565	711	140	332	621	542	109	66	175	590	3851		4045
Pb-tot	kg	331	720	175	358	564	505	106	39	82	72	2952		3100
Cu-tot	kg	1241	1591	363	694	1059	772	288	101	247	195	6550		6880
Zn-tot	kg	3850	4193	1067	1721	3195	2522	532	262	342	725	18409		19334
Fe-dis	kg	16081	15464	3104	6661	10510	8033	2471	1586	2038	3000	68948		72416
As-dis	kg	308	923	210	686	741	508	184	100	155	185	3999		4201
Cd-dis	kg	7.3	5.4	1.5	2.3	5.8	3.8	1.1	0.7	0.6	1.2	29.67		31
Cr-dis	kg	298	214	39	64	114	100	46	31	31	68	1006		1056
Mn-dis	kg	3388	7773	1312	5207	2968	1577	2305	1552	2210	5969	34261		35985
Hg-dis	kg	3.3	2.9	0.5	0.9	1.7	1.5	0.5	0.4	0.2	0.3	12.03		13
Ni-dis	kg	381	374	67	209	329	295	60	40	111	537	2403		2524
Pb-dis	kg	48	63	13	27	47	62	9	5	12	9	295.0		310
Cu-dis	kg	722	516	94	231	370	265	104	63	69	95	2529		2656
Zn-dis	kg	1971	1141	277	586	866	653	183	117	129	243	6166		6476
N-NH$_3$	t	55	95	23	143	63	28	25	10	12	25	480		504
N-NO$_3^-$	t	670	523	97	210	467	343	67	25	77	155	2634		2767
N-NO$_2^-$	t	17.8	18.1	3.8	12.0	12.9	7.0	2.9	1.1	3.4	5.0	83.9		88
N-tot	t	887	815	160	421	664	485	121	57	128	266	4005		4207
DIN	t	743	637	124	365	542	378	95	36	93	185	3198		3358
P-tot	t	52.8	63.8	12.2	23.2	25.5	27.1	8.3	3.7	5.3	6.9	228.8		240
P-PO$_4^{3-}$	t	19.2	17.4	3.8	7.3	8.1	4.5	1.6	0.9	0.8	1.1	64.7		68
Atrazine	kg	3.9	2.1	0.7	1.5	3.3	1.9	0.6	0.6	[c]	0.7	15.3	1.09	16.7
Desethyla-trazine	kg	7.3	2.2	0.6	0.9	1.9	1.3	0.3	0.2		0.4	15.0		16.3
Simazine	kg	2.2	3.2	0.5	1.1	3.5	1.1	1.3	0.5		1.6	15.0		16.3
Terbutyl-azine	kg	26	67	12	46	78	35	9.5	8.0		18	300		327
SPCBs	g	10	109	43	43	58	121	5.3	1.6	[c]	16	407		443
HCB	g	70	141	38	327	89	198	14	4.0		43	924		1006
Aroclor 1254+1261	g	133	848	303	386	475	995	56	16		147	3359		3658
ΣPCDD/PCDFs	g		1.70	1.34			2.20			[c]	0.31	5.54[d]	2.38	13.17[e]
TEQ	g		0.018	0.016			0.020				0.0029	0.057[d]		0.135[e]

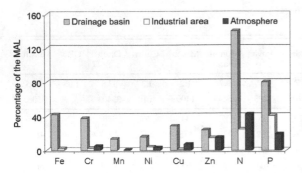

Fig. 62.2   Percentage ratio of the estimated loads for 1999 from the drainage basin, industrial area and atmosphere, with respect to the MAL limits. Data refer to total concentrations. Values for Fe and Mn are not available for the atmosphere and industrial area, respectively.

because the concentration of these compounds was generally below the analytical limit of detection (LOD). The estimates of annual loads are the main outcome of the DRAIN project, and represent a set of reference data for the framework of actions addressed to the management and safeguarding of the Venice Lagoon ecosystem.

To show the contribution of the drainage basin to the pollution of the lagoon, for the species considered by the specific Italian legislation (Ministero dell' Ambiente, 1999), Figure 62.2 compares the DRAIN project estimates with the Maximum Allowable Load (MAL). The latter comprises all the possible pollutants inputs to the lagoon. The loads from both the

industrial area of Porto Marghera (Magistrato alle Acque, 2000b) and the atmospheric inputs estimated for the same year 1999 (Magistrato alle Acque, 2000a) are also compared. The drainage basin appears as the major source for all the considered contaminants, particularly for heavy metals that, however, rank well below the MAL. The amounts of nitrogen and phosphorus are, however, higher than the MAL. In particular, the nitrogen delivered by the drainage basin alone exceeds the MAL (3000 t yr^{-1}).

By considering even the nutrient load estimates for the other main sources (ASPIV, 2000a; ASPIV, 2000b, Regione del Veneto, 2000), it can be recognized that the conditions needed to fulfil the MAL limits for nitrogen and phosphorus are far from being achieved (Table 62.3).

Apart from a discussion – which seems to be necessary – on the usefulness of the imposed limits and on their reliability, these results identify the need for a reduction of the nutrient delivery as the main target of actions for the improvement of the environmental quality of the lagoon. Nevertheless, an effective control of pollution in the lagoon ecosystem deserves a more detailed knowledge of the characteristics of pollutant sources as well as of the spatial and temporal variability of the delivered load.

Concerning the drainage basin, the mechanisms of formation and transfer of the load are intrinsically different among the tributaries, due to their varying hydraulic characteristics, as well as the type and importance of the contamination sources within the drained surfaces. Moreover, the load varies with time,

Table 62.3   *Estimated load of total nitrogen and phosphorus to the lagoon from the main sources (year 1999).*

	N-tot (t/y)	P-tot (t/y)
Drainage basin	4207	240
Industrial area	740	122
Atmosphere	1277	57
Treatment plants	435	57
Venice and islands	620	83
Total estimate	7279	559
Maximum Allowable Load (MAL)	3000	300

in response to the tributary hydrology, land use, and production cycles in the catchment. Due to the relative importance of the northern tributaries in terms of discharge, it is expected that the corresponding sector of the lagoon receives a higher pollutant load. This is particularly evident during floods: these streams are in fact characterized by more frequent and intense events, which markedly increase the amount of sediment and contaminants transferred to the lagoon. This difference is better emphasized by the dispersion diagram of Fig. 62.3, which shows the clear correlation (r = 0.94) between the monthly load of suspended particle and the corresponding rainfall, at the basin scale, for the year 1999. As a reference, the discharge hydrograph for the entire drainage basin is also shown in the figure. In April, July and November, the higher inflow causes an increase in runoff and, in turn, of suspended sediment transport. The storms of April and November, in particular, caused relevant floods in all the tributaries, which raised significantly the monthly values of the sediment load. On the contrary, in the dry season (from December to March) the lower surface runoff and the limited energy that characterises the base flow prevented the mobilisation of materials. Regardless of the differences related to the hydrological factors and the characteristics of drainage, the results obtained for the entire drainage basin can be assumed to be representative of the behaviour of the single tributaries.

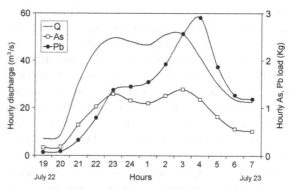

Fig. 62.4   Dese River: discharge hydrograph and hourly loads of Pb and As during the flood event of 22–23 July 1999.

The dramatic increase in streamflow energy that takes place during a flood considerably enhances the sediment transport capacity in the hydrographic network. This implies a specific increase in the concentration of the species that are predominantly associated with the solid phase, including the majority of heavy metals and micro-organic pollutants. The contemporary peak of both discharge and concentration determines a strong increase in the load of these species. One single intense flood can, therefore, greatly affect the monthly load transferred from a tributary and, by extension, from the whole drainage basin. Among the monitored tributaries, the Osellino Canal is the more impressive. Due to its flash flood regime, which is related to the sub-basin morphology and the immediate input of the runoff of Mestre urban district, this tributary delivered more than the 40 per cent of the SPM annual load in only 10 days of flood flows.

By comparison, for dissolved nutrients and the heavy metals which are mainly transported in the dissolved form (such as As), the increase of the load is substantially proportional to increases in discharge alone. Figure 62.4 shows the example of the variation of Pb and As loads during the flood event of July 1999 in the Dese River (Zonta *et al.*, in press b).

This flood was generated by a typical summer storm, which had a return period of two years. The base discharge of about 7 m³ s⁻¹ increased to a maximum of 51 m³ s⁻¹, i.e. by a factor of 7.3. The load of As showed a similar increase, up to a factor of 8.5, in the

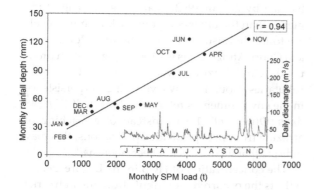

Fig. 62.3   Dispersion diagram between SPM monthly load and monthly rainfall depth, for the year 1999.
The discharge hydrograph of the entire drainage basin is also shown as a reference.

correspondence of the flood crest. In comparison, the load of Pb, which is mainly associated with the particulate fraction, increased up to 45-fold.

Besides the fundamental role of floods in determining the load, the second important aspect to be considered is the characterisation of the tributaries on the basis of their degree of contamination by the different determinants. For each tributary, the annual loads were therefore divided by the annual freshwater volume, obtaining a normalised concentration. The latter reflects the average water quality of the tributary on an annual basis, independently from both discharge and sub-basin surface. This procedure allows both the determination of possible harmful conditions permanently affecting a stream and the comparison of pollution levels between the tributaries. The distribution of the normalised concentrations is shown in Fig. 62.5 (a–i) for the principal determinants.

The higher suspended sediment transport occurring in the tributaries of the northern basin is shown by the SPM values in Fig. 62.5a. Unlike the other northern tributaries, the Silone Canal shows a very low SPM concentration. A predominant fraction of its discharge is in fact derived from the spring system of the Sile River, which carries relatively clean waters. Since it is intimately associated with the particulate fraction, the average concentration of Pb (Fig. 62.5a), as well as Fe (not shown), follows the SPM distribution. The values obtained for the other metals clearly reflect the presence of specific sources in the related sub-basins. For instance, Osellino and Lusore show the highest average concentration for most of the considered metals (Pb, Cu, As, Cd and Zn – Fig. 62.5a–c), due to the contamination by urban and industrial sources. In fact, the first receives the runoff from the urban district of Mestre, whereas the lower reach of the second crosses the industrial area of Porto Marghera. Particularly significant are the high As contamination in the Lusore, which is at least twice that of the other tributaries, and the high level of Cd in the Osellino. The concentration of both Mn and Ni of the Cuori Canal are by far the highest found (Fig. 62.5d), and this is largely due to a high amount of their dissolved form. This feature may be related to the abundance of these two metals in the peat characterising the lower part of the sub-basin area. Release processes from the soil to the water

course may therefore be expected as a consequence of the changes in physico-chemical conditions (Kharkhordin and Atroshchenko, 1998; Tipping *et al.*, 2003), thereby enhancing the dissolved concentration in the water column. Naviglio Brenta and secondly Taglio Nuovissimo show, instead, the highest value of Cr (Fig. 62.5e), mainly as a consequence of the geochemical fingerprint of the sediments from the Brenta River which are characterized by higher Cr background levels. Furthermore, unlike the other tributaries, the metal in these two streams is predominantly associated with the solid phase.

The concentration of total nitrogen is rather similar in all the tributaries, with the exception of Lusore and Cuori. The latter, in particular, has nearly twice the value found in the other streams, due to the high level of both nitrate and organic nitrogen (Fig. 62.5f). The abundant agricultural activities in the sub-basin of the Cuori Canal is therefore highlighted. Despite its relatively high nitrate yield, the total nitrogen in the Lusore is instead determined by the ammonia, that reaches the greatest level in the drainage basin (Fig. 62.5g). The highest values of total phosphorus are found in Dese, Osellino and Lusore: for these tributaries, as well as Silone, orthophosphate ranges from two-to fivefold the values of the other streams. The contamination from herbicides is consistent with the data about their use in the Veneto Region (Masone, 1998). The concentration of terbutylazine (TBZ, Fig. 62.5h), which is presently used as a generic herbicide in replacement of atrazine after this was banned by law in 1992, is an order of magnitude higher than the other compounds. Maximum levels are found in Lusore, Naviglio Brenta, Montalbano and Cuori. Atrazine – as well as its degradation product desethylatrazine – was still detected in all the tributaries, though at low levels. Since its persistence in the environment is relatively short (half-life = 48 days, Otto *et al.*, 1996), instead of a leaching of atrazine from past pesticide applications, this may indicate that its use has not been abolished.

The concentration of HCB in the Lusore clearly reflects the occurrence of inputs from the industrial area, being about an order of magnitude higher than the other tributaries. Concerning PCBs (Fig. 62.5i), concentrations are markedly different among the tributaries. The highest values are found in the

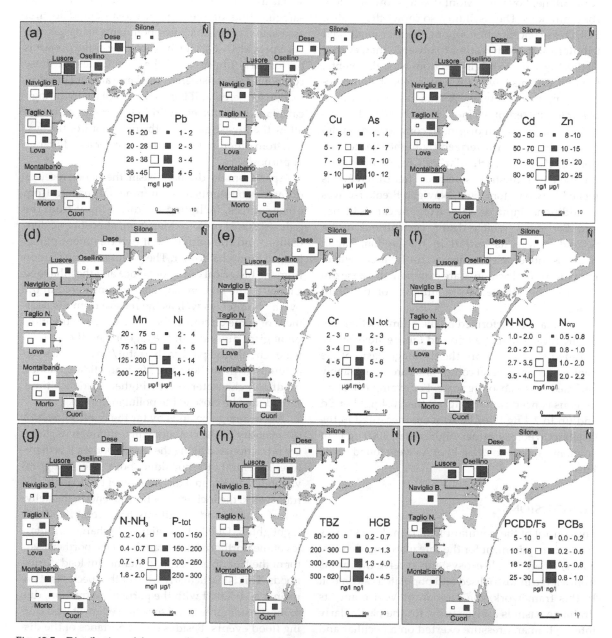

Fig. 62.5   Distribution of the normalised concentrations among the tributaries for the principal determinants. The size of the squares indicate the relative magnitude of contaminant concentrations measured. Heavy metal values given refer to total concentrations.

Osellino and Taglio Nuovissimo, while the more rural Silone, Lova and Montalbano show very low concentrations. The Osellino also shows the greatest PCDD/Fs concentration among the four monitored tributaries. It is interesting to observe that, even if Taglio Nuovissimo originates from Naviglio Brenta, the two streams have different levels of some specific compounds. Naviglio Brenta has greater values of nitrate and terbutylazine, which are related to pollution of agricultural origin. In comparison, Taglio Nuovissimo is characterized by higher concentrations of HCB and particularly PCBs, which can be ascribed to industrial sources. This difference is related to two factors. First, Naviglio Brenta receives an additional input of water from the canals Pionca and Serraglio, which drain agricultural areas. Second, before its outfall in the lagoon, Taglio Nuovissimo crosses a heavy traffic-area, where minor industrial activities are common: this zone can therefore be affected by a number of pollutant sources.

The analysis performed by the normalisation procedure reveals that Osellino and Lusore are the most polluted streams within the drainage basin of the Venice Lagoon. Several compounds are in fact highly enriched on an absolute scale. For example, instantaneous concentrations higher than 0.4 $\mu g l^{-1}$ of Cd and up to 2.4 ng l^{-1} of PCBs were measured in the Osellino, whereas values higher than 3000 $\mu g\ l^{-1}$ of ammonia and up to 20 $\mu g\ l^{-1}$ of As were found in the Lusore.

CONCLUSIONS

The control of the pollutant load from land sources is a basic requirement for the safeguarding of coastal zones, where fragile ecosystems receive the inputs from a large industrialised and populated catchment. In this framework, the Lagoon of Venice and its drainage basin is a case study, for the particularly intense human pressure exerted on a peculiar and complex coastal environment.

The DRAIN project represents a considerable improvement in the knowledge of the actual contribution of the drainage basin to the comprehensive pollutants budget in the Venice Lagoon. The estimates of the annual load of heavy metals and nutri-

ents, which are the main outcome of the project, constitute a reference data set for the management and safeguarding of the lagoon ecosystem. The first assessment of organic micropollutants loads has also been provided. This further outcome of the project is particularly important if the harmfulness of these compounds is considered. The large amount of information gathered and the complementary hydrological analysis performed within the project activities allow the investigation of the behaviour of the system from different perspectives and for a wide range of purposes.

Apart from the drainage basin, the lagoon receives different pollutants of environmental concern from other sources, including the direct input from the industrial area of Porto Marghera, atmospheric deposition, and the outfall from two treatment plants located along its border. The pollutant contribution from the historical centre of Venice and the inhabited islands of the lagoon may also be considered in the overall budget, as well as the release from the sediment of some circumscribed areas. Due to the morphological diversity of the lagoon and the general water circulation pattern, harmful/toxic species can accumulate in shallow areas (salt and tidal marshes, mudflats) characterized by higher residence time, should they be close to the pollutant inputs.

The drainage basin is by far the main source for the majority of the contaminants, although its delivery is fractionated along the lagoon border. It is therefore essential to consider both the temporal variability and spatial distribution of this load. In this framework, the incidence of flood events in the delivery of matter and pollutants has been effectively highlighted. Due to their higher discharge, tributaries of the northern basin transfer to the northern sector of the lagoon a relevant contaminant load during floods. This is particularly true for the species that are mainly associated with the particulate fraction, due to the considerable increase of the sediment load during flood events. Besides the importance of the discharge in the overall freshwater delivery and the incidence of floods in each tributary, the level of contamination for the different species must be ascertained, in order to recognise the occurrence of critical quality conditions. By considering the annual average concentration, a relevant variability of several species among the monitored tributaries is demonstrated. In particular, the analysis performed pinpointed the Osellino and Lusore canals as crucial examples, as a consequence of the effects of urban and industrial sources.

Thanks to the DRAIN project outcomes, the contribution of the drainage basin in the pollution of the Venice Lagoon has been defined. The acquired knowledge is fundamental to the framework of the integrated management of the lagoon ecosystem. Two related issues should be urgently considered with the same detail, in the direction of an actual improvement of the quality of the lagoon water. The first is the assessment of the fate of pollutants deriving from the different sources in the lagoon environment. The second concerns an accurate census of pollutant sources within the drainage basin, with particular attention being paid to the priority hazardous compounds.

REFERENCES

ASPIV. 2000a. *Impianto di depurazione di Fusina.* Relazione annuale 1999, 62pp.

ASPIV. 2000b. *Impianto di depurazione di Campalto.* Relazione annuale 1999, 50pp.

Avanzi, C., Fossato, V., Gatto, P., Rabagliati, R., Rosa Salva, P., and A. Zitelli. 1984. *Ripristino, conservazione ed uso dell'ecosistema lagunare veneziano.* Venice: Comune di Venezia Ed., 199pp.

Collavini, F., Bettiol, C., Zaggia, L., and R. Zonta. In press. Pollutant loads from the drainage basin to the Venice Lagoon (Italy). *Environment International.*

Cucchini, E. 1928. *Le acque dolci che si versano nella Laguna di Venezia.* In *Scritti sugli apporti di acqua dolce nella Laguna di Venezia.* Venice: Servizio Idrographico e Mareografico Nazionale (1992), 208pp.

Cossu, R., and E. De Fraja Frangipane. 1985. *Stato delle Conoscenze sull'Inquinamento della Laguna di Venezia.* Venice: Consorzio Venezia Nuova, Servizio Informativo, Vols. I–IV.

Francia, C., and F. Juhasz. 1993. The Lagoon of Venice, Italy. In *Coastal Zone Management, Selected Case Studies.* Paris: OECD, Chapter 5, pp. 111–28.

Kharkhordin, I. L., and F. G. Atroshchenko. 1998. Simulation of heavy metals migration in peat deposits. *Ecological Chemistry* 8, 1, 37–43.

Littlewood, I. J. 1992. Estimating contaminant loads in rivers: a review. *Institute of Hydrology Report* 117. Wallingford (UK): Institute of Hydrology.

Magistrato alle Acque. 2000a. *Programma generale delle attività di approfondimento del quadro conoscitivo di riferimento per gli interventi ambientali: Progetto 'Orizzonte 2023'.* Valutazione preliminare dello stato attuale dell'ecosistema lagunare veneziano. Consorzio Venezia Nuova – Thetis S.p.A., Rapporto Tecnico, 722pp.

Magistrato alle Acque. 2000b. *Relazione sulle caratteristiche degli scarichi idrici dell'area di Porto Marghera.* Dati relativi al 1999, 112pp.

Masone, M. 1998. *Fitofarmaci e fattori di pressione ambientale.* ANPA – Agenzia Nazionale per la protezione dell'Ambiente. Rapporto tecnico interno. RTI-INT 13.

Ministero dell'Ambiente. 1999. *Decreto Ministeriale del 09/02/99.* Carichi massimi ammissibili complessivi di inquinanti nella laguna di Venezia. Gazzetta Ufficiale Italiana n.35 del 12/02/99.

Otto, S., Vicari, A., Zanin, G., and P. Catione. 1996. Aspetti ecotossicologici e stima del rischio ambientale. In *Atti del Convegno 'Il diserbo delle aree extra agricole', Padova, 12 Dicembre,* pp. 97–133.

Regione Del Veneto. 2000. *Piano per la prevenzione dell'inquinamento e il risanamento delle acque del bacino idrografico immediatamente sversante nella laguna di Venezia.* Piano Direttore 2000.

Tipping, E., Smith, E. J., Lawlor, A. J., Hughes, S., and P. A. Stevens. 2003. Predicting the release of metals from ombrotrophic peat due to drought-induced acidification. *Environmental Pollution* 123, 239–53.

Webb, B. W., Phillips, J. M., Walling, D. E, Littlewood, I. G., Watts, C. D., and G. J. L. Leeks. 1997. Load estimation methodologies for British rivers and their relevance to the LOIS RACS(R) programme. *Sci. Tot. Environ.* 194/195, 379–89.

Zonta, R., Bettiol, C., Collavini, F., Costa, F., Fagarazzi, O. E., Zaggia, L., and A. Zuliani. 2001. *Protocolli metodologici.* DRAIN project Report No.11B, 62pp.

Zonta, R., Costa, F., Collavini, F., and L. Zaggia. In press a. Objectives and structure of the DRAIN project: an extensive study of the delivery from the drainage basin of the Venice Lagoon (Italy). *Environment International.*

Zonta, R., Collavini, F., Zaggia, L., and A. Zuliani. In press b. The effect of the transport of suspended sediments and contaminants: a case study from the estuary of the Dese River (Venice Lagoon Italy). *Environment International.*

Zuliani, A., Zaggia, L., Collavini, F., and R. Zonta. In press. Freshwater discharge from the drainage basin to the Venice Lagoon (Italy). *Environment International.*

Scientific paper

63 • Trace metal fluxes in the Venice Lagoon

G. CAPODAGLIO, W.R.L. CAIRNS, A. GAMBARO, G. TOSCANO AND C. TURETTA

SUMMARY

Experiments were carried out to assess the contribution of aerosol and remobilization from sediment on the content of trace metals in the Lagoon of Venice. The sampling of aerosol was made at four sites by selecting the wind direction to identify the hypothetical sources of metals. The samples were fractionated to six aerodynamic dimensions and metal concentrations determined in each fraction. The results emphasize that crustal elements (Al, Fe, Mn) present a bimodal distribution whilst anthropogenic elements (Pb, V, Cd) are associated preferentially with fine particles. The highest mean total concentration for a larger part of the trace elements was at Moranzani (downwind of the industrial area) but all the sites presented comparable concentrations of anthropogenic metals, showing a diffuse distribution of these elements. There is evidence that a significant flux of metals can reach the lagoon by a long range mechanism. Benthic fluxes were determined at two sites known for their high sediment metal concentrations (Canale delle Tresse and Campalto) by benthic chamber experiments. Results show a release of metals from sediment for some elements (Mn, Cd and Zn) while the flux was nil for some others (Fe and Cu), emphasizing the necessity for more studies to better understand processes controlling remobilization. Comparing results, it can be concluded that the atmosphere represents an important source of Cu, Fe, Pb and V, while the remobilization of sediments contributes significantly more to the concentration of Zn, Mn and Cd.

INTRODUCTION

Many sources contribute to the input of pollutants into the Lagoon of Venice. In particular, the contribution of industrial activities has been intensively studied in the past (Donazzolo *et al.*, 1984; Martin *et al.*, 1994; Frignani *et al.*, 1997; Scarponi *et al.*, 1998; Moret *et al.*, 2001). Recently, the contribution of other sources, such as the catchment basin have been investigated by a specific programme (DRAIN) and lagoon inlet exchanges were examined in a 2000-3 CORILA research programme project. However, little information is available about (1) the contribution of the atmosphere to the transport of pollutants; and (2) the mobility of contaminants in sediments. Changes in the chemical characteristics of water or physical processes can re-mobilize contaminants present in sediments, producing secondary pollution (Zago *et al.*, 2000; Simpson *et al.*, 1998). The atmosphere is a major pathway where anthropogenic organic compounds and trace elements are cycled in the environment and air masses can be an important source of contamination (Gambaro *et al.*, 2003; Guieu *et al.*, 1997). Trace elements chemically toxic to humans, such as Hg, Pb and Cd, are emitted into the atmosphere from anthropogenic sources by combustion and industrial processes. They contribute to the worsening of the air quality and to increasing problems related to human health (Guieu *et al.*, 1997; Chow *et al.*, 2001). The aim of this study was to investigate previously insufficiently assessed processes contributing to the input of pollutants into the waters of the Lagoon of Venice. Attention was particularly focussed on (1) the role of aerosols; and (2) the remobilization of pollutants contained in sediments as sources of contaminants for the lagoon environment.

EXPERIMENTAL DESIGN AND METHODS

Aerosol fluxes

Aerosol sampling was carried out at four sites in the Venice area based on the preferential wind directions

Flooding and Environmental Challenges for Venice and its Lagoon: State of Knowledge, ed. C. A. Fletcher and T. Spencer.
Published by Cambridge University Press, © Cambridge University Press 2005.

for the transport of aerosols to the Venice Lagoon. By selecting the wind direction and sampling site position, it is possible to make hypotheses about sources affecting aerosol composition. At sampling sites 1, 3 and 4 the selected wind direction was from the NE quadrant, while for site 2 the direction was from the SE (Fig. 63.1). For all the sites the minimal wind speed at time of sampling was 1m s^{-1}. Site 1 was located close to the industrial area of Porto Marghera, collecting aerosols influenced by urban and industrial sources. Site 2 was located at the Lido Inlet, to collect marine aerosol and aerosol transported by long range mechanisms. Site 3 was located in the Euganean Hills, *c.* 70 km from Venice, where aerosol composition is less affected by local sources of inorganic pollutants. Site 4 was located close to Venice Airport and downwind from one of the most heavily used motorways in Italy. Five campaigns were performed between July 2001 and October 2002, each

one lasting about 10-15 days. Samples were collected on cellulose membranes cleaned with acidified water before use. Sampling was performed by a six-stage cascade impactor mounted on a high volume pump equipped with a PM10 size-selective inlet (Chan *et al.*, 2000). Cellulose filters used for sampling of aerosol were weighed before and after sampling, to assess the amount of particulate matter collected; they were mineralized in an acid mixture using a microwave digestion system. Elemental analysis was carried out by ICP-SFMS (Element 2, Finnigan Mat). The accuracy and repeatability of the analytical method were controlled using a certified reference material (urban particulate matter SRM 1648, NIST). Blanks deriving from filters and handling were estimated for each campaign by exposing cleaned membranes for a few minutes without activating the pump. Quantification was carried out by the multiple standard addition method.

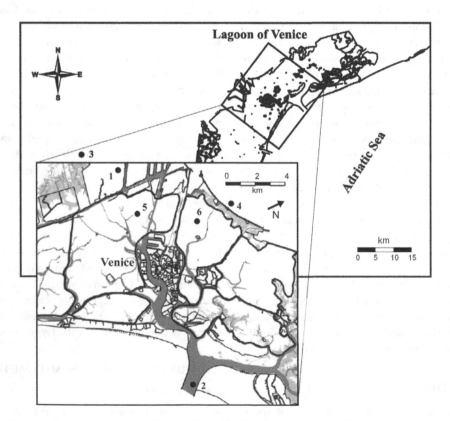

Fig. 63.1   Site locations for aerosol sampling (1–4) and benthic chamber experiments (5,6).

Benthic fluxes

Experiments were carried out using benthic chambers in two contaminated sites (Fig. 63.1), Canale delle Tresse (site 5) near the industrial area of Porto Maghera, and in the Campalto area (site 6) near the causeway, in October 2002 and May 2003 respectively. Water samples were collected over approximately 50-60 hours every 3-4 hours from 90L chambers. The samples were taken by means of a peristaltic pump and filtered by a cartridge filter (0.20 μm). Two sediment cores were collected by piston corer in June 2003 (one at each site) to extract pore water and to determine the benthic fluxes by an independent procedure. Cores were sealed and placed into a glove box conditioned with nitrogen to eliminate oxygen. They were then divided into slices of 1 cm length. The sediment was centrifuged to extract the pore water which was then quickly filtered in an inert atmosphere and frozen until analysis. Total dissolved metal concentrations were determined by Inductively Coupled Plasma-Sector Field Mass Spectrometry applying the procedure previously reported (Turetta et al., 2003). Cd was determined in low-resolution mode ($m/\Delta m$=300) using a micro-concentric nebulizer and a desolvation system (Aridus, Cetac) to reduce oxide interferences. Cu, Fe, Mn and Zn were determined in medium resolution mode ($m/\Delta m$=3000) by direct introduction using a microconcentric nebulizer and a Teflon spray chamber. Quantification was carried out by a calibration curve obtained by multiple-standard additions to one aliquot of coastal certified reference material (CASS-4).

Calculation of fluxes

The fluxes of metals in the lagoon through the aerosol phase were calculated using the following equation:

$$Fa = \frac{3600 * C * \overline{V} * \overline{t}}{24}$$ Eq. 63.1

where Fa is the metal flux in $\mu g/m^2/day$, C is the concentration in $\mu g/m^3$, $\overline{V}$ is the mean wind speed in m s^{-1}, $\overline{t}$ is the daily mean sampling time in hours. Mean wind speed and mean sampling time were esti-

mated by data measured during the sampling experiments. The benthic fluxes were estimated by detecting the changes of concentration inside the chambers during the experiments. However, the calculation of fluxes on the basis of point-to-point variations contributes to high variability in the results. This calculation might work for data with a very low experimental error but is unsuitable for trace components subject to a relatively high uncertainty. Therefore, to reduce the effect of variability, fluxes were calculated by plotting pmol of metals contained inside the chamber during the experiment as a function of time and calculating the linear regression of these experimental data, by the equation:

$$\overline{Fb} = \frac{R}{S}$$ Eq. 63.2

where $\overline{Fb}$ is the mean benthic flux during the experiment, R is the slope of the regression line in pmol/h and S is the surface covered by the chamber (3600 cm^2). Positive fluxes result when concentrations increase with time. In this event, material released from the sediment and/or from the particulate matter to the water phase is inferred. Conversely, negative fluxes result when concentrations in water decrease with time.

Diffusive fluxes were estimated by measurements of gradient concentrations at the water/sediment interface by

$$Fd = D\frac{\Delta C}{\Delta x}$$ Eq. 63.3

where Fd is the diffusive flux, D is the diffusion coefficient that can be estimated to be about 10^{-6}cm^2/s, ΔC is the difference of concentration between the pore water and the overlying bottom water in pmol/cm^3 and Δx is the distance in cm. It represents an over estimation of benthic flux because it does not consider the flux from the water to the sediment.

RESULTS AND DISCUSSION

Trace elements in aerosol

The granulometric distribution of aerosol (Fig. 63.2) shows that the particle size distribution assume a bi-

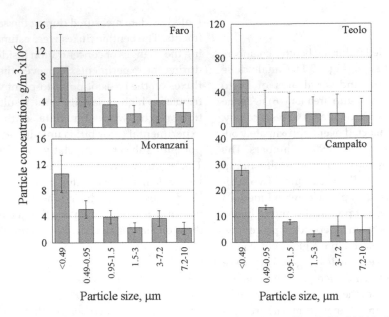

Fig. 63.2   Mean size particle distribution at the four sites. (a) Marghera (site 1 (Fig. 63.1)). (b) Lido (site 2). (c) Teolo (site 3). (d) Tessera (site 4).

modal distribution for the four sites. The finest particles (0–0.49 μm) are always the most abundant, with a second mode in the 3 0–7.2 μm fraction. The atmospheric concentration of crustal trace elements, such as Fe, Al and Mn, shows a bi-modal distribution (Fig. 63.3) reaching the highest concentrations in particles with a diameter greater than 3 μm (34, 40 and 0.75 mg/g for Fe, Al and Mn respectively). This distribution is quite typical for elements whose origin is principally related to natural sources (Chow, 1995). The abnormally high Al concentration at Moranzani (industrial area of Marghera) can be explained by the presence of industrial plants to treat bauxite. Elements related to anthropic sources (for example V, Pb, Cd) are associated preferentially with fine particles with an exponential distribution (figure 3), the highest concentration is in particles smaller than 1.5 μm (65, 680 and 450 μg/g for Cd, Pb and V respectively). Therefore, it is concluded that fine particles make the largest contribution to the transport of micropollutants and significantly affect the content of metals in the lagoon environment. Measurements showed that the lead isotopic composition in particles with different dimensions was significantly different,

suggesting that they may have a different origin (Rosman *et al.*, in preparation). The industrial activities present at site 1 (Moranzani) are responsible for the highest mean total concentration determined for this site. However, comparable concentrations of anthropogenic metals were reached at all sites, emphasising the diffuse distribution of these elements.

The mean daily fluxes of metals reaching the Lagoon of Venice by aerosol are reported in Table 63.1. By comparing the values reported here with atmospheric deposition of trace metals estimated by direct measurements and vertical distribution in salt marsh deposits, we assess that 1-5% of the total aerosol reaching the lagoon area is deposited on the lagoon water surface (Cochran *et al.*, 1998; Rossini *et al.*, submitted).

Benthic fluxes

The concentration of dissolved metals (Cd, Cu, Fe, Mn, Zn), O_2 concentration (mg/l) and salinity (psu) are reported in Figs. 63.4–63.6. During the two experiments sub-oxic conditions were reached in the benthic chambers. O_2 saturation at the beginning of the

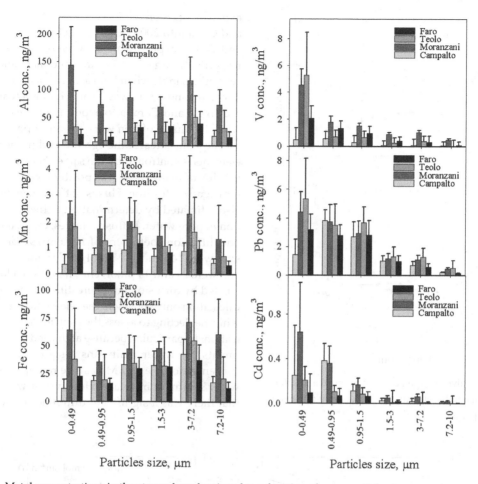

Fig. 63.3   Metal concentrations in the atmosphere fractioned as a function of mean particle size.

Table 63.1   *Mean daily metal fluxes determined at the sampling sites between July 2001 and October 2002.*

Site	Flux, mg/m²/day						
	Al	Fe	Mn	Cd	Cu	Pb	V
Faro	11.7	9.3	2.0	0.055	0.36	0.55	0.31
Teolo	11.1	8.2	2.6	0.020	0.67	0.57	0.21
Moranzani	39.1	26.1	4.8	0.101	0.78	1.07	1.17
Campalto	21.7	1.7	16.6	0.043	0.53	0.68	0.12

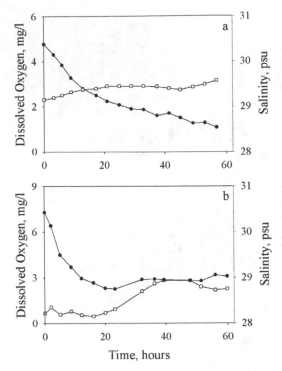

Fig. 63.4 Dissolved oxygen concentration (•) and salinity (□) inside the benthic chamber experiments at (a) Tresse and (b) Campalto.

experiments were 60% and 85% in the Tresse 2002 and Campalto 2003 cases respectively while at the end of the experiments they were 13% and 40% respectively (Fig. 63.4). The iron and manganese concentrations in the chambers are reported in Fig. 63.5. The manganese flux was 19 and 12 pmol/cm^2/h for the Tresse and Campalto experiments respectively, while the fluxes of iron were –3 and –7 pmol/cm^2/h for the same sites. Although the cycling of the two elements is controlled by the same chemical process, the flux of manganese was positive while it was nil (or negative) for iron. Fluxes of these elements were also estimated by determination of their concentration in pore water. Diffusive fluxes, calculated using one diffusion coefficient of 10^{-6} cm^2/s, for the two sites were 140 and 170 pmol/cm^2/h for manganese and were 24 and 14 pmol/cm^2/h for iron. Fluxes calculated by diffusion are overestimated because this calculation only considers the release from sediment, while neglecting to assess the contribution from deposition. In particular, because at the end of the experiments the water maintains a significant oxygen concentration, Fe and Mn can be re-depleted at the water sediment interface. However, it was evident that there was a significant difference between the

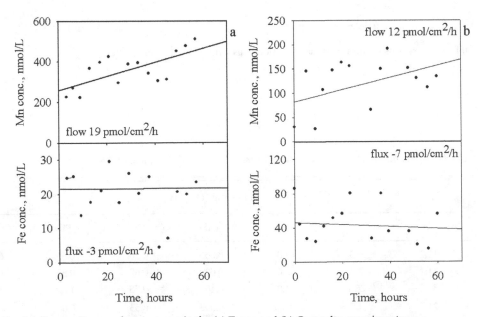

Fig. 63.5 Benthic fluxes of iron and manganese in the (a) Tresse and (b) Campalto experiments.

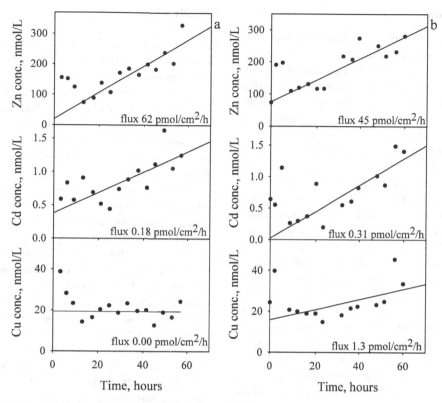

Fig. 63.6   Benthic fluxes of Cu, Cd and Zn in the (a) Tresse and (b) Campalto experiments.

fluxes of manganese and iron as a consequence of their different standard redox potential.

The concentrations of dissolved Cd, Cu and Zn for both the experiments showed an initial increase arising from remobilization during the settling of the chambers and was followed by a negative flux until about 15 hours into the experiment (Fig. 63.6); this was more evident in the Campalto experiment. The zinc and cadmium concentrations were then characterized by mean positive fluxes until the end of the experiments. For the two experiments the fluxes were 62 and 45 pmol/cm^2/h for Zn and 0.18 and 0.31 pmol/cm^2/h for Cd. The flux of Zn was c. 2 orders of magnitude higher than for Cd. However, if we examine it in relation to the differences in concentration in the dissolved phase, the mean zinc concentration was about one hundred times higher than cadmium. Therefore, the effect in terms of the increase in relative concentration is similar.

As for the other elements, the copper concentration showed an initial increase (positive flux), resulting from the sediment resuspension during the chamber settling, followed by a negative flux; after about 10 hours the flux became practically nil until the end of the experiment. The mean concentration was 19 nmol/l and 23 nmol/l for Tresse and Campalto respectively. During the experiment carried out in Campalto, the mean benthic flux was positive (1.3 pmol/cm^2/h) but only weakly significant, considering the repeatability of data (the increase was c. 5% of the mean concentration). Although, the processes controlling cadmium, zinc and copper distribution are always related to biological activity and organic complexation, the mobility of the latter seemed quite different with respect to the first two elements.

CONCLUSIONS

The study has confirmed the importance of local sources of trace elements transported by aerosols but it has also been highlighted that marine aerosol and long range transport (related to fine particles) contributions are not negligible. The results obtained in the two contaminated areas show that trace elements can be remobilized from the sediment to the water by resuspension or from sub-oxygenation of the water. Some preliminary conclusions can be drawn from the data. Despite the differences in concentrations observed in the two sites studied, the benthic flux of total dissolved Cd, Zn, Fe, Mn and Cu does not seem to be particularly affected by seasonal processes. Sediment resuspension affects the mobility of a large part of the metals while the mobility due to chemical and/or diffusive processes seems quite different, even for elements with similar geochemical characteristics. Furthermore, by studying the organic speciation of these trace elements it may be possible to better clarify whether the sediments are a source of metals and the mechanisms, since benthic fluxes may be relevant for the biogeochemical cycles of the trace elements studied in the lagoon.

Comparing the results of metals transported by aerosol and published data on deposition, we can make the approximate estimation that 1-5% of metals transported by aerosols contribute to the lagoon water composition. From the estimated fluxes at the water/sediment interface and at the atmosphere/water interface, it can be concluded that the atmosphere represents a more important source for Cu, Fe, Pb and V, while the remobilization of sediments contributes significantly more to the concentration of Zn, Mn and Cd.

ACKNOWLEDGEMENTS

This work was supported by CORILA under the Project 'Role of Aerosol and Secondary Pollution to the Chemical Contamination of the Lagoon of Venice' and by the National Research Council of Italy (CNR). The authors thank M. Frignani for the helpful discussions and sediment sampling, the staff of ARPAV – Meteorological Centre of Teolo and of EZI – Industrial Zone Agency of Porto Marghera and V. Zampieri for their sampling support.

REFERENCES

Chan, Y. C., Vowles, P. D., McTainsh, C. H., Simpson, R. W., Cohen, D. D., Bailey, G. M., and G. D. McOrist. 2000. Simultaneous collection of airborne particulate matter on several collection substrates with a high-volume cascade impactor. *Atmospheric Environment* **34**, 2645–51.

Chow, J. C. 1995. Measurement methods to determine compliance with ambient air quality standards for suspended particles. *Journal of the Air and Waste Management Association* **45**, 320–82.

Cochran, J. K., Frignani, M., Salamanca, M., Bellucci, L. G., and S. Guerzoni. 1998. Lead-210 as a tracer of atmospheric input of heavy metals in the northern Venice Lagoon. *Marine Chemistry* **62**, 15–29.

Chow, J. C., and J. G. Watson. 2001. Zones of representation for PM10 measurements along the US/Mexico border. *The Science of the Total Environment* **276**, 49–68.

Donazzolo, R., Orio, A. A., Pavonoi, B., and G. Perin. 1984. Heavy metals in sediments of the Venice Lagoon. *Oceanologica Acta* **7**, 25–32.

Frignani, M., Bellucci, L. G., Langone, L., and H. Muntau. 1997. Metal fluxes to the sediments of the northern Venice Lagoon. *Marine Chemistry* **58**, 275–92.

Gambaro, A., Manodori, L., Piazza, R., Moret, I., and P. Cescon. 2003. Distribution of 'particulate'- and 'vapour'-phase polychlorobiphenyls and polycyclic aromatic hydrocarbons in the atmosphere of the Venice Lagoon. *Organohalogen Compounds* **61**, 474–7.

Guieu, C., Chester, R., Nimmo, M., Martin, J.-M., Guerzoni, S., Nicolas, E., Mateu, J., and S. Keyse. 1997. Atmospheric input of dissolved and particulate metals to the northwestern Mediterranean. *Deep-Sea Research* **44**, 655–74.

Martin, J. M., Huang, W. W., and Y. Y. Yoon. 1994. Level and fate of trace-metals in the Lagoon of Venice (Italy). *Marine Chemistry* (1995) **48**, 343.

Moret, I., Piazza, R., Benedetti, M., Gambaro, A., Barbante, C., and P. Cescon. 2001. Determination of polychlorobiphenyls in Venice Lagoon sediments. *Chemosphere* **43**, 559–65.

Rosman, K. J. R., Howearth, L., Gambaro, A., and G. Capodaglio. In preparation. Lead isotopic ratios in the Lagoon of Venice atmospheric aerosols.

Rossini, P., Guerzoni, S., Molinaroli, E., Rampazzo, G., De Lazzari, A., and A. Zancanaro. Submitted. Atmospheric bulk deposition to the Lagoon of Venice: part I, fluxes of metals, nutrients and organic contaminants. *Environ. Internat.*, special issue.

Scarponi, G., Turetta, C., Capodaglio, G., Toscano, G., Barbante, C., Moret, I., and Cescon, P. 1998. Chemometric studies in the Lagoon of Venice, Italy. 1. The environmental quality of water and sediment matrices. *J. Chem. Inf. Comput. Sci.* **38**, 552–62.

Simpson, S. L., Apte, S. C., and G. E. Batley. 1998. Effect of short term resuspension events on trace metal speciation in polluted anoxic sediments. *Environmental Science & Technology* **32**, 620–5.

Turetta, C., Cozzi, G., Barbante, C., Capodoglio, G., and P. Cescon. In press. Trace elements determination in seawater by ICP-SFMS coupled with a micro-flow nebulization/desolvation system. *Analytical Bioanal.*

Zago, C., Capodaglio, G., Ceradini, S., Ciceri, G., Abelmoschi, L., Soggia, F., Cescon, P., and G. Scarponi. 2000. Benthic fluxes of cadmium, lead, copper and nitrogen species in the northern Adriatic Sea in front of the River Po outflow, Italy. *The Science of the Total Environment* **246**, 121–37.

64 • Sediment contamination assessment of the Venice canal network (Italy)

R. ZONTA, L. ZAGGIA, F. COLLAVINI, F. COSTA AND M. SCATTOLIN

INTRODUCTION

The pollution of coastal sediments is a problem of worldwide concern. Contaminants from different sources accumulate in lakes, rivers, and coastal zones, particularly in sediment deposits characterized by fine-grained particles and/or high organic carbon content. Such material may significantly impair the quality of the overlying waters, to the extent that aquatic life is damaged. It is estimated that about 10 per cent of the sediment underlying surface waters in the United States is sufficiently contaminated with toxic species to pose potential risks to human and wildlife food chains (US EPA, 1998).

Urban sources are widely recognized as one of the main factors in the overall budget of man induced pollutants, impairing the quality of air, soil and water. Mage *et al.* (1996) analysed data from the air of 20 megacities (over 10 million people), and always found at least one major pollutant exceeding the health protection guidelines given by the World Health Organization. Metal contamination in the soil of different cities in the world is shown by Salvagio Manta *et al.* (2002). As far as an urban stream or waterbody is concerned, the quality of the receiving water and underlying sediment are affected as well, depending on the efficiency of the facilities for pollution control. Nriagu and Pacyna (1988) discussed the worldwide contamination by trace metals, indicating domestic wastewater effluents among the major sources of these species in aquatic ecosystems. In an exhaustive investigation on the sediment contamination in the Chicago River, Collier and Cieniawski (2003) found high metal and microorganics concentrations; in particular, PAH and PCB values were as high as 721 and 76 mg kg^{-1}, respectively. PCB 'hot spots' were also identified by Tolosa

et al. (1997) near sewage outfalls from highly industrialized and populated cities of the Mediterranean Sea (Barcelona, Marseille, Nice and Naples).

Polluted sediments are difficult to handle, since both dredging and disposal must be done in an environmentally sound manner (OSPAR, 1998). For this reason, criteria for the assessment of the nature and extent of contamination in sediments have been developed in recent years (US EPA, 1992; Stortelder, 1995; Förstner and Calmano, 1998). A number of actions have also been planned to identify pollution sources (US EPA, 1987) and to prevent further contamination (US EPA, 1998). In this context, the management of the sediment in the historical centre of Venice represents a peculiar case. The city is located in the central part of a shallow water lagoon, which is connected to the Adriatic Sea by three inlets, and is crossed by a branching network of canals (known as 'rii').

Organic and inorganic matter from untreated domestic and commercial sewage effluents, as well as from the erosion of building surfaces and paved areas, reaches the canals causing a progressive silting. Periodic dredging operations have been performed since ancient times to permit navigation and to maintain acceptable hygienic conditions. From the early 1960s, however, the dredging was interrupted, except for some minor interventions, and a layer of sediment up to 1 m thick accumulated in the network. A specific law of the Italian Parliament (n. 139/1992) and the successive agreement achieved among the Ministry of Public Works, and the regional and city governments (Chiozzotto *et al.*, 1993) started, in 1994, a wide series of engineering interventions for the maintenance of the canals and the improvement of the urban utilities. Besides a comprehensive activity of remediation and maintenance dredging, the

Flooding and Environmental Challenges for Venice and its Lagoon: State of Knowledge, ed. C. A. Fletcher and T. Spencer.
Published by Cambridge University Press, © Cambridge University Press 2005.

works include the restoration of building foundations, the introduction of a number of septic tanks, and the renewal of the culvert system (Gardin, 1999).

Due to the high level of heavy metals (Zucchetta, 1983) and hydrophobic organic chemicals accumulated in the sediment, the Ministry of the Environment imposed a specific technical protocol (Ministero dell'Ambiente, 1993) for the classification of sediment prior to dredging. On the behalf of the city government, our research group sampled and classified the entire canal network in the period 1994–2000 (project 'Integrato Rii'), in order to assess the degree of contamination of materials and to identify appropriate disposal sites.

AIMS AND OBJECTIVES

The main objective of the study was to determine the contaminant distribution in the Venice canal sediments. This has been a unique chance to study the past and recent history of contamination in the network. In fact, the sediment deposit recorded the evolution of the suspended matter composition over a period of at least thirty years. At the same time, several additional investigations were performed within different interrelated research projects: (1) formation of authigenic metal-sulphides in the canal sediments (Zaggia and Zonta, 1997); (2) rate of sulphurization-oxidation processes during simulated sediment resuspension under controlled conditions (Zaggia *et al.*, 1997) and release of chemical species from the sediment during wet dredging operations (Collavini *et al.*, 2000); (3) water circulation, sediment transport and deposition fluxes of particulate matter and heavy metals (Zaggia *et al.*, 2000; Collavini *et al.*, 2004); (4) estimate of sediment and heavy metal accumulation rates in an inner canal, by means of inventories of radionuclides (Zonta *et al.*, in press); (5) monitoring of the concentration of nutrients (Arizzi Novelli *et al.*, 1998), heavy metals and microbiological parameters (Zonta *et al.*, 2004) in the water column of selected areas of the network; (6) study of the grain-size and heavy metal distributions in the sediment of the Grand Canal (Zonta *et al.*, in preparation).

These investigation have been helpful in the interpretation of data resulting from the classification activity, allowing to relate concentrations found in the sediment with the pollution sources, and to study potential routes by which contaminants are delivered to the system. The information acquired has provided a comprehensive description of the functioning of the canal network and the environmental processes governing the behaviour and fate of contaminants.

THE STUDY SITE

The historical centre of Venice (Fig. 64.1) has a population of about 66 000 inhabitants, additional presences of about 10 000 overnight stay tourists and 20 000 day tourists, plus a net daily flux of 46 000 commuters. The primary economic activities in the city include commerce and tourism; craftsman productions and small boatyards are also present.

Due to its peculiar urban structure, Venice has never been provided with a main sewage system, and untreated effluents from a large number of inputs are discharged into the canals. However, the number of septic tank systems has significantly increased in the last decade, as new criteria have been introduced with the restoration of the buildings.

The city covers a surface area of 6311 km², including the ship terminal (Marittima) and the car-park area (Tronchetto). The canal network (excluding the 0.120 km² of the historic naval arsenal) and the Grand Canal account for 0.433 and 0.244 km², respectively. The canal network, about 40 km long, has a mean width of 10 m (with a range from 3 m to *c.* 50 m). If regularly dredged, the mean depth of the canals should be around 2 m. The average tide excursion is 80 cm and 30 cm under spring and neap tide conditions respectively. The hydrodynamics is governed by the tide, which propagates from the Lido inlet toward the mainland border and generates an approximate water circulation from SE to NW during a rising tide (Umgiesser and Zampato, 2001). Water exchanges on the western side of the city are mainly driven by the Grand Canal, which is directly connected to the Lido inlet by first order tidal channels. The flow on the eastern side is, instead, substantially directed from south to north during a rising tide. Opposite flows are typically observed in the ebb phase, although significant differences in the circulation can be induced by meteorological conditions.

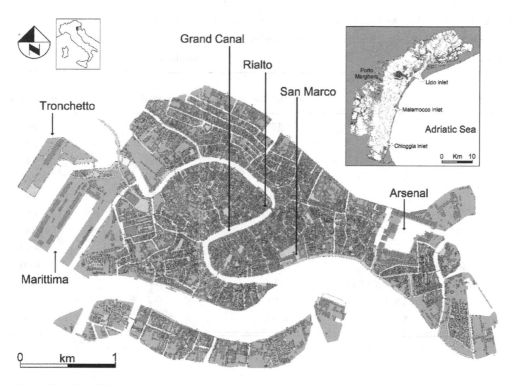

Fig. 64.1   Map of the city of Venice; the entire lagoon is shown in the inset.

Due to the intricate morphology of the canal net-work, a complex circulation pattern characterises the system. Current speeds are generally low, from a few cm s^{-1} in the more confined branches up to *c.* 50 cm s^{-1} in the larger canals, in high hydrodynamic conditions. The mean salinity of the canal water is *c.* 30 PSU (Practical Salinity Unit); it is largely conditioned by the salinity of the surrounding lagoon areas, which in turn are affected by the tide and meteorological conditions. Temporary variations are induced by tidal circulation and the freshwater inputs from the urban effluents; their relative importance increases towards the inner parts of the network.

METHODS

Determinants and sediment classification

The specific protocol of the Ministry of the Environment establishes the sampling methodologies,

analyses and criteria for the classification of the sediment prior to dredging and disposal. Materials are classified according to concentration limits (Table 64.1) for twelve chemicals: As, Cd, Cr, Cu, Hg, Ni, Pb, and Zn as total; total hydrocarbons (TH); polycyclic aromatic hydrocarbons (PAHs); polychlorinated biphenyls (PCBs); organochlorine pesticides (OCs). Four classes of contamination are identified which correspond to increasingly strict constraints for the disposal. When only one of the considered species exceeds the limit between two classes, the sediment is categorized in the higher class. A tolerance of 10 per cent is permitted when only one species determines the class.

Non-polluted sediment (class 'A') can be used for restoration of the lagoon morphology (salt marsh reconstruction, reclamation works). Moderately polluted sediment ('B') can also be used for restoration but only when ensuring a permanent confinement against the release of contaminants in the lagoon.

Table 64.1 *Concentration limits for the classification of sediment. The corresponding classes are shown.*

Determinants	Limits and classes		
	A mg/kg d.w.	B mg/kg d.w.	C mg/kg d.w.
As	15	25	50
Cd	1	5	20
Cr	20	100	500
Cu	40	50	400
Hg	0.5	2	10
Ni	45	50	150
Pb	45	100	500
Zn	200	400	3000
Total hydrocarbons (TH)	30	500	4000
Polycyclic aromatic hydrocarbons (PAHs)	1	10	20
Polychlorinated biphenyls (PCBs)	0.01	0.2	2
Organochlorine pesticides (OCs)	0.001	0.02	0.5

Polluted sediment ('C') can only be used for reclamation of elevated areas that are never in contact with the lagoon water or subjected to erosion or dispersion by infiltration. Finally, highly polluted sediment (class '>C') must be disposed of in controlled landfills outside the lagoon.

Sampling

The entire canal network was sampled at 50 m intervals, by means of a 'plexiglas' syringe corer (diameter 6 cm), collecting the sediment at a distance of 1 m from the bank, alternatively on the right and the left side of the canal. The length of the retrieved sediment cores varied between 15 and 100 cm, depending on the thickness of the deposits. According to the ministerial protocol, sediment cores longer than 50 cm were halved and identified as upper and lower samples. In total, 775 upper and 292 lower samples were collected; their length was 40 ± 7 cm (mean ± standard deviation) and 39 ± 5 cm, respectively. Samples could not be collected from twenty-five sites as an insufficient amount of sediment was consolidated over the channel bed. These sites were located in canal sections

in the vicinity of the confluence with either the Grand Canal or the larger tidal channels surrounding the city. For the analyses required by the ministerial protocol, samples were accurately homogenized and divided in different aliquots; a small amount of each sample was archived for future analyses.

During the project, further sampling events were performed for different purposes. About forty sediment cores were collected in representative sites to investigate the distribution of heavy metal concentrations (including Fe and Mn) along vertical profiles. The superficial sediment layer (20 cm depth) of some interesting zones of the network was also sampled in order to map total concentrations of N and P. Finally, sediment cores were collected at c. one hundred sites for the measurement of the redox potential along the vertical profile.

Chemical analyses

Chemical analyses were performed by a small group of quality-accredited laboratories, following the Italian standard procedures set by CNR-IRSA (National Research Council of Italy – Water Research

Table 64.2   *Coefficients of variation (CV) resulting from the analysis of five to eight replicated samples in ten sites of the canal network*

	As	Cd	Cr	Cu	Hg	Ni	Pb	Zn	TH	PAHs	PCBs	OCs
	0.03	0.03	0.03	0.02	0.04	0.02	0.03	0.02	0.02	0.05	–	–
	0.03	0.03	0.04	0.05	0.05	0.03	0.03	0.02	0.04	0.07	0.08	0.05
	0.04	0.04	0.05	0.05	0.07	0.03	0.03	0.04	0.04	0.07	0.08	0.08
	0.05	0.04	0.06	0.06	0.08	0.04	0.04	0.03	0.04	0.08	0.10	0.12
	0.05	0.05	0.06	0.07	0.09	0.05	0.06	0.05	0.05	0.11	0.10	0.13
CV	0.05	0.11	0.06	0.07	0.09	0.05	0.07	0.05	0.07	0.12	0.13	0.14
	0.06	0.12	0.10	0.07	0.09	0.05	0.08	0.06	0.08	0.14	0.15	0.17
	0.09	0.13	0.11	0.08	0.11	0.09	0.08	0.10	0.08	0.15	0.16	0.22
	0.11	0.14	0.12	0.08	0.12	0.11	0.09	0.11	0.11	0.17	0.20	0.22
	0.11	0.17	0.15	0.09	0.16	0.11	0.11	0.14	0.17	0.17	0.29	0.32
Mean	0.06	0.09	0.08	0.06	0.09	0.06	0.06	0.06	0.07	0.11	0.14	0.16

Institute). The considered PAH congeners were: naphthalene, acenaphthilene, acenaphthene, fluorene, phenanthrene, anthracene, fluoranthene, pyrene, benzo(a)anthracene, chrysene, benzo(b)fluoranthene, benzo(K)fluoranthene, benzo(a)pyrene, dibenzo(a,h)anthracene, benzo(g,h,i)perylene, indeno (1,2,3,-cd)pyrene. The concentration of PCBs was determined using Aroclor 1248, 1254, 1260 standards. Finally, OCs were measured as the sum of DDT, DDD, DDE, HCH congeners, and HCB, aldrin, dieldrin, methoxychlor. All concentrations were referred to the sample dry weight, which was determined by weight loss at 105 °C.

In order to ensure a high quality standard for analytical data, a Quality Control procedure (Ibe and Kullenberg, 1995; US EPA, 1995) was applied to the laboratory practices. This included the use of field and analytical replicate samples, reference materials, and interlaboratory comparisons. The uncertainty affecting chemical data was tested by collecting and analysing from five to eight replicated samples in ten sites of the network, which were selected to represent a wide range of concentration levels. The obtained coefficients of variation (CV = standard deviation / mean value) are shown in Table 64.2. The resulting variability, which may be mainly ascribed to the sampling bias, was fairly low. The higher CV data for PCB and OC distributions is likely to have been due to their low concentrations.

ORGANIC MATTER CONTENT AND GRAIN-SIZE ANALYSES

The organic matter content was determined in all the collected samples by loss on ignition at 550°C. Redox potential measurements were performed following a methodology (Argese *et al.*, 1992) that involves the use of combined Pt electrodes in sealed 'plexiglas' containers. Grain-size distribution and structural parameters were determined on composite samples representative of each single canal. These samples were obtained by consolidating equal aliquots of the material collected from each sampling site of the canal. Some interesting canals were investigated in greater detail, by means of measurements in each of the sampling sites. An overall number of more than 300 analyses were performed. Grain-size was measured by means of a laserbeam analyzer (Microtrac, Leeds and Northrup, USA), obtaining particle-size distribution in 15 dimensional intervals in the diameter range d < 300 μm. Geotechnical parameters, including Atterberg limits, specific gravity of the sample and of its solids, were also determined on the same samples, following the Italian standard procedures (CNR-UNI 10013 and 10014).

Table 64.3 *Mean value and basic statistical data of some descriptive parameters of the sediments. The grain size is expressed as a volumetric percentage of particle diameters in five dimensional classes.*

	Water content	Organic matter	Plastic index	Redox potential	Sand 300 >d > 125 μm	Fine sand 125 >d > 62 μm	Rough silt 62 >d > 16 μm	Fine silt 16 >d > 4 μm	Clay 4 >d > 0.7 μm
	%	%		mV	%	%	%	%	%
Mean	47.0	16.8	30	-205	7.4	9.5	33.8	34.4	15.0
Minimum	33.2	8.5	4	-269	0	2.5	22.8	21.1	5.5
Maximum	61.9	24.5	58	-137	32.6	20.3	44.3	43.8	22.4
σ	5.5	2.7	9	28	5.8	3.2	3.2	4.8	3.3
CV%	12%	16%	30%	14%	78%	34%	10%	14%	22%

RESULTS

Sediment characteristics

Table 64.3 reports the average value and basic statistics for some of the descriptive parameters of the collected samples. The sediment accumulated in the Venice canals is a strongly reduced and organic-rich material, characterized by a high plasticity. The water content is generally high and quite frequently is above the limit of liquidity. The low CV per cent values for most of the determined parameters indicate that the sediment of the canals has rather homogeneous characteristics.

A large proportion of particles were in the silt and clay interval (62 > d > 0.7 μm, Fig. 64.2). Moreover, due to the high organic content of the sediment, the occurrence of mineral-organic aggregates (Johnson, 1974; Rashid, 1985) is to be expected, and the true grain-size spectrum is possibly finer than those measured in the samples. The sand fraction generally accounts for only 15 per cent of the total, and has a greater spatial variability than the finer components. In particular, higher amounts of sand particles were found in the larger canals, as well as in channel margins where the weathering and erosion of building surfaces supply the bulk of the coarser fraction.

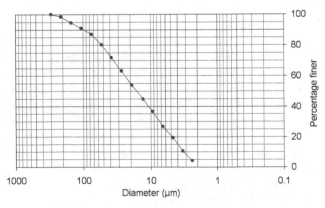

Fig. 64.2   Average grain-size distribution in the upper sediment layer of the canals.

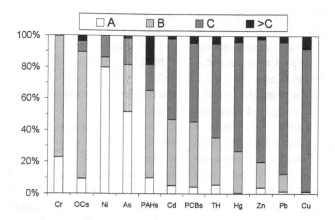

Fig. 64.3   Histogram of the percentage distribution of the entire sample population (1067 samples) in the four classes, with respect to the 12 analysed chemical species.

On comparison with data obtained for the 292 samples collected from the lower layer, it can be seen that the sediment characteristics do not show significant variations with the depth.

Contamination assessment

On the basis of the described classification criteria, the material ranks in the 'C' and '>C' classes, and therefore can only be disposed in specific dump sites, which need necessary standards. In fact, only 2 (0.2 per cent) of the 1067 analysed samples (775 from the upper and 292 from the lower layers) belong to the

'B' class, whereas 729 (68.3 per cent) and 336 (31.5 per cent) rank in 'C' and '>C' classes, respectively. None of the collected samples ranks in the 'A' class. The histogram of Fig. 64.3 shows the percentage distribution of the entire sample population in the four classes, for each of the 12 analysed species. The sediment is only slightly contaminated by Cr, OCs, Ni, and As, which have little or no influence on the classification. On the contrary, 99 per cent of the samples rank in the 'C' or '>C' classes because of the high content of Cu only. Lead, Zn, and Hg also have considerable effect on the classification of samples in the two higher classes (from 88 to 73 per cent, respec-

Table 64.4   *Mean value and basic statistics of the concentration distribution in the upper sediment layer (775 samples), for the 12 chemical species considered by the study.*

	As mg/kg d.w.	Cd mg/kg d.w.	Cr mg/kg d.w.	Cu mg/kg d.w.	Hg mg/kg d.w.	Ni mg/kg d.w.	Pb mg/kg d.w.	Zn mg/kg d.w.	TH mg/kg d.w.	PAHs mg/kg d.w.	PCBs mg/kg d.w.	OCs mg/kg d.w.
mean	16.6	6.6	30.8	245	3.7	35.9	222	889	1384	15.3	0.595	0.108
maximum	159	99	113	8285	46	261	5300	8860	18339	302	34.8	3.45
minimum	1.0	0.3	7.4	13	0.4	9.8	20	53	15	0.02	0.001	0.0001
σ	13.4	6.1	15.0	369	3.5	18.4	329	722	1549	31.7	1.584	0.273
CV%	81%	92%	49%	150%	95%	51%	148%	81%	112%	207%	266%	252%

tively). The concentration of PAHs is generally low, 65 per cent of samples ranking in the 'A' and 'B' classes. However these compounds are responsible for the highest number of samples to exceed the threshold for the '>C' class (18 per cent).

Concentration levels in the sediment are quite high for the majority of the investigated chemical species (Table 64.4), and are comparable-to or even greater-than those found in sediments near the industrial area of Porto Marghera, which are commonly considered as the most polluted of the Venice Lagoon (Donazzolo *et al.*, 1984; Marcomini *et al.*, 1999; Frignani *et al.*, 2001; Bellucci *et al.*, 2002).

These levels of contamination have led some authors to assume the prevailing influence of external sources, and in particular the industrial area (Zucchetta, 1983; Salizzato *et al.*, 1998; Peruzzi *et al.*, 2000). However, the spatial distribution of contaminants in the whole network and the concentration trends in the different sub-systems of interlinked canals clearly show the existence of diffuse and point sources within the city itself. In other terms, the urban centre appears to be mainly responsible for the pollution of its canal network.

The contribution of atmospheric deposition to the sediment contamination is very low when compared with the effect of these sources. Table 64.5 shows a comparison between the annual loads delivered to the canal network from the atmosphere and the annual amount of contaminants accumulated in sediment. The first were estimated by referring to the whole city surface, the annual fluxes being measured at a gauge station located within the historical centre (Rossini *et al.*, 2001). The second value was instead estimated on the basis of the mean concentrations of Table 64.4, assuming an average sedimentation rate of 2 cm/y in the whole canal network, according to the results of a study on a test canal (Zonta *et al.*, in press).

Even if we assume that the atmospheric load is entirely transferred to the canal sediment, it is of minor importance in terms of the total accumulation. Its contribution only accounts for a percentage varying from 0.3 per cent (PCBs) up to 5.2 per cent (Ni).

Two European cities can be usefully compared with Venice: Delft in the Netherlands and Stockholm in Sweden. Figure 64.4 shows the ratios between the mean concentrations found in Venice sediments (upper layer, as in Table 64.4) and in the two cities. Data for Delft (Kelderman *et al.*, 2000) refer to the mean values determined in the canals of the inner city, which are the most polluted. Among those investigated by Lindström *et al.* (2001) in Stockholm, the site characterized by the highest metal concentration in the sediment was instead selected. Unfortunately, As and micro-organics were not analysed in Stockholm. Data shown in Fig. 64.4 display higher accumulation of Hg, TH, and particularly Cd, OCs, PCBs in the sediment of the Venice canals. The concentration of the other species is rather similar in the three cases, with the exceptions of Cr – that is particularly enriched in Stockholm – and Pb. The latter is higher in the two comparison cities probably because of the large effect of vehicular

Table 64.5 *Annual loads delivered to the canal network from the atmosphere and annual amount of pollutants accumulated in the sediments. The percentage ratio between the two estimates is also shown.*

	As kg	Cd kg	Cr kg	Cu kg	Hg kg	Ni kg	Pb kg	Zn kg	PAH kg	PCBs kg	OCs kg
Atmospheric annual load	3.5	1.8	12	38	0.3	17	38	259	2.2	0.02	0.03
Annual accumulation	151	60.2	296	2235	32.8	328	2026	8112	140	5.47	1.00
Percentage ratio	2.3%	3.0%	3.9%	1.7%	0.9%	5.2%	1.9%	3.2%	1.6%	0.3%	3.2%

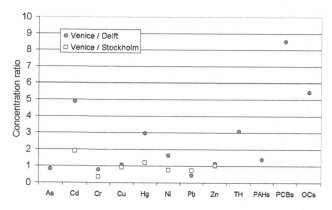

Fig. 64.4   Ratio between the average concentrations measured in the sediment of the Venice canals (upper layer) and those found in the inner city of Delft and in the most polluted site in Stockholm, for the twelve determinants. As and micro-organics were not analysed in Stockholm.

traffic, which is limited to boats in Venice. It is worth-while to observe that around 65 per cent of Zn and Pb, and even 85 per cent of Cu in the sediment of Delft was ascribed by Kelderman and co-authors to originate from outside the inner city via the Rijn-Schie canal, which is a main connection between the river Rhine and the Delft canals. Taking this into account decreases the difference in Pb concentration between sediments in Venice and Delft, whereby it increases the difference in Cu and Zn levels.

The high concentrations found in the Venice canals are caused by several interlinking factors. The first is the absence of a main sewage system: the untreated effluents carry a considerable input of metals and micro-organics – as well as organic matter and nutri-ents – into the network. Once delivered to the canals, the effluents also undergo an abrupt change of salinity, which causes a rapid precipitation of fine particles and contaminants, enhancing the effectiveness of sediment deposition and accumulation processes. Another fac-tor is the slack hydrodynamics characterising the net-work, which also favours the accumulation of materials on the canal bottom. The silting-up and the consequent decrease of the canals cross-section area determine a further reduction in the hydrodynamics. Sediment particles are therefore preferentially retained by the system, causing local increases in sedimenta-tion rate (Zonta *et al.*, in press). Finally, the peculiar

structure of the historical centre of the city magnifies the effects of urban pollution. In fact, the receiving canal network is only 10 per cent of the city surface area and it concentrates contaminants delivered by wastewaters and the runoff from horizontal and ver-tical urban surfaces.

Identification of sources

A general decrease of the contamination over time is observed for all the species of concern, by compar-ing concentrations found in the upper and lower lay-ers of the sediment in the 292 long-core sites (Fig. 64.5). In particular, for organic compounds the con-centration ratio between the two layers ranges from 2 to 4.3 for TH and OCs, respectively. As indicated by metal concentrations along the vertical profile in additional cores collected in the framework of this study, the actual reduction of contamination is even greater than the results shown in Fig. 64.5. In fact, the simple halving of the cores, requested by the min-isterial protocol, does not reflect a true separation between 'past' and 'present' records. As demon-strated by an investigation of the chronology of sed-iment deposits, this decrease started in the 1970s (Zonta *et al.*, in press).

Apart from these important findings, the data can be exploited to identify contaminant sources

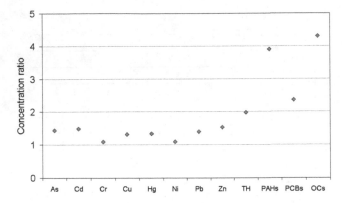

Fig. 64.5   Ratio between concentrations of the lower and upper sediment layers of the 292 sites where cores longer than 50 cm were collected.

within the canal network. However, this task is quite difficult mainly due to the large number of inputs as well as the relative importance of the diffuse contributions. Therefore, the signal from different unrecognizable sources may overlap, and the contribution of a single contaminant can be masked in the sediment of a specific location. Moreover, some point sources are currently discontinued, even though their fingerprint is quite detectable in the sedimentary record.

According to US EPA (1987) approaches, the final aim of this kind of study is to identify a series of suspected sources rather than proof of actual sources. A Geographic Information System was therefore implemented to investigate the spatial distribution of contaminants by means of thematic maps. Appropriate concentration ranges were selected with a resolution higher than that established by the classification criteria of the ministerial protocol. Four examples of significant species are described here to show the relationships between the concentration levels and the presence of point and diffuse sources within the network. Despite the presence of a few isolated hot spots, the spatial distribution of Zn (Plate 25a) is basically 'uniform', suggesting a predominance of diffuse sources for this metal. Zinc is, in fact, commonly found in the urban environment, and is released mainly from paints and pigments, roof coverage and drainpipes, as well as household goods (US

Department of Health and Human Services, 1994; Rose *et al.*, 2001; Bergbäck *et al.*, 2001).

The same typology of sources (US Department of Health and Human Services, 2002) is responsible for the input of Cu (Plate 25b), which is also released by tap water systems and electric cables (Sörme *et al.*, 2001). Nevertheless, some sectors of the network show higher contamination levels, indicating the presence of point sources. For example, three to six-fold the average Cu concentration values were found in a large sector of the northern network (white arrow in Plate 25b). This sector is also contaminated by other metals, such as As and Zn whose distributions correlates with that of Cu (r = c. 0.80).

Point sources are also responsible for the relevant accumulation of Hg in some sites of the network (Plate 26a). Mercury in the urban environment is released principally by dental amalgam, pharmaceuticals and pesticides, as well as light-bulbs, thermometers, batteries and combustion of fossil fuels (US Department of Health and Human Services, 1999; Sörme *et al.*, 2001). In particular, medical wastes are widely recognized as potential point sources of this element (Svidén and Jonsson, 2001). Therefore, it is not surprising that high values of Hg were found in the vicinity of the two hospitals (white arrows in Plate 26a). However, the highest concentrations of Hg were found in two canals of the inner network (red arrow in Plate 26a), with values up to 46 mg kg^{-1}. This location also contains high levels of other compounds, particularly, As, Pb and TH, and they are clearly correlated (r = 0.65 – 0.92). The suspected source for this contamination is the printing office of a local newspaper, which was discontinued in the 1980s. The contamination from PCBs (Plate 26b) is substantially confined to the eastern network, where large boatyards are located. Concentrations as high as 34.8 and 40.2 mg kg^{-1} were measured in the upper and lower sediment layer, in a site close to the historic naval arsenal (red arrow). Due to the ban of these compounds, the sources should be considered as discontinued (US Department of Health and Human Services, 2000).

CONCLUSIONS

The continuous input of organic and inorganic materials and associated contaminants from different sources, including domestic effluents, commercial activities and the runoff of urban surfaces, has led to the accumulation of a strongly reduced and polluted sediment in the Venice canal network. Copper, lead and zinc are the species which mainly contributed to environmentally harmful levels of contamination, as established by the criteria of Italian Ministry for the Environment. Due to the peculiarity of the investigated system, some levels are quite high if compared with other urban environments, as well as with the most polluted areas of the lagoon, close to the industrial area of Porto Marghera. However, it is important to recognize that the concentration in the sediment refers to the history of at least three decades of contamination, and some sources are now much lower or even discontinued. By comparing the concentrations in the upper and lower layers of the sedimentary column, a general decrease is observed for the more recent period, particularly for micro-organic compounds. This reduction is ascribed to the effect of protection measures such as the ban of PCBs and DDT, the replacement of lead as petrol additive, the use of more environmentally sound materials.

A large proportion of the sediment investigated in this study has been removed by the dredging interventions performed recently, which induces a greater water movement in the canal network, allowing for a more effective water renewal with the tide. Thanks to the restoration and improvement of the sewage systems, the amount of matter and the load of associated contaminants delivered to the network is expected to undergo a progressive diminution in the future. This, in turn, will imply a lower rate of the canal silting and a general improvement of the water quality in the canals. Nowadays, the research is therefore addressed to define suitable methodologies for the monitoring of the environmental conditions in the network (Collavini et al., 2004; Zonta et al., 2004).

Acknowledgements

We are grateful to the team who assisted us in this research. Special thanks are due to: the technicians of the CNR Institute (G. Magris, M. Meneghin, R. Ruggeri, F. Salviati, F. Simionato, G. Zamperoni) for the field and laboratory work; R. Pini and M. Scatena (CNR-ISE, Pisa) for the grain-size analyses; M. Botter (CNR-ISMAR) for her precious support in the data treatment.

References

Argese, E., Cogoni, G., Zaggia, L., Zonta, R., and R. Pini. 1992. Study of the redox state and grain-size of sediments in a mud flat of the Venice Lagoon. *Environmental Geology and Water Sciences* **20**, 1, 35–42.

Arizzi Novelli, A., Collavini, F., Zonta, R., and L. Zaggia. 1998. Dissolved nitrogen and phosphorous species in a highly contaminated urban ecosystem: the Venice canal network. In Proceedings of the VII International Congress of Ecology (INTECOL), Florence, Italy, 19–25 July, p. 19.

Bellucci, L. G., Frignani, M., Paolucci, D., and M. Ravanelli. 2002. Distribution of heavy metals in sediments of the Venice Lagoon: the role of industrial area. *Science of the Total Environment* **295**, 35–49.

Bergbäck, B., Johansson, K., and U. Mohlander. 2001. Urban metal flows – a case study of Stockholm. *Water, Air, and Soil Pollution: Focus* **1**, 3–24.

Chiozzotto, E. et al. 1993. *La manutenzione urbana della città di Venezia*. Venice: Comune di Venezia., 151pp.

Collavini, F., Zonta, R., Arizzi Novelli, A., and L. Zaggia. 2000. Heavy metals behaviour during resuspension of the contaminated anoxic sludge of the Venice canals. *Toxicological and Environmental Chemistry* **77**, 171–87.

Collavini, F., Coraci, E., Zaggia, L., Zaupa, A., Zonta, R., and A. Zuliani. 2004. Environmental characterisation of the Venice canals: first results of the ICARO project. In Proceedings of the 37th CIESM Congress, Barcelona, Spain, 7–11 June, p. 182.

Collier, D., and S. Cieniawski. 2003. *October 2000 and August 2002, Survey of Sediment Contamination in the Chicago River, Chicago, IL.* Chicago: US EPA Great Lakes National Program Office, 46 pp.

Donazzolo, R., Orio, A. A., Pavoni, B., and G. Perin. 1984. Heavy metals in sediments of the Venice Lagoon. *Oceanologica Acta* 7, 25–32.

Förstner, U., and W. Calmano. 1998. Characterisation of dredged materials. *Water Science and Technology* 38, 11, 149–57.

Frignani, M., Bellucci, L. G., Carraro, C., and S. Raccanelli. 2001. Polychlorinated biphenyls in sediments of the Venice Lagoon. *Chemosphere* 43, 567–75.

Gardin, P. 1999. La ripresa dello scavo dei rii. In *Venezia la città dei rii.* Venice: UNESCO – Insula Eds., pp. 91–105.

Ibe, A. C., and G. Kullenberg. 1995. Quality assurance / quality control (QA/QC) regime in marine pollution monitoring programmes: the GIPME perspective. *Marine Pollution Bulletin* 31, 4–12, 209–13.

Johnson, R. G. 1974. Particulate matter at the sediment–water interface in coastal environments. *Journal of Marine Research* 32, 313–30.

Kelderman, P., Drossaert, W. M. E., Zhang Min, Galione, L. S., Okonkwo, L. C., and I.A. Clarisse. 2000. Pollution assessment of the canal sediments in the city of Delft (The Netherlands). *Water Research* 34, 3, 936–44.

Lindström, M., Jonsson, A., Brolin, A. A., and L. Hakanson. 2001. Heavy metal sediment load from the city of Stockholm. *Water, Air, and Soil Pollution: Focus* 1, 103–18.

Mage, D., Ozolins, G., Peterson, P., Webster, A., Orthofer, R., Vanderweerd, V., and M. Gwynne. 1996. Urban air pollution in megacities of the world. *Atmosphere Environment* 30, 5, 681–6.

Marcomini, A., Bonamin, V., Degetto, S., and A. Giacometti. 1999. Occurrence of organochlorine pollutants in three dated sediment cores from the lagoon of Venice. *Organohalogen Compounds* 43, 373–82.

Ministero dell'Ambiente. 1993. *Criteri di sicurezza ambientale per gli interventi di escavazione, trasporto e reimpiego dei fanghi estratti dai canali di Venezia.* Art. 4, comma 6, legge 330/91, Technical protocol, 23pp.

Nriagu, J. O., and J. M. Pacyna. 1988. Quantitative assessment of worldwide contamination of air, water and soils by trace metals. *Nature* 333, 134–9.

OSPAR. 1998. *Guidelines for the Management of Dredged Material.* Ministerial Meeting of the OSPAR Commission, Sintra 22–23 July, 98/14/1-E, Annex 43 (OSPAR Reference Number: 1998-20), 32pp.

Peruzzi, F., Bianchi, M., Bortoli, A., Marzollo, A., Carrera, F., and H. Muntau. 2000. Presenza e distribuzione di composti organoclorurati nei rii del centro storico di Venezia. In *La ricerca scientifica per Venezia.* Venice: Il Progetto Sistema Lagunare Veneziano, Istituto Veneto di Scienze, Lettere ed Arti Ed., vol. II(I), pp. 49–54.

Rashid, M. A. 1985. *Geochemistry of Marine Humic Compounds.* Berlin: Spinger-Verlag, 349pp.

Rose, S., Crean, M. S., Sheheen, D. K., and A. M. Gahzi. 2001. Comparative zinc dynamics in Atlanta metropolitan region stream and street runoff. *Environmental Geology* 40, 983–92.

Rossini, P., De Lazzari, A., Guerzoni, S., Molinaroli, E., Rampazzo, G., and A. Zancanaro. 2001. Atmospheric input of organic pollutants to the Venice Lagoon. *Annali di Chimica* 91, 491–501.

Salizzato, M., Pavoni, B., Volpi Ghirardini, A., and P. F. Ghetti. 1998. Sediment toxicity measured using Vibrio Fischeri as related to the concentrations of organic (PCBs, PAHs) and inorganic (metals, sulphur) pollutants. *Chemosphere* 36, 14, 2949–68.

Salvagio Manta, D., Angelone, M., Bellanca, A., Neri, R., and M. Sproveri. 2002. Heavy metals in urban soils: a case study from the city of Palermo (Sicily), Italy. *The Science of the Total Environment* 300, 229–43.

Sörme, L., Bergback, B., and U. Lohm. 2001. Century perspective of heavy metal in urban areas. A case study in Stockholm. *Water, Air, and Soil Pollution: Focus* 1, 197–211.

Stortelder, P. B. M. 1995. The management of contaminated sediments, an overview. *European Water Pollution Control* 5, 5, 8–15.

Svidén, J., and A. Jonsson. 2001. Urban metabolism of mercury turnover, emissions and stock in Stockholm 1795–1995. *Water, Air, and Soil Pollution: Focus* 1, 179–96.

Tolosa, I., Readman, J. W., Fowler, S. W., Villeneuve, J. P., Dachs, J., Bayona, J. M., and J. Albaiges. 1997. PCBs in the western Mediterranean. Temporal trends and mass balance assessment. *Deep-Sea Research Part II* 44, 3–4, 907–28.

Umgiesser, G., and L. Zampato. 2001. Hydrodynamic and salinity modeling of the Venice channel network with coupled 1-D–2-D mathematical models. *Ecological Modelling* 138, 75–85.

US Department of Health and Human Services. 1994. *Toxicological Profile for Zinc.* Atlanta, GA: US Department of Health and Human Services, 259pp.

US Department of Health and Human Services. 1999 *Toxicological Profile for Mercury*. Atlanta, GA: US Department of Health and Human Services, 676pp.

US Department of Health and Human Services. 2000. *Toxicological Profile for Polychlorinated Biphenyls (PCBs)*. Atlanta, GA: US Department of Health and Human Services, 948pp.

US Department of Health and Human Services. 2002 *Toxicological Profile for Copper*. Atlanta, GA: US Department of Health and Human Services, 295pp.

US EPA. 1987. *An Overview of Sediment Quality in the United States*. EPA-905/9-88-002. Washington DC: US EPA, 202pp.

US EPA. 1992. *Sediment Classification Methods Compendium*. EPA 823-R-92-006. Washington DC: US EPA, 228pp.

US EPA. 1995. *QA/QC Guidance for Sampling and Analysis of Sediments, Water, and Tissues for Dredged Material Evaluations. Chemical Evaluations*. EPA 823-B-95-001. Washington DC: US EPA, 297pp.

US EPA. 1998. *EPA's Contaminated Sediment Management Strategy*. EPA-823-R-98-001. Washington DC: US EPA, 131pp.

Zaggia, L., and R. Zonta. 1997. Metal-sulphides formation in the contaminated anoxic sludge of the Venice canals. *Applied Geochemistry* 12, 527–36.

Zaggia, L., Collavini, F., and R. Zonta. 1997. Sulphurization and oxidation processes in the Venice canals sludge. In Proceedings of the Meeting on Environmental Sedimentology (IAS-SEPM), Venice, Italy, 27–29 October, pp. 78–9.

Zaggia, L., Zonta, R., Collavini, F., and C. Bettiol. 2000. Study of particulate matter and heavy-metal fluxes in the Venice canal network by large volume filtration and sediment traps. In Proceedings of the Int. Conf. on Heavy Metals in the Environment. Ann Arbor, Michigan, USA, August, contribution no. 1401, 4pp.

Zonta, R., Scattolin, M., Botter, M., and F. Collavini. 2004. Water quality of the Venice canal network: results of an intensive monitoring. In Proceedings of the 37th CIESM Congress, Barcelona, Spain, 7–11 June, p. 255.

Zonta, R., Cochran, J. K., Collavini, F., Costa, F., Scattolin, M., and L. Zaggia. In press. Sediment and heavy metal accumulation in a small canal, Venice, Italy. *Aquatic Ecosystem Health & Management*.

Zonta, R., Zaggia, L., Collavini, F., and M. Botter. In preparation. Heavy metals in the sediment of the Grand Canal (City of Venice, Italy).

Zucchetta, G. 1983. Determinazione dei metalli pesanti nei sedimenti di canali e rii di Venezia. *Atti e Memorie dell'Ateneo Veneto* 21, 2, 91–109.

65 • Water quality in the channels of Venice: results of a recent survey

C. DABALÀ, N. CALACE, P. CAMPOSTRINI, M. CERVELLI, F. COLLAVINI, L. DA ROS,
A. LIBERTINI, A. MARCOMINI, C. NASCI, D. PAMPANIN, B. M. PETRONIO,
M. PIETROLETTI, G. POJANA, R.TRISOLINI, L. ZAGGIA AND R. ZONTA

INTRODUCTION

The historical centre of Venice is crossed by a 40 km long canal network, which covers about the 10 per cent of the city surface area. Most urban wastewaters are discharged directly into this network, determining the accumulation of particulate matter and associated pollutants in the canal bottom (Zonta *et al.*, in press and chapter 64). Periodical dredging is therefore required to prevent the canals from silting up, as well as an effective control for water quality. The sediment of the network shows a high concentration of the pollutant species which are typical of the urban environment, such as Cu, Pb, Zn and PAHs (Zonta *et al.*, 2004).

A specific monitoring system was developed between 1996 and 1999 within the project WATERS (Water data Acquisition in real Time for coastal Ecosystems Research and Services), in the framework of the European Union Programme Life '96. The system consisted of the semi-continuous acquisition of physico-chemical variables in the water column by means of multiparametric probes installed on boats that travel daily and systematically over the whole canal network. In 2001-2, the system was employed, on the behalf of Magistrato alle Acque – Consorzio Venezia Nuova, to monitor the major canals of both the city network and the surrounding lagoon. A preliminary investigation on the levels of contamination in the water column was also performed, by analysing total heavy metals (Fe, Mn, Pb, Cu, Zn, Hg), nitrogen and phosphorus species and microbiological parameters. Additional investigations were carried out within an integration project, managed by CORILA for Magistrato alle Acque – Consorzio Venezia Nuova, in 2002, in order to study the effects of anthropic pollution by means of different investi-

gation approaches (Fig. 65.1). The resuspension of bottom sediments by motorboats was studied in a test canal affected by heavy traffic. The presence in the water of 'endocrine-disruptor' substances, which interfere with an organism's hormonal system (in reducing fecundity and/or fertility, in abnormally elevating plasma vitellogenin and intersex gonads) was also investigated, being related to civil and industrial waste. In particular, the detection of vitellogenin-like proteins was studied as a biomarker to assess the estrogenic effect in bivalves (*Mytilus galloprovincialis*). Further biomarker and genetic measurements in *M. galloprovincialis* were used to classify sampling stations, with regard to the presence of a specific or generic (i.e. not related to a specific contaminant) stress. Finally, another study concerning humic matter was performed, to evaluate the natural capability of the environment to remove and deactivate the effects of certain organic and inorganic pollutants.

EXPERIMENTAL DESIGN, METHODS AND RESULTS

Sediment resuspension by boat traffic

Organic and inorganic matter is delivered to the canals in the historical centre of Venice from untreated wastewaters and urban runoff. The physico-chemical conditions of the water column as well as the slack hydrodynamics favour the deposition on the canal bottoms of fine-grained sediments. These are associated with different kinds of pollutants (heavy metals, hydrocarbons and nutrients). The high availability of organic matter in the sediment promotes the metabolic activity of anaerobic bacterial pools, which induces strongly reduced conditions (Zaggia and

Flooding and Environmental Challenges for Venice and its Lagoon: State of Knowledge, ed. C. A. Fletcher and T. Spencer.
Published by Cambridge University Press, © Cambridge University Press 2005.

☐ Resuspension effects.

◇ Estrogenic substances: Site 1 - S. Giuliano; Site 2 - Fusina; Site 3 - Malamocco.

○ Biological effects of the pollution: Site A - Salute; Site B - San Tomà; Site C - Tre Archi;
    Site D - Treporti.

☆ Humic substances.

Fig. 65.1   Sampling sites in the lagoon (upper) and city (lower) of Venice.

Zonta, 1997). Due to resuspension, pollutant remo-
bilisation from the sediment anoxic environment to
the more oxidized water column is to be expected
(Gambrell *et al.*, 1980; Müllerand Schleichert, 1977;
Petersen *et al.*, 1997; Collavini *et al.*, 2000).

The aim of the study was the observation of the
resuspension processes by motorised boat traffic, and
their incidence on the concentration of some chemical species in the water. The investigation was performed in close conjunction with a concurrent
research project named ICARO (managed by Insula
S.p.A.), which was mainly focussed on the description
of the canal hydrodynamics and the study of transport and accumulation mechanisms governing the
sediment and associated pollutants (Collavini *et al.*,
2004). Two measurement surveys (18–20 December
2001 and 3–4 April 2002) were carried out in a test
canal, the Rio dei Mendicanti, which is directly connected to the lagoon at its northern end. Its southern
reach is linked with the Grand Canal via a series of
inner canals. A main measurement section (14 m wide
and 3 m deep) was selected, where time series of the
hydrodynamic parameters (current speed and direction) and physical-chemical variables (salinity, temperature, dissolved oxygen, pH, redox potential and

turbidity) were acquired. Additional measurements
were performed along the vertical profile and water
samples were collected to determine the concentration of suspended particulate matter (SPM), heavy
metals in total and dissolved phase (Fe, Mn, Zn, Cu,
Pb), and dissolved species of nitrogen and phosphorus. During each field survey, the samples for SPM
analyses were collected every 30 minutes for 48 hours;
for metals and nutrients, sampling was performed
hourly over a 24-hour period.

The first field surveys occurred under spring tide
conditions. The average SPM concentration was consistent with typical winter values (10-20 mg l[-1]), when
the presence of phytoplankton is at its minimum. The
occurrence of SPM concentration peaks were mainly
related to the intensity of boat traffic and to the low
water level in the canal. A direct relationship between
SPM concentration and current velocity was not
observed under medium to high tide conditions,
because the current field in the canal is generally too
weak to cause a significant resuspension of bottom
material. Four resuspension events were distinguishable (Fig. 65.2). Two were in correspondence with the
diurnal high tide (labelled 'H'), whereas the other
two occurred in the high current speed transient

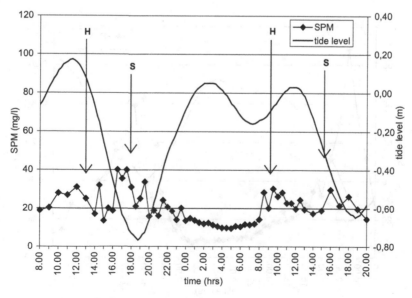

Fig. 65.2   SPM concentration during the first survey (18–19 December 2001). Four resuspension events (H, S) were
identified (see text for discussion).

(labelled 'S'), which characterises the late ebb phase of the spring tide. In the first case, the resuspension was induced by intense boat traffic occurring in the morning period (0900 – 1300); these effects were evident despite the high water level. By comparison, in the second case, the increased SPM concentration was caused by the current shear-stress, which was enhanced by the decrease of the water level.

In the second survey, which was undertaken under neap tide conditions, the current field in the canal was very weak. Here minimum water levels also corresponded to the period of greater traffic. This characteristic enhanced the phenomena that cause sediment resuspension, leading to SPM concentration values higher than 100 mg l^{-1} (Fig. 65.3). In the afternoon, when the boat traffic gradually diminished and the water level increased, the concentration dropped to the typical base values (10–20 mg l^{-1}) measured in the previous monitoring survey.

The behaviour of total heavy metals during the two field surveys substantially followed the trend of the SPM. Species that are present in the water at lower concentrations (Zn, Cu and Pb) were affected by greater time variability, as well as the dissolved fraction of all the analysed metals. An example of the trend of heavy metals is given in Figs. 65.4–65.5, which show total and dissolved concentrations of Cu in the two surveys. The comparison with SPM trend shows that the higher values of total concentrations for both metals correspond to the periods characterized by sediment resuspension. In the second survey metal concentrations were generally higher but a clear rise of metals in the dissolved fraction related to resuspension events was not observed. This suggests that metal release from particulate matter which is mobilized from the canal bottom is not effective. As observed by studying the effects of wet dredging of a test canal (Collavini *et al.*, 2000), resuspended matter undergoes fast re-deposition. As a consequence, the residence time in the water column may not be sufficient to permit the oxidation of the reduced compounds, and eventual metal release.

Unlike total heavy metals, dissolved species of nitrogen and phosphorus did not show dependence on resuspension events. These compounds are mainly delivered to the canals by the direct discharge of untreated wastewater, which is particularly relevant in the inner network. Therefore, the concentration level of these species can reflect the origin of water masses. Ammonia and orthophosphate, in particular,

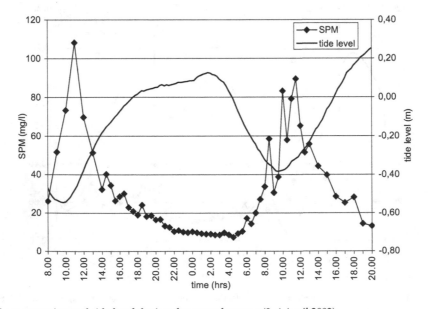

Fig. 65.3   SPM concentration and tide level during the second survey (3–4 April 2002).

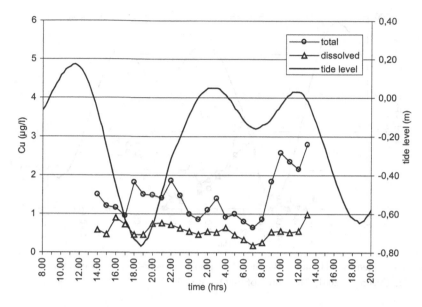

Fig. 65.4   Total and dissolved Cu concentrations and tide level during the first survey (18–19 December 2001).

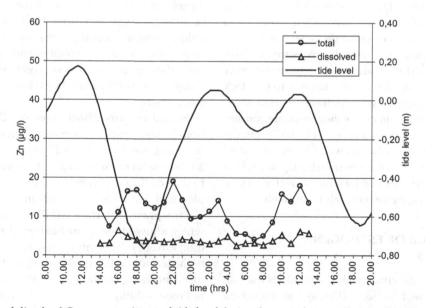

Fig. 65.5   Total and dissolved Cu concentrations and tide level during the second survey (3–4 April 2002).

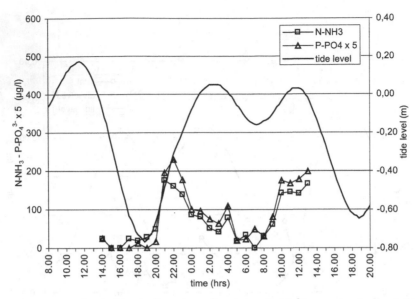

Fig. 65.6   Ammonia nitrogen (N-NH$_3$) and orthophosphate phosphorus (P-PO$_4^{3-}$) concentrations and tide level during the first survey (18–19 December 2001). The values of P-PO$_4^{3-}$ are ×5.

are useful in the determination of water circulation, as well as in the first assessment of water quality in the investigated system. During the two field surveys, the concentration of these two species (Figs. 65.6, 65.7) rapidly increased in correspondence with the beginning of the flood phase, when the measurement section was affected by water coming from the inner canals. Under neap tide conditions, due to the slack dynamics, the effects of wastewater effluents in determining high ammonia and orthophosphate concentrations are particularly evident (Fig. 65.7). On the contrary, after the flow reversal, values decreased because of the transit of water from the lagoon, which is generally characterized by a higher water quality and, in particular, by lower levels of nutrients.

OCCURRENCE OF ESTROGENIC SUBSTANCES

The environmental concern about chemicals that can alter endocrine functions (EDCs) has increased remarkably over recent years due to the finding of EDCs in the aquatic environment and the potential hazard of these chemicals to both aquatic (Körner, 2001) and human life (Commission of the European Communities, 1999). Reduced fecundity and/or fertility, abnormally elevated levels of plasma vitellogenin and intersex gonads are the most commonly observed effects produced by EDCs on aquatic wildlife species (Jobling, 1998). All these effects can cause an increase in the feminisation process that is very dangerous for the survival of species and ecosystems, as they may ultimately lead to a loss of biodiversity.

Spatial and time distributions of EDCs concentrations in the Venice Lagoon were determined in four sampling sessions (from December 2001 to May 2002) by solid phase extraction followed by high performance liquid chromatography separation coupled with ion trap-mass spectrometry detection via electrospray interface (SPE-HPLC-ESI-IT-MS), which allowed the identification of natural (estradiol, estrone) and synthetic estrogenic compounds, both steroidal (ethynylestradiol) and non-steroidal (benzophenone, bisphenol-A, nonylphenol, nonylphenol monoethoxylate carboxylate) (Pojana *et al.*, in press a, b). The sampling stations were selected in order to evaluate the possible main EDC sources for the Lagoon of Venice. Site 1 (S. Giuliano, Fig. 65.1) was located near the mouth of the Osellino river,

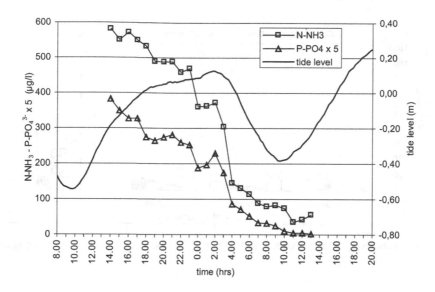

Fig. 65.7   Ammonia nitrogen (N-NH$_3$) and orthophosphate phosphorus (P-PO$_4^{3-}$) concentrations and tide level during the second survey (3–4 April 2002). The values of P-PO$_4^{3-}$ are ×5.

which carries both treated and untreated sewage from the mainland, as well as agricultural runoff. Site 2 (Fusina, Fig. 65.1) was located near the outlet of a large sewage treatment plant (c. 320 000 equivalent inhabitants), receiving both municipal and industrial sewage. Site 3 (Malamocco, c. 4 m deep, Fig. 65.1), located close to the Lido island, is affected by municipal sewage from the nearby residential area. The remaining 22 stations (average depth: 1 m), located in the inner canals of Venice, receive municipal untreated sewage from the historical centre of Venice.

The overall spatial distribution of EDC water concentrations clearly indicated that all selected EDCs are widespread contaminants of the Venice Lagoon waters. Moreover, both municipal and industrial sources were found to contribute significantly to overall EDC contamination, which appears to be redistributed inside the central lagoon by hydrodynamic processes. While synthetic non-steroidal analytes were recorded in the 1–1040 ng l⁻¹ range, steroidal EDCs (estradiol, ethynylestradiol) concentrations were lower, at 1–125 ng l⁻¹. The measured individual EDC concentrations permit the assessment of the environmental exposure of the Venice Lagoon waters to EDCs in terms of possible endocrine effects (e.g., estrogenic potency). This can be done by using estradiol equivalents (EEQs), the EDC concentrations expressed as estradiol, calculated by estradiol equivalency factors (EEFs) reported in literature (Gutendorf and Westendorf, 2001; Körner et al., 1999; Lagler, 2001; Körner et al., 2000; Thomas et al., 2001; Routledge and Sumpter, 1996; Lagler et al., 2002). The average EEQs in the four sampling sessions for the selected stations are presented (Fig. 65.8). Synthetic steroids (ethynylestradiol) accounted for a mean 77 per cent of total EEQ values, while natural steroids accounted for the remaining 23 per cent (with estradiol accounting for c. 22 per cent). Although synthetic non-steroidal EDCs were systematically found in the lagoon waters, and displayed remarkable peak concentrations, they only accounted for a negligible portion of the total EEQ (<0.1 per cent) because of the low EEFs. By comparing literature effect concentrations and measured EEQs, the Venice Lagoon waters have a significant capability for inducing effects on the endocrine system of aquatic organisms (Harries et al., 1997; Seki et al., 2002).

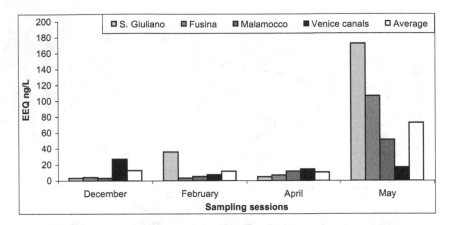

Fig. 65.8   Average estradiol equivalent concentrations (EEQs, ng l⁻¹) found at the examined sites during the four sampling sessions.

BIOLOGICAL EFFECTS OF POLLUTION: MONITORING OF ESTROGENIC COMPOUND EFFECTS ON ORGANISMS

In this study, a spatial and temporal survey at three sites (site A: Salute, site B: San Tomà, site C: Tre Archi; Fig. 65.1) located in the 'canals' of the Venice historical centre and at a reference site (site D: Treporti; Fig. 65.1) was undertaken to evaluate estrogenic and general stress effects on mussels (*Mytilus galloprovicialis*). Mussels were sampled in the Venice urban area, where raw sewage is discharged directly into the water without treatment.

Vitellogenin (Vg) is a major energy-rich lipophosphoprotein for developing embryos in vertebrates and invertebrates. In bivalves, it is suggested that Vg-like protein is synthesized within the female gonads under the control of the estrogen receptor pathway, and then transported via hemolymph to the oocytes. In contrast, it is absent (or present at very low level) in the male individual, and its production is stimulated by the presence of estrogenic compounds. At present it has not been developed as a biomarker of estrogenic effects on the lagoon mussel species *Mytilus galloprovincialis* but there is information about other bivalves, especially clams and other mussel species. The alkali-labile phosphate (ALP) assay is considered to be an indirect method for studying the

effect/presence of estrogenic compounds in vertebrates and invertebrates (such as bivalves), and it has already been used in some studies (Blaise *et al.*,1999; Gagné and Blaise, 2000; Gagné *et al.* 2001a, 2001b, 2002, 2003). In this study the alkaline-labile phosphate (ALP) assay was proposed as an indirect method for the detection of Vg-like proteins in the hemolymph of mussels. According to this procedure, ether-extracted lipophosphoproteins are subjected to an alkali treatment in order to release labile phosphates. Finally, the level of total phosphates is determined by a colorimetric phosphomolybdenum method. This method has been shown to be significantly correlated with other assays, such as immunoassays for trout Vg and the presence of Vg bodies in oocytes (Blaise *et al.*, 1999). In addition, the general health condition of the mussels was evaluated by means of a battery of biomarkers: metallothionein (MT) content following the method of Viarengo *et al.* (1997); micronucleus (MN) formation by the procedure of Bolognesi *et al.* (1996); condition index (CI) according to Lucas and Beninger (1985); and survival in air (SOS) as described in Eertman *et al.* (1993).

The results showed a significant increase of ALP levels in male mussels collected in the urban area during the reproductive period (April 2002), in comparison with the ones from the reference site. In the

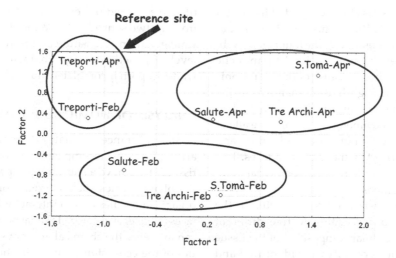

Fig. 65.9   The Principal Component Analysis (PCA) of data (biomarkers) produced a two dimensional pattern which explained 73 per cent of the total variance. Spatial separation of sampling sites is shown. The meaning of the three circles: one is related to the separation of the reference site from the urban area sites; the other two show the separation between seasons (February and April samplings).

female animals, collected in the same period in the urban area, the ALP values increased, but not significantly. The presence of chemicals with estrogenic effect was probably able to induce the expression of the Vg-like protein in the male, while in the female, where this protein may normally be expressed during the gametogenesis, the parallel process could be masked. Moreover in April, the total protein content was similar in all the samples. This could indicate the presence of chemical compounds able to increase the phosphorilation state of proteins involved in the vitellogenesis. A general impairment of health condition was also confirmed in mussels collected in the urban area by the other biomarker results (MT, MN, CI and SOS). The Principal Component Analysis (PCA), (Fig. 65.9), performed on all biomarker data clearly distinguished the Venice historical centre sites from the reference one, evidencing a decreasing stress gradient from S. Tomà>Tre Archi>Salute>Treporti.

BIOLOGICAL EFFECTS OF POLLUTION: BIOLOGICAL MONITORING THROUGH GENETIC DAMAGE MARKERS

The blue mussel (*Mytilus galloprovincialis*) from four sites in the Venice Lagoon, i.e. three inside (site A: Salute, site B: San Tomà, site C: Tre Archi. Fig. 65.1) and one outside (site D: Treporti; Fig. 65.1) the urban area, was investigated through electrophoretic analysis of proteins and flow cytometric assay of haemolymph. The electrophoretic analysis was performed on ten polymorphic enzymatic loci, extracted from the adductor muscle and the digestive gland. Statistic elaboration of the data allowed the estimation of: (1) the actual heterozygosity (the estimate of the genetic variability); (2) the Hardy-Weinberg equilibrium (the theoretical distribution of gene frequencies given by the expansion of a squared binomial); (3) the fixation indexes indicating the levels of inbreeding, the genetic drift and the differentiation among populations (Wright, 1978; Nei, 1987); and (4) the genetic distance (Nei, 1987).

A heterozygosity lower than the one expected according to the Hardy-Weinberg equilibrium was found in the samples (as already reported in mussel genus by Koehn *et al.*, 1976). This condition should have been homogeneous among the four lagoon sub-populations as they are supposed to be panmictic and continuously mixed because of the planktonic larval stage and the high hydrodynamism of the lagoon basin. On the contrary, differences in the heterozigosity were found among the loci and the alleles depending on the station under consideration and this was also confirmed by the fixation indexes. This phenomenon suggests different environmental influences at the four stations.

Flow cytometry was used to characterize the mussel haemolymph according to five different parameters: rough cellular composition of the tissue (two sub-populations of cells were identified and assumed to represent ialinocytes and granulocytes); nuclear DNA content in haemocytes; coefficient of variation in DNA content in normal diploid cells, frequency and type of abnormal distribution of nuclear DNA values. Although with different frequencies and intensity, altered patterns of the haemolymph cytometric parameters were found in the individuals coming from all the analysed stations and were presumably induced by the genotoxic action of the environment. Concerning the three urban stations, Salute was characterized by the lowest variation in parameters resembling a condition close to a 'normal' one. S. Tomà and Tre Archi followed Salute in the ranking.

The position of Treporti, initially chosen as the 'reference site', was instead considered as an outgroup site. In fact, the cytometric parameters in the samples of this station generally showed a degree of variation higher than that one of Salute. Furthermore, and conversely from what happened in the three urban sites, at Treporti ialinocyte-like rather than granulocyte-like haemocytes seem involved in the reaction to environmental stress. The hydrodynamics of the lagoon basin, and previous results coming from investigations on environmental genetic damage (Battaglia *et al.*, 1980: Libertini, 2000), suggests the existence of two different origins of pollutant in the Venice Lagoon: industrial wastes from the Marghera harbour (affecting mainly Tre Archi and to a lesser extent Salute) and urban wastes (acting on Tre Archi and S. Tomà). Thus the reported ranking fits the expected pattern of diffusion of pollutants from the two sources in the Venice Lagoon. The most similar stations are Salute and S. Tomà, the highest level of genetic damage characterizes Tre Archi, whilst Treporti is completely outside the urban effect.

ANALYSIS OF HUMIC SUBSTANCES

Humic substances are natural organic compounds arising from the decomposition of plant and animal tissues. They have a substantial capacity to complex dissolved species such as metal ions and cationic organic molecules and to interact with mineral surfaces (Haitzer *et al.*, 1998; Takahashi *et al.*, 2002). They can influence the chemical and physical characteristics of the ecosystem, can act in the transport and removal of both water-soluble and -insoluble species and can control the bioavailability of metals acting as nutrients or toxicants.

The objective of this research was to study the nature and the concentration of humic compounds, both dissolved and in particulate form, in channel water in the historic centre of Venice, considering also the amount of particulate, its organic and inorganic fractions and the ratio between organic fraction and humic compounds present in particulate. In this way it should be possible to evaluate the role that humic compounds have in this complex environment. In order to study the dependence on seasonality, three sets of superficial water samples were collected (23 November 2001, 4 February 2002, 7 May 2002) in twenty stations placed in the area of urban pollution and representative of a large area (Fig. 65.1). Each sample, divided into two sub-samples and utilized for humic acid recovery, had a volume of 20 litres, while in order to study particulate composition we collected two 1-litre samples. All samples were analysed within one week of collection. The humic and fulvic acids, both dissolved and present in particulate forms, were obtained according to the Malcolm and Thurman procedure (Thurman and Malcolm, 1981) and characterized using spectroscopic and thermogravimetric techniques. Elemental analysis was also performed. The amount of particulate matter was determined by filtering water samples on pre-weighed cellulose-acetate filters. The filters were washed with 500 ml of deionized water to remove salts; the filters were then dried on silica gel and re-weighed. Total suspended material (TSM) was determined by the weight change of the filters. Organic and inorganic fractions in particulate were obtained by putting pre-weighed porcelain crucibles containing the filters and the particulate matter in a muffle furnace at 700 °C for three hours. The increase in weight of the crucibles is reported as mineral suspended material (MSM), because the filter and organic matter are lost during the firing. The difference between TSM and MSM is reported as organic suspended material (OSM).

Humic compounds present in water, both dissolved and particulate, consisted exclusively of fulvic acids. The concentration of dissolved fulvic acids ranged between 0.2 to 0.8 mg l^{-1}. The ratios between signals (FTIR, C13-NMR) characterizing structural units of fulvic acids of various samples differed, underlining different humification levels. Moreover, the higher aliphatic carbon region in comparison with the aromatic ones, and the absence of ramified chains, indicates that they represent young materials. Consequently, they were rich in some functional groups that are responsible for metal complexation.

Particulate composition was variable and seemed to be influenced both by the nature of the channels and by the season. The particulate organic matter concentration was found to be constant in time. Nevertheless the composition of particulate matter (inorganic and organic fractions) changed in time and in February the smallest percentage of organic matter was observed, pointing to an enrichment of inorganic fraction in particulate matter. The fulvic acids present in particulate are less homogeneous than the dissolved ones; moreover the humification rate of organic matter is variable. Generally values of the first sampling were the highest, indicating that in summer the humification process is faster because of higher temperatures. When the humification rate is high, the fraction of molecules having high molecular weight increases. In this case, it may be hypothesized that the humified process is more advanced and/or that the formerly humified material present in the water is adsorbed onto the particulate materials. The high number of functional groups present in the structures of fulvic acids, both dissolved and particulate, make these compounds able to bind metal ions. We found an ash content of fulvic acids, both dissolved and particulate,

Table 65.1 Metal concentration (µg l^{-1}) bound in stable forms to fulvic acids.[1]

	St.5 (3°)	St.7 (1°)	St.7 (2°)	St.7 (3°)	St.12 (3°)	St.15 (1°)	St.15 (2°)	St.15 (3°)
Ni	0.034	0.13	0.067	0.099	0.24	0.38	0.085	0.40
Cr	0.034	0.043	0.074	0.10	0.13	0.20	0.042	0.35
Pb	0.11	0.040	0.082	0.034	0.12	0.089	0.046	0.29
Cu	0.17	0.25	0.16	0.38	0.55	0.78	0.18	0.96
Zn	0.23	0.92	0.29	0.36	1.18	0.89	0.96	1.61
Ti	0.090	0.19	0.28	0.49	0.62	0.79	0.27	1.98
Mn	0.019	0.033	0.085	0.15	0.11	0.042	0.043	0.36
V	0.011	0.0078	0.034	0.051	0.063	0.039	0.026	0.26
Cd	0.0009	0.0012	0.0011	0.0014	0.0030	0.0040	0.0029	0.0096
Mo	0.058	0.015	0.097	0.15	0.046	0.12	0.14	0.12
As	0.0026	0.0042	0.012	0.052	0.013	0.0098	0.012	0.044
Sn	0.017	0.024	0.031	0.090	0.056	0.24	0.062	
Co	0.0032	0.0036	0.0066	0.0089	0.0085	0.013	0.006	

[1] Relative standard deviations are less than 3%.

Table 65.2   *Correlation matrix: significant coefficients (p<0.050) are in bold.*

	PFA	Cr	Pb	Cu	Ti	Cd	Ni	Zn	Mn	V	Mo	As	Sn	Co
PFA	1.00													
Cr	**0.92**	1.00												
Pb	**0.83**	**0.85**	1.00											
Cu	**0.82**	**0.95**	**0.73**	1.00										
Ti	**0.97**	**0.98**	**0.88**	**0.91**	1.00									
Cd	**0.93**	**0.94**	**0.88**	**0.86**	**0.97**	1.00								
Ni	**0.74**	**0.90**	0.66	**0.97**	**0.84**	**0.83**	1.00							
Zn	**0.72**	**0.71**	0.63	**0.73**	**0.75**	**0.83**	**0.77**	1.00						
Mn	**0.93**	**0.85**	**0.82**	0.70	**0.91**	**0.84**	0.57	0.59	1.00					
V	**0.97**	**0.85**	**0.86**	**0.73**	**0.92**	**0.89**	0.67	**0.76**	**0.92**	1.00				
Mo	0.31	0.33	0.07	0.25	0.33	0.32	0.16	-0.02	0.35	0.10	1.00			
As	0.67	0.55	0.35	0.46	0.61	0.47	0.28	0.20	**0.79**	0.57	0.63	1.00		
Sn	-0.14	0.13	-0.30	0.33	-0.02	-0.04	0.41	-0.05	-0.31	-0.34	0.36	-0.07	1.00	
Co	-0.42	-0.15	-0.54	0.05	-0.29	-0.35	0.11	-0.26	-0.45	-0.58	0.27	-0.11	**0.87**	1.00

ranging between 15 and 20 per cent. Ash nature is not attributable to silica, as evidenced by the FTIR spectra of fulvic acids. Preliminary analyses carried out on some fulvic acid samples have shown the presence of metals (Ni, Cr, Pb, Cu, Zn, Ti, Mn, V, Cd, Mo, As, Sn, Co) bound in stable form to fulvic acid structures. Their concentration depends on the nature of metals and changes both with the sampling station and with sampling time.

Generally a correlation between metal concentration (μg l^{-1}) and fulvic acid concentration (μg l^{-1}) in particulate matter was observed, indicating the scavenging role of humic compounds. Moreover some metal concentrations are cross-correlated, suggesting a common origin.

CONCLUSIONS

The results of the study on resuspension indicate that the mobilization of sediment induced by boat traffic is an effective process in the transport of materials within the canal network of Venice. It is also responsible for the resuspension of considerable amounts of pollutants associated with the particulate fraction. On the contrary, there is no evidence of a significant increase of metals in the dissolved phase during resuspension events, suggesting the scarce importance of release processes from the sediment to the water column. This is also true for nitrogen and phosphorus species. Ammonia and orthophosphate levels are useful parameters, both in water quality assessments and in tracing water circulation patterns in the system.

Observed levels of estrogenic substances do not show significant differences between sites. Even if the mean values are low, the results underline the presence of significant peaks of concentrations in the waters of Venice and adjacent areas of the lagoon. This presence in the environment could be related to the significant alteration of the production of the Vg-like protein measured in the male mussel *Mytilus galloprovincialis*, while in the female, where this protein may normally be expressed during the gametogenesis, the parallel process could be masked. The level of stress shown by biomarkers and genetic measurements decreases from the stations of the historical centre to the outer sites of the lagoon.

The humic substances analysed on the particulate component display a wide concentration range, related to different parameters. The portion with lower molecular weight (younger humic material) is predominant. The analysis confirms the *scavenging* role of humic compounds, which is strongly dependent on the metal under consideration.

ACKNOWLEDGEMENTS

Measurements and the evaluations were performed by different groups lead by the following people: resuspension effects: F. Collavini, L. Zaggia, R. Zonta (ISMAR-CNR); estrogenic substances: A. Marcomini, G. Pojana (University of Venice); biological effects of pollution: C. Nasci, L. Da Ros, D. Pampanin, R. Trisolini, M. Cervelli, A. Libertini (ISMAR-CNR); and humic substances: B. M. Petronio, N. Calace, M. Pietroletti (University of Rome). The authors thank Ecotema srl for their support in sampling.

REFERENCES

Battaglia, B., Bisol P., and E. Rodinò. 1980. Experimental studies on some genetic effects of marine pollution. *Helgoländer Meeresunters* **33**, 587–95.

Blaise, C., Gagné, F., Pellerin, J., and P. D. Hansen. 1999. Determination of vitellogenin-like properties in Mya arenaria hemolymph (Saguenay Fjord, Canada): a potential biomarker for endocrine disruption. *Environ. Toxicol.* **14**, 455–65.

Bolognesi, C., Rabboni, R., and P. Roggieri. 1996. Genotoxicity biomarkers in M. Galloprovincialis as indicators of marine pollutants. *Comp. Biochem. Physiol.* **113C**, 2, 319–23.

Collavini, F., Zonta, R., Arizzi Novelli, A., and L. Zaggia. 2000. Heavy metals behaviour during resuspension of the contaminated anoxic sludge of the Venice canals. *Toxicological and Environmental Chemistry* **77**, 171–87.

Collavini, F., Coraci, E., Zaggia, L., Zaupa, A., Zonta, R., and A. Zuliani. 2004. Environmental characterisation of the Venice canals: first results of the ICARO project. In Proceedings of the 37th Congress of CIESM, Barcelona, Spain, 7–11 June, p. 182.

Commission of the European Communities. 1999. *Communication from the Commission to the Council and the European Parliament.* COM (1999) 706 final. Brussels.

Eertman, R. H. M., Wagenvoort, A. J., Hummel H., and A. C. Smaal. 1993. Survival in air of the blue mussel Mytilus edulis L. as a sensitive response to pollution-induced environmental stress. *J. Exp. Mar. Biol. Ecol.* **170**, 179–95.

Gagné, F., and C. Blaise. 2000. Organic alkali-labila phosphates in biological materials: a generic assay to detect vitellogenin in biological tissue. *Environ. Toxicol., Technical methods,* 243–7.

Gagné, F., Blaise, C., Lachance, B., Sunuhara, G., and H. Sabik. 2001a. Evidence of coprostanol estrogenicity to the freshwater mussel Elliptio complanata. *Environ. Poll* **115**, 97–106.

Gagné, F., Blaise, C., Salazar, M., Salazar, S., and P. D. Hansen. 2001b. Evaluation of estrogenic effects of municipal effluents to the freshwater mussel Elliptio complanata. *Comp. Biochem. Physiol.* **128C**, 213–25.

Gagné, F., Blaise, C., Pellerin, J., and S. Gauthier-Clerc. 2002. Alteration of the biochemical properties of female gonads and vitellins in the clam Mya arenaria at contaminated sites in the Saguenay Fjord. *Mar. Env. Res.* **53**, 295–310.

Gagné, F., Blaise, C., Pellerin, J., Pelletier, E., Douville, M., Gauthier-Clerc, S., and L. Viglino. 2003. Sex alteration in soft-shell clams (Mya arenaria) in an intertidal zone of the Saint Lawrence river (Quebec, Canada). *Comp. Biochem. Physiol.* **134C**, 189–98.

Gambrell, R. P., Rashid, A. K., and W. H. Patrick. 1980. Chemical availability of mercury, lead, and zinc in Mobile Bay sediment suspensions as affected by pH and oxidation-reduction conditions. *Environ. Sci. Technol.* **14**, 431–6.

Gutendorf, B., and J. Westendorf. 2001. Comparison of an array of in vitro assay for the assessment of estrogenic potential of natural and synthetic estrogens, phytoestrogens and xenoestrogens. *Toxicology* **166**, 79–89.

Haitzer, M., Höss, S., and W. Traunspurger. 1998. *Chemosphere* **37**, 1335.

Harries, J. L., Sheahan, D. A., Jobling, S., Matthiessen, P., Neall, P., Sumpter, J. P., Tylor, T., and N. Zaman. 1997. Estrogenic activity in five United Kingdom rivers detected by measurement of vitellogenesis in caged male trout. *Environ. Toxicol. Chem.* **16**, 534–42.

Jobling, S., Nolan, M., Tyler, C. R., Brighty, G., and J. P. Sumpter. 1998. Widespread sexual disruption in wild fish. *Environ. Sci. Technol.* **32**, 2498–506.

Koehn, R. W., Milkman R., and J. B. Mitton. 1976. Population genetics of marine pelecypods. IV. Selection, migration and genetic differentiation in the blue mussel Mytilus edulis. *Evolution* **30**, 2–32.

Körner, W., Hanf, V., Schuller, W., Kempter, C., Metzger, J., and H. Hagenmaier. 1999. Development of a sensitive E-screen assay for quantitative analysis of estrogenic activity in municipal sewage plant effluents. *The Science of the Total Environment* **225**, 33–48.

Körner, W., Bolz, U., Sïumuth, W., Hiller, G., Schuller, W., Hanf, V., and H. Hagenmaier. 2000. Input/output balance of estrogenic active compounds in a major municipal sewage plant in Germany. *Chemosphere* **40**, 1131–42.

Körner, W., Spengler, P., Bolz, U., Schuller, W., Hanf, V., and J. W. Metzger. 2001. Substances with estrogenic activity in effluents of sewage treatment plants in southwestern Germany. 2. Biological analysis. *Environ. Toxicol. Chem.* **20**, 2142–51.

Lagler, J. 2001. Development and application of in vitro and in vivo reporter gene assays for the assessment of (xeno-) estrogenic compounds in the aquatic environment. Unpublished Thesis, Wageningen University, Netherlands.

Lagler, J., Dennekamp, M., Vethaak, A. D., Brouwer, A., Koeman, J. H., Van der Burg, B., and A.J. Murk. 2002. Detection of estrogenic activity in sediment-associated compounds using in vitro reporter gene assay. T*he Science of the Total Environment* **293**, 69–83.

Libertini, A. 2000. *Individuazione di anomalie genomiche in organismi dell'ecosistema lagunare, per mezzo della citofluorimetria in flusso.* Final report of the Project 'Orizzonte 2023'. Venice: Ministero dei Lavori Pubblici – Magistrato alle Acque. Venezia

Lucas, A., and P. G. Beninger. 1985. The use of physiological condition indices in marine bivalve aquaculture. *Aquaculture* **44**, 187–200.

Müller, D., and U. Schleichert. 1977. Release of oxygen-consuming and toxic substances from anaerobic sediments by whirling-up and aeration. In H. L. Golterman (ed.), *Proc. Interactions Between Sediments and Freshwater*, Intern. Symposium, Amsterdam: Sept. 1976, pp. 415–22.

Nei, M. 1987. *Molecular Evolutionary Genetics.* New York: Columbia University Press.

Petersen, W., Willer, E., and C. Willamowski. 1997. Remobilization of trace elements from polluted anoxic sediments after resuspension in oxic water. *Water, Air and Soil Pollut.* **99**, 515–22.

Pojana, G., Busetti, F., Collarin, A., Bonfà, A., and A. Marcomini. In press a. Determination of natural and synthetic estrogenic compounds in coastal lagoon waters by HPLC-Electrospray-Ion Trap-Mass Spectrometry. *International Journal of Environmental Anal. Chem.*

Pojana, G., Bonfà, A., Busetti, F., Collarin, A., and A. Marcomini. In press b. Estrogenic potential of the Venice lagoon waters. *Environ. Toxicol. Chem.*

Routledge, E. J., and J. P. Sumpter. 1996. Estrogenic activity of surfactants and some of their degradation products assessed using a recombinant yeast screen. *Environmental Toxicology and Chemistry* **15**, 3, 241–8.

Seki, M., Yokota, H., Matsubara, H., Tsuruda, Y., Maeda, M., Tadokoro, H., and K. Kobayashi. 2002. Effect of ethinylestradiol on the reproduction and induction of vitellogenin and testis-ova in medaka (Oryzias latipes). *Environ. Toxicol. Chem.* **21**, 1692–8.

Takahashi, Y., Kimura, T., and Y. Minai. 2002. *Geochim.Cosmochim. Acta* **66**, 1.

Thomas K. V., Hurst M. R., Matthiessen, R., and M. Waldock. 2001. Characterization of estrogenic compounds in water samples collected from United Kingdom estuaries. *Environ. Toxicol. Chem.* **20**, 10, 2165–70.

Thurman, E. M., and R. L. Malcom. 1981. Preparative isolation of aquatic humic substances. *Environ. Sci. Technol.* **15**, 463.

Viarengo A., Ponzano, E., Pondero, F., and R. Fabbri. 1997. A simple spectrophotometric method for metallothionein evaluation in marine organisms: an application to Mediterranean and Antartic molluscs. *Mar. Environ. Res.* **44**, 69–84.

Wright, S. 1978. *Evolution and the Genetics of Populations.* Chicago: University of Chicago Press.

Zaggia, L., and R. Zonta. 1997. Metal-sulphides formation in the contaminated anoxic sludge of the Venice canals. *Appl. Geochem.* **12**, 527–36.

Zonta, R., Scattolin, M., Botter, M., and F. Collavini. 2004. Water quality of the Venice canal network: results of an intensive monitoring. In Proceedings of the 37th Congress of CIESM, Barcelona, Spain: 7–11 June.

Zonta, R., Cochran, J. K., Collavini, F., Costa, F., Scattolin, M., and L. Zaggia. In press. Sediment and heavy metal accumulation in a small canal, Venice, Italy. *Aquatic Ecosystem Health & Management.*

66 • Fusina integrated project: a global approach to waste water treatment and reuse in the central area of the Venice Lagoon

R. CASARIN, F. STRAZZABOSCO, P. ROSSETTO AND G. ZANOVELLO

INTRODUCTION

This chapter describes the integrated project that the Veneto Region has recently approved to regulate the municipal and industrial pollution that threatens the central area of the Venice Lagoon and to optimize water management both for drinking and non drinking purposes. The project is an important element of the 'Master Plan 2000' to reduce and control the pollution of the Venice Lagoon, regarding all sources (point and diffuse) and means of pollution transport to the lagoon, and considering all the possible sources of environmental impact: urban settlements, agriculture, industry, atmospheric deposition. This Master Plan includes state of the art techniques and technologies and represents the first example in Italy of an integrated approach to an environmental problem, designed in a proactive, instead of a reactive, framework. The general strategy for surface waters begins with preventative actions, to reduce potential loads, then moves on to reduction measures. An increase in the self purification capacity of the rivers discharging into the lagoon is planned as a final measure although, if really necessary, the diversion of the final flows from the lagoon may be considered.

The Fusina integrated project (FIP) is particularly significant because it acts not only to reduce pollution but also in respect of pollution prevention, through the reuse of discharge waters. A detailed environmental impact assessment (EIA) was undertaken to choose the best solution for the final discharge of the treated waters.

REGIONAL AND NATIONAL PLANNING TO REDUCE THE POLLUTION OF THE VENICE LAGOON

The lagoon drainage basin extends over 2038 km^2 and supports a population of more than one million inhabitants. The Master Plan 2000 aims to reduce the annual nutrient loads to control macroalgae and the risk of environmental crisis. The legally acceptable load for nitrogen is 3000 tonnes yr^{-1}, while for phosphorus at 300 tonnes yr^{-1}. Over the past ten years nutrient loads have been reduced by pollution control measures and they are set to fall further and to fulfil quality targets. The plan aims also to reduce the concentration of micropollutants in water and sediments to levels that ensure the protection of humans from adverse effects associated with the consumption of fish and shellfish.

The Fusina integrated project deals with the most densely populated part of the lagoon basin. It is based on simple concepts: use water twice; reduce sewage flow; treat all waters to higher standards; and disperse the residual outflow into the sea, away from the lagoon. The target is to dramatically reduce loads and the mass discharge of micropollutants and eutrophizing nutrient compounds in the lagoon, but with sustainable costs. Specifically, the Master Plan 2000 involves:

- extensive restoration of the old combined municipal sewage networks to reduce the spurious inflow of groundwater by 50 per cent;
- separation of industrial wastewater collection and treatment from municipal wastewater sewage and treatment;
- upsizing and renewal of the main municipal Waste Water Treatment Plant (WWTP) in Fusina

to improve reliability, in both dry and in wet seasons. Most rainwater will be stored and treated by a mixture of biological processes and chemically enhanced primary treatments and wetlands;

- WWTP effluent refinement in a huge constructed wetland, which will produce a large amount of fresh water for non-drinking reuse. This flow will substitute the current industrial withdrawal of good quality water from the Sile River which will be more properly saved and put into the main drinking water supply scheme of the Central Veneto region;
- a new pipeline to supply low cost non drinking water to the Marghera industrial zone (1000 hectares with four power plants that use large quantities of cooling water and many chemical industries that use process and washing water);
- a new sewage system dedicated to the collection and transfer of all industrial waters (currently discharged directly into the lagoon) to a centralized post-treatment plant. The industrial process waters will first be pre-treated at source then pumped to the central plant. The rainwater runoff coming from industrial areas will be stored locally then gradually sent to the central plant. The polluted infiltration waters, drained along the impermeable barriers that separate the industrial areas from the lagoon, will be sent to a dedicated section of the central plant;
- a post-treatment plant aimed at controlling and reducing residual pollution by industrial sewage flows. Due to the high variability of the inflow characteristics and to the high quality standards required, this plant will be composed of big storage tanks and many treatment devices (physical, chemical and biological). A complex pilot plant will address the choice of the best technology and the way to use it; and
- a long discharge pipe across the lagoon to disperse the residual outflow with its low chemical and organic residual levels into the Adriatic Sea.

To protect the lagoon waters, very strict legal limits (Table 66.1) and targets for nutrient compounds and micropollutants have been set by the Environment Ministry (general limits) and by the Veneto Region (after the EIA) for the Fusina WWTP.

GLOBAL IMPACT ASSESSMENT

The Environmental Impact Assessment (EIA) was also used as a tool to identify ways to improve the FIP performance. The main considerations in the EIA were:

- reduction of sewage water discharge by prevention actions and by industrial reuse of treated waters;
- centralized management and post-treatment of industrial flows, runoff and seepage groundwaters; and
- low-cost improvement in water treatment using constructed wetlands.

The EIA, developed in compliance with the requirements of the Regione Veneto (regional law 10/99) was carried out in parallel with the preliminary design of the FIP, so enabling the optimization of the design solutions, not only for what concerns the technical and economic aspects but also with respect to the preservation of the environment and landscape, and taking account of any archaeological aspects. The main optimisation aspects concerned:

1) analysis of three alternatives for the location of the residual plant discharge. Such alternatives had been already identified within the Master Plan 2000; and
2) identification and evaluation of the critical aspects of the areas located next to the lagoon border where the wetlands for the refinement of the treatment plant outflow are to be constructed.

The main study guidelines were:

- to carry out the environmental analysis, including the comparison of the different strategies identified within the Master Plan 2000 for the management of the residual discharge from the treatment plant;
- to consider the Lagoon of Venice and the neighbouring marine coastal area as a whole when assessing the environmental impact of the residual plant outflow;
- to take account of the effects of construction of the mobile barriers at the lagoon inlets;
- to consider the possible impacts generated by the construction of a discharge pipe across the lagoon

Table 66.1   *Target limits for Fusina WWTP final discharge.*

Parameter	Units	General limits		FIP limits
		Environment Ministry limits for the lagoon (DM July 1999)	Environment Ministry limits for the sea (DLgs 258/2000)	limits for the sea (after EIA 2002)
Total suspended solids (TSS)	mg l^{-1}	35	35	10
Chemical oxygen demand (COD)	mg l^{-1}	120	125	125
Biochemical oxygen demand (BOD5)	mg l^{-1}	25	25	25
Total nitrogen (N)	mg l^{-1}	5[1]	10[2]	10[2]
Total phosphorus (P)	mg l^{-1}	1	1	1
Chromium (Cr)	mg l^{-1}	0.1	2	2
Cadmium (Cd)	mg l^{-1}	1	20	5
Mercury (Hg)	mg l^{-1}	0.5	5	3
Lead (Pb)	mg l^{-1}	10	200	50
Arsenic (As)	mg l^{-1}	1	500	50
Cyanide	mg l^{-1}	5	500	5
Dioxine (TE)	mg l^{-1}	0.5		20
IPA	mg l^{-1}	1		5

[1] Average yearly limit as Master Plan 2000.
[2] The average monthly limit is 20 mg l^{-1}.

both on the lagoon ecosystem and on the archaeological sites found along its path; and

- to consider the planning, environmental, naturalistic and socio-economic aspects of the areas where the treatment wetlands are to be constructed;

In particular, for the management and siting of the treatment plant residual discharge, an EIA was carried out for three different possibilities in the Master Plan 2000. The first option considered locating the final outflow within the lagoon, next to the Malamocco inlet, so that the treated wastewaters would quickly flow out to sea. The second option considered the residual outflow being discharged into the Brenta river, and from here flowing into the Adriatic Sea just south of the lagoon. The third option considered an underground pipe across the lagoon and direct discharge into the sea, c. 10 km offshore. The discharge would have to meet the quality limits

for the sea (law 152/99) and have negligible impact – also under extreme conditions – on coastal and lagoon waters. The CRUP model, developed by the Venice Water Authority (consisting of a finite elements two-dimensional hydrodynamic model coupled to a dispersive model) was used to evaluate the impacts of the three different options on the lagoon waters. The CORMIX model, developed by the EPA, was used to evaluate the impact on the marine waters. The environmental impact analyses revealed the following:

- The first option (discharge within the lagoon) would have a positive impact on the quality of the lagoon waters and ecosystems, due to the significant abatement of the nutrient and micropollutant loads with respect to the present state, and to the location of the discharge in an area characterized by much higher water exchange rates.

However, large areas adjacent to the lagoon border would need to be converted into treatment wetlands for the refinement of the Fusina residual outflow in order to ensure that the lagoon quality limits imposed by the 'Ronchi-Costa' decrees were met;

- The second option (discharge into the Brenta river) had the main advantage of taking the residual pollutant load completely out of the lagoon, resulting in an even greater positive impact on lagoon water quality and ecosystems. It was characterized, however, by a negative impact on the Brenta waters and animal communities, which was particularly relevant during the long lasting low discharge periods which often occur in the lower river course;
- The third option (discharge into the Adriatic Sea) had the same positive impacts on the lagoon already highlighted for the previous option, while the low pollution levels of the outflow would have negligible impacts on seawater quality.

The final choice after the EIA was to take the residual Fusina WWTP outflow directly offshore into the Adriatic Sea.

TECHNICAL FEATURES OF FIP

Currently, the most important point pollution source that significantly affects lagoon water quality

is the Fusina WWTP discharge. Its sewage network covers a wide part of the lagoon watershed with 300 000 equivalent inhabitants (EI). The nitrogen load (361 tonnes N yr^{-1}) discharged by the plant is equivalent to half of the point source urban inputs to the lagoon from the catchment area and corresponds to one-third of the target reduction expected by the year 2013. The key intervention of 'Master Plan 2000' addresses the Fusina WWTP as an opportunity to build an integrated and reliable new generation sanitation system. The specific technical objectives are:

- to achieve the full control of all the local point sources of pollution (urban, industrial and runoff) with highly reliable systems and structures;
- to dramatically reduce the mass discharge of nutrients and micropollutants, not only in terms of overall concentration reduction but also through saving water; and
- to optimize fresh water management by greater re-use of treated waters.

In this context, a general water treatment and reuse plan has been devised. It is based on minimization of the residual pollution discharge to the sea; maximization of the industrial reuse of treated waters; and the saving of fresh water through reduced withdrawal from the Sile river. The water saved in this way will be used more appropriately for drinking water supply.

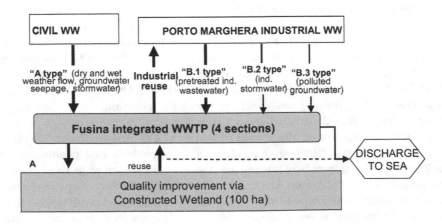

Fig. 66.1   Fusina wastewater treatment, reuse and discharge.

Components and fluxes

The scheme of the Fusina plant is straight forward (Fig. 66.2): urban sewage flow ('A') will be treated in a traditional enhanced biological plant, followed by additional polishing in a constructed free water surface (FWS) wetland. The treated water will be pumped to the industries of Porto Marghera, substituting the current supply from the Sile River. Industrial sewage water ('B') will be carried by dedicated pipes and conduits to Fusina to be further treated to comply with the sea discharge limits. This is a sort of 'water use open cycle' that does not accumulate salts and pollutants in reused waters and reduces by 50 per cent the final discharge volume.

The 'A' flow (85 000 m³ d⁻¹ in dry weather, up to 288 000 m³ d⁻¹ in wet weather) comes from urban sites that are normally drained by combined sewage networks and also carry large quantities of groundwater seepage and stormwater. The 'B' flow has three components: 'B1' (43 200 m³ d⁻¹ in winter; 57 400 m³ d⁻¹ in summer) is pretreated water from industrial processes; 'B2' (up to 17 200 m³ d⁻¹) is the polluted runoff water that will gradually be discharged by the first flush tanks, to be built in the industrial sites (the maximum number of admitted sewage overflows is five times per year for industrial areas so local storage tanks are necessary to limit rainwater treatment capacity at Fusina WWTP to a maximum of 30 per cent of average daily B1 flow); 'B3' (from 3000 to 5500 m³ d⁻¹) is the polluted groundwater flow that will be drained behind the impermeable linings of the Porto Marghera industrial canals (under construction). The key action is to centralize the management and treatment of all polluted waters of the whole area in a single, integrated system.

The biological plant

The biological plant for 'A' flows is based on a traditional biological reactor system with high redundancy in its main components. The physical pretreatment can manage up to 12 000 m³ h⁻¹ and the biological reactor can treat up to 8000 m³ h⁻¹ for long periods.

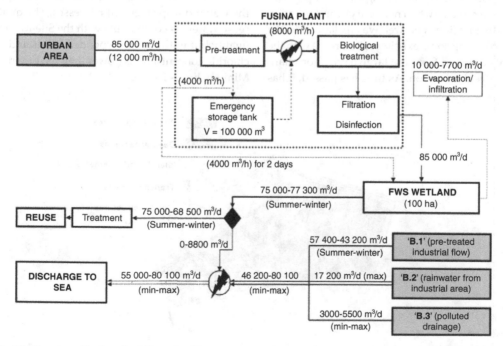

Fig. 66.2   Medium-term Fusina plant flow chart.

The latter is composed of four plug flow reactors with variable nitrification/denitrification volume (with a total volume of 69 000 m³) and the final sedimentation tanks are dimensioned for 8000 m³ h⁻¹ continuous flow. It has large storage and equalization capacities too (100 000 m³, i.e. more than one day of average flow). A chemical enhanced primary treatment (CEPT) section is dedicated to stormwater pretreatment (up to 4000 m³ h⁻¹) before overflowing into the wetland. It also has a protective filtration section and UV disinfection before treated water is sent to the quality improvement wetland (which will be a renaturalized site open to visitors). The existing anaerobic digesters will be revised to treat sludge at a higher temperature (55 °C thermophilic digesters) to increase the biogas production.

The constructed wetland

The peculiarity of 'A' flow treatment is the large constructed wetland (100 ha) which assures the reuse quality standards (Table 66.2) at sustainable post-treatment costs. It will be built in a reclaimed area close to Fusina. Further polishing treatments are planned prior to reuse, including chemical-physical corrective treatment (when necessary), sand filtration and UV disinfection. The area was reclaimed in the 1960s by covering some existing saltmarshes with the sediments dredged from the large navigation channels of the Venice Lagoon. As time has passed, it has regained some of its original natural features. At present it is mostly covered with reeds and characterized by several ecologically valuable bird species, both nesting and wintering. The design of the constructed wetland aims to create a pleasant natural area with various open water bodies and vegetated areas, functioning also as an aquatic park and recreation area. Many experiences abroad, some of which are over ten years old, have had positive results in terms of social benefits together with treatment efficiency and functionality. Wide open water bodies constitute the treatment basins and the connected areas covered by aquatic vegetation enable many waterfowl species to rest and nest. They are also usable by other vertebrate species typical of wetlands, such as reptiles and amphibians.

Reuse treatments and pipes

The treated water will be reused mainly by power station cooling systems (48 840 m³ d⁻¹ in summer) and by chemical factories in Porto Marghera (from 25 000 to 30 000 m³ d⁻¹). Hence withdrawal from the Sile River will be reduced by as much as 75 000 m³ d⁻¹. This water will meet good quality standards, due to the wetland improvement processes. The quality of reuse waters is comparable with the Sile River water quality but with lower suspended solids and higher chloride concentrations (lower than the Environment Ministry limits for reuse).

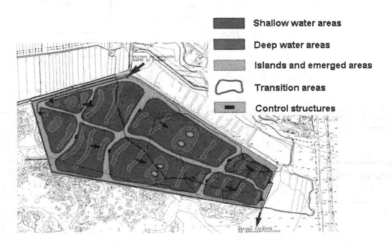

Fig. 66.3   Cassa di Colmata A – preliminary design of the constructed wetland.

Table 66.2   Quality standards for water reuse.

Parameter	Units	Environment Ministry limits	FIP reuse standards	Sile River water quality
Temperature	°C		14 – 15	14.01
pH		6 – 9.5	7 – 8	7.8
Total suspended solids (TSS)	mg l^{-1}	10	< 3	12
Chemical oxygen demand (COD)	mg l^{-1}	50	10 – 12	10.9
Biochemical oxygen demand (BOD5)	mg l^{-1}	10	2 – 5	2.5
Chloride (Cl)	mg l^{-1}	250	130 – 150	7.9
Total nitrogen (N)	mg l^{-1}	15	4 – 7	3.5
Total phosphorus (P)	mg l^{-1}		0.6	0.4
Aluminium (Al)	µg l^{-1}	1000	50 – 150	105
Cadmium (Cd)	µg l^{-1}	5	0.1 – 0.3	< 0.1
Chromium (Cr)	µg l^{-1}	100	1 – 2	0.7
Iron (Fe)	µg l^{-1}	2000	50 – 150	80
Lead (Pb)	µg l^{-1}	100	1 – 3	0.5
Mercury (Hg)	µg l^{-1}	1	0.2 – 0.3	< 0.1
Nickel (Ni)	µg l^{-1}	200	10 – 20	1.8
Copper (Cu)	µg l^{-1}	1000	5 – 7	5.4
Zinc (Zn)	µg l^{-1}	500	50 – 150	12.8
Escherichia coli	ufc/100 ml	10	< 2	300

Industrial flows post-treatment

The 'B' flows post-treatment will be calibrated by complex calibration devices. A large storage volume (75 000 m^3) will assure quality control and mixing before treatment. The planned processes are:

- post-denitrification (mainly for concentrated flows coming from cooling towers);
- flocculation/settlement; and
- filtration/microfiltration (aimed to reach a suspended solids (SS) concentration lower than 1 mg l^{-1}, to fulfil the legal limits of most micropollutants).

The process calibration device will be operated mainly for the treatment testing of polluted groundwaters that contain iron (Fe), manganese (Mn), fluoride, hydrocarbons and chlorinated solvents (HCl).

Sea discharge pipe

The selected option for the location of the final discharge of the treated plant effluent (direct discharge into the sea 10 km from the Lido shoreline, 2 km beyond the lagoon water exchange zone) is expected to have positive impacts on the lagoon. Moreover, transferring the residual pollutant load to the Adriatic Sea seems compatible with the preservation of the coastal environment. To evaluate the impacts, a model analysis was performed using the CORMIX model, developed by EPA, focussing on the abatement of the pollutant concentrations along the main dispersion direction of the discharge plume. Taking account of the seabed geometry, winds and currents, climate and the discharge location, the main modelling aims were:

- to identify the area affected by significant increases in pollutant concentrations with respect to the present conditions, paying special attention to the possible impacts on the many offshore mussel farms located along the coast and on bathing; and
- to make sure that the discharge location is far enough from the lagoon inlets to prevent the discharged treated wastewaters from coming back into the lagoon.

Different model simulations were performed reproducing the most frequent wind and water stratification conditions, including 'extreme' events. All the simulations considered an average treated wastewater discharge of 100 000 m³ d⁻¹, equal to the peak plant discharge (the medium term average discharge goes from a minimum of 55 000 m³ d⁻¹ in winter dry weather to a maximum of 80 100 m³ d⁻¹ in summer wet weather) and pollutant concentrations equal to the maximum admitted by Italian law. A pumping station located in the middle of the pipe at Malamocco will be able to regulate the discharge to the sea. The model simulations confirmed that:

- the distance of the discharge location from the coastline prevents impacts on both mussel farming and bathing and not even a small proportion of the discharged pollutants come back into the lagoon; and
- the concentration of discharged pollutants in seawater is appreciable only in a limited area adjacent to the discharge site, as a result of dilution processes providing a quick abatement of the concentrations on moving away from the discharge site.

FIP OPERATION AND ENVIRONMENTAL QUALITY MANAGEMENT

The Fusina integrated project (FIP) aims to optimize (in terms of safety, flexibility, reliability) sanitation systems and water use in the most complex and sensitive area of the lagoon basin by:

- centralized control of point pollution sources (urban, industrial, runoff);
- strong reduction of fresh water withdrawal through reuse of treated water up to 75 000 m³ d⁻¹;

- dramatic reduction of macro- and micro-pollution using advanced high efficiency techniques supported by process calibration devices; and
- minimum environmental impact of final discharge (large constructed wetland, residual outflow to the sea).

To achieve these goals detailed modelling and monitoring of the residual pollution dispersion in the Adriatic Sea is necessary. The monitoring programme for the coastal waters is being set up to be performed before, during and after the new discharge pipe is put into operation. A 'routine' activity programme will be developed together with a special programme, to be activated when the exceedance of the predefined threshold values (on water, sediment, biota) is revealed by routine monitoring. Monitoring will be supported by a mathematical model specifically developed and calibrated for local conditions; this will be able to simulate the dispersion of the treated outflow and its effects on the marine ecosystem under different wind and current conditions. The model consists of a hydrodynamic-dispersive module, a resuspension and sediment transport module, a eutrophication module and an ecotoxicology module. The use of such a model, constantly fed with field data, will enable prediction and prevention of situations potentially able to affect the trophic conditions of the recipient water body and to create an ecological risk assessment for the plant and animal communities and for human health (fish or seafood consumption).

TIMESCALE AND FINANCE NEEDED FOR THE FIP

The estimated cost of the FIP is €280 000 000. Currently the programme is financed partly (€126 000 000) by the Regional Government of the Veneto. At the moment, an official proposal for project financing made by a group of societies is under examination. If the result of this examination is positive, it will need about five years for the first new sections of FIP to become operational.

REFERENCE

'Piano per la prevenzione dell'inquinamento e il risanamento delle acque del bacino idrografico immediatamente sversante nella laguna di Venezia – Piano Direttore 2000' – Deliberazione del Consiglio regionale 1° marzo 2000, no. 24 – Bollettino Ufficiale della Regione del Veneto n. 64 – 14/07/2000.

Part VII · Venice and its Lagoon: synthesis and prospect

67 • Venice and the Venice Lagoon: communication, uncertainty and decision making in an environmentally complex system

J. DA MOSTO, T. SPENCER, C. A. FLETCHER AND P. CAMPOSTRINI

INTRODUCTION

The 'Venice problem' dates back to c. 1797, the end of the Serenissima Republic, when the State of Venice lost its independence and came under French, then Austrian and finally Italian rule. From that time, the city entered an age when it would be overshadowed by considerations of its odd and awkward aspects as a 'continental appendix' rather than a robust system, the fruit of ingenious human intervention and innovation through the ages, which permitted the lagoon and unique city to persist. The 'problem' manifests itself in the difficulty in making progress on understanding the Venice system, in arriving at decisions, and in then implementing those decisions.

In modern times, the 'Venice problem' was brought into even sharper focus by the dramatic floods of November 1966. But that event was merely a more extreme expression of the habitual problems facing the city. Issues raised in the publications which emerged in the aftermath of that disaster (e.g. Olbici, 1967; UNESCO, 1969; AAVV, 1970) are strikingly similar to the questions that this volume addresses, nearly four decades later and despite a significant research effort since that time. This apparent lack of clarity with regard to the state of Venice can be broken down into a number of separate components. Such a disaggregation is necessary in order to overcome the idea that just because Venice is unique – an extraordinary city and cultural heritage in the midst of a lagoon distinguished by its biodiversity and unusual tidal characteristics (compared with the rest of the Mediterranean) – its problems are equally so. Rather, the uniqueness of Venice really lies in the *combination* and *number* of issues that need to be considered in order to understand and manage the system. Much can be learnt in Venice from other areas of the world, with regard to specific phenomena, research techniques and remedies, just as Venice itself can share experience with other places.

Modern research techniques, ranging, for example, from lagoon-wide, repeat monitoring from satellites, to highly sensitive instrumentation for the analysis of micropollutant mobilization from point samples and mapping of genomic variability in fish (e.g. Gallini *et al.*, 2004), are helping to deepen our understanding of the system at physical, biological and chemical levels. Advances in computational power and mathematical modelling are providing support in identifying and defining cause and effect relationships between environmental parameters and facilitating the integration of various knowledge threads, ultimately for the support of decision making. However, progress in recognizing, addressing and resolving the problems of safeguarding Venice is by no means linear. At a scientific level, it must be observed that improved techniques of investigation bring with them further levels of enquiry and the recognition of ever greater complexity in the system itself. At an organizational level, the co-ordination of research activities, monitoring programmes, institutional roles and access to information is a perennial issue. And with regard to policy making, the 'polymorphism' of the city and its lagoon, given its long history of man/nature interaction, and the vital underlying natural dynamics that maintain the system, demand the wider participation of a well-informed public in decision making. A companion booklet (Fletcher and Da Mosto, 2004) to this publication has already gone some way towards filling the gap between debate amongst experts and the concerns and interest of the general public. This comes after decades of polarization generated by the debate as to whether or not to implement the MOSE barrier

Flooding and Environmental Challenges for Venice and its Lagoon: State of Knowledge, ed. C. A. Fletcher and T. Spencer.
Published by Cambridge University Press, © Cambridge University Press 2005.

system and the growth of misconceptions about the lagoon system.

In Venice especially, the public needs to be better informed about the multitude of factors determining the health of the lagoon and the preservation of the city in terms of the actual issues, the uncertainties embedded in interpreting data, and the complexities of the interrelationships between human, environmental and physical factors. Policy makers also need to be likewise informed, so that they can adopt measures which permit flexible responses in the face of the uncertainties surrounding environmental futures and changing social, cultural and political forces. In this context, the scientific community needs to participate in the simplification of their findings for the purposes of information dissemination and, in turn, collaborate with policy makers in order to ensure that the inherent complexities of the system (including those areas of uncertain knowledge) are accommodated in a scenario of long-term sustainable development (Vellinga and Lasage, chapter 68).

SYSTEM COMPLEXITY

The Venice Lagoon is a remarkably complex system, with physical, chemical and biological interactions taking place across a range of space and time scales, incorporating interactions between terrestrial and aqueous, marine and freshwater environments plus the wet/dry tidal cycles. Lagoonal environments are transitional environments – ecotones – between marine processes on the one hand and terrestrial processes on the other. Both these sets of processes have their own natural variability. When taken together, the range of possible lagoonal states and evolutionary pathways is thus extremely broad, even in a 'pristine' system.

An important physical parameter is relative water level. This is determined partly by geological subsidence, which continues as a persistent process over millennia. However, especially in the built areas, subsidence also results from the weight of buildings, relatively recent land reclamation and groundwater extraction. All these impacts have varied in their effects and persistence over the last 500 years. Other determinants of water level include eustasy (on centennial/decadal timescales) and short-term climate

patterns and weather events (on decadal and interannual timescales as well as on a daily and seasonal basis). Each of these phenomena also has a characteristic spatial signature. Thus long-term subsidence can be thought of as a lagoon-wide process but short-term events may show quite different patterns in different parts of the city and areas of the lagoon. The *bora* wind, for example, can cause a 20cm differential in water level between the northern lagoon and the town of Chioggia in the south of the lagoon. Overlying these physical processes are a whole series of ecological processes which themselves have varying time-space dynamics. There are, for example, long-term changes in salt marsh and seagrass meadow extent which appear in part to be related to changes in water level (on decadal and centennial timescales) but the system is also prone to short-term shocks, such as the anoxic crises of the 1980s and 1990s.

A particular analytical difficulty is created by the absence of a reference state for Venice given that it is both a highly altered system and a system with a long and on-going history of alteration. Thus natural system space–time variability is overlain by a whole series of human interventions which have their own rhythms, from the long-term and persistent to the large scale, sudden and recent. Man has constantly intervened in the 'functionality' of Venice, starting with the defence system offered by the lagoon's hidden channels which grounded invading enemies. This was tempered by the deviation of the main rivers away from the lagoon to abate complete siltation of the system, in an iterative and lengthy process lasting nearly three centuries. These activities were followed by construction of the jetties at the inlets in the nineteenth and early twentieth centuries, to promote natural scouring of the navigation channels for port traffic. Interventions since the latter half of the twentieth century have gone further, incorporating larger scale infrastructural requirements within the lagoon, such as extensive land reclamation for industry, housing, agriculture, airport development and the re-use of dredged materials; further excavation and alignment of the main navigation channels to service heavy industry at Marghera (which has increased tidal exchange and led to widespread erosion by tidal currents); closing off areas of the lagoon to tidal expansion for aquaculture; and the introduction of mechanical clam fishing (to the detriment of the lagoon bed ecology). As a result, it can be argued that many of the natural dynamics of the lagoon system that supported the millennial symbiosis of man and the environment of Venice have been overrun by changes made in the last few decades. The choice now seems to be between either a reliance on engineering solutions to regulate the physical relationship between the city's development and water levels, with uncertain consequences for the environment, or attempting to restore the lagoon's natural dynamics at the expense of certain economic activities (notably the scale of port traffic which, in turn, determines the dimensions of the main channels and inlets). Whatever is decided and done to manage the system, it is always in the context of an 'artificialized' system. As mentioned above, no pristine, natural reference state exists for Venice and the choice of state must, therefore, inevitably be a political one.

This unique cultural social model, Venice and its lagoon, is nonetheless dependent on its environmental heritage. Natural resources and biodiversity need to be preserved as the Venice Lagoon is the largest coastal wetland in Italy and has unique ecological characteristics in the context of the Mediterranean basin. This is mainly because it is one of a very few tidal lagoons in the region and has distinctive bioclimatic features in a transitional band between Mediterranean and Atlantic conditions (Smart and Vinals, 2004). Ornithological surveys have shown that Venice is one of the most important wetlands in Italy for wintering, migrant and breeding wildfowl. Fish species and populations are also an important factor, not least in terms of supporting the local economy and associated traditional practices, as is the rich endemic flora, especially on the saltmarshes (*barene*). At the present time, the lagoon is threatened by marked environmental degradation. Firstly the loss of two-thirds of the saltmarshes since 1900 has affected the hydro-geomorphological status of the lagoon by reducing resistance to incoming tides and the erosive forces of tidally-generated currents. In ecological terms, the *barene* and associated structures are important fish nurseries, bird feeding sites and, more generally, act as the 'lungs' of the lagoon. Water quality is variable in the lagoon, affected by pesticide

and fertilizer inputs (agricultural runoff from the catchment); atmospheric deposition of pollutants from the industrial and urbanized hinterland; remobilization and transport of heavy metal and other pollutants in lagoon sediments from former industrial activity; and urban waste disposal and sewage treatment. Tidal exchanges and biological activity (affecting the solubility and binding capacity of micropollutants, for example) also contribute to the fact that at any one point in time, the system is a snapshot of all these different impacts, taking place across a range of space and time dimensions. Some impacts can be seen as almost instantaneous but many are likely to show lagged responses. Understanding is not helped, either, by continuing interventions in the lagoon, as, for example, in the construction of 'dredge islands' in an attempt to substitute natural saltmarsh, lost through wave attack and *in situ* drowning.

With the escalating impacts of humans in the lagoon, it is a struggle to tease apart the separate components given the 'shifting baseline' problem, well known in ecology (e.g. Pauly, 1995; Jackson *et al.*, 2001). Recent monitoring initiatives notably MELa (chemical and biological features) and DRAIN regarding catchment area inputs (Zirino, chapter 55) and near-continuous current measurements at the inlets (Gačić *et al.*, chapter 49, and now being extended to measuring sediment exchanges) are beginning to show the kind of detailed temporal and spatial sampling programmes that are required to appreciate this complexity. But more time is needed to build up a picture of inter-annual variability at the lagoon-wide scale. In addition, specific and continued monitoring of areas such as the artificial marshes made from dredged material is needed so that the success of these remedial measures can be assessed against well-defined goals. Monitoring is also needed to observe the subtle changes in ecology of the lagoon arising from both natural fluctuations but also from the introduction of alien species, either intentionally or otherwise, by mankind.

KNOWLEDGE DIFFICULTIES

To facilitate the development of a sound understanding of the lagoon and its dynamics, good experimental design, careful field sampling and mathematical modelling must be combined with the development of integrated models. Modelling, in turn, needs to be based on more formal inter-comparison of models (numerous separate initiatives have been described in this volume alone) including the development of more transparent 'databanks' and increased support for model calibration and validation. The development and testing of predictive models would be particularly valuable. Furthermore, the stubborn problems of communication between different disciplines, where different researchers work in different ways and employ different 'languages', must not be overlooked.

Beyond purely scientific issues, current understanding has been hampered by a monitoring system of overlapping and fragmented institutional responsibilities between various administrations (and even among departments of a single entity), although Penna *et al.* (chapter 56) show that the situation is improving. Thus, for example, whilst pollution abatement in the lagoon watershed is the responsibility of the Regional Administration (Regione Veneto), lagoon water quality is policed by the anti-pollution service of the Venice Water Authority (Magistrato alle Acque, an agency of central government) and control of fishing activities in the lagoon is the responsibility of the Provincial Administration (Provincia di Venezia). One tide gauge network is run by a branch of central government, the Agency for the Protection of the Environment and for Technical Services, while analysis and forecasting of storm surges and issuing flood warnings is undertaken by Centro Previsioni e Segnalazioni Maree, a body of the local municipality (Comune di Venezia), which has its own tide gauges. Maintenance and dredging of the inner canals in the urban areas in and around the lagoon is under the jurisdiction of the Municipality of Venice (Comune) but the larger canals in the lagoon and its open waters are under the control of the Magistrato alle Acque. One consequence of the piecemeal institutional governance of the city and lagoon is the interruptions this often causes to the timely and wide circulation of data (studies conducted for public sector bodies often have to pass through a long approvals and audit process before being made available to the research community and general public) and studies undertaken by the Magistrato alle Acque, through the Consorzio Venezia Nuova, within the applied context of designing and planning the barrier system have not been made readily available, in spite of the fact that this information could be of considerable value to furthering scientific understanding of lagoon processes.

WHAT IS NEEDED?

Important lessons can be learned from the involvement of the international community in the preservation and restoration of Venice's architectural and artistic heritage, latterly under the umbrella of UNESCO as the 'Association of Private Committees for the safeguarding of Venice'. Their successful role in catalysing the work of the national and local institutions with direct responsibility for safeguarding heritage could be transferred to the scientific arena. An international collaboration to produce an ongoing 'audit' of the state of knowledge on the lagoon system, involving Venetian, Italian and other experts, is proposed, alongside the coordination and integration activities carried out at a local level through CORILA. This would range from assessing the coordination of research activities through to distilling and channelling final results to those that need them. Some (and only some) of the questions that might be considered by such a permanent 'forum' include: measures to facilitate the freer flow of data; model inter-comparison studies and monitoring of model predictions; capacity building for public consultation processes; and how to better link scientific research, policy-making and policy implementation by improving lines of communication between stakeholders, decision-makers and scientists. Research questions change over time. The flood of 1966, not surprisingly, provoked a major effort in the field of tide forecasting. In ecological studies, eutrophication in the late 1980s and early 1990s led to major water quality monitoring initiatives and stricter pollution control regulations for the catchment area as a result of concerns over nutrient enrichment. Now, new technical tools as well as the demands from changing social, cultural and political attitudes (notably European legislation, e.g. Habitat and Water Framework Directives) have moved attention to biogeochemical cycling and functioning at the ecosystem level (Solidoro *et al.*, chapter 57). Ultimately, the future of Venice is governed by concerns linked to global environmental change (sea level rise, surge frequency and hence flooding of the city) and the associated timescale to be considered when evaluating alternative policy options and intervention measures. Overall the Venice 'case study' could also be a challenging test ground for furthering the integration of policy making, social sciences and natural sciences in the area of public perception, risk assessment, risk management and how best to deal with scientific uncertainties, particularly those generated by near-future global environmental change (e.g. Webster *et al.*, 2001; Webster *et al.* 2002).

ACKNOWLEDGEMENTS

This chapter draws on materials generated by the Cambridge Project 'Flooding and Environmental Challenges for Venice and its Lagoon' and in particular from the reports arising from the Project workshops held in 2002 and the final discussion session at the International Meeting in 2003:

1) Creating a Forum for Discussion: Workshops Summary Report (Cambridge 9–20 September 2002). http://ccru.geog.cam.ac.uk/events/venice 2003/workshops.html
2) State of Knowledge: General Discussion: Conclusions and Way Forward. (Cambridge 14–17 September 2003). http://ccru.geog.cam.ac.uk/ events/venice2003/programme.html

We wish to record our immense gratitude to all those who participated in these discussions and helped develop our thinking.

REFERENCES

AAVV. 1970. *Concorso di idee su opere di difesa dall'aqua alta nella laguna di Venezia*. Venice: Consiglio Nazionale delle Richerche / Fondazione Giorgio Cini.

Fletcher, C. A., and J. Da Mosto. 2004. *The Science of Saving Venice*. Turin: Umberto Allemandi & C.

Gallini, A., Zane, L., and P. M. Bisol. 2004. Isolation and characterisation of microsatellites in Zosterisessor ophiocephalus. In P. Campostrini (ed.), *Scientific Research and Safeguarding of Venice CORILA Research Programme 2001–2003* Volume II. Venice: CORILA.

Jackson, J. B. C., Kirby, M. X., Berger, W. H., Bjorndal, K. A., Botsford, L. W., Bourque, B. J., Bradbury, R. H., Cooke, R., Erlandson, J., Estes, J. A., Hughes, T. P., Kidwell, S., Lange, C. B., Lenihan, H. S., Pandolfi, J. M., Peterson, C. H., Steneck, R. S., Tegner, M. J., and R. R. Warner. 2001. Historical overfishing and the recent collapse of coastal ecosystems. *Science* **293**, 629–38.

Olbici, G. 1967. *Venezia fino a quando*. Venice: Marsilio Editori.

Pauly, D. 1995. Anecdotes and the shifting baseline syndrome of fisheries. *Trends in Ecology and Evolution* **10**, 430.

Smart, M., and M. J. Vinals. 2004. *The Lagoon of Venice as a Ramsar Site*. Venice: Provincia di Venezia – Assessorato alla Caccia, Pesca e Polizia Provinciale.

UNESCO. 1969. *Rapporto su Venezia*. Vencie: Mondadori.

Webster, M., Babiker, M. H., Mayer, M., Reilly, J. M., Hamisch, J., Hyman, R., Sarofim, M. C., and C. Wang. 2001. *Uncertainty in Emissions Projections for Climate Models*. MIT Joint Program on the Science and Policy of Global Change, Report 79. Cambridge, MA: MIT.

Webster, M., Forest, C., Reilly, J., Babiker, M., Kicklighter, D., Mayer, M., Prinn, R., Sarofim, M., Sokolov, A., Stone, P., and C. Wang. 2002. *Uncertainty Analysis of Climate Change and Policy Response*. MIT Joint Program on the Science and Policy of Global Change, Report 95. Cambridge, MA: MIT.

68 • Venice, an issue of sustainability

P. VELLINGA AND R. LASAGE

SUMMARY

Conservation of monuments and protection against sea-level rise are only two of the many challenges facing Venice. Reversing demographic trends and revitalising the structure of the economy are equally relevant. In fact, the limited capability to adjust to 'external' physical and socio-political conditions is the critical issue for Venice. Externally funded measures to protect and conserve the natural and cultural heritage will not be sufficient to ensure longer-term sustainability. Imperative is the introduction of regulatory and economic instruments aimed at sustainable use of all forms of capital: natural, social, cultural and economic. Research into the socio-economic, the demographic and the political system in parallel to on-going research into the bio-geo-physical system will be necessary to support innovation aimed at a more sustainable situation for Venice and its lagoon.

INTRODUCTION

Venice and its lagoon are a source of inspiration for many of us; artists, writers, architects and historians, professionals and amateurs alike. Ecologists and engineers also enjoy the beauty and the complexity of the interaction of culture and nature that have made Venice and its lagoon what it is today. Over the years, numerous analyses and plans have been made to conserve the monuments and protect the city. However, implementation is slow and incomplete. Meanwhile the historical beauty and the ecological values continue to weaken because of sea-level rise, mass tourism and overexploitation of the lagoon.

Most of the solutions proposed and developed are technical and conservation oriented. Such plans and measures will be more successful when they are part of a more comprehensive socio-economic (intervention) approach. This chapter demonstrates that sustainability analysis provides such a wide approach. Sustainability here includes all three pillars: ecological sustainability, social sustainability and economic sustainability (United Nations, 1992).

This paper explores how the sustainability concept could help to describe and address the complexities of Venice. It starts with a discussion of major trends and developments to be expected. Next, it identifies the most relevant conditions for ecological, social and economic sustainability. Finally, it discusses a number of illustrative measures aimed at keeping up the ecological, the social and the economic capital of the city and its lagoon. Important remaining questions form an agenda for further research. Of course, all steps do require a shared understanding about trends, conditions for sustainability and the trade-offs. Sustainability as a concept can help to structure and guide the process of analysis, framing, discussion and decision-making.

TRENDS AND DEVELOPMENTS

Sea-level rise and human pressure on the water system

Both natural and anthropogenic trends of relative sea-level rise affect Venice and its lagoon. Geological analysis supported by archaeological data reveals a long-term eustatic sea-level rise in the order of 11-12 cm per century (Brambati *et al.*, 2003; Tsimplis, 2002). This rate has been relatively constant at least for the last few thousand years. However, the recent dataset from 1960 to 1994 suggests that the sea-level rise stagnated in the Northern Adriatic (Tsimplis,

Flooding and Environmental Challenges for Venice and its Lagoon: State of Knowledge, ed. C. A. Fletcher and T. Spencer. Published by Cambridge University Press, © Cambridge University Press 2005.

2002) while the rate increased up to 20 mm a year in the eastern Mediterranean. These temporal and local variations should be considered as 'noise' on the longer term, as there is insufficient evidence to interpret this observation as a new trend. The long-term eustatic sea-level rise alone accounts for an almost 1m rise in sea level since the foundations for Venice were laid in the twelfth to fourteenth centuries. There is no specific reason to assume this Mediterranean basin wide sea-level rise will not continue over the decades and centuries to come.

On top of eustatic sea-level rise, there is (regional) subsidence of the land, partly from natural geological origin and partly driven by human activities such as groundwater withdrawal. The natural geological component is relatively small when compared to the basin wide contribution; it is in the order of 4 cm per century and it varies somewhat over time (Marzocchi and Mulargia, 1996). The human component became very significant in the twentieth century. Groundwater withdrawal started in the 1930s and reached a peak between 1950 and 1970. The maximum subsidence of 17 mm yr^{-1} was recorded between 1968 and 1969. After recognising the detrimental effects, the withdrawal of groundwater was stopped and a small rebound effect of 2 cm in the historical city occurred. The anthropogenic contribution to subsidence lies around 9 cm

over the previous century (Brambati et al., 2003). There is no reason to assume that the Venetian authorities will allow for such activities in the future. The total elevation loss of the land relative to the sea level in the last century was in the order of 23 cm.

The major new phenomenon for the future is climate-induced sea-level rise. The Intergovernmental Panel on Climate Change (IPCC) of the United Nations projects an additional rise of sea level in the range of 9 to 90 cm by the end of 2100 (Church et al., 2001). The wide range has to do with the careful scientific analysis and the uncertainty about emissions of greenhouse gasses and uncertainty regarding the regional and global response of the oceans where thermal expansion will occur as a result of a global average rise in temperature. Uncertainty about the response time of the oceans is also a reason; there is a lag in the increase in temperature and the thermal expansion of the sea. Once the average global temperature has risen by, say, 2 °C the major part of the response will come in the first two to three centuries but the process of adjustment continues for many centuries thereafter. All contributions to relative sea-level rise locally for Venice are illustrated in Fig. 68.1. As the effects of climate change are still difficult to quantify precisely, the future range of sea-level rise is rather broad. Figure 68.2 presents the ranges at a

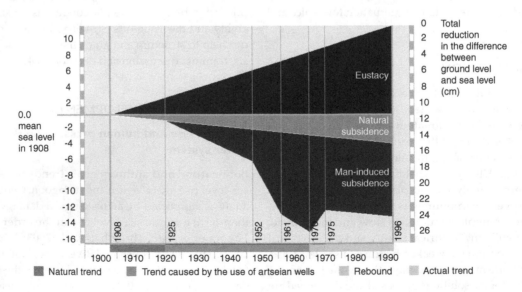

Fig. 68.1   Eustasy and subsidence (Venice Water Authority, 2002).

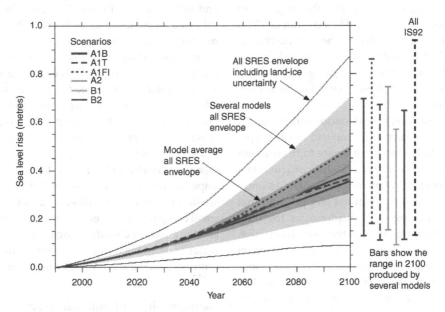

Fig. 68.2   Projected sea-level rise (Church *et al.*, 2001).

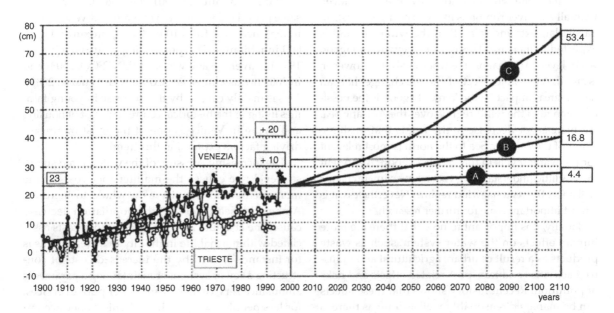

Fig. 68.3   Sea-level rise scenarios and comparison with measured levels in the twentieth century (Cecconi *et al.*, 1998).

global scale as expressed by the IPCC (Church *et al.*, 2001). It is important to realize that the longer-term regional eustatic rise in sea level in the order of 12 cm per century should be added to the climate change projections. The Committee of International Experts in their report about Venice (Collegio di esperti di livello internazionale, 1998) have used scenarios as illustrated in Fig. 68.3.

In Fig. 68.3 sea-level rise scenarios and measured levels of the last century are combined in one graph. The star marks in the record of Venice indicate recent measurements (Cecconi *et al.*, 1998). Scenario A (4.4 cm rise in the next 110 years) is based on the assumption that long-term eustatism has stopped and that the long-term geological subsidence is the only process contributing to relative sea-level rise. Scenario B (16.8 cm for the next 110 years) is based on the assumption that long-term eustatism and long-term geological subsidence will continue as they have done so over the last few hundred years. Scenario C (53.4 cm for the next 110 years) includes the mean value of additional climate change-induced sea-level rise as projected by the IPCC.

Climate change does not only affect sea level. Rainfall and river run-off is likely to change as well; in particular, extreme rainfall in the winter period is expected to increase. This can be relevant for lagoon water levels. The magnitude when analysed, however, is small. A possible effect of climate change that can have a great impact on the Venice Lagoon is the possible change in the wind regime over the Adriatic Sea. An increase or decrease in the winds that push the water of the Adriatic Sea to the north is most relevant for the water levels in the Venice Lagoon. At present there is little understanding yet on how climate change will affect the regional wind regime. Nevertheless, it is a most important issue for further research.

Finally, it is important to note that there are more human effects on the water system, such as waste products as a result of urban, agricultural and industrial activities and transport-related effects and risks of pollution. Recent history shows that these effects can be managed reasonably well as long as there is sufficient public awareness, monitoring activity and related policy response. For the future this issue is important but it can be dealt with in a way that has proved to work in many other places in the world. In particular for Venice it is important that agricultural and industrial activities and practices in and around the basin are adjusted in such a way that the environmental burden continues to decrease. To reduce the human and urban waste pressure on the lagoon and the adjacent coastal area it is most important that a sewage system is introduced in the city of Venice. The effects and risks of sea transport on the lagoon system are quite significant. It is important to study these effects in detail and to reduce the effects and the risks where possible. The transport channels affect the lagoon sediment and water system and enhance the impact of the tidal current from the sea increasing the erosion of the lagoon (Ravera, 2000). This effect is small compared to the effects of eustatic sea-level rise and the effects of climate change, giving an increase in water level of between 16 and 60 cm in the coming century (Cecconi *et al.*, 1998; Church *et al.*, 2001).

Demography and political structure

The population of the historic centre of Venice is declining, from nearly 175 000 inhabitants in 1950 to 65 500 inhabitants in 2001 (Fig. 68.4). During the same period the population of mainland Venice (terra ferma) increased from 100 000 to approximately 180 000 inhabitants. The population is also ageing; in 1991 the average age was 45.5 (UNESCO, 2003) and in 2003 it was almost 48 (Comune di Venezia, 2003). This is mainly caused by the decrease in young families living in the historical centre. The cost of housing and living in Venice is one of the most important driving factors explaining this trend.

The increasing pressure of the water as a result of sea-level rise causes the maintenance costs of the monuments and the city as a whole to increase. Along with sea-level rise, the population decrease causes the absolute and per capita public costs associated with normal life in the city to increase. This is for instance caused by the increased costs of constructing a sewer system, or maintaining a paramedic ambulance service, banks and other public services, for less people. The old city of Venice alone accommodates 70 churches, which all need resources for their maintenance. The lagoon also needs to be maintained and these costs are paid for by fewer inhabitants then in the past.

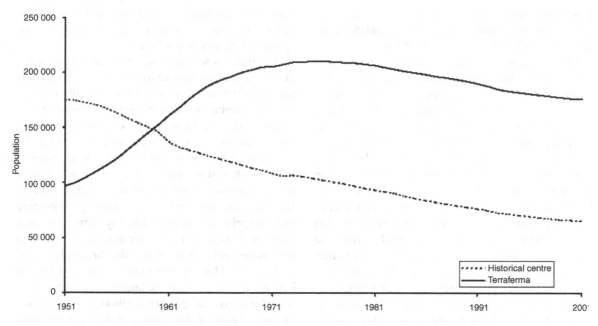

Fig. 68.4   The historical evolution of the population of Venice (Comune di Venezia, 2003).

These are clear signs of decreasing vitality. In parallel, the profile of the group of people deciding over policies and interventions in Venice and their time frame reference is becoming increasingly different from the profile of the group of people with an interest in Venice as a World Heritage Site. There are many layers of institutions in some way responsible for Venice, including four formal Italian layers of government, the EU and the UN. After the flood in 1966 three political committees were established to help the area. Together with the many dedicated committees for local, regional and interregional water management, environment, development etc., there are some 30 democratically elected committees involved in decision-making regarding policies and measures for Venice. Research in this issue has been carried out by Musu (2001) who concludes that plans for solving problems almost never become operational in a satisfactory way because of the structure of government of Venice. There is no transparency in governability, accountability and legitimacy of the different committees.

Economic structure

Employment in the Venice region has decreased from the 1980s as a result of the decline in the industrial sector, which no longer has the leading role in the economic development of the city but still employs 30% of the inhabitants (Musu, 2001). The harbour and its activities also employ 30% of the inhabitants. The impact of tourism on the Venetian economy has consistently increased and contributes to 35% of economic activities in the historical centre (Musu, 2001). More than 10 million people visit Venice each year, most of them visiting the historical centre on day trips. This is a structural transformation in tourist demand from residential tourism to day tripping. This makes tourism, as presently organised, an activity producing on the one hand high economic value, but on the other hand producing environmental pressures and potentially damage to both the quality of life of Venetians and to the physical structures of the city. A concern is that the city will become increasingly dependent on a single economic activity that also negatively influences the quality of life in Venice.

The externalities of tourism consist of physical congestion, urban degradation, environmental degradation and small-scale crime (Musu, 2001).

Other economic activities in and near the historical centre are shipyards and glass production. In the historical city, more than 50 per cent of the production is of artisan fabric, such as blown and handmade glass. These products are sold to tourists, so this is not a source of differentiation but it reinforces the tourism monoculture (Musu, 2001). In the Venice region economic activities consist of fish farming in the lagoon, industrial, chemical and oil sector in the Porto Maghera areas, as well as the port and container terminal. The economy of the port is changing from an industrial port (oil and chemical products) to a port of transit for containers. This transition decreases the potential impact of a shipping accident. In summary: there are important elements of economic vitality in the direct vicinity of the historical centre. However, the historical centre itself is increasingly dependent on mass tourism.

THREE PILLARS OF SUSTAINABILITY

Sustainable development is development that meets the needs of the present without compromising the ability of future generations to meet their needs. The World Bank distinguishes three development dimensions, or systems, relevant to sustainable development. The socio-cultural system, the economic system and the environmental system (Munasinghe, 1996). Each system has its own characteristics and its own requirements for maintenance. Sustainable development would then require a balanced maintenance of each system. In the communication of the European Union on urban development (EU, 2004) the three dimensions of sustainable development are used to arrive at strategies for keeping the cities of Europe habitable. In the following paragraphs the issues that play a role in Venice are described for the three dimensions.

Social sustainability

Social sustainability applies to the following subjects for the inhabitants of Venice: security; safety and health; demography; and democracy and culture (identity, social fabric / social divide). These have to be taken into account when thinking about the development of the city in the future.

Approaches for social sustainable development may consist of incentives to restore a demographic balance which will make the city a better place to live in for its inhabitants. One of the measures to be considered is the development of low cost or subsidized housing schemes such that young people would be encouraged to live in Venice. The introduction of incentives to restore the demographic balance should be accompanied with measures to protect and stimulate local economic and cultural activity supporting and strengthening the city's identity. Simultaneously, measures need to be taken to ensure the safety of the inhabitants and visitors against flooding and ecological hazards. The ongoing plan to construct mobile barriers at the inlets to the lagoon, in combination with the many ecological restoration projects carried out in and around the lagoon, provide a good basis for physical and ecological safety and well-being.

Economic sustainability

For sustainable economic development the policy goals should be defined for the longer term, in contrast to the presently dominating short-term tourist industry character of the economy. The tourist industry is not a very sustainable basis for the protection of specific monuments and will involve over-exploitation of the historic urban and lagoon area as a whole. Sustainable use of all resources, cultural and natural, instead of pure protection, is a very powerful and, in the longer-term, a more viable concept. Sustainable use means use while maintaining quality. This concept has been developed on the basis of experience in agricultural and fishery systems. Nevertheless it is equally valid for a historic centre such as Venice. An important concept in sustainable use is internalising external cost. For Venice this could imply that visiting tourists pay a fee which generates enough resources to maintain and sustain the entire cultural and natural setting. The same concept applies to the fisheries in the lagoon. How can the sustainable use of the natural resources be ensured? The external cost should be included in the cost of fish harvested in the lagoon. The concept of sustainable use equally applies to industrial and container port activities. The fees from these activities should be such that possible damage to the lagoon and maintenance costs are internalized in the prices of these activities.

To develop the Venetian economy in a sustainable way through the concepts listed above means that some constraints will have to be taken into account (Musu, 2001):

- compatibility between the different possible types of activities and polymorphism of the production structure;
- adequate levels of employment and economic growth for the local population;
- respect for local culture and prospects for growth (competence, innovation and research); and
- respect for the environment.

Within these constraints, economic activities need to be developed. The universities and other knowledge producing institutes in and around the city could be involved, for instance, in developing strategies to make tourism more sustainable. They can also create additional income through activities that generate less pollution, like the science and technology park.

Ecological sustainability

Ecological health of the lagoon, including interaction with the Adriatic Sea, is a pre-condition for all other activities in the region. The algae bloom in the lagoon and the related decay and odour problems in the late 1980s reminded the city how important this ecological health is to all people living in and visiting Venice. Maintenance of landscape and biodiversity in the lagoon and surrounding region is equally important. To achieve a good quality of the lagoon ecosystem, reduction and control of the input of human waste, nutrients and chemicals into the lagoon to sustainable levels is a *sine qua non*. This holds true for all urban, recreational, industrial and agricultural activities including the control of ecological risks related to industrial port activities and shipping and connecting all the buildings in the old city to a sewer system.

Ecological and economic sustainability go hand in hand when it comes to fisheries in the lagoon and the surrounding Adriatic Sea. The externalities need to be internalised in the costs of the fish and the car-

rying capacity of the system needs to be taken into account in deciding the amount of fish that is produced in the lagoon. Access to the natural habitats of the lagoon is important for the appreciation of nature and the understanding of the long-term natural-social interaction in and around Venice. These activities can contribute to the broadening of the tourist activities in the region besides the old city. For sustainability reasons, tourist access should be organized in such a way that natural habitats are maintained and not disturbed.

The hydraulic and ecological exchange between lagoon and open sea is one of the most important pillars of the ecological system of the lagoon and its wider surroundings. This permanent exchange of water, sediments, nutrients, biomass and aquatic species provides the basis for the ecologically rich lagoon and wetland system. This system can adjust itself to slow changes in climate and sea level conditions. The Venice Lagoon has a long history of human interventions. Over many centuries inflowing rivers have been diverted, tidal channels have been adjusted and the beach and tidal entrances have been protected with moles and walls and additional sand replenishments. In fact, the present lagoon is a genuine product of social-natural co-evolution. Restoration of the inflow of fresh water including minerals and sediments may help to revitalize the lagoon marsh system. The mobile barrier project can play a decisive role in protecting the city and the lagoon. As such it can be seen as a continuation of the historic co-evolution. Its function can be seen as guiding and slowing down the process of the takeover of the lagoon by the Adriatic Sea. Without the mobile barriers the Adriatic would slowly swallow the lagoon and the city. A more radical and ecological disruptive alternative would be protection of the lagoon and city through a permanent barrier. The concept of the mobile barrier system in its current form can extend the lifetime of the present open lagoon system for at least another 100 to 200 years, even when sea-level rise accelerates as a result of climate change. To serve over such a time period the technical system will have to be updated every, say, 50 years. As part of such an update the design of mobile barriers can be tuned to the then observed and projected sea-level rise. Without the mobile barrier system the lagoon and the city will be flooded more and more frequently and step by step the rising Adriatic will transgress across the lagoon. The other alternative, a permanent barrier, if installed would destroy the lagoon system, because the water would become fresh and the tidal movement would disappear, leading to dramatic ecological changes. Moreover Venice would lose the major pillar of its identity. Given the scenarios of sea-level rise, the mobile barrier system offers Venice and its lagoon an additional 100 to 200 years of ecologically and economically sound open access to the sea.

RESEARCH CONSIDERATIONS

Socio-political research

Research can be an important instrument to enhance sustainability. With regard to the socio-political aspects and related sustainability issues, the following themes can be mentioned:

Research and evaluation of current governance structures of Venice, the way decisions are taken and where responsibilities lie; in other words, multi-level governance, analysis and design. In what ways can current governance structures and practices be restructured such that their functionality and effectiveness is improved? How can such local and regional institutions better interact with national and international institutions and interest groups?

Another major research question is how international institutions and interest groups can play a role in the future of Venice. What can we learn from the (sustainable and unsustainable) management and governance structures of other world heritages, natural as well as cultural heritages?

The question of which incentives would be suitable for Venice to stimulate certain age-groups and professions to come and live in the historic centre is also relevant for the longer term vitality of the city. What measures can be identified that would help to restore the demographic balance and would help to diversify and revitalize the economy?

Finally, analysis of ongoing trends and monitoring social and political dynamics is an important part of the research agenda.

Economic research

Sustainable use of natural and cultural resources is generally recognised as a better concept than protection and isolation. However, much present day activity and research is focussed on protection and isolation. Sustainable use requires important changes in governance and 'ownership' of the resources at stake. Research can help to explore and create the political basis for such changes.

An important research theme is the identification of policy instruments, including economic instruments that help avoid over-exploitation of cultural and natural resources. Economic instruments such as user fees are an example of how to manage access to Venice and pay for its maintenance.

Economic research can also help to monitor and explore developments in the economic structure of Venice and its surroundings. To what extent can the structure of economic activities be influenced and what are the most promising instruments to help broaden the economic base of Venice?

Bio-geo-physical research

The two research themes discussed above focus on the social, political, economic and cultural aspects of Venice and its lagoon. As we often speak about co-evolution in this chapter, it is equally important to explore the natural processes and the effect of human activity on these natural processes.

Particularly relevant for Venice and the lagoon is research into the mechanisms and trends of sea-level rise, including the effects of climate change. This includes research on the effects of climate change on storm surges that determine the levels and frequencies of high water in Venice.

An equally important research theme is lagoon dynamics, including the hydraulic, sediment and biological regimes and changes in these regimes. The goal of research should be to better understand the dynamics and the driving forces for changes in the lagoon habitats and landscape. This type of research should consider simultaneously the natural processes (including the role of rivers in lagoon sedimentation processes) and the effects of human activities including urban, industrial, recreational, transport and agricultural activities.

REFERENCES

Brambati, A., Carbognin, L., Quaia, T., Teatini, P., and L. Tosi. 2003. The Lagoon of Venice: geological setting, evalution and land subsidence. *Episodes* **26**, 264–8.

Cecconi, G., Canestrelli, P., Corte, C., and M. Di Donato. 1998. Climate record of storm surges in Venice. Paper presented at RIBAMOD workshop – Impact of Climate Change on Flooding and Sustainable River Management, Wallingford, 26–27 February 1998.

Church, J. A., *et al.* 2001. *Changes in sea level, Climate Change 2001 – The Scientific Basis. Contribution of Working Group I to the Third Assessment Report of the Intergovernmental Panel on Climate Change.* Houghton, J.T. *et al.* (eds.), Cambridge: Cambridge University Press, pp. 637–93.

Collegio di esperti di livello internazionale. 1998. *Report on the Mobile Gate Project for the Tidal Flow Regulation at the Venice Lagoon Inlets.* Venice.

Comune di Venezia. 2003. www.comune.venezia.it/statistica/ popola1.asp.

EU. 2004. *Communication from the Commission to the Council, the European Parliament, the European Economic and Social Committee and the Committee of the Regions; Towards a Thematic Strategy on the Urban Environment.* Brussels, 56 pp.

Marzocchi, W., and F. Mulargia. 1996. Scale analysis to sort the different causes of mean sea level changes: an application to the northern Adriatic Sea. *Geophysical Research Letters* **23**, 1119–22.

Munasinghe, M. 1996. *Environmental Impacts of Macroeconomic and Sectoral Policies.* Washington DC: The World Bank.

Musu, I. 2001. *Sustainable Venice: Suggestions for the Future.* Dordrecht, The Netherlands: Kluwer Academic Publishers, 269 pp.

Ravera, O. 2000. The Lagoon of Venice: the result of both natural factors and human influence. *Journal of Limnology* **59**, 19–30.

Tsimplis, M. N. 2002. Sea level in the Mediterranean Sea: the contribution of temperature and salinity changes. *Geophysical Research Letters* **29**.

UNESCO. 2003. www.unesco.org/culture/tourism/html_eng/cities3.shtml.

United Nations. 1992. *Rio Declaration on Environment and Development – Declaration of the UN Conference on Environment and Development.* United Nations, Doc.A/CONF.151/26/Rev.1, Report of the UNCED, vol 1, New York.

Venice Water Authority. 2002. *Measures for the Protection of Venice and its Lagoon.* Venice.

Part VIII · Appendices

A1 • An overview of the main findings of the Third Assessment Report of the IPCC

V. FERRARA, D. GAUDIOSO AND A. RAUDNER

INTRODUCTION

Planetary climate change is linked to variabilities at all scales, from the global to the local. Land, ocean and atmosphere interactions are particularly active in coastal areas. The understanding of the vulnerability of the Venice coastal area depends in part upon the scientific information made available by the Inter-governmental Panel on Climate Change (IPCC) over the last decade. Confidence in the credibility of this information, and its continual revisions, is needed both by society and by decision – and policy – makers. The objective of this appendix is to describe the organization, structure and activity of the IPCC, the five yearly reports of which are considered the best source of information on climate change at the international level.

THE MISSION OF THE IPCC

The Intergovernmental Panel on Climate Change (IPCC) was established in 1988 by the World Meteorological Organization (WMO) and the United Nations Environment Programme (UNEP) as an intergovernmental body to assess the existing scientific knowledge on the causes and impacts of climate change, as well as mitigation strategies. It is the largest effort of its kind and has been taken as a reference for similar assessment processes in other fields, such as biodiversity or water issues.

The IPCC does not carry out research itself nor does it monitor climate related data or other relevant parameters. Its role is 'to assess on a comprehensive, objective, open and transparent basis the scientific, technical and socio-economic information relevant to understanding the scientific basis of risk of human-induced climate change, its potential impacts and options for adaptation and mitigation'. It bases its assessment mainly on peer reviewed and published scientific/technical literature.

Being an organization at the interface between science and policy, the IPCC is thought to fulfil a twofold purpose: it should provide credibility to the scientific community and is intended to feed scientific and technical information into the political negotiation and implementation process (Bolin, 1994). Therefore, the IPCC process regularly incorporates political actors in the processes of evaluation, discussion and approval, with different degrees of influence on the assessment process.

In order to understand the intergovernmental organizational design of the IPCC and the governmental approval mechanism, we can refer to the following sentence by Bert Bolin, the first chairman of the IPCC: 'Right now, many countries, especially developing countries, simply don't trust assessments in which their scientists and policymakers have not participated. Don't you think credibility demands global participation?' (Schneider, 1991, 25). Since its launch, more and more governments have participated in the plenary sessions of the IPCC where the final documents must be approved. Whereas the first session was attended by representatives from only 30 countries, at the ninth session in 1995 their number totalled 117. Subsequent participation in the plenary sessions varied between 80 and 110 countries (Siebenhüner, 2003).

THE ORGANIZATIONAL STRUCTURE

The IPCC has three Working Groups and a Task Force:

• Working Group I assesses the scientific aspects of the climate system and climate change;

Flooding and Environmental Challenges for Venice and its Lagoon: State of Knowledge, ed. C. A. Fletcher and T. Spencer. Published by Cambridge University Press, © Cambridge University Press 2005.

- Working Group II assesses the vulnerability of socio-economic and natural systems to climate change, negative and positive consequences of climate change, and options for adapting to it;
- Working Group III assesses options for limiting greenhouse gas emissions and otherwise mitigating climate change; and
- the Task Force on National Greenhouse Gas Inventories, responsible for the IPCC National Greenhouse Gas Inventories Programme.

The Panel meets in plenary sessions about once a year. It accepts/approves/adopts IPCC reports, decides on the mandates and work plans of the Working Groups and the Task Force, the structure and outlines of its reports, the IPCC Principles and Procedures, and the budget. The Panel also elects the IPCC Chair, the IPCC Bureau and the Bureau of the Task Force on National Greenhouse Gas Inventories. The IPCC Bureau meets two to three times per year and assists the IPCC Chair in planning, co-ordinating and monitoring progress in the work of the IPCC.

The IPCC is managed by the IPCC Secretariat, which is hosted by WMO in Geneva and supported by UNEP and WMO. In addition, each Working Group and the Task Force has a Technical Support Unit. These Technical Support Units are supported by the government of the developed country co-chair of that Working Group or Task Force and hosted by a research institution in that country. A number of other institutions provide in-kind support for IPCC activities. The official interaction between scientists and policy makers is restricted to well-defined stages of the assessment process. Regular plenary sessions of the panel consisting of the representatives of national governments provide the forum for the constitution of the Bureau. Governmental delegates elect the Bureau members on the basis of nominations from a nominating committee.

It is the scientists of the Bureau who develop an outline of the report, the topics of the working groups and the division of labour among them. Suggestions are taken to the national governments at the plenary sessions, where a final decision is taken. Then they select the authors and reviewers based on the principles of scientific expertise and geographic representation. Coordinating lead authors and lead authors are chosen by the co-chairs and vice chairs of each working group, following consideration of nominations from governments and participating organizations, and other experts. Since Bureau members have to found their decisions on their knowledge of publications and works, their decisions usually build to a large extent on informal contacts and communications between the scientists. Governments can nominate candidates in this phase but it remains the responsibility of the Bureau to select the most appropriate authors. On the other hand, the preparation of the chapters and the first round of peer review takes place entirely at the scientific level, as they are carried out exclusively by scientific experts. Governments enter the process once again in the second round of review when their comments to the revised drafts are solicited. Finally, they have a crucial role in the approval of the summary for policy makers and the synthesis report as here their agreement is required. Government representatives are entitled to attend plenary sessions of the working groups and discuss and approve the documents presented to them in a line-by-line procedure. Their comments and requests for amendments, however, have to be based on published papers in the scientific literature. In summary, thus far, the influence of national governments is largely restricted to the nomination and election of Bureau members. This approval mechanism grants them rather limited influence in the preparation of the documents.

MAIN ACTIVITIES AND PRODUCTS

Since its beginning, the IPCC has produced three major assessment reports, concluded in 1990, 1995 and 2001.

The first report, a four-volume report, confirmed the scientific basis for concern about climate change and inspired governments to establish the Intergovernmental Negotiating Committee, which adopted the UN Framework Convention on Climate Change in 1992.

The second report was adopted in 1995 and published in April 1996. It consists of four volumes: three volumes were prepared by the working groups and the fourth volume was a synthesis of scientific technical information relevant to interpreting Article 2

– the objective – of the UNFCCC. This report contributed to the negotiations which lead to the adoption of the Convention's Kyoto Protocol one year later at COP 3.

The Third Assessment Report, like its predecessors was a comprehensive and up to date assessment of the policy relevant scientific, technical and socio economic dimensions of climate change (IPCC, 2001). It concentrated on new findings since 1995, paying greater attention to the regional (in addition to the global) scale, and including non-English literature as far as possible. It consisted of:

- Climate Change 2001: the scientific basis – contribution of Working Group I to the Third Assessment Report;
- Climate Change 2001: impacts, adaptation and vulnerability – contribution of Working Group II to the Third Assessment Report;
- Climate Change 2001: mitigation – contribution of Working Group III to the Third Assessment Report; and
- Climate Change 2001: synthesis report of the Third Assessment Report.

The Synthesis Report addressed a range of policy relevant scientific and technical questions based on submissions from governments and identified by the IPCC in consultation with the subsidiary bodies of the UNFCCC. It synthesised and integrated the information contained in the Third Assessment Report and drew upon all previously approved IPCC reports.

The IPCC has decided to continue to prepare comprehensive assessment reports and has agreed to complete its Fourth Assessment Report in 2007. Considerations about scope and outline are about to start.

KEY CONCLUSIONS OF THE THIRD ASSESSMENT REPORT

The key 'political' conclusions of the Third Assessment Report (TAR) pointed out the following relevant aspects:

- Climate change is not just an environmental issue, but is part of the larger challenge of sustainable development;

- The Earth's climate system has demonstrably changed on both global and regional scales since the pre-industrial era, with some of these changes attributable to human activities. There is new and stronger evidence that most of the observed warming of the past 50 years is attributable to human activities;
- Recent regional changes in climate, particularly increases in temperature, have already affected hydrological systems and terrestrial and marine ecosystems in many parts of the world. The rising socio-economic costs related to weather damages and to regional variations in climate suggest increasing vulnerability to climate change;
- Projected climate change will have beneficial and adverse effects on both environmental and socio-economic systems, but the larger the changes and rate of change in climate, the more the adverse effects predominate. The impacts of climate change will fall disproportionately upon developing countries and poor people in all countries and thereby exacerbate inequalities in health status and access to adequate food, clean water and other resources;
- The stabilization of atmospheric CO_2 concentrations at 1990 levels (350 ppm) requires a reduction of current CO_2 emissions of at least 60 per cent. The stabilization of atmospheric CO_2 concentrations at 450, 650 or 1000 ppm would require global anthropogenic CO_2 emissions to drop below the year 1990 levels within a few decades (for 450 ppm), within about a century (for 650 ppm) and within about 2 centuries (for 1000 ppm). Eventually CO_2 emissions would need to decline to a very small fraction of current emissions;
- Adaptation has the potential to reduce adverse effects of climate change and can often produce immediate ancillary benefits, but will not prevent all damages. Adaptation is a necessary strategy at all scales to complement climate change mitigation efforts. Together they can contribute to sustainable development objectives; and
- Mitigation costs estimates vary for many reasons, and are of the order of a few GNP percentages. The TAR suggests substantial opportunities for lowering mitigation costs.

The key 'technical' conclusions of the Third Assessment Report (TAR) of the Intergovernmental Panel on Climate Change (IPCC) can be summarized as follows:

1) The Earth's climate system has changed, globally and regionally:
   - the Earth has warmed 0.6 ± 0.2 °C since 1860 with the last two decades being the warmest of the last century;
   - the increase in surface temperatures over the twentieth century for the Northern hemisphere is likely to be greater than that for any other century in the last 1000 years;
   - precipitation patterns have changed with an increase in heavy precipitation events in some regions; and
   - sea level has risen 10-20 cm since 1900; most non-polar glaciers are retreating; and the extent and thickness of Arctic sea ice is decreasing in summer.

2) Emissions of greenhouse gases and aerosols due to human activities continue to alter the atmosphere in ways that are expected to affect the climate:
   - concentrations of atmospheric greenhouse gases and their radiative forcing have continued to increase as a result of human activities;
   - anthropogenic aerosols (mostly originated from fossil fuel and biomass burning) are short-lived and mostly produce negative radiative forcing;
   - the radiative forcing from the increase in anthropogenic greenhouse gases since the pre-industrial era is positive (warming) with a small uncertainty range; that from the direct effects of aersols is negative (cooling) and smaller; whereas the negative forcing from the indirect effects of aerosols (on clouds and the hydrologic cycle) might be large but is not well quantified; and
   - natural factors (solar variation and volcanic aerosols) have made small contributions to radiative forcing over the past century.

3) There is new and stronger evidence that most of the warming observed over the last 50 years is attributable to human activities:
   - the Second Assessment Report (SAR) (IPCC, 1995) concluded: 'Most of the observed warming of the last 50 years is attributable to human activities';
   - since the SAR, progress has been made in reducing uncertainty, particularly with respect to distinguishing and quantifying the magnitude of responses to different external influences;
   - detection and attribution studies consistently find evidence for an anthropogenic signal in the climate record of the last 35 to 50 years; most of these studies find that, over the last 50 years, the estimated rate and magnitude of warming due to increasing concentrations of greenhouse gases alone are comparable with, or larger than the observed warming;
   - furthermore, most model estimates that take into account both greenhouse gases and sulphate aerosols are consistent with observations over this period;
   - the best agreement between model simulations and observations over the last 140 years has been found when all of the above and natural forcing factors are combined; and
   - these results show that the forcings included are sufficient to explain the observed changes, but do not exclude the possibility that other forcings may also have contributed.

4) Carbon dioxide, surface temperatures, precipitation and sea level are all projected to increase globally during the twenty-first century because of human activities:
   - all IPCC projections show that the atmospheric concentration of carbon dioxide will increase significantly during the next century in the absence of climate change policies;
   - climate models project that the Earth will warm 1.4 to 5.8 °C (2.5 to 10.8 °F) between 1990 and 2100, with most land areas warming more than the global average;
   - precipitation will increase globally, with increases and decreases locally, with an increase in heavy precipitation events over most land areas;
   - sea level is projected to increase 8–88 cm between 1990 and 2100; and

- models project an increase in extreme weather events e.g. heat waves, heavy precipitation events, floods, droughts, fires, pest outbreaks, mid-latitude continental summer soil moisture deficits and increased tropical cyclone peak wind and precipitation intensities.

5) Observed changes in regional climate over the past 50 years have affected biological and hydrological systems in many parts of the world:
   - there has been a discernible impact of regional climate change, particularly increases in temperature, on biological systems in the twentieth century (bird migration patterns and egg laying, migration of plants, insects and animals);
   - coral reefs are adversely affected by rising sea surface temperatures;
   - changes in marine systems, particularly fish populations, have been linked to large-scale climate oscillations; and
   - changes in stream flow, floods and droughts have been observed.

6) There are preliminary indications that some human systems have been affected by recent increases in floods and droughts. The rising socio-economic costs related to weather damage and to regional variations in climate suggest increasing vulnerability to climate change:
   - extreme weather or climatic events cause substantial, and increasing, damage;
   - the fraction of weather-related losses covered by insurance varies considerably by region; and
   - climate-related health effects are observed.

7) Projected changes in climate will have both beneficial and adverse effects on water resources, agriculture, natural ecosystems and human health but the larger the changes in climate the more the adverse effects dominate:
   - socio-economic sectors (e.g. agriculture, forestry, fisheries, water resources and human settlements), terrestrial and aquatic ecological systems and human health, which are all vital to human development and well-being, are all sensitive to the magnitude and rate of climate change, as well as to changes in climate extremes and variability; and

- while there are some positive effects of climate change e.g. increased agricultural productivity at mid and high latitudes for small increases in temperature and reduced winter mortality, most of the impacts are adverse, particularly in response to an increase in extreme weather events, with most natural systems and most people being adversely affected by climate change. Projections include:
   - a decrease in water availability in many water scarce regions, particularly in the sub-tropics;
   - a decrease in agricultural productivity for almost any increase in temperature in most tropical and sub-tropical regions;
   - an increase in heat stress mortality and the number of people exposed to vector-borne (e.g. malaria and dengue) and water-borne (e.g. cholera) diseases;
   - a widespread increase in the risk of flooding for tens of millions of people due to increased heavy precipitation events and sea-level rise;
   - some natural systems may undergo significant and irreversible damage, in particular glaciers, coral reefs and atolls, mangroves, and polar and alpine systems; an increased risk of extinction of some more vulnerable species and loss of biodiversity;
   - adaptation is a necessary strategy to complement climate change mitigation efforts; and
   - developing countries, and the poor within them, are the most vulnerable because they lack the financial, technical and institutional resources to adapt.

8) There are many technological options to reduce near-term greenhouse gas emissions and opportunities for lowering costs, but barriers to the deployment of climate friendly technologies need to be overcome:
   - stabilization of the atmospheric concentrations of greenhouse gases will require emissions reductions in all regions;
   - lower greenhouse gas emissions will require different patterns of energy resource development and production (trend towards de-carbonisation) and increases in end-use efficiency;
   - significant technical progress has been made in the last five years and at a faster rate than

expected (e.g. wind turbines, elimination of industrial by-products, hybrid engine cars, fuel cell technology and underground carbon dioxide storage);

- half of the projected increase in global emissions between now and 2020 could be reduced with direct benefits (negative costs), while the other half at less than $100 per tC;
- realizing these reductions involves supporting policies/overcoming barriers, increased R & D and effective technology transfer;
- some reductions in emissions can be obtained at no, or negative, costs by exploiting no regrets opportunities:
- reducing market or institutional imperfections;
- ancillary benefits e.g. local and regional air quality improvements;
- revenues from taxes or auctioned permits can be used to reduce existing distortionary taxes through revenue recycling;
- forests, agricultural lands and other terrestrial ecosystems offer significant carbon sequestration potential globally i.e. up to 200 GtC over the next 50 years;
- in the absence of trading, Annex B costs of complying with the Kyoto Protocol, range from 0.2 – 2%, whereas with full Annex B trading the costs are reduced to 0.1 – 1% (these models only consider energy-related emissions of carbon dioxide). These costs could be further reduced by using sinks, the Clean Development Mechanism, multiple greenhouse gases, efficient tax recycling, ancillary benefits and induced technological change; and
- costs of stabilizing carbon dioxide increase moderately from 750 ppm to 550 ppm, but with a larger increase going from 550 ppm to 450 ppm due to the premature retirement of capital stock.

KEY UNCERTAINTIES IN THE THIRD ASSESSMENT REPORT

Despite significant progress made by the TAR in understanding climate change and the human response to it, there remain important areas where further work is required, in particular (IPCC, 2001):

- the detection and attribution of climate change;
- the understanding and prediction of regional changes in climate and climate extremes;
- the quantification of climate change impacts at the global, regional and local levels;
- the analysis of adaptation and mitigation activities;
- the integration of all aspects of the climate change issues into strategies for sustainable development; and
- comprehensive and integrated investigations to support the judgment as to what constitutes 'dangerous anthropogenic interference with the climate system'.

Some criticisms had been raised in the past, especially towards the Summaries for Policymakers, since they do not convey information about the levels of uncertainty of the different findings. It has been argued that a thorough understanding of the uncertainties is essential to the development of good policy decisions (NAS, 2001). This comment is partially addressed in the Summaries for Policymakers of WG I and II by assigning qualitative confidence levels (i.e. virtually certain, likely, unlikely) that represent the authors' collective judgment in the validity of a conclusion based on observational evidence, modelling results, and theory that they have examined.

Uncertainty in the projections of climate change

In particular, the TAR has been criticized for lacking quantitative estimations of the uncertainties of the projections of future climate change (Reilly *et al.*, 2001). Although the preparation of climate change scenarios has received particular attention in the preparation of the TAR, the final outcome is not completely satisfactory, essentially due to lack of time in developing a consistent approach throughout the Report.

Projections of future climate change in the Second Assessment Report

For the Second Assessment Report, the IPCC had developed in 1992, a range of long term emissions scenarios, IS92a–f, of future greenhouse gas and aerosol precursor emissions based on assumptions concerning population and economic growth, land use, technological changes, energy availability and fuel mix during the period 1990 to 2100 (IPCC, 2000). Through understanding of the global carbon cycle and of atmospheric chemistry, these emissions were used to project atmospheric concentrations of greenhouse gases and aerosols and the perturbation of natural radiative forcing.

For the midrange IPCC emissions scenario IS92a, assuming the 'best estimate' value of climate sensitivity and including the effects of future increases in aerosols, models used in the SAR projected an increase in global mean surface temperature relative to 1990 of c. 2 °C by 2100. This estimate is about one third lower than the 'best estimate' in 1990. This is primarily due to lower emissions scenarios (particularly for CO_2 and CFCs), the inclusion of the cooling effect of sulphate aerosols, and improvements in the treatment of the carbon cycle. Combining the lowest IPCC emission scenario (IS92c) with a 'low' value of climate sensitivity and including the effects of future changes in aerosols concentrations leads to a projected increase of c. 1 °C by 2100. The corresponding projection for the highest IPCC scenario (IS92e) combined with a high value of climate sensitivity gives a warming of c. 3.5 °C.

Average sea level is expected to rise as a result of thermal expansion of the oceans and melting of glaciers and ice sheets. For the IS92a scenario, assuming the 'best estimate' values of climate sensitivity and of ice melt sensitivity to warming, and including the effects of future changes in aerosols, models project an increase in sea level of c. 50cm from the present to 2100. This estimate is approximately 25 per cent lower than the best estimate in 1990 due to the lower temperature projection, but also reflects improvements in the climate and icemelt models. The lowest sea-level rise projection is c.15 cm from the present to 2100. The corresponding projection for the highest emission scenario IS92e combined with high climate and icemelt sensitivities gives a sea level rise of c. 95 cm from the present to 2100.

PROJECTIONS OF FUTURE CLIMATE CHANGE IN THE THIRD ASSESSMENT REPORT

In 1995, the IS92 scenarios were evaluated. The evaluation recommended that significant changes (since 1992) in the understanding of driving forces of emissions and methodologies should be addressed. These changes in understanding relate to, amongst other factors, the carbon intensity of energy supply, the income gap between developed and developing countries and to sulphur emissions. This led to a decision by the IPCC Plenary in 1996 to develop a new set of scenarios, the SRES (Special Report on Emission Scenarios) scenarios (IPCC, 2000).

These SRES scenarios assist in climate change analysis, including climate modelling and the assessment of impacts, adaptation and mitigation. They exclude only outlying 'surprise' or 'disaster' scenarios in the literature. The possibility that any single emissions path will occur as described in the scenarios is highly uncertain. There are no assigned probabilities of occurrence for any of the scenarios. The set of SRES emissions scenarios is based on an extensive assessment of the literature, yielding six alternative modelling approaches. They cover a wide range of the main driving forces of future emissions, from the demographic to technological and economic developments and include the range of emissions of all relevant species of greenhouse gases and sulphur and their driving forces.

The SRES scenarios cover most of the range of carbon dioxide, other GHGs and sulphur emissions found in the recent literature and SRES scenario database. Their spread is similar to that of IS92 scenarios for CO_2 emissions from energy and industry as well as total emissions but represents a much wider range for land use change. The six scenarios cover wide and overlapping emission ranges. The range of GHG emissions in the scenarios widens over time to capture the long term uncertainties reflected in the literature for many of the driving forces; after 2050 it widens significantly as a result of different socio economic developments.

The SRES scenarios extend the IS92 range toward higher emissions (SRES maximum of 2538 GtC compared to 2140 GtC for IS92), but not towards lower emissions. The lower bound for both scenario sets is approximately 770 GtC. The range of emissions of HFCs in the SRES scenario is generally lower than in the earlier IPCC scenarios. Because of new insights about the availability of alternatives to HFCs as replacements for substances controlled by the Montreal Protocol, initially HFC emissions are generally lower than in previous IPCC scenarios. Sulphur emissions are generally below the IS92 range, because of structural changes in the energy system as well as concerns about local and regional air pollution. These reflect sulphur control legislation in Europe, North America, Japan, and more recently other parts of Asia and other developing regions. Although global emissions of SO_2 for the SRES scenarios are lower than in the IS92 scenarios, uncertainty about SO_2 emissions and their effect on sulphate aerosols has increased compared to the IS92 scenarios because of diverse regional patterns of SO_2 emissions in the scenarios. After initial increases over the next two to three decades, global sulphur emissions in the SRES scenarios decrease, consistent with the findings of the 1995 IPCC scenario evaluation and recent peer-reviewed literature.

Projections of future climate change in the TAR are obtained using the SRES scenarios, and a simple climate model whose climate sensitivity and ocean heat uptake are calibrated to each of seven complex climate models. The climate sensitivity used in the simple models range from 1.7 to 4.2 °C, which is comparable to the commonly accepted range of 1.5 to 4.5 °C. For the atmosphere ocean general circulation model (AOGCM) experiments for the end of the twenty first century (2071 to 2100) compared with 1961 to 1990, the mean warming for SRES scenario A2 is 3 °C with a range of 1.3 to 4.5 °C, while for SRES scenario B2 the mean warming is 2.2 °C with a range of 0.9 to 3.4 °C.

This later generation of models projects globally averaged surface temperature to increase by 1.4 to 5.8 °C over the period 1990 to 2100, about two to ten times larger than the central value of observed warming during the last 100 years. For the periods 1990 to 2025 and 1990 to 2050, the projected increases are 0.4 to 1.1 °C and 0.8 to 2.6 °C, respectively. These results are for the full range of 35 SRES scenarios, based on a number of climate models.

Temperature increases are projected to be greater than those in the SAR, which were c. 1.0 to 3.5 °C, based on six IS92 scenarios. The higher projected temperatures and the wider range are due primarily to lower projected sulphur dioxide emissions in the SRES scenarios relative to the IS92 scenarios, because of the structural changes in the energy system, as well as concerns about local and regional air pollution. The projected rate of warming is much larger than the observed changes during the twentieth century and is very likely to be without precedent during at least the last 10 000 years, based on paleoclimatic data.

As already mentioned, no assigned probabilities of occurrence are available for any of the scenarios. This is the main reason why no confidence level can be assigned to the estimation of future changes in temperature and sea level, which are not even comparable to previous results obtained by the IPCC, for instance, in the SAR.

RECENT DEVELOPMENTS

A more rigorous probabilistic approach, which assigns a probability distribution function to the mean global temperature change on the basis of uncertainties in the values of parameters, has been proposed by different authors (Webster *et al.*, 2001; Webster *et al.*, 2002; Wigley and Raper, 2001). In particular, Wigley and Raper (2001) conclude that, in the most complete and realistic case, the 90 per cent confidence interval of the 1990–2100 mean temperature increase is 1.7–4.9 °C, while the median value is 3.1 °C.

RECENT TRENDS IN CLIMATE RESEARCH

Two years after its publication, the IPCC's Third Assessment Report (TAR), and particularly the contribution of Working Group I, still represents an admirable summary of research activities on climate issues, both in terms of scientific credibility, completeness and social and political relevance.

Significant progress has been made in some areas, such as reconstruction of paleoclimatic records, understanding of the direct and indirect roles of aerosols in radiative forcing, regional models, detection and attribution of climate impacts and stabilization scenarios.

Increased evidence has been drawn up concerning some of the most divisive conclusions of the TAR, such as the biological response to recent global warming or the possible occurrence of abrupt climate changes.

Biological response to recent global warming

Two US research groups have attempted to take the discussion a stage further by searching for a climate fingerprint in the overall patterns in studies of a wide range of plants and animals. Parmesan and Yohe present a meta-analysis of studies of more than 1700 species, and find that there have been significant range shifts averaging 6.1 km per decade towards the poles, and that spring has advanced by 2.3 days per decade (Parmesan and Yohe, 2003). With 'very high confidence' (in IPCC terms), this means that climate change is already affecting living systems. Root *et al.* also detect a temperature-related 'fingerprint' in species from insects to mammals, and grasses to trees (Root *et al.*, 2003). The observed changes are most marked at high latitudes and high altitudes, where the largest changes have been predicted.

Abrupt climate changes

Recent scientific evidence shows that major and widespread climate changes have occurred abruptly in the past, especially when the climate system was being forced to change most rapidly. The abrupt changes of the past are not fully explained yet, and climate models typically underestimate the size, speed, and extent of these changes. Greenhouse warming might thus result in large, abrupt and unwelcome regional or global climate events which cannot be predicted with confidence, and climate surprises are to be expected. A report prepared in 2001 by the US National Research Council for the US Global Change Research Program provides a description of the current state of knowledge in the field and identifies priorities for further research (NRC, 2001).

REFERENCES

Bolin, B. 1994. Science and policy making. *Ambio* **23**, 25–9.

IPCC. 1995. *Climate Change 1995: The Scientific Assessment.* Cambridge: Cambridge University Press.

IPCC. 2000. *Emission Scenarios*. Special Report of the IPCC Working Group III. Geneva: IPCC (World Meteorological Organization), 27pp.

IPCC, 2001. *Climate Change 2001 – Synthesis Report of the IPCC Third Assessment Report*. Cambridge: Cambridge University Press.

NAS, 2001. *Climate Change Science: an Analysis of Some Key Questions*. Committee on the Science of Climate Change Division on Earth and Life Studies National Research Council. Washington, DC: National Academy Press.

NRC, 2001. *Abrupt Climate Change: Inevitable Surprises*. Washington, DC: National Research Council, US Global Change Research Program.

Parmesan, C., and G. Yohe. 2003. A globally coherent fingerprint of climate change impacts across natural systems. *Nature* **421**, 37–42.

Reilly, J., Stone, P. H., Forst, C. E., Webster, M. D., Jacoby, H. D., and R. G. Prinn. 2001. Uncertainty and climate change assessment. *Science* **293**, 430–3.

Root, T. L., Price, J. T., Hall, K. R., Schneider, S. H., Rosenzweig, C., and J. A. Pounds. 2003. Fingerprints of global warming on wild animals and plants. *Nature* **421**, 57–60.

Schneider, S. H. 1991. Three reports of the Intergovernmental Panel on Climate Change. *Environment* **33**, 25–30.

Siebenhüner, B. 2003. *The Changing Role of Nation States in International Environmental Assessments – The Case of the IPCC*. Global Governance Working Paper No. 7. Potsdam, Berlin: Oldenburg, The Global Governance Project.

Webster, M. D., Babiker, M. H., Mayer, M., Reilly, J. M., Harnisch, J., Hyman, R., Sarofim, M. C., and C. Wang. 2001. *Uncertainty in Emissions Projections for Climate Models*. Report 79. MIT Joint Program on the Science and Policy of Global Change. Cambridge, MA: MIT, 29pp.

Webster M. D., Forest, C., Reilly, J., Babiker, M., Kicklighter, D., Mayer, M., Prinn, R., Sarofim, M., Sokolov, A., Stone, P., and C. Wang. 2002. *Uncertainty Analysis of Climate Change and Policy Response*. Report 95. MIT Joint Program on the Science and Policy of Global Change. Cambridge, MA: MIT, 25pp.

Wigley, T. M. L., and S. C. B. Raper. 2001. Interpretation of high projections for global-mean warming. *Science* **293**, 451–4.

A2 · Special Laws for Venice

During Venice's time as a republic (13th century–1797), there were carefully specified and vigilantly applied laws for the lagoon and mainland domains. This attention dwindled from the nineteenth century onwards and consequently the lagoon's defining characteristics and underlying dynamics altered significantly. Only following the 1966 disaster was care taken again for the special characteristics of the lagoon and its inseparability from safeguarding measures for the unique city.

While ordinary laws set the duties of the State and local authorities as regards territorial management, the Special Law of 1973 established that Venice is of pre-eminent national interest and gave the State authority to determine specific objectives and associated financing for measures. The main objectives of the first Special Law were:

- safeguarding of the environment (landscape, historical, archaeological, artistic features);
- protection of the hydraulic and hydrogeological equilibrium;
- regulation of watercourses (natural and artificial) feeding into the lagoon;
- reduction, followed by regulation, of tide levels;
- coastal defence works; and
- pollution protection.

Another iteration of the Special Law in 1984 had a slightly different orientation. It committed the State to the design, experimentation and execution of works to:

- re-establish the hydrogeological equilibrium of the lagoon;
- arrest and invert the process of degradation;
- protect the *insulae*; and

- protect the urban settlements from the exceptionally high tides, 'also via measures at the inlets (...), with the characteristics of experimentability, reversibility and gradualness'.

The Law also allowed the Public Works Ministry to identify a single operating agency to guarantee the 'unitary nature' of the interventions for safeguarding the lagoon. Consorzio Venezia Nuova, made up of a group of Italian construction and engineering firms, was nominated as the executive body – essentially a private company with public objectives. The 1984 Law also established the Comitatone (see below) to oversee the implementation of safeguarding objectives and budgetary allocations among the various institutional bodies. Then in 1992 the Third Special Law made it a requirement to obtain the opinions of the Veneto regional administration and those of the Venice and Chioggia municipalities for safeguarding strategies and measures. Its main other points are:

- adjustment and reinforcement of the long breakwaters at the three lagoon inlets;
- local defences from high waters for built areas;
- restoration of lagoon morphology;
- halting the process of the lagoon's deterioration;
- coastal defences;
- substitution of petrol tanker traffic in the lagoon; and
- opening up of the closed acquaculture areas to tidal expansion.

As well as setting out the safeguarding objectives, each Special Law also set aside State funds for fulfilling them. Since there has not been a Special Law since 1992, Parliament has meanwhile set aside funds

Flooding and Environmental Challenges for Venice and its Lagoon: State of Knowledge, ed. C. A. Fletcher and T. Spencer.
Published by Cambridge University Press, © Cambridge University Press 2005.

for safeguarding Venice in its annual budget. But in 2003, for the first time, no provisions were made for Venice in the budget, nor again in 2004. Funding for the mobile barriers, however, has continued via the Strategic Objectives, a measure introduced to fund major infrastructural projects such as this and a bridge from the Italian mainland to Messina, Sicily.

A3 • Institutional framework and key organizations

Management of Venice and its lagoon is notoriously complex. Indeed, the lack of progress on implementation of a robust plan for safeguarding measures has been attributed to overlapping and fragmented institutional responsibilities among various administrations and even within departments of a single entity.

Venice is the capital of the Veneto Region in northern Italy and of one of the region's seven provinces. National ministries and regional, provincial, and local authorities, as well as the Venice City Council, are all responsible for various aspects of the administration of the city of Venice, its lagoon, and the surrounding areas. For example, the land that drains into the Venice Lagoon, an area of almost 1,900 km², is administered by 100 local authorities. The 1973 Special Laws assigned specific responsibilities to various authorities concerned with safeguarding Venice and the lagoon. The Italian State has responsibility for the physical safeguarding and restoration of the hydrogeological balance in the lagoon. The Veneto Region is responsible for the abatement of pollution, especially from the drainage basin. And the city council of Venice and the town of Chioggia at the southern end of the lagoon are responsible for urban conservation and maintenance, as well as activities aimed at promoting socio-economic development.

The Venice Water Authority (Magistrato alle Acque), which was founded in 1501, has responsibility under the Italian Ministry of Public Works for ensuring the survival of Venice, its lagoon, and the living species that inhabit it, and for protecting the lagoon from both natural events and human impacts. In 1992, a general plan backed by law to implement measures designed to safeguard Venice was launched by the Venice Water Authority. The plan is two-pronged in its approach. The first prong relates to the protection of the city from high water levels and flood damage by improving its physical defences by, for example, raising embankments, and also by reinforcing the coastline to protect it from storm damage. The second prong aims at improving the environmental quality of the lagoon through a series of interventions that tackle the factors causing the deterioration of the lagoon's ecosystem, and that also reconstruct and maintain the natural environment of the lagoon.

COMMISSIONE DI SALVAGUARDIA – SAFEGUARDING COMMISSION FOR VENICE

Instituted by the First Special Law (1973), this Committee was granted authority to express its discretionary opinion on all building works as well as territorial transformations and modifications planned by private and public bodies everywhere within the Venice Lagoon boundary. There are about 20 members, including UNESCO and the National Research Council (CNR), presided over by the President of the Region. To some extent it overlaps with the Town Council's land use planning remit and it has been accused of causing bureaucratic delays. Decision-making parameters are dominated by aesthetic rather than technical considerations.

'COMITATONE'

The Second Special Law (1984) instituted a mixed committee of government ministers and local authorities, known as the Large Committee or 'Comitatone'. It decides strategy, coordination and control of the implementation of all measures to safeguard Venice and the lagoon, especially how to divide the budget. The committee is chaired by the President of the Council of Ministers (the Italian Prime Minister) and consists of the heads of five ministries, their executive branches and the various local administrations, including the President of the Venice Water Authority (Secretary), the Minister for Infrastructure and Transport, the Minister of the Environment, the Minister of Cultural Heritage, the Minister of Transport and Navigation, the Minister of Universities and Scientific and Technological Research, the President of the Veneto Region, the Mayor of Venice, the Mayor of Chioggia and two representatives of the many other local authorities bordering the lagoon.

MAGISTRATO ALLE ACQUE (MAV) – VENICE WATER AUTHORITY
www.magisacque.it

The Venice Water Authority is a technical agency of the Ministry for Infrastructure and Transport with direct and primary responsibility for the safeguarding, security and hydraulic protection of a large area spread across a number of regions (Veneto, Friuli and Lombardy). It was established in 1907 but the name dates back to an organ of the Venetian Republic of the same name, founded in 1501. MAV duties cover five national river catchments (Adige, Brenta-Bacchiglione, Piave, Livenza, Tagliamento), one international catchment area (Isonzo), and the lagoons of Venice, Marano and Grado. With regard to the Venice Lagoon, MAV carries out systematic monitoring activites and oversees the planning and execution of safeguarding measures via Consorzio Venezia Nuova, its executive agency. A few technical offices have remained within MAV, notably the Anti-pollution Service, which publishes a monthly digest of lagoon water quality according to a series of physical-chemical characteristics at five monitoring stations.

CONSORZIO VENEZIA NUOVA (CVN)
www.salve.it

The Consorzio Venezia Nuova is the executive agency of the Ministry for Infrastructure and Transport – Venice Water Authority (as established by the Second Special Law) responsible for planning and implementing the measures to safeguard Venice and its lagoon, delegated by law to the State.

The Second Special Law established that a single body could take on responsibility for the interventions as a whole, on behalf of and controlled by the Venice Water Authority, and on the basis of a general plan of interventions defined and approved by the 'Comitatone' (and therefore by the institutions represented on the Committee) and by Parliament. Hence CVN is required to prepare and implement an 'integrated plan' to tackle the various aspects of the physical and environmental safeguarding of the lagoon ecosystem in a unitary and organic fashion and with a systematic approach.

Measures to monitor and improve the quality of water and sediment in the lagoon – most recently known as the MELa Programmes 1 and 2 – are also the responsibility of the State and carried out by CVN, in collaboration with the Regione, ARPAV and university research departments.

REGIONE VENETO – REGIONAL ADMINISTRATION
www.regione.veneto.it

Pollution abatement within the lagoon watershed area is the responsibility of the Veneto Region, which has established a framework programme of measures to monitor and reduce pollution in the drainage basin, known as the Master Plan 2000. Pollution abatement and prevention measures programmed by the Veneto Region for the drainage basin to reduce the loads arriving in the lagoon are now closely coordinated with lagoon water quality control measures (the remit of MAV/CVN).

The Regione is also responsible for drawing up the territorial plan for the conservation and development of the Venice environmental and settlement system (PALAV). It covers 16 municipalities from three provinces (Padua, Treviso and Venice). This was requested under the 1973 Special Law – it took over a decade to formulate, underwent several iterations and was finally approved in 1995.

PROVINCIA DI VENEZIA – PROVINCIAL ADMINISTRATION
www.provincia.venezia.it

The Provincial Administration does not participate in the Comitatone, except on a consultative basis, although it oversees certain key elements of Venice's natural resources, along with about 40 other municipalities within its domain. It has some land use management and environmental protection responsibilities, notably regulating fishing activities in the lagoon as well as hunting licences and it runs the environmental police department. It oversees the protection of flora and fauna, natural parks, organizes waste disposal at the provincial level and controls effluents, emissions and noise pollution. It is responsible for certain restructurings and restorations financed by the Special Law (notably schools and the island of San Servolo); territorial planning including landscape protection and environmental resources; planning and environmental education for pollution prevention and control. A small area of the Venice Lagoon falls within the domain of the Province of Padua.

COMUNE DI VENEZIA – VENICE MUNICIPALITY
www.comune.venezia.it

The Venice Municipality governs the main urban areas of the lagoon – the historic centre of Venice, Lido, the islands of Murano, Burano, Mazzorbo and Torcello, and the coastal strip of Pellestrina and San Pietro in Volta in the southern lagoon, as well as Mestre and Marghera. (Some nine other municipalities have territories within the lagoon and are duly subjected to the special legislation for Venice.) In the context of the Special Law, the Comune must look after Venice's historical, cultural, architectural and environmental heritage while also addressing the socio-economic factors that determine the city's identity and wellbeing – everything from stimulating productive activities and controlling tourism to assisting young couples with obtaining affordable housing and providing crèches for babies. Specifically, Special Law funds are used by the Town Council for:

- acquisition, restoration and refurbishment of buildings for residential, social and cultural, commercial and artisan uses – considered essential to maintaining the socio-economic identity of lagoon settlements;

- basic infrastructure (street lights, utilities, etc.) as well as bridges, embankments and canals within the municipal domain, i.e. the 45km inner canal network (whereas the larger canals are within the MAV remit along with the rest of the lagoon open waters);

- subsidies for restoration and maintenance works for private buildings, including for example the installation of lifts in tall palaces (pending approval of the Safeguarding Commission and the local branch of the cultural heritage ministry); and

- acquisition of areas to be converted to productive activities and the associated infrastructural requirements.

The Comune set up a mixed company – Insula S.p.A. – for the integrated management of canal dredging, raising street levels, revision of utility pipelines, underground maintenance works, etc. (see below). The tide forecasting office (CSPM – see below) is also part of the Comune. Executives from the Comune set up a working group in 1999 to review the Environmental Impact Study of the mobile barriers.

OTHER KEY ORGANIZATIONS IN VENICE AND THE VENETO

Agenzia Regionale per la Prevenzione e Protezione Ambientale del Veneto (ARPAV) – Veneto Regional Agency for Environmental Protection
www.arpa.veneto.it

The Veneto Regional Agency for Environmental Protection (ARPAV) is essentially the technical branch of the regional administration. Its tasks can be summarized as:

- environmental protection and monitoring;
- weather forecasting, monitoring and statistical elaborations;
- organization and management of the regional information system for environmental monitoring and environment related epidemiology;
- environmental education and information services; and

- technical and scientific services for environmental impact assessments and evaluation of environmental damage.

Agenzia per la protezione dell'ambiente e per i servizi tecnici (APAT) – National agency for environmental protection and technical services
www.apatvenezia.it

The Hydrographic Office has been operating in the Lagoon since 1907 within Magistrato alle Acque but recently changed to be run from central government. It manages a network of 52 tide gauge stations for the systematic measurement of tide level and related parameters, such as wind direction and wind speed, atmospheric pressure, precipitation and wave-height.

Autorità Portuale di Venezia – Venice Port Authority
www.port.venice.it

The Port of Venice is situated right on the axis of east-west European traffic and connected to prosperous NE Italy as well as linking the newer EU states with the rest of the territory. It is the largest economic sector, besides tourism, for the Venice Province and accounts for about 18 000 jobs in the area. Organization of the Port was transformed in 1996 when the Port Authority became a public entity responsible for the promotion, planning and management of its commercial and industrial activities. The total port area is 2045 ha and annual traffic is growing markedly, especially as regards passenger and cargo traffic.

Consorzio per la gestione del Centro di Coordinamento delle Attività di Ricerca inerenti il Sistema Lagunare di Venezia (CORILA) – Consortium for Coordination of Research Activities Concernine the Venice Lagoon System
www.corila.it

CORILA is an association of Ca'Foscari University, University IUAV of Venice, the University of Padua,

National Institute of Oceanography and Experimental Geophysics (OGS) and Italy's National Research Council. A non-profit organization, it is overseen by the Ministry of Education, Universities and Research (MIUR). It was founded in 1999 to coordinate and manage research on Venice and the Venice Lagoon and thereby provide decision support information to policy makers and public administrations dealing with Venice. Activities are organised within the framework of three-year research programmes, which in turn are divided into thematic areas and research lines. The principal thematic areas are: economics; architecture and cultural heritage; environmental processes; data management and dissemination. The first research programme (2000–3) cost c. €10.8 million of which nearly 60 per cent was funded by the Special Law via the Ministry for Research. Co-financing was provided by other administrations as well as the research departments and other partners themselves. The second research programme (2004–7) has just under €6m from the Special Law plus co-financing.

Centro Previsioni e Segnalazioni Maree – Tidal Forecasting and Early Warning Centre
www.comune.venezia.it/maree/

Founded in 1980, this organization within the Venice Municipality is responsible for the study and forecasting of storm surge events and for alerting the city in case of flood events. Observation of sea level and meteorological parameters is carried out through a monitoring network (11 stations) that gives a real-time view of marine and weather conditions in the Venice Lagoon and along the Adriatic coast. All stations measure sea level and some also collect meteorological parameters: air pressure, humidity, wind velocity and direction, waves and air temperature.

Consiglio Nazionale delle Ricerche (CNR) – National Research Council
www.cnr.it

The Italian National Research Council is an authoritative public organization in the field of scientific and technological research. Founded in 1923, it is composed of dozens of separate institutes, each with a particular specialization. In response to the increasing worldwide concern for the survival of Venice, CNR established the Institute for the Study of the Dynamics of Large Masses in 1969, now incorporated within the Marine Sciences Institute (ISMAR). Created first as a laboratory, it has spread from basic research in oceanography and geology to applied research. The other Venice-based branch of CNR-ISMAR is the Institute for Marine Biology, established in 1946 as the National Centre of Talassographic Studies, which focuses on pure and applied biological oceanography, marine and lagoon biology.

Consorzio per la Ricerca e la Formazione (COSES) – Consortium for Research and Training
www.provincia.venezia.it/coses/

COSES was established in 1967 by the municipal and provincial administrations to carry out analyses, studies and projects to support public sector activities – essentially via market research, data collection and statistical analyses concerning the urban and regional economy, building sector and housing, commercial distribution, tourism, culture, teaching and education, immigration, demographics, transport (especially water) and urban planning.

Insula S.p.A.
www.insula.it

Founded in 1997 by the Venice Town Council (52 percent share) together with Vesta (waste management), Enel.Hydro (electricity), Italgas (gas) and Telecom Italia (telephones), Insula's mission concerns urban maintenance and, more precisely, measures such as clearing canals of accumulated silt, restoration of canal walls, foundations and façades of buildings lining canals, restoring bridges, rationalization of urban subsoil (utility lines and sewer system), maintenance and renovation of paving, raising of footpaths above the level of medium-high tides.

Project integration and works coordination (all parties involved in the operations work side by side, not least the public utilities, who are Insula's partners) is therefore essential to this complicated process to minimize inconvenience, while also boosting the

efficiency, in terms of economies of scale, in such a delicate urban environment.

Istituto Veneto di Scienze Lettere ed Arti
www.istitutoveneto.it

The Istituto Veneto di Scienze, Lettere ed Arti was founded by Napoleon Bonaparte 'to collect discoveries, and to perfect the arts and sciences'. Its current mission is to increase, promulgate and safeguard the sciences, literature and arts, bringing together outstanding figures from the world of scholarship. The Institute also supports special research projects that concern Venice and the Veneto, and which are addressed by the international community. Together with various universities and the National Research Council, it has also set up specialized centres for research into environmental questions, into philological and literary aspects of the language of the Veneto, and into hydrology, meteorology and climatology. It runs a programme to integrate and share environmental data among all the major institutions and research bodies.

Università Ca' Foscari di Venezia
www.unive.it

With 4 faculties and 19 departments, the University dates back to the late nineteenth century and its students account for a significant proportion of the local population. It covers many areas of chemistry and environmental sciences; in the area of economics it carries out specific environmental economics studies; in the area of mathematics and IT it has developed and has access to models and data management instruments; and significant contributions are also made in the field of law.

Università IUAV di Venezia – University IUAV of Venice
www.iuav.it

Founded in 1926, it is an international reference point for architecture, architectural history, design, and restoration as well as for town and land use planning. It also has laboratories for construction science and analysis of ancient materials. There are about

8000 enrolled students, 209 tenured professors and 240 contract professors, plus support and technical personnel.

Università degli Studi di Padova – Padua University
www.unipd.it

Founded in 1222, it was the first university in the world to award a degree to a woman, Elena Lucrezia Cornaro Piscopia, in 1678 (in philosophy). It has 13 faculties and 62 departments and is a world leader in hydraulic engineering, biology, agricultural sciences, chemistry, mathematics and many other branches of science – not to mention centuries of tradition in law and medicine – matched by first-class facilities and instrumentation.

UNESCO Office in Venice – Regional Bureau for Science in Europe (ROSTE)
www.unesco.org

Following the disastrous floods of 1966 in Venice and Florence and the Italian Government's invitation for UNESCO to contribute, the Liaison Office for the Safeguarding of Venice was established in 1973 on the occasion of the UNESCO International Campaign for the Safeguarding of Venice. In 1988 the UNESCO Scientific Co-operation Bureau for Europe (SC/BSE) was relocated from Paris to Venice and renamed as Regional Office for Science & Technology for Europe (ROSTE). In 2002, UNESCO established a single office in Venice with the mandate to achieve UNESCO's and Member States' goals in the fields of science and culture. The UNESCO Office in Venice actively promotes, sponsors and convenes international scientific and cultural events in Europe and in the Mediterranean region. A unifying theme for UNESCO is its contribution to peace and human development in an era of globalization, through education, the sciences, culture and communication.

International Private Committees for the Safeguarding of Venice

Following the appeal launched by the Director General of UNESCO in 1966, over 50 private organizations were established in a number of countries to collect and channel contributions to restore and preserve Venice. Over the years, the International Private Committees have worked closely with the Superintendencies of Monuments and Galleries of Venice, through UNESCO, to identify and address priority needs. Since 1969, they have funded the restoration of more than 100 monuments and 1000 works of art, provided laboratory equipment and scientific expertise, sponsored research and publications and awarded innumerable grants for craftsmen, restorers and conservators to attend specialist courses in Venice. Expenditure by the Private Committees for the five-year period 1996–2000 was well in excess of €5 million (c. £3 million).

A4 • Committees of the Cambridge Project 2001–2004

PROJECT MANAGEMENT TEAM

Dr Pierpaolo Campostrini – Director, CORILA, Venice

Frances Clarke – President, Venice in Peril, UK

Jane Da Mosto – Researcher, CORILA, Venice

Dr Caroline Fletcher – Venice Fellow, Churchill College, Cambridge, and Cambridge Coastal Research Unit, University of Cambridge

Professor Peter Guthrie – Centre for Sustainable Development, Department of Engineering, University of Cambridge

Paul Richens – Vice Master, Churchill College, Cambridge

Anna Somers Cocks – Chairman, Venice in Peril, UK

Professor Robin Spence – Director, Cambridge University Centre for Risk in the Built Environment

Dr Tom Spencer (Chair) – Director, Cambridge Coastal Research Unit, University of Cambridge

INTERNATIONAL MEETING ADVISORY COMMITTEE

Professor Sir Herman Bondi – Churchill College, Cambridge

Sir John Boyd – Master, Churchill College, Cambridge

Ing. Roberto Casarin – Environment and Public Works Officer, Regione Veneto

Professor Paolo Costa – Mayor of Venice

Dr Stefano Della Sala – Environmental Affairs Manager, Venice Port Authority

Professor Bruno Dolcetta – President, Insula, Venice

Professor David Fisk – Department of Civil and Environmental Engineering, Imperial College, University of London

Dr Filippo Giorgi – Abdus Salaam International Centre for Theoretical Physics, Trieste

Professor Sir John Houghton – Inter-governmental Panel for Climate Change (IPCC) and UKMC, England

Professor Deborah Howard – Department of History of Art, University of Cambridge

Professor David Ingram – Chair, Committee for Inter-disciplinary Environmental Studies (CIES), University of Cambridge

Professor Robert Mair – Department of Engineering, University of Cambridge

Ing. Giovanni Mazzacurati – Managing Director, Consorzio Venezia Nuova, Venice

Dr Howard Moore – Director, UNESCO-ROSTE, Venice

Sir Alan Muir Wood – past President, Institution of Civil Engineers, UK

Ing. Maria Giovanna Piva – President, Magistrato alle Acque, Venice

INTERNATIONAL MEETING SCIENTIFIC PROGRAMME COMMITTEE

Dr Ida Brøker – Danish Hydraulics Institute, Hørsholm

Dr Pierpaolo Campostrini – Director, CORILA, Venice

Professor Paolo Cescon – Department of Environmental Sciences, Ca' Foscari University, Venice

Dr Giovanni Cecconi – Engineering Services, Consorzio Venezia Nuova, Venice

Jane Da Mosto – Researcher, CORILA, Venice

Professor Luigi D'Alpaos – Department of Hydraulic, Marine and Geotechnical Engineering, Padua University

Flooding and Environmental Challenges for Venice and its Lagoon: State of Knowledge, ed. C. A. Fletcher and T. Spencer. Published by Cambridge University Press, © Cambridge University Press 2005.

Professor Trevor Davies – Dean, School of Environmental Sciences, University of East Anglia

Dr Caroline Fletcher – Venice Fellow, Department of Geography, University of Cambridge

Professor Marino Folin – Rector, University Institute of Architecture of Venice (IUAV), Venice

Dr Roberto Frassetto – IGBP and CNR-Marine Sciences Institute, Venice

Dr Miro Gačić – National Institute of Oceanography and Geophysics, Trieste

Professor Don Harleman – Department of Civil and Environmental Engineering, MIT

Dr Stephen Malcolm – Centre for Environment, Fisheries and Aquaculture Science, Lowestoft

Professor Antonio Marcomini – Department of Environmental Sciences, Ca' Foscari University, Venice

Dr Sandro Rabitti – CNR-Marine Sciences Institute, Venice

Professor Keith Richards – Department of Geography, University of Cambridge

Professor Andrea Rinaldo – Department of Hydraulic, Marine and Geotechnical Engineering, Padua University

Dr Agustín Sánchez-Arcilla – CIIRC, Universitat Politècnica de Catalunya, Barcelona

Professor Robin Spence – Centre for Risk in the Built Environment, University of Cambridge

Dr Tom Spencer – Cambridge Coastal Research Unit, University of Cambridge

Professor Alberto Tomasin – Department of Mathematics, Ca' Foscari University, Venice

Dr Georg Umgiesser – CNR-Marine Sciences Institute, Venice

Professor Pier Vellinga – Institute for Environmental Studies, Free University of Amsterdam, The Netherlands

Arch. Ettore Vio – Proto della Basilica di S. Marco, Venice

Index

Italicised page numbers indicate an illustration.

Flooding and Environmental Challenges for Venice and its Lagoon: State of Knowledge, ed. C. A. Fletcher and T. Spencer.
Published by Cambridge University Press, © Cambridge University Press 2005.